Spatial Data Analysis in Ecology and Agriculture Using R

Spatial Data Analysis in Ecology and Agriculture Using R

Spatial Data Analysis in Ecology and Agriculture Using R
Second Edition

Richard E. Plant

Departments of Plant Sciences and Biological and Agricultural Engineering
University of California, Davis

CRC Press
Taylor & Francis Group
Boca Raton London New York

CRC Press is an imprint of the
Taylor & Francis Group, an **informa** business

CRC Press
Taylor & Francis Group
6000 Broken Sound Parkway NW, Suite 300
Boca Raton, FL 33487-2742

First issued in paperback 2020

ISBN 13: 978-0-367-73232-5 (pbk)
ISBN 13: 978-0-815-39275-0 (hbk)

Library of Congress Cataloging-in-Publication Data

Names: Plant, Richard E., author.
Title: Spatial data analysis in ecology and agriculture using R / by Richard E. Plant.
Description: Second edition. | Boca Raton, Florida : CRC Press, [2019] | Includes bibliographical references and index.
Identifiers: LCCN 2018040387| ISBN 9780815392750 (hardback : alk. paper) | ISBN 9781351189910 (e-book)
Subjects: LCSH: Agriculture--Statistical methods. | Spatial analysis (Statistics) | R (Computer program language)
Classification: LCC S566.55 .P53 2019 | DDC 338.1072/7--dc23
LC record available at https://lccn.loc.gov/2018040387

Visit the Taylor & Francis Web site at
http://www.taylorandfrancis.com

and the CRC Press Web site at
http://www.crcpress.com

To Kathie, Carolyn, Suzanne, and Eleanor, with all my love

Contents

Preface to the First Edition

This book is intended for classroom use or self-study by graduate students and researchers in ecology, geography, and agricultural science who wish to learn about the analysis of spatial data. The book originated in a course entitled "Spatial Data Analysis in Applied Ecology" that I taught for several years at UC Davis. Although most of the students were enrolled in Ecology, Horticulture and Agronomy, or Geography, there was a smattering of students from other programs such as Entomology, Soil Science, and Agricultural and Resource Economics. The book assumes that the reader has a background in statistics at the level of an upper division undergraduate applied linear models course. This is also, in my experience, the level at which ecologists and agronomists teach graduate applied statistics courses to their own students. To be specific, the book assumes a statistical background at the level of Kutner et al. (2005). I do not assume that the reader has had exposure to the general linear model or modern mixed-model analysis.

The book is intended for those who want to make use of these methods in their research, not for statistical or geographical specialists. It is always wise to seek out a specialist's help when such help is available, and I strongly encourage this practice. Nevertheless, the more one knows about a specialist's area of knowledge, the better able one is to make use of that knowledge. Because this is a user's book, I have elected on some occasions to take small liberties with technical points and details of the terminology. To dot every *i* and cross every *t* would drag the presentation down without adding anything useful.

The book does not assume any prior knowledge of the R programming environment. All of the R code for all of the analyses carried out in this book is available on the book's companion website, https://psfaculty.plantsciences.ucdavis.edu/plant/sda.htm. One of the best features of R is also its most challenging: the vast number of functions and contributed packages that provide a multitude of ways to solve any given problem. This provides a special challenge for a textbook, namely, how to find the best compromise between exposition via manual coding and the use of contributed package functions, and which functions to choose. I have tried to use manual coding when it is easy or there is a point to be made, and save contributed functions for more complex operations. As a result, I sometimes have provided "homemade" code for operations that can also be carried out by a function from a contributed package.

The book focuses on data from four case studies, two from uncultivated ecosystems and two from cultivated ecosystems. The data sets are also available on the book's companion website. Each of the four data sets is drawn from my own research. My reason for this approach is a conviction that if one wants to really get the most out of a data set, one has to live with it for a while and get to know it from many different perspectives, and I want to give the reader a sense of that process as I have experienced it. I make no claim of uniqueness for this idea; Griffith and Layne (1999), for example, have done it before me. I used data from projects in which I participated not because I think they are in any way special, but because they were available and I already knew them well when I started to write the book.

For most of my career, I have had a joint appointment in the Department of Biological and Agricultural Engineering and a department that, until it was gobbled up in a fit of academic consolidation, bore the name Agronomy and Range Science. Faculty in this second department were fortunate in that, as the department name implies, we were able to work in both cultivated and uncultivated ecosystems, and this enabled me to include two of each in the book. I was originally motivated to write the book based on my experiences working in precision agriculture. We in California entered this arena considerably later than our colleagues in the Midwest, but I was in at the beginning in California. As was typical of academic researchers, I developed methods for site-specific crop management, presented them to farmers at innumerable field days, and watched with bemusement, as they were not adopted very often. After a while I came to the realization that farmers can figure out how best to use this new technology in the ways that suit their needs, and indeed they are beginning to do so. This led to the further realization that we researchers should be using this technology for what we do best: research. This requires learning how to analyze the data that the technology provides, and that is what this book is about.

I have been very fortunate to have had some truly outstanding students work with me, and their work has contributed powerfully to my own knowledge of the subject. I particularly want to acknowledge those students who contributed to the research that resulted in this book, including (in alphabetical order) Steven Greco, Peggy Hauselt, Randy Horney, Claudia Marchesi, Ali Mermer, Jorge Perez-Quezada, Alvaro Roel, and Marc Vayssières. In particular, Steven Greco provided Data Set 1, Marc Vayssières provided Data Set 2, and Alvaro Roel provided Data Set 3. I also want to thank the students in my course who read through several versions of this book and made many valuable suggestions. In particular, thanks go to Kimberley Miller, who read every chapter of the final draft and made many valuable comments. I have benefited from the interaction with a number of colleagues, too many to name, but I particularly want to thank Hugo Firpo, who collected the data on Uruguayan rice farmers for Data Set 3, Tony Turkovich, who let us collect the data for Data Set 4 in his fields, Stuart Pettygrove, who managed the large-scale effort that led to that data set and made the notes and data freely available, and Robert Hijmans, who introduced me to his raster package and provided me with many valuable comments about the book in general. Finally, I want to thank Roger Bivand, who, without my asking him to do so, took the trouble to read one of the chapters and made several valuable suggestions. Naturally, these acknowledgments in no way imply that the persons acknowledged, or anyone else, endorses what I have written. Indeed, some of the methods presented in this book are very *ad hoc* and may turn out to be inappropriate. For that, as well as for the mistakes in the book and bugs in the code that I am sure are lurking there, I take full responsibility.

Davis
California

Preface to the Second Edition

I am grateful for the opportunity to write a second edition of this book. R is very dynamic, and there have been sufficient changes to degrade the book's usefulness for learning R. There have also been major advances on several fronts in spatial data analysis. The most dramatic changes have been in the analysis of spatiotemporal data. These have been sufficient for me to completely revise that chapter. Major advances have also been made in the application of Bayesian methods to spatial data. The use of the generalized additive model has been rapidly gaining ground in ecology. Finally, but not least importantly, the use of packages associated with the tidyverse of Hadley Wickham and his colleagues have made graphical analysis much simpler. I have incorporated all of these, although, for reasons I have elaborated in Chapter 2, I have decided to continue to use the traditional R graphics to construct the figures in this book. Of course this means that some things have had to be deleted to make room. Most prominent has been the removal of the section on principal components analysis. The substitution of a section on the generalized additive model has prompted me to switch the order of this chapter and the one on linear models. I have placed the section on principal components analysis on the books companion website, https://psfaculty.plantsciences.ucdavis.edu/plant/sda2.htm in a section called "Additional Topics." Additional material to accompany the book can be accessed at https://psfaculty.plantsciences.ucdavis.edu/plant/additionaltopics.htm. I hope to add discussions of other relevant topics there as well. I have also had the opportunity to correct numerous errors in the first edition; I hope I have not introduced too many new ones into the second. In addition to the people I acknowledged in the first edition, I want to thank Meili Baragatti, James Graham, and Andrew Latimer for their thoughtful reviews of the first edition. I would also like to thank my editors, John Sulzycki and Alice Oven, for their guidance and assistance. They have been a true pleasure to work with.

Davis
California

Author

Richard E. Plant is a professor emeritus of Plant Sciences and Biological and Agricultural Engineering at the University of California, Davis. He is the co-author of *Knowledge-Based Systems in Agriculture* and is a former editor-in-chief of *Computers and Electronics in Agriculture* and current associate editor of *Precision Agriculture*. He has published extensively on applications of crop modeling, expert systems, spatial statistics, remote sensing, and geographic information systems to problems in crop production and natural resource management.

1

Working with Spatial Data

1.1 Introduction

The last decades of the twentieth century witnessed the introduction of revolutionary new tools for collecting and analyzing ecological data at the landscape level. The best known of these are the global positioning system (GPS) and the various remote sensing technologies, such as multispectral, hyperspectral, thermal, radar, and LiDAR imaging. Remote sensing permits data to be gathered over a wide area at relatively low cost, and the GPS permits locations to be easily determined on the surface of the earth with sub-meter accuracy. Other new technologies have had a similar impact on some areas of agricultural, ecological, or environmental research. For example, measurement of apparent soil electrical conductivity (EC_a) permits the rapid estimation of soil properties, such as salinity and clay content, for use in agricultural (Corwin and Lesch, 2005) and environmental (Kaya and Fang, 1997) research. The "electronic nose" is in its infancy, but in principle it can be used to detect the presence of organic pollutants (Capelli et al., 2014) and plant diseases (Wilson et al., 2004). Sensor technologies such as these are beginning to find their way into fundamental ecological research.

The impact of these technologies in agricultural crop production has been strengthened by another device of strictly agricultural utility: the yield monitor. This device permits "on the go" measurement of the flow of harvested material. If one has spatially precise measurements of yield and of the factors that affect yield, it is natural to attempt to adjust management practices in a spatially precise manner to improve yield or reduce costs. This practice is called *site-specific crop management*, which has been defined as the management of a crop at a spatial and temporal scale appropriate to its natural variation (Lowenberg-DeBoer and Erickson, 2000).

Technologies such as remote sensing, soil electrical conductivity measurement, the electronic nose, and the yield monitor are precision measurement instruments at the landscape scale, and they allow scientists to obtain landscape-level data at a previously unheard-of spatial resolution. These data, however, differ in their statistical properties from the data collected in traditional plot-based ecological and agronomic experiments. The data often consist of thousands of individual values, each obtained from a spatial location either contiguous with or very near to those of neighboring data. The advent of these massive spatial data sets has been part of the motivation for the further development of the theory of spatial statistics, particularly as it applies to ecology and agriculture.

Although some of the development of the theory of spatial statistics has been motivated by large data sets produced by modern measurement technologies, there are many older data sets to which the theory may be profitably applied. Figure 1.1 shows the locations of a subset of about one-tenth of a set of 4,101 sites taken from the Wieslander survey

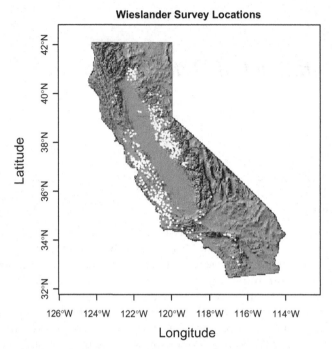

FIGURE 1.1
Locations of survey sites taken from the Wieslander vegetation survey. The white dots represent one-tenth of the 4,101 survey locations used to characterize the ecological relationships of oak species in California oak woodlands. The hillshade map was downloaded from the University of California, Santa Barbara Biogeography Lab, http://www.biogeog.ucsb.edu/projects/gap/gap_data_state.html, and reprojected in ArcGIS.

(Wieslander, 1935). This was a project led by Albert Wieslander of the US Forest Service that documented the vegetation in about 155,000 km^2 of California, or roughly 35% of its area. Allen et al. (1991) used a subset of the Wieslander survey to develop a classification system for California's oak woodlands, and subsequently Evett (1994) and Vayssières et al. (2000) used these data to compare methods for predicting oak woodland species distributions based on environmental properties.

Large spatial data sets like those arising from modern precision landscape measuring devices or from large-scale surveys such as the Wieslander survey present the analyst with two major problems. The first is that, even in cases where the data satisfy the assumptions of traditional statistics, traditional statistical inference is often meaningless with data sets having thousands of records because the null hypothesis is almost always rejected (Matloff, 1991). Real data never have *exactly* the hypothesized property, and the large size of the data set makes it impossible to be close enough. This would seem to make these data sets ideal for modern nonparametric methods associated with "Big Data." This approach, however, is often defeated by the second problem, which is that the data records from geographically nearby sites are often more closely related to each other than they are to data records from sites farther away. Moreover, if one were to record some data attribute at a location in between two neighboring sample points, the resulting attribute data value would likely be similar to those of both of its neighbors, and thus would be limited in the amount of information it added. The data in this case are, to use the statistical term, *spatially autocorrelated*. We will define this term more precisely and discuss its consequences in greater detail in Chapter 3, but it is clear that these data violate the assumption of both traditional statistical analysis and modern nonparametric analysis that data values or errors are mutually independent. The theory of spatial statistics concerns itself with these sorts of data.

Data sets generated by modern precision landscape measurement tools frequently have a second, more ecological, characteristic that we will call "low ecological resolution." This is illustrated in the maps in Figure 1.2, which are based on data from an agricultural field. These three thematic maps represent measurements associated respectively with soil properties, vegetation density, and reproductive material. The figures are arranged in a rough sequence of influence. That is, one can roughly say that soil properties influence vegetative growth, and vegetative growth influences reproductive growth. These influences,

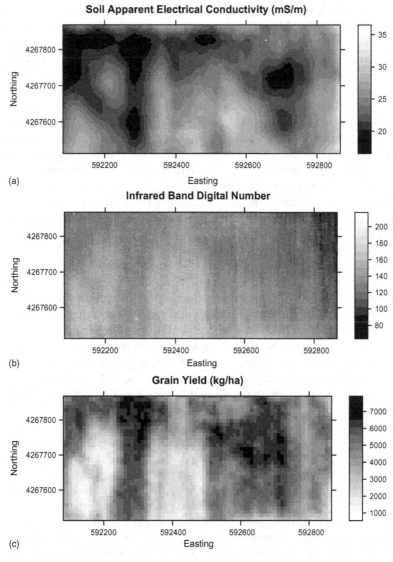

FIGURE 1.2
Spatial plots of three data sets from Field 2 of Data Set 4, a wheat field in Central California: (a) gray-scale representation of apparent soil electrical conductivity (mS m⁻¹); (b) digital number value of the infrared band of an image of the field taken in May, 1996; and (c) yield map (kg ha⁻¹).

however, are very complex and nonlinear. Three quantities are displayed in Figure 1.2: soil apparent electrical conductivity (EC_a) in Figure 1.2a, infrared reflectance (IR digital number) in Figure 1.2b, and grain yield in Figure 1.2c. The complexity of the relationship between soil, vegetation, and reproductive material is reflected in the relationship between the patterns in Figures 1.2a, 1.2b, and 1.2c, which is clearly not a simple one (Ball and Frazier, 1993; Benedetti and Rossini, 1993). In addition to the complexity of the relationships among these three quantities, each quantity itself has a complex relationship with more fundamental properties like soil clay content, leaf nitrogen content, and so forth. To say that the three data sets in Figure 1.2 have a very high spatial resolution but a low "ecological resolution" means that there is a complex relationship between these data and the fundamental quantities that are used to gain an understanding of the ecological processes in play. The low ecological resolution often makes it necessary to supplement these high spatial resolution data with other, more traditional data sets that result from the direct collection of soil and plant data.

The three data sets in Figure 1.2, like many spatial data sets, are filled with suggestive visual patterns. The two areas of low grain yield in the southern half of the field in Figure 1.2c clearly seem to align with patterns of low infrared reflectance in Figure 1.2b. However, there are other, more subtle patterns that may or may not be related between the maps. These include an area of high EC_a on the west side of the field that may correspond to a somewhat similarly shaped area of high infrared reflectance, an area of high apparent EC_a in the southern half of the field centered near 592,600E that may possibly correspond to an area of slightly lower infrared reflectance in the same location, and a possible correspondence between low apparent EC_a, low infrared reflectance, and low grain yield at the east end of the field. Scrutinizing the maps carefully might reveal other possibilities. The problem with searching for matching patterns like these, however, is that the human eye is famously capable of seeing patterns that are not really there. Also, the eye may miss subtle patterns that do exist. For this reason, a more objective means of detecting true patterns must be used, and statistical analysis can often provide this means.

Data sets like the ones discussed in this book are often unreplicated, relying on observations rather than controlled experiments. Every student in every introductory applied statistics class is taught that the three fundamental concepts of experimental design, introduced by Sir Ronald Fisher (1935), are randomization, replication, and blocking. The primary purpose of replication is to provide an estimate of the variance (Kempthorne, 1952, p. 177). In addition, replication helps to ensure that experimental treatments are interspersed, which may reduce the impact of confounding factors (Hurlbert, 1984). Some statisticians might go so far as to say that data that are not randomized and replicated are not worthy of statistical analysis.

In summary, we are dealing with data sets that may not satisfy the traditional assumptions of statistical analysis, that often involve relations between measured quantities for which there is no easy interpretation, that tempt the analyst with suggestive but potentially misleading visual patterns, and that frequently result from unreplicated observations. In the face of all these issues, one might ask why anyone would even consider working with data like these, when traditional replicated small plot experiments have been used so effectively for over a hundred years. A short answer is that replicated small plot experiments are indeed valuable and can tell us many things, but they cannot tell us everything. Often phenomena that occur at one spatial scale are not observable at another, and sometimes complex interactions that are important at the landscape scale lose their importance at the plot scale. Richter et al. (2009) argue that laboratory experiments conducted under highly standardized conditions can produce results with little validity beyond the specific environment in which the experiment is conducted, and this same argument can also be applied to small plot field experiments. Moreover, econometricians and other scientists for whom replicated experiments are often impossible have developed powerful statistical methods for analyzing and drawing inferences from observational data. These methods have much to recommend them for ecological and agricultural use as well. The fundamental theme of this book is that methods of spatial statistics, many of them originally developed with econometric and other applications in mind, can prove useful to ecologists and crop and soil scientists. These methods are certainly not intended to replace traditional replicated experiments, but when properly applied they can serve as a useful complement, providing additional insights, increasing the scope, and serving as a bridge from the small plot to the landscape. Even in cases where the analysis of data from an observational study does not by itself lead to publishable results, such an analysis may provide a much-improved focus to the research project and enable the investigator to pose much more productive hypotheses.

1.2 Analysis of Spatial Data

1.2.1 Types of Spatial Data

Cressie (1991, p. 8) provides a very convenient classification of spatial data into three categories, which we will call (1) geostatistical data, (2) areal data, and (3) point pattern data. These can be explained as follows. Suppose we denote by D a site (the *domain*) in which the data are collected. Let the locations in D be specified by position coordinates (x, y), and let $Y(x, y)$ be a quantity or vector of quantities measured at location (x, y). Suppose, for example, the domain D is the field in Figure 1.2. The components of Y would then be the three quantities mapped in the figure together with any other measured quantities. The first of Cressie's (1991) categories, *geostatistical data*, consists of data in which the components of $Y(x, y)$ vary continuously in the spatial variables (x, y). The value of Y is measured at a set of points with coordinates (x_i, y_i), $i = 1, ..., n$. A common goal in the analysis of geostatistical data is to interpolate Y at points where it is not measured.

The second of Cressie's (1991) categories consists of data that are defined *only* at a set of locations, which may be points or polygons, in the domain D. If the locations are polygons, then the data are assumed to be uniform within each polygon. If the data are measured at a set of points then one can arrange a mosaic or lattice of polygons that each contain one or more measurement points and whose areal aggregate covers the entire domain D. For this reason, these data are called *areal data*. In referring to the spatial elements of areal data, we will use the terms *cell* and *polygon* synonymously. A *mosaic* is an irregular arrangement of cells, and a *lattice* is a regular rectangular arrangement.

It is often the case that a biophysical process represented by areal data varies continuously over the landscape (Schabenberger and Gotway, 2005, p. 9). Thus, the same data set may be treated in one analysis as geostatistical data and in another as areal data. Figure 1.3 illustrates this concept. The figure shows the predicted values of clay content in the same field as that shown in Figure 1.2. Clay content was measured in this field by taking soil cores on a rectangular grid of square cells 61 m apart. Figure 1.3a shows an interpolation of the values obtained by kriging (Section 6.3.2). This is an implementation of the geostatistical model. Figure 1.3b shows the same data modeled as a lattice of discrete

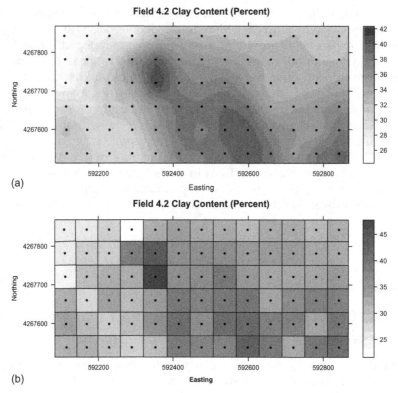

FIGURE 1.3
Predicted values of soil clay content in Field 2 of Data Set 4 using two different spatial models: (a) interpolated
values obtained using kriging based on the geostatistical model and (b) polygons based on the areal model. The
soil sample locations are shown as black dots.

polygons. In both figures, the actual data collection locations are shown as black dots. It
happens that each polygon in Figure 1.3b contains exactly one data location, but in other
applications, data from more than one location may be aggregated into a single polygon,
and the polygons may not form a regular lattice.

The appropriate data model, geostatistical or areal, may depend on the objective of the
analysis. If the objective is to accurately predict the value of clay content at unmeasured
locations, then generally the geostatistical model would be used. If the objective is to determine how clay content and other quantities influence some response variable such as yield,
then the areal model may be more appropriate. In many cases, it is wisest not to focus on
the classification of the data into one or another category but rather on the objective of the
analysis and the statistical tools that may be best used to achieve this objective.

The third category of spatial data consists of points in the domain D that are defined
by their location, and for which the primary questions of interest involve the pattern of
these locations. Data falling in this category, such as the locations of members of a species
of tree, are called *point patterns*. For example, Figure 1.6 below shows a photograph of oak
trees in a California oak woodland. If each tree is considered as a point, then point pattern
analysis is concerned with the pattern of the spatial distribution of the trees. When spatial
data are collected at irregularly located points, as is the case with the Wieslander data,
knowledge of their pattern may be helpful in determining whether the sampling points
have introduced a bias into the data. The primary focus of this book is on geostatistical

and areal data, which we will jointly refer to as *georeferenced* data. We do, however, occasionally discuss point pattern data analysis, primarily as a means of better understanding georeferenced data.

1.2.2 The Components of Spatial Data

Georeferenced data have two components. The first is the *attribute component*, which is the set of all the measurements describing the phenomenon being observed. The second is the *spatial component*, which consists of that part of the data that at a minimum describes the location at which the data were measured and may also include other properties such as spatial relationships with other data. The central problem addressed in this book is the incorporation into the analysis of attribute data of the effect of spatial relationships. As a simple example, consider the problem of developing a linear regression model relating a response variable to a single environmental explanatory variable. Suppose that there are two measured quantities. The first is leaf area index, or LAI, the ratio between total leaf surface area and ground surface area, at a set of locations. We treat this as a response variable and represent it with the symbol Y. Let Y_i denote the LAI measured at location i. Suppose that there is one explanatory variable, say, soil clay content. Explanatory variables will be represented using the symbol X, possibly with subscripts if there are more than one of them. We can postulate a simple linear regression relation of the form

$$Y_i = \beta_0 + \beta_1 X_i + \varepsilon_i, \ i = 1, ..., n. \tag{1.1}$$

There are several ways in which spatial effects can influence this model. One very commonly occurring one is a reduction of the *effective sample size*. Intuition for this concept can be gained as follows. Suppose that the n data values are collected on a square grid and that $n = m^2$ where m is the number of measurement locations on a side. Suppose now that we consider sampling on a $2m$ by $2m$ grid subtending the same geographic area. This would give us $4n$ data values and thus, *if these values were independent*, provide a much larger sample and greater statistical power. One could consider sampling on a $4m$ by $4m$ grid for a still larger sample size, and so on *ad infinitum*. Thus, it would seem that there is no limit to the power attainable if only we are willing to sample enough. The problem with this reasoning is that the data values are *not* independent: values sampled very close to each other will tend to be similar. If the values are similar, then the errors ε_i will also be similar. This property of similarity among nearby error values violates a fundamental assumption on which the statistical analysis of regression data is based, namely, the independence of errors. This loss of independence of errors is one of the most important properties of spatial data, and will play a major role in the development of methods described in this book.

1.2.3 Spatial Data Models

In the previous two sections, we have defined georeferenced data as data that contains an attribute component and a spatial component and for which a value can be assigned to any location in the domain D. Section 1.2.1 described one model for a geographic system that might generate such data, namely, the areal data model in which every point in the domain D is contained in a polygon, and the mosaic or lattice of these polygons covers the entire domain. This representation is called the *vector data model* (Lo and Yeung, 2007, p. 87).

It works well for systems in which large contiguous area have uniform or nearly uniform values (such as data defined by state or province) and less well for systems in which the attribute values vary continuously in space.

The second commonly used model in the analysis of georeferenced data is the *raster model* (Lo and Yeung, 2007, p. 83). In this model, the data are represented by a grid of raster cells, each of which contains one attribute value. The data represented in Figure 1.2 may be considered as exemplary of raster data. Much of the analysis in this book will be carried out using the vector model, but we will also on numerous occasions use the raster model, which more accurately represents georeferenced data that vary continuously in space. It is wise to develop a facility with both models in order to be able to select the one most appropriate for the given situation and analytical objective.

1.2.4 Topics Covered in the Text

The topics covered in this book are ordered as follows. The statistical software used for almost all of the analysis in the book is R (R Development Core Team, 2017). No familiarity with R is assumed, and Chapter 2 contains an introduction to the R package. The property of spatial data described in the previous paragraph, that attribute values measured at nearby locations tend to be similar, is called *positive spatial autocorrelation*. In order to develop a statistical theory that allows for spatial autocorrelation, one must define it precisely. This is the primary topic of Chapter 3. That chapter also describes the principal effects of spatial autocorrelation on the statistical properties of data. Having developed a precise definition of autocorrelation, one can measure its magnitude in a particular data set. Chapter 4 describes some of the most widely used measures of spatial autocorrelation and introduces some of their uses in data analysis.

Beginning with Chapter 5, the chapters of this book are organized in the same general order as that in which the phases of a study are carried out. The first phase is the actual collection of the data. Chapter 5 contains a discussion of sampling plans for spatial data. The second phase of a study is the preparation of the data for analysis, which is covered in Chapter 6. This is a much more involved process with spatial data than with non-spatial data, first because the data sets must often be converted to a file format compatible with the analytical software; second, because the data sets must be georegistered (i.e., their locations must be aligned with the points on the earth at which they were measured); third, because the data are often not measured at the same location or on the same scale (this is the problem of *misalignment*); and fourth, because in many cases the high volume and data collection process make it necessary to remove outliers and other "discordant" data records. After data preparation is complete, the analyses can begin.

Statistical analysis of a particular data set is often divided into exploratory and confirmatory components. Chapters 7 through 9 deal with the exploratory component. Chapter 7 discusses some of the common graphical methods for exploring spatial data. Chapter 8 describes the use of linear methods, including multiple regression and the general linear model. These methods are here used in an exploratory context only since spatial autocorrelation is not taken into account. The discussion of regression analysis also serves as a review on which to base much of the material in subsequent chapters.

Chapter 9 continues the application of non-spatial methods to spatial data, in this case focusing on nonparametric methods, that is, methods that do not make assumptions about the structure of the random component of the data. These include generalized additive models, recursive partitioning, and random forest. Chapters 8 and 9 serve as a link between the exploratory and confirmatory components of the analysis. The focus of Chapter 10 is the effect of spatial autocorrelation on the sample variance. The classical estimates based on independence of the random variables are no longer appropriate, and in many cases an analytical estimate is impossible. Although approximations can be developed, in many cases bootstrap estimation provides a simple and powerful tool for generating variance estimates. Chapter 11 enters fully into confirmatory data analysis. One of the first things one often does at this stage is to quantify, and test hypotheses about, association between measured quantities. It is at this point that the effects of spatial autocorrelation of the data begin to make themselves strongly felt. Chapter 12 describes these effects and discusses some methods for addressing them.

In Chapter 12, we address the incorporation of spatial effects into models for the relationship between a response variable and its environment through the use of mixed model theory. Chapter 13 describes autoregressive models, which have become one of the most popular tools for statistical modeling. Chapter 14 provides an introduction to Bayesian methods of spatial data analysis, which are rapidly gaining in popularity as their advantages become better exploited. Chapter 15 provides a very brief introduction to the study of spatiotemporal data. Chapter 16 describes methods for the explicit incorporation of spatial autocorrelation into the analysis of replicated experiments. Although the primary focus of this book is on data from observational studies, if one has the opportunity to incorporate replicated data, one certainly should not pass this up. Chapter 17 describes the summarization of the results of data analysis into conclusions about the processes that the data describe. This is the goal of every ecological or agronomic study, and the methods for carrying out the summarization are very much case dependent. All we can hope to do is to try to offer some examples.

1.3 The Data Sets Analyzed in This Book

Every example in this book is based on one of four data sets. There are a few reasons for this arrangement. Probably the most important is that one of the objectives of the book is to give the reader a complete picture, from beginning to end, of the collection and analysis of spatial data characterizing an ecological process. A second is that these data sets are not ideal "textbook" data. They include a few warts. Real data sets almost always do, and one should know how to confront them. Also, sometimes they yield ambiguous results. Again, this can happen with real data sets and one must know what to do when it does happen.

Two of the data sets involve unmanaged ecosystems and two involve cultivated ecosystems (ecologists should also study the cultivated systems and agronomists should also study the natural systems). All four data sets come from studies in which I have participated. This is not because I feel that these studies are somehow superior to others that I might have chosen, but rather because at one time or another my students and collaborators and I have lived with these data sets for an extended period, and I believe one must live with a data set for a long time to really come to know it and analyze it effectively. The book is not a research monograph, and none of the analytical objectives given here is an actual current research objective. Their purpose is strictly expository. In that context, although the discussion of each method has the ostensible purpose of furthering the stated research objective, the true purpose is to expose the methods to the reader. This does have one beneficial side effect. In real life, the application of a particular method to a particular data set sometimes does not lead to any worthwhile results. That sometimes happens with the analyses in this book, and I hope this gives the reader a more realistic impression of the process of data analysis.

The present section contains a brief description of each of the four data sets. A more thorough, "Materials and Methods" style description is given in Appendix B. The data sets, along with the R code to execute the analyses described in this book, are available on the book's website http://psfaculty.plantsciences.ucdavis.edu/plant/sda2.htm. The data are housed in four subfolders of the folder *data*, called *Set1*, *Set2*, *Set3*, and *Set4*. In addition to a description of the data sets, Appendix B also contains the R statements used to load them into the computer. These statements are therefore not generally repeated in the body of the text.

Two other data folders, which are also provided on the website, appear in the R code: *auxiliary* and *created*. Readers should be able to generate these files for themselves, but they are made available just in case. The folder *created* contains files that are created as a result of doing the analyses and exercises in the book. The folder *auxiliary* contains data downloaded from the Internet. Large, publicly available data sets play a very important role in the analysis of ecological data, and one of the objectives of this book is to introduce the reader to some of these data sets and how they can be used.

1.3.1 Data Set 1: Yellow-Billed Cuckoo Habitat

This data set was collected as a part of a study of the change over a 50 year period in the extent of suitable habitat for the western yellow-billed cuckoo (*Coccyzus americanus occidentalis*), a California state listed endangered species that is one of California's rarest birds (Gaines and Laymon, 1984). The yellow-billed cuckoo, whose habitat is the interior of riparian forests, once bred throughout the Pacific Coast states. Land use conversion to agriculture, urbanization, and flood control have, however, restricted its current habitat to the upper reaches of the Sacramento River. This originally meandering river has over most of its length been constrained by the construction of flood control levees. One of the few remaining areas in which the river is unrestrained is in its north central portion (Figure 1.4).

Data Set 1 Site Location

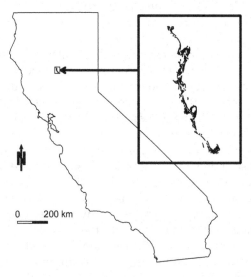

FIGURE 1.4
Location of study area of yellow-billed cuckoo habitat in California, in the north-central portion of the Sacramento River.

The data set consists of five explanatory variables and a response variable. The explanatory variables are defined over a polygonal map of the riparian area of the Sacramento River between river-mile 196, located at Pine Creek Bend, and river-mile 219, located near the Woodson Bridge State Recreation Area (although metric units are used exclusively in this book, the river-mile is a formal measurement system used to define locations along a river). This map was created by a series of geographic information system (GIS) operations as described in Appendix B.1. Figure 1.5 shows the land cover mosaic in the northern end of the study area as measured in 1997. The white areas are open water and the light gray areas are either managed (lightest gray) or unmanaged (medium gray) areas of habitat unsuitable for the yellow-billed cuckoo. Only the darkest areas contain suitable vegetation cover. These areas are primarily riparian forest, which consists of dense stands of willow and cottonwood. The response variable is based on cuckoo encounters at each of the 21 locations situated along the river.

Our objective in the analysis of Data Set 1 is to test a model for habitat suitability for the yellow-billed cuckoo. The model is based on a modification of the California Wildlife Habitat Relationships (CWHR) classification system (Mayer and Laudenslayer, 1988). This is a classification system that "contains life history, geographic range, habitat relationships, and management information on 694 species of amphibians, reptiles, birds, and mammals known to occur in the state" (https://www.wildlife.ca.gov/Data/CWHR). The modified CWHR model is expressed in terms of habitat patches, where a *patch* is defined as "a geographic area of contiguous land cover." Photographs of examples of the habitat types used in the model are available on the CWHR website given above. Our analysis will test a habitat suitability index based on a CWHR model published by Laymon and Halterman (1989). The habitat suitability model incorporates the following explanatory variables: (1) patch area, (2) patch width, (3) patch distance to water, (4) within-patch ratio of high vegetation to medium and low vegetation, and (5) patch vegetation species. The original discussion of this analysis is given by Greco et al. (2002).

Data Set 1 Land Cover, Northern End, 1997

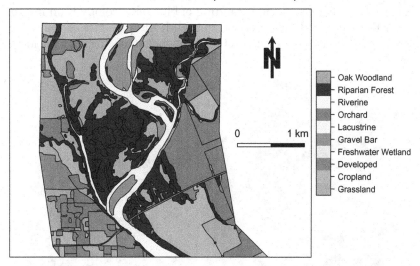

FIGURE 1.5
Land use in 1997 at the northern end of the region along the Sacramento River from which Data Set 1 was collected. The white areas are open water (riverine = flowing, lacustrine = standing). The light gray areas are unsuitable; only the charcoal colored area is suitable riparian forest.

The scope of inference of this study is restricted to the upper Sacramento River, since this is the only remaining cuckoo habitat. Data Set 1 is the only one of the four data sets for which the analysis objective is predictive rather than explanatory. Moreover, the primary objective is not to develop the best (by some measure) predictive model but rather to test a specific model that has already been proposed. As a part of the analysis, however, we will try to determine whether alternative models have greater predicative power or, conversely, whether any simpler subsets of the proposed model are its equal in predictive power.

1.3.2 Data Set 2: Environmental Characteristics of Oak Woodlands

Hardwood rangelands cover almost four million hectares in California, or over 10% of the state's total land area (Allen et al., 1991; Waddell and Barrett, 2005). About three-quarters of hardwood rangeland is classified as oak woodland, defined as areas dominated by oak species and not productive enough to be considered timberland (Waddel and Barret, 2005, p. 12). Figure 1.6 shows typical oak woodland located in the foothills of the Coast Range. Blue oak (*Quercus douglasii*) woodlands are the most common hardwood rangeland type, occupying about 500,000 ha. These woodlands occupy an intermediate elevation between the grasslands of the valleys and the mixed coniferous forests of the higher elevations. Their climate is Mediterranean, with hot, dry summers and cool, wet winters. Rangeland ecologists have expressed increasing concern over the apparent failure of certain oak species, particularly blue oak, to reproduce at a rate sufficient to maintain their population (Standiford et al., 1997; Tyler et al., 2006; Zavaleta et al., 2007; López-Sánchez et al., 2014). The appearance in California in the mid-1990s of the pathogen *Phytophthora ramorum*, associated with the condition known as sudden oak death, has served to heighten this concern (Waddel and Barrett, 2005).

The reason or reasons for the decline in oak recruitment rate are not well understood. Among the possibilities are climate change, habitat fragmentation, increased seedling predation resulting from changes in herbivore populations, changes in fire regimes, exotic plant and animal invasions, grazing, and changes in soil conditions (Tyler et al., 2006). One component of an increased understanding of blue oak ecology is an improved characterization of the ecological factors associated with the presence of mature blue oaks. Although changes in blue oak recruitment probably began occurring coincidentally with

FIGURE 1.6
Oak woodlands located in the foothills of the Coast Range of California. (Photograph by the author).

the arrival of the first European settlers in the early nineteenth century (Messing, 1992), it is possible that an analysis of areas populated by blue oaks during the first half of the twentieth century would provide some indication of environmental conditions favorable to mature oaks. That is the general objective of the analysis of Data Set 2. This data set consists of records from 4,101 locations surveyed as a part of the Vegetation Type Map (VTM) survey in California, carried out during the 1920s and 1930s (Wieslander, 1935; Jensen, 1947; Allen et al., 1991a; Allen-Diaz and Holzman, 1991), depicted schematically in Figure 1.1. Allen-Diaz and co-workers (Allen-Diaz and Holzman, 1991) entered these data into a database, and Evett (1994) added geographic coordinates and climatic data to this database. The primary goal of the analysis of Data Set 2 is to characterize the environmental factors associated with the presence of blue oaks in the region sampled by the VTM survey. The original analysis was carried out by Vayssières et al. (2000).

The intended scope of inference of the analysis is that portion of the state of California where blue oak recruitment is possible. As a practical matter, this area can be divided roughly into four regions (Appendix B.2 and Figure 1.1): the Coast Range on the west side of the Great Central Valley, the Klamath Range to the north, the Sierra Nevada on the east side of the valley, and the Traverse Ranges to the south. The analysis in this book will be restricted to the Coast Range and the Sierra Nevada, with the Klamath Range and the Traverse Ranges used as validation data.

1.3.3 Data Set 3: Uruguayan Rice Farmers

Rice (*Oriza sativa* L.) is a major Uruguayan export crop. It is grown under flooded conditions. Low dykes called *taipas* are often used to promote even water levels (Figure 1.7). Rice is generally grown in rotation with pasture, with a common rotation being 2 years of rice followed by 3 years of pasture. Data Set 3 contains measurements taken in 16 rice fields in eastern Uruguay (Figure 1.8) farmed by twelve different farmers over a period of three growing seasons. Several locations in each field were flagged and measurements were made

FIGURE 1.7
A typical Uruguayan rice field. The low dykes, called *taipas*, are used to promote more even flooding. (Courtesy of Alvaro Roel).

Data Set 3 Field Locations

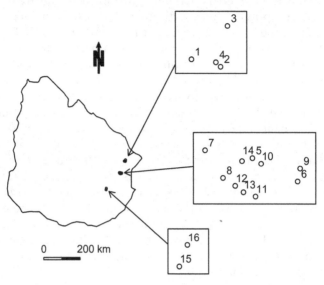

FIGURE 1.8
Map of Uruguay, showing locations of the 16 rice fields measured in Data Set 3. Insets are not drawn to scale.

at each location of soil and management factors that might influence yield. Hand harvests of rice yield were collected at these locations. Regional climatic data were also recorded.

Some of the farmers in the study obtained consistently higher yields than others. The overall goal of the analysis is to try to understand the management factors primarily responsible for variation in yields across the agricultural landscape. A key component of the analysis is to distinguish between management factors and environmental factors with respect to their influence on crop yield. We must take into account the fact that a skillful farmer, or one with access to more resources, might manage a field not having conditions conducive to high yield in a different way from that in which the same farmer would manage a field with conditions favorable to high yield. For example, a skillful farmer growing a crop on infertile soil might apply fertilizer at an appropriate rate, while a less skillful farmer growing a crop on a fertile soil might apply fertilizer at an inappropriate rate. The less skillful farmer might obtain a better yield, but this would be due to the environmental conditions rather than the farmer's skill. Conversely, a farmer, skillful or otherwise, who could not afford to purchase fertilizer might not apply it at a rate sufficient to maximize yield. The original analysis was carried out by Roel et al. (2007). The scope of the study is intended to include those rice farmers in eastern Uruguay who employ management practices similar to those of the study participants. For simplicity, we assume that the farmers' objective is to maximize yield rather than profit.

1.3.4 Data Set 4: Factors Underlying Yield in Two Fields

One of the greatest challenges in site-specific crop management is to determine the factors underlying observed yield variability. Data Set 4 was collected as a part of a four-year study initiated in two fields in Winters, California, which is located in the Central Valley.

This was the first major precision agriculture study in California, and the original objective was simply to determine whether or not the same sort of spatial variability in yield existed in California as had been observed in other states. We will pursue a different objective in this book, namely, to determine whether we can establish which of the measured explanatory variables influenced the observed yield variability. Data from one of the fields (designated Field 2) is shown in Figure 1.2.

In the first year of the experiment, both fields were planted to spring wheat (*Triticum aestivum* L.). In the second year, both fields were planted to tomato (*Solanum lycopersicum* L.). In the third year, Field 1 was planted to bean (*Phaseolus vulgaris* L.) and Field 2 was planted to sunflower (*Helianthus annuus* L.), and in the fourth year Field 1 was planted to sunflower and Field 2 was planted to corn (*Zea mays* L.). Each crop was harvested with a yield monitor equipped harvester so that yield maps of the fields were available. False color infrared aerial photographs (i.e., images where infrared is displayed as red, red as green, and green as blue) were taken of each field at various points in the growing season. Only those from the first season are used in this book. During the first year, soil and plant data were collected by sampling on a square grid 61 m in size, or about two sample points per hectare. In addition, a more densely sampled soil electrical conductivity map of Field 2 was made during the first year of the experiment.

As was mentioned in the discussion of Data Set 2, the climate in California's Central Valley is Mediterranean, with hot, essentially rain-free summers and cool, wet winters. Spring wheat is planted in the late fall and grown in a supplementally irrigated cropping system, while the other crops, all of which are grown in the summer, are fully irrigated. Both fields were managed by the same cooperator, a highly skilled farmer who used best practices as recommended by the University of California. Thus, while the objective of the analysis of Data Set 3 is to distinguish the most effective management practices among a group of farmers, the objective of the analysis of Data Set 4 is to determine the factors underlying yield variability in a pair of fields in which there is no known variation in management practices.

The original analysis of the data set was carried out by Plant et al. (1999). The scope of the study, strictly speaking, extends only to the two fields involved. Obviously, however, the real interest lies in extending the results to other fields. Success in achieving the objective would not be to establish that the same factors that influence yield in these fields were also influential in other fields, which is not the case anyway. The objective is rather to develop a methodology that would permit the influential factors in other fields to be identified.

1.3.5 Comparing the Data Sets

The four data sets present a variety of spatial data structures, geographical structures, attribute data structures, and study objectives. Let's take a moment to compare them, with the idea that by doing so you might get an idea as to how they relate to your own data set and objectives. Data Sets 2 and 3 have the simplest spatial structure. Both data sets consist of a single set of data records, each of which is the aggregation of a collection of measurements made at a location modeled as a point in space. Data Set 2 contains quantities that are measured at different spatial scales. This data set covers the largest area and presents the greatest opportunity to incorporate publicly available data into the analysis. The spatial structure of Data Set 1 is more complex in that it consists of data records that are modeled as properties of a mosaic of polygons. Data Set 1 is the only data set that includes this polygonal data structure in the raw data. Data Set 4 does not include any polygons, but it does include raster data sets. In addition, Data Set 4 includes point data sets measured at different locations. Moreover, the spatial points that represent the data locations in different point data sets actually represent the aggregation of data collected over different areas. This combination of types of data, different locations of measurement, and different aggregation areas leads to the problem of *misalignment* of the data, which will be discussed frequently in this book, but primarily in Chapter 6.

Moving on to geographic structure, Data Set 4 is the simplest. The data are recorded in two fields, each of size less than 100 ha. We will use Universal Transverse Mercator (UTM) coordinates (Lo and Yeung, 2007, p. 43) to represent spatial location. If you are not familiar with UTM coordinates, you need to learn about them. In brief, they divide the surface of the earth into 120 *Zones*, based on location in the northern or southern hemisphere. Within each hemisphere, UTM coordinates are based on location within 60 bands running north and south, each of width six degrees of longitude. Within each zone, location is represented by an *Easting* (x coordinate) and a *Northing* (y coordinate), each measured in meters. Locations in Data Sets 1 and 3 also are represented in UTM coordinates, although both of these data sets cover a larger geographic area than that of Data Set 4. Depending on the application, locations in Data Set 2 may be represented in either degrees of longitude and latitude or UTM coordinates. This data set covers the largest geographic area. Moreover, the locations in Data Set 2 are in two different UTM Zones.

1.4 Further Reading

Spatial statistics is a rapidly growing field, and the body of literature associated with it is growing rapidly as well. The seminal books in the field are Cliff and Ord (1973) and, even more importantly, Cliff and Ord (1981). This latter monograph set the stage for much that was to follow, providing a firm mathematical foundation for measures of spatial autocorrelation and for spatial regression models. Ripley (1981) contains coverage of many topics not found in Cliff and Ord (1981). It is set at a slightly higher level mathematically but is nevertheless a valuable text even for the practitioner. Besides Cliff and Ord (1981), the other seminal reference is Cressie (1991). This practically encyclopedic treatment of spatial statistics is at a mathematical level that some may find difficult, but working one's way through it is well worth the effort. Griffith (1987) provides an excellent introductory discussion of the phenomenon of spatial autocorrelation. Upton and Fingleton (1985) provide a wide array of examples for applications-oriented readers. Their book is highly readable while still providing considerable mathematical detail.

Among more recent works, the two monographs by Haining (1990, 2003) provide a broad range of coverage and are very accessible to readers with a wide range of mathematical sophistication. Legendre and Legendre (1998) touch on many aspects of spatial data analysis. Griffith and Lane (1999) present an in-depth discussion of spatial data analysis through the study of a collection of example data sets. Waller and Gotway (2004) discuss many aspects of spatial data analysis directly applicable to ecological problems. Lloyd (2010) gives an introduction to spatial statistics for those coming from a GIS background. Chun and Griffith (2013) provide a very nice, brief undergraduate-level introduction to spatial statistics. Oyana and Margai (2015) provide a data-centric treatment of the subject.

There are also a number of collections of papers on the subject of spatial data analysis. Two early volumes that contain a great deal of relevant information are those edited by Bartels and Ketellapper (1979) and Griffith (1989). The collection of papers edited by

Arlinghaus (1995) has a very applications oriented focus. The volume edited by Longley and Batty (1996), although it is primarily concerned with GIS modeling, contains a number of papers with a statistical emphasis. Similarly, the collection of papers edited by Stein et al. (1999), although it focuses on remote sensing, is broadly applicable. The volume edited by Scheiner and Gurevitch (2001) contains many very useful chapters on spatial data analysis. Fotheringham and Rogerson (2009) have assembled papers by primary contributors on a number of topics in spatial analysis into a single handbook.

The standard introductory reference for geostatistics is Isaaks and Srivastava (1989). Webster and Oliver (1990, 2001) provide coverage of geostatistical data analysis that is remarkable for its ability to combine accessibility with depth of mathematical treatment. Another excellent source is Goovaerts (1997). Journel and Huijbregts (1978) is especially valuable for its historical and applied perspective. Pielou (1977) is an excellent early reference on point pattern analysis. Fortin and Dale (2005) devote a chapter to the topic. Diggle (1983) provides a mathematically more advanced treatment.

A good discussion of observational versus replicated data is provided by Kempthorne (1952). Sprent (1998) discusses hypothesis testing for observational data. Gelman and Hill (2007, Chapters 9 through 10) discuss the use of observational data in the development of explanatory models. Bolker (2008) and Zuur et al. (2007) are good sources for analysis of ecological data.

For those unfamiliar with GIS concepts, Lo and Yeung (2007) provide an excellent introduction. In addition to Lo and Yeung (2007), which we use as the primary GIS reference, there are many others who provide excellent discussions of issues such as map projections, Thiessen polygons, spatial resampling, and so forth. Prominent among these are de Smith et al. (2007), Bonham-Carter (1994), Burrough and McDonnell (1998), Clarke (1990), and Johnston (1998). Good GIS texts at a more introductory level include Chang (2008), Demers (2009), and Longley et al. (2001).

2

The R Programming Environment

2.1 Introduction

2.1.1 Introduction to R

The analyses discussed in this book are, with a few minor exceptions, carried out in the R programming environment (R Core Development Team, 2017). R is an open source program, which means, among other things, that it is freely distributed and that in addition to the executable files, the source code is made available to all. A full description of open source software is given on the website of the Open Source Initiative at http://www.opensource.org/docs/osd. R, like some other well-known statistical software packages, is primarily *command driven*, which means that analyses in R must generally be carried out by typing a series of statements (we will throughout this book use the terms *command* and *statement* synonymously). R runs on Macintosh, Windows, and Linux platforms. This book was written using R version 3.4.3 on a Windows platform. The book is not intended to be a complete introduction to R, since many such texts are already available. Some of these are listed in Section 2.8. No prior experience with R is assumed, and indeed one of the goals of this book is to introduce R to those in the applied ecology research community who are not familiar with it. The reader who is completely unfamiliar with R is encouraged to read one or more of the introductory R books listed in Section 2.8 along with this book. Section 2.1.2 provides information on how to download R and get set up to use it.

Unlike some programming environments and statistical software packages, R is designed to be used interactively. With most programming languages and with many command-driven statistical packages, you run in *batch mode*. This means that you enter a series of statements, execute them all at once, and receive output. Although the final product of an R session is sometimes run like this, in many cases you will use R by typing one or a few lines, observing the response, and then typing more lines based on this response. This is the *interactive mode*. Moreover, R will frequently execute commands and give you no response at all. Instead, you will have to type a statement or statements requesting certain specific results from the computations.

Aside from being freely available and having freely available source code, another major feature of an open source program like R is that it is supported by a large user community. When you install R, you are installing the base package, which contains some fundamental software. Many (several thousand!) other contributed packages have been written and are maintained by people in the R community. The fact that R is open source means that anyone can write and contribute a package or modify an existing package for his or her own use. In this book, for example, we will make considerable use of the contributed packages sf (Pebesma 2017)[1] and sp (Bivand et al., 2013b). Will see how to install contributed packages in the next section.

Finally, one feature of R, well known within the R community, is that, given any task that you want to perform, there are almost always at least a half dozen different ways to do it. There is generally no effort expended in this book to teach you more than one of them, and that one is usually the one I consider simplest and most effective.

2.1.2 Setting Yourself Up to Use This Book

Although it is possible for you to type statements into R based on the printed statements in this book, that is not the intent, and moreover, most of the code would not work because this book does not include every statement used in the analysis. Instead, the R files, as well as the data, should be downloaded from the book's companion website, http://psfaculty. plantsciences.ucdavis.edu/plant/sda2.htm (don't forget the 2, or else you will get the material from the first edition, which is different). One of the great things about a programming environment like R, as opposed to a menu-driven application, is that code can be reused after slight modification. I hope that the code in this book will serve as a template that enables you to write programs to analyze your own data.

To set up the files on the disk, you need to first set up the directory structure. In the book, we will use the term "folder" and "directory" synonymously. In the Windows world they are called folders, and in the R world they are called directories. Suppose you want to store all of your materials associated with R in a folder on the *C:* drive called *rdata*. First open a window on the *C:* drive and create a new folder called *rdata*. Suppose you want to store the material associated with this book in a subfolder called *SDA2* (for "spatial data analysis, second edition"). In Windows, open the *rdata* folder and create a new folder called *SDA2*. Now you can download the data from the companion website. When you visit this site, you will see a link for data and one for R code. Clicking on each of these links will enable you to download a zip file. These zip files may be uncompressed into two folders. One contains the four data sets, each in a separate subfolder, together with auxiliary and created data. The other contains the R code. The R code is arranged in subfolders by chapter and within each folder by section within the chapter. If you wish, you may also download two other folders that contain data created during the course of working through the book as well as auxiliary data from other sources on the web. You are encouraged, however, to generate yourself as much of these data as you can.

In order to do serious R programming, you will need an editor. There are a number of editors available for R that provide improved functionality over the built-in R editor. Two very good ones that I use are TINN-R (TINN stands for This Is Not Notepad) (https:// sourceforge.net/projects/tinn-r/) and RStudio (http://www.rstudio.com/). I use them both, generally TINN-R for shorter jobs and RStudio for longer projects. You can run them both on your computer at the same time, running two separate implementations of R. This is often useful for comparing, modifying, and debugging code. Also, I have found that TINN-R uses less memory than RStudio. This doesn't usually matter, but sometimes it does. Nevertheless, for purposes of consistency we will use RStudio throughout the book.

As you work though this book, you should run the code and see the results. In some cases, particularly those that involve sensitive numerical calculations where round-off

---

[1] We will make many multiple references to contributed packages. Therefore, to avoid overwhelming the reader with repeated citations, I will generally follow the practice of citing R packages only on their first use in the book.

errors may differ from computer to computer, or those involving sequences of pseudorandom (i.e., random appearing—these are discussed further in Section 2.3.2) numbers, your results may be somewhat different from mine.

2.2 R Basics

If you have RStudio ready to go, open it and let's begin. You don't need to open R because it will open within RStudio. The RStudio window should contain four panes. If it only contains three, click *File->New File->R Script* and the fourth will appear. Reading clockwise from the upper left, they should say something like "Unititled1," "Environment," "Files" or "Packages," and "Console." These are called, respectively, the *Script, Workspace, Plot/Help,* and *Console.*

Upon starting RStudio, in the Console window you will see some information (which is worth reading the first time you use R) followed by a prompt that, by default, looks like this:

```
>
```

The prompt signals the user to enter statements that tell R what to do. Like most programs, you can exit RStudio by selecting *File -> Exit.* You can also exit from RStudio by typing q() or quit() at the command prompt in the Console. The parentheses indicate that q() is a function. It is a general property of R that once a function is identified unambiguously, all further identifying letters are superfluous, and quit() is the only base function that begins with the letter *q.* To stop R from executing a statement, hit the *Esc* key (see Exercise 2.1) or the Stop sign in the upper right corner of the Console. Another very useful function for the beginner is help.start(). If you type this at the command prompt (or in the script window as discussed below), you will obtain in the RStudio Plot/Help window a list of information useful for getting started in R. Don't forget to include the parentheses.

R programs are a series of *statements*, each of which is either an *assignment* or an *evaluation.* Here is a sequence of statements.

```
> # Assign the value of 3 to w
> w <- 3
> w # Display the value of w
[1]  3
```

Like all listings in the book, this shows the actual execution of the statements in the R environment. We will display code listings using courier font, with prompts and input in normal font and R output in bold. These code listings are copied directly from the R screen. When typing statements into RStudio, you should of course not type the prompt character >. Also, to repeat, when running the R programs in this book you should use the code downloaded from the companion website http://psfaculty.plantsciences.ucdavis.edu/plant/sda2.htm rather than copying the statements from the text, because not all of the statements are included in the text, only those necessary to present the concept being discussed. In fact, you should open the file 2.2.r right now so you can see the actual code. In RStudio, click *File->Open* and navigate to this file. It will appear in the Script window.

Note that in RStudio the first line and part of the third line are displayed in green. The first line in the code is a *comment*, which is any text preceded by a # symbol. Everything in the line after the comment symbol # is ignored by R. We will emphasize comments by italicizing them. Comments are very useful in making long code sequences understandable. They can be placed anywhere in a line, so that for example in line 3 above, a short comment is placed after the statement that causes the display of the value of w. Line 2 of the code above is an *assignment* statement, which creates w and assigns to it the value 3. All assignment statements in this book will have the form A <- B where A is created by the assignment and B is the value assigned to A. The symbol in between them is a < followed by a – to create something that looks like an arrow. There are other ways to make an assignment, but to avoid confusion we will not even mention them. Line 3 is an *evaluation* statement. This implicitly and automatically invokes the R function print(), which prints the result of the evaluation. In this book we will follow the practice of indicating a function in the text body by placing parentheses after it.

Now click and drag to select the entire set of three lines in the Script window (they should turn blue) and then click on the Run icon at the top of the Script window (it looks like a box with a green arrow and should be about the fifth icon from the left). The three lines should show up in the Console (the window below the Script window), together with the fourth line showing the value of w. The Console contains the actual running implementation of R. The symbol **[1]** in line 4 is explained below. Also, the Workspace window (upper left right corner) should show that w has been assigned the value 3. Finally, the bottom line of the Console contains another command prompt, showing that it is ready to accept the next commands.

R is an *object oriented* language. This means that everything in R is an object that belongs to some class. Let's see the class to which the object w belongs. Repeat the process of the previous paragraph to execute the following statement.

```
> class(w) # Display the object class of w
[1] "numeric"
```

This is an evaluation statement that generates as a response the class to which the object w belongs, which is numeric. Now execute the statement ?class to see in the Plot/Help window (lower right corner) an explanation of the class() function. Much of R is actually written in R. To see the source code of the function class(), execute class (without the parentheses). The source code of class() is not especially revealing. To see some actual source code, execute sd to generate the code of sd(), a function that computes the standard deviation. Even if you are new to R, you should recognize some things that look familiar in this function.

When executing an assignment statement, R must determine the class of the object to which the assignment is made. In the example above, R recognizes that 3 is a number and accordingly w is created as a member of the class numeric. Another commonly used class is character, which is indicated by single or double quotes around the symbol.

```
> # Assign z a character value
> z <- "a"
> z
[1] "a"
> class(z)
[1] "character"
```

The quotes are necessary. Although R recognizes the symbol 3 as a number, it does not recognize the symbol a without the quotes as a character. Let's try it.

```
> z <- a
Error: object 'a' not found
```

Since R is an object-oriented language, everything in R is an object that is a member of some class.

```
> class(class)
[1] "function"
```

As a programming language, R has many features that are unusual, if not unique. For this reason, experienced programmers who have worked with other languages such as C++ sometimes have problems adjusting. However, R also has much in common with these and other languages. Like many programming languages, R operates on *arrays*. These are data structures that have one or more indices. In this book, arrays will generally be *vectors* (one dimensional arrays) or *matrices* (two dimensional arrays, see Appendix A.1). We will, however, have occasion to use a three-dimensional array, which can be visualized as a stack of matrices, so that it looks like a Rubik's cube. Some programming languages, such as C++ and Fortran, require that the size of an array always be declared before that array has values assigned to it. R does not generally require this.

One way of creating an array is to use the function c() (for *concatenate*).

```
> # Example of the concatenate function
> z <- c(1, 7, 6.2, 4.5, -27, 1.5e2, 7251360203, w, 2*w, w^2, -27/w)
> z
 [1]     1.0              7.0     6.2     4.5     -27.0
 [6]   150.0     7251360203.0     3.0     6.0       9.0
[11]    -9.0
```

The array z has a total of 11 elements. The sixth is expressed in the assignment statement in exponential notation, and the last four involve the already defined object w, whose value is 3. When an array is displayed in R, each line begins with the index of the first element displayed in that line, so that for example the sixth element of z is 150. The object w is itself an array with one element, which is the reason that the line 4 in the code sequence at the start of this section begins with [1].

If no specific indices of an array are indicated, R acts on the entire array. We can isolate one element, or a group of elements, of an array by identifying their indices using square brackets.

```
> z[6]
[1] 150
> z[c(6,8)]
[1] 150   3
```

In the second statement the concatenate function c() is used inside the brackets to identify two particular indices. Specifying the negative of an index removes that element.

```
> # Remove the 6th element of z
> z[-6]
 [1]               1.0    7.0    6.2    4.5    -27.0
 [6] 7251360203.0    3.0    6.0    9.0     -9.0
> z[-(6:11)]
[1] 1.0    7.0    6.2    4.5    -27.0
```

In R the expression m:n is interpreted as the sequence beginning at *m* and incremented by plus or minus 1 until *n* is reached or the next number would pass *n*.

If the class of the array cannot be determined from the right-hand side of an assignment statement, then the array must be constructed explicitly.

```
> # This causes an error
> v[1] <- 4
Error in v[1] <- 4: object 'v' not found
> # This works
> v <- numeric()
> v[1] <- 4
> v[2] <- 6.2
> v
[1] 4.0 6.2
```

Now we will assign the value of v to z and display the value of z in one statement.

```
> # Assign v to z and display the result
> print(z <- v)
[1] 4.0 6.2
```

The statement z <- v executes normally because the class of z can be determined: it is the same as the class of v. R allows us to display the result of the assignment statement without having to type the object on a subsequent line by including the assignment statement itself as an argument in the function print().

Here are examples of some R vector and matrix operations. First we create and display a vector w with components 1 through 30.

```
> print(w <- 1:30)
 [1]  1  2  3  4  5  6  7  8  9 10 11 12 13 14 15 16 17 18 19 20
[21] 21 22 23 24 25 26 27 28 29 30
```

Similarly, we can execute

```
> print(z <- 30:1)
 [1] 30 29 28 27 26 25 24 23 22 21 20 19 18 17 16 15 14 13 12 11 10 9
[23]  8  7  6  5  4  3  2  1
```

and get a decreasing sequence. Other sequences can be generated by the function seq(). Thus, we could have executed w <- seq(1,30) and gotten the same result as that above. To learn more about the function seq(), execute ?seq. The resulting Help screen shows us several things. One is that seq() is one of a group of related functions. A second is that seq() has several arguments. Each of these arguments has a name and some have default values. Actually, in the case of seq(), *all* of the arguments have default values, but this is not always true. When an argument has a default value, if you do not specify a different value to override the default, the function will execute with the assigned default value.

Thus, if you just type seq(), with no arguments assigned, you get the number 1, since this is the default of the from and the to arguments. If they are not specified, arguments are assigned in the order that they appear in the help file, so that if you type seq(1,30,2) you will get a sequence from 1 to 30 by 2. This is the same as specifying seq(from = 1, to = 30, by = 2). If you specify the name of an argument, this overrides the order in the Help screen, so that seq(to = 30, from = 1) gives the same result as seq(1,30).

Now let's return to arrays. Indices in arrays of more than one dimension are separated by a comma. The first index of a matrix (an array of two dimensions) references the row, and the second index references the column. Appendix A.1 contains a discussion of matrix algebra. The reader unfamiliar with this subject is encouraged to consult the appendix as needed. We now create the matrix *a* and assign values.

```
> print(a <- matrix(data = 1:9, nrow = 3))
     [,1]  [,2]  [,3]
[1,]    1     4     7
[2,]    2     5     8
[3,]    3     6     9
> a[1,]
[1] 1 4 7
> 2 * a[1,]
[1] 2 8 14
```

Note that by default R assigns values to a matrix by columns. The statement a[1,] causes an evaluation of the first row of the matrix a; since no column is specified, all are evaluated. The statement 2 * a[1,] demonstrates the application of an operation (multiplication by 2) to a set of values of an array. If no specific index is given, then an R operation is applied to all indices. The object created by the operation 2 * a[1,] is printed but not saved. To retain it, it is necessary to use an assignment statement, for example b <- 2 * a[1,]. If you are typing statements like these in which there is a lot of repetition from one statement to the next, a very convenient feature of R is that if you press the *up* arrow while in the Console, R will repeat the last statement executed. Try it! As was mentioned in Section 2.1, R runs in the interactive mode. Running lines of code from the RStudio script window is much more like the batch mode, and when you are programming you might often find it convenient to work directly in the Console.

One useful function for creating and augmenting matrices is cbind() (short for *column bind*).

```
> w <- c(1,3,6)
> z <- (1:3)
> cbind(w, z)
     w z
[1,] 1 1
[2,] 3 2
[3,] 6 3
```

The names w and z are preserved in the matrix output. It is possible to assign names to both rows and columns of a matrix using the function dimnames(). R functions like cbind() that match pairs of vectors generally return a value even if the numbers of elements of the two vectors are not equal. The elements of the smaller vector are recycled to make up the difference. There is function rbind() that performs a similar function by rows.

Finally, let's try installing the contributed package sf mentioned in Section 2.1. If you are working in RStudio, simply go to the Plot/Help window and click the Packages tab.

Then click *Install*. You will be prompted for the packages to install. We will be needing the packages sf, rgdal, raster, and maptools very soon, so go ahead and install these. You should be able to see them in the Packages tab. However, this does not load the packages into your R session. To load, for example, the sf package you must type library(sf). Many packages depend on the presence of other packages. Occasionally you may get an error message that reads "there is no package called 'XXX' " where "XXX" is the name of some package. This probably resulted from a problem in downloading, so all you have to do is go back and download this package. You only *install* a package once, but you have to *load* it using library() every time you use it. Also, you can check the dependencies of a package. Try the following.

```
> library(maptools)
> library(tools)
> package_dependencies("maptools")
$maptools
[1] "sp"      "foreign"    "methods"    "grid"    "lattice"    "stats"
[7] "utils"   "grDevices"
```

The package maptools depends on a lot of other packages. In particular, it depends on sp, which is the fundamental package for much of the spatial work in this book. Therefore, if you load maptools, you automatically also load, among others, sp.

This may be a good time to take a break. Recall that one way of exiting R and RStudio is to type and execute q(). This time, try typing Q(). It doesn't work. R is *case sensitive*. When you type q(), or exit in any other way, you get a message asking if you want to save the workspace image. If you say yes, then the next time you start R all of the objects to which you have assigned values (in this case w, z, and a) will retain these values, so that you don't have to recompute them the next time you run a program R. This can be very convenient, but it is a double-edged sword. I have found that if I retain my workspace, I soon forget what is in it and as a result I am much more likely to create a program with bugs. You can always clear your workspace in RStudio with the menu selection *Session -> Clear workspace*, and it is not a bad idea to do this whenever it is convenient.

2.3 Programming Concepts

2.3.1 Looping and Branching

In order to use R at the level of sophistication required to implement the material covered this book, it will be necessary to write computer code. Computer programs often contain two types of expression that we will use extensively, *iteration* statements and *conditional* statements. Iteration, which is also known as *looping*, is the repetition of a set of instructions, often modified by an index. As an example, we can create the same vectors w and z as in the previous section, but using a for loop in the same way as conventional programming languages like C++ and Fortran.

```
> w <- numeric(30)
> z <- numeric(30)
> for(i in 1:30){
+    w[i] <- i
```

```
+    z[i] <- 30 - i +1
+}
> w
 [1]  1  2  3  4  5  6  7  8  9 10 11 12 13 14 15 16 17 18 19 20 21
[22] 22 23 24 25 26 27 28 29 30
```

The *index* i is explicitly incremented from 1 to 30 and the array elements indexed by i are assigned values individually, as opposed to the code in the previous section, in which the array values are all assigned in the same operation. This code also demonstrates *grouped expressions* in R. A grouped expression is an expression consisting of a collection of individual statements that are evaluated together. A grouped expression in R is delimited by brackets {}, so that no statement in the group is carried out until all statements have been entered. In the code above, line 3 ends in a left bracket. This indicates the start of the group. The + signs at the beginning of the next three lines indicates that R is withholding evaluation until the right bracket in line 6, which indicates the end of the expression. We will follow the convention of indenting grouped expressions and putting the closing brackets on separate lines. This makes it easier to distinguish the beginning and the end of each expression. RStudio and TINN-R do this automatically.

The second construct common to computer programs is the conditional statement, also known as *branching*. Conditional statements evaluate an expression as either true or false and execute a subsequent statement or statements based on the outcome of this evaluation (Hartling et al., 1983, p. 64). Here is an example in R that, although the bracket placement looks a little different from what one might expect, will otherwise be familiar to any traditional programmer.

```
> w <- 1:10
> for (i in 1:10){
+   {if(w[i] > 5)
+       w[i] <- 20
+   else
+       w[i] <- 0
+   }
+ }
> w
 [1]  0  0  0  0  0 20 20 20 20 20
```

The conditional if–else statement says that if the expression in brackets evaluates as TRUE, then the first statement or set of statements is executed; otherwise the second. The inner set of brackets is placed in front of the if term to ensure that the entire statement is executed correctly. This is the part that might look a little unusual to an experienced programmer, and it is not strictly necessary, but is good defensive programming. Because R operates in the interactive mode, if a line of code appears to be complete, R will execute it before going on to the next line. Therefore, R may think you are done typing an expression when you are not. In the present example, the expression evaluates correctly without the extra brackets, but I generally include them to be on the safe side.

As with looping, many branching operations in R can be carried out without the explicit use of an if statement. The code below carries out the same operation as that above.

```
> w <- 1:10
> w > 5
 [1] FALSE FALSE FALSE FALSE FALSE  TRUE  TRUE  TRUE  TRUE  TRUE
```

```
> w[w > 5]
[1]  6  7  8  9 10
> w[w > 5] <- 20
> w[w <= 5] <- 0
> w
[1]  0  0  0  0  0 20 20 20 20 20
```

Unlike many languages, R can evaluate a conditional expression contained in a subscript, so that the expression w[w > 5] in line 2 is interpreted as "all elements of the array *w* whose value is greater than 5." The expression w > 5 returns a logical vector, each of whose elements evaluates to the truth or falsehood of the expression for that element of w. Again unlike many languages, R can correctly interpret operations involving mixtures of arrays and scalars. Thus, the assignment w[w > 5] <- 20 assigns the value 20 to each element satisfying the conditional expression w[i] > 5. A word of caution is in order here. This code sequence does not work in this way for all values of the replacement quantities. See Exercise 2.5.

2.3.2 Functional Programming

Many things in R that do not look like functions really are. For example, the left square bracket in the expression w[i] is a function. To see an explanation, type ?"[" in the Console. In this context, the lesson of the last two code sequences of the previous section is that an operation involving looping or branching may be carried out more directly in R by using a function. This feature is not unique to those two operations. R supports *functional programming*, which permits the programmer to replace many instances of looping and branching with applications of a function or operation. This is hard on old-school programmers (like me) but frequently makes for much more concise and legible code.

Many programming operations require repetition and in other languages involve a for loop. In R, these operations often make use of variants of a function called apply(). These can be illustrated by application of one member of the apply() family, the function tapply() (short for *table apply*). We will illustrate functional programming by performing a fairly complex task: we generate a sequence of 20 random appearing numbers from a standard normal distribution, divide them into four sequential groups of five each, and then compute the mean of each group.

The first step is to generate and display a sequence of *pseudorandom numbers*, that is, a sequence of numbers that, although it is deterministic, has many statistical properties similar to those of a sequence of random numbers (Rubinstein, 1981, p. 38; Venables and Ripley, 2002, p. 111). The actual numbers generated are members of a sequence of integers, each of which determines the value of the next. The first number in the sequence is called the *seed*. Since the sequence is deterministic and since each number determines the next, setting the value of the seed establishes the entire sequence. Controlling this value, which is done by the function set.seed(), ensures that the same sequence of numbers is generated each time the program is executed. This is useful for debugging and comparison purposes. To control the first number, one includes a numeric argument in the function set.seed(), such as set.seed(123). The value of the argument of set.seed() does not generally matter; I use 123 because it is easy to type. The sequence of pseudorandom integers is transformed by the generating function so that the resulting sequence of numbers approximates a sequence of random variables with a particular distribution. To generate an approximately normally distributed sequence with mean zero and variance 1, one uses the generating function rnorm() as in line 3 below. Functions rpois(), runif(), rbinom(), etc. are also available.

```
> # Generate a sequence of 20 pseudorandom numbers
> set.seed(123) # Set random seed
> print(w <- rnorm(20), digits = 2)
 [1] -0.56 -0.23 1.56 0.07  0.13 1.72 0.46 -1.27 -0.69 -0.45
[11]  1.22  0.36 0.40 0.11 -0.56 1.79 0.50 -1.97  0.70 -0.47
```

Sometimes print() gives you more precision than you want. You can limit the number of significant digits printed using the argument digits (this does not affect the computation, only the printing). In working through the code, remember that a complete description of a function can be obtained by typing that function preceded by a question mark (e.g., ?set. seed) at the R statement prompt.

The next step is to create an index array that has a sequence of five ones, then five twos, then five threes, and then five fours. We do this initially in two steps, although ultimately we will squeeze it down to one. First, we use the function rep() to create a replicated sequence z of five subsequences running from 1 to 4.

```
> print(z <- rep(1:4, 5))
 [1] 1 2 3 4 1 2 3 4 1 2 3 4 1 2 3 4 1 2 3 4
```

Next, we use the function sort() to sort these in increasing order creating the index for the four sequences.

```
> print(z <- sort(z))
 [1] 1 1 1 1 1 2 2 2 2 2 3 3 3 3 3 4 4 4 4 4
```

The statement z <- sort(z) indicates that we replace the old object z with a new one rather than creating a new object. We can combine these two steps into one by using the statement z <- sort(rep(1:4, 5)). It is more efficient to nest functions this way, but care must be taken not to overdo it or the code may become difficult to understand.

The R function tapply(w, z, ftn,...) applies the function whose name is ftn to values in array w as indexed by the values in array z. In our application, w is the array w of pseudorandom numbers, z is the array z of index values, and ftn is the function mean(). Thus the mean is computed over all elements of w whose index is the same as those of the of the elements of the vector z having the values 1, then over all w values whose index is the same as those of the vector z whose value is 2, and so forth.

```
> # Apply the mean function to w indexed by z
> tapply(w, z, mean)
          1           2           3           4
 0.19357026 -0.04431897 0.30790173 0.10934219
```

Note that the third argument is just the name of the function, without any parentheses. The actual operation takes only four lines, and is very clean and easy to understand. Here they are in sequence.

```
> set.seed(123)
> w <- rnorm(20)
> z <- sort(rep(1:4, 5))
> tapply(w, z, mean)
```

This could actually be compressed into just two lines:

```
> set.seed(123)
> tapply(rnorm(20), sort(rep(1:4, 5)), mean)
```

but this is an example of the risk of making the code less intelligible by compressing it. The earlier chapters of this book will avoid compressing code like this, but the later chapters will assume that the reader is sufficiently familiar with R syntax to put up with a bit of compression.

Code carrying out the same operation with explicit looping and branching would be much longer and more obscure. That is the upside of functional programming. The biggest downside of functional programming is that it makes use of a large number of separate functions, often housed in different contributed packages. This makes the learning curve longer and enforces continued use of R in order not to forget all the functions and their actions. With traditional languages one can stop programming for a while, take it up again, and remember with ease what a *for* loop or an *if-then-else* statement does. It is because of this downside that I decided to include an R "thesaurus" as Appendix C.

2.4 Handling Data in R

2.4.1 Data Structures in R

For our purposes, the fundamental object for storing data in R is the *data frame*. Venables et al. (2006, p. 27) state that "A data frame may for many purposes be regarded as a matrix with columns possibly of differing modes and attributes. It may be displayed in matrix form, and its rows and columns extracted using matrix indexing conventions." In describing the contents of a data frame we will use two standard terms from GIS theory: *data field* and *data record* (Lo and Yeung, 2007, p. 76). A data field is a specific attribute item such as percent clay content of a soil sample. When data are displayed in tabular form, the data fields are ordinarily the columns. A data record is the collection of the values of all the data fields for one occurrence of the data. When data are displayed in tabular form, the data records are ordinarily the rows.

Data frame properties may best be illustrated by two examples. The first is a simple data frame from Data Set 1 that contains the results of a set of surveys at 21 locations along the Sacramento River to identify the presence and estimate the number of individuals of a population of western yellow-billed cuckoos (Section 1.4.1). In the next section, we show how to read this data set into an R data frame called data.Set1.obs. Once this is done, one can use the function str() (for *structure*) to display the structure of the object.

```
> str(data.Set1.obs)
'data.frame' :    21 obs. of   5 variables:
 $ ID:         int  1 2 3 4 5 6 7 8 9 10...
 $ PresAbs   : int  1 0 0 1 1 0 0 0 0 0...
 $ Abund     : int  16 0 0 2 2 0 0 0 0 0...
 $ Easting   : num  577355 578239 578056 579635 581761...
 $ Northing  : num  4418523 4418423 4417806 4414752 4415052...
```

The components of the data frame and their first few values are displayed. These are: ID, the identification number of the location; PresAbs, 1 if one or more birds were observed at this location and 0 otherwise; Abund, the estimated total number of birds observed; Easting; the UTM Easting in Zone 10N of the observation site; and Northing, the UTM Northing.

Data frames can contain much more complex information. In the code below, the object data.Set4.1 is a data frame containing the point sample data from Field 1 of Data Set 4, one of the two fields in the precision agriculture experiment (Section 1.4.4).

```
> class(data.Set4.1)
[1] "data.frame"
> names(data.Set4.1)
 [1] "ID"        "Row"       "Column"    "Easting"   "Northing"
 [6] "RowCol"    "Sand"      "Silt"      "Clay"      "SoilpH"
[11] "SoilTOC"   "SoilTN"    "SoilP"     "SoilK"     "CropDens"
[16] "Weeds"     "Disease"   "SPAD"      "GrainProt" "LeafN"
[21] "FLN"       "EM38F425"  "EM38B425"  "EM38F520"  "EM38B520"
> data.Set4.1$EM38F425[20:30]
 [1] 73 NA 74 63 69 79 74 77 NA 75 79
> data.Set4.1[20:30,25]
 [1] 94 86 80 69 86 96 94 98 89 86 81
> data.Set4.1[25,25]
[1] 96
> data.Set4.1$RowCol[1:5]
[1] 1a 1b 1c 1d 1e
86 Levels: 10a 10b 10c 10d 10e 10f 11a 11b 11c 11d 11e 11f 12a... 9f
>
```

The call to class(), as usual, produces the class of the object. The call to the function names() produces the set of names of each data field in the data frame, displayed as an array. Once again, the numbers in square brackets at the beginning of each line correspond to the position in the array of the first element in the row, so that, for example, Northing is the fifth element and Sand is the seventh. Thus, if the data frame is considered as a matrix, the Sand data field forms the seventh column. The contents of a column may be accessed using the $ operator, which identifies a named data field in a data frame, in the example above rows 20 through 30 of EM38B425 are displayed. These same contents may also be accessed as in the next statement, in which the data frame is treated as a matrix. The symbol NA, which stands for "not available," refers to a data record with a missing or null value. The contents of the twenty-fifth row (i.e., twenty-fifth data record) of the twenty-fifth column (i.e., twenty-fifth data field) may also be accessed in matrix fashion as is done with the statement data.Set4.1[25,25]. A data frame can also hold non-numeric information such as the alphanumeric code for the combined row and column number of the sample points in the sample grid data.Set4.1$RowCol[1:5].

The last line of R output in this code listing refers to *levels*. We can see where this comes from by applying the class() function to two data fields in data.Set4.1.

```
> class(data.Set4.1$GrainProt)
[1] "numeric"
> class(data.Set4.1$RowCol)
[1] "factor"
```

The values of the data field RowCol are alphanumeric. The function read.csv(), which was used to read the data and which is described in the next section, automatically assigns the class factor to any data field whose values are not purely numerical. Venables et al. (2006, p. 16) define a *factor* as "a vector object used to specify a discrete classification (grouping) of the components." The collection of all individual values is called the *levels*. Any allowable symbol, be it numerical, alphabetical, or alphanumeric, may be converted into a factor:

```
> print(w <- c("A", "A", "A", "B", "B", "B"))
[1] "A" "A" "A" "B" "B" "B"
> class(w)
[1] "character"
> print(z <- as.factor(w))
[1] A A A B B B
Levels: A B
> class(z)
[1] "factor"
```

Note that the quotation marks are absent from the elements of z; this is because these elements are not characters but factors. The use of the function as.factor() is an example of *coercion*, in which an object is converted from one class to another. In R, functions that begin with "as" generally involve coercion.

In addition to the data frame, a second type of data storage object that we shall have occasion to use is the *list*. Technically, a data frame is a particular type of list (Venables et al., 2006), but viewed from our perspective these two classes of object will have very little in common, so forget that you just read that. Again quoting Venables et al. (2006, p. 26), "A list is an ordered collection of objects consisting of objects known as its components." Lists can be created using *constructor functions*; here is an example:

```
> list.demo <- list(crop = "wheat", yield = "7900", fert.applied =
+    c("N", "P", "K"), fert.rate = c(150, 25, 15))
> list.demo[[1]]
[1] "wheat"
> list.demo$crop
[1] "wheat"
> list.demo[[4]]
[1] 150  25  15
> list.demo[[4]][1]
[1] 150
```

This shows the use of the simple constructor function list() to create a list. To access the *n*th component of a list one uses double square brackets (lines 3, 7, and 9). A list can contain character string components, vector components, and other types of components as well. The $ operator can be used to identify named components just as it can with a data frame (line 5). If a component itself has elements, then these elements can themselves be singled out (line 9).

Finally, suppose we want to construct a data frame from scratch. The constructor function used for this purpose is, not surprisingly, data.frame(). We can use elements of an existing data frame, or create new values. Here is a simple example.

```
> data.ex <- data.frame(Char = data.Set4.1$RowCol[1:3],
+    Num = seq(1,5,2))
> data.ex
  Char Num
1   1a   1
2   1b   3
3   1c   5
```

2.4.2 Basic Data Input and Output

In addition to carrying out assignments and evaluating expressions, a computer program must be able to input and output data. A good deal of the data that we will be using in this book is spatial data with a very specialized format, and we will defer discussion of the specifics of spatial data input and output until later. However, we introduce at this point some of the basic concepts and expressions. This material is specific to Windows, but if you work in another environment, you should be able to translate it. One important concept is the *working directory*. This is the directory on the disk that you tell R to use as a default for input and output. Suppose, for example, that you store all of your R data for a project in a single directory. If you make this your working directory, then R will automatically go to it by default to read from or write to the disk. The function setwd() is used to set a particular directory (which must already exist) as the working directory. Directory names must be enclosed in quotes, and subdirectories must be indicated with a *double* backslash (or forward slash). For example, suppose you have set up the folders in Windows as described in Section 2.1.2. To establish the Windows directory *C:\rdata\SDA2\data* as the working directory you would enter the following statement.

```
> setwd("c:\\rdata\\SDA2\\data")
```

Don't forget the quotation marks! If you prefer to put your project directory in your *My Documents* folder, you can do that too. For example, on my computer the *My Documents* folder is a directory called *C:\documents and settings\replant\my documents*. To establish a subfolder of this folder as a working directory I would enter the following.

```
> setwd("c:documents and settings\\replant
+    \\my documents\\rdata\\SDA2\\data")
```

R is case sensitive, but Windows is not, so character strings passed directly to Windows, such as file and folder names, are not case sensitive. Thus, this statement would work equally well.

```
> setwd("c:documents and settings\\replant
+    \\my documents\\rdata\\sda2\\data")
```

Once a working directory has been set, R can access its subdirectories directly. This feature is convenient for our purposes in that it permits the input and output statements in this book to be independent of specific directory structure. In this book, we will not display the calls to setwd(), since my working directory will be different from yours.

All attribute data and point sampling data used in this book are stored in *comma separated variable* or csv files. These have the advantage that they are very simple and can be accessed and edited using both spreadsheet and word processing software. Suppose you want to input a csv-formatted file named *file1.csv* in the directory *C:\rdata\SDA2\data\newdata*, and that *C:\rdata\SDA2\data* has already been set as the working directory. You would type read.csv("newdata\\file1.csv") to read the file. Often the file will have a *header*, that is, the first line will contain the names of the data fields. When this is the case, the appropriate expression is read.csv("data\\newdata\\file1.csv", header = TRUE).

The headers are input into R and are therefore are case sensitive.

Writing a data frame as a csv file is equally simple. Suppose we have a data frame called my.data, and we want to write this to the file *file2.csv* in the subdirectory *C:\rdata\SDA2\data\newdata*. Suppose again that *C:\rdata\SDA2\data* has already been set as the working directory. We would then type the statement write.csv(my.data, "newdata\\file2. csv"). Don't forget the double slashes and the quotation marks.

We will do all attribute data input and output using csv files because this keeps things simple. R is, however, capable of reading and writing many other file formats. The package foreign contains functions for some of these, and others may be found using the help. search() and RSiteSearch() functions described in Section 2.7.

2.4.3 Spatial Data Structures

There are a number of spatial data structures in contributed R packages, but among the most useful for our purposes are those in the sf, raster, and sp packages. We will present a highly simplified description of these data structures starting with sf, then moving on to sp, and finally to raster. The sf package includes functions that allow one to work with the vector model of spatial data (see Section 1.2.3), so that it can model point data, line data, and polygon data. A very complete description is given by Pebesma (2017). Although line data are very useful in some applications, we will have no call in this text to use them, so we will focus on point and polygon data. The term "sf" stands for "simple features," where the term "feature" is used in the sense of a GIS representation of a spatial entity (Lo and Young, 2007, p. 74). Without going into too much detail, the geometric definition of a "simple feature" is something whose boundary, when drawn, can be represented by a set of points in a plane connected by straight lines. This is generally true of the vector data that are likely to arise from ecological applications.

Referring back to Section 2.4.1, an sf object combines a data frame to store the attribute data with a list to store the information for the geometric coordinates. The easiest way to introduce the sf concept is to construct an sf object that describes a set of point data. We will begin with the data frame data.Set1.obs that we just saw in Section 2.4.1. It contains data associated with yellow-billed cuckoo observations in Data Set 1. If you have set up your directory structure as described in Section 2.1.2, this file would be located in the file *C:\rdata\SDA2\data\set1\set1\obspts.csv*. If you have executed the statement setwd("c:\\ rdata\\SDA2\\data"), then to load the data into R as you would execute the following statement, taken from the file *Appendix\B.1 Set1 Input.r*.

```
> data.Set1.obs <- read.csv("set1\\obspts.csv", header = TRUE)
```

Now that the data frame is loaded, let's look again at its class and structure.

```
> class(data.Set1.obs)
[1] "data.frame"
> str(data.Set1.obs)
'data.frame' :     21 obs. of 5 variables:
 $ ID       : int  1 2 3 4 5 6 7 8 9 10 ...
 $ PresAbs  : int  1 0 0 1 1 0 0 0 0 0 ...
 $ Abund    : int  16 0 0 2 2 0 0 0 0 0 ...
 $ Easting  : num  577355 578239 578056 579635 581761 ...
 $ Northing : num  4418523 4418423 4417806 4414752 4415052 ...
```

The last two lines are the data fields Easting and Northing, which contain the geometric data. To create an sf object, we first need to load the sf package. If you have not installed sf on your computer, you need to do so now. If you the missed instructions the first time, go back to the end of Section 2.2 to learn how to do this. Once the package is installed on your computer, you load it into R for that session by using the function library().

```
> library(sf)
```

We will generally not explicitly include the call to the function library() in the text when we employ a function from a particular package, but you always need to load that package first. To create an sf object, we specify that these are the coordinates of the points.

```
> data.Set1.sf <- st_as_sf(data.Set1.obs, coords = c("Easting",
    "Northing"))
```

This uses the function st_as _ sf() to create the sf object data.Set1.sf.
  Let's see what the object data.Set1.sf looks like

```
> class(data.Set1.sf)
[1] "sf"             "data.frame"

> str(data.Set1.sf)
Classes 'sf' and 'data.frame':      21 obs. of 4 variables:
 $ ID       : int  1 2 3 4 5 6 7 8 9 10 ...
 $ PresAbs  : int 1 0 0 1 1 0 0 0 0 0 ...
 $ Abund    : int 16 0 0 2 2 0 0 0 0 0 ...
 $ geometry :sfc_POINT of length 21; first list element: Classes 'XY',
 'POINT', 'sfg' num [1:2] 577355 4418523
 - attr(*, "sf_column")= chr "geometry"
 - attr(*, "agr")= Factor w/ 3 levels "constant","aggregate",..: NA NA
   NA
 .. - attr(*, "names")= chr "ID" "PresAbs" "Abund"
```

The object data.Set1.sf is still a data frame, but it has a geometry object attached to it. This is represented in the geometry data field:

```
> str(data.Set1.sf$geometry)
sfc_POINT of length 21; first list element: Classes 'XY', 'POINT',
 'sfg' num [1:2] 577355 4418523
```

The last task remaining to create an R spatial object that we can work with is to assign a map projection. Projections are handled in the package sf in two ways. The first is by using the PROJ.4 Cartographic Projections library originally written by Gerald Evenden (http://trac.osgeo.org/proj/). To establish the projection or coordinate system, we use the function st_crs().

```
> st_crs(data.Set1.sf) <- "+proj=utm +zone=10 +ellps=WGS84"
```

Here "crs" stands for *coordinate reference system*. The projection is specified in the argument, and therefore must be known to the investigator. In this case, it is UTM Zone 10 using the WGS84 ellipsoid. Experienced programmers may find this statement a bit odd in that an assignment is made to a function rather than a function assigning a value to a variable. This is an example of a *replacement function*, which is described in The R Language Definition under Manuals in the R Project website, http://www.r-project.org/. If you don't see what the issue is, or if this is too much information, don't worry about it.

To see what happened, we can type the same function but without any assignment. In this case, st_crs() retrieves the coordinate system.

```
> st_crs(data.Set1.sf)
$epsg
[1] 32610

$proj4string
[1] "+proj=utm +zone=10 +datum=WGS84 +units=m +no_defs"

attr(,"class")
[1] "crs"
```

The data field epsg refers to the epsg code. This is a set of numeric codes created by the European Petroleum Survey Group. If you know your projection (for example UTM 10N), you can go to the site http://www.spatialreference.org/ and search for the corresponding EPSG code. The function st_crs accepts these codes as well as a PROJ4 character string, so you could use

```
> st_crs(data.Set1.sf) <- 32601
```

and get the same result.

To introduce the use of spatial features to deal with polygon data, we will create an sf object describing the boundary of Field 2 of Data Set 4. This is the field shown in Figures 1.2 and 1.3. (This is a simple set of polygon data because there is only one polygon; in Section 5.3.4 we create an sf object with multiple polygons.) The creation of a polygon describing the boundary of an area to be sampled requires a set of geographic coordinates describing that boundary. These can be obtained by walking around the site with a GPS or, possibly, by obtaining coordinates from a georeferenced remotely sensed image, either a privately obtained image or a publicly available one from a database such as Google Earth® or Terraserver®. If the boundary is not defined very precisely, one can use the function locator() (see Section 7.3) to obtain the coordinates quite quickly from a map or image. In the case of Field 2 the coordinates were originally obtained using a GPS and modified slightly for use in this book. They are set to an even multiple of five and so that the length of the field in the east-west direction is twice that in the north-south direction.

```
> N <- 4267873
> S <- 4267483
> E <- 592860
> W <- 592080
> N - S
[1] 390
> E - W
[1] 780
```

We first put the boundary values into a matrix coords.mat that holds the coordinates describing the polygon. By default, matrices in R are created by columns, so that in this case the first four members of the sequence form the first column and the second four members form the last column. Because the region is rectangular, the coordinates of its corners are specified by four numbers. In the creation of a polygon with n vertices, a total of $n + 1$ coordinate pairs are assigned, with the first coordinate pair being repeated as the last. Thus, in this case five vertices are specified.

```
> print(coords.mat <- matrix(c(W, E,E, W,W, N,N, S,S, N),
+    ncol = 2))
        [,1]      [,2]
[1,]  592080  4267873
[2,]  592860  4267873
[3,]  592860  4267483
[4,]  592080  4267483
[5,]  592080  4267873
```

Since the geometry of an `sf` object is specified as a list, we use the constructor function `list()` to create a list containing the boundary points.

```
> coords.lst <- list(coords.mat)
```

The next step is to tell the system that we are creating a polygon rather than a set of points whose first is the same as the last. This is done by creating an `sfc` object, which is called a *list-column*.

```
> coords.pol = st_sfc(st_polygon(coords.lst))
```

Next, we once again apply the function `st_sf` to create the `sf` polygon object describing the boundary. We assign the single cell in the polygon an attribute z with the value 1.

```
> Set42bdry = st_sf(z = 1, coords.pol))
```

Once again, we must assign a coordinate system to the boundary using `st_crs()`. This time we will use the EPSG code.

```
> st_crs(Set42bdry) <- 32601
```

Section 2.6.2 contains a more thorough introduction to the construction of maps in R using `sf` objects. At this point, however, it is a good idea to take a look at the polygon we have created. We can do this with a simple call to the function `plot()`.

```
> plot(Set42bdry)
```

A plot of the boundary should appear in the Plot/Help window of Rstudio.

We need to save the boundary that we have created as an ESRI shapefile for later use. This is the first of many such objects that we will need to save. The output of `sf` objects is done via the function `st_write()`. As in Section 2.1.2, let us assume that you have set your working directory using `setwd()`. Suppose you have created a subfolder of this directory called *created*. The function call would then be

```
> st_write(Set42bdry, "created\\Set42bdry.shp")
```

**Writing layer `Set42bdry' to data source `created\Set42bdry.shp' using
driver `ESRI Shapefile'
features: 1
fields: 1
geometry type: Polygon**

The function st_write() detects the extension *.shp* in the filename and appropriately creates a shapefile.

The input of data into sf objects is carried out using the function st_read(). We can illustrate this by reading the shapefile containing the land cover data shown in Figure 1.5. This is stored in the shapefile *landcover.shp*. By the way, if you are unfamiliar with ESRI shapefiles, they actually consist of at least three separate files, each with the same filename and a different extension. We will follow the usual convention of referring to only one file, but be aware that for these files to function, all of them must be present. The correct function call is

```
> data.Set1.landcover <- st_read("Set1\\landcover.shp")
```

and the R response is not reproduced here to save space. By the way, in assigning names to objects, I tend to be quite verbose, with the idea that this helps to avoid confusion as to what these objects represent. The feature of RStudio that it automatically fills in the names of variables after a few key strokes comes in handy here. Anyway, we can examine the structure of data.Set1.landcover in the usual way.

```
> str(data.Set1.landcover)
Classes 'sf' and 'data.frame':  1811 obs. of  7 variables:
 $ SP_ID      : Factor w/ 1811 levels "0","1","10","100",..: 1 2 924 ...
 $ Area       : num  34748 21338 183208 9626 3889...
 $ Perimeter  : num  717 834 2128 1014 335...
 $ VegType    : Factor w/ 10 levels "a","c","d","f",..:2 9 2 9 5 8 5  ...
 $ HeightClas : Factor w/ 4 levels "0","h","l","m":1 4 1 3 1 1 1 1 4  ...
 $ CoverClass : Factor w/ 5 levels "0","d","m","p",..: 1 2 1 2 1 1 1  ...
 $ geometry   :sfc_POLYGON of length 1811; first list element: List of 1
  ..$ : num [1:17, 1:2] 576493 576691 576704 576708 576697...
  ..- attr(*, "class")= chr "XY" "POLYGON" "sfg"
 - attr(*, "sf_column")= chr "geometry"
 - attr(*, "agr")= Factor w/ 3 levels "constant","aggregate",..: NA NA NA
NA NA NA
  ..- attr(*, "names")= chr "SP_ID" "Area" "Perimeter" "VegType"...
```

We can obtain nice plots of all of the attributes of data.Set1.landcover with a simple call to plot(). In general, the results of these calls to plot() are not shown, but the calls are included in the code.

We now move on to the sp package. Although sf may someday replace sp, as of the time of the writing of this book, a complete reliance on sf is not possible. One reason for this is that the raster package, discussed below, works with sp. A second reason is that we will also make extensive use of the spdep and maptools packages (Bivand et al., 2013; Bivand and Piras, 2015), and these work with sp objects.

To begin with the sp package, we will use the coercion function as() to coerce data. Set1.landcover into an sp object.

```
> data.Set1.landcover.sp <- as(data.Set1.landcover, "Spatial")
> str(data.Set1.landcover.sp)
Formal class 'SpatialPolygonsDataFrame' [package "sp"] with 5 slots
  ..@ data       :'data.frame': 1811 obs. of 6 variables:
  .. ..$ SP_ID    : Factor w/ 1811 levels "0","1","10","100",..: 1 2
924 1035 1146 1257 1368 1479 1590 1701...
  .. ..$ Area     : num [1:1811] 34748 21338 183208 9626 3889...
  .. ..$ Perimeter : num [1:1811] 717 834 2128 1014 335...
  .. ..$ VegType  : Factor w/ 10 levels "a","c","d","f",..: 2 9 2...
  .. ..$ HeightClas: Factor w/ 4 levels "0","h","l","m": 1 4 1 3 1...
  .. ..$ CoverClass: Factor w/ 5 levels "0","d","m","p",..: 1 1 2 1...
  ..@ polygons  :List of 1811
  .. ..$:Formal class 'Polygons' [package "sp"] with 5 slots
  .. .. .. ..@ Polygons:List of 1
  .. .. .. .. ..$:Formal class 'Polygon' [package "sp"] with 5 slots
  .. .. .. .. .. .. ..@ labpt  : num [1:2] 576596 4420706
  .. .. .. .. .. .. ..@ area   : num 34748
  .. .. .. .. .. .. ..@ hole   : logi FALSE
  .. .. .. .. .. .. ..@ ringDir: int 1
  .. .. .. .. .. .. ..@ coords : num [1:17, 1:2] 576493 576691 576704
576708 576697...
  .. .. .. ..@ plotOrder: int 1
  .. .. .. ..@ labpt    : num [1:2] 576596 4420706
  .. .. .. ..@ ID       : chr "1"
  .. .. .. ..@ area     : num 34748

          *          *          *
```

You get a lot of output! Only the very beginning is shown here. The object data.Set1.
landcover.sp is a SpatialPointsDataFrame. The data in this object are contained in *slots*.
Without going into any more detail than is necessary, the SpatialPointsDataFrame
itself contains five slots, only the first two of which (data and polygons) are shown here.
The data slot contains the attribute data. To access the data in a slot one can use the function
slot(). The slots themselves are objects that are members of some class.

```
> class(slot(data.Set1.landcover.sp, "data"))
[1] "data.frame"
```

To display the first ten elements of the VegType data field, we can type

```
> slot(data.Set1.landcover.sp, "data")$VegType[1:10]
 [1] c v c v g r g a v l
Levels: a c d f g l o r v w
```

Since slot(data.Set1.landcover.sp,"data") is a data frame, the $ operator may
be appended to it to specify a data field, and then square brackets may be appended to that
to access particular data records. A shorthand expression of the function slot() as used
above is the @ operator, as in the following.

```
> data.Set1.landcover.sp@data$VegType[1:10]
 [1] c v c v g r g a v l
Levels: a c d f g l o r v w
```

The polygons slot contains a list with the geometrical information. It is in the form of a list, with one list element per polygon. Each element of the list is a Polygon object, which itself has slots. Most of these we can ignore, but we will be using two, the labpt slot and the ID slot. Only these are shown here.

```
> data.Set1.landcover.sp@polygons[[1]]
An object of class "Polygons"
Slot "Polygons":
[[1]]
An object of class "Polygon"
Slot "labpt":
[1] 576595.8 4420706.2
           *         *       *
Slot "ID":
[1] "1"
           *         *       *
```

We will use these in later chapters to check that attribute values are at their correct locations. Since the data are in lists, we can use lapply() to extract them. The function lapply() is one of the family of apply() functions discussed in Section 2.3.2. Type ?lapply to see an succinct description of it. We can obtain the ID values as follows.

```
> lapply(data.Set1.landcover.sp@polygons, slot, "ID")
[[1]]
[1] "1"

[[2]]
[1] "2"
           *         *       *
```

In this case, the ID values are in the order of the polygons, but one cannot guarantee that this is always so. Extracting the label points is trickier because they are each a pair of coordinates.

```
> y <- lapply(data.Set1.landcover.sp@polygons, slot, "labpt")
> data.Set1.landcover.sp.loc <- matrix(0, length(y), 2)
> for (i in 1:length(y))
+    data.Set1.landcover.sp.loc[i,] <- unlist(y[[i]])
> data.Set1.landcover.sp.loc
            [,1]      [,2]
  [1,] 576595.8 4420706
  [2,] 576710.6 4420652
  [3,] 577163.5 4420675
           *         *       *
```

To complete our discussion of sp objects we will mention the function coordinates(). This is the sp equivalent of the function st_crs() described above, and, like st_crs(), it can work as a replacement function if it is on the left side of the assignment arrow or it can retrieve the coordinate system in place otherwise.

```
> data.Set1.sp <- data.Set1.obs
> coordinates(data.Set1.sp) <- c("Easting", "Northing")
> proj4string(data.Set1.sp) <- CRS("+proj=utm +zone=10 +ellps=WGS84")
```

There is one subtle but important difference between the sp function coordinates()
and the sf function st_crs(). The function st_crs() creates a new sf object, while the
function coordinates() converts an existing data frame into an sp object.

To work with raster data, we will usually use the raster package (Hijmans, 2016). There
are three fundamental objects in this package: the RasterLayer, the RasterBrick, and
the RasterStack. A RasterLayer represents a single layer of raster data, such as the
infrared reflectance values shown in Figure 1.2b. When working with a grid of remotely
sensed values, however, often one works with data from more than one spectral band. As
described in Section 1.3.4, the data use to create this figure come from a multispectral sensor (Lo and Young, 2007, p. 294) and include three bands, one at the infrared level, one at
the red level, and one at the green level. For this reason, these data cannot be represented
by a single RasterLayer. The RasterBrick accommodates multilevel data sets such
as this by incorporating multiple layers. The data we use is stored in a *tiff* file and can be
loaded into R very easily using the function brick.

```
> # Read the three bands of a tiff image as a raster stack
> library(raster)
> data.4.2.May <- brick("set4\\Set4.20596.tif")
> data.4.2.May
class         : RasterBrick
dimensions    : 621, 617, 383157, 3 (nrow, ncol, ncell, nlayers)
resolution    : 1.902529, 1.902529 (x, y)
extent        : 591965.6,  593139.5, 4267139, 4268321 (xmin, xmax, ymin, ymax)
coord. ref.   : NA
data source   : c:\rdata\Data\Set4\Set4.20596.tif
names         : Set4.20596.1, Set4.20596.2, Set4.20596.3
min values    :            0,            0,            0
max values    :          255,          255,          255
```

There is no coordinate reference system because none has yet been assigned. We can assign
one using the function projection().

```
> projection(data.4.2.May) <- CRS("+proj=utm +zone=10 +ellps=WGS84")
> data.4.2.May
class         : RasterBrick
dimensions    : 621, 617, 383157, 3 (nrow, ncol, ncell, nlayers)
resolution    : 1.902529, 1.902529 (x, y)
extent        : 591965.6,  593139.5, 4267139, 4268321 (xmin, xmax, ymin, ymax)
coord. ref.   : +proj=utm +zone=10 +ellps=WGS84
data source   : c:\aaRdata\Book\Second Edition\Data\Set4\Set4.20596.tif
names         : Set4.20596.1, Set4.20596.2, Set4.20596.3
min values    :            0,            0,            0
max values    :          255,          255,          255
```

You can take a look at all three layers of the RasterBrick using the plot() function.

```
> plot(data.4.2.May)
```

To access one layer of the brick, use double square brackets. For instance, to display only
the first, infrared layer, enter the following.

```
> plot(data.4.2.May[[1]])
```

The `RasterBrick` object permits us to work with an object consisting of multiple layers, but it has an important property. If we look again at Figure 1.2 in its entirety, we can see that it consists of multiple layers coming from multiple data files. Each of these files contains gridded data, but the data from different files. A `RasterBrick` must contain data from a single file. To deal with data from multiple files we use the `RasterStack` object. We can build a raster stack as follows. Suppose we already have a `RasterLayer` object created to hold the soil electrical conductivity data shown in Figure 1.2a. This layer is created using the kriging interpolation procedure described in Section 6.3 to create an object called `EC.grid`. The procedure for doing this will be described in that section. This object is converted into a `RasterLayer` object called `EC.ras` using the `raster` function `interpolate()`.

```
> EC.ras <- interpolate(grid.ras, EC.krig)
```

Our objective is to combine this with the infrared layer of the raster brick `data.4.2.May` to create a raster stack.

Although the two raster objects describe data from the same geographical area, since they were created using different methods they may not exactly overlap. We can check this quickly by determining whether they have the same extent, using the raster function of that name.

```
> extent(EC.ras)
class      : Extent
xmin       : 592077
xmax       : 592867
ymin       : 4267513
ymax       : 4267873
> extent(data.4.2.May)
class      : Extent
xmin       : 591965.6
xmax       : 593139.5
ymin       : 4267139
ymax       : 4268321
```

The extents, and thus the grids, are indeed different. In order to create a stack, we must make them line up. We do this using the function `resample()`, which is named after a common GIS operation in which data on one grid is "resampled" to another grid (Lo and Young, 2007, p. 154).

```
> IR.4.2.May <- resample(data.4.2.May[[1]], EC.ras)
```

Now we can create the `RasterStack` object.

```
> data.4.2.May.stack <- stack(IR.4.2.May, EC.ras)
> plot(data.4.2.May.stack)
```

As a contributed package, `raster` may change at some point in the future, so it is good to have another package as a fallback. A good one is the package `rgdal` (Keitt et al., 2017). For purposes of illustration, this function is occasionally used to read in image data, for example in the code in Sections 1.1 and 6.4.2.

2.5 Writing Functions in R

Like most programming languages, R provides the user with the capacity to write functions. As an example demonstrating the basics of writing an R function, we will create one that computes a simplified "trimmed standard deviation." Most statistical packages contain a function that permits the computation of a trimmed mean, that is, the mean of a data set from which a certain number or fraction of the highest and lowest values have been removed. The R function mean() has an optional argument trim that allows the user to specify a fraction of the values to be removed from each extreme before computing the mean. The function sd() does not have a similar argument that would permit one to compute the standard deviation of a trimmed data set. The formula used in our example is not standard for the "trimmed standard deviation" and is intended for illustrative purposes only. See Wu and Zuo (2008) for a discussion of the trimmed standard deviation.

Given a vector z and a fraction trim between 0 and 0.5, the standard deviation of the trimmed data set can be computed as follows. First we sort the data in z.

```
> z <- c(10,2,3,21,5,72,18,25,34,33)
> print(ztrim <- sort(z))
 [1]  2  3  5 10 18 21 25 33 34 72
```

Next we compute the number to trim from either end, based on the value of trim. We use the coercion function as.integer() to obtain an integer value.

```
> trim <- 0.1
> print(tl <- as.integer(trim * length(ztrim)))
 [1] 1
```

Next we trim the values, first the low end and then the high end.

```
# Trim from the low end
> print(ztrim <- ztrim[-(1:tl)])
 [1]  3  5 10 18 21 25 33 34 72
```

Next we trim from the high end.

```
> # Trim from the high end
> th <- length(ztrim) - tl
> print(ztrim <- ztrim[-((th + 1):length(ztrim))])
 [1]  3  5 10 18 21 25 33 34
```

Finally, we compute the standard deviation.

```
> print(zsd <- sd(ztrim))
 [1] 11.91563
```

Now we will create a function sd.trim() that carries out the entire sequence of steps. The syntax is

```
> # Function to compute the trimmed standard deviation
> sd.trim <- function(z, trim){
```

```
+ # trim must be between 0 and 0.5 or this won't work
+ # Sort the z values
+    ztrim <- sort(z)
+ # Calculate the number to remove
+    tl <- as.integer(trim * length(ztrim))
+ # Remove the lowest values
+    ztrim <- ztrim[-(1:tl)]
+ # Remove the highest values
+    th <- length(ztrim) - tl
+    ztrim <- ztrim[-((th + 1):length(ztrim))]
+ # Compute the standard deviation
+    zsd <- sd(ztrim)
+    return(zsd)
+ }
```

The function is defined by function(arguments) expression. The *arguments* represent one or more objects whose values are explicitly named in the statement invoking the function. The *expression* is the step or sequence of steps that carries out the specified calculations. In this book, the last statement of any function with more than one statement will generally be the *return* statement, which specifies the vector of values to return. If no return statement exists, the last object computed is returned. In the case of a function with only one assignment statement, the return statement is superfluous. An example of such a function is given later in this section The application of a user-defined function proceeds just like the calculation of any other function.

```
> sd.trim(z, 0.1)
[1] 11.91563
```

We will be writing a large number of functions in this book. It is important to note at the very outset, however, that the code we display will deviate from good programming practice in one very important way. In an effort to keep the code simple and concise, we will generally not include the error trapping facilities that would be built into good commercial software, or into a good contributed R package. For example the code in sd.trim() computes the standard deviation of a data set after a fraction *trim* of the largest and smallest values have been removed. The code does not include a test to determine whether $0 \leq trim \leq 0.5$; such a test would be an absolute necessity in software that was intended for general use. Error trapping and resolution is a complex process (Goodliffe, 2007, Ch. 6) and to do it properly would dramatically increase the length and complexity of the code. Instead, to borrow a phrase from Ed Post (Post, 1983), our error trapping philosophy will be "you asked for it, you got it."

Recall that R is an object-oriented programming language. In addition to the class structure described in Section 2.2, a feature common among object-oriented programming languages is *polymorphism*. This refers to the ability of functions to accept arguments of different classes and to adjust their evaluation depending on the class of the argument or arguments. This feature can be illustrated using the function diag(). When the argument of this function is a matrix, diag() returns the diagonal of this matrix.

```
> a <- matrix(c(1,2,3,4), nrow = 2)
> a
     [,1]  [,2]
[1,]    1     3
```

```
[2,]    2    4
> diag(a)
[1] 1 4
```

When its argument is a vector, diag() returns a diagonal matrix whose diagonal is the components of the vector.

```
> v <- c(1,2)
> v
[1] 1 2
> diag(v)
     [,1] [,2]
[1,]    1    0
[2,]    0    2
```

Polymorphism is an incredibly useful feature in simplifying the coding of R programs involving different kinds of data structures. For example, the function plot() knows how to plot many kinds of objects and will generally create the most appropriate plot for a given object class (see Exercise 2.10). The first argument of a polymorphic functions like plot() determines which version of the function is used.

One way in which R differs from some other object-oriented programming languages is that R implements a feature called *lexical scoping* (Venables and Ripley, 2000, p. 63). Without going into details, an important feature of lexical scoping can be illustrated by creating a rather strange function called to.the.n() to evaluate x^n.

```
> to.the.n <- function(w) w^n
> n <- 3
> to.the.n(2)
[1] 8
```

The function is strange because it does not include the value of n as one of its arguments. Since the value of n is not available as one of the arguments of to.the.n(), R looks for a value outside the scope of the function. The value 3 has been assigned to a variable n prior to the call to the function, and R uses that value. The function better.to.the.n(), which does include n in its argument, uses this value and ignores (actually, never looks for) the value 3 that is assigned to n outside the function but is not used.

```
> better.to.the.n <- function(w, n) w^n
> n <- 3
> better.to.the.n(2,4)
[1] 16
```

This is a vastly oversimplified treatment of lexical scoping. We will not have much need to make use of it in this book, and a more detailed discussion is given by Venables et al. (2006, p. 46) and also in the R project FAQs (http://www.r-project.org/). The main reason for mentioning it is to warn those used to programming in some other languages of its existence. Also, if you are not careful, lexical scoping can give you some very nice rope with which to hang yourself. Specifically, if you forget to specify some variable in a function and a variable by that name happens to be hanging around in your workspace, perhaps from a completely different application, R will happily plug that value into the computation and give you a completely incorrect result. A bug like this can be very hard to track down. For this reason, it is better to pass all values used by the function through the argument list unless there is good reason to do otherwise. In Section 5.3.1, lexical scoping is used to call a function from inside another function. This makes the code simpler.

2.6 Graphics in R

R has a very powerful set of graphics tools, and in this section we describe some of these tools using some simple examples. Our primary objective is data visualization. Data visualization, like presentation graphics, involves displaying of features of the data in graphical form. However, unlike presentation graphics, whose purpose is to convey already identified features of the data to an audience not necessarily familiar with the data themselves, the purpose of data visualization is to allow the analyst to gain insight into the data by identifying its properties and relationships (Haining, 2003, p. 189). The human brain instinctively searches for and perceives patterns in visual data. In order for the process of data visualization to be successful, the encoding of the data into a graphical object must be done in such a way that the decoding of the object by the human eye is effective (Cleveland, 1985, p. 7). This means that the visual patterns in the graphical object must not induce "optical illusions" in which the eye perceives patterns differently from the actual arrangement of the data.

The R programming environment, both in its base and contributed packages, provides an incredibly powerful and flexible array of tools for constructing graphical images. Every figure in this book except the photographs (R is not *that* good) was constructed using R. The base R graphics system embodies both the strengths and the challenge of the R programming environment. The strengths are the flexibility and the power to generate almost any type of graphical image one can desire. The challenge is that the graphics software is controlled by a function-based, command-driven programming environment that can sometimes seem discouragingly complex.

At the outset, one thing must be emphasized. As with everything else, the R programming environment provides several alternative methods for generating the graphical images displayed here. Not all of these alternatives will be presented in this book, although in some cases we will describe more than one if there is a good reason to do so. Readers who wish to learn more about other alternatives are advised to begin with Murrell (2006) and then to proceed to other sources listed in Section 2.8. One package in particular, however, deserves special mention: the package ggplot2 (Wickham, 2009). This package, and the function ggplot() that it contains, provide a very powerful alternative to the function plot(). In particular, ggplot() is so simple to learn that it practically overcomes the complexity of use issue described in the previous paragraph. At least in my opinion, however, plot() is stronger in the creation of presentation graphics such as the figures in this book. This is partly due to the excellent description by Murrell (2006) of how to control every aspect in the presentation of a figure. Therefore I have decided to use each of these plotting functions for what I consider their strengths. The figures, which are presentation graphics, are made with traditional graphics such as plot(). In every case where quick and effective data visualization is appropriate, however, I have included in the code a call to ggplot(). Both plot() and ggplot() are described in the next section.

2.6.1 Traditional Graphics in R: Attribute Data

The R presentation graphics presented in this book are based on two fundamental packages, a "traditional" system implemented in the base package and a trellis graphics system implemented in the `lattice` package (Sarkar, 2008). Trellis graphics is a technique originated by Cleveland (1993) that seeks to organize multivariate data in a manner that simplifies the visualization of patterns and interactions. We focus on `lattice` because it is the foundation for the spatial plotting functions discussed in this book. Our discussion references figures in other chapters. The code to produce these figures is in the section containing the figure, rather than in this section. All the plots in this book were created by executing the R code using TINN-R as the editor and running R as a separate application, so that it produces a separate window that can be printed or saved. If you create these plots in RStudio, they will appear in the Plot/Help window and will generally look very different.

Functions in R that produce a graphical output directly are called *high level* graphics functions. The standard R traditional graphics system has high level functions such as `pie()`, `barplot()`, `hist()`, and `boxplot()` that produce a particular type of graph. Some of these will be described later on. The fundamental high level workhorse function of the standard presentation graphics system, at least for us, will be the function `plot()`. When applied to attribute data (i.e., with no spatial information), the output of `plot()` is a plot of an array Y against an array X. We will introduce `plot()` using Figure 3.6a in Section 3.5.1, which is one of the simplest figures in the book. It is a plot of an array called `error.rate` on the abscissa against an array called `lambda` on the ordinate. After working through this plot, you will have the basic information needed to construct a plot in R. You may wish to skim the remainder of the section, which concerns more advanced plots, on first reading. If you do so, stop skimming when it gets to the material covering the `ggplot2` package.

A simple statement to generate a scatterplot is the command

```
> plot(lambda, error.rate)
```

We have not yet discussed the *error rate*, which is a quantity that is explained in Section 3.5.1. For now, we are only interested in the plot of this quantity (again, you should open the code file *3.5.1.r* to follow this discussion). If you try this with arrays generated in Chapter 3, you will get a plot that resembles Figure 3.6a but has some important differences. There is no title, the abscissa and ordinate labels are smaller, and instead of the Greek letter λ and the words "Error Rate," the axes contain the names of the objects, `lambda` and `error.rate`, themselves. These will need adjusting. The R graphics system does, however, automatically do several things very well. The plotting region is clearly demarcated by four scale lines, with the tick marks on the outside, and, as a default, having an approximately square shape (Cleveland, 1985, p. 36).

Some of the information needed to control the appearance of the graph can be included directly in the call to the high level function `plot()`. We can specify a title. Type the following.

```
> plot(lambda, error.rate, main = "Error Rate")
```

That gets us a title. Your title may look different from that of Figure 3.6a because the publisher of the book has altered the graphics to save space. The next step is the axis labels. The ordinate is easy:

```
> plot(lambda, error.rate, main = "Error Rate", ylab = "Error Rate")
```

The abscissa requires one more step. This is to use the function `expression()` within the call to `plot()`. To see a complete list of the things `expression()` can do, type `?plotmath`. From this we see that to plot a Greek letter for the ordinate label we can type the statement

```
> plot(lambda, error.rate, main = "Error Rate", ylab = "Error Rate",
+    xlab = expression(lambda))
```

This almost gets us to where we want to be, but the labels are still too small. The general indictor in R standard graphics for size is `cex`. This is a factor by which the size of a label is multiplied when creating a graph; the default is a value of 1. The graph in Figure 3.6a is created with the statement

```
> plot(lambda, error.rate, xlab = expression(lambda),
+    ylab = "Error Rate", cex.lab = 1.5) # Fig. 3.6a
```

This solves the label size problem, but creates another problem: the abscissa label may be crowded against the side of the graphics window. We need to increase the size of the area surrounding the graph, which is called the *figure margin* (actually, this is an oversimplification; see Murrell, 2006, Section 3.1 for a full explanation). There are a number of options one can use to control this. All must be set not as an argument of `plot()` but rather as an argument of the function `par()`, which controls the R plotting environment. A very simple method is to use the argument `mai`, which gives the margin size in inches of the left, right, top, and bottom margins, respectively, in inches (yes, inches!). For our application the statement

```
> par(mai = c(1,1,1,1))
```

will do the trick. This statement is used prior to almost all traditional plots in this book and will no longer be explicitly mentioned. Entering the statement and then the call to `plot()` above produces Figure 3.6a. Settings made in a high level or low level plotting function are applied only to that instance of the function, but settings made in a call to the function `par()` are retained until they are changed or the R session is stopped without saving the workspace. Again, in the version that you see of Figure 3.6a the size and font of the title and axis labels may have been changed in the process of setting the book up for publication. For this reason, the figure may look different from the one you get as output.

Now we will move on to more advanced plots. Figure 3.6b provides another example of the use of the function `expression()`, this time for the abscissa label. The quantity plotted on the abscissa is $\bar{Y}$, the sample mean, as a function of λ. This requires two actions on the character: putting a bar over it and italicizing it. From `?plotmath` we see that the code in `expression()` to produce a bar over the Y is `bar(Y)` and the code to italicize is `italic()`, so we can combine these.

```
> plot(lambda, mean. Y, xlab = expression(lambda),
+    ylab = expression("Mean"~italic(bar(Y))),
+    ylim = c(-0.1, 0.1), cex.lab = 1.5) # Fig. 3.6b
> title(main = "Mean of Sample Means", cex.main = 2)
```

The tilde ~ character in the function `expression()` inserts a space. To separate two symbols without inserting a space, use an asterisk, as in the code for Figure 3.6c. Note that the parentheses have quotes around them.

```
> plot(lambda, sem.Y, xlab = expression(lambda),
+     ylab = expression("Mean s{"*italic(bar(Y))*"}"),
+     ylim = c(0,0.5), cex.lab = 1.5) # Fig. 3.6c
> title(main = "Mean Estimated Standard Error", cex.main = 2)
```

The code sequence introduces another concept, the *low level* plotting function. The function `title()` is a low level plotting function. It adds material to an existing graphic rather than creating a new graphic as `plot()` does. The philosophy of traditional R graphics is to start with a basic graph and then build it up by using functions to add material. A common way (but not the only way, as we will see below) to add material is by using low level plotting functions.

Now let's turn to a more complex attribute plot, that of Figure 3.1. This is a plot of a cross section of the components of the artificially generated surface shown in Figure 3.2a (which we will get to in a moment). The figure shows three line segment plots: (a) the trend $T(x,y)$, (b) the trend plus autocorrelated component $T(x,y)+\eta(x,y)$, and (c) the trend plus autocorrelated component plus uncorrelated component $Y(x,y) = T(x,y)+\eta(x,y)+\varepsilon(x,y)$. The components T, η, and ε are represented in the R code by `Yt`, `Yeta`, and `Yeps`, respectively, and Y is represented by `Y`. The initial plot is constructed with the following statement.

```
> plot(x, Y, type = "l", lwd = 1, cex.lab = 1.5, #Fig. 3.1
+     ylim = c(-0.5,2.5), xlab = expression(italic(x)),
+     ylab = expression(italic(Y)))
```

The argument `type = "l"` specifies that the graph will be a line. The argument `lwd` controls the line width. This is set at 1, which is the default value. The argument `ylim = c(-0.5,2.5)` controls the abscissa limits. A similar specification of the argument `xlim` is not needed in this particular case because the automatically generated ordinate values are adequate. The `plot()` statement produces the axes, labels, and the thin curve representing $Y(x,y) = T(x,y)+\eta(x,y)+\varepsilon(x,y)$. Now we will build the rest of the graphic by adding material. First, we add the curve representing $T(x,y)+\eta(x,y)$ with the following statement.

```
> lines(x, Yt, lwd = 3)
> lines(x, Yt + Yeta, lwd = 2)
```

The argument lwd = 3 specifies that the line width will be three times the default width. Now it is time to add the legend. There is a function `legend()` that will do this for us (Exercise 2.8), but for this plot we will use the `text()` function. It takes a bit of trial and error to locate the correct placement, but finally we come up with the statements

```
> text(x = 12.5, y = 0.25, "T(x, y)", pos = 4)
> lines(c(10,12.5),c(0.25,0.25),lwd = 3)
```

to produce the first line of the legend, and similar statements for the rest of the legend. The argument pos = 4 causes the text to be placed to the right of the location specified in the argument. The default is to place the center of the text at this point.

As our final, and most complex, example of an attribute graph, we will discuss Figure 5.14b. First, we issue a plot() statement to generate the basic graph.

```
> plot(data.samp$IRvalue, data.samp$Yield, # Fig. 5.14b
+     xlab = "Infrared Digital Number", ylab = "Yield (kg/ha)",
+     xlim = c(110, 170), ylim = c(2000, 7000),
+     main = "Transect 2", pch = 16,
+     cex.main = 2, cex.lab = 1.5, font.lab = 2)
```

Because we want the axes of this figure to match those of Figure 5.14a and 5.14c, we control the scale of the axes using the arguments xlim = c(110, 170), ylim = c(2000, 7000). The argument pch = 16 generates black dots. The next step is to plot a regression line fit to these data. The least squares fit is obtained with the statement

```
> Yield.band1 <- lm(Yield ~ IRvalue, data = data.samp)
```

The line is added to the plot with the statement

```
> abline(reg = Yield.band1)
>
```

The next step is to add data points and regression lines for the second transect. The data for this transect is stored in the data frame data.samp2 in the data fields IRvalue and Yield. The data points are added to the graphic with the statement

```
> with(data.samp2, points(IRvalue, Yield, pch = 3))
```

The regression line is then added with the statements

```
> Yield. IR2 <- lm(Yield ~ IRvalue, data = data.samp2)
> abline(reg = Yield. IR2, lty = 3)
```

Here the argument lty = 3 generates a dotted line (lty stands for "line type"). The other scatterplots and regression lines are added with similar statements. The next step is to add the mean values. The following statement produces the large black dot located on the solid line that fits all the data.

```
> points(mean(data.samp$IRvalue), Y.bar,
      + pch = 19, cex = 2)
```

Next, a symbol of the form $\bar{Y}$ is added using the statement

```
> text(144, 4600, expression(bold(bar(Y))))
```

The other means are added similarly. The subscripts are inserted using square brackets, for example in the statement

```
> text(139, 5500, expression(bold(bar(Y)[2])))
```

Finally, the legends are added. The first legend statement has this form:

```
> legend(155, 6000, c("All Data", "Transect 2"),
+    pt.cex = 1, pch = c(19,3), y.intersp = 1,
+    title = "Sample")
```

The other legend statement is similar. Once again, the axis labels and header in the figures in the book may be different from those you produce due to the publisher's saving of space in the figures.

Before taking up the application of the function plot() to spatial data, we briefly discuss the function persp(), which serves as a sort of segue between attribute plots and spatial plots. This function is used to produce the "wireframe" three-dimensional graphs of Figure 3.3. We will not have much occasion to use these wireframe graphs, so we do not spend much time discussing their use. Here is the statement that generates Figure 3.2a.

```
> x <- (1:20) + 1.5
> y <- x
> Y.persp <- matrix(4*Y, nrow = 20)
> persp(x, y, Y.persp, theta = 225, phi = 15,
+    scale = FALSE, zlab = "Y") #Fig. 3.2a
```

Here theta and phi are the angles defining the viewing direction; theta gives the azimuthal direction and phi the colatitude, that is, the angle of elevation from which the view is seen.

As you can see, the function plot() gives you lots of control, but at the price of lots of complexity. Now let's look at ggplot2(), which in some sense is the opposite. Calls to ggplot2() can be done in many ways, but here we will use the form described by Wickham and Grolemund (2017). The "gg" in "ggplot" stands for "grammar of graphics" (Wilkinson, 2005), an organizational structure for graphic presentation on which ggplot() is based. The quickest way to introduce ggplot() is with an example. This one comes from Section 7.3, which involves the exploration of Data Set 2, the blue oak data (you may want to go back and read Section 1.3.2 to get an idea of the data set). At this point, you should load the code file *7.3.r* and run it along with me.

We would like to get an idea of the relationships between blue oak presence/absence and environmental factors. We know for a start that blue oak is influenced directly by elevation. We can try plotting blue oak presence (*QUDO* = 1) or absence (*QUDO* = 0) against *Elevation*. So we try this.

```
> ggplot(data = data.Set2) +
+    geom_point(aes(x = Elevation, y = QUDO))
```

The first line is the call to ggplot() itself. It sets the coordinate system (the default is Cartesian) and identifies the data source. The next line adds a layer to the plot in the form of a "geom." A geom is a type of plot, and the function geom _ point() specifies a scatter plot. In this case, as you can see, the plot does not provide much information, because the values of QUDO are so dense. We can get a better picture of the dependence of *QUDO* on *Elevation* using a different geom.

```
> ggplot(data = data.Set2) +
+    geom_smooth(aes(x = Elevation, y = QUDO))
```

This smooths the data using a generalized additive model (Section 9.2) and provides a good indication of how blue oaks vary with elevation. What might be causing this variation? One possibility is mean annual temperature. For a nicer plot, we first convert QUDO into a character. Then we can plot mean annual temperature, *MAT*, against *Elevation* while using color to display oak presence or absence. We can also add a smooth curve summarizing the relationship between *MAT* and *Elevation*.

```
> data.Set2$QF <- as.character(data.Set2$QUDO)
> ggplot(data = data.Set2) +
+    geom_point(aes(x = Elevation, y = MAT, color = QF)) +
+    geom_smooth(aes(x = Elevation, y = MAT))
```

We can add as many geoms as we wish to the plot by adding more calls to the appropriate function.

Another climatic quantity that might influence blue oak presence or absence is annual precipitation. We can easily examine the relationship between *MAT* and *Precip*.

```
> ggplot(data = data.Set2) +
+    geom_point(aes(x = Precip, y = MAT, color = QF))
```

There are many more geoms that can be accessed (lines, boxplots, histograms, etc.), and these will be demonstrated when appropriate, especially in Chapter 7. There is much more that can be done with ggplot(), but with this simple introduction you should have enough to use it effectively in data visualization.

2.6.2 Traditional Graphics in R: Spatial Data

Let us now turn to the use of the function plot() to create maps of spatial data. Before we begin, we need to distinguish between the plotting of sf objects (i.e., spatial features) and that of sp objects (see Section 2.4.3). Because of polymorphism (see Section 2.5), the same command plot() can be used with objects of both classes and R will respond appropriately. For any polymorphic R function, the class is determined by the first argument. I have found that the plotting of sf objects is most useful for initial quick looks at the data, and that the sp version of plot() is most effective for plotting presentation graphics, such as those that are used in this book. The code, therefore, will occasionally contain, especially in Chapter 7, calls to the sf plot() function, but all of the figures in the book are produced using the sp version.

We will begin by discussing Figure 3.4a, which is relatively simple and illustrates some of the basic ideas. Once again, after this basic plot, we will discuss more complex plots, and you may want to skim this on first reading. As usual, the graphic is initiated with an application of the function plot(), in this case, used to generate the map of the boundary, which is contained in the SpatialPolygonsDataFrame bdry.spdf.

```
> plot(bdry.spdf, axes = TRUE) #Fig. 3.4a
```

When applied to an sp object, the function plot() cannot be used to generate axis labels (Bivand et al., 2013b, p. 62), but if desired, these can be added later with the function title(). The next step is to add the symbols for the value of the plotted quantity, in this case the detrended sand content.

```
> plot(data.Set4.1, add = TRUE, pch = 1,
+    cex = (data.Set4.1$Sand / 15))
```

Attribute values are represented in graphs created using the function plot() by controlling the appearance of the plotted symbols. Usually one controls either their size, through the argument cex; their form, through the argument pch; or their color, through the argument col. In our case, we use the argument cex to control the size. The argument cex = (1 + data.Set4.1$Sand / 15 assigns a vector to cex with the same number of values as the number of data values to be plotted, and this is used to enable the size of the symbol to represent the data value. The value assigned is obtained by trial and error until the symbols "look right." The other elements of the map are then added with the title() and legend() functions.

```
> title(main = "Field 4.1 Sand Content", cex.main = 2,
+    xlab = "Easting", ylab = "Northing", cex.lab = 1.5)
> legend(592450, 4270600, c("15", "30", "45"), pt.cex = 1:3, pch = 1,
+    y.intersp = 1.5, title = "Percent Sand")
```

Now let's take on a more complex spatial plot, the map of the survey location of Data Set 1 displayed in Figure 1.4. This consists of a map of the state of California with the location of the study area identified and an insert that shows the study area on a smaller map scale. The map of California is obtained from the maps package (Becker et al., 2016).

```
> library(maps)
> data(stateMapEnv)
> cal.map <- map("state", "california",
+    fill=TRUE, col="transparent", plot = FALSE)
```

The map is converted to a SpatialPolygons object using the maptools function map2SpatialPolygons().

```
> cal.map <- map("state", "california",
+    fill=TRUE, col="transparent", plot = FALSE)
> cal.poly <- map2SpatialPolygons(cal.map, "California")
> proj4string(cal.poly) <- CRS("+proj=longlat +datum=WGS84")
```

The Data Set 1 shapefile is then read. It is stored on disk in UTM Zone 10N coordinates, so the rgdal function spTransform() is used to convert it to WGS84 longitude and latitude.

```
> data.Set1.sp <- readShapePoly("set1\\set1data.shp")
> proj4string(data.Set1.sp) <- CRS("+proj=utm +zone=10 +ellps=WGS84")
> data.Set1.wgs <- spTransform(data.Set1.sp,
+    CRS("+proj=longlat +datum=WGS84"))
```

Now we are ready to begin the plot. First, we plot the map of California and add the title and the study area map at the California scale. Because we are not going to be using axes in this map, we do not make our usual call to the function par().

```
> plot(cal.poly, axes = FALSE) # Fig. 1.4
> title(main = "Set 1 Site Location",cex.main = 2)
> plot(data.Set1.wgs, add = TRUE)
```

Next, we build a small box around the study site map within California.

```
> lines(c(-122.15,-121.9), c(39.7,39.7))
> lines(c(-122.15,-122.15), c(39.7,39.95))
> lines(c(-122.15,-121.9), c(39.95,39.95))
> lines(c(-121.9,-121.9), c(39.7,39.95))
```

Then we build a large box to hold the second map of the study area, as well as an arrow to connect the two boxes.

```
> lines(c(-118,-113), c(36.3,36.3), lwd = 3)
> lines(c(-118,-113), c(41.9,41.9), lwd = 3)
> lines(c(-118,-118), c(36.3,41.9), lwd = 3)
> lines(c(-113,-113), c(36.3,41.9), lwd = 3)
> arrows(-118, 39.82, -121.85, 39.82, lwd = 3, length = 0.1)
```

Now we use the sp function `SpatialPolygonsRescale()` to add a scale bar. The radius of the earth at the equator is approximately 40,075 km, and we use this to create a scale bar of the appropriate length for the latitude at which it is displayed. The code to display the scale bar is modified from Bivand et al. (2013, p. 65).

```
> deg.per.km <- 360 / 40075
> lat <- 34
> text(-124, lat+0.35, "0")
> text(-122, lat+0.35, "200 km")
> deg.per.km <- deg.per.km * cos(pi*lat / 180)
> SpatialPolygonsRescale(layout.scale.bar(), offset = c(-124, lat),
+    scale = 200 * deg.per.km, fill=c("transparent","black"),
+    plot.grid = FALSE)
```

Bivand et al. use the function `locator()` to place the scale bar. This has the advantage of being quick, but the disadvantages that it must be repeated each time the plot is created and that the objects are never quite in the same place in two different versions of the map. The alternative, shown here, is to play around with the numbers until the map looks right. The north arrow is placed similarly.

```
> SpatialPolygonsRescale(layout.north.arrow(), offset = c(-124, 36),
+    scale = 1, plot.grid = FALSE)
```

Now it is time to insert the second map of the study area. The code for this is modified from that used to create Figure 1.5 of Murrell (2006). First, we use the `par()` function to reset the coordinates of the figure region. These are set relative to the default values 0, 1, 0, 1 (horizontal and vertical). Then we indicate that a new plot will be created with a call to the function `plot.new()`. We create the ranges of the plot based on the bounding box of the `SpatialPolygons` object, create a plot window with these specifications, and plot the map within this plot window.

```
> par(fig = c(0.5, 0.94, 0.3, 0.95), new = TRUE)
> plot.new()
> xrange <- slot(data.Set1.wgs, "bbox")[1,]
> yrange <- slot(data.Set1.wgs, "bbox")[2,]
> plot.window(xlim = xrange, ylim = yrange)
> plot(data.Set1.wgs, add = TRUE, axes = FALSE)
```

Once again, a bit of trial and error is required to get the values right in the function par().

Finally, let us consider the plotting of maps that involve an image such as an aerial photograph. Figure 5.7, which depicts a map of weed infestation level in Field 4.2 over an image of the infrared band of an aerial photograph of the field, provides a good example. Creating these maps is easily done using the function image(), which is part of the traditional graphics package. Here is the relevant code to draw the map.

```
> image(data.May.tiff, "band1", col = greys, # Fig. 5.7
+   xlim = c(592100,592900), ylim = c(4267500,4267860),
+ axes = TRUE)
> plot(bdry.spdf, add = TRUE)
> data.samp$hiweeds <- 1
> data.samp$hiweeds[data.samp$Weeds == 5] <- 2
> plot(data.samp, add = TRUE, pch = 1, cex = data.samp$hiweeds)
> title(main = "High Weed Locations", xlab = "Easting",
+   ylab = "Northing", cex.main = 2, cex.lab = 1.5)
```

The high weed infestation level locations are depicted by larger circles, and the polygonal sampling boundary is added to the map as well, all using the standard concept of R traditional graphics: start with a basic graphic and then build on it by adding material.

2.6.3 Trellis Graphics in R, Attribute Data

Trellis graphics are based on ideas developed by Cleveland (1993). Although they can be used to generate almost any of the plots produced by traditional graphics, they are particularly useful for the exploration of multivariate data. They are important for our use because the function spplot(), which is the plotting workhorse of the sp package, works with trellis graphics, and also because many of the plots generated by the package nlme that we will be using for regression analysis use this system. Trellis graphics are implemented in R in the package lattice (Sarkar, 2008). The use of trellis graphics is illustrated by Figure 4.3. This figure displays the comparison of three Moran correlograms, one created with a distance-based spatial weights matrix, one with a path-based matrix, and one with a spatial weights matrix based on Thiessen polygons. Moran's I values are computed at five spatial lags using each of these methods in Section 4.5.1. The values are stored in three arrays, I.d, I.p$res[,1], I.t$res[,1] respectively, and we start with these arrays. The three plots in Figure 4.3 are distinguished based on a *grouping variable* (also called a conditioning variable). Trellis graphics produces a separate plot for each value of the grouping variable.

The easiest way to introduce the use of the grouping variable is to display the code for Figure 4.3. First, we construct a data frame as follows. We start by concatenating the three arrays.

```
> moran.cor <- c(I.d, I.p$res[,1], I.t$res[,1])
```

Next, we create the grouping variable, an index of the method used to create the Moran's *I* values.

```
> cor.type <- sort(rep(1:3, 5))
> cor.type
[1] 1 1 1 1 1 2 2 2 2 2 3 3 3 3 3
```

Next, we convert this into a factor and add labels identifying the method associated with each index value.

```
> cor.index <- factor(cor.type,
+    labels = c("Distance Based", "Path Based", "Thiessen Polygon"))
```

Finally, we create a data frame with the Moran's *I* and index variables, and add a data field representing the number of lags, which varies from one to five for each of the three plots.

```
> mor.frame <- data.frame(moran.cor)
> mor.frame$index <- cor.index
> mor.frame$lag <- rep(1:5, 3)
```

The final data frame looks like this.

```
> head(mor.frame, 2)
     moran.cor          index lag
1 0.20666693 Distance    Based   1
2 -0.06646138 Distance   Based   2
> tail(mor.frame, 2)
     moran.cor          index lag
14 -0.06585552 Queen's   Case   4
15 -0.16373871 Queen's   Case   5
```

Here we have used the functions `head()` and `tail()` to display the first and last records of the data frame. Now we are ready to create the plot. The trellis graphics analog of `plot()` is `xyplot()`, and that is what we will use. First, we display the code.

```
> xyplot(moran.cor ~ lag | index, data = mor.frame, type = "o",
+    layout = c(3,1), col = "black", aspect = c(1.0),
+    xlab = "Lag", ylab = "Moran's I",
+    main = "Moran Correlograms") # Fig. 4.3
```

The first argument is a *formula*. Formulas will also find use in linear regression, but they play an important role in trellis graphics. In the formula `moran.cor ~ lag | index`, `moran.cor` is the *response* and `lag | index` is the *expression* (Venables and Ripley, 2000, p. 56). As used in the `lattice` package, the response identifies the quantity to be plotted on the abscissa and the expression identifies the quantity to be plotted on the ordinate. The expression is in turn subdivided by a | symbol. The quantity to the left of this symbol is to be plotted on the ordinate. The quantity to the right is the grouping variable. The plots are grouped according to this variable as shown in Figure 4.3.

The second argument, `data = mor.frame`, identifies the data source for the plot. Among the remaining arguments, only `layout` and `aspect` are unfamiliar. The argument `layout` indicates the arrangement of the three *panels* of the lattice plot, that is, the three individual plots. The argument `aspect` indicates the aspect ratio (height to width) of each panel. A very important difference between trellis graphics and traditional graphics is that trellis graphics completely lacks the concept of creating a base graph using a high level function like `plot()` and then adding to it using low level functions like `lines()`

and text(). Once a plot is created, it cannot be augmented. There is a function called update() that permits graphs to be incrementally constructed, but they are replotted rather than augmented.

2.6.4 Trellis Graphics in R, Spatial Data

Now let us move on to the function spplot(). We will begin with Figure 5.1b. This is a plot of two quantities, the yield in kg ha^{-1} of the artificial population used to test various sampling schemes, and the yield data that were interpolated to produce the artificial population. The code to produce Figure 5.1b is as follows. The two data sets are contained in two SpatialPointsDataFrame objects, called pop.data and data.Set4.2.sp. The former is the artificial population, and the latter is the actual yield data with extreme values removed. If we were using the function plot(), we could create the plot with the first data set and then add the second data set using the either the function points() or the function plot() with the argument add = TRUE. With the lattice-based function spplot(), however, we must set everything up first and then use a single call to the plotting function. Therefore, we must combine the two data sets into one. This operation is carried out with the sp function spRbind(), which combines the two data sets and effectively does an rbind() operation (see Section 2.2) on the attribute data.

```
> data.Pop <- spRbind(pop.data, data.Set4.2.sp[,2])
```

In the statement above, we have taken advantage of the fact that although data.Set4.2.sp is a SpatialPolygonsDataFrame, its attribute data can be accessed as if it were an ordinary data frame (Section 2.4.3). The graph is then plotted with the following statement:

```
> greys <- grey(0:200 / 255)
> spplot(data.Pop, zcol = "Yield", col.regions = greys,
+    xlim = c(592500,592550), ylim = c(4267750,4267800),
+    scales = list(draw = TRUE), # Fig. 5.1b
+    xlab = "Easting", ylab = "Northing",
+    main = "Actual Yield and Artificial Population")
```

We get a warning message about duplicated row names that can be ignored. The color scale greys is restricted to the range 0–200/250 in order to avoid plotting white or nearly white dots, which would be invisible.

Now let us turn to Figure 1.5, which shows land use in a portion of the survey area of Data Set 1. This figure displays categorical data and includes a north arrow and scale bar rather than a set of axis scales. The land use types are in a code, so we first create a set of labels corresponding to this code.

```
> levels(data.Set1.cover$VegType)
 [1] "a" "c" "d" "f" "g" "l" "o" "r" "v" "w"
> data.Set1.cover@data$VegType <-
+    factor(as.character(slot(data.Set1, "data")$VegType),
+    labels = c("Grassland", "Cropland","Developed",
+    "Freshwater Wetland", "Gravel Bar", "Lacustrine", "Orchard",
+    "Riverine", "Riparian Forest", "Oak Woodland"))
```

As with the previous application of spplot(), we now create the elements that we are going to use in the final plot. These are the north arrow, the scale bar, and the text accompanying the scale bar.

```
> north <- list("SpatialPolygonsRescale", layout.north.arrow(),
+     offset = c(579800,4420000), scale = 600)
> scale <- list("SpatialPolygonsRescale", layout.scale.bar(),
+     offset = c(579400,4419000), scale = 1000,
+     fill = c("transparent", "black"))
> text1 <- list("sp.text", c(579400, 4419200), "0")
> text2 <- list("sp.text", c(580300, 4419200), "1 km")
> map.layout = list(north, text1, text2, scale)
```

The shades of gray are chosen so that water is displayed as white, unsuitable habitat as light gray, and suitable habitat as dark gray. The actual call to function spplot() is then quite short.

```
> greys <-
+     grey(c(180, 190, 140, 240, 150, 250, 130, 170, 30, 250) / 255)
> spplot(data.Set1.cover, "VegType", col.regions = greys,
+     main = "Land Use, Northern End, 1997",
+     xlim = c(576000,580500), ylim = c(4417400,4421000),
+     sp.layout = map.layout)
```

The function spplot() can also be used to create multiple panel plots. Figure 3.9 is an example. This figure displays the distribution of values of the solution Y of the equation $Y = \mu + \lambda W(Y - \mu) + \varepsilon$, where W is the spatial weights matrix, for three values of the autocorrelation coefficient λ. The plot is created using the following code, which is based on code in the sp Gallery https://edzer.github.io/sp/.

```
> greys <- grey(0:255 / 255)
> spplot(Y.data, c("Y0", "Y4", "Y8"),
+     names.attr = c("0", "0.4", "0.8"),
+     col.regions = greys, scales = list(draw = TRUE),
+     layout = c(3,1), xlim = c(1,23)) #Fig. 3.9
```

The second argument of spplot() is the name(s) of the data field(s) of the SpatialPolygonsDataFrame object Y.data that are to be plotted; since there is more than one data field in this case, spplot() plots them each in a separate panel. The third argument provides a means for labeling the individual panels. The value of xlim is set to leave empty space on either side of the plots so that they do not run together.

So which is better, plot() or spplot()? Before taking this on, I should also note that you can also plot spatial objects with ggplot(). Again, I have found this to be best for quick maps in data exploration (along with the sf version of plot()). As far as the question that begins the paragraph, I don't think that there is a single correct answer, so the book provides numerous examples that you can use to learn both. The function spplot() is better for figures that show multiple panels, such as Figure 3.9. In some cases I have found that the capacity of plot() to build graphs by successive applications of low level functions can be used to advantage.

2.6.5 Using Color in R

In order to keep the cost of the book as low as possible, all of the figures are rendered in gray scale. Color versions of a selection of the figures are provided on the book's companion website, http://psfaculty.plantsciences.ucdavis.edu/plant/sda2.htm. The figures on the website are examples where interpretation is enhanced through the use of color, and code is provided for all figures. It is important to have a facility with color images, both because this format is often available for publications and because even when it is not, color images are often much more useful both for exploration and presentation than gray scale ones. Fortunately, R makes it very easy to create attractive color images. Before discussing these, it is useful to review some concepts of color cartography. Although we commonly create colors using the red—green—blue (RGB) color model, in which each color is a mixture of intensities of these colors, an alternative color model called the hue- value- saturation model is also useful (Lo and Yeung, 2007, p. 260; Tufte, 1990, p. 92). In this model, hue is the color property associated with the spectral wavelength or mixture of wavelengths (the "color"), value is the degree of lightness or darkness (the "shade"), and saturation is the degree of richness. Hue is the most visually "interesting" property, and differences in hue are easily perceived when placed next to each other. For this reason, hue is generally considered the best means with which to represent categorical quantities. Value is considered the most effective means to represent numerical quantities because we can more easily perceive ordinal differences in values than in hues. These are not hard-and-fast rules but rather guidelines, and indeed we will have occasion to violate them below.

The base R package provides a number of color palettes in the form of functions: rainbow(), heat.colors(), terrain.colors(), topo.colors(), and cm.colors() (for cyan-magenta). Each of these requires an argument n, which indicates the number of colors in the palette. Use ?rainbow to see information on all of them. As an example, Figure 1.1 includes a topographical map of California, so one is tempted to try either topo.colors() or terrain.colors() on it. The statement

```
> image(ca.elev, col = topo.colors(10), axes = TRUE)
```

produces a map that is not particularly attractive, at least to my eye. The statement

```
> image(ca.elev, col = terrain.colors(10), axes = TRUE)
```

is better, but the lower elevations are brown and the upper elevations are green, which, although it accurately reflects the color of vegetation in California during the summer, is not the form in which elevation is typically displayed. We can reverse the colors by creating a vector col.terrain using this statement.

```
> col.terrain <- terrain.colors(10)[10:1]
> image(ca.elev, col = col.terrain, axes = TRUE)
```

This renders the lower elevations as green and the upper elevations as brown, but creates a green background. The background is the highest color value, so we can obtain a white background with the statement

```
> col.terrain[10] <- "white"
```

This creates an intuitive elevation map. We can then locate the survey points with a nice, contrasting color like red.

```
> plot(Set2.plot, pch = 16, cex = 0.5, add = TRUE, col = "red")
```

Note here that our use of brown and green to represent altitude violates the guideline that an interval scale quantity (elevation) should be represented by color value and not hue. The very use of the term "terrain colors" indicates that our eye is used to seeing maps in which green and brown represent terrain, and therefore can easily interpret an elevation map using this color scale.

   We would usually like to plot "quantitative" quantities using different values of the same hue. We can create these different values by mixing hues with white using the function `colorRampPalette()` as described by Bivand et al. (2013, p. 79). The code to do this for Figure 3.9 is a very simple modification of that given in the previous section.

```
> rw.colors <- colorRampPalette(c("red", "white"))
> spplot(Y.data, c("Y0", "Y4", "Y8"),
+     names.attr = c("0", "0.4", "0.8"),
+     col.regions = rw.colors(20), scales = list(draw = TRUE),
+     layout = c(3,1))
```

The function `rainbow()` produces different hues, which may be used for categorical variables. Let's try them on Figure 1.4, which displays land-use categories.

```
spplot(data.Set1.cover, "VegType", col.regions = rainbow(10),
   main = "Land Use, Northern End, 1997",
   xlim = c(576000,580500), ylim = c(4417400,4421000),
   sp.layout = map.layout)
```

This produces a map that satisfies the theory of different hues for different categories. There are, however, two problems. The first is that the map looks pretty garish. The second is that, although by coincidence the river water is blue, none of the other classes have colors that correspond to what we think they should look like. Moreover, there is nothing to distinguish the suitable class, riparian, from the others. We can make a more attractive and intuitive map by using hues that correspond to our expectations and distinguishing between them using values. One of the most popular R color systems is the package RColorBrewer (Neuwirth, 2014). This provides a set of palettes, and the paper of Brewer et al. (2003) provides a discussion of their appropriate use. We can use the package to modify the map of Figure 1.4 as follows. We will make the water shades of blue, the "natural" vegetation shades of green (except for riparian forest, the cuckoo habitat), the cropped vegetation and developed areas shades of purple, the gravel bar tan, and the riparian forest yellow (to make it stand out). To do this we will use two of the RColorBrewer palettes, BrBG, which ranges from brown to blue to blue-green, and PiYG, which ranges from purple to green.

```
> brown.bg <- brewer.pal(11,"BrBG")
> purple.green <- brewer.pal(11,"PiYG")
> color.array <- c(
+     purple.green[8], #Grassland
```

```
+     purple.green[4], #Cropland
+     purple.green[3], #Developed
+     brown.bg[7], #Freshwater Wetland
+     brown.bg[4], #Gravel Bar
+     brown.bg[8], #Lacustrine
+     purple.green[2], #Orchard
+     brown.bg[7], #Riverine
+     "yellow", #Riparian Forest
+     purple.green[11]) #Oak Woodland
> spplot(data.Set1.cover, "VegType", col.regions = color.array,
+     main = "Land Use, Northern End, 1997",
+     xlim = c(576000,580500), ylim = c(4417400,4421000),
+     sp.layout = map.layout)
```

Good color selection in cartography is, of course, an art.

2.7 Continuing on from Here with R

As you continue to work with R, you will find two things to be true: first, there is always more to learn; and second, if you stand still you will fall behind. To deal with the first issue, we must address the question of where one goes to continue learning. Of course, Google is always a good option. When I have a specific question, I can almost always find the answer there or on the specialized site https://rseek.org/. Beyond that, when you are starting out with R one of the best things you can do is to subscribe to the R-help mailing list, together with any of the special interest mailing lists that may be appropriate (e.g., R-SIG-Geo for spatial data analysis and R-SIG ecology for ecological data analysis). These are available via the website https://www.r-project.org/mail.html. You will see that they are all quite active. At first you may feel a bit overwhelmed by technical questions that you don't understand, but don't despair! You will gradually find these to be a great source for answers to questions that you didn't realize you had. Another good source is the R Bloggers site https://www.r-bloggers.com/.

A good deal of information is available directly from R. In addition to the help.start() function, the function help.search() uses fuzzy matching to search R for documentation matching the character string entered in the argument of the function (in quotes); use ?help. search to obtain details. The function R.SiteSearch() searches for matching words or phrases in the R-help mailing list archives, R manuals, and help pages. Finally, many R packages contain one or more *vignettes*, which provide documentation and examples. A list of available vignettes is accessible by typing vignette(), and specific vignettes can then be displayed by typing vignette("name") where name is the name of the vignette.

Now for the challenges. R is very dynamic. It is driven entirely by its users, and these users are constantly making changes an improvements in the software that they have contributed. New packages appear all the time, and old ones are modified. The discussion lists described in the previous paragraphs are a good way to keep up with these changes. One consequence of this dynamism is that you will occasionally make a call to a function and get a warning that it has been *deprecated*. This means that the maintainer of the package has stopped supporting the function, generally because it has been replaced by something better, but that the maintainer is still making it available to give users a time to switch. Ordinarily the announcement will contain a reference to the new function, and you will just have to learn how to use it.

2.8 Further Reading

There are many excellent books describing the R programming environment and its use in statistical analysis; indeed, the number of such books is increasing exponentially, and we will only mention a subset. Historically, R was created as an open source version of the language S, developed by Bell Labs (Becker et al., 1988). The S language evolved into a commercial product called S-Plus. Good entry points to R are Venables et al. (2006, 2008). Spector (1994) also provides a good introduction to the R and S-Plus languages, and Crawley (2007) provides almost encyclopedic coverage. Venables and Ripley (2002) is the standard text for an introduction to the use of R in statistical analysis, and Heiberger and Holland (2004) also provide an excellent, and somewhat more detailed, discussion. For those moving from other languages, Marques de Sá (2007) can be useful as it provides a sort of Rosetta stone between R and other statistical packages. Rizzo (2008) discusses a broad range of topics at a slightly higher level. Leisch (2004) provides a discussion of S4 classes, used in the sp package, in R.

As an open source project, R makes all of its source code available for inspection. Some R functions are written in R, and these can be accessed directly by typing the name of the function, as was illustrated in Section 2.2 with the function sd(). Many R functions, however, are programmed in a compiled language such as C++. Ligges (2006) provides an extensive discussion on how to retrieve source code for any R function.

The best source for R graphics is Murrell (2006). Sarkar (2008) is an excellent source for trellis graphics. Bivand et al. (2013) provide an excellent introduction to graphics with the sp functions. Wickham and Grolemund (2017) is the standard printed reference for ggplot(), although you can learn most of what you need to know just by googling "ggplot."

For those unfamiliar with GIS concepts, Lo and Yeung (2007) provide an excellent introduction. In addition to Lo and Yeung (2007), which this book uses as the primary GIS reference, there are many others that provide excellent discussions of issues such as map projections, Thiessen polygons, spatial resampling, and so forth. Prominent among these are de Smith et al. (2007), Bonham-Carter (1994), Burrough and McDonnell (1998), Clarke (1990), and Johnston (1998). Good GIS texts at a more introductory level include Chang (2008), Demers (2009), and Longley et al. (2001).

Exercises

All of the exercises are assumed to be carried out in the R programming environment, running under Windows.

2.1 Venables and Ripley (2002) make the very cogent point that one of the first things to learn about a programming environment is how to stop it from running away. First type i <- 1:20 to create an index. Then type the statement while (i < 5) i <- 2. What happens? Why does this happen? To stop the execution, either hit the <Esc> key, or, if you are running RStudio, click on the little Stop sign that appears in the upper right corner of the Console. This is how you "bail out" of any R execution in Windows.

2.2 Like all programming languages, mathematical operators in R have an *order of precedence*, which can be overridden by appropriate use of parentheses. (a) Type the following expressions and observe the result:

```
3 * 4 ^ 2 + 2
(3 * 4) ^ 2 + 2
-1:2
-(1:2)
```

(b) Type 1:3 - 1 and observe the result.

(c) Type the statement w <- 1:10 to create a vector w. Type the statement w > 5. The R symbol for equality used in a test is the double equal sign ==. Type w == 5. The R symbol for inequality used in a test is !=. Type w != 5.

2.3 (a) Using a single R statement, create a matrix A with three rows and four columns whose elements are the row number (i.e., every element of row 1 is 1, every element of row 2 is 2, and every element of row 3 is 3) (hint: make sure you read the Help file ?matrix). Display the matrix, then display the second row of the matrix, then display every row but the first.

(b) In a single statement using the function cbind(), create a matrix B whose first column is equal to the first column of A and whose second column is five times the first column of A.

(c) Create a data frame C from the matrix B and then give the fields of C the names "C1" and "C2."

(d) Create a list L whose first element L1 is the vector w, whose second element L2 is the matrix B, and whose third element L3 is the data frame C. Display L$L3$C2. Now display the same object using the [[]] and []operators (*note*: in R nomenclature, these operators are denoted [[and [. To look them up in Help, type ?"[" and ?"[[".

2.4 (a) Use the function c() to create a character vector w whose elements are the number 6, 2.4, and 3. Then create a vector z whose elements are the characters "*a*," "*b*," and "*c*." Note that in R characters are always enclosed in quotes.

(b) Use the function c() to create a vector z that is the concatenation of the vectors w and z created in part (a). Evaluate the vector z. What has happened to the elements of the vector w?

(c) Apply the function as.numeric() to the vector z to recover the numerical values of the vector w of part (a). What happens to the components of the vector z?

2.5 In Section 2.3.1, we created a vector w whose elements were 1:10, and then we used two code sequences to convert the first five of these elements to 0 and the second five to 20. Repeat this code exactly, but instead of converting the second five elements to 20, attempt to convert them to 3. Do the two code sequences give the same result? If not, create a second code sequence that gives the same results as the first for and values of the two replacement quantities 20 and 3.

2.6 (a) In Section 2.3.2, we used the functions rep() and sort() to create a sequence of five ones, five twos, five threes, and five fours. In this exercise we will use another approach. First, use the constructor function numeric() to create a vector z of 20 zeroes.

(b) Use the *modulo* operator %% to assign a value of 1 to every fifth component of the vector z. The modulo operation u %% v evaluates to u mod v, which, for integer values of u and v is the remainder of the division of u by v. Thus, for example 6 mod 5 = 1. The only numbers i such that i mod 5 = 0 are 0, 5, 10, etc., and therefore the components of the vector z with these indices are assigned the value 1. To read about an operator using special characters, you must place it in quotes, so you type `?"%%"`.

(c) Next, use the R function `cumsum()` to compute the index vector. For a vector z, the ith component of $cumsum(z)$ is $\sum_{k=1}^{i} z_k$.

2.7   (a) R has some built-in numbers, including `pi`. Read about the function `print()`, and then evaluate pi to 20 significant digits.

   (b) Read about the function `sin()`. Read about the function `seq()` and create a vector w whose components run from 0 to 6 in steps of 0.25 (i.e., 0, 0.25, 0.5,0.75, ..., 6.0) Now set n equal to 0.5 and create a vector z1 by evaluating $sin(n\pi w)$.

   (c) Create a new function `sin.npi(w)` defined as `sin(n * pi * w)`. Do *not* pass n in the argument of the function; instead, use lexical scoping to determine it. Create a vector z2 by the statement `z2 <- sin(n * pi * w)`. Use the function `cbind()` to create an array z whose columns are z1 and z2, and visually compare them to see that they are equal.

   (d) Read about the function `all.equal()` and use this to test the equality of z1 and z2.

   (e) Create a function `e.m.sin(w)` defined as `exp(-sin.npi(w))` (note that because of lexical scoping you do not have to pass the function `sin.npi(w)` as an argument). Repeat parts (b) and (c) with this new function.

2.8   (a) R contains many built-in data sets. These are accessed using the function `data()`. Read about this function in the usual way by typing `?data`. Then type `data` to read the list of available data sets. One of the data sets, `UKgas`, which gives quarterly gas consumption in the UK between 1960 and 1986. Type `help(UKgas)` to read about this data set.

   (b) We would like to compute and plot the mean UK gas consumption by year. The `UKgas` data set is a member of the `ts` (time series) object class. Although we will occasionally deal with time series in this book, we will not use this object class. Therefore we will convert `UKgas` to a matrix `UKg`. Use the function `matrix()` to accomplish this (hint: you can ignore the year column in the `ts` object in this conversion).

   (c) Try typing `mean(UKg)` to compute the yearly mean. That clearly doesn't work. Instead, we will use the function `apply()`. Type `?apply` to read about this function and then use it to compute a vector `UKgm` of mean annual gas consumption. Check your results by computing the mean consumption in one particular year using a different method.

   (d) Create a data frame `UKg.df` from the matrix `UKg`. Then add a field `mean` from `UKgm` that contains the mean consumption. Finally, add a field `year` that contains the year. Give these fields the names *Winter, Spring, Summer, Fall, Mean,* and *Year*.

   (e) Repeat part (c) computing the mean gas consumption over years by quarter.

2.9 (a) Use the function plot() to plot a graph of mean UK gas consumption against year for the years 1960 through 1986 as computed in Problem 2.8.

(b) For these data, plot() automatically plots points. Type the correct statement to plot a line graph instead, with the title "Mean UK Gas Consumption," the abscissa label "Year," and the ordinate label "Mean Consumption."

(c) Read about the function points() and then add points to indicate the gas consumption in the winter of each year. This causes a problem. Adjust the scale of the ordinate, find the maximum consumption over all quarters, and replot the graph with an ordinate scale a bit larger than this.

(d) Plot the data points for all the seasons. Use the argument pch to distinguish seasons.

(e) R has a function legend() that we will discuss in part (f), but for now we will make a key by hand. In the upper left, at about $w = 1965$, $z = 1000$, plot a single point using the same value of pch as you used for the Winter data. Next to this, use the function text() to insert the word "Winter." Make sure you read about the argument pos of the function text() and specify it correctly to locate the text to the right of the symbol. Repeat this process to insert the appropriate symbol and text for the other seasons at appropriate locations.

(f) Now read about the function legend() and redraw the plot using this function in place of the handmade legend of part (e).

2.10 One of the consequences of polymorphism is that there may be more than one version of the same function available. To see how you can find all available versions of a function, first load the raster package by typing library(raster). Next type getAnwhere("plot"). Next, type raster::plot and graphics::plot. The double colon is used to specify membership of a function in a class. Now type library(sf). The plotting function in the sf package is called plot _ sf(), but the package will also understand the command plot(). Type ?plot. Try typing ?plot.sf and ?plot.raster to see another way of seeing the help file for a specific instance of a function.

2.11 Field 1 of Data Set 4 has a trapezoidal shape. The UTM coordinates of the western, southern, and northern boundaries are approximately 592025, 4270452, and 4271132, respectively. The Easting at the northern boundary is 592470, and at the southern boundary 592404. Create an sf polygon object describing the boundary of this field and save it as an ESRI shapefile named *Set4.19697bdry.shp*.

3

Statistical Properties of Spatially Autocorrelated Data

3.1 Introduction

Waldo Tobler made famous his "first law of geography," which states that "everything is related to everything else, but near things are more related than distant things" (Tobler, 1970, p. 236), but he was obviously not the first person to realize this. Awareness of the need to take into account the effect of temporal and spatial proximity when carrying out statistical analyses goes back at least as far as Student (1914). Nowadays, data that obey the first law of geography are said to be *spatially autocorrelated*. The prefix "auto" comes from the Greek word αυτό, meaning "self." Thus, to say a feature is spatially autocorrelated means etymologically that its attribute values are correlated with attribute values of the same feature at nearby locations. Different authors define the term *spatial autocorrelation* differently. Anselin and Bera (1998, p. 241) provide a concise verbal definition: "spatial autocorrelation can be loosely defined as the coincidence of value similarity with locational similarity." For example, one of the quantities recorded in the Weislander survey illustrated in Figure 1.1 is mean annual precipitation. Nearby locations would probably tend to have similar mean annual precipitation levels. Anselin and Bera (1998) also provide a more formal definition as follows: a nonzero spatial autocorrelation exists between attributes of a feature defined at locations i and j if the covariance between feature attribute values at those points is nonzero. If this covariance is positive (i.e., if data with attribute values above the mean tend to be near other data with values above the mean), then we say there is *positive* spatial autocorrelation; if the converse is true, then we say there is *negative* spatial autocorrelation. Positive autocorrelation is much more common in nature, but negative autocorrelation does exist, for example, in the case of conspecific allelopathy, the tendency of some plants to inhibit the nearby growth of other plants of the same species (Rice, 1984). Nevertheless, in this book whenever we use the term spatial autocorrelation we will mean *positive* spatial autocorrelation.

The areal data discussed in this book generally consist of sets of attribute values $Y(x, y)$ together with the coordinates x and y describing the location of each attribute value. The locations (x, y) are fixed and not random, but the attribute values Y are assumed to have a random component. Errors in measuring the location of the phenomenon are incorporated into the uncertainty in the value of the phenomenon at each location. The coordinates themselves are assumed to be measured without error. For example, Figure 1.3 shows the values of clay content in a farmer's field. Error in this case would represent an inaccurate measurement of clay content, but the locations are assumed to be measured accurately.

Consider now the process of modeling a data set such as that of Figure 1.3. A statistical model includes a *random variable*, that is, very roughly speaking, a quantity that is sampled and whose values are distributed according to some probability distribution. A more rigorous discussion of random variables is provided by Larsen and Marx (1986, p. 104). A random variable that is measured at a set of locations is called a *random field* (Besag, 1974, Cressie, 1991, p. 8). Any particular set of measurements of the random field (e.g., the set of clay content measurements illustrated in Figure 1.3), is called a *realization* of the random field (or of the random variable). The random variable may not actually be measurable at every point in the domain. For example, if we are measuring tree yield in an orchard, this quantity is not measurable at locations where there is no tree. The manner in which we deal with this situation will depend on the class of spatial data model (geostatistical or areal, see Section 1.2.1) that we employ in the analysis.

Each data record in every spatial data set is identified as having a location specified by an (x, y) coordinate pair or by a polygon. For the clay content data of Figure 1.3, for example, this location is the point at which the measurement is made. Figure 1.5 shows a spatial mosaic whose cells are polygons defined by land cover class. Observations were made in these cells of the presence or absence of the western yellow-billed cuckoo, with no more than one observation site in each cell. The observations were made by playing a recording of the bird's call and listening for a response. In this case, there is no natural choice for the precise location of the coordinate pair describing the observation site, and it is simply identified as being in the polygon.

3.2 Components of a Spatial Random Process

3.2.1 Spatial Trends in Data

It is often convenient to model a spatial random field as the sum of a collection of separate components, each with its own properties. Cressie (1991, p. 112) presents a model for spatial variability in which the data consist of the sum of four components with differing scales of variability. Burrough and McDonnell (1998, p. 134) describe a slightly simpler system that will suit our needs. In this system, the data are represented as consisting of the sum of three components. These are (1) a "structural," or "deterministic" component that may consist of a trend, of large scale variation, or both; (2) a spatially autocorrelated random process; and (3) an uncorrelated random variable representing uncorrelated random variation and measurement error. Displaying the (x, y) coordinates explicitly, one can write this as

$$Y(x, y) = T(x, y) + \eta(x, y) + \varepsilon(x, y). \tag{3.1}$$

Here $T(x, y)$ represents the deterministic trend, $\eta(x, y)$ represents the spatially autocorrelated random process, and $\varepsilon(x, y)$ represents the uncorrelated random variable. This idea is shown schematicallly in Figures 3.1 and 3.2 (code is also included to create Figure 3.1 using ggplot()). If, for example, $Y(x, y)$ represents mean annual precipitation at the point (x, y), then we might imagine that there is some underlying, large-scale variability that may be modeled as the trend $T(x, y)$ (e.g., due to elevation, slope, and aspect). There are smaller scale components that, taken together, may be modeled as the autocorrelated

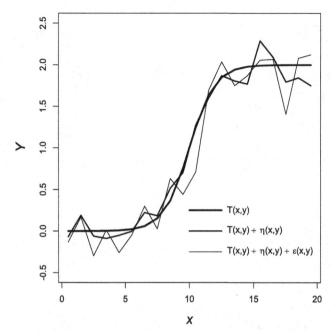

FIGURE 3.1
Plot of a cross section of the variable $Y(x, y)$ shown in perspective in Figure 3.2a.

random variable $\eta(x, y)$ (e.g., due to small scale topographic features and low altitude air movement), and there are still other random components that taken together may be modeled as uncorrelated variability and measurement error $\varepsilon(x, y)$ (Haining, 2003, p. 185). It is important to emphasize that in dealing with real data, any decomposition of the form of Equation 3.1 cannot be unique. To quote Cressie (1991, p. 114), "one person's deterministic trend may be another person's correlated error structure." Keep this in mind as you read the following discussion.

In order to separate a spatial data set into components, one must first estimate the trend term $T(x, y)$ and, if it is large enough to warrant attention, subtract it from the data. The trend is often estimated using either a linear regression model (Appendix A.2) or the *median polish* technique, although a generalized additive model can also be very effective (Dormann et al., 2007, see Exercise 9.3). To use a linear regression model, one fits the data with a regression function in which the predictor variables are functions of the coordinates x and y. The simplest such function is a first-degree model of the form

$$Y_i = \beta_0 + \beta_1 x_i + \beta_2 y_i + \varepsilon_i, \tag{3.2}$$

where Y_i is the measured value at the location (x_i, y_i), $i = 1, ..., n$. Other more complex models, either linear regression models with higher degree terms in the coordinates or nonlinear regression or generalized additive models, may also be used (Unwin, 1975). The parameters of the regression model may be estimated using ordinary or nonlinear least squares, depending on the model. One must exercise caution in interpreting the results of least squares with polynomial terms since the regression may be ill-conditioned (i.e., have numerical properties that lead to an inaccurate solution) (Haining, 2003, p. 326).

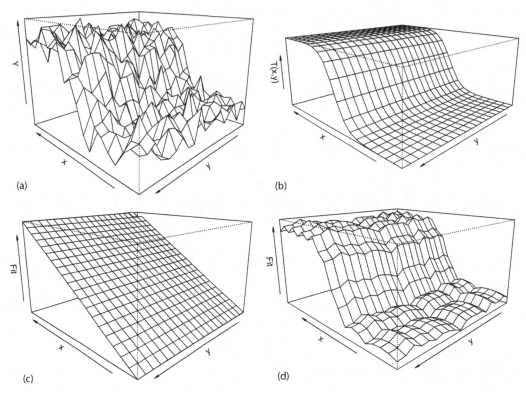

(a)   (b)

(c)   (d)

FIGURE 3.2
(a) Perspective plot of a random field $Y(x, y)$ made up of a trend surface, an autocorrelated random component, and an uncorrelated random component; (b) trend $T(x, y)$ of the random field; (c) linear regression estimate of $T(x, y)$; (d) median polish estimate of $T(x, y)$.

To illustrate the process of fitting a trend surface, we will employ a made-up example consisting of a random field $Y(x, y)$ satisfying Equation 3.1 with the following components. The deterministic component $T(x, y)$ is a logistic function of the form

$$T(x, y) = \frac{ae^{-cy}}{1 + e^{-cy}}. \tag{3.3}$$

where a has the value 2 and c has the value 1. Figure 3.2a shows a perspective plot of the random variable $Y(x, y)$ whose cross section is shown in Figure 3.1. Figure 3.2b shows the actual trend surface component $T(x, y)$ First, we estimate the trend $T(x, y)$ using linear regression with Equation 3.2. The R function most commonly used for ordinary linear regression is lm(). This function is discussed in Appendix A.2. The R package spatial (Venables and Ripley, 2002, p. 420) contains a function surf.ls() that can also be used to fit a trend surface by least squares, but for this simple case doing it directly with lm() is easier and more transparent. The values of $Y(x, y)$ are stored in a data frame called Y.df. The statements that create this data frame are in the code that accompanies this book. The code to compute the estimated trend and display a perspective plot on a 20 by 20 grid is the following.

```
> model.lin <- lm(Y ~ x + y, data = Y.df)
> coef(model.lin)
        (Intercept)            x             y
        -0.435929572 0.143930575 0.000212603
> trend.lin <- predict(model.lin)
> Y.lin <- matrix(trend.lin, nrow = 20)
> Fit <- 4 * Y.lin
> persp(x, y, Fit, theta = 225, phi = 15, scale = FALSE)
```

The first argument in `lm()` is the regression formula (see Appendix A.2), and the second is the source of the data. The estimated regression coefficients are $b_0 = -0.436$ $b_1 = 0.144$, and $b_3 = 0.0002$. The function `predict()` gives the predicted values of the linear model at the data points. The next two lines of code generate the trend surface shown in Figure 3.2c. The function `persp(x, y, Y, theta, phi, scale...)` displays a perspective plot of Y as a function of x and y. The arguments `theta` and `phi` give the azimuth and latitude from with the plot is viewed, and `scale` is a logical argument that, if true, allows each axis to be scaled independently. In lieu of this, we simply multiply the z component by four (obtained by trial and error) to exaggerate it.

The second approach to fitting a trend surface, median polish, is a nonparametric method, that is, one in which no assumptions are made about the distribution of the error terms, in the way that they are with linear regression. Median polish is an iterative procedure originally described by Tukey (1977) and discussed by Emerson and Hoaglin (1983) and Cressie (1991, p. 185). At each iteration, one alternately subtracts row and column medians from the data set. The R function that performs median polish is `medpolish()`. The code to carry out the operations is

```
> Y.trend <- matrix(Y.df$Y, nrow = 20)
> model.mp <- medpolish(Y.trend)
> Y.mp <- Y.trend - model.mp$residuals
> Fit <- 4 * Y.mp
> persp(x, y, Fit, theta = 225, phi = 15, scale = FALSE)
```

The median polish trend surface of the simulated data set is shown in Figure 3.2d. In this example, the median polish fit more accurately represents the trend surface, and because of the logistic shape of this surface, median polish would probably still be more accurate even if higher order powers of x and y were included in the regression in Equation 3.2.

The choice of method for trend estimation depends on the application, and, to some extent, on the trend surface. Median polish frequently provides a more accurate fit $\hat{T}(x, y)$ of the trend, which is useful for visualization and, if the trend surface is complex, subtracting this estimate from $Y(x, y)$ provides a more accurate prediction $\hat{\eta}(x, y) + \hat{\varepsilon}(x, y)$ of the random quantity represented as the sum $\eta(x, y) + \varepsilon(x, y)$ in Equation 3.1. On the other hand, the parameter values generated by fitting a regression model can provide useful summary information, such as whether the trend is stronger in the x or y direction, or whether it has a strong quadratic component or interaction, etc. The parameter values are also useful in comparing trends of different attributes of a data set.

We can compare the methods using a real data set. Field 1 of Data Set 4 has the shape of a trapezoid about twice as long in the north-south direction as in the east-west direction (see Figure 3.4a below). The contents of the file *Set4.196sample.csv* are loaded into the data frame `data.Set4.1` using the code in Appendix B.4. There is a strong north-south trend in sand content. The data for plotting with the function `persp()` must be in the form of a matrix, so we make use of the *Row* and *Column* data used to identify the sample locations (Appendix B.4) to construct a matrix using a simple `for` loop.

```
> Sand <- matrix(nrow = 13, ncol = 7)
> for (i in 1:86){
+    Sand[data.Set4.1$Row[i], data.Set4.1$Column[i]] <-
+        data.Set4.1$Sand[i]
+ }
```

The percent sand values range up to about 45%. Rather than using the `scale` argument of the function `persp`, we simply scale the figure manually by multiplying the row and column numbers of the sample locations by 3, again obtained by trial and error.

```
> North <- 3 * 1:13
> West <- 3 * 1:7
> persp(North, West, Sand, theta = 30, phi = 20,
+    zlim = c(0,45), scale = FALSE) # Fig. 3.3a
```

The resulting perspective plot is shown in Figure 3.3a. Because of the curvature of the plot in the *y* (north-south) direction, it appears that a linear regression model fit should include quadratic terms, at least in this direction. We again use the function `lm()` to fit the trend.

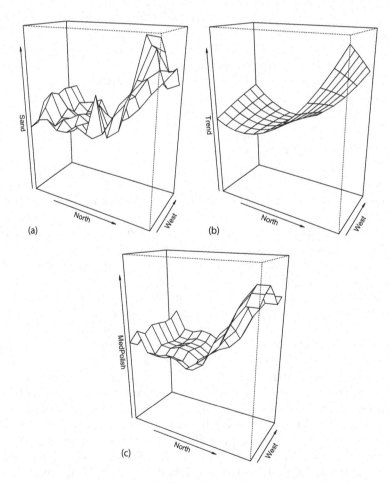

FIGURE 3.3
Perspective views of the sand data of Field 1 of Data Set 4: (a) actual data; (b) trend surface estimated by linear modeling including second-order terms; (c) trend surface estimated by median polish.

```
> trend.lm <- lm(Sand ~ Row + Column + I(Row^2) +
+   I(Column^2) + I(Row*Column), data = data.Set4.1)
```

The expressions such as I(Row^2) are used to indicate the evaluation of a function such as Row^2. The fitted model is $\hat{T}(x,y) = 22.42 + 1.29x - 1.95y - 0.21xy + 0.28y^2$; the x^2 coefficient is very small and has been omitted. This function captures the main feature of the trend: its rapid increase in the y direction at the south end of the field. The slight increase in the x (east-west) direction in the north end as well as the interaction may well be spurious, but in any case, they are not pronounced (Figure 3.3b). The median polish approximation (Figure 3.3c) appears to fit the data better than the linear model in the y direction and about the same in the x direction (where there is little if any trend).

One of the main reasons for computing the trend often is to remove it. Figure 3.4a shows a bubble plot of Field 4.1 in which the size of the circle is proportional to the sand content.

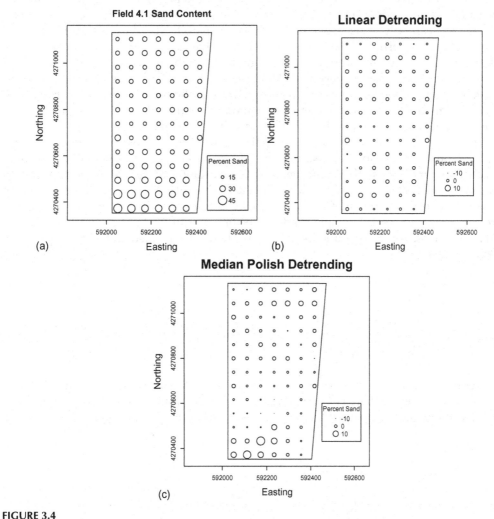

FIGURE 3.4
(a) Thematic bubble maps of sand content data of Field 1 of Data Set 4: (actual data); (b) data after trend removal using the linear model; (c) data after trend removal using median polish. The size of the circles represent the sand content values.

Equation 3.1 indicates that the quantity $\hat{\eta}(x, y) + \hat{\varepsilon}(x, y) = Y(x, y) - \hat{T}(x, y)$ (here as well as generally in this book the hats are used to indicate estimated values) should contain little if any trend; this quantity is referred to as the *detrended data*. Figure 3.4b and c show the detrended sand content data based on linear model and on median polish, respectively. A desirable property of these detrended data is to be *stationary*. This concept is discussed in the next section.

3.2.2 Stationarity

As stated in Section 3.1, the data described in this book are conceptualized as being a realization of a random field, that is, of a set of random numbers each of which is associated with a spatial location. To say that a data set is a *realization* of a random field means that nature has assigned, according to some well-defined law, values to the random variables that make up this data set. In theory, the assignment could be carried out again according to the same law to produce a different set of values that would be a different realization. Generally, one can only observe one realization of a spatial data set, and thus if one is to infer any of the properties of the law by which the values are assigned, one must use the measurements at different locations of that one realization, analogous to how one uses replications in a controlled experiment (Haining, 1990, p. 33). If these measurements at multiple locations are to be considered as if they were replications, then the law by which they are assigned must not vary from one location to another. A spatial random process whose properties do not vary by location is said to be *stationary*.

We will use a coin-tossing simulation to gain an intuitive idea of stationarity. As with spatial stationarity described as location invariance in the previous paragraph, a time series is stationary if its statistical properties are invariant in time. We begin by simulating a sequence of 30 tosses, each of which has a probability 0.5 of landing heads. Since the probability is independent of time, the process is stationary. Instead of simulating the entire set of coin tosses as a single random vector, we will use a for loop to explicitly simulate each toss. We use the function rbinom(), which is included in the base R package (try ?rbinom).

```
> set.seed(123)
> n.tosses <- 30
> head <- numeric(n.tosses)
> for (i in 1:n.tosses) head[i] <- rbinom(1,1,0.5)
> head
 [1] 0 1 0 1 1 0 1 1 1 0 1 0 1 1 0 1 0 0 0 1 1 1 1 1 1 1 1 1 0 0
```

In the sequence of tosses, every value has an equal probability of turning up heads (one) or tails (zero). By chance there is a very long run of heads towards the end of the sequence. Now let us make each toss dependent on the previous toss.

```
> set.seed(123)
> n.tosses <- 30
> head <- numeric(n.tosses)
> p <- 0.5
> for (i in 1:n.tosses){
+   head[i] <- rbinom(1,1,p)
+   p <- ifelse(head[i] > 0, 0.8, 0.2)
```

```
+ }
> head
   [1] 0 0 0 1 0 0 0 1 1 1 0 0 0 0 0 1 1 1 1 0 1 1 1 0 0 0 0 0 0 0
```

This sequence is also stationary. It is true that, for example, in this data set the probability that the fifth toss will be a head is 0.8, because the fourth toss was a head, while the probability that the sixth toss will be a head is 0.2, because the fifth toss was a tail. However, the same rule for determining the outcome of the toss, a probability of 0.8 that it will be the same as the previous toss, applies for all tosses, and this independence of time of the toss is what determines stationarity in a time series.

Next, suppose that the probability that a toss lands heads starts at a low value gradually increases as the sequence proceeds.

```
> set.seed(123)
> n.tosses <- 30
> head <- numeric(n.tosses)
> p <- 0.3
> for (i in 1:n.tosses){
+    head[i] <- rbinom(1,1,p)
+    p <- p + 0.02
+ }
> head
   [1] 0 1 0 1 1 0 0 1 1 0 0 1 0 0 1 0 1 1 1 0 0 1 1 0 1 1 1 1 1 1
```

This sequence is not stationary, because the probability of heads depends on position in the sequence.

It would be very hard to determine based on the data which of the sequences in these three examples is stationary and which is not. Moreover, if in the second example the probabilities of head and tail were 0.99 and 0.01 instead of 0.8 and 0.2, it could well be the case that the data would consist of a sequence of, for example, only heads followed by a sequence of only tails. Nevertheless, this would be a stationary process. We can recall in this context the comment of Cressie (1991) quoted in Section 3.2.1 that one person's deterministic trend is another person's correlated error structure. We will have to confront this issue again and again over the course of our analyses.

Now we return to spatial processes. For the present, we define stationarity only for areal data such as that represented in Figure 1.3b. The most useful definition of stationarity for areal data is *second-order stationarity*. A random field Y defined on a finite mosaic consisting of n locations is second order stationary if (Cliff and Ord, 1981, pp. 142–143, see also Anselin, 1988, p. 42, Isaaks and Srivastava, 1989, p. 222)

$$E\{Y_i\} = \mu,$$

$$\text{var}\{Y_i\} = \sigma^2, \tag{3.4}$$

$$\text{cov}\{Y_i, Y_j\} = \sigma^2 cor\{(x_i, y_i) - (x_j, y_j)\},$$

where μ and σ^2 are independent of location i, (x_i, y_i) and (x_j, y_j) are the position vectors of locations i and j, and $cor(x, y)$ is a correlation function depending only on the relative locations of (x_i, y_i) and (x_j, y_j), and not on their specific positions. A stronger definition of stationarity, called *strict stationarity*, requires that the independence of location expressed in Equations 3.4 hold for all moments of the distribution. If the distribution is normal, then second-order stationarity implies strict stationarity since the normal distribution is defined by its first two moments, the mean and the variance, but this is not true for probability distributions in general. However, second-order stationarity ensures that two of the most important statistical properties, the mean and the variance, do not vary with position. Our original motivation for requiring stationarity was to use multiple measurements as replications. In this context, second-order stationarity is sufficiently strong to serve the purpose, but not so strong as to exclude data sets that we want to be able to consider stationary.

Sometimes there are obvious reasons why spatial data are not stationary. Perhaps the most common example is if the data involve species dispersal. By definition, the interesting part of these data is generally how their pattern changes in time, and a population that is stationary is not dispersing. Another example is if there is an obvious gradient in some important quantity. For example, the oak distribution data in Data Set 1 are unlikely to be stationary since the trees are heavily influenced, either directly or indirectly, by elevation, which is on a spatial gradient. The exploratory methods described in Chapters 7 through 9 do not require spatial stationarity since they are non-spatial in nature, but the spatially based methods described in Chapters 13 and after do assume stationarity in some sense. In the case of regression methods, however, often it is not stationarity of the data that is required but rather stationarity of the residuals. Sometimes the residuals of a properly parameterized regression model will be stationary (or at least close enough) even if the data are not. Sometimes they are not, however, and so some means of testing for stationary can be very important.

There is an extensive literature on testing for stationarity in time series (e.g., Elliot et al., 1996). Ultimately, however, stationarity is a modeling assumption and, since there are many ways in which a process can be non-stationary, there is no test that can cover all of them. There has been less comparable development of tests of stationarity for spatial data than for time series. There are two fundamental differences between time series data and spatial data that make the notion of testing for stationarity of spatial data less meaningful (Haining, 2003, p. 47). The first is that time is characterized by a flow (and hence a dependency) in one direction. One can distinguish between past, present, and future: one can specify that events in the present or past cannot depend on the future. There is no equivalent concept with spatial data. Second, spatial data $Y(x, y)$ exist (in this book) in two dimensions, and there is no guarantee that the structure of spatial data be the same in one dimension as it is in the other. If the dependency has the same structure independent of direction, the data are said to be *isotropic*, otherwise the data are *anisotropic*. If you rub your hand along a piece of corduroy fabric, the resistance depends on whether you rub with the cords or perpendicular to them, so it is anisotropic. Plain fabric does not offer a different resistance depending on direction; it is isotropic.

The properties of stationarity and isotropy in spatial data must always be presented as assumptions, which can be examined in an exploratory way but cannot be subjected to rigorous hypothesis testing. As stated succinctly by ver Hoef and Cressie (2001, p. 299), "these assumptions are impossible to test, because it is impossible to go back in time again and again and generate the experiment each time to check whether each experimental unit has the same mean value or whether the correlation is the same for all pairs of plots that are at some fixed distance from each other. However, any gross spatial trends in the residuals...would cause suspicion that they are not stationary." A data set like the sand content data discussed in Section 3.2.1 obviously is not stationary, although the residuals of the detrended data (i.e., the sum $\hat{\eta}(x, y) + \hat{\varepsilon}(x, y)$) might be expected to satisfy the assumption of stationarity. There are some statistical tools that can be used to explore the stationarity of spatial data. One of these is the local Moran's *I*, discussed in Section 4.5.3, and another is geographically weighted regression, discussed in Section 4.5.4.

3.3 Monte Carlo Simulation

In many cases, the properties of a statistical model cannot be worked out analytically. *Monte Carlo simulation* is a very useful tool for estimating these properties numerically. Monte Carlo simulation has a number of definitions (Ripley, 1981, Besag and Clifford, 1989, Manly, 1997), but a very general one is given by Rubinstein (1981, p. 11) as "a technique, using random or pseudorandom numbers, for solution of a model." The statistical definition of an experiment is an action that can in principle be replicated an arbitrary number of times (Larsen and Marx, 1986, p. 14). We have made a point of distinguishing between a replicated experiment and an observational study, so at the risk of being pedantic we will put the word "experiment" in quotes when used in the statistical context just given. In practice "experiments" are generally only carried out once or a few times. The idea of a Monte Carlo simulation is to carry out a simulated "experiment" many (e.g., 1000 or 10,000) times, to observe the properties of the resulting distribution of outcomes, and to compute statistics characterizing this distribution. It is usually no problem for R to generate the pseudorandom numbers (see Section 2.3.2) needed to simulate the "experiment" many times.

The R function `replicate()` can be used to carry out Monte Carlo simulations. Consider the "experiment" consisting of tossing a fair coin $n = 20$ times and recording the number of heads. The probability of the coin turning up heads on any one toss is $p = 0.5$. According to standard statistical theory (Larsen and Marx, 1986, p. 96), the number of heads in 20 tosses follows a binomial distribution with a mean $np = 10$ and a variance $np(1 - p) = 5$. Let us implement and display the results of a Monte Carlo simulation of this "experiment." The actual "experiment" is placed in a function called `coin.toss()` that generates the number of heads as a pseudorandom number from a binomial distribution.

```
> coin.toss <- function (n.tosses, p){
+    n.heads <- rbinom(1,n.tosses, p)
+ }
```

The function takes two arguments, the number of tosses n.tosses and the probability of heads p. Note the lack of a `return` statement. As mentioned in Section 2.5, when there is no `return` statement, the last quantity computed, which in this case is n.heads, is returned.

Now we are ready to carry out the Monte Carlo simulation.

```
> set.seed(123)
> n.tosses <- 20
> p <- 0.5
> n.reps <- 10000
> U <- replicate(n.reps, coin.toss(n.tosses, p))
```

The function `replicate()` generates a vector U whose elements are the number of heads in each of the 10,000 replications of the experiment.

```
> mean(U)
[1] 9.9814
> var(U)
[1] 4.905345
> hist(U, cex.lab = 1.5, # Fig. 3.5
+     main = "Number of Heads in 20 Tosses",
+     cex.main = 2, xlab = "Number of Heads")
```

The mean and variance of U are close to their theoretically predicted values of 10 and 5, respectively, and a histogram of U indicates that it has the approximately normal distribution (Figure 3.5) that is expected based on central limit theorem (Larsen and Marx, 1986, pp. 206, 322). This theorem says, roughly speaking, that, given a set of random variables from any distribution, not necessarily normal, the sample mean of this set is approximately normally distributed, with the approximation becoming better as the sample size increases.

In the simple example of the coin tossing "experiment" just carried out, the statistical properties of the sampling distribution (i.e., its mean, variance, etc.) could be computed analytically, but for many "experiments," particularly those involving spatial statistics, this is not the case. In such circumstances, Monte Carlo simulation may be the only possibility for obtaining insight into the properties of the data. Even in those cases where a closed form solution is possible, a Monte Carlo simulation often provides insight and intuition not so easily available from analytical calculations.

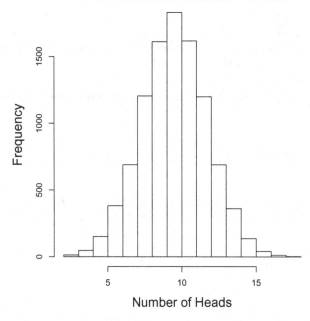

Number of Heads in 20 Tosses

FIGURE 3.5
Histogram of the results of 10,000 Monte Carlo simulations of the tossing of a fair coin 20 times in succession and counting the number of heads.

3.4 A Review of Hypothesis and Significance Testing

Spatial autocorrelation often affects the outcome of significance tests. For this reason, we will spend some time reviewing this procedure. The review is informal and intuitive; for a more formal discussion see Muller and Fetterman (2002, p. 4) and Schabenberger and Pierce (2001 p. 22). Suppose we have measured a set of values $\{Y_1, Y_2, ..., Y_n\}$ of a normally distributed random variable Y whose mean μ and variance σ^2 are unknown. In this book, both the random variable and its values are denoted with an upper case Latin letter. Other texts frequently denote the values of the random variable with lowercase letters, but we do not follow this practice because the symbols x and y are used to indicate spatial coordinates. We can write the expression for the Y_i as

$$Y_i = \mu + \varepsilon_i, \ i = 1, ..., n, \tag{3.5}$$

where the ε_i are independent, identically distributed random variables drawn from a normal distribution with mean zero and variance σ^2. We wish to test the null hypothesis that the value of the mean μ is zero against the alternative that it is not,

$$H_0 : \mu = 0,$$
$$H_a : \mu \neq 0. \tag{3.6}$$

When the population from which the values Y_i are drawn is normal, this hypothesis can be tested using the Student t statistic. This statistic is defined as follows. The sample mean $\bar{Y}$ is given by

$$\bar{Y} = \frac{1}{n} \sum_{i=1}^{n} Y_i. \tag{3.7}$$

If the Y_i are independent, then the variance of $\bar{Y}$ is given by

$$\text{var}\{\bar{Y}\} = \frac{\sigma^2}{n} \tag{3.8}$$

(Larsen and Marx, 1986, p. 321). Moreover, the sample variance, defined by

$$s^2 = \frac{1}{n-1} \sum_{i=1}^{n} (Y_i - \bar{Y})^2, \tag{3.9}$$

is an unbiased estimator of σ^2. Therefore s^2 / n is an unbiased estimator of the variance of $\bar{Y}$. The square root of this quantity is called the *standard error* and is given by

$$s\{\bar{Y}\} = \frac{s}{\sqrt{n}}. \tag{3.10}$$

With these quantities defined, the Student t statistic is given by

$$t = \frac{\bar{Y} - \mu}{s\{\bar{Y}\}}. \tag{3.11}$$

TABLE 3.1

The Four Possible Outcomes of the Test
$H_0 : \mu = 0$ against the Alternative $H_a : \mu \neq 0$

	H_0 True	H_0 False
Don't reject H_0	OK	Type II error
Reject H_0	Type I error	OK

The distribution of this statistic under $H_0 : \mu = 0$ in (3.5) is called the Student t distribution.

One way to carry out the test is to compare the value of the t statistic computed in Equation 3.11 with the percentiles of the Student t distribution. If H_0 is true, then the value of t should be small. According to the standard theory (Larsen and Marx, 1986, p. 343), if we define $t_{\alpha/2}$ as the $\alpha/2$ percentile of the Student t distribution, then $P\{|t| > t_{\alpha/2}\} = \alpha$, i.e., the probability that the magnitude of the t statistic will exceed the $\alpha/2$ percentile, is equal to α. There are four possible outcomes of this test, which are shown in Table 3.1. Two of these represent an error, that is, a failure to draw the correct conclusion from the test. In particular, the test will with probability α result in a Type I error, that is, a rejection of the null hypothesis when it is true (Table 3.1). It is important to emphasize that the most appropriate interpretation of a failure to reject the null hypothesis is *not* to conclude that the null hypothesis is true and that $\mu = 0$. We have not proven that $\mu = 0$; we have simply not been able to demonstrate with adequate certainty that it is not zero. The most appropriate way to verbalize a failure to reject the null hypothesis that $\mu = 0$ is to say that, based on our available evidence, we cannot distinguish the value of μ from zero.

There are two approaches to the test of H_0 (Schabenberger and Pierce, 2001, p. 22). The first is to fix α at some pre-assigned value, say, 0.1 or 0.05, and determine based on the value of the percentile of the t distribution relative to this fixed value whether or not to reject H_0. The second approach is to report the probability of observing a statistic at least as large in magnitude as that actually observed (i.e., the p value) as the *significance* of the test. Manly (1997, p. 1) advocates the latter policy, pointing out that "To avoid the characterization of belonging to 'that group of people whose aim in life is to be wrong 5% of the time' (Kempthorne and Doerfler, 1969), it is better to regard the level of significance as a measure of the strength of evidence against the null hypothesis rather than showing whether the data are significant or not at a certain level." Nevertheless, the magic number 0.05 is sufficiently ingrained in applied statistics that it is impossible to completely avoid using it. In addition, we will see that it provides a convenient means to measure the effect of spatial autocorrelation on the results of the test.

If, for whatever reason, the null hypothesis H_0 is not rejected, there is the possibility that this decision is incorrect. Failure to reject the null hypothesis when it is false is called a Type II error (Table 3.1). The *power* of the test is defined as the probability of correctly rejecting H_0 when it is false, that is, $power = 1 - \Pr\{Type\ II\ error\}$.

As an illustration of the use of R in a test of a null hypothesis such as that in Equation 3.6, we generate a sample of sample size $n = 10$ from a standard normal distribution.

```
> set.seed(123)
> Y <- rnorm(10)
> Y.ttest <- t.test(Y, alternative = "two.sided")
> Y.ttest$p.value
 [1] 0.8101338
```

It happens that this time the p value is about 0.81. We can modify this to carry out a Monte Carlo simulation of the test. The default value α of the function `ttest()` is 0.05, and indeed the fraction of Type I errors is close to this value.

```
> set.seed(123)
> ttest <- function(){
+    Y <- rnorm(10)
+    t.ttest <- t.test(Y,alternative = "two.sided")
+    TypeI <- t.ttest$p.value < 0.05
+ }
> U <- replicate(10000, ttest())
> mean(U)
[1] 0.0485
```

There is an alternative method for carrying out hypothesis and significance tests that we shall occasionally employ. This is called, depending on the author, either a *permutation test* or a *randomization test*. We will use the former term, although by doing so we probably put ourselves in the minority, in order to avoid confusion with the *randomization assumption*, which is discussed in Chapter 4. Permutation tests are nonparametric, so that they do not depend on the population fitting any parameterized distribution, in the way that the t test depends on the normal distribution (see Exercise 3.1). They are simple both to understand and to carry out. For a reason that will soon become apparent, we will not use a permutation test to test the null hypothesis of Equation 3.6. Instead, we introduce this form of hypothesis test using a *two-sample* test, in which we test the hypothesis that two independent samples come from the same probability distribution (Manly, 1997, p. 97). We first draw two random sequences $Y_1 = \{Y_{1i}\}$, $i = 1,...,5$ and $Y_2 = \{Y_{2i}\}$, $i = 1,...,5$ of five values each. Each sequence is drawn from a standard normal distribution (since this test is nonparametric, the parameters could be drawn from any distribution, but the normal is a convenient one). The difference d between the means of the two sequences is small but not negligible.

```
> set.seed(123)
> print(Y1 <- rnorm(5))
[1] -0.56047565 -0.23017749 1.55870831 0.07050839 0.12928774
> print(Y2 <- rnorm(5))
[1] 1.7150650 0.4609162 -1.2650612 -0.6868529 -0.4456620
> print(d <- mean(Y1) - mean(Y2))
[1] 0.2378892
```

We will test the hypothesis that the expected values of the parent distributions of the two samples are the same. The basic idea is to repeatedly create new vectors Y_1' and Y_2' by rearranging (i.e., permuting) the elements of Y_1 and Y_2, and comparing the mean of the differences $d' = Y_1' - Y_2'$ with that of $d = Y_1 - Y_2$. If Y_1 and Y_2 come from the same population then, if we repeat this permutation many times, the difference d between the means of Y_1 and Y_2 should not have an extreme value in the distribution of means of permuted arrays. If it does have an extreme value, then it is likely that Y_1 and Y_2 do not come from the same population.

The first step of the procedure is to concatenate the two sequences into a single sequence of 10 elements. Then we rearrange these in a random order value and finally we separate the reordered array into two new sub-arrays of five elements each. These are the test arrays Y_1' and Y_2'. We use the R function `sample()` to implement the permutation. Read about this function using ?sample. By default sample(x,...) returns an array of the same length as x of randomly sampled values of the elements of x. One of the optional arguments of sample() is replace, which indicates sampling with replacement if it has the value TRUE. By default the value of replace is FALSE.

```
> Y <- c(Y1, Y2)
> Ysamp <- sample(Y,length(Y))
> print(Yprime1 <- Ysamp[1:5])
[1] -0.68685285 0.46091621 1.71506499 -0.44566197 0.07050839
> print(Yprime2 <- Ysamp[6:10])
[1] -1.2650612 1.5587083 -0.2301775 -0.5604756 0.1292877
> print(dprime <- mean(Yprime1) - mean(Yprime2))
[1] 0.2963386
```

The difference $d' = Y_1' - Y_2'$ in this case happens to be about the same magnitude as d. Now we create a function perm.diff() that carries out the permutation and computation of d'

```
> perm.diff <- function(Y1, Y2){
+    Y <- c(Y1, Y2)
+    Ysamp <- sample(Y,length(Y))
+    Yprime1 <- Ysamp[1:5]
+    Yprime2 <- Ysamp[6:10]
+    dprime <- mean(Yprime1) - mean(Yprime2)
+ }
```

We can now use the function replicate() to generate a set of differences between the means of permutations of the elements of Y_1 and Y_2. We start with nine permutations and add the observed difference d as the tenth.

```
> set.seed(123)
> U <- replicate(9, perm.diff(Y1, Y2))
> sort(c(d,U))
 [1] -0.64644597 -0.61387764 -0.20122761 -0.08870639 -0.08856700
 [6]  0.15530214  0.23788923  0.29633862  0.86680338  1.14863761
```

The observed d occupies the seventh position among the differences in permutations. To get a more precise evaluation, we can generate a p value by running a large number (say, 10,000) permutations and counting the number in the "tail," that is, the number of values that are farther away from the median than is d. Doubling this number (to account for both tails) and dividing by 10,000 yields a p value.

```
> set.seed(123)
> U <- replicate(9999, perm.diff(Y1, Y2))
> U <- c(U, d) # Add original observation to get 10,000
> U.low <- U[U < d] # Obtain diff. values less than d
> {if (d < median(U)) # Is d in the upper or lower tail?
+    n.tail <- length(U.low)
+  else
+    n.tail <- 10000 - length(U.low)
+ }
> print(p <- 2 * n.tail / 10000) # Two tail test
[1] 0.714
```

The elements of the vector U are the 9,999 differences between means of random permutations of the values in Y_1 and Y_2; the original value d is the ten thousandth. The statement U.low <- U[U < d] is used to determine how many values of U are less than the difference between the original samples. If this value is either very small or very large, then d is judged sufficiently extreme to justify rejection of the null hypothesis. In the example, about 71% of the randomized rearrangements have differences farther away from the middle value than the observed value of 0.238 (to three decimal places) so the p value computed by the permutation test is $p = 0.71$. This is not far from the value computed by a t test.

```
> t.test(Y1, Y2, "two.sided")$p.value
[1] 0.7185004
```

Having carried out the two sample permutation test, we can see why a one sample permutation test would not work. Rearranging the order of a single sample does not affect the sample mean, so there is nothing to test (Manly, 1997, p. 17). There are other nonparametric tests that can be used for a one sample test, the simplest of which is the *sign test* (Crawley, 2007, p. 300), but we will not pursue these here.

There has been considerable discussion over the years about the scope of inference of permutation tests and how it compares with the scope of inference of traditional parametric tests such as the t-test. Our summary of this discussion is based on comments by Romesburg (1985) and Edgington (2007). One of the assumptions underlying parametric tests such as the Student t-test is that the data are a random sample of the population, and in many cases this assumption is violated. Permutation tests, on the other hand, do not require this randomness assumption. Fisher (1935) originally developed the permutation test as a means of demonstrating the robustness of the Student t-test when used with non-normal data. Romesburg (1985, p. 22) asserts that because of the close correspondence between the results of parametric tests and permutation tests, a parametric test may be considered as an approximation of a permutation test. In this context, the assumption of randomness can, according to Romesburg (1985), be relaxed for the parametric test as well. On the other hand, Edgington (2007) and others have pointed out that, strictly speaking, permutation tests apply only to the data sets on which they are based, and cannot with statistical validity be extended to a wider population. In other words, the results of an analysis using parametric statistics can be extended to the entire population from which the sample is drawn, but require assumptions of random sampling that are almost never satisfied in practice. The results of a permutation test, on the other hand, do not depend on assumptions of random sampling but cannot be extended to the entire population from which the sample is drawn. What does one do in the face of this dilemma? First, when both parametric and permutation tests can be applied, the best approach is to use them both and compare the results. When one does not have a theoretical justification for extending one's scope of inference, one must be as careful as possible to do everything practical to justify this extension in scope, that is, to make sure that the data set is not selected from the population in a way that makes it a biased sample either because the selection process was biased or because the scope of inference is extended too far. It is useful to recall a quote of G.E.P. Box (1976): "Since all models are wrong, the scientist must be alert to what is importantly wrong. It is inappropriate to be concerned about mice when there are tigers abroad." It is, however, entirely appropriate to do everything possible to ensure that no tigers are inadvertently released.

3.5 Modeling Spatial Autocorrelation

3.5.1 Monte Carlo Simulation of Time Series

We begin the discussion of the simulation of spatial data with a simpler topic: the simulation of time series data. We will use time series to illustrate the effects of positive autocorrelation on the outcome of hypothesis tests. Much of the theory of spatial statistics grew out of the analogy with time series (Whittle, 1954; Bartlett, 1935, p. 18) and because of the one-dimensional and directional nature of time series data, many of the most important concepts are more intuitive in that domain. Simulation through the use of artificially generated data sets has the advantage of permitting the analysis of data whose distributional properties are known. We will use Monte Carlo simulation to explore the effect of temporal autocorrelation on the outcome of a test of the null hypothesis that an observed data set is drawn from of a population with mean zero.

In Section 3.4, we presented the results of a simulation using uncorrelated data. The simulated error rate, that is, the fraction of times a Type I error occurred, was 0.0485.

Now we will explore the effect of autocorrelation on the outcome of the *t*-test. The autocorrelation takes the form of a first-order autoregressive time series in the error terms (Kendall and Ord, 1990, p. 56). The model is initially written as

$$Y_i = \mu + \eta_i$$
$$\eta_i = \lambda \eta_{i-1} + \varepsilon_i, \, i = 1, 2, \ldots \qquad (3.12)$$

where $-1 < \lambda < 1$ and the ε_i are independent normally distributed random variables with mean zero and variance σ^2. The η_i are the autoregressive error terms. The model is specified in this form to maintain consistency with the spatial model that will be discussed in the next section.

The equations are easiest to work with if we substitute $\eta_i = Y_i - \mu$ and $\eta_{i-1} = Y_{i-1} - \mu$ into the second of Equations 3.12 to get

$$Y_i - \mu = \lambda(Y_{i-1} - \mu) + \varepsilon_i, \, i = 1, 2, \ldots \qquad (3.13)$$

If $\lambda = 0$ then this reduces to Equation 3.5. The time series is initiated by specifying Y_1 and setting $\eta_0 = 0$ and $\varepsilon_1 = 0$. We then have

$$Y_2 = \mu + \lambda(Y_1 - \mu) + \varepsilon_2$$
$$Y_3 = \mu + \lambda(Y_2 - \mu) + \varepsilon_3$$
$$= \mu + \lambda^2(Y_1 - \mu) + \lambda\varepsilon_2 + \varepsilon_2 \qquad (3.14)$$
$$Y_4 = \mu + \lambda(Y_3 - \mu) + \varepsilon_4$$
$$= \mu + \lambda^3(Y_1 - \mu) + \lambda^2\varepsilon_2 + \lambda\varepsilon_3 + \varepsilon_4.$$

and so forth. Since $-1 < \lambda < 1$, $\lambda^j \to 0$ as j increases, so that the influence of the initial term declines with each time step.

We will simulate this process with the value of μ set to zero so that the null hypothesis is true. Since $\mu = 0$, it follows that Equation 3.13 becomes

$$Y_i = \lambda Y_{i-1} + \varepsilon_i, \ i = 1, 2, ..., \tag{3.15}$$

and this is how the simulation is programmed. The first simulation, with λ set at 0.4, involves a sample of size 10. The initial value is set at $Y_1 = 0$, which is a fixed, non-random number. This may cause the data to be non-stationary for the first few values of i. This effect is sometimes called an "initial transient," and to avoid it, twenty values of Y_i are generated and the first ten are discarded.

```
> lambda <- 0.4 #Autocorrelation term
> set.seed(123)
> Y <- numeric(20)
> for (i in 2:20) Y[i] <- lambda * Y[i - 1] + rnorm(1)
> Y <- Y[11:20]
```

There is an R function `arima.sim()` that accomplishes the same thing as this code without using a for loop. The call to the function in this case would be `Y <- arima.sim(list(ar=lambda),n = 10, n.start = 10)`. The for loop is used explicitly to emphasize the connection with Equation 3.15.

We can now carry out a *t*-test of the null hypothesis $H_0 : \mu = 0$.

```
> Y.ttest <- t.test(Y,alternative = "two.sided")
> #Assign the value 1 to a Type I error
> TypeI <- as.numeric(Y.ttest$p.value < 0.05)
> Ybar <- mean(Y)
> Yse <- sqrt(var(Y) / 10)
> c(TypeI, Ybar, Yse)
[1] 0.0000000 0.2790157 0.3028951
```

We concatenate into a single vector an indicator variable for a Type I error (1 = Type I error, 0 = H_0 not rejected), the sample mean $\bar{Y}$ and standard error $s\{\bar{Y}\}$.

Now we are ready to combine this into a Monte Carlo simulation.

```
> set.seed(123)
> lambda <- 0.4
> ttest <- function(lambda){
+    Y <- numeric(20)
+    for (i in 2:20) Y[i] <- lambda * Y[i - 1] + rnorm(1, sd = 1)
+    Y <- Y[11:20]
+    Y.ttest <- t.test(Y,alternative = "two.sided")
+    TypeI <- as.numeric(Y.ttest$p.value < 0.05)
+    Ybar <- mean(Y)
+    Yse <- sqrt(var(Y) / 10)
+    return(c(TypeI, Ybar, Yse))
+ }
```

We use the function replicate() to carry out the simulation as described in Section 2.6.

```
> U <- replicate(10000, ttest(lambda))
> mean(U[1,]) # Type I error rate
[1] 0.1936
> mean(U[2,]) # Mean value of Ybar
[1] 0.003212074
> mean(U[3,]) # Mean est. standard error
[1] 0.3128384
> sd(U[2,]) # Sample std. dev. of Ybar
[1] 0.5028021
```

The return value of the function ttest() is c(TypeI, Ybar, Yse). This is interpreted by R as a column vector, and thus the output of replicate() is a $3 \times 10,000$ matrix (i.e., each column of the matrix is one replication).

The fraction of Type I errors in the 10,000 simulation runs is 0.194, well above the nominal error rate of 0.05. The effect of the positive autocorrelation has been to make the t-test dramatically more "liberal" (sometimes called "anti-conservative"), that is, to reject the null hypothesis more often than the theory predicts. Figure 3.6a shows a plot of the Type I error rate of the hypothesis test of Equation 3.6 applied to the model of Equation 3.15 as a function of λ. The error rate increases dramatically with increasing λ.

Why does this happen? The outcome of the test is based on the value of the t statistic in Equation 3.11. From this equation, we can see that since the null hypothesis specifies $\mu = 0$ there are two possible explanations for an increased Type I error rate: either the magnitude of the numerator $\bar{Y}$ is overestimated (i.e., the estimator $\bar{Y}$ is biased), or the denominator $s\{\bar{Y}\}$, the standard error, underestimates the true standard deviation of $\bar{Y}$ (or both). Based on Figure 3.6b, the numerator $\bar{Y}$ appears to be estimated as approximately zero. However, comparing Figure 3.6c and 3.6d indicates that although the value of $s\{\bar{Y}\}$ as computed in Equation 3.10 stays approximately constant and near its theoretical value of $1/\sqrt{10} = 0.316$, the actual standard deviation of $\bar{Y}$ increases dramatically as λ increases from zero.

It is not hard to show that the estimate $\bar{Y}$ of the population mean is unbiased whether or not the errors are autocorrelated. Indeed, from Equation 3.7, since ε_i has mean 0 and variance σ^2, we have

$$E\{\bar{Y}\} = E\left\{\sum_{i=1}^{n}\frac{Y_i}{n}\right\} = E\left\{\frac{1}{n}\left[\sum_{i=1}^{n}\mu + \lambda(Y_{i-1} - \mu) + \varepsilon_i\right]\right\}$$

$$= \frac{n\mu}{n} + E\left\{\sum_{i=1}^{n}\lambda(Y_{i-1} - \mu)\right\} + \frac{1}{n}\sum_{i=1}^{n}E\{\varepsilon_i\}$$

$$= \mu + \lambda\sum_{i=1}^{n}E\{Y_{i-1} - \mu\} + 0$$

$$= \mu + \lambda \times 0 + 0$$

$$= \mu$$

(3.16)

The variance of the mean, on the other hand, is (Haining, 1988)

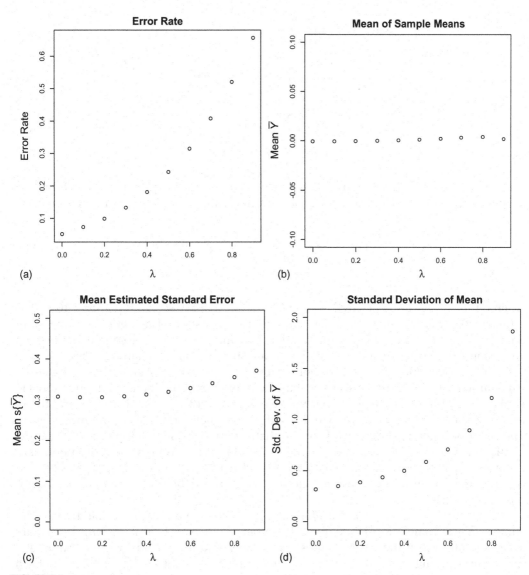

FIGURE 3.6

Results of Monte Carlo simulations of an autoregressive time series model with parameter λ for $0 \leq \lambda \leq 0.8$. Plotted against values of λ are: (a) Type I error rate (fraction of times a Type I error is made) for of a test of the null hypothesis $H_0 : \mu = 0$ against the alternative $H_a : \mu \neq 0$; (b) mean value of $\bar{Y}$ over the 10,000 simulations; (c) mean value of $s\{\bar{Y}\}$ over the 10,000 simulations; (d) standard deviation of $\bar{Y}$ over the 10,000 simulations.

$$\mathrm{var}\{\bar{Y}\} = \mathrm{var}\left\{\frac{1}{n}\sum_i Y_i\right\}$$

$$= \frac{1}{n^2}\sum_i \mathrm{var}\{Y_i\} + \frac{2}{n^2}\sum_i \mathrm{cov}\{Y_i, Y_{i-1}\} \tag{3.17}$$

$$= \frac{\sigma^2}{n} + \frac{2}{n^2}\sum_i \mathrm{cov}\{Y_i, Y_{i-1}\}.$$

Therefore, if $\mathrm{cov}\{Y_i, Y_{i-1}\} > 0$ then $\mathrm{var}\{\bar{Y}\} > \sigma^2/n$ and therefore the quantity σ^2/n in Equation 3.8 underestimates $\mathrm{var}\{\bar{Y}\}$. Moreover, the expected value of s^2 defined in Equation 3.9 is

$$E\{s^2\} = \sigma^2 - \frac{2}{n(n-1)}\sum_i \mathrm{cov}\{Y_i, Y_{i-1}\}. \tag{3.18}$$

(Haining, 1988). Therefore, if $\mathrm{cov}\{Y_i, Y_{i-1}\} > 0$ then the quantity $s\{\bar{Y}\}$ of Equation 3.10 underestimates the standard deviation of $\bar{Y}$ through a combination of two effects on this standard deviation: first, because $s\{\bar{Y}\}$ in Equation 3.10 underestimates the square root of σ^2/n, and second, because σ^2/n, in Equation 3.8 underestimates $\mathrm{var}\{\bar{Y}\}$ (Haining, 1988).

This effect of temporal autocorrelation, that the quantity $s\{\bar{Y}\} = s/\sqrt{n}$ underestimates the true standard deviation of $\bar{Y}$, means that the denominator of the t statistic in Equation 3.11 is smaller than it should be to properly take into account the variability of $\bar{Y}$, and therefore the t statistic is larger than it would be if the random variables Y_i were uncorrelated. The effect of inflating the value of the t statistic is to increase the Type I error rate, the fraction of times that the test exceeds the threshold for rejection of the null hypothesis. If, for example, the size of α is fixed at $\alpha = 0.05$ then the actual fraction of tests carried out in which the null hypothesis is rejected will actually be higher than 0.05, as demonstrated above.

Another way of interpreting this effect is that when data are autocorrelated, each data value provides some information about the other data values near it. When the data are independent, each of the n data values carries only information about itself and, in statistical terms, brings a full degree of freedom to the statistic (Steel et al., 1997, p. 24). Each autocorrelated data value, however, brings less than a full degree of freedom, and thus the effective sample size, which is in the denominator of Equation 3.10, is not n but rather a value less than that.

3.5.2 Modeling Spatial Contiguity

The results of the previous section indicate that the positive autocorrelation among the random variables Y_i in the time series of Equation 3.13 has the effect of increasing the variance $\mathrm{var}\{\bar{Y}\}$. It is reasonable to expect that the same effect may prevail when the random variables are spatially autocorrelated.

The simplest spatial analog for Equation 3.12 is a spatially autocorrelated random variable on a square $m \times m$ lattice. By direct analogy with Equation 3.12 one can write

$$Y_i = \mu + \eta_i$$

$$\eta_i = \lambda \left(\sum_{j=1}^{n} w_{ij}\eta_j \right) + \varepsilon_i, \; i = 1, 2, \dots, n. \tag{3.19}$$

Here $n = m^2$, w_{ij} measures connection strength between Y_i and Y_j, $\varepsilon_i \sim N(0, \sigma^2)$, and as before λ measures the overall autocorrelation strength.

By analogy with time series models, in which one speaks of a time lag, the term $\Sigma w_{ij}\eta_j$ is referred to as a *spatial* lag (Anselin and Bera, 1998). As an example, consider the case $m = 3$. The lattice for this case is shown in Figure 3.7. The matrix $W = [w_{ij}]$ in Equation 3.19 is called the *spatial weights matrix*. One simple W matrix can be constructed by letting

$$w_{ij} = \begin{cases} 1 \; \textit{if the cells share a boundary of length} > 0 \\ 0 \; \textit{otherwise} \end{cases} \tag{3.20}$$

The matrix W as defined in Equation 3.20 describing the lattice of Figure 3.7 is

$$W = \begin{bmatrix} 0 & 1 & 0 & 1 & 0 & 0 & 0 & 0 & 0 \\ 1 & 0 & 1 & 0 & 1 & 0 & 0 & 0 & 0 \\ 0 & 1 & 0 & 0 & 0 & 1 & 0 & 0 & 0 \\ 1 & 0 & 0 & 0 & 1 & 0 & 1 & 0 & 0 \\ 0 & 1 & 0 & 1 & 0 & 1 & 0 & 1 & 0 \\ 0 & 0 & 1 & 0 & 1 & 0 & 0 & 0 & 1 \\ 0 & 0 & 0 & 1 & 0 & 0 & 0 & 1 & 0 \\ 0 & 0 & 0 & 0 & 1 & 0 & 1 & 0 & 1 \\ 0 & 0 & 0 & 0 & 0 & 1 & 0 & 1 & 0 \end{bmatrix}, \tag{3.21}$$

as can be verified by inspection. Note that the diagonal elements w_{ii} are all zero. That is, a cell does not share a spatial connection with itself. This is the usual convention in defining the spatial weights matrix.

The spatial weights matrix W determines how the spatial relationships between spatial objects (polygons or points) are modeled. It is evident that the assignment of values to the

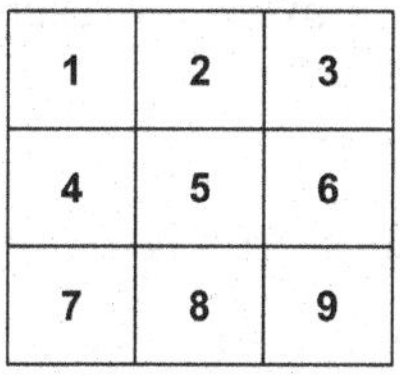

1	2	3
4	5	6
7	8	9

FIGURE 3.7
Numbering of a square 3×3 lattice.

components w_{ij} of W plays an important role in the modeling of the spatial structure of the data and therefore in the statistical analysis of this structure. There are two aspects of this assignment of values to consider: determining which if any of the w_{ij} are assigned a value of zero, indicating no spatial connection; and determining what number to assign to the nonzero values of w_{ij}.

There is a considerable body of literature on this issue, with many proposed methods for assigning values. In discussing these methods, it is most convenient to separate them initially into two categories. The first category consists of methods for describing relationships among cells such as those of Figure 1.3b, whose boundaries are explicitly defined, and the second consists of collections of methods for describing relationships among spatial points such as those of Figure 1.3a, for which the cell boundaries are not explicitly specified. We discuss four methods of determining the w_{ij} for data falling into the first category, the explicitly defined mosaic of cells. These methods are the rook's case, queen's case, distance, and a method proposed by Cliff and Ord (1981). We will then discuss two methods for determining the w_{ij} for the second category of spatial data, the set of points with no explicitly defined boundaries. These are Euclidean distance and distance threshold.

Beginning with the case of explicitly defined boundaries, the first and simplest choice for W is to assign a positive value to contiguous cells and a value of zero to non-contiguous cells. This requires a definition of contiguity. Probably the most common definition is that cells are contiguous only if they are actually in contact with each other (this is called *first order contiguity*). The *rook's case* contiguity rule specifies that the elements of W are nonzero only for those w_{ij} representing cells whose adjoining boundaries have length greater than zero. In the *queen's case*, those elements of W representing cells whose corners are touching are also assigned nonzero values. The connection to the allowable moves of chess pieces is apparent (to chess players). For the benefit of non-chess players, the rook can only move laterally or forward and back, so that a rook located in lattice cell 5 of Figure 3.7 could only move in the directions of cells 2, 4, 6, and 8. A queen can also move diagonally, so the queen could also move in the direction of cells 1, 3, 7, and 9. In general, we will use the term *rook's case* to refer to any system of assigning values to W in which only boundaries with length greater than zero have nonzero weight, and we will use the *queen's case* to refer to any system of assigning weights in which all boundaries between adjacent cells, whether of zero or greater length, have nonzero values.

When values are assigned according to either rook's case or queen's case contiguity, the simplest rule for assigning a numerical value is to let $w_{ij} = 1$ for contiguous cells and $w_{ij} = 0$ for noncontiguous cells. This is called the *binary* weights assignment. While this has the virtue of simplicity and does simplify certain computational formulas, there is an alternative assignment rule that is also commonly used. This is to assign positive values to contiguous cells in such a way that the row sum $\Sigma_j w_{ij}$ always equals 1. This coding of this weights matrix, called *row normalized*, may be represented as

$$w_{ij} \begin{cases} > 0 \text{ if the cells share a boundary of length} > 0 \\ = 0 \text{ otherwise} \end{cases}$$

(3.22)

$$\sum_{j=0}^{n} w_{ij} = 1.$$

With this coding, the W matrix of Figure 3.7 has the form

$$
W = \begin{bmatrix}
0 & 1/2 & 0 & 1/2 & 0 & 0 & 0 & 0 & 0 \\
1/3 & 0 & 1/3 & 0 & 1/3 & 0 & 0 & 0 & 0 \\
0 & 1/2 & 0 & 0 & 0 & 1/2 & 0 & 0 & 0 \\
1/3 & 0 & 0 & 0 & 1/3 & 0 & 1/3 & 0 & 0 \\
0 & 1/4 & 0 & 1/4 & 0 & 1/4 & 0 & 1/4 & 0 \\
0 & 0 & 1/3 & 0 & 1/3 & 0 & 0 & 0 & 1/3 \\
0 & 0 & 0 & 1/2 & 0 & 0 & 0 & 1/2 & 0 \\
0 & 0 & 0 & 0 & 1/3 & 0 & 1/3 & 0 & 1/3 \\
0 & 0 & 0 & 0 & 0 & 1/2 & 0 & 1/2 & 0
\end{bmatrix}, \quad (3.23)
$$

so that all rows sum to 1. Although this rule does not have a strong intuitive justification, it turns out to have some important advantages (Anselin and Bera, 1998, p. 243). A disadvantage, both conceptually and computationally, is that under this rule W is generally not symmetric, that is, for many combinations of i and j $w_{ij} \neq w_{ji}$. However, every row normalized matrix is *similar* to a symmetric matrix (Ord, 1975), that is, they have the same eigenvalues (Appendix A.1). It is often the case that this property is all that is required to carry out computations associated with spatial analysis. Because of its statistical properties, and because certain operations are greatly simplified by it, we will frequently use this rule in our analyses. There are other rules for assignment besides binary and row normalized; references for these are given in Section 3.7.

In this book, we will only develop models for spatial contiguity using either the rook's case or queen's case rules. However, for purposes of exposition we describe two other rules. These address a problem with the rook's case and queen's case rules, which is that they are based strictly on topological properties, with no reference to geometry. Topological properties are those properties that are invariant under continuous transformation of the map. That is, they are properties that do not change when the map is rotated and distorted as if it were made of rubber. The fact that two polygons share a boundary is a topological property. Geometric properties do vary with continuous distortion; the property that one polygon is north of another, or larger than another, is a geometric property.

Cliff and Ord (1981, p. 15), citing Dacey (1965), refer to the problem with the rook's case and queen's case rules as the problem of "topological invariance." That is, the strength of connection is determined only by the topology of the map, and not by either its geometry or by other factors that might influence the strength of connection between polygons. The potential influence of geometry is shown in Figure 3.8. Under rook's case, contiguity the lattice in Figure 3.8a has the same spatial weights matrix as that in Figure 3.8b. The two figures are not topologically equivalent, but they could be made so by poking a very small hole in the center of the lattice of Figure 3.8a. In this case, the two figures would have the same spatial weights matrix under queen's case contiguity as well. More generally, factors that might influence the degree of contiguity between two polygons include barriers. For example, two polygons that are depicted as contiguous on a map might actually be separated by a river or highway that renders direct transit between them difficult.

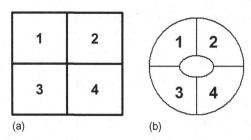

FIGURE 3.8
Under the rook's case contiguity rule, maps (a) and (b) have the same spatial weights matrix. (Modified from Cliff and Ord, 1991, Figure 1.6, p. 15. Used by permission of Pion Limited, London, UK.)

One fairly simple way to define contiguity geometrically is to let $w_{ij} = d_{ij}$, where d_{ij} is some measure of the distance between cell i and cell j. A common choice is to let d_{ij} be the Euclidean distance between centroids of the cells. While this rule does have a certain intuitive appeal, it is not necessarily the best choice. Chou et al. (1990) compared the results of statistical tests in which weights matrices based on this distance rule were used with tests using weights matrices based on some topological form of contiguity rule such as rook's or queen's case. They found that the tests based on the contiguity rules were much more powerful. The tests were carried out on data from an irregular mosaic with greatly varying cell sizes, and the authors' intuitive explanation for the difference in power is that larger cells, which tend to have reduced w_{ij} values due to the greater distance of their centroids from those of other cells, in reality tend to exert the greatest influence on the values of neighboring cells. Manly (1997) points this out as well, and suggests using the inverse of the distance d_{ij}^{-1} instead. Cliff and Ord (1981) suggest the use of the rule $w_{ij} = b_{ij}^{\beta}/d_{ij}^{\alpha}$, where b_{ij} is the fraction of the common border between cell i and cell j that is located in the perimeter of cell i, and α and β are parameters that are chosen to fit the data. This rule would better distinguish between the structures in Figure 3.8a and b. As we have said, however, in this book we will employ only the rook's case or queen's case rules for polygonal data.

We turn now to the specification of weights for the second type of spatial data, in which the location data consist of a finite number of points in the plane as in Figure 1.3a. Since there are no explicit boundaries, a specification by topology, such as the rook's case or queen's case, is impossible. The only means available for specification of weights is geometric. One possibility is to use some form of distance metric such as the inverse distance discussed in the previous paragraph. A second method, which is the one we will use, is to establish a geometric condition that, if satisfied, defines two locations i and j to be neighbors, in which case the value of w_{ij} is greater than zero. The nonzero values of w_{ij} in this formalism are determined similarly to those of the topologically defined neighboring lattice cells already discussed, for example, using a definition analogous to Equation 3.20 or 3.22. In this context, one can either make the condition for a nonzero w_{ij} depend on distance between points being less than some threshold value, or one can declare the k closest points to point i to be neighbors of this point. When the locations of the data form a rectangular set of points such as that defined by the centers of the cells in Figure 1.3b, the threshold distance metric, for an appropriately chosen threshold, generates a weights matrix identical to that of the rook's case (Equation 3.20, see Exercise 3.3). The k nearest neighbors approach does not, because points near the boundary are assigned more neighbors than would be the case under a corresponding lattice or mosaic topology.

3.5.3 Modeling Spatial Association in R

Spatial association is modeled in R by means of the spatial weights matrix W discussed in Section 3.5.2. The representation of this matrix is broken into two parts, matching the two part process of defining the matrix described in that subsection. The first part identifies the matrix elements that are greater than zero (signifying spatial adjacency), and the second assigns a numerical value to adjacent elements. Because spatial weights matrices tend to be very large and sparse (i.e., the vast majority of their terms are zero, as can be seen in Equations 3.21 and 3.23), W is not stored explicitly as a matrix in the spdep package. Instead, it is represented by a *neighbor list*, also called an nb object, which can be converted into a spatial weights matrix. A neighbor list describing a rectangular lattice like the one in Figure 3.7 can be generated using the spdep function cell2nb(). The sequence of commands is as follows.

```
> library(spdep)
> nb3x3 <- cell2nb(3,3) #Rook's case by default
# Examine the structure of the neighbor list
> str(nb3x3)
List of 9
 $ : int [1:2] 2 4
 $ : int [1:3] 1 3 5
 $ : int [1:2] 2 6
 $ : int [1:3] 1 5 7
 $ : int [1:4] 2 4 6 8
 $ : int [1:3] 3 5 9
 $ : int [1:2] 4 8
 $ : int [1:3] 5 7 9
 $ : int [1:2] 6 8
 - attr(*, "class")= chr "nb"
 - attr(*, "call")= language cell2nb(nrow = 3, ncol = 3)
 - attr(*, "region.id")= chr [1:9] "1:1" "2:1" "3:1" "1:2" ...
 - attr(*, "cell")= logi TRUE
 - attr(*, "rook")= logi TRUE
 - attr(*, "sym")= logi TRUE
```

The order of cell indices is left to right and top to bottom as in Figure 3.7, so neighbors of cell 1 are cells 2 and 4, and so forth. The argument type of the function cell2nb() can be used to specify rook's case or queen's case contiguity. Since neither is specified in the listing, the default, which is rook's case, is used. The neighbor list does not specify the numerical values of the nonzero spatial weights.

Assignment of weights is accomplished by the function nb2listw(), which creates a listw object by supplementing the neighbor list object with spatial weight values. The function nb2listw() has an argument style, whose value can take on a number of character values, including "B" for binary, "W" for row normalized, and several others not described in this book. The default value is "W", which is invoked in the example here since no value is given. Only a few lines of the response to the call to function str() are shown, but these indicate that the elements of W33 match those in Equation 3.23.

```
> W33 <- nb2listw(nb3x3)
> str(W33)
List of 3
 $ style : chr "W"
```

```
$ neighbours:List of 9
  ..$ : int [1:2] 2 4
   *    *    *    DELETED    *    *    *
$ weights:List of 9
  ..$ : num [1:2] 0.5 0.5
  ..$ : num [1:3] 0.333 0.333 0.333
   * * * DELETED    *    *    *
- attr(*, "call")= language nb2listw(neighbours = nb3x3)
```

Let us now use the `raster` and `sp` objects to simulate the spatial autocorrelation in Equation 3.19. Recall that in the case of the temporal model of Equation 3.12, the autocorrelation coefficient λ was required to satisfy $-1 < \lambda < 1$ to keep the terms Y_i from blowing up. The analogous restriction in the case of Equation 3.19 depends on the form of the spatial weights matrix W. For the binary form of Equation 3.20, the restriction is $-\frac{1}{4} < \lambda < \frac{1}{4}$. To see why, note that if $\lambda > \frac{1}{4}$ in Equation 3.19, then the sum of terms on the right-hand side of Equation 3.19 will tend to be larger than the left-hand side, which will cause the Y_i to increase geometrically in magnitude. For the row standardized coding of Equation 3.22, the corresponding restriction is $-1 < \lambda < 1$.

The spatial model (Equation 3.19) is easier to deal with if the equations are written in matrix notation (Appendix A.1). Let Y be the vector of the Y_i (i.e., $Y = \begin{bmatrix} Y_1 & Y_2 & \ldots & Y_n \end{bmatrix}'$ where the prime denotes transpose), and similarly let μ be a vector, each of whose components is the scalar μ, let η be the vector of the η_i (i.e., $\eta = \begin{bmatrix} \eta_1 & \eta_2 & \ldots & \eta_n \end{bmatrix}'$), and similarly let ε be the vector of the ε_i. The second of equations (Equation 3.19) can be written

$$\eta_1 = \mu + \lambda \left(\sum_{j=1}^{n} w_{1j}\eta_j \right) + \varepsilon_1$$

$$\eta_2 = \mu + \lambda \left(\sum_{j=1}^{n} w_{2j}\eta_j \right) + \varepsilon_2 \tag{3.24}$$

$$\cdots\cdots$$

$$\eta_n = \mu + \lambda \left(\sum_{j=1}^{n} w_{nj}\eta_j \right) + \varepsilon_n.$$

Thus equations (Equation 3.19) may be written in matrix form as

$$Y = \mu + \eta$$
$$\eta = \lambda W \eta + \varepsilon. \tag{3.25}$$

The term $W\eta$ is the spatial lag in this case (Anselin, 1988, p. 22). However, because the lag is applied to the error in Equation 3.25, this particular model is called the *spatial error model* (see Chapter 13).

Haining (1990, p. 116) describes the following method of simulating this model. Subtracting $\lambda W \eta$ from both sides of the second of equations (Equation 3.25) yields $\eta - \lambda W \eta = \varepsilon$, or $(I - \lambda W)\eta = \varepsilon$. This equation may be inverted to obtain

$$\eta = (I - \lambda W)^{-1} \varepsilon. \tag{3.26}$$

We can generate a vector $Y = \{Y_1, Y_2, ..., Y_n\}$ of autocorrelated random variables by generating a vector of $\varepsilon = \{\varepsilon_1, \varepsilon_2, ..., \varepsilon_n\}$ of normally distributed pseudorandom numbers, premultiplying μ by the matrix $(I - \lambda W)^{-1}$ to obtain η, and adding μ in Equation 3.25 to obtain Y. The R package spdep implements this solution method. We will demonstrate it using a square grid generated by the function cell2nb(). Because cells on the boundary of the lattice only have a neighbor on one side, they have asymmetric interactions. This can cause a non-stationarity analogous to the initial transient in the time series model of Equation 3.12, discussed in Section 3.5.1. In that example, the first ten values of Y_i were discarded to eliminate or at least reduce initial transient effects. In a similar way, we will reduce the boundary effect by generating a larger region than we intend to analyze and deleting the values Y_i of lattice cells near the boundary. We will generate code to implement equation (Equation 3.25) with $\mu = 0$ (so that the null hypothesis is true) for a square grid of 400 cells plus two cells on the boundary that will be dropped. Since $\mu = 0$, we have $Y = \eta$, and therefore equations (Equation 3.25) become $Y = \lambda WY + \varepsilon$.

In code below the side of the region actual region computed is set at 24 and then two cells are removed from each boundary to create the final display. The grid cells are placed at the center of cells running from 0 to 24.The function expand.grid() (see Exercise 3.4) is used to generate a data frame Y.df of coordinates of the centers of the lattice cells.

```
> library(spdep)
> Y.df <- expand.grid(x = seq(0.5,23.5),
+    y = seq(23.5, 0.5, by = -1))
```

Notice the second argument to the function expand.grid(), which is y = seq(23.5, 0.5, by = -1). This sequences y down from 23.5 to 0.5 and ensures that the generated values of y are matched with the correct grid cell. The next step is to create a data field Y in Y.df with the implementation of Equation 3.26. First we assign a value to λ and create the square grid. This is again done using cell2nb(). Given the neighbor list created in this way, the function invIrM() generates the matrix $(I - \lambda W)^{-1}$.

```
> lambda <- 0.4
> nlist <- cell2nb(24, 24)
> IrWinv <- invIrM(nlist, lambda)
```

Since no value of type is specified in the arguments cell2nb(), the default rook's case neighbor list is generated. Similarly, since no style argument is specified in the arguments of invIrM(), the default, which is row normalized spatial weights, is used. The R operator for matrix multiplication (Appendix A.1) is %*%, so that the generation of the vector λ and multiplication by $(I - \lambda W)^{-1}$ is carried out as follows.

```
> eps <- rnorm(24^2)
> Y.df$Y <- IrWinv %*% eps
```

The final step is to remove the boundary cells from the lattice.

```
> Y.df <- Y.df[(Y.df$x > 2 & Y.df$x < 22
+    & Y.df$y > 2 & Y.df$y < 22),]
```

Figure 3.9 contains thematic maps of the values of the spatial lattices created using a grid of raster cells whose attribute values are the data generated above using expand.grid(). The figures show the spatial patterns if Y for three values of the autocorrelation parameter λ. The pattern for $\lambda = 0.8$ appears to have more of a tendency for high values to

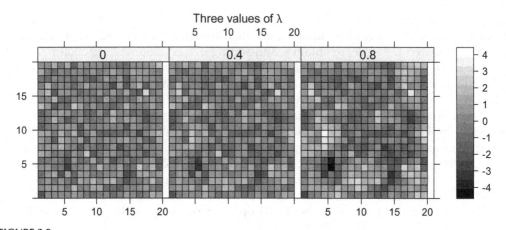

FIGURE 3.9

Gray-scale maps of values of Y in a spatial autoregression model for three values of λ.

be near other high values and low values to be near low values than does the pattern for $\lambda = 0$. As these figures show, however, it is not always easy to identify the level of spatial autocorrelation by eye.

The construction of the grids in Figure 3.9 provides an opportunity to discuss some aspects of the use of spatial objects from the `raster` and `sp` classes and of the process of coercion. If you are not too familiar with these classes, it would be a good idea to review them in Section 2.4.3. The three sets of values of `Y.df` computed above are defined on a square 20 by 20 grid. We can create such a grid using the creation function `raster()`. This was used in Section 2.4.3 to read a file from the disk, but it can also be used to create a grid directly.

```
> library(raster)
> Y.ras <- raster(ncol = 20, nrow = 20, xmn = 0, xmx = 20, ymn = 0,
+   ymx = 20,crs = NULL)
```

This creates a 20 by 20 `rasterLayer` grid. By default the grid has WGS84 coordinates, and the argument `crs = NULL` suppresses this. The next step is to use the coercion function `as()` to coerce this into a `SpatialPolygons` object.

```
> Y.sp <- as(Y.ras, "SpatialPolygons")
```

A `SpatialPolygons` object is, in GIS terminology, an object that has spatial data but no attribute data (Section 2.4.3). The attribute data is contained in the data frame `Y.df` created above. We use the constructor function `SpatialPolygonsDataFrame(Sp, data, match.ID)` to create the `SpatialPolygonsDataFrame` object `Y.spdf`.

```
> Y.spdf <- SpatialPolygonsDataFrame(Y.sp, Y.df, FALSE)
```

The first argument specifies the `SpatialPolygons` object, the second argument specifies the data frame, and the third argument specifies whether the attribute data records are to be matched with the polygons by matching their row names. In our case, these row names are different, and so it is vitally important to verify that the order of the data records in `Y.df` is the same as the order of the polygons in `Y.sp` (Exercise 3.4). This is sufficiently important that we will devote a subsection to it in Section 3.6 when we work with real data, and as preparation for this you should do the exercise.

In order to examine the effect of spatial autocorrelation on the test of the null hypothesis $\mu = 0$ we will use functions from the spdep package to carry out a Monte Carlo simulation experiment. Formally, the null and alternative hypotheses are the same as Equation 3.6,

$$H_0 : \mu = 0,$$
$$H_a : \mu \neq 0,$$

(3.27)

where μ is a parameter in Equation 3.25. In the simulations, the value of μ is set to zero so that the null hypothesis is true. The results of Monte Carlo experiments in Section 3.4.1 indicated that an increase in the value of the autocorrelation term λ in a time series led to an increase in the Type I error rate. We will see whether the same phenomenon is observed in the case of spatially autocorrelated data by carrying out a series of Monte Carlo simulations of hypothesis tests of models of the form (Equation 3.25) with increasing values of λ. The code for these tests does not require that the locations of the lattice cells be specified explicitly; these locations are implicit in the spatial weights matrix W describing them. The coding is therefore slightly different from that just given. We use cell2nb() to generate a neighbor list of a 14 by 14 grid of cells, generate the random numbers and then remove the outer two cells to remove edge effects. Then the function invIrM() is applied to the neighbor list, again with the default values. We again use the defaults to create a row normalized rook's case spatial weights matrix.

```
> lambda <- 0.4
> nlist <- cell2nb(14,14)
> IrWinv <- invIrM(nlist, lambda)
```

Since we will use the function replicate() to run the simulations, we must create a function that we call ttest() to generate each individual simulation.

```
> ttest <- function(IrWinv){
+ # Generate a vector representing the autocorrelated data
+ # on a 14 by 14 grid
+    Y.plus <- IrWinv %*% rnorm(14^2)
+ # Convert Y to a matrix and remove the outer two cells
+    Y <- matrix(Y.plus, nrow = 14,
+        byrow = TRUE)[3:12,3:12]
+    Ybar <- mean(Y)
+ # Carry out the test and return the outcome and the mean
+    t.ttest <- t.test(Y,alternative = "two.sided")
+    TypeI <- t.ttest$p.value < 0.05
+    return(c(TypeI, Ybar))
+ }
```

Here are the results of the simulation.

```
> set.seed(123)
> U <- replicate(10000, ttest(IrWinv))
> mean(U[1,])
[1] 0.1876
> mean(U[2,])
[1] -0.0009136907
> sqrt(var(U[2,]))
[1] 0.1594549
```

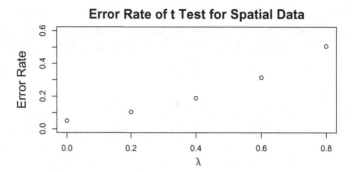

FIGURE 3.10

Type I error rate as a function of λ in 10,000 Monte Carlo simulations of a spatial autoregression model.

In the listing, the autocorrelation term `lambda` is set to 0.4, and the error rate is inflated.

Figure 3.10 shows the Type I error rate as a function of the value of λ used in the simulation. At the value $\lambda = 0$, the error rate is approximately 0.05. As in the case of time series, the error rate increases with increasing λ. The spatial analog of Equation 3.17, describing the effects of autocorrelation on the variance of $\overline{Y}$, is

$$
\begin{aligned}
\mathrm{var}\{\overline{Y}\} &= \mathrm{var}\left\{\frac{1}{n}\sum_i Y_i\right\} \\
&= \frac{1}{n^2}\sum_i \mathrm{var}\{Y_i\} + \frac{2}{n^2}\sum_{i \neq j}\mathrm{cov}\{Y_i, Y_j\} \\
&= \frac{\sigma^2}{n} + \frac{2}{n^2}\sum_{i \neq j}\mathrm{cov}\{Y_i, Y_j\}.
\end{aligned}
\tag{3.28}
$$

and a similar analog exists for Equation 3.18, describing the effect of autocorrelation on $E\{\sigma^2\}$. As with temporally autocorrelated data, so with spatially autocorrelated data the t statistic is larger than it would be if the random variables Y_i were uncorrelated, so that the Type I error rate is inflated. To repeat, when data are spatially autocorrelated, each data value provides some information about the other data values near it, so that the effective sample size less than the number n of data records.

3.6 Application to Field Data

3.6.1 Setting Up the Data

Our primary descriptor of spatial relationships is the spatial weights matrix. The elements of this matrix depend on the rules used to define what it means to be a spatial neighbor. The choices for neighbor status on which we focus are the rook's case and queen's case for polygons (Figure 1.3b); and threshold distance and k nearest neighbors for points (Figure 1.3a). In this section we examine, using a real data set, the sensitivity of the results of spatial analyses to the way this matrix is constructed. The data we will use are detrended percent sand content in Field 1 of Data Set 4 (abbreviated Field 4.1). The data, detrended using linear regression and median polish, are shown in Figure 3.4. These figures indicate that the detrended sand content data may be spatially autocorrelated in this field. Both methods tend to leave low detrended values near the center and higher values in the northern and southern ends.

We will fit the spatial error model of Equation 3.25 to the sand data that have been detrended using the linear model (Figure 3.4b). There are other models besides the spatial error model that may fit these data better, and no attempt is made at this point to justify the use of the model; this issue is taken up in Chapter 13. For the present, we simply use this as an example of one type of model that might be used for spatially autocorrelated data. We first test methods for modeling spatial relationships between point data such as those displayed in Figure 3.4b. The first method that we will use to generate the nonzero values of a spatial weights matrix is based on a threshold distance between neighboring points, and the second is based on selecting the k nearest neighbors of each point. We begin with the former. We identify the locations using the UTM coordinates, which are in the data fields *Northing* and *Easting*. Because these are large numbers that, when squared, disrupt the accuracy of the linear regression, we first create new coordinates by subtracting the minimum values. The input of the data frame data.Set4.1 is described in Appendix B.4.

```
> data.Set4.1$x <-data.Set4.1$Easting - min(data.Set4.1$Easting)
> data.Set4.1$y <-data.Set4.1$Northing - min(data.Set4.1$Northing)
```

Next, we fit the trend surface and use the function predict() described in Section 3.2.1 to create this surface. The trend is subtracted from the Sand data field to create the detrended sand content SandDT.

```
> trend.lm <- lm(Sand ~ x + y + I(x^2) +
+    I(y^2) + I(x*y), data = data.Set4.1)
> data.Set4.1$SandDT <- data.Set4.1$Sand - predict(trend.lm)
```

Next, the data frame is converted into a spatial points data frame using the function coordinates(). The function dnearneigh() from the spdep package is used to create a neighbors list in which points within 61 m (the point to point distance of the sample locations in the field) are considered neighbors.

```
> coordinates(data.Set4.1) <- c("x", "y")
> nlist <- dnearneigh(data.Set4.1, d1 = 0, d2 = 61)
```

The neighbor list is used to construct a listw object representing the spatial weights matrix, and then the function errorsarlm() is applied to generate the estimated value of λ in Equation 3.25.

```
> W <- nb2listw(nlist, style = "W")
> Y.mod <- errorsarlm(SandDT ~ 1, data = data.Set4.1, listw = W)
> print(Y.mod$lambda, digits = 4)
   lambda
0.4005
```

The estimate is $\hat{\lambda} = 0.4005$. When this operation is repeated using the argument style = "B" in the first line, the estimate is $\hat{\lambda} = 0.1180$. Recall that the limiting values of λ for stationarity are $-1 < \lambda < 1$ for the binary W and $-1/4 < \lambda < 1/4$ for the binary W, so we expect the estimated value $\hat{\lambda}$ for the binary W to be about one-fourth that of the row normalized W.

Next, we try an implementation of the k nearest neighbors rule using the function knearneigh(). For a rectangular grid such as this, the four nearest neighbors are the only locations within the point to point distance threshold. Note that this function does not generate the neighbor list directly and has to be called as an argument of the function knn2nb().

```
> nlist <- knn2nb(knearneigh(data.Set4.1, k = 4))
> W <- nb2listw(nlist)
> Y.mod <- errorsarlm(SandDT ~ 1, data = data.Set4.1, listw = W)
> print(Y.mod$lambda, digits = 4)
   lambda
0.3299
```

The estimate is $\hat{\lambda} = 0.3299$. The estimate for binary spatial weights is $\hat{\lambda} = 0.0824$. In this example there is a considerable difference, possibly due to the difference between the weights matrices for points on the boundary.

One of the questions that will be of interest is the extent to which the method used to detrend the data influences the results. Referring to Equation 3.1, $Y(x, y) = T(x, y) + \eta(x, y) + \varepsilon(x, y)$, different detrending methods will apportion the values of $Y(x, y)$ in this equation to the three components on the right-hand side in different ways. To repeat the comment of Cressie, "one person's deterministic trend may be another person's correlated error structure" (by the end of this book you will be tired of reading this). In Exercise 3.8, you are asked to repeat these calculations for data that have been detrended using median polish and for data that have not been detrended at all.

We turn now to polygonal data. Since the data themselves are point measurements, the creation of polygons to represent these data is somewhat arbitrary, particularly since the field is not rectangular (Figure 3.4). The most natural choice is to create Thiessen polygons. Given a set of points $P_1, P_2, ..., P_n$ in the plane, Thiessen polygons (also called Voronoi polygons or Dirichlet cells) have the property that every polygon contains one of the points P_i as well as every point in the region that is closer to P_i than to any other point in the set $P_1, P_2, ..., P_n$ (Ripley, 1981, p. 38; Lo and Yeung, 2007, p. 333). When the set of points forms a regular grid, Thiessen polygons form a square or rectangular lattice. R has the capacity to construct Thiessen polygons using functions in the packages tripack (Renka et al., 2011), deldir (Turner, 2011), and spatstat (Baddeley and Turner 2005).

We will first use the point pattern analysis package spatstat to create a ppp file, which is a spatstat point format representing a point pattern data set in the two-dimensional plane. This file is created using the constructor function ppp(), which takes as arguments a vector of the x coordinates of the points, a vector of the y coordinates, and an argument owin that defines the bounding box (see Section 2.4.3) of the region.

```
> library(spatstat)
> W <- 592000
> E <- 592500
> S <- 4270300
> N <- 4271200
> cell.ppp <- ppp(data.Set4.1$Easting, data.Set4.1$Northing,
+       window = owin(c(W, E), c(S, N)))
```

Next, we use the spatstat function dirichlet() to create a spatstat object of class tess (for "tessellation").

```
> thsn.tess <- dirichlet(cell.ppp)
```

Next, we follow the process we used in Section 3.5.3. We use a coercion function from maptools to coerce the tess object into a SpatialPolygons object using the coercion function as(). Then we use the constructor function SpatialPolygonsDataFrame() to create the SpatialPolygonsDataFrame object thsn.spdf.

```
> library(maptools)
> thsn.sp <- as(thsn.tess, "SpatialPolygons")
> thsn.spdf <- SpatialPolygonsDataFrame(thsn.sp,
+    slot(data.Set4.1, "data"), FALSE)
```

Before we can move on, we must make sure that the ordering of the spatial data matches the ordering of the attribute data. That is the subject of the next subsection.

3.6.2 Checking Sequence Validity

In the last section and in Exercise 3.4, we alluded to an issue that is sufficiently subtle, and whose consequences are sufficiently dire, that it is worthwhile to emphasize it by devoting an entire subsection. Look at the two lines of code just above. The first line creates the object thsn.sp, which describes the polygon structure and contains spatial data but no attribute data. It was created by forming Thiessen polygons around the set of point coordinates in data.Set4.1, and these polygons are arrayed in an order determined by the creation process, independent of the values in the ID field in data.Set4.1. Now look at the following line of code. This assigns attribute data to the polygons. There is no guarantee that the order of the attribute data matches the order of the polygons, and if it does not, then the attribute data will be assigned to the wrong polygons. Therefore, in these situations one must always check to make sure that the polygon order matches the attribute data order.

Because it is simpler to work with sf (spatial features) objects, we will use coercion to move into this domain.

```
> library(sf)
> thsn.geom.sf <- st_as_sf(thsn.sp)
> str(thsn.geom.sf)
Classes 'sf' and 'data.frame':      86 obs. of  1 variable:
 $ geometry:sfc_POLYGON of length 86; first list element: List of 1
  ..$ : num [1:5, 1:2] 592081 592081 592000 592000 592081 ...
  ..- attr(*, "class")= chr "XY" "POLYGON" "sfg"
 - attr(*, "sf_column")= chr "geometry"
 - attr(*, "agr")= Factor w/ 3 levels "constant","aggregate",..:
  ..- attr(*, "names")= chr
```

Although there are no identification numbers, the 86 polygons are arranged in a particular order, namely, the order assigned to them when they were created. Again, we coerce thsn.spdf into an sf object and look at its structure.

```
> thsn.sf <- st_as_sf(thsn.spdf)
> str(thsn.sf)
```

```
Classes 'sf' and 'data.frame':        86 obs. of  27 variables:
 $ ID       : int 1 2 3 4 5 6 7 8 9 10 ...
 $ Row      : int 1 1 1 1 1 1 1 2 2 2 ...
 $ Column   : int 1 2 3 4 5 6 7 1 2 3 ...
 $ Easting  : num 592051 592112 592173 592234 592295 ...
 $ Northing : num 4271104 4271104 4271104 4271104 4271104 ...
...
```

Now we come to the important part. The data records in the data frame data.Set4.1 are indexed by the data field ID. The polygons in thsn.sf were created based on the values of Easting and Northing in this data frame. Therefore, the values of thsn.sf$ID should match the order of records of this data frame. With 86 records this is easy to check.

```
> thsn.sf$ID
 [1]  1  2  3  4  5  6  7  8  9 10 11 12 13 14 15 16 17 18 19 20 21 22
[23] 23 24 25 26 27 28 29 30 31 32 33 34 35 36 37 38 39 40 41 42 43 44
[45] 45 46 47 48 49 50 51 52 53 54 55 56 57 58 59 60 61 62 63 64 65 66
[67] 67 68 69 70 71 72 73 74 75 76 77 78 79 80 81 82 83 84 85 86
```

If there are hundreds of data records, it might be less easy to visually, but we can do it with the function all.equal().

```
> all.equal(1:length(thsn.sf$ID), thsn.sf$ID)
[1] TRUE
```

We can also check for correct alignment visually. First, we plot the attribute *ID* values.

```
> plot(thsn.spdf, pch = 1, cex = 0.1)
> text(coordinates(thsn.spdf),
+       labels=as.character(thsn.spdf@data$ID))
```

Next, we plot the polygon *ID* values and see that they match.

```
> plot(thsn.spdf, pch = 1, cex = 0.1)
> text(coordinates(thsn.spdf),
+       labels=lapply(thsn.spdf@polygons, slot, "ID"))
```

Now that we have assured ourselves that there is no misalignment of geometry and attribute data, we can move on to the spatial autocorrelation tests. See Exercise 3.8 for some practice on this topic.

3.6.3 Determining Spatial Autocorrelation

There are four cases we will consider in estimating the parameter λ of Equation 3.25 for the Thiessen polygons, namely, the possible combinations of rook's case and queens case contiguity rules and binary and row normalized spatial weights. The function used to carry out the estimation is again errorsarlm.

```
> nlist <- poly2nb(thsn.spdf,
+   row.names = as.character(thsn.spdf$ID), queen = FALSE)
> W <- nb2listw(nlist, style = "W")
> Y.mod <- errorsarlm(SandDT ~ 1, data = thsn.spdf, listw = W)
> print(Y.mod$lambda, digits = 4)
   lambda
0.4033
```

The corresponding estimate for the binary rook's case W is $\hat{\lambda} = 0.1182$. For queen's case contiguity, the estimates are $\hat{\lambda} = 0.2009$ for the row normalized W and $\hat{\lambda} = 0.0407$ for the binary W. Thus, with this data set there is little difference between the estimates for different contiguity rules. Not surprisingly, the rook's case contiguity rule gives a similar estimate to that of the corresponding nearest neighbor distance rule for point data. Which estimate is "better"? We don't know if any of them are any good, because the model itself could be in error. We will leave the discussion of this issue until we take up autoregressive models in Chapter 13.

3.7 Further Reading

The fundamental source for almost all of the material discussed in this chapter is Cliff and Ord (1981). The seminal paper on the modeling spatial autocorrelation is generally considered to be that of Whittle (1954). This paper explicitly draws on the analogy of spatial processes with time series, using the notion of a line transect in space as a conceptual bridge. Bartels (1979) also provides a discussion of the relationship between time series data and spatial data.

Davis (1986) provides an extensive discussion of the properties of trend surfaces and develops a method for testing their significance. The difficulty in distinguishing between responses to a trend and a spatial error structure is amplified by Sokal et al. (1993). They point out that apparent spatial autocorrelation may be due to the response of the mean to a deterministic regional trend. They refer to this as "spurious spatial autocorrelation."

There are a number of references that provide a more detailed discussion of creating a spatial weights matrix W (Cliff and Ord, 1981; Anselin, 1988; Anselin and Bera, 1998; Tiefelsdorf, 2000). Getis and Aldstadt (2004) describe a method for creating W using local spatial statistics, which are discussed in Section 4.5.3. Several papers discuss the assignment of spatial weights in various contexts. For example, Tiefelsdorf et al. (1999) describe the effects of various choices for the structure of W on the exact distribution of Moran's I. Case and Rosen (1993) use a combination of spatial and attribute data to assign weights between states to model budget development. Lacombe (2004) analyzes different methods for generating the matrix in the context of county welfare policy.

All of the artificial spatial data sets in this book are generated using the method of Haining (1990, p. 116), which is simple and easy to understand. This is not, however, the only way to generate random spatial data. Griffiths (1988, Ch. 9) provides an extensive discussion of the simulation of spatially autocorrelated data. The geoR package (Ribeiro and Diggle, 2016) contains the function grf(), which develops a Gaussian random fields simulation of autocorrelated data (Wood and Chan, 1994).

Finally, G. E. P. Box's more famous quote is "all models are wrong; some models are useful," which is from Box (1979).

Exercises

3.1 To test the sensitivity of the t-test to skewed data, carry out a Monte Carlo simulation of the test using data generated from a normal and a lognormal distribution. Be careful about the value of μ for the lognormal distribution.

3.2   The Monte Carlo simulation should not be applied blindly. Consider the coin tossing experiment conducted in Section 3.2.2. Look up the function `binom.test()` and then carry out a simulation of 50 tosses of a fair coin using the following function:

```
coin.toss <- function (n.tosses){
  z <- rbinom(1,n.tosses,0.5)
  n.heads <- sum(z)
  b <- binom.test(n.heads, n.tosses,0.5,"two.sided",0.95)
  TypeI <- b$p.value < 0.05
}
```

Use the function `replicate()` to run this function 10,000 times and compute the fraction of Type I errors. Is it close to 0.05? Why not? HINT: explore the possible values that b$p.value can take on.

3.3   Using the `spdep` functions described in this chapter, generate a `listw` object for a four by four square lattice. Use the function `listw2mat()` to generate a row normalized spatial weights matrix. Use `apply()` to show that the rows sum to 1.

3.4   Use the function `expand.grid()` to generate a three by three set of points matching the lattice of Figure 3.7. Then use the `spatstat` functions discussed in this chapter to generate Thiessen polygons. Use the function `str()` to check whether the numbering of these polygons matches Figure 3.7. How do you have to order the x and y sequences in `expand.grid()`?

3.5   (a) Use the function `read.csv()` to read Data Set 2 (don't forget to set your working directory). From this data set, create a small data set by selecting only those sites with longitude values between $-123.2463°$ and $-123.0124°$ and latitude values between $39.61633°$ and $39.95469°$; (b) Use the statement `?coordinates` to read about the function `coordinates()`. Then use this function to create a `SpatialPointsDataFrame` by assigning data fields to be the coordinates of the data set; (c) Plot the locations of the data set using the `plot()` function; (d) Use the function `str()` to explore the structure of the `SpatialPointsDataFrame`.

3.6   (a) Use `knearneigh()` to create a spatial weights matrix and use the function `errorsarlm()` to estimate the value of λ in a spatial error model for the data set created in Problem 3.5; (b) Give a physical interpretation of the meaning of the terms in the spatial lag model (Equation 3.25) as applied to the data in part (a). Does this model make sense?

3.7   Use Monte Carlo simulation to compute the Type I error rate for a 10 by 10 square data set created with an autocorrelation value of λ equal to 0.6. Compute the sample mean and sample standard deviation of the replicates.

3.8   It may help with the idea of ID misalignment to see what happens when data are misaligned with polygons. Create a data set identical to Data Set 4.1 of Section 3.6.2 but with order of the ID values reversed so that they run from 86 to 1 and repeat the analyses of that section on this data set.

4

Measures of Spatial Autocorrelation

4.1 Introduction

The previous chapter introduced the definition of positive spatial autocorrelation as a non-zero covariance between spatial proximity and attribute value proximity (i.e., nearby values are more similar than they would be if they were arranged randomly). This definition imposes the need for a means to measure, based on a sample of data values, the covariance between nearby points and to decide whether or not this covariance is consistent with a random spatial arrangement of values. We have to define a statistic and establish a null hypothesis concerning the value of that statistic when no spatial autocorrelation exists, and an alternative hypothesis concerning the value when spatial autocorrelation does exist. Consider the soil sand content data of Section 3.6. In that section, a model called the spatial error model was fit to the detrended sand content data of Field 4.1 of Data Set 4 (denoted Field 4.1). The parameter λ of the model was estimated in this process. One could measure and test autocorrelation using this parameter, and indeed we shall have occasion to do something like this in Chapter 13. However, the validity of this test requires that the data satisfy the spatial error model. In many applications there is considerable advantage to having a statistic that does not depend on a particular model of the autocorrelation structure. In this chapter, we introduce a collection of such statistics that are used to measure the strength of spatial autocorrelation. After some preliminary discussion in Section 4.2, Section 4.3 discusses tests for spatial autocorrelation of categorical data and Section 4.4 discusses tests for spatial autocorrelation of quantitative data. Section 4.5 discusses measures of autocorrelation structure within a data set. Each of these sections is concerned with areal data. Section 4.6 provides a brief discussion of measures of autocorrelation of continuous (geostatistical) data.

4.2 Preliminary Considerations

4.2.1 Measurement Scale

Figure 4.1a shows a plot of the numerical value of weed infestation level in Field 4.1. These data were obtained by having an expert agronomist walk through the field and at each sample location assess the weed infestation level on a scale of 1 = no infestation to 5 = complete weed cover. Figure 4.1b shows a map of detrended sand content in the same field. This is a modification of Figure 3.4b in which the actual numerical values of the data are shown.

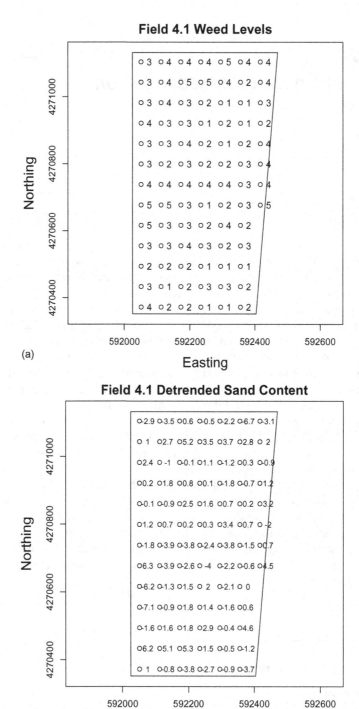

FIGURE 4.1
(a) Location map of numerical values of *Weeds* in Field 1 of Data Set 4; (b) Location map of numerical values of detrended sand content in Field 1 of Data Set 4.

These data sets are superficially similar, but there is an important difference. The sand content values in Figure 4.1b are represented by numbers, and some mathematical operations on these numbers are meaningful; for example, the difference between a detrended sand content of 2% and one of 1.5% is the same as the difference between 0% and −0.5%. The weed levels in Figure 4.1a are rankings, and although they, like the data representing sand content, are expressed as numbers, the interpretation of these numbers is quite different. The weed level rating (5, 4, 3, 2, 1) could also be represented, for example, with the sequence (10, 6.2, 0, −3, −25), and there would be no difference in interpretation. An analogous change in the sand content data, however, would completely change the interpretation.

The differences between these data types can be understood through the concept of *measurement scale* (not to be confused with spatial scale). A measurement can be defined as "the assignment of numerals to objects or events according to rules," and a scale in this context is, in the most general sense, a "means of dealing with aspects of these objects" (Stevens, 1946). Measurement scales may be classified according to whether or not operations such as equality comparison, ordering, and arithmetic are meaningful. The classification of measurement scales has been discussed by many statisticians, but the most generally recognized classification, which we will use in this book, is due to Stevens (1946). In this classification a measured quantity falls into one of the following four categories.

Nominal scale data. These are data for which the measurement defines membership in a category, but there is no implied relationship among categories. Familiar examples include soil type (as represented by set of numbers, one for each soil type), telephone number, and postal code. The only meaningful mathematical operation that can be carried out with nominal scale data is equality comparison: any two data values can be either equal or not equal. Thus, for example, if the number 27 represents the soil type *Rincon silty clay loam*, then all data records assigned the number 27 would be this soil type. A special case of nominal scale data is *binary data*, in which there are two categories. These categories can be represented by the number zero and one. There is an implied order in the relationship in that one is larger than zero, but because there are only two possible values the grouping of data by order is mathematically equivalent to grouping by equality comparison.

Ordinal scale data. These are data for which a rank order can be established, but differences between ranks do not have meaning. The weed infestation level data shown in Figure 4.1a are an example. In addition to equality comparison, an ordering can be established, but an operation on data values such as subtraction of one from another has no meaning. Thus, for example in the case of the weed level classes of Figure 4.1a, one cannot infer that the difference between a level 3 and a level 2 weed level is the same as the difference between weed levels 2 and 1.

Interval scale data. These are data for which the operations of addition and subtraction have meaning but, because no absolute datum is established, multiplication and division do not. Temperature measured on either a Celsius or Fahrenheit scale, is an example. We cannot say that 70°C is "twice" 35°C because if we convert the measurement units to °F, for example, the resulting temperatures (158°F and 95°F, in this case) do not have the same numerical relationship. The detrended sand content data are interval scale data.

Ratio scale data. These are data in which a datum exists, so that the operations of multiplication and division produce meaningful results. Population density is an example, as is the original sand content data set of Figure 3.4a. It is meaningful to say that if one location has a 20% sand content and a second location has a 40% sand content, then the sand content at the second location is twice that at the first.

From the perspective of measurement of spatial autocorrelation, there is no need to distinguish between interval and ratio scale data, and indeed we shall see that ordinal scale data are often, although not always, handled in the same way as well (this is most justifiable if the ordinal scale data take on values that are in some sense "analogous" to interval or ratio scale values). Thus, we will generally distinguish primarily between nominal (or "categorical") data and "quantitative" data. The term "quantitative" subsumes the categories of interval, ratio, and sometimes ordinal data.

4.2.2 Resampling and Randomization Assumptions

By carrying out statistical analysis of the data, we implicitly assume that these data represent a realization of a random field, that is, a set of random variables whose realized values are mapped onto (in our case) a two-dimensional surface. It turns out that the statistical properties of the sample depend on how this realization takes place. Following Cliff and Ord (1981, p. 14), we will allow four different possibilities. These are defined according to two dichotomous criteria. The first criterion is the method by which values would be assigned under the null hypothesis to the n locations (polygons or points) in successive hypothetical "experiments" (if you have forgotten why "experiments" is in quotes, see Section 3.3). The second criterion is the distribution of the population from which the values in the sample are drawn, which for all of our examples will be either binary or normal.

Regarding the first criterion, one way to carry out the assignment of values to locations is in each "experiment" to draw n random samples with replacement from the population and assign one of these values to each location. This is called the *resampling assumption*. The second way is to draw n samples from the population and then to randomly reassign these fixed numbers one to each location in each "experiment." This is called the *randomization assumption*. Let us look at each of these more closely.

> *Resampling assumption.* Under this assumption, the n attribute values Y_i (e.g., the values of sand content in Figure 3.4a) in the sample are obtained from n independent drawings from a population having a given distribution. The null hypothesis states that each cell is assigned a value equal to a random number drawn from the population independently of the other cell assignments. For example, if time could be reset and another hypothetical drawing of sand content values in Figure 3.4a could be generated, they would be drawn from the same population as the first but would be independent of the values from the previous sample. If the data are assumed to be drawn from a normal population, they are sometimes said to follow a *normality* assumption (Cliff and Ord, 1981, p. 14). If the distribution is binomial, then the probability that a cell has the value one is p and the probability that a cell has the value zero is $q = 1 - p$. This case is sometimes called the *free sampling* assumption, and can be considered as sampling *with* replacement from a population with $p \times n$ ones and $q \times n$ zeroes.

> *Randomization assumption.* If the data are assumed to follow a resampling assumption as defined in the previous paragraph, then each realization will in general produce a completely different attribute data set. Thus, the values assigned to the polygons or points will generally be different from one realization to the next.

In the case of binary data under the resampling assumption, the number of ones and zeroes will generally be different in each realization since they are produced by n independent draws from a binomial distribution.

In contradistinction to this, the randomization assumption considers only the n attribute values observed in the sample. For example, in the hypothetical second drawing of detrended sand content data in Figure 3.4a, the values would be the same in each realization but the locations of these values would be randomly rearranged and independent of the locations from the previous drawing. In the case of normally distributed data the observed sample is considered to be one of the $n!$ (this is read as "n factorial" and is equal to $n \times (n-1) \times (n-2) \times \ldots \times 2 \times 1$) possible arrangements in which the n observed numbers could be assigned to the n cells. The randomization occurs in the assignment to the n geographical locations. In the case of binary data, if k of the n sites have the value one, then the null hypothesis is that the observed pattern is a random arrangement of k ones and $n-k$ zeroes, considering that all $\binom{n}{k}$ (this is read "n choose k" and is equal to $n!/k!(n-k)!$) possibilities are equally likely. This special binary case is sometimes called *nonfree sampling*, and can be considered as sampling *without* replacement from a population with $p \times n$ ones and $q \times n$ zeroes. Unfortunately, the terms used to characterize the assumptions mostly begin with N or R, and this can cause confusion. Table 4.1 can be used for ready reference.

One must decide which assumption makes more sense for a particular data set. Goodchild (1986, p. 23) expresses the opinion that the resampling assumption "is more reasonable in most contexts." In general, unless otherwise stated, we will assume that the data follow the resampling assumption. There is, however, one very important exception. The very useful permutation-based hypothesis tests described in Section 4.3 require the randomization assumption, and when we use a permutation test we will be implicitly assuming that the data follow this assumption. As usual, the most sensible course, where possible, is to carry out the analysis under both assumptions and, if they are substantially different, try to understand why.

4.2.3 Testing the Null Hypothesis

We are now in a position to address the question posed at the start of this chapter, namely, whether or not the data in a set of observations may be considered to be spatially autocorrelated based on their observed spatial distribution. This question is formulated in terms of a hypothesis test. The null hypothesis is that the data are randomly assigned to locations according to one of the two assumptions described in Section 4.2.2, the resampling assumption or the randomization assumption. We compute a test statistic and the probability of observing a value of this statistic at least as extreme as the one we actually observe (the p value). If this probability is very low, then we reject the null hypothesis that the data are randomly arranged by location. The evidence in this case indicates that the data display some level of spatial autocorrelation.

TABLE 4.1

Terminology Used to Characterize the Various Assumptions about the Behavior of the Data under the Null Hypothesis of Zero Autocorrelation

Assumption	Resampling N	Randomization R
Binary Data	Free Sampling	Nonfree Sampling
"Numerical" Data	Normality	Randomization

Referring to the notation given above, a very general form for a statistic to be used in a test of the null hypothesis of zero autocorrelation is the *Mantel* statistic (Mantel, 1967; Mantel and Valand, 1970; Anselin, 1995), given by

$$\Gamma = \sum_{i=1}^{n} \sum_{j=1}^{n} w_{ij} c_{ij} \qquad (4.1)$$

Here w_{ij} is the ij^{th} element of the spatial weights matrix and represents the contiguity between locations i and j, and c_{ij} is a quantity that indicates how much statistical weight is assigned to the pair ij. Different ways of computing c_{ij} characterize different autocorrelation statistics. We will initially restrict ourselves to a few special forms that lead to widely used measures of autocorrelation. Cliff and Ord (1981) show that each of the forms of Γ we consider, under both the resampling and randomization assumptions, is asymptotically normally distributed as n becomes very large. Because of this, one method of testing the null hypothesis is to assume that the random quantities are normally distributed and use statistical tests based on this assumption, such as the t-test. However, there may be a concern that the actual sample size is not sufficiently large that asymptotic theory is justified. In the next sections, we will look at the application of various forms of Mantel statistics to specific cases.

4.3 Join–Count Statistics

Join–count statistics may be used to test the null hypothesis of no spatial autocorrelation for nominal data. The properties of the distribution of the join–count statistic were worked out by Moran (1948) and are discussed in detail by Goodchild (1986, p. 37). We restrict discussion to binary data, which is the most common case (for the more general, "multicolor" case, see Cliff and Ord, 1981, p. 19). Figure 4.2 shows the data of Figure 4.1a with weed levels 4 and 5 combined into one class called "High" and the remaining values combined into the class "Low." The data are represented by two colors, black (actually, dark gray so the borders are visible) and white (actually, light gray). Define a *join* to be a nonzero contiguity between spatial objects (points or polygons) i and j as defined by a nonzero spatial weights matrix element w_{ij}. In the case of a mosaic of polygons, a join is a boundary under the contiguity rule used in the analysis (rook's case or queen's case, see Section 3.5.2) between two cells. To avoid constant repetition, we will from now on refer to the objects as polygons, recognizing that what is said also applies to points if an appropriate spatial weights matrix is used.

Assume the two attribute values are denoted B, for black, and W, for white (be careful not to confuse this use of W with that of the spatial weights matrix). Define J_{rs} as the sum of weighted joins between color r and color s, where $r, s = B, W$, and the weight of the join between polygon i and polygon j is the value of the element w_{ij} of the spatial weights matrix W. We will use the statistics J_{BB}, J_{WW}, and J_{BW} to test the null hypothesis that the data have zero autocorrelation. A two-sided test would imply that the null hypothesis of zero auto-correlation is being tested against the alternative that the data are either positively or negatively autocorrelated, but the analyst does not know which. This situation is unlikely to arise in practice, since ordinarily the physical or biological characteristics of the data would favor either one alternative or the other. Therefore, a one-sided alternative is generally used.

Cliff and Ord (1981, p. 11) give formulas for computing J_{BB}, J_{WW}, and J_{BW}. Let $Y_i = 1$ if polygon i is black and 0 if it is white. Then the formulas for J_{BB} and J_{BW} are

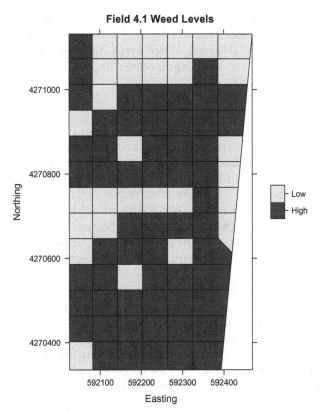

FIGURE 4.2
Thematic map of a binary interpretation of the weed levels in Field 1 of Data Set 4. Values of *Weeds* greater than or equal 4 are classified as High and values between 1 and 3 are classified as Low.

$$J_{BB} = \frac{1}{2}\sum_i\sum_j w_{ij}Y_iY_j,$$

$$J_{BW} = \frac{1}{2}\sum_i\sum_j w_{ij}(Y_i - Y_j)^2,$$

(4.2)

where the identity $w_{ii} = 0$ is required for all i. The value of J_{WW} is $J - J_{BB} - J_{BW}$ where J is the total number of weighted joins. In order to test parametrically the null hypothesis of zero autocorrelation, one must compute the expected value and variance of the statistic under the resampling (free sampling in the case of binary data) or randomization (nonfree sampling) assumption. These are worked out by Cliff and Ord (1981, p. 19). The complexity of the general formulas greatly outweighs their usefulness, and so we will not show them but instead refer the interested reader to the cited reference.

The distribution theory of join–count statistics is discussed by Cliff and Ord (1981, p. 36). Under the free sampling model the probabilities that a cell has the value one or zero are distributed binomially, whereas under the nonfree sampling model they follow a hypergeometric distribution. The distributions are asymptotically normal as n approaches infinity (Cliff and Ord, 1981, p. 51–54). Cliff and Ord (1981, p. 63) carry out Monte Carlo simulations to show that the normal approximation is reasonable even for moderate values of n.

For any one map there are three join–count statistics that can be used to test the null hypothesis of zero autocorrelation, namely, J_{BB}, J_{WW} and J_{BW}. We will focus on the *BB* and *WW* joins, since these are available in R. When invoking the asymptotic normality of the statistic, the join–count test is carried out as a *t*-test using the statistic

$$t = \frac{J_{rr} - E\{J_{rr}\}}{\sqrt{Var\{J_{rr}\} / n}},$$ (4.3)

where $E\{J_{rr}\}$ and var$\{J_{rr}\}$ are the expectation and variance under the null hypothesis of zero autocorrelation using the formulas given by Cliff and Ord (1981).

Let us apply these formulas to the data represented in Figure 4.2. The contents of the file *Set4.196sample.csv* are loaded into the data frame data.Set4.1 using the code in Appendix B.4. This creates a spatial feature file thsn.sf, which is then coerced in to a SpatialPolygonsDataFrame called thsn.sp (see Section 2.4.3). Thiessen polygons can be created as described in Section 3.6. The polygons in Figure 4.2, which are topologically equivalent but match the nonrectangular boundary of the field, were created in ArcGIS and are in the file of auxiliary data on the book's website.

The R function joincount.test() in the package spdep (Bivand et al., 2011) can be used to compute the expected values as well as the significance level under the normal approximation of the nonfree sampling model for *BB* and *WW* joins. We will take this opportunity to utilize two functions, although they make the code slightly more complex. For the test of the data in Figure 4.2, we first use the function lapply() to create a list (Section 2.4.1) in which the logical test *Weeds* ≥ 4 is successively applied to each data record. The function unlist() then converts this list into a vector. The second statement creates a data field HiWeeds in the Thiessen Polygons, coverts the results of the logical test into a vector of factors, and creates labels for the factors.

```
> HiWeeds <- unlist(lapply(data.Set4.1$Weeds,
+    function(x)(as.numeric(x >= 4))))
> thsn.sp@data$HiWeeds <- factor(HiWeeds,
+    labels = c("High", "Low"))
```

(For practice, try implementing the two steps lapply() and unlist() separately and observing the output.) Next, we create a spatial weights matrix using the methods discussed in Section 3.5.2 and apply the function joincount.test(). In this implementation, we use a rook's case binary contiguity rule.

```
> nlist <- poly2nb(thsn.sp,
+    row.names = as.character(thsn.sp$ID), queen = FALSE)
> W <- nb2listw(nlist, style = "B")
> joincount.test(thsn.sp$HiWeeds, W)
        Join count test under nonfree sampling
data:   thsn.sp$HiWeeds
weights: W

Std. deviate for High = 2.9134, p-value = 0.001787
alternative hypothesis: greater
sample estimates:
Same colour statistic     Expectation            Variance
            70.00000         59.82271            12.20276
```

```
         Join count test under nonfree sampling
data:    thsn.sp$HiWeeds
weights: W

Std. deviate for Low = 3.4465, p-value = 0.0002839
alternative hypothesis: greater
sample estimates:
Same colour statistic        Expectation           Variance
            25.000000          15.218057           8.055317
```

The alternative tested is that the population parameter is greater than the observed value. The function returns the results for *BB* joins (value 1) and *WW* joins (value 0), and both indicate rejection of the null hypothesis, although with slightly different *p* values.

Permutation tests of join–count can be carried out using the spdep function joincount. mc(). The full output is not shown, only the line with the *p* values.

```
> set.seed(123)
> joincount.mc(thsn.sp$HiWeeds, W,1000,
+ alternative = "greater")
      Monte-Carlo simulation of join-count statistic
Join-count statistic for High = 70, rank of observed statistic = 997,
p-value = 0.003996

Join-count statistic for Low = 25, rank of observed statistic = 999,
p-value = 0.001998
```

Simulations with different random number seeds will produce slightly different results, but in general they should closely approximate the results obtained with joincount.test(). In this case, the permutation tests of both J_{BB} and J_{WW} once again indicate rejection of the null hypothesis. Table 4.2 shows the results of *t* tests and permutation tests for the data in Figure 4.2 for binary and row normalized spatial weights matrices. There is quite a bit of difference among the tests, with the binary queen's case contiguity rule being the most conservative.

For any data set, three different join–count statistics, J_{BB}, J_{WW}, and J_{BW} are available to test a single null hypothesis, and it is very possible to draw completely different conclusions depending on which statistic is tested. This is especially true with relatively small data sets (Exercise 4.1). Which statistic should be used to interpret the significance level? Cliff and Ord (1981, p. 63) report on extensive Monte Carlo simulations under a wide variety of conditions. As is generally the case in the comparison of test statistics, the comparison by Cliff and Ord focuses on the power of the test using each statistic (Section 3.4).

TABLE 4.2

Results (*p* values) of Tests of the Null Hypothesis of Zero Autocorrelation for the Maps of Figure 4.2

	Rook's Case				Queens's Case			
	Binary		Row Norm.		Binary		Row Norm.	
	t test	permut.	*t* test	permut.	*t* test	permut.	*t* test	permut.
J_{WW}	0.002	0.004	<0.001	0.001	0.037	0.054	<0.001	0.001
J_{BB}	<0.001	0.002	<0.001	0.001	0.007	0.011	<0.001	0.001

The conclusions of Cliff and Ord may be summarized as follows. Both the randomization test and the normal approximation appear to apply reasonably well, so an observed large difference in the results from the functions `joincount.test()` and `joincount.mc()` should be interpreted as a sign of possible problems with the data, such as a failure to satisfy the assumptions of normality, or possibly an indication that `joincount.mc()` should be evaluated again with a larger number of permutations. Second, the color with the larger number of cells generally produces the more accurate result, so this p value should be used to interpret significance. This would be J_{WW}, the statistic for low weed level, in our case. A final conclusion (Cliff and Ord 1981, p. 63) is that the statistic J_{BW} measuring different colored joins often has a higher power than statistics measuring joins of the same color. There is a more complex `spdep` function `joincount.multi()` that permits the testing of *BW* joins as well as spatial objects with more than two colors. Finally, the functions `joincount.test()` and `joincount.mc()` both have the capability to extend the join–count statistic for same color joins to maps with more than two colors.

4.4 Moran's I and Geary's c

The Moran's I (Moran, 1948) and Geary's c (Geary, 1954) statistics were both developed to test the null hypothesis of zero autocorrelation with interval or ratio data. Although not developed for use with ordinal data, these statistics often work better than the alternatives with this measurement scale as well. As with the join–count statistic, the distribution theory of I and c is extensively discussed by Cliff and Ord (1981) as well as by Goodchild (1986) and Upton and Fingleton (1989). Both I and c can be expressed as special cases of the general Mantel statistic introduced in Equation 4.1 and defined by the particular choice of the attribute similarity measure c_{ij}. Moran's I is written

$$I = \frac{n}{S_0} \frac{\sum_i \sum_j w_{ij}(Y_i - \bar{Y})(Y_j - \bar{Y})}{\sum_i (Y_i - \bar{Y})^2}, \tag{4.4}$$

where $S_0 = \sum_i \sum_j w_{ij}$. The similarity measure c_{ij} in Equation 4.1 can therefore be written

$$c_{ij} = \frac{n}{S_0} \frac{(Y_i - \bar{Y})(Y_j - \bar{Y})}{\sum_i (Y_i - \bar{Y})^2}, \tag{4.5}$$

showing that Moran's I bears a strong conceptual resemblance to the Pearson correlation coefficient (Appendix A). Geary's c may be written

$$c = \frac{n-1}{2S_0} \frac{\sum_i \sum_j w_{ij}(Y_i - Y_j)^2}{\sum_i (Y_i - \bar{Y})^2}, \tag{4.6}$$

so that the similarity measure in this case is

$$c_{ij} = \frac{n-1}{2S_0} \frac{(Y_i - Y_j)^2}{\sum_i (Y_i - \bar{Y})^2}. \tag{4.7}$$

Thus, the similarity measure underlying Geary's c is a squared difference, analogous to the variogram measure, which will be discussed in Section 4.6.1, for data on a spatial continuum.

The similarity of appearance of Moran's I to the Pearson correlation coefficient r might lead one to believe that the I statistic would display similar behavior to r. However, while the behavior of I is somewhat analogous to that of r, there are important differences. The primary difference is that the expected value of I under the null hypothesis of zero spatial autocorrelation is not zero but rather $-1/(n-1)$ (Cliff and Ord, 1981, p. 42). Nevertheless, increasingly positive values of I are associated with increasingly strong positive autocorrelation, and negative values with negative autocorrelation. The expected value of Geary's c under the null hypothesis of zero spatial autocorrelation is one. A value of c between zero and one indicates positive autocorrelation, and a value $c > 1$ indicates negative spatial autocorrelation (note from the form of Equation 4.6 that c cannot be negative).

Moran's I and Geary's c are both asymptotically normally distributed under the normality assumption as n increases (Cliff and Ord, 1981, p. 47). Thus, as with the join–count statistics, the null hypothesis of zero autocorrelation may be tested by computing the expected value and variance of the statistic under normality and then testing via a t-test. Alternatively, one may employ a permutation test under the randomization assumption.

In summary, the possible ways that we have discussed that one may carry out a test of the null hypothesis of zero spatial autocorrelation using the Moran's I or the Geary's c statistic are as follows. First, that data may be modeled as points or polygons. Second, the data may be assumed to be distributed under the null hypothesis according to the normality assumption or the randomization assumption. Third, one may compute the value of the statistic and its variance under either assumption and compute the p value according to asymptotic normality, or one may carry out a permutation test (in which case the randomization assumption is implicitly made). Finally, one can carry out any of these calculations using any one of the forms for the spatial weights matrix W. Thus, a number of p values can be computed.

We can compare some of these methods using calculations involving the percent silt data discussed in Field 4.1. We use the silt data because it has little or no trend (this will be shown in Section 4.6.1) and there are issues (these will be discussed at the end of this section) with applying these tests to detrended data. Here is the code for the Moran test and the Geary test for distance based, rook's case point data under the normality assumption.

```
> nlistk <- knn2nb(knearneigh(data.Set4.1, k = 4))
> W.kW <- nb2listw(nlistk, style = "W")
> W.kB <- nb2listw(nlistk, style = "B")
> Silt <- data.Set4.1@data$Silt
> moran.test(Silt, W.kW, randomisation = FALSE, alternative = "greater")
        Moran's I test under normality
data:  Silt
weights: W.kW
Moran I statistic standard deviate = 2.2547, p-value = 0.01208
alternative hypothesis: greater
sample estimates:
Moran I statistic         Expectation           Variance
        0.371990105          -0.011764706        0.006482012
```

TABLE 4.3

Values and Variances of Moran's I and Geary's c for the Field 4.1 Silt
Data Under the Resampling Assumptions Using Row-Normalized (W)
and Binary (B) Spatial Weights Matrices for I and c Using Rook's Case
and Queen's Case Contiguity Rules

| | Point Data | | | Polygon Data | | | |
| | Binary | | | Rook's Case | | Queen's Case | |
	Value	p.		Value	p.	Value	p.
I_d	0.372	0.000	I_B	0.364	0.000	0.326	0.000
I_k	0.369	0.000	I_W	0.371	0.000	0.330	0.000
c_d	0.624	0.000	c_B	0.599	0.000	0.636	0.000
c_k	0.606	0.000	c_W	0.623	0.000	0.670	0.000

Note: For the point data, the subscript d refers to d distance based weights and
the subscript k refers to k nearest neighbor weights and only the binary
data are shown. Results under the randomization assumption are virtually
identical and are not shown.

Table 4.3 shows results of the test of the null hypothesis of zero autocorrelation against the
alternative of positive autocorrelation using the point location model. Results are shown
for the four combinations of neighbor relationship (distance = 61 m and four nearest neighbors) and weight coding (binary coded and the row-normalized). Permutation tests displayed very similar results (not shown). All of the tests for point data indicate significant
spatial autocorrelation in the data.

The I and c statistics may be used for ordinal as well as ratio and interval scale data. Cliff
and Ord (1981, p. 15) point out that for ordinal data either the multicolor join–count statistic or
the I or c statistic may be used, but that latter are preferable because they preserve the ordering
of the relationship. Moreover, neither statistic is very sensitive to departures from normality.

Since two statistics are available, Moran's I and Geary's c, the natural question arises as
to which should be preferentially used. Upton and Fingleton (1989, p. 170) point out that
because of the form of Equations 4.4 and 4.6, c would be expected in general to be more
sensitive to differences in values of Y while I would be more sensitive to extreme Y values.
Cliff and Ord (1973, p. 45) suggest that I is less affected by the distribution of the data than
is c. Cliff and Ord (1981, Chapter 6) carry out a detailed comparison of I and c focusing on
two methods of comparison: an analytical method and Monte Carlo simulation.

The null and alternative hypotheses considered by Cliff and Ord (1981, p. 167) in their
comparison of the two statistics are $H_0 : I = -1/(n-1)$ versus $H_a : I \neq -1/(n-1)$ and $c = 1$
versus $c \neq 1$. For these tests, Cliff and Ord compute the *asymptotic relative efficiency*, which
is (roughly speaking) the limit as n approaches infinity of the power of two tests. They
show that for the binary weights matrix the asymptotic relative efficiency c is always less
than that of I. Perhaps somewhat more convincingly, Cliff and Ord (1981, p. 175) report on a
series of Monte Carlo simulations that indicate that the statistical power of I is greater than
that of c. They conclude that I is to be preferred over c. Griffith and Layne (1999, p. 15) provide additional comparison and also conclude that the Moran's I is generally preferable,
although corroboration of Moran's I results by the computation of Geary's c is desirable. By
displaying the mathematical relationship between the two, Griffith and Layne (1999, p. 15)
show that a difference between the two might be observed in cases of highly irregular lattice cell sizes and outliers in the Y_i values. As always, the wise course is to compute both

statistics. If they indicate similar properties, one breathes a sigh of relief and moves on. If they indicate dissimilar properties, one must try to determine the reason for this, and, ultimately, report the dissimilarity in any discussion of the analysis.

Two final notes of caution. First, if there is a substantial trend in the data, then the method chosen for detrending the data may have an effect on the results of a test for autocorrelation (see Exercise 4.3). In this context, referring to Equation 3.1, $Y(x,y) = T(x,y) + \eta(x,y) + \varepsilon(x,y)$, with detrended data, the test for autocorrelation in Equation 4.4 is not made on the values of the Y_i but rather on $Y_i - \hat{T}(x_i, y_i)$, where $\hat{T}(x,y)$ is an estimate of the trend $T(x,y)$. This may disrupt the results. Second, the tests described in this section do not apply to data that have been transformed via linear regression because the residuals of the regression are inherently correlated (see Section 13.2). This also means that if the trend estimate $\hat{T}(x,y)$ has been obtained by regression, then its residuals are autocorrelated. In this context, the issues associated with testing detrended data and transformed data are similar. Valid tests for such data do exist and will be described in Section 13.2. In general, in dealing with data that has been subjected to some form of transformation, it is wise to carry out tests using both the functions described in this section and those in Section 13.2 and compare the results.

4.5 Measures of Autocorrelation Structure

4.5.1 The Moran Correlogram

The autocorrelation statistics described in Sections 4.3 and 4.4 describe the overall level of autocorrelation of a spatial data set, but they say nothing about the structure of this autocorrelation. The *spatial correlogram* (Cliff and Ord, 1981, Chapter 5) describes the manner in which spatial autocorrelation changes with increasing lag distance between locations. For the polygon and point data with which we are concerned, distance is measured via the spatial weights matrix W. Locations i and j that are directly contiguous, as reflected in a nonzero term w_{ij} in the spatial weights matrix W, are said to have *first-order contiguity*. If locations i and j are first-order contiguous and a location k is not first-order contiguous with location i but is so with location j, then i and k have *second-order contiguity*. Higher orders of contiguity are defined similarly. To compute a spatial correlogram one must first fix the manner in which one models the spatial relationship between data locations with higher levels of contiguity.

One simple way to define higher-order contiguity is to base it on the elements of the spatial weights matrix (Bivand et al., 2013b, p. 269). We can define (somewhat informally) a path between two locations to be a sequence of nonzero elements of the spatial weights matrix that connects them. For example, suppose again that locations i and j are first-order contiguous and that location k is not first-order contiguous with location i but is with location j, so that i and k have *second-order contiguity*. If the locations are polygons and the spatial weights matrix is constructed using the rook's case contiguity rules, then the path is a sequence of polygons, each of which share a common nonzero boundary. One can then say that two locations are at a *lag distance h* if the shortest rook's path between these cells passes through $h - 1$ intervening cells (Haining, 2003, p. 79). For example, consider the map of nine cells in Figure 3.7, and assume rook's case contiguity. Polygons 1 and 2 are at a lag distance of one, since they share a common boundary. Polygons 1 and 5 are at a lag distance of two, since they do not share a nonzero boundary, and they do share a boundary with polygon 2.

Polygons 1 and 6 are separated by a lag distance of three, since these are not at lag distance two and they share a boundary with at least one polygon at lag distance 2.

For each lag distance h, determined either using a distance or a contiguity rule, one can define a spatial weights matrix $W(h)$ using binary weights, row-normalized weights, or any other type of weights, and then compute a Moran's I statistic $I(h)$ via $W(h)$. This function $I(h)$ is called the *Moran correlogram*. It gives a measure of the change in correlation structure as distance between lattice cells is increased.

We will compare the Moran correlograms for three different methods of construction of $W(h)$ using the detrended sand content data from Section 3.6. First, we create a distance based $W(h)$ using the function dnearneigh() with lag spacings of 61 m as described above. We generate the Moran correlogram using the spdep function sp.correlogram(neighbours, var, order, method, style,...). The argument method = "I" specifies a Moran correlogram, and style = "W" specifies a row normalized spatial weights matrix.

```
> nlist <- dnearneigh(data.Set4.1, d1 = 0, d2 = 61)
> I.d <- sp.correlogram(nlist, Silt, order = 5,
+    method = "I", style = "W")
```

Next, we construct a correlogram based on Thiessen polygons and a rook's case spatial weights rule. The Thiessen polygons in this section are constructed with the functions in the R package spatstat using the same coding as is described in Section 3.5.3. As such, they are rectangular rather than trapezoidal in shape, but they are topologically equivalent to the trapezoidal figure shown in Figure 4.2.

The attribute data are then added using the following statement.

```
> thsn.spdf <- SpatialPolygonsDataFrame(thsn.sp,
+    data.Set4.1@data, FALSE)
```

As always, we must check to see that each attribute data record is correctly matched with the appropriate polygon. We do this by first comparing the attribute data *ID* values.

```
> thsn.sf <- st_as_sf(thsn.spdf)
> all.equal(1:length(thsn.sf$ID), thsn.sf$ID)
[1] TRUE
```

See Section 3.6.2 for a visual check for misalignment.

Having verified that the attribute data is correctly aligned with the spatial data, we can construct the correlogram.

```
> nlist <- poly2nb(thsn.sp, queen = FALSE)
> I.r <- sp.correlogram(nlist, Silt, order = 5,
+    method = "I", style = "W", randomisation = TRUE)
```

Using similar code, we also construct a correlogram with queen's case contiguity. Figure 4.3 shows three Moran correlograms. They are similar, and all display a weak negative correlation at more than one spatial lag. This is a common occurrence and is due to the patchiness of the autocorrelated data (Sokal and Oden, 1978; Fortin and Dale, 2005, p. 127). The value of the spatial lag at which I is no longer significantly positive can be used as an indication of the range of autocorrelation of the data. Cliff and Ord (1981, Ch. 5) provide further discussion of the interpretation of the correlogram.

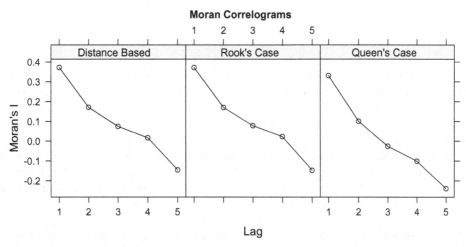

FIGURE 4.3

Moran correlograms for the detrended sand data of Field 1 of Data Set 4. Left to right are a correlogram of point data; a correlogram of Thiessen polygon data under rook's case contiguity, and a correlogram of Thiessen polygon data under queen's case contiguity.

4.5.2 The Moran Scatterplot

Another very useful graphical tool based on the Moran's *I* statistic is the *Moran scatterplot* (Anselin, 1996). This is a bivariate plot that takes advantage of the fact that the Moran's *I* statistic can be represented as the slope of a regression line. By plotting this regression line together with the coordinate pairs that it regresses, one can identify potential outliers and gain insight into the local structure of the data. To derive the regression equation, we observe that if W is row normalized then $S_0 = n$ so Equation 4.4 becomes

$$I = \frac{\sum_i \sum_j w_{ij}(Y_i - \bar{Y})(Y_j - \bar{Y})}{\sum_i (Y_i - \bar{Y})^2}. \tag{4.8}$$

We can rewrite Equation 4.8 as

$$I = \frac{\sum_i (Y_i - \bar{Y}) \sum_j w_{ij}(Y_j - \bar{Y})}{\sum_i (Y_i - \bar{Y})^2} \tag{4.9}$$

or

$$I = \frac{\sum_i (Y_i - \bar{Y})(Z_i - \bar{Z})}{\sum_i (Y_i - \bar{Y})^2} \tag{4.10}$$

where:

$$Z_i = \sum_j w_{ij} Y_j \qquad (4.11)$$

Equation 4.11 may be written in matrix form as $Z = WY$, and comparison of Equation 4.11 with Equation 3.19 indicates that Z represents a spatial lag in Y.

Equation 4.10 represents the slope of the regression line of Z on Y (Equation A.25 in Appendix A) with a zero intercept term. This indicates that the Moran's I may be considered as the slope of the regression of the spatial lag WY on Y. It follows that a scatterplot of WY against Y should be a very useful way to visualize the spatial distribution of Y. Anselin (1996) calls such a plot a *Moran scatterplot*. In particular, the Moran scatterplot reveals the extent to which the global I statistic effectively summarizes the spatial structure.

We can construct a Moran scatterplot using the spdep function moran.plot(). As with any plotting function, the act of creating the plot is a side effect of the evaluation of the function. Many plotting functions do not return a meaningful value, but moran.plot() does. It returns a matrix whose columns are the results of diagnostic tests by the function influence.measures() for each data record (X_i, Y_i). These diagnostics are discussed in Appendix A.2.3. We will create an object to hold this matrix, and, as a side effect, print the Moran scatterplot (Figure 4.4). We do not run into the same issues using detrended data that we did in Section 4.4, so for purposes of comparison with Figure 3.4, we will use the detrended sand content.

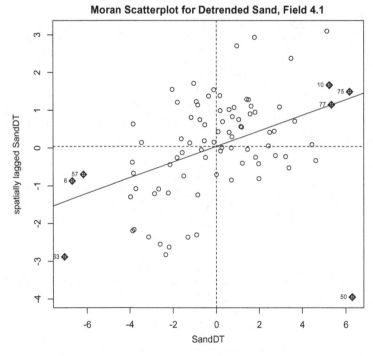

FIGURE 4.4
Moran scatterplot for the detrended sand content data of Figure 3.4b.

```
> trend.lm <- lm(Sand ~ x + y + I(x^2) +
+   I(y^2) + I(x*y), data = data.Set4.1)
> data.Set4.1$SandDT <- data.Set4.1$Sand - predict(trend.lm)
> SandDT <- data.Set4.1@data$SandDT
> SandDT.mp <- moran.plot(SandDT, W) # Fig. 4.4
> title(main = "Moran Scatterplot for Detrended Sand, Field 4.1")
```

The matrix of influence measure tests is in the slot is.inf. We can first determine which data records test positive for at least one influence measure by summing across the rows of the matrix. Any row with a positive sum indicates at least one positive test.

```
> print (inf.rows <- which(rowSums(SandDT.mp$is.inf) > 0))
 6 10 50 57 63 75 77
```

Seven data records test positive for at least one influence measure. These are also identified in Figure 4.4. Let's see how many measures are violated by each record.

```
> SandDT.mp$is.inf[inf.rows,]
    dfb.1_  dfb.x dffit cov.r cook.d    hat
6    FALSE  FALSE FALSE  TRUE  FALSE   TRUE
10   FALSE  FALSE FALSE  TRUE  FALSE  FALSE
50   FALSE   TRUE  TRUE  TRUE   TRUE   TRUE
57   FALSE  FALSE FALSE  TRUE  FALSE   TRUE
63   FALSE  FALSE FALSE  TRUE  FALSE   TRUE
75   FALSE  FALSE FALSE  TRUE  FALSE   TRUE
77   FALSE  FALSE FALSE  TRUE  FALSE  FALSE
```

Examining Figure 4.4 indicates that of these potentially influential points, data record 50 is the most distinctive. This data record has an exceptionally high value relative to its neighbors (see Figure 4.5a below). The location of point 50 in the extreme lower right of the plot is consistent with the following interpretation of the Moran scatterplot (Anslein, 1996, p. 117). Points in the upper right and lower left quadrants correspond to positive autocorrelation, with the upper right quadrant containing high-valued points with high-valued neighbors and the lower left quadrant containing low-valued points with low-valued neighbors. The upper left quadrant (low-valued points with high-valued neighbors) and the lower right quadrant (high-valued points with low-valued neighbors) correspond to pockets of negative spatial autocorrelation. Data record 50 is an extreme case of a high-value with low-valued neighbors.

4.5.3 Local Measures of Autocorrelation

The Moran scatterplot discussion in Section 4.5.2 indicates that the Moran's *I* itself summarizes the contribution of many spatial lag pairs in the same way that a regression line summarizes many pairs of response and predictor variables. The local autocorrelation structure may, however, be quite different in some areas from that described by a global statistic such as the Moran's *I*. For example, the region around data record 50 in the detrended sand data of Field 4.1 is characterized by negative spatial autocorrelation, although the overall detrended sand content displays a high level of positive spatial autocorrelation. Spatial structure exists at many scales and one might expect that subregions of a greater whole could exhibit a local autocorrelation structure very different from that characterized by the single statistic that describes the entire region.

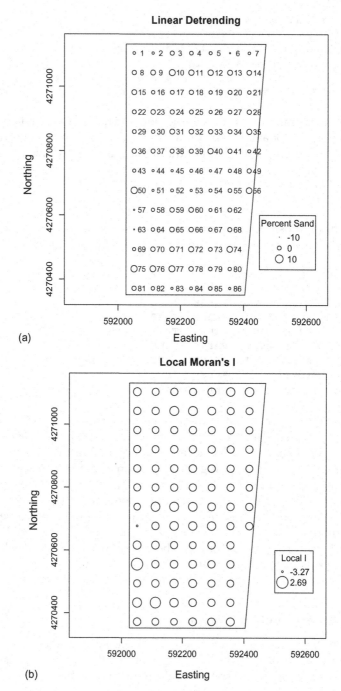

FIGURE 4.5
(a) Bubble plot of detrended percent sand content of Field 1 of Data Set 4 (same data as Figure 3.5b), shown with sample location ID numbers; (b) Local Moran's I for the data of Figure 4.5a.

The first widely used explicitly local measures of spatial autocorrelation statistics are attributed to Getis and Ord (1992, 1996), although in a prior paper Chou et al. (1990) actually proposed a special case of these statistics. Getis and Ord (1992) discussed a pair of statistics they called $G_i(d)$ and $G_i*(d)$. Here i is the index of the polygon element and d is a measure of distance from element i. In the original paper, d is assumed to be a Euclidean distance, but it could equally be a path distance similar to that used in computing the correlogram. The statistic $G_i(d)$ is defined as

$$G_i(d) = \frac{\sum_{j=1}^{n} w_{ij}(d)Y_j}{\sum_{j=1}^{n} Y_j},$$ (4.12)

where the sums explicitly exclude the $j = i$ term. In the original paper, Getis and Ord (1992) only consider a binary weights matrix, but Ord and Getis (1995) discuss other forms. The special case discussed by Chou et al. (1990), which they called a spatial weights index, consists of the G_i statistic with the first-order binary spatial weights matrix. The statistic $G_i*(d)$ is defined identically to $G_i(d)$ except that the summation explicitly includes the $j = i$ term. Both of these statistics apply only to ratio scale quantities (i.e., quantities possessing a natural zero value, see Section 4.2.1). In the later paper, however, Ord and Getis (1995) extend the statistics to interval and, potentially, ordinal scale quantities.

Anselin (1995) defines another class of statistics that he calls *local indicators of spatial association*, or LISA statistics. Like the Getis-Ord statistics, LISA statistics are defined for each cell of an areal data set and serve as an indication of spatial clustering of similar values. In addition, however, LISA statistics have the property that their sum, when taken over all locations, equals the value of the corresponding global statistic. The most important LISA statistic is the local Moran's I, defined as

$$I_i = (Y_i - \bar{Y})\sum_{j=1}^{n} w_{ij}(Y_j - \bar{Y}).$$ (4.13)

Anselin (1995, p. 99) shows that $\Sigma_i I_i = I$, where I is given in Equation 4.4. LISA statistics, therefore, provide an indicator of how areas in the region contribute to the value of the global statistic.

The spdep package contains functions to compute both the G statistics and the local I. We will focus on the local I. We can again use the detrended sand data since we are considering local relationships. The function localmoran() computes the local I at each location.

```
> SandIi <- localmoran(SandDT, W)[,1]
```

Anselin (1995, p. 102) points out that the local I statistic is positive in regions of clustering of similar values and negative in regions of clustering of dissimilar values. We can plot the local I, but we should be careful to properly scale the plot. Section 2.6.2 contains an extensive discussion of the use of the function plot() for spatial data, but we give here a brief discussion of scaling. The argument cex of the function plot() is used to control the size of point symbols. The "nominal" value is 1, and we will use a range of 0 to 3 in our map. We can scale the plotting of the local I values as follows. Let cex_{min} and cex_{max} be the minimum and maximum values we wish to plot, and let I_{min} and I_{max} be the minimum and maximum local I values. Then the appropriate scaling is

$$cex = cex_{min} + \frac{I_i \times (cex_{max} - cex_{min})}{I_{max} - I_{min}}. \tag{4.14}$$

Figure 4.5a shows the linearly detrended sand content data of Field 4.1. This is the same map as Figure 3.4b, but this time the data location ID numbers are shown. Figure 4.5b shows a bubble plot of the corresponding local Moran's I values. As was the case with the Moran scatterplot analysis of the previous section, the data at location 50 stands out as being very different from its neighbors. Removal of location 50 changes the value of the global Moran's I from 0.207 to 0.317.

Anselin (1995) describes two primary uses of local statistics, to detect "hot spots," or local areas of clustered extreme values, and to identify individual outliers. The trend analysis described in Section 3.2.1 detects nonstationarity due to trends, but may not be effective in detecting other forms of nonstationarity. It is possible in principle to test the null hypothesis of zero local autocorrelation for each of these statistics (Ord and Getis, 1995, Anselin, 1995), but probably their greatest use is exploratory rather than confirmatory.

4.5.4 Geographically Weighted Regression

The ordinary least squares (OLS) regression model with a single explanatory variable is (Appendix A.2, Equation A.22)

$$Y_i = \beta_0 + \beta_1 X_i + \varepsilon_i, \, i = 1, ..., n. \tag{4.15}$$

Geographically weighted regression (GWR) (Fotheringham et al., 1997, 2002) modifies Equation 4.15 to let the regression coefficients be functions of position:

$$Y_i = \beta_0(x_i, y_i) + \beta_1(x_i, y_i) + \varepsilon_i, \, i = 1, ..., n. \tag{4.16}$$

The method can easily be extended to more than one explanatory variable, in which case it becomes more convenient to write the regression equations in matrix notation (Appendix A). In particular, the equation for the least squares estimate of the OLS regression coefficients $b = [b_0 \, b_1 \, ... \, b_{p-1}]'$, where the prime denotes transpose, is Equation A.35

$$b = (X'X)^{-1}X'Y, \tag{4.17}$$

where Y is the vector whose components are the Y_i and X is a matrix, called the design matrix, whose elements are values of the explanatory variables X_i (see Appendix A.2).

In this notation, the geographically weighted regression model incorporates local variation in the estimated regression coefficients $b_k(x_i, y_i)$ by incorporating a geographic weighting matrix $G(x_i y_i)$. This is a diagonal matrix (see Appendix A.1) whose diagonal elements incorporate the spatial distance d_{ij} between the location (x_i, y_i) of the explanatory variable X_i and the location (x_j, y_j) of the response variable Y_j.

The modified form of Equation 4.17 used in GWR is written

$$b(x_i, y_i) = (X'G(x_i, y_i)X)^{-1}X'G(x_i, y_i)Y. \tag{4.18}$$

This is not a single equation but rather a system of matrix equations that is evaluated at each location.

As d_{ij} increases, the explanatory variable is expected to have less influence over the response variable, so the elements of G should reflect this. There is more than one form available for G, but a very common one is the *Gaussian* form, in which case the elements are $g_{jj}(x_i, y_i) = \exp(-[d_{ij} / h]^2)$. The parameter h, which determines the width of the normal shaped curve, is called the *bandwidth* and must be estimated from the data. One way to estimate the bandwidth is by using cross validation, in which one iteratively solves the regression equations and chooses h to minimize the quantity

$$Q = \sum_{i=1}^{n} \left[Y_i - \hat{Y}_{-i}(h) \right]^2, \tag{4.19}$$

where $\hat{Y}_{-i}(h)$ is the predicted value of Y_i resulting from a regression where the date record (X_i, Y_i) is not used.

GWR is particularly useful for the exploration of stationarity in the data. If there is a large variation in the estimated regression coefficients $b_k(x_i, y_i)$, then this can be taken as indicating that the data are not stationary. The method is implemented in R in the package spgwr (Bivand and Yu, 2017). We first illustrate it using an artificial data set on a 20 by 20 grid.

```
> data.df <- expand.grid(x = seq(1,20),
+    y = seq(20,1, by = -1))
```

We create two variables X and Y and convert the grid into a SpatialPointsDataFrame.

```
> set.seed(123)
> data.df$X <- rnorm(400)
> data.df$Y <- data.df$X + 0.1*rnorm(400)
> coordinates(data.df) <- c("x", "y")
```

Taking advantage of the fact that we can use data.df as if it were a data frame, we can compute the OLS regression coefficients.

```
> coef(lm(Y ~ X, data = data.df))
(Intercept)            X
0.0004473601 1.0028106263
```

We know that data are stationary because there is no dependence on position. Let's see what we get from a geographically weighted regression analysis. The first step is to compute the bandwidth via cross-validation using Equation 4.19. This is done using the spgwr function gwr.sel().

```
> library(spgwr)
> Y.bw <- gwr.sel(Y ~ X, data = data.df)
```

Next, we plug this bandwidth plus the formula and data source into the function gwr().

```
> Y.gwr <- gwr(Y ~ X, data= data.df, bandwidth = Y.bw)
```

One of the data fields of the object Y.grw is denoted SDF. This is a Spatial PointsDataFrame with, among other things, the estimated regression coefficients. There are 400 of them, one for each location. Let's look at the range.

```
> range(Y.gwr$SDF$X)
[1] 0.9932662 1.0107629
```

The range is very small.

Fotheringham et al. (2000, p. 114) recommend the following as an indication of nonstationarity. If the data are stationary, then the regression coefficients should not be affected by random rearrangements of the data. The standard deviation of the regression coefficients is one measure of this effect. Therefore, one can carry out a permutation test of (say) 99 random rearrangements of the X_i and compute the standard deviation of each rearrangement. Using the original data as the hundredth rearrangement, it should not be an extreme value. To carry out this test, we embed the code sequence above into a function and replicate it 99 times.

```
> demo.perm <- function(){
+     data.test <- data.df
+     data.test@data$Y <- sample(data.df@data$Y,
+         replace = FALSE)
+     Y.bw <- gwr.sel(Y ~ X, data = data.df)
+     Y.gwr <- gwr(Y ~ X, data = data.test,
+         bandwidth = Y.bw)
+     return(sd(Y.gwr$SDF$X))
+ }
> set.seed(123)
> U <- replicate(99, demo.perm())
```

You will get a lot of output! You can ignore it, because the important thing is the object U that comes at the end (after a long time). Here is the result of the test.

```
> print(Y.sd <- sd(Y.gwr$SDF$X), digits = 3)
[1] 0.00384
> length(which(U >= Y.sd)) / 100
  [1] 0.98
```

Ninety-eight percent of the permutations have a standard deviation at least as large as the data. This is a strong indication that the data are stationary.

Now let's try this same procedure with the sand data (not detrended) from Field 4.1. We expect gwr() to produce an indication of nonstationarity. This uses the interpolated yield data file *set4.1yld96ptsidw.csv* that contain 1996 yield data interpolated to the locations at which the data in the file *data.Set4.1* were sampled. It is created in Exercise 6.5, but is also available in the *created* folder on the book's website. We will use this to regress yield against sand concentration, which, due to its strong trend, we expect to be nonstationary. The code for the regression is analogous to that given above and is not shown.

```
> Sand.perm <- function(){
+     data.test <- data.Set4.1
+     data.test@data$Sand <- sample(data.Set4.1@data$Sand,
+         replace = FALSE)
+     Sand.bw <- gwr.sel(Yield ~ Sand, data = data.test)
```

```
+    Sand.gwr <- gwr(Yield ~ Sand, data = data.test,
+        bandwidth = Sand.bw)
+    return(sd(Sand.gwr$SDF$Sand))
+ }
> set.seed(123)
> U <- replicate(99, Sand.perm())
> print(Sand.sd <- sd(Sand.gwr$SDF$Sand), digits = 3)
[1] 143
> length(which(U >= Sand.sd)) / 100
[1] 0
```

We do indeed see a strong indication of nonstationarity. None of the permutations have a standard deviation as large as the observed data.

It is important to remember (Section 3.2.2) that stationarity is not a property that can be tested statistically, and that one person's nonstationarity is another person's correlated error structure. Dealing with apparent nonstationarity and trends is one of the most subjective and tricky aspects of spatial data analysis. The best practice is to try different alternative models for the trend $T(x,y)$ of Equation 3.1 and try to gain an understanding of how these alternatives influence the outcome of the analysis. Whatever you do, don't just rely on GWR as an easy way out of the problem.

4.6 Measuring Autocorrelation of Spatially Continuous Data

4.6.1 The Variogram

Although ecological landscape data is almost always more accurately modeled as varying continuously in space, all of the statistics discussed so far have assumed an areal model. This is mostly a matter of convenience because many of the statistical methods described in the following chapters are based on this model. In attempting to characterize the data as completely as possible, however, it makes sense to try to represent them using the model that most accurately characterizes their natural structure. The *variogram* is the most common means of characterizing the spatial structure of data that vary continuously across the landscape. In this section, we describe the construction of the variogram. Variogram analysis is associated with kriging, a widely used interpolation method for spatial data, and in this context it has been discussed in some excellent texts at both the intermediate (e.g., Isaacs and Srivastava, 1989; Panatier, 1996) and advanced (e.g., Cressie, 1991) level. Chapter 6 includes a brief introductory discussion of kriging interpolation.

The discussion of the variogram will use the soil sample data from Field 4.1. We begin by taking up again the *spatial lag*, which represents the difference between the locations of two measurements. In geostatistics, the spatial lag is traditionally denoted h. The spatial lag h is a vector quantity since it has both a magnitude and a direction. Figure 4.6 shows a plot of the sample points of Field 4.1 together with three different lag vectors. The distance between sample points is 61 m in both the north–south and east–west directions. Therefore, the vectors shown in the figure are $h_1 = (61,0)$, $h_2 = (61,61)$, $h_3 = (0,183)$. Throughout this discussion, we assume that the data are isotropic (i.e., independent of direction; see Section 3.2.2). The theory can be extended to more general cases (Cressie, 1991, p. 61; Isaacs and Srivastava, 1989, p. 96; Panatier, 1996, p. 53). In the isotropic case, the variogram properties

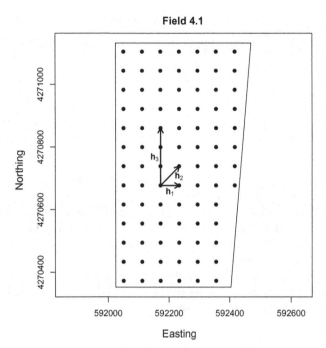

FIGURE 4.6
Sample points of Field 4.1, showing three different spatial lags.

depend only on the magnitude of h, and not on its direction. Because of this, we use the symbol h to denote the magnitude of the vector h as well as the vector itself.

In the isotropic case, we define the *variogram* to be the following function of the spatial lag (Cressie, 1991 p. 58):

$$\gamma(h) = \frac{1}{2}\,\text{var}\{Y(x+h)-Y(x)\},\qquad(4.20)$$

where Y is the measured quantity and x is a position vector with coordinates (x, y). Because of the factor ½, the correct term for $\gamma(h)$ is the *semivariogram*. However, because the R functions for computing this quantity generally use the term "variogram," so will we. Whether or not one includes the ½ will turn out not to matter too much. The variogram is generally estimated by the *experimental variogram* $\hat{\gamma}(h)$, which is defined as follows. Let $m(h)$ be the number of lag vectors that have a magnitude h. Then (Isaacs and Srivastava, 1989, p. 60; Panatier, 1996, p. 38),

$$\hat{\gamma}(h) = \frac{1}{2m(h)}\sum_{i=1}^{m(h)}[Y(x_i+h)-Y(x_i)]^2,\qquad(4.21)$$

where the sum is taken over all sample points separated by a lag vector of magnitude h. Cressie (1991, citing Matheron, 1963) points out that $\hat{\gamma}(h)$ of Equation 4.21 is a method of moments estimator (Larsen and Marx, 1986, p. 267) for $\gamma(h)$ of Equation 4.20. The reason for the 2 in the denominator is that the sum counts each vector pair twice, once in each direction of the vector h.

In the case of a regularly spaced grid of data locations, the grid spacing is often referred to as the *fundamental lag*. For example, in Figure 4.6, if the fundamental lag is 61 m, then the sample points shown as separated by lag vector h_1 are in the group separated by one lag, and so are the same two points separated by the vector in the opposite direction. Although some of the lag vectors, such as h_1 and h_3 in Figure 4.6, have integer magnitude, others, such as h_2, do not. As the magnitude of h increases one arrives at a situation where there are relatively few lag vectors that have the same magnitude (i.e., $m(h)$ becomes small relative to the number of lags), but there are groups of lag vectors that have almost the same magnitude. For this reason, the calculation in Equation 4.21 is generally modified by dividing the lag vectors into *lag groups*, each of which contains all of the lag vectors whose magnitude lies within a specified interval, and carrying out the summation over these lag groups (Panatier, 1996, p. 30). When we need to represent these lag groups explicitly, we will denote the kth group by $H(k)$, defined by those points separated by a distance $kh_0 \leq h < (k+1)h_0$ for some fixed h_0. For example, in the case of the data of Field 4.1 we might set $h_0 = 61\,\text{m}$.

There are several R packages containing functions that may be used to compute the variogram. The computations described here use the package gstat (Pebesma, 2004). We begin by computing and plotting the variogram of the silt content of Field 4.1. The code to compute a variogram is as follows.

```
> library(gstat)
> Silt.vgm <- variogram(Silt ~ 1, data.Set4.1, cutoff = 600)
```

The argument Silt ~ 1 in the function variogram() is a *formula* and indicates that the variogram of Silt is to be computed without reference to any other variables. The second argument provides the data set. The third argument, cutoff, has a couple of uses, but the main one is to exclude any spatial lags larger than the value provided. If no value of cutoff is provided, variogram() will use a default (see ?variogram). Figure 4.7a plots the experimental variogram of the soil silt content of Field 4.1. This is almost a "textbook" plot of spatial data. The solid curve is the variogram model, computed using the function fit.variogram() using the following statement.

```
> Silt.fit <- fit.variogram(Silt.vgm, model = vgm(1, "Sph", 700, 1))
```

The first argument of this function provides the experimental variogram. The second argument specifies the model. This model is specified by the function vgm(). Let's see what each of the arguments of vgm() represents, starting with the second, the class of model (we will get to the first argument below). The argument "Sph" signifies a spherical model. The spherical variogram model has the form (Isaaks and Srivastava, 1989, p. 374; Pebesma, 2001, p. 38)

$$\gamma_{sph}(h) = \begin{cases} 1.5\dfrac{h}{a} - 0.5\left(\dfrac{h}{a}\right)^3 & : h \leq a. \\ 1 & otherwise \end{cases} \tag{4.22}$$

The spherical model is one of a class of variogram models that are *positive definite*. These models are commonly used in kriging interpolation because they ensure that the kriging variance will be positive (Isaacs and Srivastava, 1989, p. 372). The fitting parameter a is called the *range*, and represents the value of h at which the $\gamma_{sph}(h)$ reaches a constant value

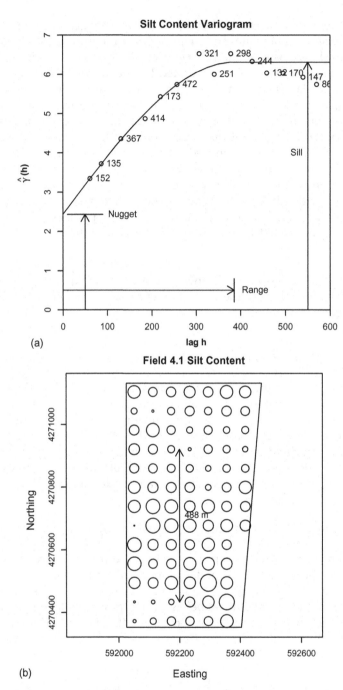

(a)

(b)

FIGURE 4.7
(a) Experimental variograms of percent silt content data of Field 4.1; (b) thematic map of percent silt content of Field 4.1, indicating that low values at the north and south ends are separated by a distance of about 450–500 m.

(i.e., "flattens out"). The range represents the distance at which attribute values cease to be spatially autocorrelated. Figure 4.7a indicates that the range for silt content is about 350 meters. Since $\gamma_{sph}(0) = 0$ and $\gamma_{sph}(h) = 1$ for $h \geq a$, the function actually fit by nonlinear least squares involves two other fitting parameters and has the form

$$\gamma(h) = b + (c - b)\gamma_{sph}(h), \, h > 0. \tag{4.23}$$

From the defining Equation 4.20, $\gamma(0)$ must equal zero. The parameter b is called the *nugget* and represents $\lim_{h \to 0} \gamma(h)$. The nugget represents the combination of sampling error and short-range variability that causes two samples apparently taken from the same location to have different values. If the nugget b is nonzero, then the variogram is discontinuous at zero. The parameter c is called the *sill*. This represents the long-range variability of the samples. These three quantities, the range, nugget, and sill, are defined not only for the spherical model, but also for many other variogram models. The arguments of the function vgm() are respectively the initial guess for the sill, the class of model, the initial guess for the range, and the initial guess for the nugget.

The numbers to the right of each experimental variogram point in Figure 4.7a are the number of points in the lag group used to compute the corresponding value of $\hat{\gamma}(h)$. The greater the number of pairs of points in a lag group, the more confidence can be placed in the computed experimental variogram value. Webster and Oliver (1990, p. 222) suggest that one hundred sample data values should be regarded as the minimum to compute an accurate variogram estimate. By this standard, the data set used in this section, with 86 values, falls just short. Webster and Oliver (1992) point out that in some cases as many as 200 data values may be necessary.

One way in which the experimental variogram of soil silt content in Figure 4.7a does not display "textbook" behavior is that it declines for very large values of the lag h. There are two reasons why this might occur. One is that the number of points in the lag group for very large lags becomes smaller, which can lead to increased variability in the variogram at these lags. Ordinarily, however, this variability includes values above as well as below the sill, while in Figure 4.7a the values of $\hat{\gamma}(h)$ are all below the sill for large h. Figure 4.7b shows that in Field 4.1 the low values of silt content are at the north ends of the field, separated by a distance of about 450–500 m. Therefore, values at this distance are actually more highly correlated than values from sample sites located closer to each other. Experimental variograms that decline with distance or oscillate in magnitude can be indicative of a patchy spatial structure.

Figure 4.8a shows experimental variogram for the "raw" (i.e., not detrended) sand content data. This variogram does not level off. The failure to reach a sill is a common manifestation of a trend. In the presence of a trend, the squared difference $(Y(x_i + h) - Y(x_i))^2$ in Equation 4.21 never reaches a constant value as h increases but rather continues to increase. The detrended variogram can be computed using the data field SandDT developed in Section 3.5, and it can also be computed directly using the function variogram() (see Exercise 4.5b). Figure 4.8b shows the experimental variogram of the detrended sand content. This experimental variogram displays an unfortunate property not uncommon with real data, namely, it does not resemble a "textbook" variogram. There is some indication of a residual trend, and the value of $\hat{\gamma}$ for the smallest lag group is not much smaller than the values for larger lags. A few attempts to fit this variogram with an automatic method were not successful. The computer program Variowin (Panatier, 1996), which is out of print but available on the internet, contains software that allows the user to easily fit a model to an experimental variogram manually.

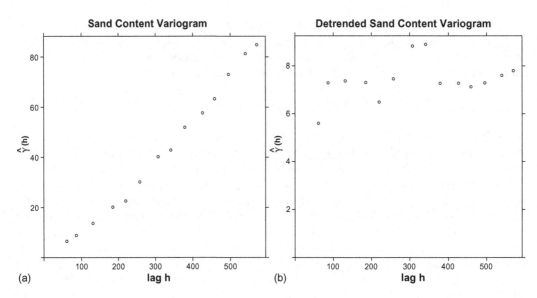

FIGURE 4.8
Experimental variograms of percent sand content data of Field 4.1. (a) Sand content data without detrending, (b) detrended sand content data.

4.6.2 The Covariogram and the Correlogram

Rather than using the variogram, it is sometimes more convenient to work with the *correlogram* $\rho(h)$. This quantity is defined in terms of the *covariogram*. For a second-order stationary random field Y, the covariogram $C(h)$ is defined as (Cressie, 1991, p. 53)

$$C(h) = \text{cov}\{Y(x), Y(x + h)\}, \tag{4.24}$$

and, provided $C(0) > 0$, the correlogram is then defined as (Cressie, 1991, p. 67).

$$\rho(h) = \frac{C(h)}{C(0)}. \tag{4.25}$$

The following discussion will gloss over some technical stationarity assumptions. Banerjee et al. (2004) provide a good discussion of these. We do not discuss them because for ecological data sets they are unlikely to be violated.

From the definition of $\gamma(h)$ (Equation 4.20), it follows that $C(0)$ is equal to the sill of the variogram $\gamma(h)$ and $\lim_{h \to \infty} C(h)$ is equal to the nugget. In other words, for a second-order stationary random field Y, the covariogram $C(h)$ and the variogram $\gamma(h)$ are related by the equation

$$\gamma(h) = C(0) - C(h). \tag{4.26}$$

As with the variogram, the covariogram and the correlogram are generally estimated over lag groups $H(k)$, where k is the index of the lag group. To avoid unnecessary confusion we explicitly note that the symbol h represents a spatial lag, and the symbol k represents the index of a lag group. The experimental covariogram is given by (Isaaks and Srivastava, 1989, p. 59; Ripley, 1981, p. 79)

$$\hat{C}(k) = n_k^{-1} \sum_{i,j \in H(k)} (Y(x_i, y_i) - \bar{Y})(Y(x_j, y_j) - \bar{Y}), \tag{4.27}$$

that is, it is the covariance of data values whose separation is in lag group k. The experimental correlogram $\hat{r}(k)$ can then be estimated by dividing by the sample variance as $\hat{r}(k) = \hat{C}(k) / s^2$.

4.7 Further Reading

Griffith (1987) provides an excellent introduction to autocorrelation statistics. Cliff and Ord (1981) provide a very thorough yet readable discussion of the resampling and randomization assumptions and of the derivation and properties of the join–count, Moran, and Geary statistics. Good entry level sources for permutation tests are Good (2001), Manly (1997), and Sprent (1998). Books at a somewhat higher level include Good (2005), Rizzo (2008), and Edgington (2007). The terminology distinguishing permutation tests is very inconsistent among authors. Some authors use the terms *randomization test* and *permutation test* synonymously, others distinguish between permutation tests in which all possible permutations are examined and randomization tests in which a random sample is drawn without replacement from the permutations, and still other authors use the term *randomization test* for something else entirely (Sprent, 1998, p. 26; Edgington, 2007, p. 20).

Haining (1990, 2003) is an excellent source for discussion of global and local measures of autocorrelation. There are many excellent geostatistics texts. Isaaks and Srivastava (1989) is universally considered to be a great resource. Good sources for Monte Carlo simulation include Ripley (1981), Bivand and Clifford (1989), Manly (1997), and Rubinstein (1981). Efron and Tibshirani (1993) is also very useful, and Rizzo (2008) provides R code. The eigenvectors and eigenvalues of the spatial weights matrix W can provide information about the spatial structure of the data. See Griffiths (1988, Ch. 3) for a discussion. It is considerably easier to understand the application of matrix theory and linear algebra to statistical problems by making use of the concept of the *subject space*, which is discussed by Wickens (1995).

Exercises

4.1 For very small samples, the join–count test can give meaningless results. Do a join–count test using k nearest neighbors with $k = 4$ on the *QUDO* data of Data Set 2 for longitude between $-121.6°$ and $-121.55°$ and latitude between $36.32°$ and $36.45°$. Plot the QUDO data using `pch = QUDO` to see what you are testing.

4.2 Using the *Weed* data from Field 1 of Data Set 4, test the null hypothesis of zero spatial autocorrelation for the full range of levels (i.e., without combining levels). Use both a multicolor join–count statistic and a Moran's I test. How do the results compare?

4.3 (a) Carry out a significance test of spatial autocorrelation using Moran's I for both the "raw" sand content data and the detrended sand content data of Field 1 of Data Set 4. What do you think causes the difference? This is an example of "spurious autocorrelation" discussed in Section 13.1. (b) Compare the results of the test of spatial autocorrelation using Moran's *I* for the detrended data using the linear and median polish detrending.

4.4   Create a Moran scatterplot for mean annual precipitation of Data Set 2 for longitude between −121° and −119° and latitude between 37° and 39°. Identify two points with extreme values. What is the interpretation of these extreme values? Use the function `plot()` to plot the location of these points. Now plot the elevation and discuss whether there might be a physical reason for the extreme local *I* values.

4.5   Using the data of Field 2 of Data Set 4, create Thiessen polygons of different sizes and average the *Yield* data over these polygons. Compute the Moran's *I* for these polygons. What is the effect of polygon size on the Moran's *I*?

4.6   (a) Compute using the function `variogram()` the variogram of detrended sand content using the `SandDT` data field computed via the call to function `lm()` in Section 3.5.

(b) Compute the variogram for sand content directly with the function `variogram()` using the formula `Sand ~ x + y + I(x*y) + I(x^2) + I(y^2)` as the first argument. Compare the result with that of part (a).

5

Sampling and Data Collection

5.1 Introduction

Spatial data consist of two components, a spatial component and an attribute component (Haining, 2003). Spatial sampling can refer to the collection of information about either of these two components. An example of the sampling of the spatial component of a data set is the use of a GPS to record locations in a landscape of point features such as trees. An example of sampling of the attribute component is the collection of soil properties such as clay content, cation exchange capacity, and so forth, according to some sampling pattern. One could also sample for both the spatial location and the attribute values. All four of the data sets used in this book conform to the second type of sampling, in which the location of the sample is specified and some attribute value or values are recorded. It is implicitly assumed in this sort of sampling that the locations of the sites are measured without error. This assumption is not always valid in real data sets, but it is a fundamental one on which much of the theory is based. Since the true location is usually close to the measured location, the usual way to take location uncertainty into account is to augment attribute uncertainty.

In the sampling plans described in this chapter, the locations at which one collects data are generated according to a rule. Sometimes the rule involves some form of randomization, although this is not the case with any of the four data sets in this book. Sometimes the rule involves the systematic selection of locations, as is the case with Data Set 4. Alternative methods of site selection, neither random nor systematic, also exist. These are sometimes imposed by the conditions under which the data are collected. Cochran (1977, p. 16) describes some of them.

1. The sample may be restricted to that part of the population that is readily accessible. The observation sites of western yellow-billed cuckoos in Data Set 1 were collected by floating down the river in a boat, playing bird calls, and listening for a response. Because of this collection method, the sites were restricted to areas near the water.

2. The sample may be selected haphazardly. The data in Data Set 2, which involve the distribution of oak trees in the foothills of the Sierra Nevada and the factors that affect this distribution, were collected during the 1930s in a partly systematic and partly haphazard manner (Wieslander, 1935).

3. The sampler may select "typical" members of a population. This was more or less the case in the selection of sample sites within the rice fields of Data Set 3.

4. The sample may consist of those members of the population who volunteer to be sampled. This is often the case in sampling campaigns involving private property owners. The data in Data Set 3 involves samples collected in the fields of a collection of Uruguayan rice farmers, all of whom agreed to have the samples collected in their fields.

To quote Cochran (1977, p. 16), "Under the right conditions, any of these methods can give useful results." This is a good thing, because at some level virtually every data set collected in the course of a field experiment or observational study can be described at least partly as having one or more of these characteristics (Gelfand et al., 2006). Agricultural researchers collect data on the farms of cooperating growers or on an experiment station. Researchers travel to sites that they can reach. Locations that appear anomalous may be, perhaps subconsciously, excluded. However, it is frequently the case that, once one has either haphazardly or for purposes of convenience selected a site, one can carry out the sampling campaign on that site according to a well-considered plan. This is the case with Data Sets 3 and 4. The locations of fields themselves were selected in consultation with the cooperating farmers, but once the fields were selected the investigators were free to collect data within each field according to a plan of their choice.

In interpreting the results of the analysis of data from a study that suffers from an imperfection, such as one of the four mentioned above, the most important issue to keep in mind is whether the imperfections have introduced bias relative to the population that the sample is supposed to represent (i.e., to the scope of inference). The yellow-billed cuckoo sample in Data Set 1 may favor areas that are more easily accessible to humans. The haphazard selection of sites to sample for oak trees in Data Set 2 may favor a certain type of site, such as those that are more accessible. Data collected from a group of participating farmers such as that in Data Set 3 may be a biased representation of a population of all farmers if only the more "progressive" farmers cooperate with researchers. When reporting these results, care should be taken to describe the conditions under which the sample was taken as precisely as possible so that the reader may form a judgment as to the validity of the sample.

In many ecological applications, the data collected by sampling at point locations on the site may be accurately modeled as varying continuously in space (for example, mean annual precipitation or soil clay content). This chapter is concerned with sampling plans for such spatially continuous data. Although the data are assumed to vary continuously by position, and the samples may be treated mathematically as if they occurred at a single point, this is obviously not really the case. Any real sample, be it a soil core, an electrical conductivity measurement, a pixel in a remotely sensed image, a yield monitor data value, a visual observation, or whatever, is taken over a finite area. The size of area over which the sample is extracted is called the *support* (Webster and Oliver, 1990, p. 29; Isaaks and Srivastava, 1989, p. 190; see Section 6.4.1). In the case of a soil core, the support may be a centimeter or two for a single core, or a meter or two in the case of a composite sample. The other measurements just mentioned typically have supports ranging from one to tens of meters. As an example, Data Set 1 includes locations of bird calls taken near a river. Precise identification of the geographic location of the source of this observation is in many cases impossible, and so the support may be quite large. It is often the case that the support of the sample is different from the size of the area that the sample is intended to represent. For example, many of the analyses described in this book assume that the data are defined on a mosaic of polygons.

Soil core data with a support of a few meters may be used to represent the value of that quantity in polygons that are each tens or hundreds of square meters in area. The distribution of a soil property averaged over a set of polygons of larger area than the support will generally be much smoother than the distribution of the same property averaged over a support of a few square meters. Thus, the polygons should not be considered as the support of these data. Isaacs and Srivastava (1989, p. 190) discuss the issue of mismatch between sample size and area of intended use of the sample data in some detail. Precise definitions of support and other related terms are given in Section 6.4.1, and the effect of this difference between the actual and represented support is discussed in Section 11.5.

In the process of sampling spatial data, the first step is to delimit the area in which to sample. This explicitly defines the *population*, the set of values from which the sample is drawn. The population could be either finite, meaning that it can be indexed by a set of N integers; it may be countably infinite, meaning that it can be indexed by a set of integers, but that set of integers is infinite; or it may be uncountably infinite, meaning that it varies continuously and cannot be indexed. In terms of sampling a spatial area, only the second and third cases generally are relevant unless N is very large. If the value of every member of the population could be determined, then the process would be a *census* and not a *sample*, so it is implicit in the use of this latter term that not every member of the population will be measured. Rather, the properties of the subset that constitutes the sample will be evaluated and used to estimate properties or characteristics of the population.

According to the use of the term given in the preceding paragraph, the population from which the sample is drawn is restricted to the region of sampling. In many if not most cases, however, the results of the sampling and of the analysis of the data from the sample are actually extrapolated to a larger population that is contained in a larger region. For example, the analysis of Data Set 3 is not intended to be restricted to the 16 participating farmers. These results are implicitly extrapolated to a wider collection of similar locations. The definition of similarity in this context, however, is more a scientific issue than a statistical one, and the extrapolation process often depends on the ecological or agronomic skills of the investigator in selecting experimental sites.

In determining a spatial sampling plan, there are a number of issues that one must address (Haining, 2003 p. 94). These include (1) determining the precise objective of the sampling program, (2) determining the spatial form of the sampling pattern, and (3) determining the sample size. Haining (1990, p. 171) describes three categories into which the objective in a spatial sampling campaign may fall.

Category I has as its objective the estimation of a global, non-spatial statistic such as the value of the mean or the probability that the mean exceeds some predetermined threshold.

Category II includes objectives that require knowledge of the variation of the quantity sampled. These include the computation of graphical summaries such as the variogram as well as the development of interpolated maps.

Category III includes objectives that involve the creation of a thematic map via classification. An example is ground-truthing the classification of vegetation types in a remotely sensed image.

Our initial comparison of methods will be based on how well they satisfy an objective in Category I, namely, the estimation of the population mean. While this objective may not arise frequently in ecological research involving spatial data, it turns out that sampling plans that are good at estimating the global mean are often good at other things as well. The problems of selecting placement and number of locations may be separated into two stages. The first stage is the selection of the form of locations' spatial patterns, and the second stage is the selection of the intensity of sampling using that pattern, that is, the number of samples. Most practical spatial sampling patterns fall into one of a few general types. Therefore, it is possible to simplify the process by choosing from a relatively small number of candidate patterns.

The method we use for comparison of sampling patterns is to construct an artificial population, run Monte Carlo simulations on the application of various spatial sampling patterns to this population, and compare the results. This approach does not imply any generality of the comparisons, because it only considers one example. The comparison is useful as a means of gaining insight into the strengths and weaknesses of the patterns, but not as a means of developing general conclusions about the effectiveness of the patterns in sampling other populations. Nevertheless, it provides a useful and intuitive device for increasing understanding of the issues associated with spatial sampling. In Section 5.2, we set up the artificial pattern and formulate the ground rules for the comparison of patterns.In Section 5.3 we develop a set of sampling plans to collect data and compare these plans according to the rules developed in Section 5.2. Section 5.4 contains a discussion of sampling for variogram estimation, and Section 5.5 describes methods for estimating the appropriate sample size based on a targeted confidence interval. Section 5.6 briefly discusses sampling when the intent is to construct a thematic map. Section 5.7 describes the concept of model-based sampling.

5.2 Preliminary Considerations

5.2.1 The Artificial Population

To test the different sampling patterns, we will create an artificial population that is somewhat idealized but is based on real data. Among all of our data sets, the yield monitor data from agricultural fields provide the best opportunity to create an artificial population whose characteristics closely resemble those of a real, continuously varying population. The data we use are taken from the 1996 yield map of Field 2 of Data Set 4, when the field was planted to wheat. Eliminating the edges of the field leaves a data set measuring 720 m long by 360 m wide. To regularize the data for sampling purposes, we create a grid of 144 by 72 points, for a total of 10,368 points on a 5m grid, and in each point we place the value of yield estimated by inverse distance weighted interpolation (Isaaks and Srivastava, 1989, p. 257; see Section 6.3.1). The resulting values are close to the original yield data, but the grid provides the regularity needed for a test of the various sampling designs. Moreover, since this is a finite population we can compute its parameters (mean, standard deviation, etc.) exactly and test the various sampling methods exactly. The creation of the artificial data set involves procedures that are not discussed until later in the book, and understanding these procedures is not essential to understanding the discussion in this chapter. For this reason, the code is not discussed, although like all of the code it is included on the book's website. The population is housed in a `SpatialPointsDataFrame` called `pop.data`.

Figure 5.1a shows a gray-scale plot of the population, and Figure 5.1b shows a close-up of the locations of the data from the artificial population and the real data on which it is based. We first compute some of the population parameters associated with the artificial

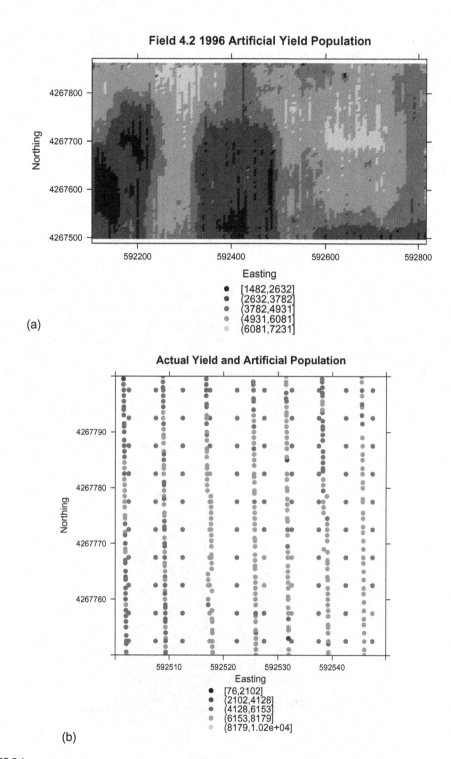

FIGURE 5.1

(a) Plot of the population of 10,368 idealized yield data values; (b) a close-up plot of the locations of a small subset of the idealized values (regular grid) together with the locations of the actual data values.

population. The summary statistics for the yield population data of Figure 5.1 are as follows (because $N = 10,368$, we can ignore the fact that sd() divides by $N - 1$ instead of N).

```
> print(pop.mean <- mean(pop.data$Yield), digits = 5)
[1] 4529.8
> print(pop.sd <- sd(pop.data$Yield), digits = 5)
[1] 1163.3
```

We can generate a plot of the histogram of the data (Figure 5.2) using the function hist().

```
> hist(data. Pop$Yield, 100, plot = TRUE,
+      main = "Histogram of Artificial Population",
+      xlab = "Yield, kg/ha", font.lab = 2, cex.main = 2,
+      cex.lab = 1.5) # Fig. 5.2
```

This reveals that the population distribution is bimodal, as might be expected from inspection of Figure 5.1a. We will see in Chapter 7 that the areas of very low yield correspond to areas of very high weed levels. Therefore, the population presumably can be considered consisting of two subpopulations, one without intense weed competition, and one with weed competition.

We must also create a boundary polygon sampbdry.sf that delimits the population. This follows exactly the same procedure as was used in Section 2.4.3 to create the

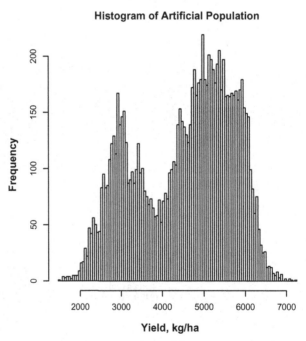

FIGURE 5.2
Frequency histogram of the yield distribution of the artificial data set shown in Figure 5.1.

boundary for the actual field. The only difference is in how the location of the boundaries is determined. For the boundary of the artificial population we extract the coordinates of the bounding box of the `SpatialPointsDataFrame` that contains the artificial population data as follows.

```
> W <- bbox(pop.data)[1,1]
> E <- bbox(pop.data)[1,2]
> S <- bbox(pop.data)[2,1]
> N <- bbox(pop.data)[2,2]
> E - W
[1] 715
> N - S
[1] 355
```

In order return to our original size, we add or subtract 2.5m as appropriate.

```
> N <- N + 2.5
> S <- S - 2.5
> E <- E + 2.5
> W <- W - 2.5
```

The remainder of the construction follows exactly the same steps as were used in Section 2.4.3.

5.2.2 Accuracy, Bias, Precision, and Variance

Sampling error is an unavoidable consequence of the fact that the data represent a sample and not a census. Suppose we are trying to use the sample mean $\bar{Y}$ to estimate the mean μ of a population. The difference $\left|\bar{Y} - \mu\right|$ is a measure of the *accuracy* of the sample (Cochran, 1977, p. 16). One source of low accuracy is a *bias* in the sample. This term will be defined more precisely later on, but intuitively it means a systematic difference between $\bar{Y}$ and μ as shown schematically in Figure 5.3c. A potential second source of low accuracy is low *precision*, as shown schematically in Figure 5.3b and 5.3d. The term refers to a large deviation of the sample means $\bar{Y}$ from the grand mean obtained by repeated applications of the sample procedure. Of course, in a real sampling campaign one does not know the true values of the population parameters, but with our artificial population we do, and so we can compare both the accuracy and the precision of the samples.

FIGURE 5.3
Four different combinations of accuracy and precision: (a) high accuracy and high precision, (b) high accuracy and low precision, (c) low accuracy due to bias and high precision, (d) low accuracy and low precision. Each x represents one application of the sampling procedure.

5.2.3 Comparison Procedures

The testing we present here employs Monte Carlo simulation (Section 3.3). The simulations generate statistics that measure the quality of the sampling plan. Two such statistical measures are used. The first is a measure of accuracy, and the second is a measure of precision. The measure of accuracy is the percent error, defined as

$$\text{Percent error} = \left| \frac{\text{Sample mean} - \text{true mean}}{\text{True mean}} \right| \times 100\%. \tag{5.1}$$

This statistic is computed for each simulation and the mean of the simulations is displayed. The measure of precision is the experimental standard error (i.e., the standard deviation of the means) computed by the simulations. This standard error will be used to compute the *relative efficiency* of two sampling schemes. We will define the relative efficiency e_{12} of estimation of the mean of two sampling schemes of the same sample size to be

$$e_{12} = \frac{se_2}{se_1}, \tag{5.2}$$

where se_i is the standard error of the mean obtained using scheme i.

5.3 Developing the Sampling Patterns

5.3.1 Random Sampling

Webster and Oliver (1990, p. 42) describe a number of simple sampling plans, including simple random sampling, geographically stratified random sampling, and grid sampling. We will evaluate five sampling plans: these three, a second form of stratified sampling, and cluster sampling. All are tested on our finite artificial population, which has a size $N = 10,368$ on a 144 by 72 grid.

The sp package contains a very convenient function spsample() for determining sample points in a spatial context. We will use the function spsample() to develop several of the sampling plans in this section. The code that generates the random sampling pattern shown in Figure 5.4 is as follows.

```
set.seed(123)
spsamp.pts <- spsample(sampbdry.sp, 32, type = "random")
```

The SpatialPolygon object sampbdry.sp describes the region in which to sample and was created in Section 5.2.1. The second argument in spsample() above specifies the sample size. The argument type is specified as "random". The function spsample() selects a random sample of the specified size from the area bounded by the polygon

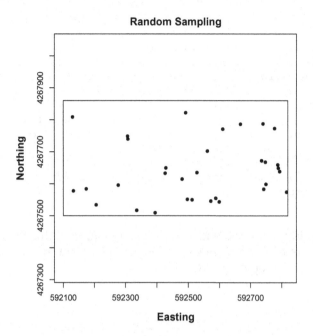

FIGURE 5.4
Thirty-two randomly sampled values from the population of size 10,368.

(Bivand et al., 2013b, p. 146). The coordinates of the sample points are contained in the `coords` slot of the `SpatialPointsDataFrame` object `spsamp.pts` and are accessible through the extractor function `coordinates()`.

Since we are dealing with a finite population defined at 10,368 locations, the sample points randomly selected by `spsample()` do not in general match the locations of the members of the population. Therefore, we will create a function `closest.point()` to determine the member of the artificial population whose location in (x, y) coordinates is closest to that of the sample point. The function `which.min()` in the fourth line returns the index of the member of the population with the smallest geographic distance to the given point.

```
closest.point <- function(sample.pt, grid.data){
   dist.sq <- (coordinates(grid.data)[,1] - sample.pt[1])^2 +
      (coordinates(grid.data)[,2] - sample.pt[2])^2
   return(which.min(dist.sq))
}
```

There is an `sp` function `spDistsN1()` that can be used to perform this task as well (Exercise 5.1).

We now create a function `rand.samp()` that computes the sample mean and the percent error (Equation 5.1) of a random sample drawn from the population. This function is rather long, so a liberal sprinkling of comments is included.

```
> rand.samp <- function (samp.size){
+ # Create the locations of the random sample
```

```
+     spsamp.pts <- spsample(sampbdry.sp, samp.size, type = "random")
+ # Extract a two column array of the x and y coords
+     sample.coords <- coordinates(spsamp.pts)
+ # Apply the function closest.point() to each row of the
+ # array sample.coords (i.e., each sample location)
+     samp.pts <- apply(sample.coords, 1, closest.point,
+         grid.data = pop.data)
+ # Each element of samp.pts is the index of the population value
+ # closest to the corresponding location in sample.coords
+     data.samp <- pop.data[samp.pts,]
+     samp.mean <- mean(data.samp$Yield)
+     prct.err <- abs(samp.mean - pop.mean) / pop.mean
+     return(c(samp.mean, prct.err))
+}
```

We are now ready to carry out the Monte Carlo simulation of this process. The function `closest.point()` remains in memory and is accessed from inside the function `rand.samp()` through lexical scoping (Section 2.5).

```
> samp.size <- 32
> set.seed(123)
> U <- replicate(1000,rand.samp(samp.size))
> print(mean(U[2,]), 3) # Equation (5.1)
[1] 0.0365
> print(sd(U[1,]), 4)
[1] 205.9
```

The mean percent error is 3.7%, and the standard deviation of the means is 205.9 kg ha^{-1}. The histogram (not shown) of the set of means is relatively normal in appearance, as it should be according to the central limit theorem (Section 3.3). One of the assumptions of the central limit theorem as stated in Larsen and Marx (1986, p. 322) is the independence of the random variables. However, the theorem can also be shown to hold in a form sufficient for our uses (Bolthausen, 1982; Guyon, 1995, p. 111; Haining, 2003, p. 274) when the data are spatially autocorrelated. Specifically, Bolthausen (1982) shows that the central limit theorem remains valid for spatially autocorrelated data provided the autocorrelation tends to zero as the distance between data values increases.

By changing the value of the sample size `samp.size` in the first line of the code sequence above, one can determine the effect of this quantity on the sample accuracy and precision. The mean percent errors of the sample of means of sizes 162 and 288 are 1.6% and 1.2%, respectively. The standard deviations of the sample of means of sizes 162 and 288 are 90 and 72 kg ha^{-1}, respectively. Even in the case of non-spatial data, when the data have some form of structure, random sampling may produce an imprecise result (Cochran, 1977, p. 89). With spatial data, it is often a poor choice. In the next subsections we will compare common spatial sampling patterns with random sampling and with each other.

5.3.2 Geographically Stratified Sampling

There are two primary problems with random sampling of a spatial region. The first is that it is time consuming to plot out a path and travel to the randomly selected points, and the second (much more serious) problem is that large portions of the region may happen to go

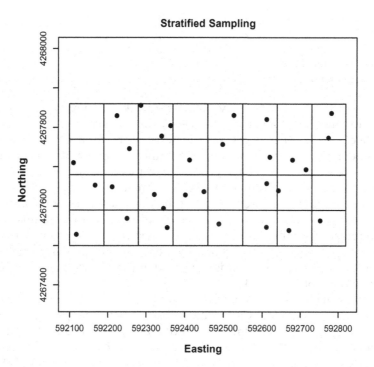

FIGURE 5.5
Stratified sampling on the basis of a subdivision of the field into 32 strata. The strata are generated internally in `spsample()` and do not conform exactly to the strata illustrated here.

unsampled, and some regions may be oversampled. For example, in the random sample illustrated in Figure 5.4, few data are collected in the western part of the field, and many of the sample points at the eastern end are so close together as to be almost duplicative. One way to increase the dispersion of the sample sites is to stratify the field geographically and to sample randomly within each stratum (Webster and Oliver, 1990, p. 43). In Figure 5.5 the field has been divided into 32 equal square strata.

A square or rectangular stratification such as that pictured in the figure is very easily constructed using the sp function `GridTopology()` (Bivand et al., 2013b, p. 48). First, we calculate `x.size` and `y.size`, the cell size in the *x* and *y* directions based on an 8 by 4 arrangement of the cells and the boundary of the sampling region as defined in Section 5.2.1.

```
> print(x.size <- (E - W) / 8)
[1] 90
> print(y.size <- (N - S) / 4)
[1] 90
```

In this case, the lattice cells are 90 m by 90 m squares. Next, we use the function `GridTopology(cellcenter.offset, cellsize, cells.dim)` to generate the geographic strata. Each of the arguments is a vector of two components, representing the *x* and *y* directions. The first argument represents the position of the center of the lower left cell relative to the point (0,0); the second argument represents the cell side length in the *x* and *y* direction, and the third argument represents the number of cells in each direction.

```
> x.offset <- W + 0.5 * x.size
> y.offset <- S + 0.5 * y.size
> samp.gt <- GridTopology(c(x.offset, y.offset),
+   c(x.size, y.size), c(8,4))
> samp.sp <- as(samp.gt,"SpatialPolygons")
```

After constructing the GridTopology object samp.gt, the next line uses the coercion function as() to convert this into a SpatialPolygons object.

For a sample of size 32, there is one sample per stratum; for a sample of size 288, there are 9 samples per stratum. The advantage of stratified sampling of autocorrelated data is that the sample does not miss any large geographic regions. Geographically stratified sampling of agricultural fields is discussed by Wollenenhaupt et al. (1994, 1997). The function spsample() accepts "stratified" as one of the possible values of the argument type. In this case, a geographic stratification is computed based on the sample size, and the appropriate number of samples are randomly located in each stratum. The stratification is carried out internally and does not perfectly match the division of the field into square regions as shown in Figure 5.5, but it is evident that the sample points are more evenly distributed than those of Figure 5.4.

Once again, we can conduct a Monte Carlo simulation with 1000 runs. The only change necessary in the previous Monte Carlo simulation test is to change the value of type in the arguments of spsample() to "stratified". For samples of size 32, 162, and 288 the mean percent errors are 2.6%, 0.8%, and 0.6% (Table 5.1). The standard deviations of the sample of means of geographically stratified samples of sizes 32, 162, and 288 are 143, 44, and 31 kg ha^{-1}, respectively, indicating an improvement in precision over that of random sampling. Using the relative efficiency defined in Equation 5.2, the results of our Monte Carlo simulation indicate that the relative efficiency in estimation of the global mean of random sampling to stratified sampling is between 32% for a sample size of 32, and 42% for a sample of size 288.

5.3.3 Sampling on a Regular Grid

Stratified random sampling alleviates the second of the two objections raised against simple random sampling, the poor coverage of large portions of the field, but it does not alleviate the first, the complexity of the sampling plan. The simplest sampling plan one could carry out is to sample the data on a regular grid. For example, we can sample the field on a 12 by 24 grid to obtain 288 sample points, a 9 by 18 grid to get 162 points, and a 4 by 8 grid

TABLE 5.1

Mean Percent Error (Equation 5.1) and Standard error of the Sample Mean Obtained in 1000 Monte Carlo Simulation Process for Each of Three Random Sampling Methods.

	Random		Geographically Stratified		Stratified by Covariate	
	Mean % error	Std. dev. of means	Mean % error	Std. dev. of means	Mean % error	Std. dev. of means
32	3.7	206	2.6	143	2.3	132
162	1.6	90	0.8	45	1.0	59
288	1.2	72	0.6	31	0.8	45

Note: For purposes of comparison, the corresponding percent errors for grid sampling with 32, 162, and 288 sample locations are 1.6%, 0.1%, and 0.3%, respectively.

to get 32 points. The function `expand.grid()` can be used to generate grid-based sampling plans. The function `spsample()` also accepts the argument `type = "regular"` to generate a regular grid (see Exercise 5.3). Here is the code to compute a 4 by 8 grid using the function `expand.grid()`.

```
> nrows <- 4
> ncols <- 8
> grid.size <- (E - W) / ncols
> grid.offset <- 0.5 * grid.size
> spsamp.pts <- expand.grid(x = seq(W+grid.offset, E,grid.size),
+    y = seq(N - grid.offset,S,-grid.size))
> samp.pts <- apply(spsamp.pts, 1, closest.point, grid.data = pop.data)
> data.samp <- pop.data[samp.pts,]
> print(abs(mean(data.samp$Yield) - pop.mean) / pop.mean, digits = 4)
[1] 0.0002843
```

The quantity `grid.offest` in line 4 of the code is computed to center the grid within the boundary of the rectangle. Figure 5.6 shows a regular square 4 by 8 grid imposed on the test population. The percent error of grid sampling for this particular experiment is 0.03% for a 32-point grid, 0.3% for a 162-point grid, and 0.09% for a 288-point grid (note that by chance the 32-point grid produces the best result). This is considerably better than the performance of the other methods (Table 5.1). Of course, as always with these numerical experiments, this result is highly dependent on the particular data set, particularly the

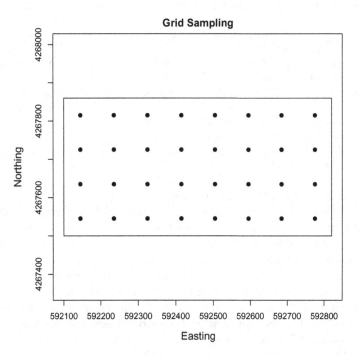

FIGURE 5.6
A regular square grid of 32 points.

very low error of the 32-point sample. Nevertheless, we shall see later that in an important way grid sampling is very close to an optimal sampling scheme.

In thinking about the possibility of carrying out a Monte Carlo simulation experiment for grid-based sampling analogous to those for simple and stratified random sampling, we can see at a glance a fundamental problem. The regularity of the grid removes the randomness of the sample points relative to one another. An element of randomness can be inserted into the sample by randomly locating the coordinates of one corner (Webster and Oliver, 1990, p. 45; Cochran 1977, p. 206). In practice, however, one ordinarily places the grid in such a way as to maximize the distance from any grid point to the field boundary, in order to minimize edge effects. In this case, there is no randomness at all, and it becomes impossible to meaningfully estimate the standard error in the same way as is done with sampling plans involving randomness in sample location. We will return to this issue in Section 5.6.

5.3.4 Stratification Based on a Covariate

The previous sampling schemes made no use of information about the properties of the site being sampled. In many cases, the investigator has information about how various aspects of the site's geography influence the response variable, and such information may be useful in increasing sampling efficiency. Indeed, in classical, non-spatial sampling, stratification is generally based on a covariate rather than on spatial location (Cochran, 1977). We can illustrate this approach for our particular site by utilizing the relationship between weed infestation level and yield. During the course of the study in Field 4.2 in 1996, it became evident that a weed infestation existed in the western part of the field that could be expected to lead to substantial yield loss (as indeed it did). We can compute the effect of weed infestation level on yield as follows. First, we use the function closest.point() to determine the yield of the member of the artificial population located closest to each of the 78 sample points.

```
> data. Set4.2 <- read.csv("set4\\set4.296sample.csv", header = TRUE)
> # Extract the coords
> sample.coords <- cbind(data. Set4.2$Easting, data. Set4.2$Northing)
> # Find the yield at the closest sample points
> samp.pts <- apply(sample.coords, 1, closest.point, grid.data = pop.
data)
> data. Set4.2$Yield <- pop.data$Yield[samp.pts]
> with(data. Set4.2, print(tapply(Yield, Weeds, mean), digits = 4))
    1    2    3    4    5
5419 5184 5286 4724 3129
```

The last line is an example of the use of the functions with() and tapply() (tapply() is described in Section 2.3.2). The function with() evaluates a function specified by its second argument in an environment constructed by the data in its first argument. Thus (ignoring the print() function) the statement is equivalent to tapply(data. Set4.2$Yield, data. Set4.2$Weeds, mean).

When stratifying by a covariate one could use either random or grid-based sampling within the strata. We will employ random sampling in order to compare this sampling plan with geographic stratification. The region of high weed infestation levels was visible on the ground, and is also apparent in the infrared aerial image taken in May at the end of the season. At this time, the crop was already senescent, but the weeds were still vegetative and therefore show up as brighter in the infrared band. Figure 5.7 shows the location of high weed levels (*Weeds* = 5), as indicated by the larger circles. The locations are

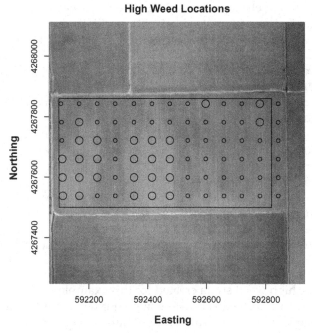

FIGURE 5.7
Locations of high weed infestation level at the sample points of Field 4.2, shown on top of the infrared band of an aerial image taken in May 1996. The image was scanned from a positive film, and there are faintly visible concentric circles in the northwest part of the image of the field and on the eastern boundary. These are Newton rings, caused by molecular interaction between the film and the scanner's glass plate.

superimposed on the infrared band of the aerial image taken in May 1996. The boundary of the sampling area is also shown. The code used to create this figure is discussed in Section 2.6.2. The first step in the process of stratified sampling based on the value of the covariate *Weeds* is to develop a boundary file that delimits the strata. At the time of the sampling campaign, the actual effect of weed level on yield was unknown. The observation of very high weed levels would, however, lead to an anticipation of yield loss, so the appropriate response would be to stratify by the high weed and low weed areas, where the high weed areas are defined as those in which the weed level has the value 5. For this application, we create a set of polygons delimiting only the high weed areas, using code similar to that used to create the boundary file, but with four individual polygons instead of one. Coordinates defining these polygons can be determined in a geographic information system or, alternatively, by using the function `locator()`.

```
> #Create boundary file
> x1 <- 592165
> x2 <- 592240
> x3 <- 592320
> x4 <- 592490
> x5 <- 592580
> x6 <- 592620
> x7 <- 592760
> y1 <- 4267690
> y2 <- 4267810
> y3 <- 4267760
```

There are four polygons defining areas of high weeds. Because they are easier to work with, we use special features to create an `sf` polygon file and then coerce this into an `sp` object. The code is a modification of that introduced in Section 2.4.3.

```
> strat1 <- matrix(c(W, y1,x1,y2,x2,y2,x2,S, W,S, W,y1), ncol = 2,
+     byrow = TRUE)
> strat2 <- matrix(c(x3,S, x3,y3,x4,y3,x4,S, x3,S), ncol = 2,
+     byrow = TRUE)
> strat3 <- matrix(c(x5,N, x6,N, x6,y2,x5,y2,x5,N), ncol = 2,
+     byrow = TRUE)
> strat4 <- matrix(c(x7,N, E,N, E,y3,x7,y3,x7,N), ncol = 2, byrow = TRUE)
> strat.pol <- st_sfc(st_polygon(list(strat1)),
+     st_polygon((list(strat2))), st_polygon(list(strat3)),
+     st_polygon(list(strat4)))
> stratbdry.sf <- st_sf(z = c(1,1,1,1), strat.pol)
> st_crs(stratbdry.sf) <- "+proj=utm +zone=10 +ellps=WGS84"
> stratbdry.sp <- as(stratbdry.sf, "Spatial")
```

Figure 5.8 shows the high weed strata created for this example.

In order to obtain an unbiased estimate using stratified sampling, the fraction of the total number of samples taken from each stratum must be equal to the relative size of that stratum in relation to the total population size (Cochran, 1977, p. 117). In our case, this means that the fraction of the samples that are carried out in the high weed stratum must

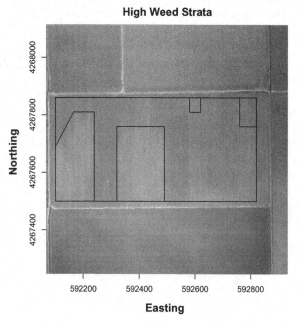

FIGURE 5.8
High weed sample strata in Field 4.2.

be equal to the fractional area occupied by this stratum. The relative size of the strata in spatial sampling can be determined from their geographic areas. It is easy to obtain the areas of the polygons of the sf objects using the function st _ area(). This returns the area of each polygon

We can also compute the total area.

```
> print(hi.area <- st_area(stratbdry.sf))
Units: m^2
[1] 39500 44200 2000 6000
> print(tot.area <- st_area(sampbdry.sf))
259200 m^2
> print(frac.hi <- as.numeric(sum(hi.area) / tot.area))
[1] 0.3537809
```

The function st _ area() returns a value in the map units, so we must use as.numeric() to coerce this into a pure number.

To carry out a Monte Carlo simulation using the artificial population, we must separate the artificial population into two subpopulations, one contained in the high weeds stratum and one in the low weeds stratum. This can be done with the function over()of the sp package (Pebesma, 2018).

```
> # over(pts, polygon) produces a data frame with NA
> # for points outside the polygon
> hiweeds <- over(pop.data, stratbdry.sp)
> head(hiweeds)
    z
1 NA
2 NA
3 NA
4 NA
5 NA
6 NA
> length(hiweeds$z)
[1] 10368
> length(which(!is.na(hiweeds$z)))
[1] 3668
```

The first argument of over() is pop.data, which contains the sample points and is a SpatialPointsDataFrame. The second argument stratbdry.sp, which defines the boundary of the high weed stratum, is a SpatialPolygons object. (By the way, if one of the functions in the argument of over() was originally created using an sf method and another was originally created using an sp method, you may get an error message. This is due to a trivial difference in these methods' use of the function proj4string() and can be corrected by simply setting projections when both are sp objects.)

When the first and second arguments respectively of a call to over() are members of these classes, the function returns a data frame whose one data field is a set of integers of the same length as the number of points in the first argument. Each element corresponds to a point. For those points that fall inside a polygon in the second argument (stratbdry.sp in this case), the value of the corresponding element of the data field is the index of the polygon in which the point falls. For those points that do not fall inside a polygon of `stratbdry.sp`, the value of the corresponding element is `NA`, as it is in this case for the first six records. Since we are aggregating all polygons together as indicating high weeds, any element of the array whose value is not `NA` corresponds to a point in the high weeds stratum. We can, therefore, create a data field `hiweeds` of the artificial population data object `pop.data`, initially assign all of its records the value zero, and then assign the value 1 to all those records whose index is not `NA`.

```
> pop.data$hiweeds <- 0
> pop.data$hiweeds[which(!is.na(hiweeds$z))] <- 1
```

We next create a function `stratified.sample()` to use in the simulation in the same way that the function `random.samp()` was used earlier. In this case, the argument `samp.size` is an array with two elements. The first is the number of samples from the low weed stratum, and the second is the number of samples from the high weed stratum. The object `subpop` is created twice, to contain the elements of both subpopulations.

```
> stratified.samp <- function(samp.size){
+ # Low weed stratum
+    subpop <- pop.data$Yield[pop.data$hiweeds == 0]
+    samp <- sample(subpop, samp.size[1])
+ # High weed stratum
+    subpop <- pop.data$Yield[pop.data$hiweeds == 1]
+    samp <- c(samp, sample(subpop, samp.size[2]))
+    samp.mean <- mean(samp)
+    prct.error <- abs((samp.mean - true.mean) / true.mean)
+    return(c(samp.mean, prct.error))
+ }
```

The quantity `frac.hi` computed above is used to calculate the sample sizes in each stratum.

```
> set.seed(123)
> sample.size <- 32
> hi.size <- round(frac.hi * sample.size, 0)
> lo.size <- sample.size - hi.size
> samp.size <- c(lo.size, hi.size)
> U <- replicate(1000, stratified.samp(samp.size))
> mean(U[2,])
[1] 0.02394189
> sd(U[1,])
[1] 132.7480
```

Figure 5.9 shows a sampling pattern for stratified random sampling with 32 sample locations.

Table 5.1 contains the mean percent error and the standard deviation of the estimated means of 1000 Monte Carlo simulations for each of the three random sampling methods we have tested. For this particular case, geographic stratification and stratification

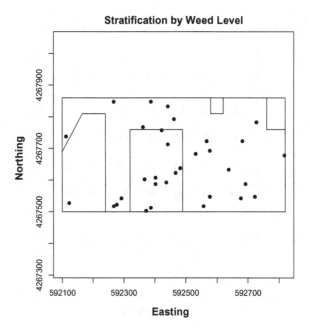

FIGURE 5.9
A sampling plan with 32 sample locations stratified by weed level.

by a covariate give similar results. This is not true in every case, but it is generally true that stratification provides better results than simple random sampling. Depending on how many strata are created in sampling by a covariate, part of the effect of this stratification may be to stratify geographically and spread the sample points more evenly.

5.3.5 Cluster Sampling

In every ecosystem, spatial variability exists on multiple scales. In Section 6.4.1, a formal model is provided for this phenomenon, but even without such a formal model one can recognize the value of gathering data on the spatial relationships of nearby locations within a site as well as relatively distant locations. Cluster sampling provides, in theory, a means of gaining precision in the estimation of short range spatial interaction effects without requiring an impossibly large number of samples. For a fixed sample size, the potentially increased local precision of cluster sampling comes at the expense of reduced coverage density of the entire site. As with stratification based on a covariate, cluster sampling can be carried out using either a random or a grid-based plan, and indeed it could be combined with any of the plans already discussed. Because its primary utility is in the improved estimation of statistical properties involving short range variability, we would not expect cluster sampling to necessarily provide improved estimates of a global statistic such as the mean. Therefore, there seems to be no particular advantage to testing a random

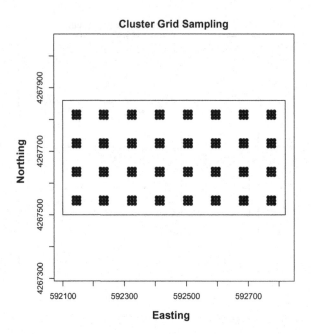

FIGURE 5.10
A cluster sampling grid consisting of 32 clusters with centers separated by 50 m, with nine sample locations in each cluster, separated by 10 m.

sampling plan, so instead we will develop a grid-based cluster sampling plan. We will be testing cluster sampling on the estimation of the mean, as with the other plans, but the real interest in this sampling plan is its performance in variogram estimation, which is discussed in Section 5.4.

Figure 5.10 shows a cluster sampling plan consisting of 32 clusters of 9 sample points each, for a total of 288 sample locations. Within each cluster, the nine sample locations are 10 m apart. The sampling plan has an error in estimation of the mean of 0.7%. This is more accurate than any of the randomization methods of Table 5.1, but for this data set it is not as accurate as the 0.3% error of the regular grid at 288 locations.

5.4 Methods for Variogram Estimation

The estimation of a single, globally based statistic such as the mean does not tell us anything about the spatial properties of the sample domain. For this we need to estimate a spatial descriptor such as one of those discussed in Chapter 4. The vast majority of the literature on this subject is devoted to estimation of the variogram (Section 4.6.1). Recall that the equation of the experimental variogram is given by (Equation 4.21),

$$\hat{\gamma}(h) = \frac{1}{2m(h)} \sum_{i=1}^{m(h)} [Y(x_i) - Y(x_i + h)]^2, \tag{5.3}$$

where h represents a lag and $m(h)$ is the number of lag vectors of lag h. As discussed in Section 4.6.1, lag vectors are generally grouped into lag groups $H(k)$ defined by those of magnitude $kh_0 \le h < (k+1)h_0$ for some fixed h_0. A reasonable size of the lag group is required to compute a good estimate $\hat{\gamma}$, and $\hat{\gamma}$ must be estimated for several lag groups, so it is evident that a lot of data values are needed to adequately estimate the variogram. Webster and Oliver (1990, 1992) state that between 100 and 200 samples should be used to generate an adequate experimental variogram.

In comparing sample plans for estimation of the variogram, it is necessary to define the standard of comparison. One possibility is to consider some direct measure of the variogram itself, such as a confidence interval. Russo (1984), for example, proposes a measure of quality based on the aggregation of the measurement accuracies of the individual values of the variogram estimate $\hat{\gamma}(h)$ at each lag h. It is generally simpler and more satisfactory, however, to consider the role of the variogram in kriging (Section 6.3.2) and to use the kriging variance (Isaaks and Srivastava, 1989; see Section 6.3.2) as the standard of comparison. The kriging variance does not depend directly on the values of the response variable $Y(x,y)$ but only on the number and locations of the sample points (x,y) and the values of the variogram $\hat{\gamma}(h)$ (Cressie, 1991, p. 315; Webster and Oliver, 1990, p. 262). The values of $Y(x,y)$ enter into the kriging variance through their influence on $\hat{\gamma}(h)$. An important special case is that in which the data are isotropic, that is, their properties relative to one another do not depend on direction (Section 3.2.2). It may be necessary to modify the grid in the case that the data are anisotropic. Cressie (1991, p. 323), and Webster and Oliver, (1990, p. 277) discuss this case.

To carry out an informal comparison of sampling plans for variogram estimation, we again use the artificial yield population of Figure 5.1a. In addition to (1) the population variogram, the following five sampling methods are tested: (2) a regular 12 by 24 grid (288 samples), (3) a regular 8 by 16 grid (128 samples), (4) a regular 6 by 12 grid (72 samples), (5) a sample of 288 randomly selected locations, and (6) the cluster sample of Figure 5.10. Figure 5.11 shows the results of the comparison of each of these with the variogram of the entire population.

In this example, the random sample (Figure 5.11d) generally performs about as well as the regular grid sample (Figure 5.11a) of the same size. The 128 and 72 point samples generally perform worse. Cluster sampling does much worse, even at small lag distances. Cressie (1991, p. 317) points out that cluster sampling is often poorly adapted to spatial data.

McBratney et al. (1981) carried out a comparison of sampling plans in accuracy of estimation of the variogram. Their results can be interpreted intuitively as indicating the optimal plan is one in which the sample points are mutually as far away from each other as possible. The absolute maximum mutual distance is achieved by using an equilateral triangular sampling grid, but a nearly equal mutual distance is achieved with a regular square grid, which is much easier to lay out and execute (McBratney et al., 1981).

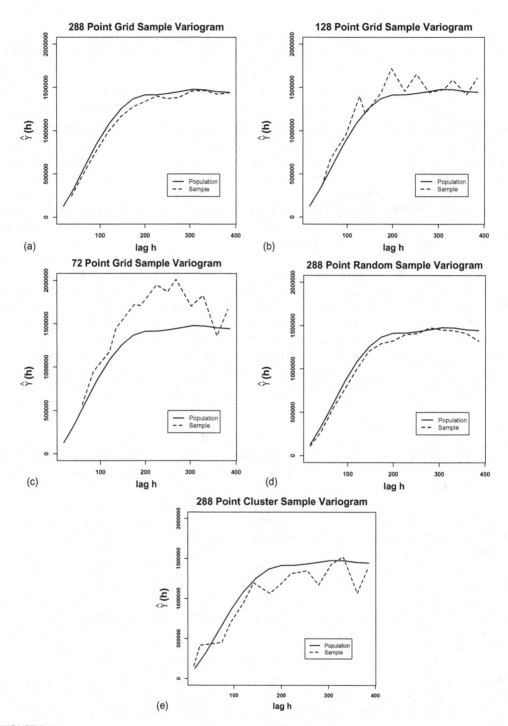

FIGURE 5.11
Plots of the comparison between the population variogram $\gamma(h)$ and the experimental variograms $\hat{\gamma}(h)$ for the artificial yield data. The variogram of the full population is shown as the solid line and that of the sample as a dashed line; (a) grid sample of size 288; (b) grid sample of size 128; (c) grid sample of size 72; (d) random sample of size 288; (e) cluster sample as in Figure 5.8.

5.5 Estimating the Sample Size

Once a sampling pattern has been established, the next step is to establish the sample size. As a practical matter, every project with which I have ever been involved used the following formula to determine the sample size:

$$\text{Number of samples} = \frac{\text{Money in the budget}}{\text{Cost per sample}} \tag{5.4}$$

Nevertheless, it is very useful to calculate an estimated sample size, if only to determine how close one can come to the ideal. Also, the process of estimating the sample size introduces a very important concept: the relationship between sample size and data variability. The number of samples necessary to estimate a parameter or test a hypothesis depends on the desired precision and the variability of the data. Many methods for estimating sample size use an independent estimate of variability, as measured by the standard error, to determine the number of samples required to achieve a given level of precision in the estimate. We consider first a simple example that illustrates some of these principles. The example is the estimation of the mean μ of a normally distributed population from a set of *independent*, identically distributed random variables drawn from the population. Assume that we want to estimate μ with a $(1-\alpha) \times 100$ percent confidence interval of width no larger than 2δ. The formula for the confidence interval is (Sokal and Rohlf, 1981, p. 148; Kutner et al., 2005, p. 1306)

$$\bar{Y} - t(1 - \alpha/2; n-1)s\{\bar{Y}\} \leq \mu \leq \bar{Y} + t(1 - \alpha/2; n-1)s\{\bar{Y}\}, \tag{5.5}$$

where $\bar{Y}$ is the sample mean, $t(1-\alpha/2; n-1)$ is the $1-\alpha/2$ percentile of the t distribution with $n-1$ degrees of freedom, and $s\{\bar{Y}\}$ is the standard error, given by

$$s\{\bar{Y}\} = \frac{s}{\sqrt{n}}, \quad s = \sqrt{\frac{\Sigma(Y_i - \bar{Y})^2}{n-1}}. \tag{5.6}$$

Therefore, if the confidence interval is to have a width no larger than 2, we must have $2\delta \geq 2t(1 - \alpha/2; n-1)s/\sqrt{n}$, which implies

$$n \geq \frac{t(1 - \alpha/2; n-1)^2 s^2}{\delta^2}. \tag{5.7}$$

Thus, an estimate of the sample variance s^2, usually drawn from a preliminary sample, is required to estimate the sample size. The important principle is that Equation 5.6, which defines a measure of the variability of the mean in terms of the sample size n, is inverted to form Equation 5.7, which expresses the sample size n in terms of the variability. One then uses some independent estimate of variability to compute the sample size.

One method of implementing inequality (Equation 5.7) to estimate sample size in developing a sampling plan is called *Stein's method* (Steel et al., 1997, p. 124). This method requires a preliminary sample of size n_0. From this sample, one computes the sample variance s_0^2 and establishes n as the smallest integer greater than or equal to $t(1 - \alpha/2; n_0 - 1)s_0^2 / \delta^2$.

There is a considerable literature on sample size estimation, much of which is discussed by Cochran (1977, Ch. 4). This literature, however, is valid for samples of independent data but may not retain its validity for spatial data.

A common but simplistic means of estimating the spatial density of a grid sample is to use range of the variogram, that is, the value of the lag h at which the variogram $\gamma(h)$ approaches its asymptote (Section 4.6.1). This represents a measure of the spatial distance between which data values are uncorrelated. As such, the sites in a sampling grid should be separated by a distance no greater than the range if some measure of the entire sampling region is to be obtained. This is a fairly blunt instrument, however. The number of data values required for a sample is also dependent on the objective of the sampling campaign. In the variogram estimation example of Section 5.4, the sample spacings for the 288, 128, and 72 point samples were 30, 45, and 60 m, respectively. Inspection of Figure 5.11 indicates that the range is about 200 m. Thus, the estimation of the variogram requires sampling at a spacing considerably smaller than the range. On the other hand, the estimation of the global mean can sometimes be done with reasonable accuracy using a larger grid spacing.

One way to determine the appropriate sample size for a spatially autocorrelated population is to estimate the *effective sample size* n_e. This quantity, which is discussed at considerable length in Chapter 10, is the size of the sample that would yield the same sample variance if the population was not autocorrelated as a sample of size n of the actual population. If one can compute an estimate $\hat{n}_e$ of n_e, then one can determine the appropriate sample size n by substituting $\hat{n}_e$ into the right-hand side of inequality (Equation 5.7). Chapter 10 provides a simple, *ad hoc* method of computing $\hat{n}_e$ using bootstrapping. Griffith (2005) describes several methods based on the theories developed in Chapters 12 and 13.

5.6 Sampling for Thematic Mapping

In many applications, the objective is not to estimate a statistic but rather to develop a thematic map of the sample region based on one or more themes (Webster and Oliver 1990, p. 273). For example, in an ecological application one may wish to develop a thematic map of a meteorological quantity such as mean annual precipitation or maximum July temperature. An example in agriculture is the so-called "management zones" of site-specific crop management (Lowenberg-DeBoer and Erickson, 2000). We will characterize the general problem of thematic mapping as devising a sampling plan such that some attribute or attributes may be classified at each location according to a finite set of categories. The main point is that we want the estimate to be as accurate as possible at all locations, as opposed to merely seeking an estimate of a global statistic or a representation of variability.

The sampled quantity Y commonly takes on values $Y(x, y)$ continuously, or piecewise continuously (i.e., allowing for jumps at some locations) as a function of position. For example, soil clay content or precipitation could be assumed to have this property. If the scale is large enough, quantities that are not continuously distributed in this way, such as species population density, can be modeled for the purpose of mapping as if they were continuously varying in geographical space. The mapping problem in this case becomes one of finding the best interpolation method to estimate values of Y at locations other than the sample locations. This is a fundamental problem of geostatistics, and is briefly discussed in Section 6.4. Interpolation is well covered in other texts (e.g., Isaaks and Srivastava,

1989; Goovaerts, 1997; Webster and Oliver, 1990; Webster and Oliver, 2001; Cressie, 1991). A number of interpolation methods are described in these texts, including inverse distance weighted interpolation, kriging, and spline interpolation. In terms of sampling plan development, kriging has the advantage of permitting the estimation of the kriging variance, which permits one to determine the sample patterns that, for example, minimize the maximum value of the kriging variance. Webster and Oliver (1990, p. 272), Cressie (1991, p. 318), and McBratney et al. (1981) discuss optimization of sample patterns in this context. The general conclusion is that under most circumstances a regular grid provides the best combination of simplicity of layout and accuracy of estimation.

Although not always the case, it frequently happens that the estimate of a particular variable is based on two or more measured quantities, that the measurement cost of the different quantities is quite different, and that the most expensive data to collect provide the most accurate estimate. For example, in estimating soil clay content the most accurate data comes from the extraction of soil cores, but apparent electrical conductivity is much cheaper and easier to measure. The question then becomes how to allocate the samples of the expensive quantity in such a way that the combination of all the measurements yields the most accurate map possible. In this case, sampling on a grid may not be the best approach for either commercial or scientific applications. Pocknee et al. (1996) provide a discussion of the drawbacks of grid sampling for this application. Probably the most commonly used method of estimation of the expensive-to-sample quantity via the cheap-to-sample quantity is cokriging (Isaaks and Srivastava, 1989, p. 400, Section 6.3.3). An alternative approach based on response surface analysis has been put forward by Lesch et al. (1995).

5.7 Design-Based and Model-Based Sampling

There are two separate approaches to the development of a sampling plan, the *design-based* approach and the *model-based* approach. In the design-based approach, also referred to as the *probability sampling* approach (Valliant et al., 2000), the population in the study region is viewed as having a fixed set of values (Haining 2003, p. 96). Sampling locations are selected according to some randomization scheme. This scheme is designed to ensure that it yields a parameter estimate with the desired statistical properties (e.g., unbiasedness or minimum variance). This reliance on a proper design is the source of the term "design-based" (Webster and Oliver 1990, p. 28). The random and stratified sampling plans discussed in Section 5.3 are examples of the design-based approach. There has been an increase in interest in using the model-based approach in spatial sampling (e.g., Haining, 2003; Griffiths, 2005), and for this reason we will give a simple example of how it might be applied to the artificial data set used in this chapter.

In the model-based approach, the population itself is considered to be one realization of a stochastic process. Other realizations are possible, and the population properties of interest, such as the mean and variance, are functions of the values of a random process. These properties are therefore themselves random variables and technically are predicted rather than estimated (only fixed parameters of a population are estimated). For this reason, model-based sampling is also called *prediction-based sampling* (Valiant et al., 2000). The process of developing a sampling plan involves developing a model for the random process generating the data and then estimating the parameters of this model.

Since the population in the study region is viewed as a random variable, there is no need to introduce randomness in the sampling pattern. Therefore, systematic sampling plans, such as the grid-based plans described in Section 5.3, can be studied statistically using a model-based formalism.

To make this distinction a bit clearer, consider the simple example in which a finite population of size N is being sampled, and the objective is to choose a sample of size n to estimate the population mean

$$\mu = \sum_{i=1}^{N} Y_i/N \tag{5.8}$$

by computing the sample mean

$$\bar{Y} = \sum_{i=1}^{n} Y_i / n \tag{5.9}$$

In the design-based approach, the population $\{Y_i, i = 1,...,N\}$ is considered as fixed, and the quantity μ is a fixed parameter. The sample $\{Y_i, i = 1,...,n\}$ is a random quantity dependent on the random selection of the n values to sample. The objective is to select a randomization process that generates a value $\bar{Y}$ that, according to some measure, optimally estimates μ. In the model-based approach, the population $\{Y_i, i = 1,...,N\}$ is viewed as a realization of a random process, and therefore $\mu(Y_1, Y_2,..., Y_N)$ defined by Equation 5.8 is a random variable. The objective of sampling is to develop a sampling plan that optimally predicts μ. Suppose, for example, that the sampling plan was to sample every tenth value of the population. Under the assumptions of the design-based approach, since the population is fixed, this sampling pattern, if applied repeatedly, would yield the same sample values and the same estimate $\bar{Y}$ each time it was applied. Under the assumptions of the model-based approach each time the sampling pattern was applied it would sample a different realization of a random process, and so the sample values and the value of $\bar{Y}$ would be random variables. There is, of course, nothing special about the mean, and the same idea can be applied to the prediction of any other function of the random variable Y.

If one has a model for the relationship between Y and some explanatory variable X, then one can use this model to improve the accuracy of prediction. Valliant et al. (2000, p. 2) introduce the concept of model-based sampling with a simple example based on the number of patients discharged per day from a hospital versus the number of beds. We will begin our discussion with an analogous presentation using the relationship between wheat yield of the artificial data set and observed weed level at the nearest sample point. In their example, Valliant et al. (2000) estimate the total number of patients discharged in 33 hospitals given a sample consisting of the total number of patients discharged in 32 of them (equivalently, they estimate the number of patients discharged in the one non-sampled hospital). We will generate a similar example using the artificial yield population from Field 4.2.

To some extent, comparing design-based and model-based plans is a matter of comparing apples and oranges. Nevertheless, we will carry out an informal comparison using 32 sample points, the same value as the minimum size of the random and grid sample methods. We use the `raster` package to manipulate raster objects in the ways described by Lo and Yeung (2007, p. 183). Here we make only the simplest use of the package's capabilities to compute a simple linear regression between weed level and May infrared image digital number. We will use the fact (Kutner et al., 2005, p. 24) that the regression line between Y and X passes through $(\bar{X}, \bar{Y})$, to estimate μ by computing the value of the regression line at $\bar{X}$). We emphasize that this is not necessarily the best way to carry out a model-based sampling plan, but it does illustrate the idea and it will enable us to demonstrate some issues associated with model-based sampling.

Figure 1.2 indicates that there is an apparently close relationship between the infrared band digital number of the May aerial image of Field 4.2 and the yield. We will base our prediction on this relationship. We start by loading the object `data.May.ras`, which contains the image band information.

```
> library(raster)
> data.May.ras <- raster("set4\\set4.20596.tif")
> class(data.May.ras)
[1] "RasterLayer"
attr(,"package")
[1] "raster"
```

By default, the function `raster()` loads band 1 of the *TIFF* file, which is the band that we want. We will use the spatial locations of the 32 grid sample points in Figure 5.6. We can apply the function `extract()` to the raster object to place the IR values of the cell containing each of the 32 sample points into the data frame `data.samp` created in Section 5.3.

```
> data.samp$IRvalue <- extract(data.May.ras, spsamp.pts)
```

Figure 5.12 shows a plot of yield vs. infrared band digital number together with the least squares regression fit. The code to generate the fit and compute the value of $\bar{Y}$ is as follows.

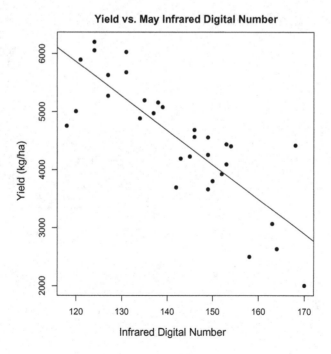

FIGURE 5.12
Linear regression model of yield vs. May IR value based on 32 sample locations.

```
> Yield.band1 <- lm(Yield ~ IRvalue, data = data.samp)
> summary(Yield.band1)
Coefficients:
            Estimate Std. Error t value Pr(>|t|)
(Intercept) 12996.28    1068.28  12.166 3.96e-13 ***
IRvalue       -59.42       7.46  -7.965 6.85e-09 ***
Multiple R-squared:  0.679,   Adjusted R-squared: 0.6683
```

The fit is reasonably good ($R^2 = 0.68$), which provides justification for the use of this model in our model-based sampling plan. The estimate of the population mean is obtained as the mean of the predicted values based on the linear regression model.

```
> print(Y.bar <- mean(predict(Yield.band1)), digits = 5)
[1] 4528.5
> print(abs(Y.bar - pop.mean) / pop.mean, digits = 3)
[1] 0.000284
```

The error in this particular example is less than that of grid sampling.

It is evident that the estimate of μ is highly dependent on accuracy of the model. This in turn depends in part on the choice of sample locations. This choice is important both for design-based sampling plans and for model-based sampling plans, but the use of a model-based plan provides an opportunity to illustrate this effect very graphically. We will consider three cases (Figure 5.13). The first is to sample seven points in a north to south transect in a high weed area; the second is to sample seven points in a north to south transect in a

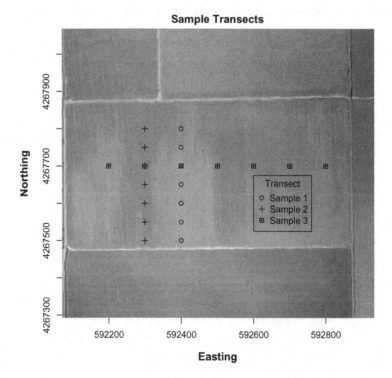

FIGURE 5.13
Three sample transects for model-based estimation of yield vs. May IR in Field 4.2.

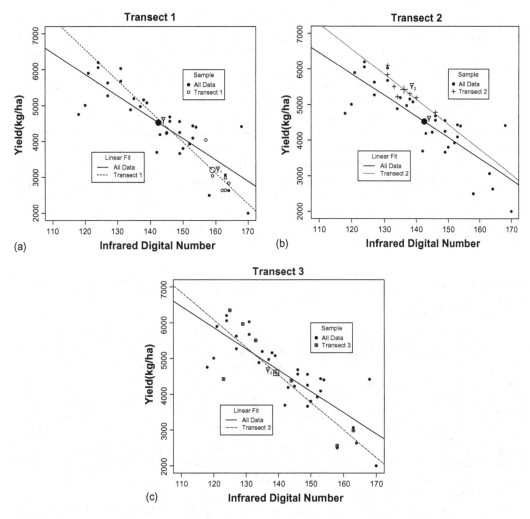

FIGURE 5.14
Regression relationships of the grid-based sample together with the models and the estimates based on the three sample transects in Figure 5.13. (a) Transect 1; (b) Transect 2; (c) Transect 3.

low weed area; and the third is to sample along an east to west transect spanning the field. The predicted yield means of the three samples are $\bar{Y}_1 = 3208$ kg/ha (29% error) for transect 1, $\bar{Y}_2 = 5424$ kg/ha (20% error) for transect 2, and $\bar{Y}_3 = 5000$ kg/ha (1.5% error) for transect 3. These results are summarized in Figure 5.14. The figure shows the data, the regression line, and the estimated mean of each of the transects. The first two transects (Figure 5.14a and b) give an incorrect representation of the yield-IR relationship because they are from regions of low and high IR value respectively, whereas the east–west transect covers the range of IR values and provides a fairly accurate representation. The estimate from the first transect (Figure 5.14a) is biased downwards, that from the second transect is biased upwards (Figure 5.14b), and that from the center transect is relatively accurate (Figure 5.14c). It is evident that care needs to be taken in selecting the sample locations. This applies equally to the results of a sampling plan using the design-based model.

The problem with both $\bar{Y}_1$ and $\bar{Y}_2$ is that they are based on samples from only a small geographic part of the field. Since the field has a strong geographic trend, this is equivalent to saying that they are based on only a small subset of the possible values of IR value and yield. Partly as a result, the regression models based on the first two sets of samples are incorrect, and since the samples are from either a higher than average range of IR values (in transect 1) or a lower than average range of IR values (in transect 2), the estimates are not robust to these incorrect models. Since the model based on the east–west transect is accurate, we cannot say anything about whether this sample is correspondingly robust to an incorrect model. We can say, however, that the most accurate estimate comes from a sample that is "representative" of the range of values of IR and yield. This concept of "representativeness" can be formalized through the property that a sample must be *balanced* (Valliant et al., 2000, p. 53), which means, roughly speaking, that each sample value represents, or is close to, about an equal fraction of the totality of values in the population.

5.8 Further Reading

Cochran (1977) is the classical reference for sampling, although it contains little in a spatial context. Valliant et al. (2013) provide an excellent recent discussion. The two books by Webster and Oliver (1990, 2001) contain a wealth of valuable material on sampling spatial data, as do Ripley (1981) and Haining (1990). Odeh et al. (1998) provide an excellent example of the use of information at multiple scales (see Chapter 6) to direct sampling. Brus (1994) describes a design-based stratified soil sampling plan. Valliant et al. (2000) provide a good introduction to model-based sampling. Olea (1984) and Lesch et al. (1995) discuss sampling plans that have model-based aspects. Edwards (2000) provides a good overview of sampling concepts for ecological data. The notion of distinguishing sampling error from nonsampling error is discussed by Biemer and Lyberg (2003). This concept goes back at least to Fisher (1935), who provides an excellent discussion of this issue.

The comparison of sampling plans in this chapter applies only to sampling a rectangular region. Van Groenigen and Stein (1998) and van Groenigen et al. (1999) provide mathematical methods for generating sampling plans on irregularly shaped regions that minimize the kriging variance. The mathematics of these plans is quite intricate, but simply by looking at the figures in the papers one can gain a good idea of how these sampling schemes relate to one that would be generated on a rectangular region. The package spcosa (Walvoort et al., 2010) uses a k-means clustering algorithm to develop sampling plans that distribute sample locations in an approximately uniform manner. The package is especially well suited for irregularly shaped regions, for which it can be used to construct compact strata of approximately equal area.

Exercises

5.1 Read about the sp function spDistsN1(). Create a new function closest.point() that uses this function.

5.2 a. Use the boundary file of Field 1 of Data Set 4 created in Exercise 2.11 and the function `spsample()` to create a regular grid sample plan with 100 sampling sites for the field. Use the function `points()` to add a plot of the sample point locations to the map.

   b. Use the function `class()` to check the object class of the sampling plan created in part (a). Use the function `str()` to display the structure of the object. Use the function `coordinates()` to display the coordinates of the first 10 sample locations in the object.

   c. Does the number of points in the sample plan equal the number you specified? Answer the question without counting the points (use the information provided by `str()`.).

5.3. Use the function `expand.grid()` to create a grid of sample points in Field 1 of Data Set 4 with the same spacing as that in Exercise 5.2. Use the function `coordinates()` to convert the object created by `expand.grid()` into a `SpatialPoints` object. Create a map showing the field boundary and the two sets of data locations, each with a different symbol.

5.4. It sometimes happens with an irregularly shaped boundary that the function `expand.grid()` creates sample locations outside the sample area boundary. Use the function `over()` to create a `SpatialPoints` object that does not include locations in the set created in Exercise 5.2 lying outside the field boundary. Create a map that shows this sample plan together with the field boundary.

5.5. Assume the 86 sample values are the entire population (i.e., $N = 86$) of clay content values in Field 1 of Data Set 4. Suppose you want to estimate the mean.

   a. Compute the error in estimating the mean based on a random sample of six clay values.

   b. Suppose you have EM38 values (which are much easier to measure) at all 86 locations, taken on April 25 from the beds (see Appendix B.4). You can collect six soil cores. Use the EM38 data to stratify the sample, creating two zones, one of high clay and one of low clay (make them the same size). Collect a random sample totaling seven samples within each zone and compute the error in estimating the mean. Remember that the strata cannot be of equal size.

   c. Compare the result of parts (a) and (b) with estimate obtained by taking a north–south transect of seven soil cores consisting of every other data location in the middle column of the data starting from sample point 4.

5.6. Suppose in the problem of Exercise 5.4 you have taken a north–south transect of seven soil cores consisting of every other data location in the middle column of the data starting from sample point 4. Use this together with the EM 38 data to construct a model-based estimate of the mean of the 86 values of clay content.

5.7 Section 5.3.1 contains a discussion of an error in the implementation of the polymorphic function `idw()` that occurs when the `gstat` and `spatstat` packages are both loaded and R tries to implement the function from the wrong package. Ordinarily R selects polymorphic functions correctly. Why does it fail in this case?

6

Preparing Spatial Data for Analysis

6.1 Introduction

Georeferenced data usually require considerable manipulation and checking before they can be subjected to statistical analysis. Point sample data, whether they are manually sampled (e.g., soil core data) or automatically sampled (e.g., yield monitor data) must often be converted from a spreadsheet or database format into a geographic information system (GIS) data file format such as the ESRI shapefile. Automatically sampled data often contain numerous outliers that must be detected and dealt with. Image data must often be geo-registered to the earth's surface. Moreover, spatial data are often *misaligned*, that is, they are not recorded at the same location and, in a sense that will be made clearer later in this chapter, they are often measured at different scales.

One of the issues that must be decided in the initial data processing phase is the projection in which to represent the data. In the northern hemisphere, the two most common choices are longitude-latitude and Universal Transverse Mercator (UTM) coordinates. This book does not provide any discussion of geographic coordinate systems. A complete discussion of this topic is given by Lo and Yeung (2007, Chapter 2). Most global positioning systems permit the user to select either longitude-latitude in the WGS 84 datum (Lo and Yeung, 2007, p. 40) or UTM. The advantage of working in UTM is that it is a conformal projection (Lo and Yeung, 2007, p. 43), that is, shapes of reasonably small areas on the ground are preserved on the map. A major disadvantage of UTM coordinates is that the UTM zones are only six degrees of longitude in width, and as a result, many geographic features span more than one zone. For example, Data Set 2 is located in UTM Zones 10 and 11. As a general rule, it is good practice when recording data in the field to record in the same projection as that of other layers in the data set, or, if these are in different native projections, to record in the same projection as that data set considered the most geographically accurate. R provides the capability to transform the projection of a data set (Section 2.4.3). Also, the functions in the sf and sp packages can correct for the spherical shape of the earth when performing distance calculations using longitude and latitude.

As was mentioned in Chapter 1, the intent of the book is to mimic the process of analysis of a real data set, using the four examples that we carry along. After four introductory chapters and a chapter on data collection, we now begin this simulated data analysis project. The objective in this chapter is to first evaluate the data sets for quality and second to put them into a form suitable for further analysis. The spatial data are assumed to be error free, so the quality evaluation focuses on the attribute data. The process of putting the data into a form suitable for further analysis, however, generally focuses on the spatial component.

Data are often represented as a table in which the columns are the *data fields* and the rows are the *data records* (Section 2.4.1, Lo and Yeung, 2007, p. 76). In this representation, a data field is a specific attribute item such as mean annual precipitation, and a data record is the collection of the values of all the data fields for one occurrence of the data. We begin in Section 6.2 with a discussion of quality control of attribute data. Sections 6.3 and 6.4 deal with the spatial component. Section 6.3 presents a brief discussion of geostatistical interpolation procedures to estimate its attribute values at different locations. Section 6.4 deals with the application of these procedures to the resolution of misalignment problems.

6.2 Quality of Attribute Data

6.2.1 Dealing with Outliers and Contaminants

"A weed is a...plant out of place" (Klingman, 1961, p. 1). Data quality control is like weed management: one must identify and deal with data values that are out of place. However, unlike weeds, it is often difficult to tell whether a data value is out of place or not, or even what it means to be out of place. In thinking about data quality, we must first consider the sources of data variability. Anscombe (1960, see also Barnett and Lewis, 1994, p. 33) identifies three sources of such variability. For our purposes, we will paraphrase Anscombe's characterization slightly. The sources are (1) *inherent variability*, the variability that would be observed in the population even if all measurements were perfectly accurate; (2) *measurement error*, error in the reading of the measurement instrument; and (3) *execution error*, which we define as the inclusion of a valid record from a population other than the one we intend to measure. There are a number of well-known cases, such as the discovery of penicillin and the discovery of the ozone hole, in which data values were initially rejected as belonging to category (2), measurement error, when indeed they belonged to category (3), observation of a different, previously unknown population, and the previously unknown population that they represented turned out to be extremely important. Occurrences such as this are often cited as reasons that data should never be completely discarded without a trace. Nevertheless, some data values warrant special attention.

In thinking about data quality, it is important to recognize the difference between *outliers* and *contaminants*. Barnett and Lewis (1994, p. 7) define an outlier to be "an observation (or subset of observations) which appears to be inconsistent with the remainder of that set of data." They define a contaminant to be a data value coming from a distribution other than the one characterizing the population being measured (i.e., a member of category (3) of the previous paragraph). An outlier might not be a contaminant (it may be an error, or it may be a valid member of the population with a one-in-a-million value), and a contaminant might not be an outlier (it may come from a different population but have a value consistent with members of the population we intended to measure).

Data quality control involves two phases: identifying data values that warrant special attention, either as potential outliers or as potential contaminants or both; and determining what action to take in response to these identified special values. These two phases should not be considered separately, because the decision of which data values to identify as special depends partly on the intended fate of these values. Values that are identified as warranting special attention will be called *discordant* values (Barnett and Lewis, 1994).

In deciding what to do with a data value that has been identified as discordant, one may consider four options: (1) include it in the analysis, possibly mentioning it in the discussion and characterizing its effect; (2) remove it from the analysis (possibly to become part of a different analysis); (3) modify its value; and (4) modify the analysis or analytical methods because of its presence. Not all special values will necessarily receive the same treatment, and indeed the treatment a data value receives depends partly on the determination of how the variable is characterized, and partly on the objectives of the data analysis. Kruskal (1960) gives a vivid if militaristic example of the importance of considering the objectives. Suppose 50 bombs are dropped on a target, and that a few go wildly astray. Suppose further that the bombs that go astray are observed to do so because their fins came loose. If we are concerned with the accuracy of the whole bombing system, then these wild bombs should be included in the analysis. If, however, the concern is with the accuracy of the bombsight, then the wild bombs should be excluded.

Most of the analyses described in this book have as their ultimate objective the determination of the relationship between observed environmental factors and some response variable such as species presence or crop yield. It is well known that certain statistics, such as the sample mean, are highly sensitive to extreme values in the data. Therefore, even if an extreme value may be the result of inherent variability in the data (the first and seemingly most legitimate of the three sources of data variability given above), including that value in the analysis may result in an unacceptably large error in the estimate of a population parameter. This error might provide justification for eliminating such a data record from the analysis (option (2) above). Removing a data record because of its effect is, however, a form of data censorship (or data snooping). It may be preferable to simply remove a certain number or fraction of the most extreme values prior to the analysis without regard to their actual magnitude. Statistics such as the *trimmed mean* are manifestations of this concept of automatic removal of the most extreme values. The trimmed mean is computed by automatically removing ("trimming") a pre-specified number or fraction of the highest and lowest data values from the computation of the mean.

It sometimes may happen that we do not want to discard an extreme value completely. This may be because doing so would change the number of observations, and we do not want to do this, or it may be because, although the value of the observation is considered too extreme, the observation's sign and the fact that the observation has a large magnitude are considered valid. A common practice in this case is to retain the data record but modify its value (option (3) above). A widely used data modification method is *Winsorization* (Barnett and Lewis, 1994, p. 78), in which the value of an observation is replaced by its nearest neighbor in an ordered list of the data. For example, in a data set consisting of the five values 3, 6, 4, 25, and 2, the data record whose value is 25 would be retained, but its value would be replaced with 6. Other similar methods are given by Barnett and Lewis (1994).

If the decision on how to proceed with the data is to adopt option (4) above and accommodate the discordant value or values, then one must determine what their effect on the analysis might be and how it can be ameliorated. A common strategy is the use of resistant statistical methods (Barnett and Lewis, 1994, Ch 3), that is, methods that are not sensitive to extreme values. For example, if we are interested in determining a measure of central tendency, then the fact that the mean is sensitive to extreme values suggests that we use the median instead.

Even if the data do not include any discordant values, they may have a different problem. Many of the spatial analysis methods described in this book assume that the data consist of random variables with a normal probability distribution and are sensitive to this assumption to some degree. If the data do not appear to be normally distributed, one may decide to transform them. Alternatively, one may wish to employ robust statistical methods, that is, methods that are not sensitive to the distributional assumption. The adoption of option (4), for example, by electing to use robust or resistant methods, represents a response to data quality problems by accommodating them and will not be discussed extensively in this book (the nonparametric methods described in Chapter 9 are, however, generally more robust than the parametric methods described in the rest of the book). The remainder of the present section will deal with options (2) and (3), in which the data values must be removed or modified in some way to make them conform to the assumptions of the statistical analysis.

6.2.2 Quality of Ecological Survey Data

Given a parameter θ of a population (for example, it might be the expected value, μ), suppose the purpose of the survey is to determine an estimate $\hat{\theta}$ (for example, if the parameter is the expected value μ, the estimate might be the sample mean $\bar{Y}$). There are two forms of error in estimating μ. The first, called *sampling error* (Biemer and Lyberg, 2003, p. 36), is the error due to the difference between the estimate and the parameter that results from the fact that not every member of the population is sampled. The second, called *nonsampling error* by Biemer and Lyberg (2003), refers to all of the other forms of error that can occur during the sampling process. Sampling error is an unavoidable consequence of the fact that the data represent a sample and not a census. Therefore, efforts to improve data quality are directed at nonsampling error. Biemer and Lyberg (2003) provide an extensive discussion of methods for data quality improvement.

The Moran scatterplot (Section 4.5.2) provides a very powerful general tool for assessing the quality of spatial data. Recall (Equations 4.10 and 4.11) that the Moran scatterplot computes a simple linear regression of Y, the measured variable, on WY, the spatially lagged value of Y. This permits regression diagnostics to be applied to the data in order to identify values that are discordant with their geographic neighbors. The spdep function `moran.plot()` employs the diagnostic statistics of the function `influence.measures()` to detect potentially discordant values. These statistics are discussed in Appendix A.2.3. Section 4.5.2 contains an example of the use of this function to identify discordant values in the detrended sand content data of Field 4.1. I am such a fan of the Moran scatterplot for this application that I am not going to discuss any alternatives, but simply present this approach in the next section.

6.2.3 Quality of Automatically Recorded Data

One of the features of modern data recording methods is that they automatically gather a very large stream of data at a high rate. Such data records warrant special attention because they often include values that would be identified as exceptional, and maybe discarded without further thought, in records collected by hand. A convenient example that typifies many of the properties of such data sets is yield monitor data in agriculture. There are a number computer programs available for quality control of yield monitor data. One excellent program is Yield Editor 2.0 (Sudduth et al., 2013). Software such as this provides an easy means to enter data directly from the yield monitor and deal with records identified as outliers or contaminants. In this section, we discuss the process of quality control in yield monitor data sets in order to develop more generally the procedure of outlier removal for automatically collected data. To give away the punch line, we are going to advocate the use of the Moran scatterplot. It is very important to emphasize that, although the discussion refers to specific events that occur in the collection of yield monitor data, analogous events occur in the collection of other forms of densely sampled point data such as remotely sensed data, soil electrical conductivity, and so forth, and the Moran scatterplot should be considered for these applications as well.

An automatically collected data stream such as that generated by a yield monitor consists of a sequence of records measured as a time series. Figure 6.1a shows a portion of the yield monitor data stream from Field 4.2. The figure shows 3,000 records out of 38,159 recorded from that field. In the general theory of outlier detection in time series, two types of outliers are commonly distinguished (Barnett and Lewis, 1994, p. 396): *additive outliers* and *innovation outliers*. An additive outlier is a "hiccup," that is, a measurement error or execution error superimposed on a single observation or set of observations and not affecting other observations outside the set. An example in combine-harvested yield monitor data is when the combine slows and turns. An innovation outlier is a measurement error or execution error that affects more than one record sequentially in the series. An example would be a change in the calibration of the sensor.

In order to properly develop the process of data quality assessment and improvement it is necessary to explicitly state one's assumptions and objective. For instance, the procedures might be very different depending on whether the objective is to discover all of the possible probability distributions underlying a data set or to exclude all data not belonging to the dominant distribution. We will assume in this section that the objective is to characterize the dominant probability distribution and to retain as many measurements of this population as we can while rejecting measurements that are suspected of containing substantial measurement error or of being members of a different population. Since we have thousands or tens of thousands of values from which to choose, we adopt the policy that any data value that can be reasonably suspected of being spurious will be deleted. This assumption is obviously very important in simplifying the analysis.

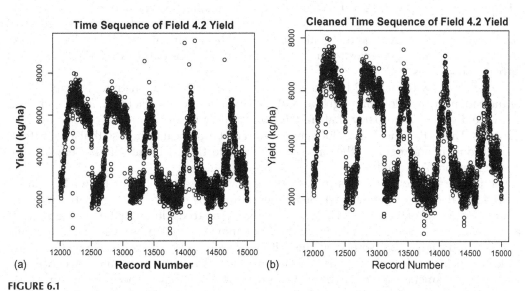

FIGURE 6.1

(a) Three thousand yield data records from Field 4.2; (b) same record interval after having eliminated data values identified as suspicious by a Moran scatterplot.

Many papers discussing elimination of outliers from data express their procedure in terms of passing the data through a series of filters, often based on the specifics of the data collection process (Edwards, 2000). In the case of yield monitor data, these filters identify and remove data satisfying criteria including the following:

1. Data recorded while or shortly after the combine header is in a raised position (Thylen and Murphy, 1996; Simbahan et al., 2004);

2. Short temporal segments of data (Simbahan et al., 2004);

3. Spatially overlapping yield records, either due to actual harvester path or to inaccurate GPS record (Blackmore and Marshall 1996; Nolan et al., 1996; Beck et al., 2001; Simbahan et al., 2004);

4. Data from less than a full swath width (Blackmore and Marshall, 1996; Blackmore and Moore, 1999; Beck et al., 2001; Colvin and Arslan, 2001);

5. Data recorded while the harvester is stopped, moving faster than a given speed, turning, accelerating, or decelerating (Murphy et al., 1995; Nolan et al., 1996; Thylen and Murphy, 1996; Blackmore and Moore, 1999; Beck et al., 2001; Colvin and Arslan, 2001; Simbahan et al., 2004);

6. Data recorded at the edge of the field (Nolan et al., 1996; Haneklaus et al., 2001);

7. Data records whose grain flow rate or moisture content are further than a given value (usually three times the standard deviation) from the mean (Beck et al., 2001; Simbahan et al., 2004);

8. Data that are outside the biological limits of the crop (Simbahan et al., 2004);

9. Data that are very different from geographically nearby values, including the results of surging in the harvester (Blackmore and Marshall 1996; Beck et al., 2001; Simbahan et al., 2004; Robinson and Metternicht, 2005).

In addition to being application specific, these criteria are somewhat ad hoc in nature. The Moran scatterplot provides a unified, theoretically justifiable means of identifying discordant values. In applying the Moran Scatterplot to yield monitor data, because we have thousands of data values and our goal is to identify the dominant probability distribution, we will discard any data records that are identified by any one of the influence diagnostics as being suspicious.

We will work with Field 4.2 from the 1996 wheat harvest to demonstrate the application of the Moran scatterplot, although we will also include code for Field 4.1. The yield data from Field 4.2 has a property that frequently occurs with automatically collected dense data sets. This is that the data include many short distance local trends, so that a value that is discordant in one region of the data set may be perfectly consistent with its neighbors in another region (Figure 6.1a).

The primary issue that arises in computing a Moran scatterplot for the yield data of Field 4.2 is the size of the data set. Since there are 38,159 data records, a spatial weights matrix relating all of them would be a 38,159 by 38,159 matrix! Although the sp objects such as the neighbor list and listw objects take advantage of the sparseness of the weights matrix (i.e., the large number of elements with value zero), a W matrix describing the entire Field 4.2 yield monitor data set is still beyond the reach of the computer. Accordingly, we apply the function moran.plot() to sub-regions sequentially. There are sp functions subset.listw() and subset.nb() that compute these objects for a subset of points, but for this application it appears more straightforward to use a simpler, brute force approach.

Specifically, we will subdivide the data set by Easting and Northing, apply the function moran.plot() to each sub-region to determine the data records that are not discordant (i.e., that we want to keep), and then reassemble these into a cleaned data set.

The raw yield data are read using code given in Appendix B.4. Since we are dealing only with point data, there is no simplification if we create an sf object and then coerce it, so we will use the sp function coordinates() to convert the code to a SpatialPointsDataFrame directly (Section 2.4.3). Because there are two fields in the data set and we would like to reuse the code for each of them, we assign the name data.Set4 to the SpatialPointsDataFrame that contains the yield data. Since we will ultimately need to reassemble a group of subsets, we first assign ID values to make this task easier.

```
> data.Set4$ID <- 1:nrow(data.Set4)
```

Based on the rectangular shape of Field 4.2 and the capacity of the computer, we will subdivide the field into 21 sub-regions arranged in three rows of seven sub-regions per row. Field 4.1 is longer in the north–south direction, so we subdivide this field into seven rows of three sub-regions per row. First, we compute a width x.size and a height y.size for the subdivisions.

```
> # For Set 4.1, x.cells = 3 and y.cells = 7
> # For Set 4.2, x.cells = 7 and y.cells = 3
> add.factor <- 4 * Field.no - 6
> x.cells <- 5 + add.factor
> y.cells <- 5 - add.factor
> x.max <- max(coordinates(data.Set4)[,1]) + 1
> x.min <- min(coordinates(data.Set4)[,1])
> y.max <- max(coordinates(data.Set4)[,2]) + 1
> y.min <- min(coordinates(data.Set4)[,2])
```

Here we are using the function coordinates() as retrieval function (see Section 2.4.3). Now we need to identify each sub-region. For Field 4.2, we can number the sub-regions from 1 to 7 in the x direction and from 1 to 3 in the y direction. The following formula will compute a coordinate x_{loc} with values ranging from 1 to 7 based on the size of the x coordinate relative to the minimum and maximum x coordinate values,

$$x_{index} = trunc\left(\frac{7[x - x_{min}]}{x_{max} - x_{min}}\right) + 1 \tag{6.1}$$

where the function trunc(x) truncates the value of x by removing its fractional part. Here is the implementation of this equation, placing the values computed in Equation 6.1 into a data field called x.index.

```
> x.index <- trunc(x.cells*(coordinates(data.Set4)[,1] -
+     x.min) / (x.max - x.min)) + 1
```

A similar operation is carried out on the y values to assign them an index of 1, 2, or 3. The next step is to combine these indices so that each of the 21 sub-regions is uniquely identified. A very simple way to do this, familiar to those who work with map algebra, is to multiply the x and y indices by different powers of 10 and then add the results of these multiplications.

```
> data.Set4$xyindex <- x.index + 10 * y.index
> print(xyindex <- unique(data.Set4$xyindex))
 [1] 31 32 33 34 35 36 37 27 17 16 15 14 13 12 11 21 22 23 24 25 26
```

The function unique() here returns all of the unique values of the sub-region index. Now that each sub-region is uniquely identified we can use a simple for loop to compute a Moran scatterplot for each sub-region. As with all irregular point sets, in creating the spatial weights matrix we must decide between using dnearneigh(), for distance-based neighbor definition, and knearneigh(), for numbers-based neighbor definition (see Section 3.6). In this case we opt for the former, since we are specifically interested in the attribute value of each data record relative to those of its nearest geographic neighbors. Some playing around is needed to get an appropriate upper distance d2. The code for this step is a bit long, so comments are inserted.

```
> for (i in 1:length(xyindex)){
+ # Separate out points in sub-region i
+     data.Set4.i <-
+         data.Set4[which(data.Set4$xyindex == xyindex[i]),]
+ # Compute the neighbor list and listw objects
+     nlist <- dnearneigh(data.Set4.i, d1 = 0, d2 = 20))
+     W <- nb2listw(nlist, style = "W")
+ # Compute the Moran scatterplot; don't print the results
+     mp <- moran.plot(data.Set4.i$Yield, W, quiet = TRUE)
+ # Create a matrix with 6 columns,
+ # one for each outlier identification statistic
+ # Convert the values TRUE and FALSE to numbers and keep those that
+ # are all zero (outlier = FALSE)
+     keep <- which(rowSums(matrix(as.numeric(mp$is.inf),
+         ncol = 6)) == 0)
+ # Concatenate these values to keep into one array
+     {if (i == 1)
+         keep.id <- data.Set4.i$ID[keep]
+     else
+         keep.id <- c(keep.id,data.Set4.i$ID[keep])}
+ }
```

The array keep.id contains the ID value for each of the data records that is *not* discordant (i.e., that is to be retained). First, we select those data records to keep.

```
> data.Set4.keep <- data.Set4[keep.id,]
```

These are not necessarily in numerical order, however. Therefore, we reorder them into increasing numerical order by ID value.

```
> data.Set4.keep <- data.Set4.keep[order(data.Set4.keep$ID),]
```

Let's see how many data records got dropped.

```
> nrow(data.Set4)
[1] 38159
> nrow(data.Set4.keep)
[1] 35343
```

Figure 6.1b shows the results of the cleaning on the same string of data records as is displayed in Figure 6.1a. Although a few questionable records have been retained (and possibly some good records have been discarded), overall the result is remarkably good. The final step is to aggregate the cleaned data and save them to a file in the folder *Created*. This exercise indicates the usefulness of the Moran scatterplot as a data cleaning tool. All of the analyses carried out in the remainder of the book that involve yield monitor data from one of the fields in Data Set 4 use data that have been cleaned in this way.

6.3 Spatial Interpolation Procedures

6.3.1 Inverse Weighted Distance Interpolation

One of the major difficulties associated with combining spatial data is the problem of *misalignment*. This occurs when data are collected at different locations or on different scales, and will be discussed in Section 6.4. It is evident, however, that if data measured at different locations are to be combined, then some data values will have to be predicted at locations other than those where they were measured. This is the process of interpolation, as this term is defined in geostatistics.

This section contains a brief introduction to three interpolation methods: inverse distance weighted interpolation, kriging, and co-kriging. In addition to these three methods, Thiessen polygons (Ripley, 1981, p. 38; Lo and Young, 2007, p. 333), which were discussed in Section 3.5, are often very useful for visualization via thematic maps (Cliff and Ord, 1981; Haining, 1990, p. 202), which permit the rapid visual examination of patterns in the data. For example, Thiessen polygons are used in Figure 3.9 to compare artificially generated data at differing levels of a spatial autocorrelation parameter. Thiessen polygons are also useful as a sort of last-resort interpolation method for data sets that are too sparse and irregular to use the methods described in this section.

We will formulate the pointwise interpolation problem as follows. Given a set of position coordinate vectors $\{x_1, x_2, \ldots x_n\}$, where each coordinate vector x_i has two components (horizontal x and vertical y), and given a set of values $Y(x_i)$, $i = 1, 2, \ldots, n$ measured at these locations, compute an interpolation of the value $Y(x)$ at a location x where the value of Y is not measured. Figure 6.2 shows a schematic example in which $n = 3$; of course in real

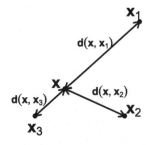

FIGURE 6.2
Schematic representation of the location of three points x_1, x_2 and x_3 at which Y is measured, and a fourth location x, at which the value of Y is to be estimated.

situations n will generally be larger than this. An interpolator $\hat{Y}(x)$ is called a *linear interpolator* if for some set of coefficients ϕ_i, $i = 1, 2, ..., n$,

$$\hat{Y}(x) = \sum_{i=1}^{n} \phi_i Y(x_i). \tag{6.2}$$

If the values $Y(x_i)$ are spatially uncorrelated, then their spatial location is irrelevant and the best (in a sense that will be made clear below) interpolator is the mean $\hat{Y}(x) = \overline{Y}$, independent of x. If, however, the values of $Y(x_i)$ are positively spatially autocorrelated, then we can expect that values of $Y(x_i)$ measured at locations close to x will be closer to the value of $Y(x)$ than values farther away. For example, in Figure 6.2, one would expect that the value of $Y(x_3)$ will be closer to that of $Y(x)$ than will the value of $Y(x_1)$, with $Y(x_2)$ somewhere in between. Thus, one would expect that increasing the value of ϕ_3 in relation to ϕ_1 and ϕ_2 in Equation 6.2 would in this case produce a more accurate interpolation.

At this point, we can note one very important property of the ϕ_i. Since the values $Y(x_i)$ in Equation 6.2 are random variables, the interpolator $\hat{Y}(x)$ is a random variable as well. An interpolator $\hat{Y}(x)$ of $Y(x)$ is *unbiased* if its expected value $E\{\hat{Y}(x_i)\}$ is equal to $Y(x)$ (Isaaks and Srivastava, 1989, p. 234). If we consider the case of uncorrelated Y values, and interpolate using the mean $\overline{Y} = \Sigma Y_i / n$, we can see that in this case $\phi_i = 1/n$ for all n, and therefore

$$\sum_{i=1}^{n} \phi_i = 1. \tag{6.3}$$

The mean is unbiased. This provides intuition for a property that turns out to be true in general, which is that in order for the interpolator $\hat{Y}(x)$ to be unbiased, the ϕ_i must satisfy Equation 6.3 (Isaaks and Srivastava, 1989, p. 234).

Inverse distance weighted (IDW) interpolation (Isaaks and Srivastava, 1989, p. 257; Lo and Yeung, 2007, p. 350) is a simple, *ad hoc* method that sometimes produces reasonably good results. Returning to Figure 6.2 for intuition, the interpolator $\hat{Y}(x)$ can be improved relative to the sample mean by reducing the weight of more distant locations like x_1 relative to closer ones like x_3, that is, by reducing the value of ϕ_i for locations with large distance values $d(x, x_i)$. We can achieve this by making ϕ_i proportional to $d(x, x_i)^{-p}$ for some $p \geq 1$. We need to scale the ϕ_i so that Equation 6.3 is satisfied, and we can do this by dividing each value of $d(x, x_i)^{-p}$ by their sum $\Sigma_{i=1}^{n} d(x, x_i)^{-p}$. This leads to the formula for the inverse distance weighted method,

$$\hat{Y}(x) = \sum_{i=1}^{n} \phi_i Y(x_i).,$$

$$\phi_i = \frac{d(x, x_i)^{-p}}{\sum_{i=1}^{n} d(x, x_i)^{-p}}. \tag{6.4}$$

There are two "control parameters" in this equation, the power p and the number of neighbors n. Varying these parameters changes the interpolator $\hat{Y}(x)$. Increasing the value of

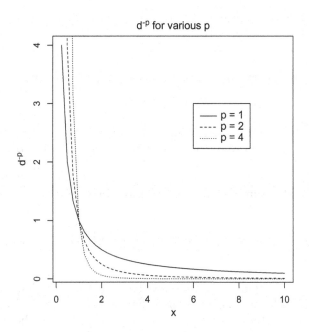

FIGURE 6.3
Curves representing d^{-p} as a function of d for three values of p. These indicate the relative rate at which the influence of data on an interpolated value drops off as a function of distance from the sampled location to the interpolated location.

n increases the number of values that contribute to the interpolation. Some interpolation software also permits the analyst to specify a maximum geographic distance $d(x, x_i)$ beyond which a value $Y(x_i)$ is not included. Figure 6.3 shows graphs of the function d^{-p} for p equal to 1, 2, and 4. Increasing the value of p reduces the influence of more distant neighbors relative to that of closer ones. The "standard" value of p in most interpolation software is 2.

In Chapter 5, we used an artificial population with 10,368 values to compare various sampling methods. We will use that same population in this section to compare interpolation methods. Using the function closest.point() from Chapter 5, the artificial yield data are sampled at the same 78 locations in Field 4.2 at which environmental data were sampled (these are shown in Figure 5.7).

```
> sample.coords <- cbind(data.Set4.2$Easting, data.Set4.2$Northing)
> # Find the yield at the closest sample points
> samp.pts <- apply(sample.coords, 1, closest.point,
+     grid.data = pop.data)
> data.Set4.2$Yield <- pop.data$Yield[samp.pts]
> coordinates(data.Set4.2) <- c("Easting", "Northing")
```

There are several packages in R that do IDW interpolation. We will use the package gstat (Pebesma, 2004), which links nicely with sp objects. First, we must create a grid of points at which to interpolate the data. We will interpolate on a 5-meter grid one-half meter inside the sample boundary.

```
> print(Left <- bbox(bdry.spdf)[1,1])
[1] 592105
> print(Right <- bbox(bdry.spdf)[1,2])
[1] 592815
> print(Top <- bbox(bdry.spdf)[2,2])
[1] 4267858
> print(Bottom <- bbox(bdry.spdf)[2,1])
[1] 4267503
> cell.size <- 5
> grid.xy <- expand.grid(Easting = seq(Left,Right,cell.size),
+    Northing = seq(Top,Bottom,-cell.size))
```

We first convert `grid.xy` into a `SpatialPoints` object and then into a `SpatialGrid` object

```
> coordinates(grid.xy) <- c("Easting", "Northing")
> gridded(grid.xy) = TRUE
```

The actual IDW interpolation can be done with the `gstat` function `idw()`. This is an R *wrapper* function, in other words, a function that performs a specialized task by calling another function that performs a more general task. In this case, the more general function is `predict.gstat()`(Rossiter, 2007). The purpose of wrapper functions is to simplify programming by making it easier to specify the arguments of the function.

```
> yield.idw <- idw(Yield ~ 1, data.Set4.2, grid.xy, idp = 2, nmax = 12)
```

The first argument is a `formula` that specifies that `Yield` is being interpolated, the second argument gives the `SpatialPointsDataFrame`, and the third and fourth arguments specify the tuning parameters p and n in Equation 6.4. Note! If you get the following message

```
Error: is.ppp(X) && is.marked(X) is not TRUE
```

it means that you have executed `library(spatstat)` and have this package, which also has an `idw()` function, hanging around. In this case, execute

```
> yield.idw <- gstat::idw(Yield ~ 1, data.Set4.2, grid.xy,
+    idp = 2, nmax = 12)
```

Figure 6.4 shows gray-scale maps of interpolated yields computed using four combinations of n and p in Equation 6.4. Figure 6.4a shows the results of what are generally considered "typical" values, $n = 12$ and $p = 2$. The round spots surrounding certain sample points, commonly called "duck eggs," are an indication that the value of $Y(x_i)$ at that x_i is sufficiently extreme to produce a "bulge" or a "hollow" in the interpolated surface. Reducing the value of n to 4 (Figure 6.4b) reduces the duck egg effect in this data set but does not eliminate it. Raising the value of p to 4 (Figure 6.4c) strongly emphasizes the nearest data location. As p becomes larger, the influence of the nearest location increases (cf. Figure 6.3), so that for large p the thematic map begins to resemble a set of Thiessen polygons. Reducing the value of p to 1 has the opposite effect (Figure 6.4d). The increased influence of more distant locations has a smoothing effect on the interpolation.

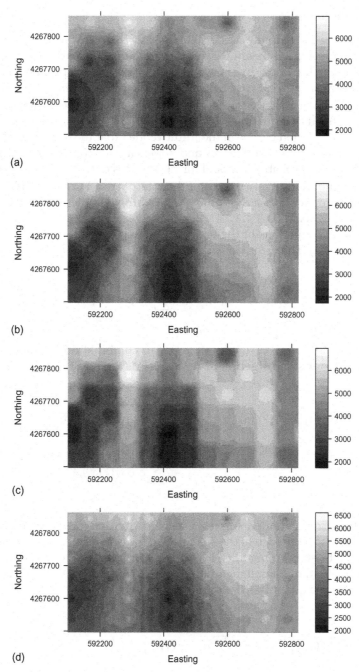

FIGURE 6.4
Thematic maps of IDW interpolation of data sampled from the artificial data set at 78 sample locations. The maps show the effect of varying the number of neighbors n and the power value p in Equation 6.4 (a) Field 4.2 interpolated yield, $n = 12$, $p = 2$; (b) Field 4.2 interpolated yield, $n = 4$, $p = 2$; (c) Field 4.2 interpolated yield, $n = 12$, $p = 4$; and (d) Field 4.2 interpolated yield, $n = 12$, $p = 1$.

6.3.2 Kriging Interpolation

The term "kriging" was originally used by Georges Matheron (Matheron, 1963) in honor of the South African mining engineer D.G. Krige, who carried out early work on this method (Krige, 1951). Those in the know, by the way, say that "kriging" is pronounced with a hard "g," like "get." There are several varieties of kriging, and we will restrict our discussion to ordinary kriging. The ordinary kriging interpolator $\hat{Y}(x)$ is a linear interpolation, which means that it is defined by Equation 6.2, similarly to the IDW interpolator. IDW interpolation is an ad hoc calculation based on the intuitive idea that the influence of a value $Y(x_i)$ should die off as the distance between x and x_i increases. Kriging interpolation is based on the assumption of a probability model for $Y(x)$, and is derived within that model with the objective of minimizing the variance of the interpolator $\hat{Y}(x)$. We will assume that Y is second-order stationary (this assumption makes things simpler but is not strictly necessary (see Cressie, 1991, p. 142).

A derivation of the equations of the ordinary kriging estimator is given by Isaaks and Srivastava (1989, p. 281). The error variance is given by

$$\text{var}\{\hat{Y}(x) - Y(x)\} = \text{var}\{\sum_{i=1}^{n} \phi_i Y(x_i) - Y(x)\}. \tag{6.5}$$

Using the formula for the variance of a linear combination of random variables, Equation 6.5 can, after some algebra, be written (Isaaks and Srivastava, 1989, p. 284)

$$\text{var}\{\hat{Y}(x) - Y(x)\} = \hat{\sigma}^2 + \sum_{i=1}^{n} \sum_{j=1}^{n} \phi_i \phi_j \, \text{cov}\{Y(x_i), Y(x_j)\}$$
$$- 2 \sum_{i=1}^{n} \phi_i \, \text{cov}\{Y(x_i), Y(x)\}. \tag{6.6}$$

To minimize the error variance, one sets to zero the derivatives of the right side of Equation 6.6 with respect to each of the ϕ_i, subject to the constraint $\Sigma \phi_i = 1$ to ensure that the interpolation is unbiased (Equation 6.3). This is a constrained optimization problem and is solved by the method of Lagrange multipliers (Appendix A.4). We introduce a Lagrange multiplier ψ so that the quantity to be minimized becomes

$$F(\phi, \psi) = \hat{\sigma}^2 + \sum_{i=1}^{n} \sum_{j=1}^{n} \phi_i \phi_j \, \text{cov}\{Y(x_i), Y(x_j)\}$$
$$- 2 \sum_{i=1}^{n} \phi_i \, \text{cov}\{Y(x_i), Y(x)\} + \psi \left(\sum_{i=1}^{n} \phi_i - 1 \right). \tag{6.7}$$

The solution is obtained by taking the partial derivatives with respect to ψ and each of the ϕ_i and setting them to zero. The result is (Isaaks and Srivastava, 1989, p. 287)

$$\sum_{j=1}^{n} \phi_j \operatorname{cov}\{Y(x_i), Y(x_j)\} + \psi = \operatorname{cov}\{Y(x_i), Y(x)\}, \; i = 1, \ldots, n$$

$$\sum_{i=1}^{n} \phi_i = 1. \tag{6.8}$$

At this point, one must introduce a model for the covariance functions in Equation 6.8. Assuming that the data are isotropic, we can write $\operatorname{cov}\{Y(x_i), Y(x_j)\} \cong \tilde{C}(h)$, where $\tilde{C}(h)$ is a model for the covariance defined in Equation 4.24 between two values separated by a distance h, and h is the distance between x_i and x_j. Although we could use the covariogram model directly, it is traditional in geostatistics to define the covariogram in terms of the variogram model. Let $\hat{\gamma}(h)$ be the experimental variogram of the data set, defined in Equation 4.21, and let $\tilde{\gamma}(h)$ be a variogram model, for example, the spherical model defined in Equation 4.22. Let us denote the nugget by $\tilde{\gamma}_0$ and the sill by $\tilde{\gamma}_\infty$. Let $\tilde{C}(h)$ denote the corresponding model for the covariogram. Then from Equation 4.26 it follows that

$$\tilde{C}(h) = \tilde{\gamma}_\infty - \tilde{\gamma}_0 - \tilde{\gamma}(h). \tag{6.9}$$

In words, the covariogram model is the sill minus the nugget minus the variogram model.

The ordinary kriging procedure is to fit the data with an experimental variogram, use Equation 6.9 to compute a covariogram model for the data, and then substitute this model into the *kriging equations*

$$\sum_{j=1}^{n} \phi_j \tilde{C}_{ij} + \psi = \tilde{C}_{i0}, \; i = 1, \ldots, n$$

$$\sum_{i=1}^{n} \phi_i = 1, \tag{6.10}$$

where $\tilde{C}_{ij}$ is the value of the covariogram model at a lag h corresponding to the distance between x_i and x_j, and $\tilde{C}_{i0}$ is the covariogram model at a lag h corresponding to the distance between x_i and x. These values of ϕ_i are then substituted into Equation 6.2 to obtain $\hat{Y}(x)$.

The first step in kriging interpolation is to compute the experimental variogram and fit it with an appropriate variogram model. Using the same data as in Section 6.3.1, we can accomplish these two steps using the gstat functions variogram() and fit. variogram().

```
> data.var <- variogram(Yield ~ 1, data.Set4.2, cutoff = 600)
> data.fit <- fit.variogram(data.var, model = vgm(150000, "Sph", 250,
+    1))
```

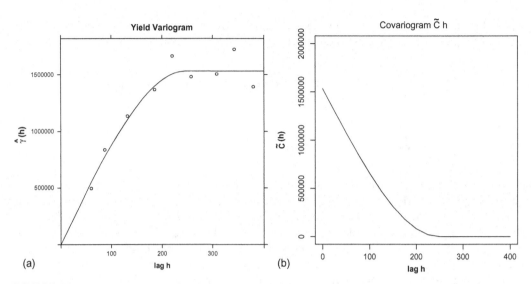

FIGURE 6.5
(a) Variogram of yield data based on 78 sample points; (b) covariogram of the same yield data as part (a).

Figure 6.5a shows the experimental variogram for the artificial yield data at the 78 sample points, together with the best fit spherical variogram model. Without going into a lot of detail, the spherical model is the one that works most often. The arguments of the function vgm() inside the function fit.variogram() are respectively, the starting estimate for the sill, the variogram model, the starting estimate for the range, and the starting estimate for the nugget. In actual practice, these were obtained from examination of the plot of the experimental variogram, but for the purposes of drawing the figure in the code the function fit.variogram() is run prior to generating the plot.

The nugget is zero, the range is 248.28m, and the sill is 1,533,695 $(\text{kg/ha})^2$. Since in our example the nugget is zero, the covariogram model is given by the sill minus the variogram model. This model is shown in Figure 6.5b. The covariogram model behaves in a similar manner to an inverse distance (Figure 6.3), with both of them tapering off to zero as the lag distance h increases. Thus, the terms $\tilde{C}_{i0}$ on the right-hand side of Equations 6.10 provide the same distance attenuation effect as the inverse distances in the IDW method. With isotropic data we have $\tilde{C}_{i0} = \tilde{C}(h)$, where h is the lag distance between point x and point x_i. The distance modeled by $\tilde{C}(h)$ is sometimes called the *statistical distance*. The role played by the terms $\tilde{C}_{ij}$ on the left-hand side of the equation is the distinguishing feature of kriging (Isaaks and Srivastava, 1989, p. 300). If two data locations are close together, they will tend to have large $\tilde{C}_{ij}$ values. When the Equations 6.10 are solved for the ϕ_i, the effect of these large $\tilde{C}_{ij}$ values is to adjust the raw statistical distances measured by $\tilde{C}_{i0}$ to take into account the effects of redundancy due to close proximity of sample locations.

Kriging is carried out in the gstat package using the wrapper function krige(). The first three arguments are the same as those of the function idw() in the previous section. The fourth argument specifies the variogram model and indeed, if this argument is not specified then krige() will compute an IDW interpolation.

```
> yield.krig <- krige(Yield ~ 1, data.Set4.2, grid.xy,
+    model = data.fit)
```

Figure 6.6 shows a gray scale plot of the kriged artificial yield data.

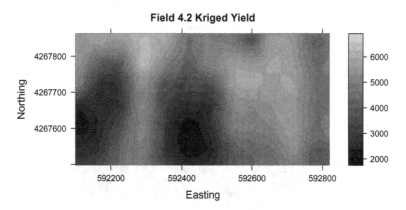

FIGURE 6.6
Thematic map of kriging interpolation of data sampled from 78 sample locations in the artificial data set.

It is important to note that although kriging provides the best linear unbiased interpolation of the data, in the sense that at any given point the interpolation has the highest precision, kriging does not provide a very good model of the *variability* of the data. Indeed, kriging, like any interpolation method, tends to smooth the data, that is, to reduce the variance of the distribution (Ripley, 1981, p. 51; Isaaks and Srivastava, 1989, p. 268). This effect can be visualized by comparing the kriged interpolation (Figure 6.6) with the original data (Figure 5.1a). Simulation methods such as Gaussian simulation (Ripley, 1987; Deutsch and Journel, 1992) are much better suited for generating a model that simulates the variability of the data. In Exercise 6.7, you are asked to compare the errors associated with IDW and kriging interpolation. The results may surprise you a bit.

6.3.3 Cokriging Interpolation

It is often the case that when trying to interpolate the values of one quantity that is expensive to measure, one also has access to data on one or more related quantities that are relatively inexpensive to measure. Consider the example of measuring clay content in Field 4.2. To obtain these data, one goes to the field with a soil coring tool and at each sample location one extracts about five or so soil cores from locations within a radius of a few meters. One then combines these into a single soil sample and takes it to the lab for analysis. There are various ways of determining sand, silt, and clay content, but all are sufficiently expensive to deter high intensity sampling. In non-saline soils, apparent soil electrical conductivity (EC_a) is related to clay content (Triantafilis et al., 2001; Corwin and Lesch, 2005; Sudduth et al., 2005). This is cheaper and easier to measure, and therefore can be sampled more intensively. Cokriging provides a means by which the correlated EC_a measurements can be used to augment the clay content measurements to obtain a more accurate interpolation.

A high density set of soil EC_a measurements is available in the file *Set4.2EC.csv*. In addition to measurement locations in WGS84 and UTM coordinates, this file contains two data fields, *ECto30* and *ECto100*. These correspond to the horizontal and vertical dipole measurements of the instrument and represent the approximate depth in centimeters to which EC_a is measured. The contents of this file are loaded into the data frame data. Set4.2EC (Appendix B.4), which is then converted into a SpatialPointsDataFrame. Figure 6.7 shows the measurement locations of the EC_a sample locations as well as the

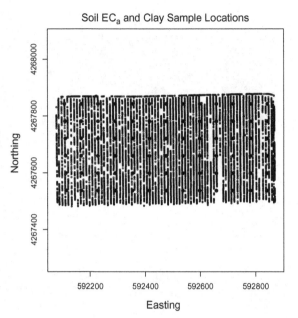

Soil EC$_a$ and Clay Sample Locations

FIGURE 6.7
Sample locations of soil EC$_a$ (small open circles) and clay content (large filled circles) for Field 4.2.

78 soil sample locations. The contents of the file *Set4.296sample.csv* are loaded into the data frame data.Set4.2 and this is converted into a SpatialPointsDataFrame. By applying the function closest.point() developed in Chapter 5 to EC$_a$, we can compare the two EC$_a$ measurements with clay content.

```
> EC.pts <-
+     apply(sample.coords, 1, closest.point, grid.data = data.Set4.2.EC)
> data.Set4.2@data$ECto30 <- data.EC@data$ECto30[EC.pts]
> data.Set4.2@data$ECto100 <- data.EC@data$ECto100[EC.pts]
```

Figure 6.8 shows a plot of clay content vs. *ECto30*, which turns out to have the closer linear relationship with clay content ($r = 0.53$). We will use this as the covariate in our cokriging interpolation of soil clay content.

The code to make Figure 6.8 is as follows.

```
> with(data.Set4.2@data, plot(ECto30, Clay, cex.main = 2,
+     main = expression(Soil~EC[a]~vs.~Clay~Content),
+     xlab = expression(EC[a]~"("*dS/m*")"), cex.lab = 1.5,
+     ylab = "Percent Clay")) # Fig. 6.8
```

There are two possibly discordant points in Figure 6.8, and this provides an opportunity to introduce the function identify(). This function takes as arguments the values on the abscissa and ordinate of the scatterplot and a value to use as a label. After invoking this function, you can use the cursor to click on a point, and it will be identified by the label. Here is the code to label the discordant points.

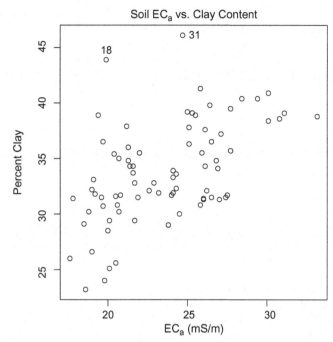

FIGURE 6.8

Scatterplot of the relationship of clay content to the variable *ECto30*, representing electrical conductivity measured to approximately 30 cm. Two possibly discordant points have been identified.

```
> identify(data.Set4.2@data$ECto30, data.Set4.2@data$Clay,
+       data.Set4.2@data$ID)
```

If you are using RStudio, you won't see anything until you click on the *Finish* button in the upper right corner of the *Plot/Help* window (after first clicking on the two discordant points). Points 18 and 31 are identified as the discordant values, so we will try removing them.

```
> data.Set4.2.cleaned <- data.Set4.2[-c(18,31),]
```

This improves the r value to 0.61, so we will go with the cleaned data.

The derivation of the cokriging equations is similar to that of the kriging equations given in the previous section, and therefore is not discussed here. The interested reader is referred to the excellent exposition in Isaaks and Srivastava (1989, Chapter 17). IDW and kriging interpolation can be implemented in the gstat package through the use of wrapper functions, but cokriging requires the use of the fundamental function predict.gstat(). As with the implementation of function polymorphism in R generally, you do not have to actually specify this function name in an R statement in order to use it. Rather, you specify the function predict() with an object from the gstat package as the first argument, and R automatically applies predict.gstat(). The first step is therefore to construct a gstat object. This is done using two applications of the gstat() constructor function.

```
> g.cok <- gstat(NULL, "Clay", Clay ~ 1, data.Set4.2)
> g.cok <- gstat(g.cok, "ECto30", ECto30 ~ 1, data.Set4.2.EC)
```

Cokriging in `gstat` requires that both the variate and the covariate have the same range. Therefore, we construct and inspect the experimental variograms of these two quantities, using the functions `variogram()` and `fit.variogram()` as with kriging. The range of *Clay* is about 400m and the range of *ECto30* is about 300m (variogram plots not shown), so we will us a range of 350m for the cokriging interpolation of the cokriging object g.cok.

```
> g.var <- variogram(g.cok)
> g.fit <- fit.lmc(g.var, g.cok, vgm(1, "Sph", 350, 1))
```

Now we are ready to carry out the cokriging interpolation. This is done a call to the polymorphic function `predict()`.

```
> g.cokrige <- predict(g.fit, grid.xy)
```

After staring at the screen for a few minutes, we decide to go out and get a cup of coffee. Upon our return (if you actually try this, it may take several hours), we find that the computer has good news and bad news for us. First, the good news.

```
Linear Model of Coregionalization found. Good.
[using ordinary cokriging]
```

The word "Good" refers to the fact that the model g.fit can indeed be used for cokriging, because the ranges have been matched correctly. Now for the bad news.

```
There were 50 or more warnings (use warnings() to see the first 50)
> warnings()
Warning messages:
1: In predict.gstat(g.fit, grid.xy):
   Covariance matrix singular at location [592100,4.26786e+006,0]:
skipping...
```

Problems like this are often caused by numerical overflow when the computer chokes on too much data. One way to determine whether that is the problem in this case is to try cokriging the clay content data in Field 4.2 by using the soil EC_a data from that same data set. You are asked to do this in Exercise 6.8. It turns out that this is indeed the problem, so we will reduce the size of the covariate data set by using the modulo function to eliminate a fraction of the data records. If you did Exercise 2.6, you learned about this function. For those who did not, the *modulo* operator u mod v, for integer values of u and v, is the remainder of the division of u by v. Thus, for example, 6 mod 5 = 1. The modulo operator in R is `%%` so that the operation u `%%` v evaluates to u mod v. If we number the data records in data. EC from 1 to n, and select only those records whose record number i satisfies i mod $k = 0$ we will eliminate $(k-1)/k$ of the records.

```
> nrow(data.Set4.2.EC)
[1] 12002
```

With 12,000 data records, we can probably eliminate nine-tenths of them and still have a useful covariate. We will reload the data file and carry out this operation.

```
> data.EC <- read.csv("Set4\\Set4.2EC.csv")
> ID <- 1:nrow(data.EC)
```

```
> mod.fac <- 10
> data.ECmod <- data.EC[which(ID %% mod.fac == 0),]
> coordinates(data.ECmod) <- c("Easting", "Northing")
```

This executes without error. Let's see what we get.

```
> g.cok <- gstat(NULL, "Clay", Clay ~ 1, data.Set4.2)
> g.cok <- gstat(g.cok, "ECto30", ECto30 ~ 1, data.ECmod)
> g.var <- variogram(g.cok)
> g.fit <- fit.lmc(g.var, g.cok, vgm(1, "Sph", 350, 1))
> g.cokrige <- predict(g.fit, grid.xy) # This works
Linear Model of Coregionalization found. Good.
[using ordinary cokriging]

> names(g.cokrige)
[1] "Clay.pred"          "Clay.var"            "ECto30.pred"        "ECto30.var"
[5] "cov.Clay.ECto30"
```

Figure 6.9 shows the cokriged and kriged clay data. They are obviously similar but not identical. Unlike the kriged artificial yield population data, we do not have a clay content population that we can use to compare the two methods. We can, however, make use of cross validation. We will discuss this method in detail in Chapter 9 in connection with

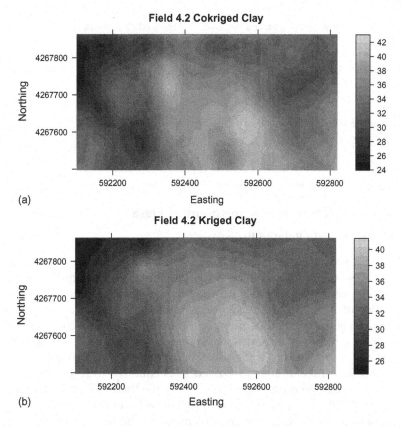

(a)

(b)

FIGURE 6.9

(a) Thematic map of clay content interpolated using ordinary kriging. (b) Thematic map of clay content interpolated using cokriging with EC_a.

generalized additive models and recursive partitioning. The basic idea is very simple. The *k-fold cross validation* procedure is to remove $1/k$ of the data records from the data set, compute the predicted values at the locations of these removed data records, and compare the predictions with the actual values. The process is repeated for successive fractions $1/k$ of the records until all records in the data set have been subjected to the procedure. One can then compute diagnostic statistics. Two of the most useful are the mean error, which indicates the bias, and root mean square error (RMSE), which indicates the variance (Section 6.2.2).

For kriging, the `gstat` library includes the wrapper function `krige.cv()`. In the application to kriging and cokriging, the most common practice is to remove and predict the data records one at a time.

```
> claykrige.cv <- krige.cv(Clay ~ 1, data.Set4.2, grid.xy,
+     model = clay.fit, nfold = nrow(data.Set4.2))
> mean(res.krige) # Mean error (bias)
[1] -0.005734252
> sqrt(mean(res.krige^2)) # RMSE (variance)
[1] 2.89842
```

In the code above, the argument `nfold = nrow(data.Set4.2)` indicates one at a time cross validation. As with cokriging itself, cokriging cross validation does not have a wrapper function. In some ways, however, the code is even easier because it uses the `gstat` object `g.cok`.

```
> claycok.cv <- gstat.cv(g.fit, nfold = nrow(data.Set4.2))
> res.cok <- claycok.cv$residual
> mean(res.cok)
[1] -0.02391489
> sqrt(mean(res.cok^2))
[1] 2.769164
```

The mean error of the cokriging residuals in this example is actually larger, while the RMSE is slightly smaller.

6.4 Spatial Rectification and Alignment of Data

6.4.1 Definitions of Scale Related Processes

Adjustments in the spatial component of a data set generally involve three primary operations. The first is georegistration, which is the alignment of the coordinates defining the spatial component of a data record with the true geographic coordinates of the corresponding location of the spatial entity. Georegistration is easiest to do in a GIS, and we will not discuss it here. The second adjustment is to transform the projection if necessary. This process can be carried out in R and is described in Section 2.4.3. The third operation, which is the focus of this section, is the correction of spatial misalignment. We will make use of the interpolation techniques discussed in Section 6.3 to resolve misalignment in spatial data. The discussion is organized in terms of the theory of scale change developed by Bierkens et al. (2000). In order to discuss concepts involving scale precisely, it is necessary to define terms precisely. Unfortunately, much of the terminology is not standardized, and indeed the term *scale* has a second meaning in cartography that is the reverse of the one used here (Lo and Yeung, 2007, p. 22). To standardize our own terminology,

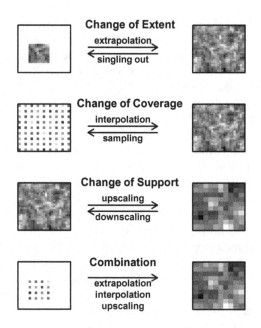

FIGURE 6.10
Schematic map of the three scale change operations defined by Bierkens et al. (2000). Redrawn from Figure 5.5 of Bierkens et al (2000, p. 9). Original figure copyright 2000, Kluwer Publishing. Used with kind permission from Springer Science and Business Media B.V.

we will use that of Bierkens et al. (2000). All of our definitions are quoted directly from their glossary on pp. 177–179. They define three scale change operations shown schematically in Figure 6.10. These are change of extent, change of coverage, and change of support.

The *extent* is the "area...over which observations are made, model outcomes are calculated, or policy measures are to be made." Suppose we make measurements in one of the rice fields of Data Set 3 and use them to represent the whole data set. This process, "increasing the extent of the research area," is called *extrapolation*. Conversely, if we use measurements made over the whole region to draw conclusions about a single field, this process of "selecting an area of smaller extent" is called *singling out*.

The term *support* is defined as "the largest...area...for which the property of interest in considered homogeneous." For example, the sand content data shown in Figure 3.4 were collected by taking four or five soil cores over an area of about 5 m^2 and compositing these cores. Thus, the support of these data is about 5 m^2. The term *scale* as we use it here is synonymous with support, with *support* being the commonly used term in the geostatistics literature. The process of moving from a smaller support (or scale) to a larger one, so that one is "increasing the support of the research area," is called *upscaling*. The opposite process, that of "decreasing the support of the research area," is called *downscaling*.

The third class of change of scale operations involves the *coverage*. Referring to the data describing Field 4.2, pixels values from an aerial image are available at any location in the field, with a support of about 3 m^2. The point sample data, on the other hand, were measured at 78 locations in the field, each with a support of about 5 m^2. The soil content value is known only at these locations, whose total support is about 400 m^2 in a 32 ha field. The ratio of "the sum of all the support units for which the average values are known and the extent" is called the *coverage*. If we want to estimate the value of, say, clay content at a location not sampled, we must interpolate it. The precise definition of *interpolation* is analogous to the geostatistical definition: the process of "increasing the coverage of the research area." The process of "selecting and observing a subset" of the support units in the extent is called *sampling*, matching the colloquial use of the term. The term *change of scale* is used somewhat loosely to refer generally to a change of one or more of these three features, support, extent, and coverage.

It should be noted that the definition given by Bierkens et al. (2000) of the term *support* leaves out one important feature. The definition should include not only the size, but also the shape of the area over which the phenomenon in question is averaged (Gotway and Young, 2002). For example, a lattice with a given number of square cells has the same average area as an irregular mosaic with the same number of cells, but their different spatial configurations may result in different values of statistics describing them, so that changing the configuration as well as the size of the subdivisions of a spatial region should be considered a change of support, even though it is neither upscaling nor downscaling.

We will not discuss in this book operations involving a change of extent. The process of extrapolation can only be meaningfully carried out in the presence of covariates, and the question of how to carry it out correctly is primarily a scientific one rather than a statistical one. The process of singling out primarily a GIS operation and is referred to in GIS terminology as "clipping," and can be done either using a GIS or functions in the package `rgeos` (Bivand and Rundel, 2017). The next two subsections contain a brief discussion of the two remaining operations, change of support, and change of coverage, as used in data set alignment.

6.4.2 Change of Coverage

One of the two change of coverage operations, sampling, is the topic of Chapter 5. The second change of coverage operation is interpolation. One of the primary concerns with the use of geostatistical interpolation to predict data at unmeasured locations is the effect of this operation on the distributional properties of the resultant data set. In exploratory analysis one is most concerned with developing data sets that can be used to visualize patterns in the data using maps and other tools of data exploration. Thus, there is little need to be concerned with the effect of the operations on the distributional properties of the data. Things are very different when one is preparing the data for parameter estimation and hypothesis testing (the phase that Tukey, 1977, p. 3, calls confirmatory analysis). In this case, changes in the distributional properties of the data induced by operations such as interpolation can have a profound effect on the statistical analysis (Anselin, 2001). This is true in particular of the smoothing effect of interpolation mentioned in Section 6.3.2.

Consider the problem of fitting a linear regression model of the form $Y = X\beta + \varepsilon$, where for example Y is a vector of crop yield values based on yield monitor data and X is a matrix of explanatory variables from data collected at sample locations. Figure 6.11 shows three forms of spatial data describing a small portion of Field 4.2. Yield is measured by a yield monitor at the locations indicated by the open circles, soil EC_a is measured at the locations indicated by the "+" signs, manual sampling was carried out at the location indicated by the * symbols, and the pixels of a remotely sensed image cover the areas indicated by the shades of gray. If one were to use IDW or kriging interpolation to predict all of these data values on a common grid, the interpolation would induce error into explanatory variables as well as into the yield data. A crucial assumption of the theory of linear regression is that the explanatory variables are measured without error (Sprent, 1969; Webster, 1997). Violations of this assumption must be handled by the *errors in variables* theory of linear regression (Johnston and DiNardo, 1997, p. 153).

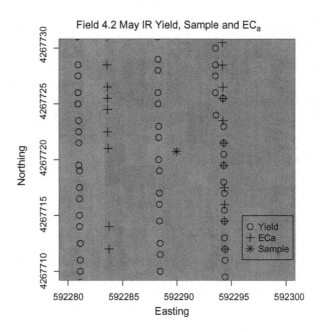

FIGURE 6.11
Spatial arrangement of the data sets from Field 2 of Data Set 4. The asterix represents a sampling location, the plus signs represent electrical conductivity measurements, the circles represent 1996 yield monitor locations, and the gray squares are pixels in the infrared band of a remotely sensed image taken in May 1996.

Fortunately, in determining the effects of environmental variables on an automatically measured response variable (such as yield in the application we discuss here, where the X values are sample data and the Y values come from a yield monitor), it is reasonable to assume that the environmental variables are measured more accurately than the response variable. Therefore, the assumptions of exact measurement of the explanatory variables required by the regression model can be considered to be satisfied if the value of the response variable is estimated at the spatial locations of the explanatory variables. In our analyses of Data Set 4, we will generally interpolate all variables to the locations of the manual sample sites. Griffin et al. (2007) provide an excellent practical discussion of protocols to follow in carrying out this procedure. One way to do this is to create buffers around each sample point and estimate the yield at each sample point as the mean of those yield monitor values falling within the buffer circle. This is called the *neighborhood mean* and is analogous to the compositing operation commonly practiced when taking soil samples. As with the size of the region over which to composite a soil sample, the size of the buffer circles is somewhat arbitrary. We will use a value of 10m so that each buffer would enclose some data values from at least two and possibly three passes of the combine harvester.

The data are loaded as described in Appendix B.4. The buffering operation can be carried out using the `spatstat` (Baddeley and Turner, 2005) function `disc()` or the `rgeos` function `gBuffer()`. We will use the former. This creates a circular `owin` object (this is a form of `spatstat` object whose details we don't need to know). We can create a disc of radius 10m centered on first sample location as follows.

```
> xy <- coordinates(data.Set4.2)[1,]
> samp.rad <- 10
> d <- disc(radius = samp.rad, centre = xy)
```

Note the spelling of the argument centre. We don't need to know the details of the owin object because we immediately coerce it into a SpatialPolygons object.

```
> buffer.10m <- as(d, "SpatialPolygons")
```

The function spRbind() joins pairs of polygons, so we will augment the object buffer.10m via a for loop that creates and adds new buffer discs one at a time.

```
> for (i in 2:nrow(data.Set4.2@data)){
+     xy <- coordinates(data.Set4.2)[i,]
+     d <- disc(radius = samp.rad, centre = xy)
+     sp.d <- as(d, "SpatialPolygons")
+     sp.d@polygons[[1]]@ID <- as.character(i)
+     buffer.10m <- spRbind(buffer.10m, sp.d)
+ }
```

The attribute data come from a different source from the spatial data, so as always we must be careful to give each disc the correct ID number before adding it. These sorts of comparisons are generally most easily done with spatial features object because of their simpler structure.

```
> library(sf)
> data.Set4.2.sf <- as(data.Set4.2, "sf")
> buffer.10m.sf <- as(buffer.10m, "sf")
> st_crs(buffer.10m.sf) <- "+proj=utm +zone=10 +ellps=WGS84"
> y <- st_contains(buffer.10m.sf, data.Set4.2.sf)
> unlist(y)
 [1]  1  2  3  4  5  6  7  8  9 10 11 12 13 14 15 16 17 18 19 20 21
[22] 22 23 24 25 26 27 28 29 30 31 32 33 34 35 36 37 38 39 40 41 42
[43] 43 44 45 46 47 48 49 50 51 52 53 54 55 56 57 58 59 60 61 62 63
[64] 64 65 66 67 68 69 70 71 72 73 74 75 76 77 78
> all.equal(data.Set4.2.sf$ID, unlist(y))
[1] TRUE
```

Figure 6.12a shows a small portion of the field with four buffer circles.

The computation of the means over the buffer circle of each of the data fields in the yield monitor data file can be carried out using the R function over() to determine the mean value of those data records in the yield file data.Yield4.2 that lie inside each buffer circle.

```
> yield.bufmean <- over(buffer.10m, data.Set4.2Yield96raw, fn = mean)
```

This usage of the function over() is as follows. The objects buffer.10m and data.Set4.2Yield96raw are SpatialPolygons and SpatialPointsDataFrame, respectively. In this case, the function over() returns a data frame with the same number of elements (length of a vector or rows of a data frame) as the first argument. If the third argument is not specified, over() returns the index of the polygon. We want to compute the mean yield over each buffer, and we accomplish this by specifying fn = mean as the third argument, which applies the function mean() over the data locations lying within each polygon in buffer.10m. It is possible, although it takes more work, to compute the same mean values with the order of arguments of over() reversed (Exercise 6.10). Since the buffers are circular, the means of the geographic coordinates will be near but generally

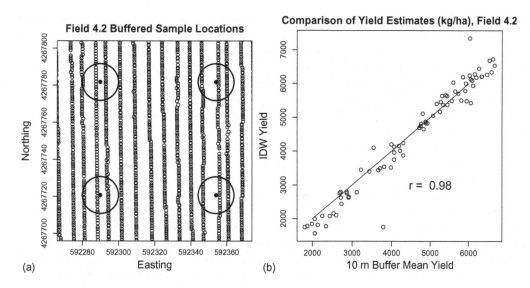

FIGURE 6.12
(a) Yield monitor sampling points of Field 4.2 in 1996 (white circles), sampling points of sparse data (black circles), and 10m buffers of the sparse sample points (large circles); (b) Scatterplot of the yield values obtained via the two methods. The solid line satisfied $y = x$.

not exactly equal to the geographic coordinates of the sample points, which are at the center of the buffer circles, but they should be close enough.

Given the issues with the effect of interpolation on the distribution of the interpolated data, it is of interest to compare the yield estimates in Field 4.2 obtained using the 10m buffers of Figure 6.12a with those obtained using interpolation. Figure 6.12b shows a plot of the two estimates. The correlation between the two estimates is $r = 0.99$. This indicates that for this data set the operation of aggregating data over a region of radius 10 meters has approximately the same smoothing effect on the estimation process as does the operation of IDW interpolation. This is not necessarily true, however, for all data sets. As always, the best procedure is to test both methods, breathe a sigh of relief, and move on if they give similar results, or try to find out what is going on if they do not. For our later use, we will create a file *yield4.296ptsidw* holding the interpolated values of yield and grain moisture and put it in the *created* folder.

```
> yield.idw.df <- data.frame(Yield = yield.idw$var1.pred,
+    Easting = coordinates(yield.idw)[,1],
+    Northing = coordinates(yield.idw)[,2])
+ write.csv(yield.idw.df, "created\\set4.2yld96ptsidw.csv")
```

In Exercise 6.10, you are asked to repeat this process for Field 1. The file you create will be important for subsequent analysis.

6.4.3 Change of Support

The process of change of support (see Figure 6.10) via interpolation is known in the GIS literature as *resampling*. Resampling for exploratory analysis involves two primary considerations. The first is how to interpolate Y values at a common set of points so that data values may best be visualized in the process of data exploration. This requires the aggregation of values over a common lattice or interpolating values at a common set of points. The second consideration involves the proper way to aggregate different data sets so that they may be compared and incorporated into statistical models. We will defer discussion of this second issue until Section 11.5, and for now focus on the issue of data exploration.

Since spatial data sets may contain thousands or tens of thousands of points, the standard forms of visualization, both spatial (e.g., thematic maps) and non-spatial (e.g., boxplots, scatterplots, etc.) may be virtually illegible unless the data are summarized to some extent. For this reason, a common practice is to predict the Y values at a set of locations having a larger support than the data. This can be done by aggregating over a mosaic of polygons or interpolating to a common set of data points. Aggregation over a mosaic can be carried out, for example, by computing the mean over each polygon of the points within it. The data may in principle be interpolated using any interpolation method, including IDW interpolation and kriging as covered in Sections 6.3.2 and 6.3.3 (Nolan et al., 1996). Where no well-defined mosaic exists a priori, we will focus on the second option, interpolation to a common grid and then sampling at a subset of grid locations.

There is, in particular, a considerable literature discussing the interpolation of yield monitor data to a regular grid, and this may serve as a model for interpolation of other similar data types. Most of this work involves a comparison between kriging and IDW interpolation. Because of the relatively large sample size and the relatively small distance between samples, one would not expect a great difference between results using different interpolation methods (Birrell et al., 1995). Indeed, comparisons between different forms of interpolation indicate little difference (Juerschik et al., 1999; Robinson and Metternicht, 2005). Kriging is more complex but has the advantage that it provides a best linear unbiased interpolation of the data at each point as well as a variance estimate (Birrell et al., 1995; Blackmore and Marshall, 1996; Thylen and Murphy, 1996; Blackmoore and Moore, 1999). The data smoothing property of interpolation methods may be problematic for statistical modeling and confirmatory analysis, but it may actually be useful for exploratory analysis since it may be possible that important patterns obscured by high data variability are more visible when this variability (i.e., noise) is reduced.

Unlike automatically sampled data sets, remotely sensed data generally consist of a pixilated grid. Each pixel contains a set of integer values (often three values in the case of multispectral sensing, and many more in hyperspectral sensing) that represent the intensity of radiation in different bands of the electromagnetic spectrum. For example, in the common case of a false color infrared image these are infrared (represented as red), red (represented as green), and green (represented as blue). If the ultimate objective is to sample image data on a subset of grid cells, it is often not necessary to place remotely sensed images onto a common grid to do so. If it is necessary to convert remotely sensed image data, or any other form of pixilated data, to a common grid, one simple operation to do so is nearest neighbor resampling (Lo and Young, 2007, p. 154).

Figure 6.13a shows a close-up view of a small portion of the remotely sensed image of Field 4.2 taken in May 1996. The "+" symbols are located at the center of each pixel and represent the location at which the band values of that pixel are considered to be correctly measured. The black dots represent the locations at which the values are to be resampled. Each pixel can be considered as a Thiessen polygon surrounding its center point, so that assigning to the resample points the value of the pixel in which they are located accomplishes nearest neighbor resampling. The sp function over() can be used to carry out this nearest neighbor resampling operation, as can the raster function resample(). Since we used over() in the last section, we will use raster() in this one. Assume that the image is loaded as the R object data.Set4.2.May (Appendix B.4) and that values of the edges of

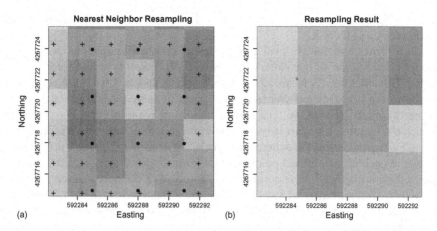

FIGURE 6.13
(a) Close-up view of a small portion of the remotely sensed image of Field 4.2 taken in May 1996. The "+" symbols are located at the center of each pixel and represent the location at which the band values of that pixel are assumed to be correctly measured. The black dots represent the locations at which the values are to be resampled. Each black dot is assigned the value of the plus sign contained in the same pixel. (b) Gray squares indicating the results of nearest neighbor resampling on the data of part (a).

the field have been assigned to R objects N, S, E, and W. The first step is to use the function raster() to read the existing image and generate a cell grid at the new cell size.

```
> library(raster)
> data.Set4.2.May <- raster("set4\\Set4.20596.tif")
> proj4string(data.Set4.2.May) <- CRS("+proj=utm +zone=10
+ellps=WGS84")
> cell.size <- 1.9
>
```

Next, we create a new raster object new.ras that defines the grid size of the resampled data.

```
> E <- 591967
> W <- 593139
> S <- 4267140.95
> N <- 4268320.95
> new.cell.size <- 3
>
> ncol.new <- (W - E) / new.cell.size
> nrow.new <- (N - S) / new.cell.size
> new.ras <- raster(ncol = ncol.new, nrow = nrow.new, xmn = E,
+   xmx = W, ymn = S, ymx = N)
> proj4string(new.ras) <- CRS("+proj=utm +zone=10 + ellps=WGS84")
```

Finally, the function resample() is used to resample the raster data.Set4.2.May to the new grid defined by new.ras, creating a raster object data.Set4.2.new.

```
> > data.Set4.2.new <- resample(data.Set4.2.May, new.ras, method =
+   "ngb")
```

The third argument specifies that nearest neighbor resampling will be used. This is discussed below. Figure 6.13b shows the results of this resampling. On my computer, neither of the plots in Figure 6.13 could be produced in RStudio because of memory limitations, but they did work in TINN-R.

In resampling from a raster grid, such as a remotely sensed image, the usual assumption is that the precise location of the value of the raster cell (pixel) is at its center. Based on the discussion in this chapter, there are at least three possible ways that one could develop a common alignment of estimated values from a set of densely measured data that includes two or more quantities. These are (1) a nearest neighbor resampling scheme to a common grid in which the estimated values at the aligned points are set to the geographically nearest data value; (2) a local aggregation scheme in which, for example, Thiessen polygons are constructed around each of the aligned points and the estimated value is set at the mean of those data values falling within the corresponding polygon; and (3) an interpolation scheme such as kriging in which values at aligned locations are interpolated from the data. The distribution of the data created from each of these methods will be different, and some evaluation method will be necessary to compare them. The raster function `resample()` provides two methods to choose from, bilinear interpolation (the default) and nearest neighbor resampling.

The cell size in the new grid in the example just provided is about one-and-one-half times the size of that of the original grid. Hijmans (2016) points out that for substantial changes in the size of the support (i.e., the cell size in this case), one of the raster functions `aggregate()` or `interpolate()` should be used to more accurately determine the appropriate attribute values of the resampled data.

There are some very serious statistical issues that must be dealt with. These have been alluded to previously, but let us gather some of them in one location. First, these dense data sets are highly autocorrelated in space, so that assumptions of independence of errors generally do not hold. Second, the relationship between, for example, remotely sensed image data and a response variable can be highly dependent on the support over which the data are aggregated (Long, 1996). This phenomenon is called the *modifiable areal unit problem* (covered in Section 11.5.1). Third, if one is using the image data to predict the response variable value, there may be substantial measurement error in both the predictor variable (e.g., image data) and the response variable, in violation of the assumptions of linear regression. Fourth, the action of georegistration, in which features in an image are brought into alignment with their geographic locations, can affect the distributional properties of the data, which compromises the conclusions of a statistical analysis (Anselin, 2001). We will not attempt to address these issues right now, but it will be necessary to keep them in mind in future chapters.

6.5 Further Reading

In addition to Lo and Yeung (2007), which we use as the primary GIS reference, there are many others that provide excellent discussions of issues such as map projections, Thiessen polygons, spatial resampling, and so forth. Prominent among these are de Smith et al. (2007), Bonham-Carter (1994), Burrough and McDonnell (1998), Clarke (1990), and Johnston (1998). Good GIS texts at a more introductory level include Chang (2008), Demers (2009), and Longley et al. (2001).

There is some literature on data quality in ecological surveys (e.g., Kanciruk et al., 1986; Edwards, 2000; Iverson, 2007). The literature on sample surveys carried out via interviews is much more extensive, however, and many of the concepts carry over directly to ecological sampling. A classic is Kish (1965). At a slightly lower level mathematically is Biemer and Lyberg (2003). Naus (1975) discusses some of the issues associated with automatic error detection and data adjustment for large data sets. Anderson et al. (1990) provide an interesting discussion of the data quality issues associated with a large, well-known study, the Framingham Heart Study. Naus (1975) provides more general information about the maintenance of database quality.

Wickham and Grolemund (2017) describe the *tidyverse*, a collection of R packages for the input, graphing, cleaning, and analysis of data sets. A data set is said to be *tidy* if it can be arranged in a set of tables whose columns represent the fields, such that within every field each record forms a row. In that context, our data sets are tidy. The graphics package `ggplot2` is one part of the tidyverse, but there are several other parts as well. Small data sets such as ours that are conveniently packaged in csv files and images may not need the tidyverse, but if your data set is a mass of notes and miscellaneous files collected without too much care for how they fit together, then you may want to look into this.

Barnett and Lewis (1994) is an excellent source for outlier detection and treatment. The papers by Anscombe (1960), Daniel (1960), and Kruskal (1960) are a part of a volume on the detection and treatment of discordant values and, despite their age, they still contain a wealth of useful knowledge. This chapter has focused on the identification of individual outliers. For a discussion of the identification of multiple outliers, see Davies and Gather (1993).

Burrough and McDonnell (1998) provide a discussion of the effect of scale on quality of spatial data. Bierkens et al (2000) provide a very systematic, almost recipe-like approach to changes of scale. There have not been very many implementations of their approach; for one example see Hauselt and Plant (2010). The effect of scale changes is greatest when the interactions among the variables are nonlinear. For discussions in an agricultural context see Cassman and Plant (1992) and Lark and Webster (2001). Odeh et al. (1998) provide an excellent example of the use of information at multiple scales to direct sampling. There is a vast literature on hierarchy theory, which provides organizing principles for spatial and temporal scales in ecosystems. The usual entry to this is O'Neill et al. (1986).

Isaaks and Srivastava (1989) is the standard introductory text for geostatistics. Journel and Huijbregts (1978) and Goovaerts (1997) provide excellent discussions that go beyond the material in Isaaks and Srivastatva (1989). Cressie (1991) is an excellent source at a higher mathematical level. Kanevski and Maignam (2004) provide a good discussion of interpolation methods, stochastic simulation, as well as neural network and support vector machine analysis, for environmental data.

Kutner et al. (2005) is a good introductory source for regression diagnostics. Belsley et al. (1980) and Fox (1991, 1997) provide a more extensive discussion.

Exercises

6.1 The square-shaped northern set of sample points in Data Set 2 is the Klamath Range. Create a map of this region, and display the map along with a map of the boundary of California, in both longitude-latitude and UTM coordinates. The region is in UTM Zone 10N.

6.2 The data field *QUDO_BA* in Data Set 2 contains measurements of the total basal area of blue oak trees at each sample location. Use the function `moran.plot()` to identify discordant values of this variable and create a map showing their location. Can you distinguish any possible source of the error based on the map?

6.3 (a) Use the function `expand.grid()` to create a rectangular grid of sample points in the (x, y) domain in which x takes on values 0, 10, 20, and 30, and y takes on the values 20, 10, and 0. Use the function `class()` to verify that this grid is a data frame. Create a third data field called `z` in this data frame and assign it numbers beginning with 1 and incrementing by 1. Load the package `maptools`. Use the function `coordinates()` to assign `x` and `y` as the coordinates of the data frame. Repeat the application of the function `class()` to the object. (b) Display the grid using the function `plot()`. Set the argument `pch` in the function `plot()` to the `z` data field. Carry out this same operation plotting the value of `z` using the function `spplot()`. (c) Use the function `gridded()` to convert the object to a `SpatialPixelsDataFrame`. Repeat the applications of `plot()` and `spplot()` of part (b).

6.4 Construct a set of Thiessen polygons built around the 78 data points of Field 2 of Data Set 4. Display the map in R along with the axes in UTM coordinates, using yield from the artificial population to determine the gray scale. Based on the discussion in Section 6.3.1, what is an example of a data set for which Thiessen polygons would be the most appropriate?

6.5 Use IDW interpolation to interpolate the yield data of Field 4.1 to the sample points and save the file as set4.1yldptsidw.csv.

6.6 The `sp` package contains the function `over()` that performs a GIS overlay operation. This can be used to clip an interpolated grid to an irregular boundary. If you have not yet created the boundary file for Field 4.1 by doing Exercise 2.11, create it now. Compute an IDW interpolation of *Clay* on a 5m grid. Apply `over()` to clip the interpolated grid to the field boundary. What class of object is the result? Use this result to create a `SpatialPointsDataFrame` of the interpolated clay content with the correct boundary. Use `spplot()` to verify your result.

6.7 The clay content data of Field 1 of Data Set 4 has a strong north–south trend, which violates the stationarity assumption. Prepare two kriging interpolations of clay content, one of which involves detrended data and other of which does not detrend the data, and compare the two. Don't forget to add the trend back onto the kriged detrended data. Notice that a spherical model will not work very well for the non-detrended data, so you will have to try something else.

6.8 Carry out IDW and kriging interpolation of the artificial yield population on the same grid as that population and compute the errors. (a) Plot the errors using `spplot()`. How do they compare visually? (b) Generate histograms of the distributions of the two error sets. (c) Compute the sums of squared errors of the two sets. (d) Can you think of circumstances under which IDW would be preferable to kriging?

6.9 The file *data, Set4.2.csv* contains two data fields representing soil EC_a: *EM38F*, the soil EC_a measured in the furrow, and *EM38B*, the soil EC_a measured in the beds. Determine which correlates better with soil clay content, and use this to carry out a cokriging interpolation of clay content with EC_a as the covariate. Compare your result with that obtained by ordinary kriging.

6.10 Repeat the computation of yield means over records located in each circular neighborhood in Figure 6.12a, but do it with a reversed order of arguments in the function `over()`.

6.11 Repeat for Field 1 of Data Set 4 the creation of a file *set4.1yld96ptsidw.csv* analogous to the one created in Section 6.4.2 doe Field 2.

7

Preliminary Exploration of Spatial Data

7.1 Introduction

"Exploratory analysis is detective work." This is how Tukey (1977, p. 1) opens the book *Exploratory Data Analysis*, and it would be impossible to write a better introductory sentence. Tukey (1977) expands the analogy between data analysis and the criminal justice system. The initial phase involves the search for evidence and is generally carried out by the police. Once the evidence has been gathered, the evaluation of the strength of that evidence is the responsibility of the judge and jury. Data analysis can also be divided into two phases (which are often carried out by the same person or team). The exploratory phase examines the properties of the data and attempts to uncover patterns and indications. The confirmatory phase tests the strength of these patterns and indications and in so doing addresses questions about the population from which the sample data were drawn.

One must be careful, however, not to press the analogy between data analysis and criminal justice too far. Ordinarily, the phases of the criminal justice process are more or less sequential, and once the case has gone before the judge and jury the investigative phase ends. With data analysis, particularly when dealing with spatial data, the process is more flexible. After an initial exploratory phase, the exploratory and confirmatory processes to some extent co-occur, with results of the confirmatory phase often motivating further exploration. The main danger that must be avoided is that of "data snooping," that is, of letting the confirmatory process be influenced inappropriately by the exploratory process through an introduced bias. Such an inappropriate influence would occur, for example, if the investigator selected for a hypothesis test subsets of the data that appeared in the exploratory phase to satisfy the hypothesis.

Exploratory data analysis (EDA) consists of a collection of tools, and the question of which tool to employ depends on the data and on the objective of the study. EDA is best introduced by means of examples. This also provides a good way to introduce the data sets that we will be studying throughout the rest of the book. In the contexts of the simulated research projects that we carry through the book, the objective in this chapter is to initiate data exploration by graphical means. In a real data analysis program, one would likely cycle through applications of graphical analysis and applications of the tools in the chapters to follow. To the extent that this is feasible we will do so in the book.

Each of the four data sets consists of a mixture of several different kinds of data, and in order to guide the initial exploration it will be useful to establish a set of categories into which these kinds fall. These categories can be arranged into a rough sequence of presumed influence (Figure 7.1). The use of association to infer influence has a long and inglorious history (Sprent, 1969, p. 29; Box et al., p. 487), and so we first emphasize that when we discuss influence, we must do so from the perspective of the ecological scientist

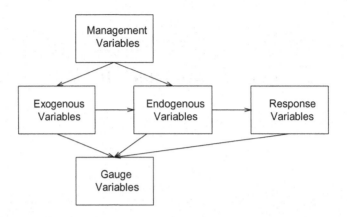

FIGURE 7.1
Classification of types of variable for use in guiding exploratory analysis. This classification provides assistance in avoiding the confounding of variables that play different roles in the ecological processes governing the behavior of the system.

rather than the data analyst. Assumptions about influence relationships must be made based on the ecologist's knowledge of process and mechanism and used to inform the analysis of the data, rather than using a blind analysis of the data to inform conclusions about influence. Failure to consider process and mechanism in the data analysis and to, for example, carry out a multiple regression in which all possible predictors are simply lumped together with no thought to their ecological relationships, will lead to conclusions that are at best confusing and at worst misleading.

I find it useful in thinking about process and influence to schematically divide the variables of the model into five categories: exogenous, endogenous, management, response, and gauge variables. To avoid unnecessary confusion, I must point out that my use of the terms *endogenous* and *exogenous*, although motivated by econometrics (Wooldridge, 2013, p. 87), is not the same as their use in this discipline. To illustrate the five categories, we will use as an example a data set describing a crop in a farmer's field, such as Data Set 3 or 4. At one end of the presumed chain of influence are *exogenous variables*, those variables that measure the influence of the environment on the processes that directly influence the values of the response variables. Soil clay content and daily high temperature are examples of exogenous variables. At the other end of the influence chain are the *response variables* (or variable; often there is only one). Crop yield is an example in Data Sets 3 and 4. In between the exogenous variables and the response variables are *endogenous variables*, which are influenced by the exogenous variables and, in turn, influence the response variables. Leaf nitrogen content in Data Set 4 is an example of an endogenous variable. In some data sets, the values of one or more of the exogenous, endogenous, or response variables may be represented by *gauge variables*. These are variables that do not directly play a role in the chain of influence, but whose values represent variables that do play such a role. The aerial images and soil electrical conductivity measurements in Data Set 4 are examples. Finally, human influence on the system may enter in the form of *management variables*. These are exogenous variables that are implemented through human intervention. Fertilizer rates and irrigation effectiveness in Data Set 3 are examples. This subdivision into categories is intended to help guide thinking, not to constrain or confuse it. The selection of categories is often a matter of taste, and, to paraphrase Cressie (1991, p. 114), one person's endogenous variable may be another person's exogenous variable.

Systems such as Data Sets 1 and 2, which do not involve human intervention to produce an end product, offer a different challenge in assessing interrelationships among variables. Although some "natural" systems may include management variables, they are not present in either of these data sets. If some form of resource were provided, such as artificial bird habitat in Data Set 1 or planting and maintenance of oak seedlings in Data Set 2, then the intensity of these interventions would be a management variable. An example of a gauge variable in either data set would be a remotely sensed quantity such as a vegetation index in aerial or satellite images. With cuckoo presence or count as the response variable in Data Set 1, there are arguably no true endogenous variables. If bird's nests could be identified, these would serve as an endogenous variable in Data Set 1. In Data Set 2, we can consider variables such as *Elevation* and *CoastDist*, the shortest distance to the Pacific coast, as exogenous variables since they do not influence blue oak presence directly.

7.2 Data Set 1

For convenience, we will repeat in our discussion of each data set some of the descriptive material that was presented in Chapter 1. Data Set 1 was collected as a part of a study of the properties of habitat suitable for the western yellow-billed cuckoo (*Coccyzus americanus occidentalis*), a California state-listed endangered species. The bird, whose habitat is the interior of riparian forests, once bred throughout the Pacific Coast states of the USA. Land use conversion to agriculture, urbanization, and flood control have, however, restricted its current habitat to the upper reaches of the Sacramento River (Figure 1.3). Greco et al. (2002) provide the following descriptive characterization of suitable habitat: "a mosaic of riparian forest vegetation consisting of willow (*Salix* spp.) and cottonwood (*Populus fremontii*) forests in combination with open-water habitats such as an oxbow lake or backwater channel. Dense vegetation less than 20 m in height is especially important for nesting while both low and high vegetation are used for foraging." Cuckoos nest in large forest patches (greater than 5–20 ha), and they are not found in parts of the patch less than 100 m wide (Laymon and Halterman, 1989), although they can cross gaps in the forest of less than 100 m (Gaines and Laymon 1984).

The Sacramento River is a naturally meandering river that has over most of its length been constrained by levees; the upper reaches are the only remaining unconstrained portion. There has been discussion of the possibility of instituting a levee setback system in which the levees would be relocated to permit some movement of the lower reaches of the river, and there is interest in estimating the effect of this policy on yellow-billed cuckoo habitat. This requires a habitat suitability model. Such a model has been developed, based on years of observation, by Laymon and Halterman (1989). The objective of the analysis of Data Set 1 is to test this model based on the combination of habitat variables and bird observation data. The data set consists of five explanatory variables (the habitat data) and a response variable (the bird observation data). A detailed description of the processes used to create the data set is given in Appendix B.1. For convenience, some of that information is summarized here.

The explanatory variables are defined over a polygonal map of the riparian area of the Sacramento River between river-mile 196, located at Pine Creek Bend, and river-mile 219, located near the Woodson Bridge State Recreation Area. This area can be visualized in the USGS Earth Explorer (https://earthexplorer.usgs.gov/) by searching for "Woodson Bridge State Recreation Area."

The explanatory variables are based on a modification of the California Wildlife Habitat Relationships (CWHR) classification system (Mayer and Laudenslayer, 1988). This is a classification system that "contains life history, geographic range, habitat relationships, and management information on 694 species of amphibians, reptiles, birds, and mammals known to occur in the state" (http://www.dfg.ca.gov/biogeodata/cwhr). The modified CWHR model is expressed in terms of habitat patches, where a *patch* is defined for the purpose of the cuckoo habitat model as "a geographic area of the following contiguous land cover categories: riparian, freshwater emergent wetland, or lacustrine." These definitions correspond to the CWHR classifications VRI, FEW, and LAC, respectively. Photographs of examples of these habitat types are available on the CWHR website. The shapefile containing the habitat patches is included in the data as the file *habitatpatches.shp*. Code for loading this and the other data files into R is given in Appendix B.1.

The original habitat suitability model of Laymon and Halterman (1989) incorporates the following predictor variables: (1) patch area, (2) patch width, (3) patch distance to water, (4) within-patch ratio of high vegetation to medium and low vegetation, and (5) patch vegetation species. Variables 1–4 were incorporated directly into the model tested here. Variable 5, patch vegetation species, was interpreted as follows (Greco et al., 2002, p. 187). As a river like the Sacramento meanders, it alters the characteristics of the land surrounding it. After the meandering of the river has exposed a formerly submerged site, the land goes through a series of successional stages (Conard et al., 1977). These generally pass from an open gravel bar to a shrub-dominated thicket to a willow/cottonwood forest to a mixed willow/cottonwood oak woodland to a mature oak woodland. From the perspective of habitat quality for the yellow-billed cuckoo, the most important aspect of this succession is the fraction of the patch dominated by the willow/cottonwood mixture. It was impossible to distinguish willow/cottonwood from oak in the aerial photographs used to create the land cover maps. Therefore, floodplain age was used as a surrogate variable for stand composition. Floodplain age was estimated by a method described by Greco et al. (2007). Portions of a patch with a high floodplain age were taken to indicate land that passed to oak dominance. Portions of a patch with a low floodplain age were taken to indicate land that was still dominated by gravel bar and shrubs. A calibration of this model indicated that habitat quality of a patch could be well represented by the ratio of the area of the patch of age less than 60 years to the total area of the patch. The optimal ratio was found to be between 0.67 and 0.8.

The habitat suitability model of Laymon and Halterman (1989) was implemented using the scheme shown in Table 7.1. Each patch was assigned a suitability score for each of the five variables, and the overall patch suitability was computed as the geometric mean of the individual scores (Greco et al., 2002). The geometric mean was used because it emphasizes patches that satisfy all of the criteria (Cooperrider, 1986). The data set in the file *set1data.shp* contains the explanatory variables. This file was constructed as described in Appendix B.1.

TABLE 7.1

Habitat Patch Suitability Model for the Yellow-Billed Cuckoo.

Suitability Class	Suitability Score	Patch Area (ha)	Patch Width (m)	Patch Distance to Water (m)	Height Class Ratio (H:H+L+M)	Floodplain Age Ratio <60
Optimum	1.0	>80	>600	<100	0.45–0.55	0.67–0.8
Suitable	0.66	40–80	200–600	<100	0.2–0.45, 0.55–0.67	0.5–0.67, 0.8–0.875
Marginal	0.33	17–40	100–200	<100	<0.2	0.375–0.5, >0.875
Unsuitable	0.0	<17	<100	>100	>0.67	<0.375

Since the classification variable *distance to water* has only one separation value, which classifies land greater than 100 m from water as unsuitable (Table 7.1), this variable is not explicitly included in the attribute data. Instead, a buffer of 100 m around water polygons was created, and land use polygons that fell outside this buffer and did not touch a polygon that overlapped the buffer were eliminated.

Since the objective of the study is to analyze a habitat suitability model for the western yellow-billed cuckoo, it is evident that the response variable must be some measure of bird population or presence. A difficulty in this regard is that the bird is, to quote Gaines (1974, p. 204), "furtive, and thus easily overlooked." The commonly used manner of surveying for the presence or absence of the yellow-billed cuckoo is to visit a site, play a tape recording of the bird's *kowlp* call, and observe whether or not there is a response (Gaines, 1974). The standard procedure (Greco, 1999, p. 232) is to play the call five to ten times at each observation point. The call is audible for a distance of about 300 m (Laymon and Halterman, 1987). Failure to elicit a response is taken to imply that no cuckoos are present at that location. The response may be simply a bird call and may not include an actual sighting. The data set used in this study is an aggregation of observations collected over a period between 1978 and 1992 (Gaines and Laymon, 1984; Laymon and Halterman, 1987; Halterman, 1991; Greco, 2002). The survey includes a total of 21 locations within the study area (Figure 7.2). There are two forms of observation data that can be used as a response variable. The first is presence/absence, that is, whether or not a cuckoo observation occurs at a location. The second is the number of observations. Although they are suggestive and should not be completely ignored, the count data are fraught with practical difficulties. One is that it is impossible to determine with certainty the number of individual birds represented by these observations. Moreover, even if this relationship could be determined, it is not clear how one would interpret the difference between a single observation of one bird, multiple observations of the same bird, and multiple observations of different birds.

A second serious problem with the interpretation of the bird count data is that it is an *extensive* property. The number of birds counted in a particular area may depend on the size of the area. An extensive property is one that depends on the size of the medium containing it, while an *intensive* property does not (Lewis and Randall, 1961). For example, in the context of ecology, population is an extensive property while population density is an intensive property. Generally speaking, an intensive property can be derived from an extensive property by dividing the extensive property by an appropriate denominator. For example, population density is derived from population by dividing by area of the region containing the population. It is easy to create misleading thematic maps by either displaying an extensive property when an intensive property would be more appropriate, or by using an inappropriate denominator to derive an intensive property from an extensive one (Kronmal, 1993; Henning, 2003, p. 195).

Although there is no guarantee that, if everything else is fixed, bird count in a patch will be proportional to patch area, this quantity is nevertheless a logical choice. Patch area is a predictor variable in the CWHR model, so choosing this as the denominator for count data would have the effect of putting this quantity on both sides of the equals sign, or, to put it another way, to enforce an assumption that population density depends on patch area. In our initial exploration and analysis of the data, we will consider only presence-absence data. We will use the count data, without normalization by patch area, in Section 8.4.4 as an example of the construction of regression models for count data, but we will have to keep in mind the weaknesses of the data set.

Following the standard procedure of Appendix B.1, data are loaded as spatial features (sf) objects and converted to spatial (sp) objects for plotting the figures in the book. As discussed

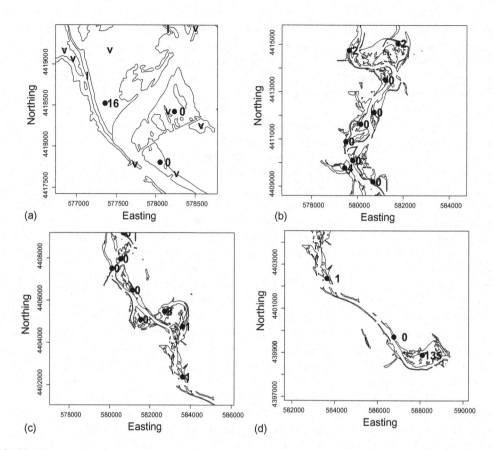

FIGURE 7.2
Habitat patches and locations of cuckoo observation points along the Sacramento River in the far northern portion of Data Set 1. The number next to each location signifies the total number of cuckoo observations in the pooled survey data. Parts (a) through (d) show regions from north to south. The letters in the Figure 7.2a are the vegetation type codes: v = riparian forest, and l = lacustrine. These are not shown in the other figures to avoid clutter.

in Section 2.6, however, we will also get a quick overview of the data using the sf function plot() as well as calls to the function ggplot(). The code for this section, as well as the other sections, includes calls to these functions, although the results are not shown.

The quick call to the sf function plot() shows that the study site is extremely long and narrow (Figure 1.3). Figure 7.2 shows the site divided into four zones of roughly equal length that include all observation points. The locations of the observation points are indicated, along with the number of birds observed at each location. The code to create Figure 7.2a is as follows. The SpatialPolygonsDataFrame data.Set1.patches contains the contents of the shapefile *habitatpatches.shp*, and the SpatialPointsDataFrame data.Set1.obs is created from the contents of the file *obspts.csv* (Appendix B.1).

```
> plot(data.Set1.patches.sp, axes = TRUE, # Fig. 7.2a
+    xlim = c(577000,578500), ylim = c(4417500,4419400))
> title(xlab = "Easting", ylab = "Northing", cex.lab = 1.5)
> points(data.Set1.obs, pch = 19, cex = 2)
> text(coordinates(data.Set1.obs)[,1] + 100,
+    coordinates(data.Set1.obs)[,2],
+    labels=as.character(data.Set1.obs$Abund), cex = 2, font = 2)
```

```
> y <- lapply(data.Set1.patches.sp@polygons, slot, "labpt")
> patches.loc <- matrix(0, length(y), 2)
> for (i in 1:length(y)) patches.loc[i,] <- unlist(y[[i]])
> text(patches.loc,
+    labels=as.character(data.Set1.patches.sp$VegType), cex=2, font=2)
```

The code using `lapply()` described in Section 2.4.3 is used to extract the attribute data values so that they can be plotted as character values in the appropriate locations. This is only done in Figure 7.2a; the other figures contain so many polygons that displaying all of their values would create too much clutter. Inspection of Figure 7.2d indicates that the southernmost observation point has an anomalously large number of sightings. Either there is something unusual about this location that makes the cuckoos really love it or there is a problem with the data.

We can get an idea of the relationships among the data by examining the small region at the northern end of the site. Figure 7.3 shows thematic maps of height class, cover class, and age class in the northern end, corresponding to Figure 7.2a. The code to create Figure 7.3a is the following. The `SpatialPolygonsDataFrame data.Set1.sp` contains the shapefile *set1.data.shp*.

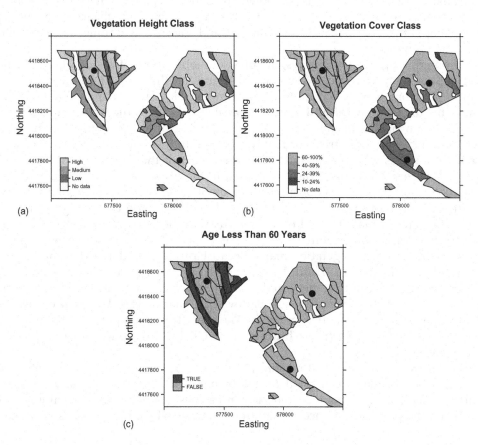

FIGURE 7.3
Geometry of the habitat patches and observation points in the northern area (Figure 7.2a): thematic maps created with the function `spplot()` distinguishing (a) vegetation size classes, (b) cover classes, and (c) age classes. The large black circles indicate patches in which cuckoo presence/absence was recorded.

```
> levels(data.Set1.sp$HtClass)
[1] "0" "h" "l" "m"
> data.Set1.sp@data$HtClass2 <-
+     ordered(as.character(data.Set1.sp@data$HtClass),
+     levels = c("0", "l", "m", "h"),
+     labels = c("No data", "Low", "Medium", "High"))
> greys <- grey(c(250, 100, 150, 200) / 255)
> obs.list = list("sp.points", data.Set1.obs, pch = 19, col = "black",
+     cex = 2)
> spplot(data.Set1.sp, "HtClass2", col.regions = greys, # Fig. 7.3a
+     scales = list(draw = TRUE), xlab = "Easting", ylab = "Northing",
+     main = "Vegetation Height Class", sp.layout = list(obs.list),
+     xlim = c(577000,578500), ylim = c(4417500,4418800))
```

Two new concepts are introduced here. The first is the *ordered factor*, which is generated by the function ordered() in the fourth line. Since *low, medium,* and *high* together make up an ordinal scale, it makes sense to plot them in this order. An ordered factor as created by the function ordered() behaves the same as an ordinary factor except that its ordinal relationships are recorded and preserved in plotting. The second new concept is the use of the argument sp.layout to overlay a point map on top of a polygon map created with the sp package function spplot(). A list called obs.list is generated that contains the type of spatial object that is to be plotted, the data set, and any special arguments to control the appearance of the plot.

Examining the figures themselves, the most obvious, and distressing, feature is that the vegetation map is truncated at the north end, so that not all of the area of the northern polygons is included in the map (compare Figures 7.2a and 7.3a). Figure 7.2 displays the data contained in the file *habitatpatches.shp*. This file was created by photointerpretation of a collection of 1997 aerial photographs, and only contains land cover class (riparian, lacustrine, or freshwater wetland, see Appendix B.1). These are the habitat patches whose area and width are a part of the habitat suitability model (Table 7.1). Figure 7.3 displays data contained in the file *set1data.shp*. The polygons in this shapefile are characterized by a combination of vegetation type, height class, cover class (this is not used in the model), and age class. Vegetation type, height class, and cover class were estimated based on the 1997 aerial photographs alone, but age class was estimated based on photointerpretation of the entire set of available aerial photos together with historical vegetation maps. Because of this, the data extent of age class was slightly less in the north end. Unfortunately, this means that the floodplain age ratios in Table 7.1 cannot be measured accurately for the northernmost observation point (the one with 16 observations in Figure 7.2a), and so this data record will have to be dropped from the statistical analysis. The other truncated patch, in the northeast of Figures 7.2a and 7.3a, with zero birds observed, is less affected, and we will leave it in the analysis.

In order to continue the exploratory analysis, we must create a spatial data set that reflects the construction of the CWHR model. This involves two fundamental steps. First, we eliminate those polygons that represent habitat patches where no observation was made and append the presence-absence data to those patches where observations were made.

Second, we compute the explanatory variables in the model as represented in Table 7.1. Elimination of the habitat patches in which there is no observation point is accomplished using the sp function over().

```
> Set1.obs <- over(data.Set1.obs, data.Set1)
```

The object data.Set1.obs is a SpatialPointsDataFrame created from the file *obspts. csv* and object data.Set1.sp is a SpatialPolygonsDataFrame created by reading the shapefile *set1data.shp*. The result of the application of the function over() with first and second arguments in these classes is the data frame Set1.obs containing the attribute data of the subset of the habitat patches in data.Set1 that contain an observation point (i.e., a point in data.Set1.obs).

Now we add the presence-absence data, the ID value, and the coordinates from data. Set1.obs. For future reference, we will also add the abundance data.

```
> Set1.obs$PresAbs <- data.Set1.obs$PresAbs
> Set1.obs$obsID <- data.Set1.obs$ID
> Set1.obs$Abund <- data.Set1.obs$Abund
> Set1.obs$Easting <- data.Set1.obs@coords[,1]
> Set1.obs$Northing <- data.Set1.obs@coords[,2]
```

This completes the first of the two steps listed above, elimination of patches with no observation site. Here is the result.

```
> names(Set1.obs) # Polygons containing an obs point
 [1] "SP_ID" "ID" "AgeID" "AgeArea"
 [5] "Age" "VegType" "PatchArea" "PatchID"
 [9] "HtClID" "HtClArea" "HtClass" "CoverClass"
[13] "PatchWidth" "HtClass2" "CoverClass2" "AgeLT60"
[17] "PresAbs" "obsID" "Abund" "Easting"
[21] "Northing"
```

We can display some of the variables used in the model as follows.

```
> with(Set1.obs, cbind(PatchID, PatchArea, PatchWidth, obsID, PresAbs))
      PatchID PatchArea PatchWidth obsID PresAbs
 [1,]     191 1257345.27    593.88     1       1
 [2,]     175  301897.91    188.28     2       0
   *    *     *    DELETED     *     *       *
[21,]      19 1058784.95    414.52    21       1
```

At this point, we again encounter the dangerous step in the analysis of spatial data discussed in Section 3.6.2. We have detached the attribute data in data.Set1.obs from the spatial data and then attached them to the records in Set1.obs. The observation point data are attached to the habitat patch data according to the order in which the records appear in the respective data sets. Linking each observation point with the correct habitat patch requires that the application of the function over() preserve the order of the records in the attribute table. In Exercise 7.1, you are asked to verify that this linkage is indeed correct.

Two of the variables in the CWHR model of Table 7.1, patch area and patch width, are already present in the object Set1.obs, as are the vegetation type code, patch ID, and value of the response variable Y (presence or absence) for each observation point. Now we must add the other variables in Table 7.1, height class ratio and floodplain age ratio. Because it is available in the data set, we will also add the cover class ratio *dense area:(medium area + sparse area)*, even though this is not in the model. Height class and cover class have the same identification value, HtClID.

We will illustrate the computation of the habitat ratios in Table 7.1 by describing the calculation of the height class ratio $HtRatio = H : (H + M + L)$, the ratio of areas of high to total vegetation. The age ratio (the fraction of the total area with vegetation less than 60 years old), and the cover class ratio (the fraction of total area with dense cover) are computed similarly. The computation of the height class ratio could be done using functional programming, but this is one of those instances where an old-fashioned for loop is probably easier and more transparent. The computations involve the object that was initially created by reading the file *set1.data.shp*, which contains all the data, rather than the just-created Set1.obs (Exercise 7.2). Recall that two objects were created, the sf, object data.Set1. sf, and the sp object data.Set1.sp. We will use the former because it has the simpler structure.

First, a data frame MLArea is created to place in the data field Area nonzero values for all those polygons in which the value of HtClass is either "m" or "l".

```
> # Create the Medium/Low data frame with ID data field
> MLArea <- data.frame(PatchID = data.Set1.sf$PatchID)
> # Initially set ML area of each patch to 0
> MLArea$Area <- 0
> # Insert the patch area if the height class is m or l
> for (i in 1:nrow(MLArea)){
+    {if ((data.Set1.sf$HtClass[i] == "m") |
        (data.Set1.sf$HtClass[i] == "l"))
+        MLArea$Area[i] <- data.Set1.sf$HtClArea[i]}}
```

The process is then repeated for the data frame HArea to hold nonzero Area values for all those polygons in which the value of HtClass is "h".

```
> HArea <- data.frame(PatchID = data.Set1.sf$PatchID)
> HArea$Area <- 0
> for (i in 1:nrow(HArea)){
+    {if ((data.Set1.sf$HtClass[i] == "h"))
+        HArea$Area[i] <- data.Set1.sf$HtClArea[i]}}
```

These data frames are then passed as denom and num respectively to the function AreaRatio() to create the height class ratio value in Table 7.1. Here is the function.

```
> AreaRatio <- function(num, denom){
+ # Add the num argument values by PatchID
+    patch.num <- aggregate(num,
+        by = list(AggID = num$PatchID), FUN = sum)
+ # Add the denom argument values by PatchID
+    patch.denom <- aggregate(denom,
+        by = list(AggID = denom$PatchID), FUN = sum)
+ # If the num and denom PatchID values are not equal, return NULL
+    ratio <- NULL
```

```
+ # Otherwise, compute the ratio (and avoid 0/0)
+     if (all.equal(patch.num$PatchID, patch.denom$PatchID)){
+         ratio <- patch.num$Area /
+             (patch.num$Area + patch.denom$Area + 0.000001)}
+     return(cbind(patch.num$AggID, ratio))
+}
```

The function `AreaRatio()` takes two arguments associated with a patch, num and denom, and computes the total area of the patch in the category num divided by the sum of the areas in num and denom. The R function `aggregate()` in the third and sixth lines does exactly what its name implies; it aggregates the contents of the data frame in its first argument by the data field in the second argument using the function defined in the third argument. The function `all.equal()` returns FALSE if all of the elements of the two arrays in the argument are not equal, in which case the function `AreaRatio()` returns NULL. Note that the argument num actually appears in both the numerator and the denominator of the ratio, since the ratio being computed is actually a faction of the total area. The value 0.000001 is added to the denominator so that a value of zero is returned in the case that the aggregated data in both num and denom are equal to zero.

The use of the function `AreaRatio()` is illustrated in the following code sequence, which computes the height class ratio in Table 7.1.

```
> Ratio <- AreaRatio(HArea, MLArea)
> HtRatio.df <- data.frame(PatchID = Ratio[,1], HtRatio = Ratio[,2])
> Set1.obs1 <- merge(x = Set1.obs, y = HtRatio.df,
+     by.x = "PatchID", by.y = "PatchID")
```

The object `Ratio` in the first line is a matrix whose first column contains the value of `PatchID` for each habitat patch and whose second column contains the height class ratio for that patch. This matrix is converted into a data frame. The function `merge(x, y, by.x, by.y)` in the third line performs what is called in GIS terminology a *join by attribute values* (Lo and Young, 2007, p. 210). That is, the function `merge()` returns a data frame in which all records in data frame x with a given value of `by.x` (which must be a data field of x) are merged (or joined) to those records in data frame y having the same value in the data field `by.y`. This creates the object `Set1.obs1`, which contains all of the data fields of `Set1.obs` plus the height class ratio.

The process is repeated for the floodplain age ratio (Exercise 7.3), creating `Set1.obs2`, and again for the cover class ratio, creating the final object `Set1.obs3`.

Now that we have created a data frame to hold the explanatory and response variables for each of the observation locations, we can begin to explore the data set. A quick check can be carried out by computing box plots of each of the explanatory variables. We can do this quickly using `ggplot()`.

```
> ggplot(data = Set1.obs3) +
+     geom_boxplot(mapping = aes(as.factor(PresAbs), PatchArea))
```

The function `geom_boxplot()` expects to see the abscissa variable as a factor, so we coerce it. The code to produce the presentation graphics boxplot for *PatchArea* is shown in Figure 7.4a.

```
> with(Set1.obs3, boxplot(PatchArea ~ PresAbs)) # Fig. 7.4a
> title(main = "Patch Area", cex.main = 2,
+     xlab = "Presence/Absence",
+     ylab = expression(Area~"("*m^2*")"), cex.lab = 1.5)
```

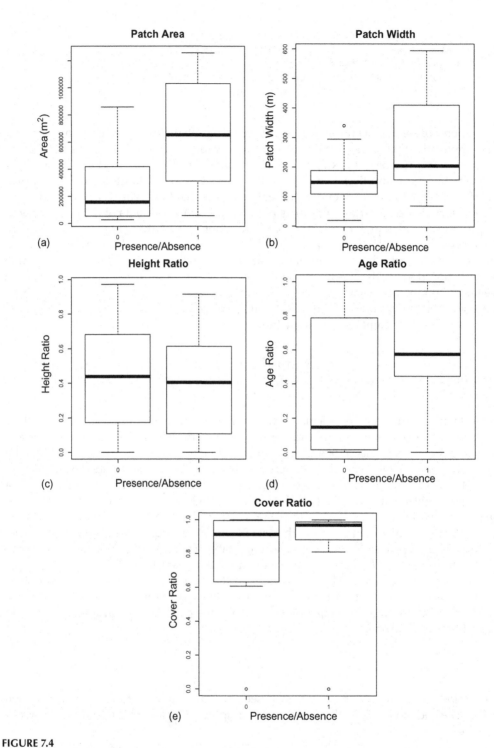

FIGURE 7.4
Boxplots showing distribution of (a) patch area, (b) patch width, (c) height ratio, (d) floodplain age ratio, and (e) cover class ratio for patches with (1) and without (0) cuckoos observed.

The other plots are created in a similar way. The heavy line inside the box indicates the median and the upper and lower extremes of the box indicate the values of the first and third quartile. The whiskers extend to the most extreme data points whose distance from the box is no more than 1.5 (this is the default value, which can be altered) times the interquartile range.

Examining these boxplots (Figure 7.4) reveals that there is considerable overlap in attribute values between occupied and unoccupied patches, and that *HtRatio*, the ratio of area of tall trees to total area, seems to play the least important role in distinguishing cuckoo presence and absence. The boxplots of *CoverRatio* are artificially compressed by one data record in which it was not measured, but this variable also does not appear to play a major role. Patch area and patch width have slightly different boxplots. The median patch area in those patches where birds are present is higher than that where birds are absent. There are also no narrow patches with birds present, but the medians of those with and without birds are fairly close. The minimum and maximum age ratios of patches with and without birds are similar, but the median age ratio of the patches with birds present is higher.

An alternative view of the data can be obtained using a pair of connected dot plots (Figure 7.5), which resemble somewhat the figure on page 49 of Tufte (1983). The graphs in Figure 7.5 are produced with the traditional graphics function `matplot()`. The function `matplot(x, y)`, where x and y are matrices, plots the columns of y against the columns of x. If there is only one argument, then this is plotted as the y variable. There are other options, which can be seen via `?matplot`. The code to produce Figure 7.5a is as follows.

```
> row.names(Set1.obs3) <- as.character(1:nrow(Set1.obs3))
> obs.trans <- data.frame(t(scale(with(Set1.obs3,
+     cbind(HtRatio, AgeRatio, CoverRatio, PatchArea, PatchWidth)))))
> n.trans <- nrow(obs.trans)
> obs.trans[n.trans + 1,] <- t(Set1.obs3$PresAbs)
> obs.YPres <- obs.trans[1:n.trans, (obs.trans[n.trans+1,] == 1)]
> obs.YAbs <- obs.trans[1:n.trans, (obs.trans[n.trans+1,] == 0)]
```

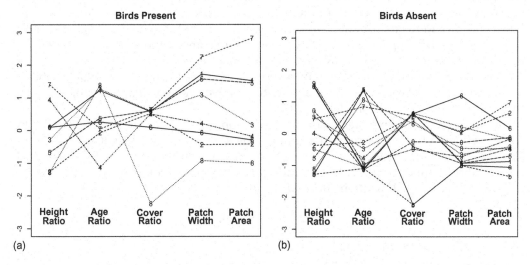

FIGURE 7.5
Connected dot plots of values of the predictor variables for observations with (a) birds present; (b) birds absent.

```
> matplot(obs.YPres, type = "o", col = "black",
+     ylim = c(-3,3), main = "Birds Present", ylab = "",
+     cex.main = 2, xaxt = "n") # Fig. 7.5a
> text(1.1,-2.5, "Height", font = 2, cex = 1.5)
> text(1.1,-2.75, "Ratio", font = 2, cex = 1.5)
     *   *   *   DELETED   *   *   *
> text(4.8,-2.5, "Patch", font = 2, cex = 1.5)
> text(4.8,-2.75, "Area", font = 2, cex = 1.5)
```

The second line applies three functions in succession. The function `scale()` centers and scales each of the explanatory data fields so that they all have zero mean and unit variance. The function `t()` creates a matrix that is the transpose of its argument, so that the data for each site is in a column. The function `data.frame()` then converts this matrix to a data frame. The next line augments the data frame by adding as the last row the presence-absence data, which is not centered and scaled. The next two lines separate the data into those where birds are present and those where birds are absent. The plot is constructed in the usual way with traditional graphics. The function `matplot()` is applied; the argument `xaxt = "n"` suppresses the *x* axis, which would consist of meaningless numbers, and instead the appropriate text is manually added.

The properties of the data in Figures 7.4 and 7.5 indicate that the explanatory data themselves may be highly correlated among themselves, and this is indeed the case.

```
> with(Set1.obs3, (cor(cbind(HtRatio, AgeRatio,
+     CoverRatio, PatchArea, PatchWidth))))
              HtRatio    AgeRatio CoverRatio PatchArea PatchWidth
HtRatio     1.0000000 -0.5313764  0.4829030 0.3103738  0.3732220
AgeRatio   -0.5313764  1.0000000 -0.2243593 0.1080622  0.1294009
CoverRatio  0.4829030 -0.2243593  1.0000000 0.5314379  0.5538397
PatchArea   0.3103738  0.1080622  0.5314379 1.0000000  0.8836189
PatchWidth  0.3732220  0.1294009  0.5538397 0.8836189  1.0000000
```

In particular (and not surprisingly), *PatchArea* and *PatchWidth* are highly correlated. However, the other important variable as indicated by Figure 7.5 is *AgeRatio*, and this is not highly correlated with either of the patch size variables. Again, there appears to be little difference in the distributions of *HtRatio* or *CoverRatio* between occupied and unoccupied patches. In comparing properties of patches with and without cuckoos present, we must bear in mind that the CWHR model is of habitat *suitability*, not of habitat *occupancy*. Thus we might expect, particularly with a very rare species, to find some suitable habitats unoccupied, but we would regard as a potential error in the model any occupied habitat that is unsuitable under the model.

The final step in the exploration process is to apply a preliminary test to the CWHR model of Table 7.1. We will construct simple 2 by 2 contingency tables (Chapter 11, see also Larsen and Marx, 1986, p. 424) of cuckoo presence/absence vs. habitat suitability/unsuitability. A function `suitability.score()` is defined to compute suitability scores for patch area and patch width in Table 7.1

```
> suitability.score <- function(x, cat.val){
+     score <- 0
+     if(x >= cat.val[3]) score <- 1
+     if(x < cat.val[3] & x >= cat.val[2]) score <- 0.66
```

```
+      if(x < cat.val[2] & x >= cat.val[1]) score <- 0.33
+      return(score)
+ }
```

This function is applied to compute the suitability scores for patch area as follows. From Table 7.1, the threshold patch areas for *Optimum*, *Suitable*, and *Marginal* are 80, 40, and 17 ha, respectively. This motivates the following code.

```
> AreaScore <- numeric(nrow(Set1.obs3))
> for (i in 1:nrow(Set1.obs3)){
+   AreaScore[i] <-
+      suitability.score(Set1.obs3$PatchArea[i] / 10000, c(17,40,80))
+ }
> Set1.obs3$AreaScore <- AreaScore
```

The same function can be used to compute the suitability scores for patch width. Similar functions are used to compute scores for height ratio and floodplain age ratio. These cannot be computed using the function `suitability.score()` because the suitability values for these variables in Table 7.1 are not monotonic (i.e., strictly increasing or strictly decreasing) (Exercise 7.4). Figure 7.6 shows a plot of the suitability score of the age ratio.

After the suitability scores are computed, the function `apply()` is next used to compute as an overall score the geometric mean of the individual scores (recall that this is used in order to emphasize the suitability of patches that score high in all explanatory variables).

```
> scores <- with(Set1.obs3, cbind(AreaScore, WidthScore,
+   AgeScore, HeightScore))
> print(Set1.obs3$HabitatScore <- apply(scores[,1:4], 1, prod)^(1/4),
```

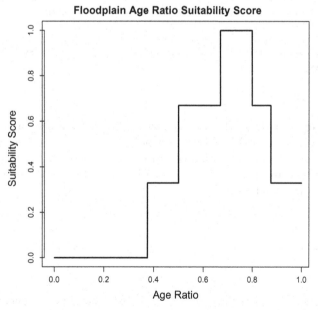

FIGURE 7.6
Plot of the habitat suitability score as a function of floodplain age ratio.

```
+    digits = 2)
 [1]  0.68 0.00 0.33 0.00 0.62 0.00 0.00 0.00 0.00 0.00 0.66 0.00
[13]  0.73 0.52 0.00 0.00 0.00 0.00 0.00 0.00 0.00
```

In this exploration, we set a very low bar: any habitat that is not unsuitable will be considered potentially suitable. Therefore, we define a habitat suitability index that takes on the value 1 if the habitat score is positive.

```
> print(Set1.obs3$HSIPred <- as.numeric(Set1.obs3$HabitatScore > 0))
 [1] 1 0 1 0 1 0 0 0 0 0 1 0 1 1 0 0 0 0 0 0 0
```

The northernmost observation point in Patch 191, for which we lack floodplain age data, is excluded.

```
> Set1.corrected <- Set1.obs3[-which(Set1.obs3$PatchID == 191),]
```

Finally, we compute and display a preliminary contingency table from the following data.

```
> print(habitat.score <- as.numeric(HabitatScore > 0))
 [1] 1 0 1 0 1 0 0 0 0 0 1 0 1 1 0 0 0 0 0 0
> print(obs <- scores$PresAbs)
 [1] 1 0 1 0 1 0 0 1 0 0 0 0 1 1 0 0 1 0 0 0
```

The contingency table indicates the occupancy of patches that are unsuitable and those that are in some measure suitable. Larger contingency tables are best done with the function `table()` (see Exercise 7.10), but we will construct this one by hand.

```
> UA <- with(Set1.corrected, which(HSIPred == 0 & PresAbs == 0))
> UP <- with(Set1.corrected, which(HSIPred == 0 & PresAbs == 1))
> SA <- with(Set1.corrected, which(HSIPred == 1 & PresAbs == 0))
> SP <- with(Set1.corrected, which(HSIPred == 1 & PresAbs == 1))
> print(cont.table <- matrix(c(length(SP),length(SA),
+    length(UP),length(UA)), nrow = 2, byrow = TRUE,
+    dimnames = list(c("Suit.", "Unsuit."),c("Pres.", "Abs."))))
        Pres. Abs.
Suit.       5    1
Unsuit.     2   12
```

There are two occupied patches (*PatchID* = 100 and 209) with a zero suitability score, and one unoccupied patch (*PatchID* = 122) with a positive suitability score. Some interpretation of this contingency table is carried out in Exercise 7.5. As a last step of preliminary analysis, you are asked in Exercise 7.6 to determine the effect on the contingency table of removing explanatory variables.

7.3 Data Set 2

Data Set 2 (Figure 1.1) consists of records from 4,101 locations surveyed as a part of the Vegetation Type Map (VTM) survey in California, carried out during the 1920s and 1930s (Wieslander, 1935; Jensen, 1947; Allen et al., 1991). At each location an 81 m^2 (9 m by 9 m) area of land was surveyed. All non-climatic variables in Table 7.2 were recorded. Allen and

TABLE 7.2

Summary of the Variables Included in Data Set 2

Variable	Type	Description
Response variables:		
QUDO	N	Presence/absence of *Quercus douglasii* (Blue oak).
QUDO_BA	R	Basal area of *Quercus douglasii* (Blue oak).
Predictor variables:		
Elevation	R	Elevation (m)
CoastDist	R	Great circle distance from coast (km)
Precip	R	Mean annual precipitation (mm)
MAT	I	Mean annual temperature (C)
JaMin	I	Mean minimum temperature in January (C)
JaMax	I	Mean maximum temperature in January (C)
JaMean	I	Mean temperature in January (C)
JuMin	I	Mean minimum temperature in July (C)
JuMax	I	Mean maximum temperature in July (C)
JuMean	I	Mean temperature in July (C)
TempR	R	Annual temperature range (C) i.e., JUMEAN–JAMEAN
GS28	R	Length of 28°F growing season (days)
GS32	R	Length of 32°F growing season (days)
PE	R	Potential evapotranspiration (mm)
ET	R	Actual evapotranspiration (mm)
SolRad6	R	Potential (cloudless) solar radiation on the average day in June (MJ/m^2)
SolRad12	R	Potential (cloudless) solar radiation on the average day in December (MJ/m^2)
SolRad	R	Yearly potential (cloudless) solar radiation (MJ/m^2)
Texture	O	Soil surface texture (from 0: rock/gravel to 5: clay; 6: no data)
AWCAvg	R	Average (over components) of available water capacity for soil map unit (mm)
PM100	N	Parent materials in 100 series: igneous intrusive rocks
PM200	N	Parent materials in 200 series: igneous extrusive, flow rocks
PM300	N	Parent materials in 300 series: igneous extrusive, pyroclastic rocks
PM400	N	Parent materials in 400 series: metamorphic rocks
PM500	N	Parent material in 500 series: soft sedimentary rocks, soft
PM600	N	Parent material in 600 series: hard sedimentary rocks

co-workers entered these data into an electronic database (Allen et al., 1991). Data Set 2 is a subset of the full VTM database and includes all locations at which at least one oak species was recorded. Evett (1994) added geographic coordinates and climatic data to this database. Records in the original database include presence-absence data for six oak species, but we will concern ourselves exclusively with one of these: blue oak (*Quercus douglasii* Hook. & Arn.) This is one of the oak species considered to be in severe population decline (Zavaleta et al., 2007). Allen et al. (1991) used TWINSPAN (Two-Way Indicator Species Analysis) (Hill, 1979a) and DECORANA (Detrended Correspondence Analysis) (Hill, 1979b) to carry out a classification based on these data. Evett (1994) used direct gradient analysis (Austin et al., 1990) to construct models of the environmental niche of all six oak species.

The VTM survey achieved fairly complete coverage of the regions of California containing oak species, although portions of the southern Sierra Nevada are not included. Because the data set developed by Allen et al. (1991) and Evett (1994) includes only those sites on which at least one oak was recorded, the question we address in our analysis of this data set is not what characterizes suitability of sites in California for blue oak, but rather what characterizes suitability for blue oak of those sites in California in which some species of oak is observed.

Pavlik et al. (1991, p. 16) describe the blue oak as an extremely drought-tolerant species that populates the foothills of the Central Valley (visible in Figure 1.1 as the large valley in the center of California) at elevations less than 1,000 m. Annual precipitation in these regions ranges from 500 to 1,000 mm and soils are generally not well developed. McClaran (1986), citing Jepson (1910), describes blue oak woodland as being characterized by its exclusivity, with a complete lack of other tree species, with the occasional exception of foothill pine (*Pinus sabiniana*). In the analysis of Data Set 1, we had to take into account the distinction between habitat suitability for the yellow-billed cuckoo and presence/absence of the cuckoo, recognizing that cuckoos may be absent from suitable habitat. In the case of Data Set 2, every site contains at least one species of oak, and therefore the geographic exclusivity of blue oak woodland renders the issue of habitat suitability versus species presence less important. An important feature of Data Set 2 is that although the data are all recorded as points, they are not all measured at the same spatial scale. The climatic variables were derived by Evett (1994) using the procedure described in Appendix B.2 and are based on climatic measurements that are made at a scale that, although it is not well defined, is considerably larger than that of the non-climatic variables.

We can get a good overview of the data relationships via the spatial features plot() function. Appendix B.2 indicates that Data Set 2 contains several quantities that may influence blue oak growth. We select for initial exploration elevation (Elevation), mean annual temperature (MAT), evapotranspiration (ET), total precipitation (Precip), and soil texture class (Texture). Here is the code to set the data up and invoke the plot function.

```
> data.Set2 <- read.csv("set2\\set2data.csv", header = TRUE)
> x <- with(data.Set2, cbind(Longitude, Latitude, QUDO, Elevation, MAT,
+    ET, Precip, Texture))
> data.plot <- data.frame(x)
> data.plot.sf <- st_as_sf(data.plot, coords = c("Longitude",
+    "Latitude"))
> st_crs(data.plot.sf) <- "+proj=longlat +datum=WGS84"
> plot(data.plot.sf)
```

As always with these initial plots, the intent is to be quick and not to generate a publication quality figure, so the plot is not displayed here. Instead, Figure 7.8 below shows nice-looking figures of the same data (the code is included in the R file but not shown here). Examination of the output of the plotted spatial features object (or of Figure 7.8) indicates that, especially in the Sierra Nevada, where the maximum elevation is higher, there is a close relationship between elevation and blue oak presence. Therefore, we begin by examining in more detail the relationship between elevation and blue oak presence and absence.

Again in preliminary exploration we can use the function ggplot() to get a quick look at the relationship between elevation and blue oak presence, and how this interacts with other environmental variables. When we only plot *QUDO* against *Elevation* we don't see much (try it!), but when we add a call to geom_smooth(), which fits a smooth curve to the data, we see that blue oak presence increases with elevation until about 400 m and then declines. We will return to this in Section 9.2. We can also visualize the interaction between elevation and other variables in their influence on blue oak presence. This is most effectively done if we introduce a factor with the same values as *QUDO*. For example, the code sequence

```
> data.Set2$QF <- as.character(data.Set2$QUDO)
> ggplot(data = data.Set2) +
+    geom_point(aes(x = Elevation, y = MAT, color = QF)) +
+    geom_smooth(aes(x = Elevation, y = MAT))
```

produces a graph that summarizes how mean annual temperature and elevation interact to influence blue oak presence. The use of the geom `geom_histogram(aes(data.`
`Set2$Elevation))` or the base graphics function `hist(data.Set2$Elevation)`
produces a histogram that indicates that the majority of data locations lie between about
250 and 1000 m, and the statement `max(data.Set2$Elevation)` reveals that the highest elevation measurement was taken at 2240 m. With this general knowledge, we can
examine the effect of elevation on blue oak presence. The first step is to aggregate the presence and absence data at 100 m intervals. We could use the R function `aggregate()` to
do this, but it is probably a case where a `for` loop is more transparent than functional programming. We first write a function that allows us to compute the number of occupied or
unoccupied sites of any oak species at any range of another variable in the data set.

```
> n.oaks <- function(var.name, oak, low, high, PresAbs){
+ length(which(var.name >= low &
+ var.name < high & oak == PresAbs))}
```

Next, we aggregate the presence and absence data.

```
> pres <- numeric(22)
> absent <- numeric(22)
> for(i in 0:22) pres[i] <- with(data.Set2,
+     n.oaks(Elevation, QUDO, i*100,(i+1)*100, 1))
> for(i in 0:22) absent[i] <- with(data.Set2,
+     n.oaks(Elevation, QUDO, i*100,(i+1)*100, 0))
```

Now we are ready to plot the fraction of occupied sites versus elevation.

```
> pa <- pres / (pres + absent)
> x <- seq(100,2200,100)
> par(mai = c(1,1,1,1))
> plot(x, pa, type = "o")
> title(xlab = "Elevation (m)", ylab = "Portion with Blue Oak",
+    cex.lab = 1.5, main = "Blue Oak Presence vs. Elevation",
+    cex.main = 2)
```

If you ran the `ggplot()` code, you have already seen that the portion of sites having a
blue oak, when aggregated over the entire data set, declines almost linearly with elevation
(Figure 7.7).

As mentioned in Section 7.1, the trees are probably not directly sensitive to elevation. It is
likely that elevation is serving as an exogenous variable in the sense of Figure 7.1, and that
the direct influence on blue oak presence or absence is one or more endogenous variables
associated with elevation. One possibility is that climate has an influence and a second,
not mutually exclusive, is that soil texture is important. Figure 7.8 shows thematic maps of
blue oak presence/absence, elevation, mean annual precipitation, mean annual temperature, and soil texture class (the pretty versions of the earlier plotted spatial features object).
The code is similar to that used to construct Figure 1.4 and is described in Section 2.6.2. The
maps indicate an apparent association between climate variables and blue oak presence.

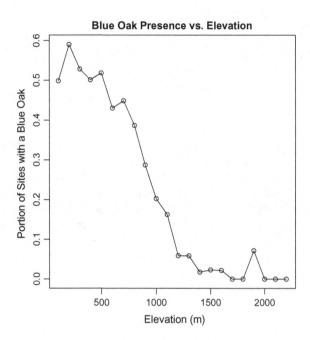

FIGURE 7.7
Plot of the fraction of sites in Data Set 2 at which a blue oak is present vs. elevation.

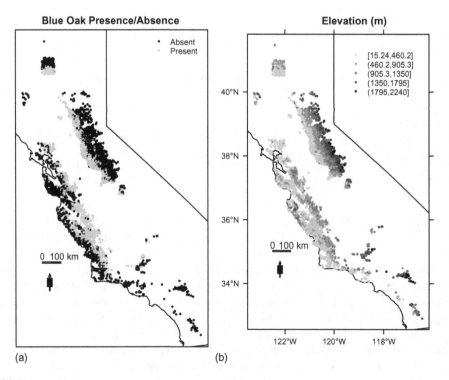

FIGURE 7.8
Thematic maps of the data set showing (a) blue oak presence/absence, (b) elevation.   (*Continued*)

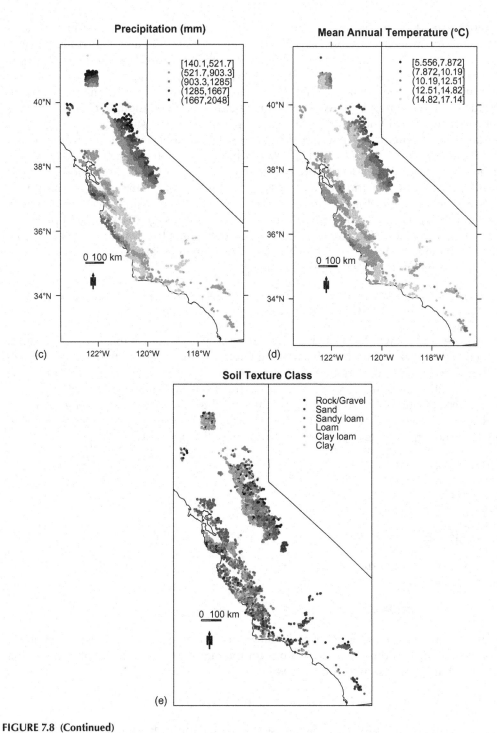

FIGURE 7.8 (Continued)
Thematic maps of the data set showing (c) mean annual precipitation, (d) mean annual temperature, (e) soil texture class at each of the sample sites.

Another issue that we can now examine is that of "secondary" spatial variables. One of the explanatory variables in Table 7.2 is *CoastDist*, the distance in kilometers from the nearest point on the coast. The maps in Figure 7.8 indicate that this might be a more appropriate east–west spatial variable than longitude, since clearly the ocean might have a strong influence. This is a situation similar to that encountered by Venables and Dichmont (2004) in their study of tiger prawn fisheries off the Australian coast. Even though it is not a climate variable per se, we will include *CoastDist* in the preliminary analysis since it may play an important role as a predictive quantity. In this context it is important to recognize that exogenous variables such as *CoastDist* and *Elevation*, while they may have great predictive value, have no explanatory value except insofar as they might lead to the identification of important endogenous variables through a process of scientific, rather than statistical, analysis.

Let's look at how these climate variables relate to each other by using a scatterplot matrix (Cleveland, 1985). The function `splom()` in the `lattice` package (Sarkar, 2008) plots one directly. We first create a data frame of the attribute data using the `data.frame()` function. Then we separate out the climate data. Not all of the climate-related data are included; we mostly focus on temperature and precipitation.

```
> climate.data <- with(data.Set2@data, data.frame(Precip, MAT, JaMin,
+     JaMax, JaMean, JuMin, JuMax, JuMean, GS32, CoastDist))
```

While we are at it, we would like to make the font of the main title of the scatterplot matrix twice the nominal size, as we have been doing with our other plots. With the `plot()` function this is done using the argument `cex.main = 2`, either in `plot()` itself or in the `title()` function. In trellis plots with `lattice` functions, we must use another approach. First, we need to find out what the term is that is used to control the plot settings. An easy way that usually works (and is often faster than wading through the documentation) is to apply the function `trellis.par.get()` to list all of the parameter settings.

```
> library(lattice)
> trellis.par.get()
$grid.pars
list()
$fontsize
$fontsize$text
[1] 12
    *    *    *    DELETED    *    *    *
$par.main.text$cex
[1] 1.2
    *    *    *    DELETED    *    *    *
```

Scrolling through a very long list of settings, we come to par.main.text$cex set at 1.2. This looks like it should govern the settings for the main text of the plot (and indeed that is exactly what it does), so we will set its value to 2.

```
> trellis.par.set(par.main.text = list(cex = 2))
```

This parameter setting is permanent (unless we change it) for the remainder of the R session. This is all right because we always want our main text to have this `cex` value, but trellis parameter settings can also be achieved for one time only by including them as arguments to the trellis plotting function. We can use this technique for some settings in our application of `splom()`.

```
> splom(climate.data, par.settings = list(fontsize=list(text=9),
+   plot.symbol = list(col = "black")), pscales = 0,
+   main = "Climate Data") # Fig. 7.9a
```

Arguments in `lattice` functions are often in the form of lists, for example, `plot.symbol = list(col = "black")`. The argument `pscales = 0` suppresses printing of scales in the diagonal boxes.

The scatterplot matrix (Figure 7.9a) reveals some very tight linear relations, but also some scatterplots, such as the one relating `JaMean` to `JuMean`, that seem to be a combination

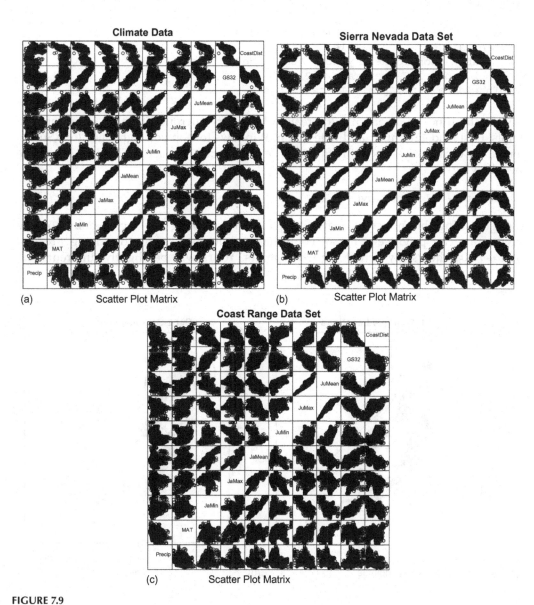

FIGURE 7.9
Scatterplot matrices of Data Set 2 climate data for (a) the full data set; (b) the Sierra Nevada subset; (c) the Coast Ranges subset.

of two or more relationships. Examination of Figure 7.8 indicates that the data sites are located in four separate groups, each of which is in a separate mountain range. The small, square northernmost group is in the Klamath Range. The two long groups running north to south are in the ranges on either side of the Central Valley. The eastern group is in the Sierra Nevada, and the western group is in the Coast Range. The small, isolated groups in the southern inlands are in the Traverse Range (http://en.wikipedia.org/wiki/Geography_of_California). The climatic conditions of these ranges are different, and it is reasonable to expect that at least some of the data relationships will be different as well. To investigate this, we will start by separating out the Sierra Nevada locations. First draw a map of the data set.

```
> plot(cal.poly, axes = TRUE) # Fig. 7.10a
> title(main = "Set 2 Sample Locations",
+    xlab = "Longitude", ylab = "Latitude",
+    cex.lab = 1.5, cex.main = 2)
> points(data.Set2, pch = 1, cex = 0.4)
```

We use the function `locator()` to create a boundary surrounding this group of data points. First read `?locator` and note that in Windows to stop you right click (in RStudio you hit the escape key—I actually find the use of `locator()` easier in TINN-R). Now type the following

```
> bdry.pts <- locator(type = "o")
```

Then, by using the mouse as a digitizing tool and clicking at selected points on the screen, one can produce a list of coordinates generated by the digitizing process. Figure 7.10a shows my actual digitized points. Yours will be slightly different. The

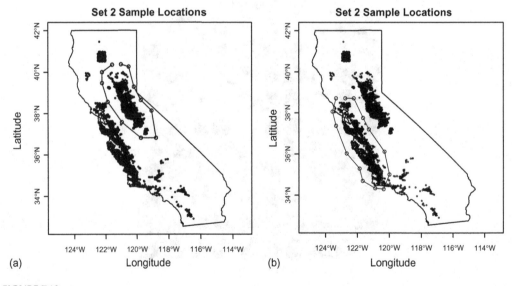

(a)

(b)

FIGURE 7.10
Boundary points drawn around the data values in (a) the Sierra Nevada and (b) the Coast Range using the function `locator()`.

argument `type = "o"` causes the digitizing to be displayed as in the figure. Note that the polygon is not closed. Next, the function `unlist()` is used to convert the list into a vector, and then in the same line the function `matrix()` converts the vector into a matrix with two columns.

```
> coords.mat <- matrix(unlist(bdry.pts), ncol = 2)
```

The polygon is closed by appending the coordinates of the first point.

```
> coords.mat <- rbind(coords.mat, coords.mat[1,])
```

The `sf` polygon object is then created using the sequence of steps described in Section 2.4.3.

```
> coords.lst <- list(coords.mat)
> coords.pol = st_sfc(st_polygon(coords.lst))
> bdry.sf = st_sf(z = 1, coords.pol)
> proj4string(bdry.spdf) = CRS("+proj=longlat +datum=WGS84")
```

A `SpatialPolygonsDataFrame` called `bdry.spdf` representing the boundary is created by coercion from the `sf` object. The data points located in the Sierra Nevada are extracted from the full data using the `sp` function `over()` to do a point in polygon operation.

```
> data.ol <- over(data.Set2,bdry.spdf)
> region.spdf <- data.Set2[which(is.na(data.ol$ID) == FALSE),]
```

This process was used to create separate data frames containing data from the Sierra Nevada and the Coast Range (Figure 7.10b) and to generate the scatterplot matrices in Figure 7.9b and c. The decision of where to establish the southern boundary of the Coast Range was somewhat subjective.

It is evident that the relationship between climatic variables is very different in the two mountain ranges. It is also evident that a very close linear relationship exists between climatic variables in the Sierra Nevada, and that the relationship is more complex in the Coast Range. If the relationship among climatic variables is different between the Sierra Nevada and the Coast Range, then it may be that the relationship between elevation and presence/absence of blue oaks may also be different. Figure 7.11 confirms that this is indeed the case. The fraction of blue oak sites declines steadily with elevation in the Sierra Nevada, while it actually increases slightly in the Coast Range.

At this point, we come to a fork in the road. The scatterplot matrix of Figure 7.9a indicates that a regression model that incorporates all of the data records will be quite complex, involving higher order terms and interactions. The scatterplot matrices of Figure 7.9b and c imply that regression models restricted to these geographic regions may be much simpler. We must therefore decide whether to analyze the entire data set or split it up and analyze regions separately. There are at least two primary arguments in favor of analyzing the entire data set. First, although the regression model may be complex, it may also represent real biophysical phenomena. For example, blue oaks apparently prefer xeric environments, but it is likely that they would not be found in completely dry areas. Therefore, one might expect that in an area that ranged from very dry to very moist locations, a regression model for blue oak presence might include a quadratic term that first increased and then decreased. The region could be split into xeric and mesic sub-regions in which the portion of sites with a blue oak was respectively increasing and decreasing, but this would not capture the biophysical process as well as a model of the entire region. In any case,

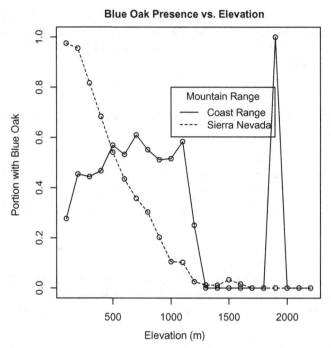

FIGURE 7.11
Plot of blue oak presence vs. elevation for the Sierra Nevada and Coast Ranges.

a spatial data set should never be split unless the two sub-regions are spatially contiguous and meaningful. A second, more practical argument in favor of analyzing the entire data set involves degrees of freedom. Splitting the data set in two means that each of the sub-regions will have only about half the number of observations, and in some cases this could have a substantial effect on the analysis by reducing the amount of available data.

Arguments in favor of a split include the following. In general, simpler models are easier to interpret and more reliable than complex ones. If the sub-regions are governed by different biophysical processes (e.g., if temperature is important in one sub-region and soil texture in the other) then two separate models may be more easily interpreted than a single model with an interaction term. Moreover, comparing and contrasting the models in the context of their geography may provide further insights into the processes they describe.

In the case of Data Set 2, with over four thousand data records, the degrees of freedom issue is not important. Examination of the scatterplots in Figure 7.9 indicates that some of the explanatory variables have a nonlinear relationship over the whole region, but the relationship is better approximated as linear over the Sierra Nevada and to a lesser extent the Coast Range by themselves. If we look at the elevation map in Figure 7.8b, we can get a sense of the geographical relationships. The elevation in the Sierra Nevada gradually increases in a more or less uniform manner from west to east. The terrain in the Coast Range is more complex. As one moves inland from the coast, elevation increases rapidly and then declines as one moves into the Salinas Valley (made famous in the works of John Steinbeck), as well as other, smaller valleys that do not show up in Figure 7.8b. Elevation rises rapidly again on the east side of the Salinas Valley, and then declines rapidly as one enters the Central Valley. The prevailing wind is from the west off the Pacific Ocean, and the terrain influences precipitation patterns. Precipitation in the Sierra Nevada generally increases with elevation. Precipitation in the Coast Ranges also tends to increase with elevation, but there is an apparent rain shadow effect on the eastern side and increased precipitation on the western side (Figure 7.8c).

It is clear that precipitation is strong associated with blue oak habitat suitability. What is not so clear right now is whether the relationship between precipitation and other climatic factors is the same in both the Sierra Nevada and the Coast Range. It does not appear to be. There may be some insight to be gained in comparing separate models for these two systems, and we can hold back the Klamath Range and the Transverse Ranges for validation purposes. Therefore, we make the decision to split the data and model the Sierra Nevada and the Coast Ranges separately. To analyze the presence-absence data, we create vectors `pa.coast` and `pa.sierra`, each housing values in groups by 100 m in elevation. Here is the code to create `pa.coast`.

```
> pres <- numeric(22)
> absent <- numeric(22)
> for(i in 0:22) pres[i] <- with(coast.sf,
+     n.oaks(Elevation, QUDO, i*100,(i+1)*100, 1))
> for(i in 0:22) absent[i] <- with(coast.sf,
+     n.oaks(Elevation, QUDO, i*100,(i+1)*100, 0))
> pa.coast <- pres / (pres + absent + 0.00001)
```

The relationship between blue oak presence/absence and climatic conditions is clearly complex, and we will delay further exploration until later chapters. The relationship among factors not included in Figure 7.9 is not very well organized. For our final exploration step of Data Set 2 in this chapter, we plot blue oak presence/absence against precipitation (Figure 7.12a) and soil texture category (Figure 7.12b). Soil texture category is an ordinal scale variable indicating the fineness of the soil particles, from 0 (rocky) to 5 (clay). Figure 7.12a is consistent with the statement of Pavlik et al. (1991, p. 16) that blue oaks are extremely drought tolerant but are out-competed by other oak species on more mesic soils. Figure 7.12b is somewhat of a surprise in the context of a statement by McDonald (1990) that blue oaks tend to be found in rocky, poorly developed soils. This apparent contradiction was also noticed by Evett (1994, p. 91).

Because soil texture is an ordinal scale variable, we plot it using a bar chart to avoid (at least for now) the idea that we might compute something like a regression line against this variable. Actually, we shall see in a later chapter that this is not always such a bad idea. In any case, the code to produce Figure 7.12b is

```
> pa <- rbind(pa.coast, pa.sierra)
> barplot(pa, beside = TRUE, # Fig. 7.12b
+    names = c("0", "1", "2", "3", "4", "5"),
+    ylim = c(0,1), xlab = "Texture Class",
+    ylab = "Percent Occupied Sites", cex.lab = 1.5,
+    legend.text = c("Coast", "Sierra"),
+    main = "Percent Blue Oak Occupation vs. Texture",
+    cex.main = 1.5)
```

Occupancy in soils in the intermediate texture class 2 is clearly different between the Coast Ranges and the Sierra Nevada, but the difference is not so obvious for the finer textured soils (classes 3, 4, and 5). Blue oak presence also declines with increasing permeability class (Figure 7.12c). However, texture class and permeability class are not closely related (Exercise 7.10).

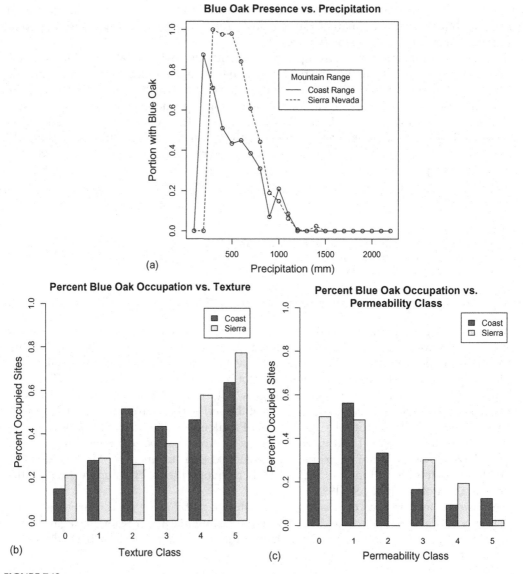

(a)

(b)

(c)

FIGURE 7.12

Plot of blue oak presence vs. (a) precipitation, (b) soil texture class, and (c) soil permeability class.

Figure 7.12a suggests that something affects the response of blue oaks to precipitation, which differs between the Sierra Nevada and the Coast Range. One possibility is temperature. We will carry out a last exploration step that allows us to introduce the function hexbin() from the package of the same name (Carr et al., 2016). This provides a convenient means to visualize bivariate frequency plots. Here is the code to generate a hexagon bin frequency plot of precipitation and mean annual temperature in the Coast Range.

```
> library(hexbin)
> plot(hexbin(coast.sf$Precip, coast.sf$MAT),
+     xlab = "Precipitation", # Fig. 7.13
```

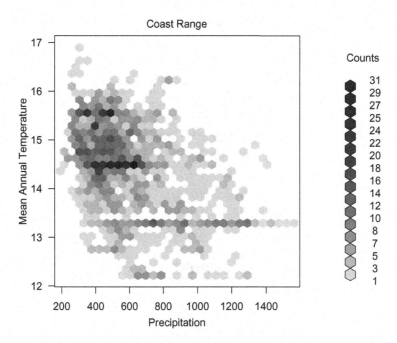

FIGURE 7.13
Hexagonal bin frequency plot of *Precip* and *MAT* in Data Set 2.

```
+    ylab = "Mean Annual Temperature",
+    main = "Coast Range")
```

Figure 7.13 shows the result. There is a bimodal distribution of *MAT* over a range of *Precip* between 200 and 700 mm. The effect is even more pronounced with the variable *JuMax* (Exercise 7.11).

In summary, we have found that blue oaks differ in their response to elevation between the Coast Range and the Sierra Nevada, with a more uniform distribution with respect to elevation in the Coast Range. Precipitation is clearly a limiting factor, with virtually no blue oak presence at sites with an annual precipitation greater than 1250 mm. The response to precipitation, however, appears to be different in the Coast Range from that in the Sierra Nevada. This may have something to do with temperature differences between the mountain ranges. Blue oaks appear to have a higher percent occupation in sites with fine textured soil, in contradiction to published anecdotal observations. It is disconcerting, however, that measured soil texture and measured permeability are not closely related.

Finally, we can use the spatstat package to analyze the spatial pattern of sample points to try to determine whether it is reasonably dispersed and does not favor any particular regions (see the discussion in Chapter 5 about spatial sampling patterns). Figure 7.14a shows a point pattern selected from the Sierra Nevada. The computation is made using the Ripley's *K* statistic (Ripley, 1977; Diggle, 1983, p. 47). This is defined as follows. Let λ be the mean number of points per unit area. Then

$$K(r) = \frac{E_p(r)}{\lambda}, \tag{7.1}$$

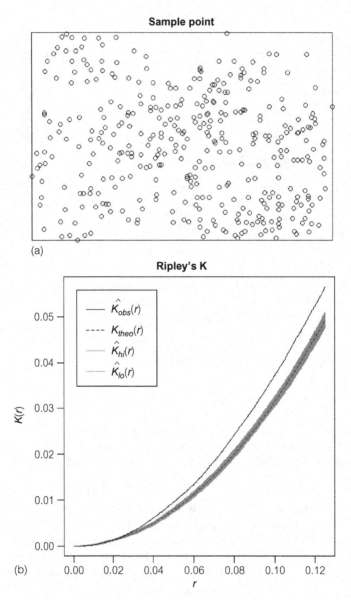

FIGURE 7.14
(a) Point pattern of a small section of sample points in the Sierra Nevada and (b) Plot of the theoretical and observed values of Ripley's K for the point pattern of Figure 7.14a.

where $E_p(r)$ is the expected number of additional points within a distance r of an arbitrarily chosen point. We can use the `spatstat` functions `Kest()` and `envelope()` to carry out this computation. Here is the code.

```
> W <- -120.8994
> S <- 38.0425
> E <- -120.1537
> samp.pts <- which(coordinates(data.Set2)[,1] <= E & coordinates(data.
Set2)[,1] >= W &
+    coordinates(data.Set2)[,2] >= S & coordinates(data.Set2)[,2] <= N)
```

```
> longitude <- coordinates(data.Set2)[samp.pts,1]
> latitude <- coordinates(data.Set2)[samp.pts,2]
> samp.ppp <- ppp(longitude, latitude, window = owin(c(W, E),c(S, N)))
> plot.ppp(samp.ppp, main = "Sample points") # Fig. 17.14a
> plot(envelope(samp.ppp, Kest), main = "Ripley's K") # Fig. 17.14b
```

Figure 7.14a shows the location of the points, which are in the central Sierra Nevada. Figure 7.14b shows a plot of the observed values of $K(r)$ as a function of r together with an envelope containing the highest and lowest simulated values obtained using a Monte Carlo simulation consisting of 99 runs of simulated data in which the same number of points are arranged at random. These are not confidence intervals in the normal sense, but they do give an indication of how different the observed value of $K(r)$ is from the theoretical value. Values of r for which the observed value of $K(r)$ is higher than the theoretical value are taken as an indication that the data are more clustered in space than randomly arranged data would be, and values of the observed $K(r)$ below the theoretical value are taken as indicating more dispersed point locations. Figure 7.14b indicates that the data are clustered at all values of r. This is interesting as it would be hard (at least for me) to draw this conclusion by eye. In Exercise 7.12, you are asked to gain further insight into Ripley's K by plotting it for the sampling patterns of Chapter 5.

7.4 Data Set 3

Data Set 3 (Appendix B.3) was collected over a period of three cropping seasons from a total of 16 rice fields in the region around Treinta y Tres, Uruguay (which is named for the 33 individuals who initiated the revolution that ultimately resulted in Uruguayan independence). The fields are all relatively large (generally about 30–50 ha). At the time of data collection, rice was typically grown in Uruguay in a five-year rotation, with two years of rice followed by three years of pasture. Many farmers do not own the land but rather rent it on a year to year basis, so that some fields were farmed by different people in different years. The fields are located in three geographic regions, northern, central, and southern (Figure 1.8). The northern region is located about 50 km to the north of the central region, and the southern region is about 75 km to its south. The data fields are given in Table 7.3. The objective of the study was to determine the practices that, by altering them, would have greatest impact on improving farmers' yields. The approach to realizing this objective was to determine the most effective practices that distinguished the farmers getting the highest yields from those getting lesser yields, taking into account field conditions.

A quick plot of UTM coordinates using ggplot() indicates that all fields in the northern region have a Northing greater than 6,340,000 m and all fields in the southern region have a Northing less than 6,280,000 m.

```
> data.Set3$SeasonFac <- factor(data.Set3$Season,
+    labels = c("2002-03", "2003-04", "2004-05"))
> ggplot(data = data.Set3) +
+   geom_point(mapping = aes(x = ID, y = Northing, color = SeasonFac))
```

The study spanned four calendar years (three Southern Hemisphere summers), and the variable *Season* represents a growing season (i.e., 1, 2, or 3). In the code we introduce the factor SeasonFac, which will serve as one of the grouping factors moving forward. Based

TABLE 7.3

Field Variables Included in Data Set 3

Variable Name	Units	Quantity Represented	Type	Scale
pH		pH	Exogenous	Ratio
Corg	%	Soil organic carbon	Exogenous	Ratio
SoilP	%	Soil P conc.	Exogenous	Ratio
SoilK	%	Soil K conc.	Exogenous	Ratio
Sand	%	Sand content	Exogenous	Ratio
Silt	%	Silt content	Exogenous	Ratio
Clay	%	Clay content	Exogenous	Ratio
Weeds		Weed level prior to herbicide	Exogenous	Ordinal
Irrig		Irrigation effectiveness	Exogenous	Ordinal
DPL	15 days	Planting date after 1 October	Management	Ratio
Cont		Level of weed control	Management	Ordinal
Farmer		Farmer ID letter	Management	Nominal
Fert	kg ha^{-1}	Total fertilizer	Management	Ratio
N	kg ha^{-1}	Fertilizer N	Management	Ratio
P	kg ha^{-1}	Fertilizer P	Management	Ratio
K	kg ha^{-1}	Fertilizer K	Management	Ratio
Var		Variety	Management	Nominal
Yield	kg ha^{-1}	Yield	Response	Ratio
Emer	Days	Emergence	Response	Ratio
D50	Days	Days after 1 Jan. to 50% flower	Response	Ordinal

on this, a `Location` data field was added to the data set that had the three values `North`, `Center`, and `South`. Let's look at examples of boxplots created using traditional and trellis graphics (the code also contains the plots created with `ggplot()`). The data frame `data.Set3` holds the contents of the file *set3.data.csv* (Appendix B.3). Figure 7.15 shows a boxplot of yield by farmer created using traditional graphics.

```
> boxplot(Yield ~ Farmer, data = data.Set3,
+    main = "Rice Yields by Farmer", xlab = "Farmer", cex.main = 2,
+    ylab = "Yield (kg/ha)", cex.lab = 1.5) # Fig. 7.15
>
```

Trellis graphics were used to create a set of box-and-whiskers plots of the yields for each farmer for the three regions (Figure 7.16), with the single grouping variable *Location*. Here is the code.

```
> trellis.par.set(par.main.text = list(cex = 2))
> trellis.device(color = FALSE)
+ labels = c("2002-03", "2003-04", "2004-05"))
> bwplot(Yield ~ Location | SeasonFac, data = data.Set3,
+    main = "Rice Yields by Year", # Fig. 7.16
+    xlab = "Location", ylab = "Yield (kg/ha)", layout = c(3,1),
+    aspect = 1)
```

Median yield in the central region is very consistent from one season to the next, although variability is somewhat higher in the first season (Figure 7.16). Yield in the south is reduced

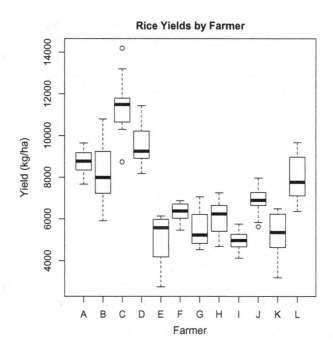

FIGURE 7.15
Box-and-whiskers plots of individual farmer yields in Data set 3, constructed using traditional graphics.

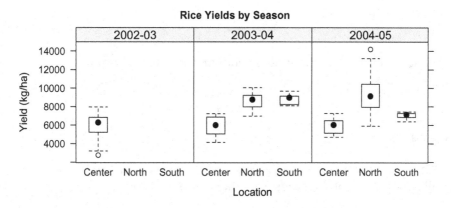

FIGURE 7.16
Trellis box-and-whiskers plots of yield by region with season number as a grouping variable.

a bit, and yield variability in the north is increased considerably in the third year. Farmers *A* through *D* are in the northern region, Farmers *E* through *K* are in the central region, and Farmer *L* is in the south. The farmers in the north obtained the highest yields in general. The farmer in the south had median yields somewhat below those of the northern farmers. The farmers in the central region tended to get much lower yields. Although he has one bad harvest, farmer *C* in the north clearly leads the pack. Farmer *L* in the south is about on par with the other three northern farmers, *A, B,* and *D*. The farmers in the center fall in behind with differing amounts of variability; farmer *I* gets consistently mediocre yields and farmers *E* and *K* have some good and some truly dreadful seasons. Farmer *J* has the highest average yield in the center.

Let us first examine in more detail the mean yield of each of the three regions in each of the three years. We can use the function `tapply()` to compute the mean yields over region and season by numbering the regions as 1 for the northern, 2 for the central, and 3 for the southern, and assigning a code to each data record identifying the region–season combination. The code for location–season combination consists of the location code plus 10 times the season code.

```
> data.Set3$LocN <- 2
> data.Set3$LocN[(data.Set3$Northing > 6340000)] <- 1
> data.Set3$LocN[(data.Set3$Northing < 6280000)] <- 3
> data.Set3$LocSeason <- with(data.Set3, Season + 10 * LocN)
> print(round(mean.yields <- tapply(data.Set3$Yield,
+     data.Set3$LocSeason, mean)))
   12   13   21   22   23   32   33
 8605 9173 5999 5900 5909 8832 7064
```

Mean yield stayed fairly consistent in the north and center and dropped substantially in the south. There was, however, only one field in each season in the south, and the fields were different. Since the differences in yield seem to be due to field differences more than season differences, we will not normalize yield over seasons.

We can also determine whether there is evidence of a second year effect, that is, whether yield tends to decline in the second year in a field in which rice is grown in two years in succession. The data field `RiceYear` indicates the year of the field in the rotation.

```
> unique(data.Set3$RiceYear)
[1] 1 2
```

Let's see which fields were in the study in both first and second years.

```
> sort(unique(data.Set3$Field[which(data.Set3$RiceYear == 1)]))
 [1]  1  2  3  4  5  6  7  8  9 10 11 12 13 14 15
> sort(unique(data.Set3$Field[which(data.Set3$RiceYear == 2)]))
[1]  3  5 12 13 14 16
```

Fields 3, 5, 12, 13, and 14 had both first and second rice years in the study. We can do a *t* test on the null hypothesis that the yields were the same in each rice year.

```
> t.test(data.Set3$Yield[which((data.Set3$Field == 3)
+     & (data.Set3$RiceYear == 1))],
+     data.Set3$Yield[which((data.Set3$Field == 3)
+     & (data.Set3$RiceYear == 2))])
        Welch Two Sample t-test
t = 1.4235, df = 43.61, p-value = 0.1617
alternative hypothesis: true difference in means is not equal to 0
95 percent confidence interval:
 -211.4933 1227.8933
sample estimates:
mean of x mean of y
  8472.84   7964.64
```

In Field 3 the second rice year yield declined, but not significantly, in Fields 5 and 12 the yield declined significantly in the second year. In Fields 13 and 14, the yields increased significantly. Here is a list of every farmer and field in each year.

```
> data.Set3$YearFarmerField <- with(data.Set3,
+     paste(as.character(Season),Farmer, as.character(Field)))
> print(tapply(data.Set3$Yield, data.Set3$YearFarmerField, mean),
+     digits = 4)
1 E 11   1 F 7   1 G 8   1 H 10   1 J 12   1 J 13   1 J 14   1 J 5   1 K 6   1 K 9
   4771    6331    5559     5995     6898     6408     7153    7404    5904    4091
 2 A 1   2 B 3   2 I 5   2 J 14   2 L 15    3 B 3    3 C 2    3 D 4   3 E 13   3 H 12
   8738    8473    4982     6742     8832     7965    11456     9542    5698    6023
3 L 16
   7064
```

Yields appeared to be more determined by the farmer than by location or season in those cases where different farmers farmed the same field in different seasons. This is most evident in the central region, in which *J* farmed Fields 5, 12, 13, and 14 in season 2, while *H* farmed Field 12 in season 4, *J* farmed Field 14 in season 3, and *I* farmed Field 5 in season 3. Of these fields, only that farmed by *J* maintained its yield in both seasons. The only field in the north farmed in two successive seasons was Field 3, which was farmed by *B* and maintained a roughly similar yield in both seasons. In summary, certain farmers appear to be much more successful in obtaining high yields than others. Can we find out the secrets to *J*'s and *C*'s success? Is it better management or better field conditions? If it is better management, what are the most influential management factors?

Our initial question is whether something in the physical environment could be affecting the regions. One possibility is terrain. Rice is grown in Uruguay as a flood-irrigated crop (Figure 1.6), and a high level of relief can affect water distribution. The SRTM (Shuttle Radar Topography Mission) digital elevation model (DEM), which has a 90 m cell size, was downloaded from the website provided by the Consultative Group on International Agricultural Research (CGIAR), http://srtm.csi.cgiar.org/. It is the tile between 50°W and 55°W and 30° and 35°S. The files in the *auxiliary* folder are included with the data. We can analyze the grid using functions from the `raster` package. We use the function `raster()` to read the file, then we assign a projection.

```
> library(raster)
> dem.ras <- raster("auxilliary\\dem.asc")
> projection(dem.ras) <- CRS("+proj=longlat +datum=WGS84")
```

The DEM is quite large, and we only need a small portion of it, so we determine the ranged of longitude and latitude values of the data and then use the function `crop()` to create a smaller grid.

```
> range(data.Set3$Latitude)
[1] -33.75644 -32.77624
> range(data.Set3$Longitude)
[1] -54.52640 -53.74849
> crop.extent <- matrix(c(-54.6,-53.7,-33.8,-32.7),
+     nrow = 2, byrow = TRUE)
> dem.Set3 <- crop(dem.ras, extent(crop.extent))
```

Next, we use the function `terrain()` to compute the slope.

```
> slope.Set3 <- terrain(dem.Set3, opt = "slope")
```

To find the slopes in the fields, we copy `data.Set3` to a file that we convert to a `SpatialPointsDataFrame` and use the function `extract()` to obtain the slope values at the field locations.

```
> Set3.WGS <- data.Set3
> coordinates(Set3.WGS) <- c("Longitude", "Latitude")
> slopes <- extract(slope.Set3, Set3.WGS)
> print(range(slopes), digits = 3)
[1] 4.91e-18 8.93e-02
```

There is a large range in slopes, with a maximum of almost 9%. An application of the function `hist()` (not shown) shows that almost all of the slope values are less than 3%. Let's see which fields are in highly sloped regions.

```
> sort(unique(data.Set3$Field[which(slopes > 0.03)]))
[1] 3 13 14
```

There is, however, only a very small correlation between slope and yield.

```
> print(cor(data.Set3$Yield, slopes), digits = 2)
[1] -0.034
```

Moreover, the fields with higher slopes do not tend to be those displaying low yields. Therefore, we do not consider terrain an important factor.

A second question is the extent to which there was a season effect. Addressing the question of a season effect allows us to use a trellis plot with two grouping variables, farmer and season. Although data were collected during three seasons, the first season's data were only collected from the central region. The box-and-whiskers plot of Figure 7.17 does not make very effective use of space, but it is informative.

```
> bwplot(Yield ~ Farmer | Location + SeasonFac, data = data.Set3,
+    main = "Rice Yields by Farmer and Season",
+    xlab = "Farmer", ylab = "Yield (kg/ha)") # Fig. 7.17
```

Once again, we see the farmer effect. Farmers *C* and *D*, who only entered the study in the last year, accounted for the greatly increased yields in that year. The only northern farmer observed in both year 2 and year 3 was *B*, whose yields were pretty consistent between years.

We can begin to examine the possibility that management actions are associated with differences in yield between the northern, central, and southern regions. We can use the function `tapply()` to get a quick look at the mean amount of fertilizer applied by region.

```
> tapply(data.Set3$Fert, data.Set3$Location, mean)
  Center North South
109.4405 158.0000 125.3200
```

The farmers in the north apply the most fertilizer, the farmers in the center apply the least, and the farmer in the south is in between. Of course, it may be that the farmers in the north apply too much fertilizer, or that the soil in the center is not suitable for higher levels of fertilizer application. This is what we are trying to determine. It is also possible that some of the farmers cannot afford more fertilizer, but we have established at the outset that we are going to ignore economic considerations.

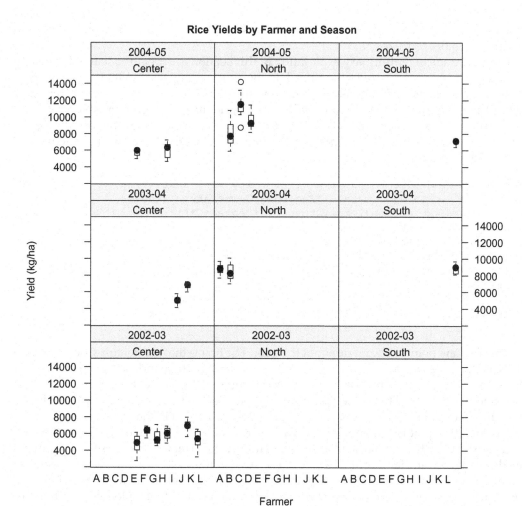

FIGURE 7.17
Trellis box-and-whiskers plots of yield grouped by farmer and season.

Using `tapply()` with other variables generates Table 7.4. In each of the management actions the farmers in the northern region managed at a higher level, the farmers in the central region managed at a lower level, and the farmer in the southern region was intermediate between the two. For example, although farmers in the north generally had lower weed problems prior to applying control measures, as indicated by a smaller average value of *Weeds*, they controlled weeds more effectively, indicated by a larger value of *Cont*. Similarly, they irrigated more effectively. Finally, although they had higher mean soil test

TABLE 7.4

Mean Values of Measured Quantities in the Three Geographic Regions of Data Set 3.

Region	Weeds	Cont	Irrig	SoilP	SoilK	Fert	N	P	K
North	2.3	4.8	4.2	6.2	0.20	158	62	72	23
Center	3.5	3.5	3.4	5.6	0.22	109	54	57	0
South	3.0	4.1	3.8	3.6	0.30	125	61	64	0

P and K values on average, the northern farmers still applied more fertilizer P and K, in addition to more fertilizer N.

The final stage of preliminary exploration in this section is to begin to search for insight as to which of the measured quantities has the greatest impact on yield. First, we consider climatic factors. Temperature, hours of sun per day, and rainfall were recorded three times a month at the Paso de la Laguna agricultural experiment station located near Treinta y Tres. The data frame `wthr.data` holds the contents of the file *set3weather. csv* (Appendix B.3). The code to plot the temperature in each year as well as the normal for that day is as follows.

```
> plot(wthr.data$Temp0203, type = "l", lty = 2, xaxt = "n",
+     ylim = c(10,27), main = "Regional Temperature Trends",
+     ylab = expression(Temperature~"("*degree*C*")"),
+     xlab = "Month", cex.main = 2, cex.lab = 1.5) # Fig. 7.18a
> lines(wthr.data$Temp0304, lty = 3)
> lines(wthr.data$Temp0405, lty = 4)
> lines(wthr.data$TempAvg, lty = 1)
> axis(side = 1, at = c(1,4,7,10,13,16,19),
+     labels = c("Oct","Nov","Dec","Jan","Feb","Mar","Apr"))
> legend(8, 15, c("2002-03", "2003-04", "2004-05",
+     "Normal"), lty = c(2,3,4,1))
```

There are a few points to be made about this plot. The first is that the default abscissa would indicate the index of each record, which is much less useful than indicating the month. For this reason, the default abscissa is suppressed through the argument `xaxt = "n"` of the function `plot()`, and the labels are added manually using the function `axis()`. In line 3, the function `expression()` is used to insert a degree symbol in the ordinate title.

The resulting plot is shown in Figure 7.18a. It is fairly clear that in each season the temperatures late in the season tended to be higher than normal, while these temperatures tended to be lower than normal earlier in the season. It is not easy, however, to visualize how much higher. Cleveland (1985, p. 276) points out that the vertical distance between two curves of varying slopes is very difficult to evaluate, and that a plot of the difference between these curves is more effective. Figure 7.18b plots the difference between each season's temperature measurements and the normal temperature. The temperature did tend to move from cooler than normal early seasons to warmer than normal late seasons, with a particularly warm late southern hemisphere fall in 2004. Differences from the normal in hours of solar irradiation did not follow any particular pattern (Figure 7.18c). None of the years were particularly dry relative to the normal (Figure 7.18d), and indeed there was heavy rainfall in the late fall in 2003.

We now consider the effect of soil-related variables. Here are the texture components of each region.

```
> Sand <- with(data.Set3, tapply(Sand, Location, mean))
> Silt <- with(data.Set3, tapply(Silt, Location, mean))
> Clay <- with(data.Set3, tapply(Clay, Location, mean))
> print(cbind(Sand, Silt, Clay), digits = 3)
  Sand Silt Clay
1 46.6 32.0 21.5
2 15.5 61.5 23.0
3 30.3 37.9 31.8
```

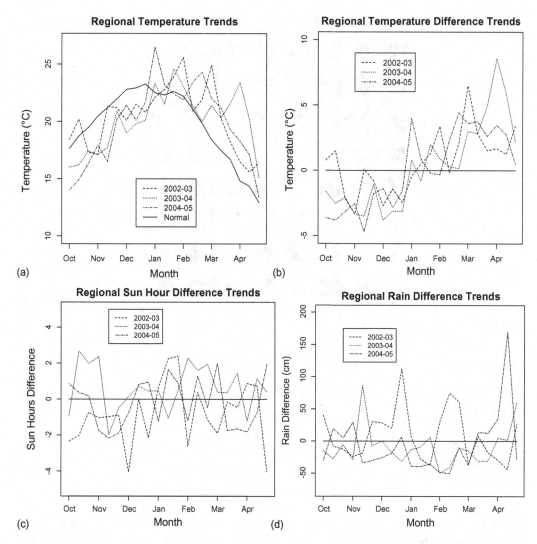

FIGURE 7.18
(a) Plot of temperatures recorded three times a month at the Paso de la Laguna experiment station, Treinta y Tres, Uruguay; (b) plot of the difference between the normal temperature and that recorded in each growing season; (c) plot of the difference between the normal number of hours of sun per day for each season; (d) plot of the difference between the normal rainfall per day for each season.

The northern region has a higher sand content on average, and the central region has a much higher silt content. A scatterplot matrix of Data Set 3 yield together with soil-related factors indicates that yield may have a positive association with sand and a negative association with silt, and possibly with pH (Figure 7.19). The scatterplot matrix provides information on relationships at the landscape scale; it is of interest to determine whether the relationship between yield and silt content "scales down" to the field level. We can make this determination using the function xyplot() from the lattice package.

Soil Data and Yield, Data Set 3

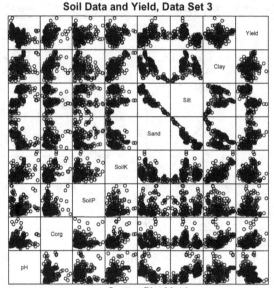

Scatter Plot Matrix

FIGURE 7.19
Scatterplot matrix of soil-related data of Data Set 3.

```
> trellis.par.set(par.main.text = list(cex = 2))
> trellis.device(color = FALSE)
> data.Set3$FieldFac <- factor(data.Set3$Field,
+    labels = c("Field 1","Field 2","Field 3",
+    "Field 4","Field 5","Field 6","Field 7",
+    "Field 8","Field 9","Field 10","Field 11",
+    "Field 12","Field 13","Field 14","Field 15",
+    "Field 16"))
> xyplot(Yield ~ Sand | FieldFac, data = data.Set3,
+    main = "Yield vs. Sand by Field") # Fig. 7.20
```

The trellis plot (Figure 7.20) indicates that there is no consistent relationship at the field level between yield and silt content. A couple of fields show a negative relationship, a couple of fields show a positive relationship, and most of the fields show no relationship. A similar lack of consistency is present in the field by field relationships with sand and pH (not shown). This is an example of a phenomenon called the *ecological fallacy*, which will be discussed in Section 11.5.2.

Finally, we examine the effect of variety. Four varieties were planted. They are coded numerically in the data set; the corresponding variety names are given by Roel et al. (2007). We can use these to compute mean yields by variety.

```
> data.Set3$Variety <- "INIA Tacuarí"
> data.Set3$Variety[which(data.Set3$Var == 2)] <- "El Pasol"
> data.Set3$Variety[which(data.Set3$Var == 3)] <- "Perla"
> data.Set3$Variety[which(data.Set3$Var == 4)] <- "INIA Olimar"
> with(data.Set3, tapply(Yield, Variety, mean))
    El Pasol  INIA Olimar INIA Tacuarí        Perla
    7064.099     7737.281     6931.632     9307.667
```

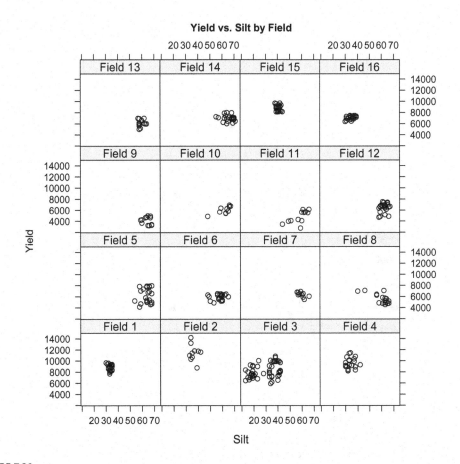

FIGURE 7.20
Trellis plot of yield vs. silt content for each of the individual fields of Data Set 3.

Obviously, there is a substantial difference in yields between the varieties. Let's look at who used which variety.

```
> with(data.Set3, unique(Farmer[which(Variety == "INIA Tacuarí")]))
[1] K I L D
Levels: A B C D E F G H I J K L
> with(data.Set3, unique(Farmer[which(Variety == "El Pasol")]))
[1] K F J G H E A L C
Levels: A B C D E F G H I J K L
> with(data.Set3, unique(Farmer[which(Variety == "INIA Olimar")]))
[1] B E H
Levels: A B C D E F G H I J K L
> with(data.Set3, unique(Farmer[which(Variety == "Perla")]))
[1] A
Levels: A B C D E F G H I J K L
```

Only one (northern) farmer used Perla, and he used El Pasol as well. The farmers in all of the regions tended to use all of the varieties. To further disentangle the data, we can plot a trellis box and whiskers plot (Figure 7.21).

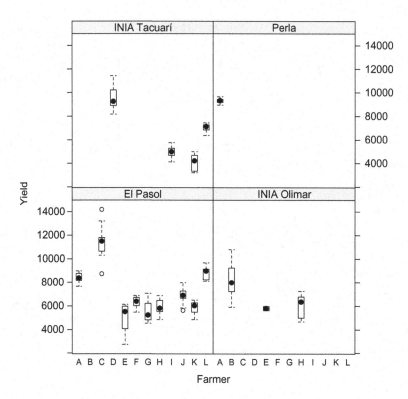

FIGURE 7.21
Trellis plot of yield by variety and farmer.

```
> bwplot(Yield ~ Farmer | Variety, data = data.Set3,
+    xlab = "Farmer") # Fig. 7.21
```

This does not appear to indicate any pattern of yield by variety.

To summarize the results we have obtained so far in the exploratory analysis of Data Set 3, we have found a pair of farmers, one in the central region and one in the north, who seem to consistently obtain better yields than the rest. Climate does not appear to have had an important influence on year-to-year yield variability in the three seasons in which data were collected, and terrain does not appear to vary much between fields. Variety does appear to have a strong association with yield, but the pattern of variety and yield is complex. In the exercises, you are asked to further explore the management variables associated with Data Set 3.

7.5 Data Set 4

Data Set 4 consists of data from the 1995–96 through the 1999 growing seasons from two fields in central California. During the 1995–1996 season the fields were planted to wheat in December 1995 and harvested in May 1996. During the remaining years, the fields were both planted to summer crops. Both fields were planted to tomato in 1997. One field, denoted Field 1, was planted to beans in 1998 and to sunflower in 1999. The other field, denoted Field 2, was planted to sunflower in 1998 and to corn (i.e., maize) in 1999.

FIGURE 7.22
Boxplots of 1996 wheat yields from the two fields in Data Set 4.

The data from 1997 through 1999 will be analyzed in Chapter 15. Prior to that chapter, we will focus exclusively on the wheat data from the 1995–96 season. The primary factor influencing economic return from wheat is grain yield, with protein content as a secondary influence. The principal objective of the analysis is to determine the principal factors underlying spatial variability in grain yield.

To create the object `data.Set4.1` describing Field 1 we first read the file *set4.196sample. csv*, which contains the values at the 86 sample locations. We then add the data field `Yield`, which contains data from the file created in Exercise 6.10 of cleaned yield monitor data interpolated to the 86 locations (see R code in Appendix B.4). Then we do the same for Field 2. As an initial exploratory step, we compare the yield distributions of the fields using a set of box-and-whisker plots. Figure 7.22 shows boxplots of yield for the two fields. Plotting of extreme values is suppressed. Typical wheat yields in this part of California range between 6,000 and 8,000 kg ha⁻¹ (Anonymous, 2008), so both fields clearly had some problem areas. The yield distribution of Field 1 is right-skewed, with a mass of relatively low values, while the distribution of Field 2 is somewhat left-skewed. As with Data Set 2, creating a similar boxplot with `ggplot()` requires that the ordinate be an ordinal scale variable.

The discussion of data exploration will focus on Field 1. The exploration of Field 2 is left to the exercises. As a first step, we examine the high spatial resolution data, looking for similarly shaped spatial patterns. For Field 1, these data consist of the yield map and gauge data derived from the three aerial images taken in December 1995, March 1996, and May 1996. Each image consists of three radiometric bands (Lo and Yeung, 2007, p. 294), infrared, red, and green. The numerical value of each band is an integer between 0 and 255 (which is $2^8 - 1$), with higher values indicating higher intensity in that band. The quantity most often used in remote sensing to estimate vegetation density is the *normalized difference vegetation index*, or NDVI, which is defined as (Tucker, 1979)

$$NDVI = \frac{IR - R}{IR + R} \tag{7.2}$$

Since healthy vegetation reflects infrared much more strongly than red, a high NDVI indicates dense vegetation, whereas an NDVI near zero indicates little or no green vegetation. Figure 7.23a shows the infrared band (IR) from the December image, the NDVI calculated from the March and May images, the IR band from the May image, and the interpolated yield. These were found by a process of trial and error to provide the most information

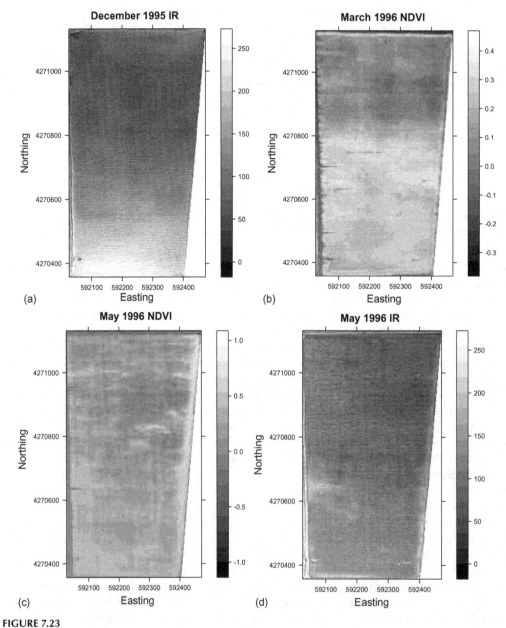

(a)　(b)　(c)　(d)

FIGURE 7.23
Plots of data from three aerial images taken of Field 4.1 during the 1995–96 season together with interpolated yield monitor data: (a) December infrared; (b) March NDVI; (c) May NDVI; (d) May infrared.　(*Continued*)

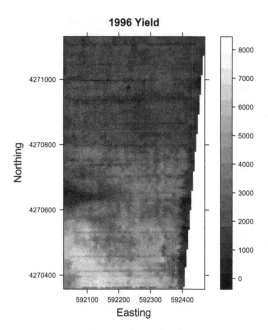

FIGURE 7.23 (Continued)
Plots of data from three aerial images taken of Field 4.1 during the 1995–96 season together with interpolated yield monitor data: (e) interpolated yield.

about the field conditions. To create this figure it was necessary to crop the December infrared image so that it was restricted to the field boundary. We saw how to do this in Section 5.4.3 using the function over(). In this section's code, we use a convenient shortcut made possible by polymorphism and the fact that the left bracket [is actually a function (Section 2.3.2). Because of this, we can write

```
f.img <- img.df[bdry.spdf,]
>
```

and the expression on the right is equivalent to img.df[!is.na(over(img.df, bdry. spdf),] (Pebesma, 2018). This makes the operation look just like the data frame equivalent.

The December infrared reflectance in Figure 7.23a is lower in the north than in the south. The aerial image was taken a few days after a heavy rainstorm (Plant et al., 1999). Water absorbs infrared radiation, so that wet soil tends to absorb this radiation while dry soil tends to reflect it. The image is an indication that the soil in the north end of the field remained wet after the south end had begun to dry. This indicates that the clay content of the soil in the northern part of the field may be higher than that in the south. This is confirmed by the soil sample data presented in Section 3.2.1. Figure 7.24 shows a map of the soil types in the field. This map was constructed by downloading a shapefile of SSURGO soil classification data from the NRCS Soil Data Mart http://soildatamart.nrcs.usda.gov and clipping this shapefile in ArcGIS with that of the field boundary. The northernmost soil type (Ca) is Capay silty clay (Andrews et al., 1972). This soil is characterized by a low permeability. The soil type in the center (BrA) is Brentwood silty clay loam, which is characterized by a moderate permeability. The soil in the south (Ya) is Yolo silt loam, which is also characterized as moderately permeable and well drained. Taken together, Figures 7.23 and 7.24 indicate

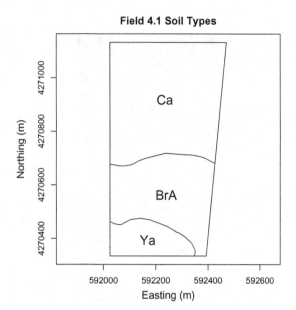

FIGURE 7.24
Soil types of Field 4.1. Ca = Capay silty clay, BrA = Brentwood silty clay loam, Ya = Yolo silt loam.

that Field 4.1 displays a pattern of variation in soil texture. Clay content and sand content have a strong negative association.

```
> with(data.Set4.1, cor(cbind(Sand, Silt, Clay)))
            Sand         Silt         Clay
Sand   1.0000000 -0.30338274 -0.93432993
Silt  -0.3033827  1.00000000 -0.05615113
Clay  -0.9343299 -0.05615113  1.00000000
```

Therefore, we will eliminate *Sand* from the analysis and instead focus on *Clay* and *Silt* as the texture components.

The southern part of the field has higher yield overall (Figure 7.23e), with the southwest corner having the highest yield. The yield trend generally appears fairly smooth except for two anomalously low areas: a triangular-shaped region on the western edge and the entire eastern edge. The March NDVI data (Figure 7.23b) generally indicates that vegetation density decreases as one moves from south to north. The May infrared data indicates this same trend, but also indicates anomalously high values in the same areas as the low areas in the yield map, namely, the triangular-shaped region and the eastern edge. In the actual study, heavy infestations of wild oat (*Avena fatua* L.) and canary grass (*Phalaris canariensis* L.) were observed in these areas. By May, the wheat was beginning to senesce, but the weeds were still green, and show up as brighter areas in the May infrared image. In summary, visual inspection of the high spatial resolution data suggests that the low yields in the northern half of the field may have been caused by aeration stress and that low yields along the eastern border and in the triangular region on the southwest border were probably caused by weed competition.

We now move to an exploration of the point sample data. In organizing the exploration, we will make use of the variable type concept illustrated schematically in Figure 7.1. Table 7.5 shows a list of variables in Data Set 4.1 together with their type

TABLE 7.5

Variables in Data Set 4

Variable Name	Quantity Represented	Type	Spatial Resolution	Comment
Sand	Soil sand content (%)	Exogenous	Low	
Silt	Soil silt content (%)	Exogenous	Low	
Clay	Soil clay content (%)	Exogenous	Low	
SoilpH	Soil pH	Exogenous	Low	
SoilTOC	Soil total organic C (%)	Exogenous	Low	
SoilTN	Soil total nitrogen (%)	Exogenous	Low	
SoilP	Soil phosphorous content	Exogenous	Low	
SoilK	Soil potassium content	Exogenous	Low	Field 1 only
Weeds	Weed level (1 to 5)	Exogenous	Low	
Disease	Disease level (1 to 5)	Exogenous	Low	
CropDens	Crop density (1 to 5)	Endogenous	Low	
LeafN	Leaf nitrogen content	Endogenous	Low	
FLN	Flag leaf N content	Endogenous	Low	Field 1 only
SPAD	Minolta *SPAD* reading	Gauge	Low	
EM38F	Furrow EM38 reading	Gauge	Low	date sampled
EM38B	Bed EM38 reading	Gauge	Low	date sampled
IR	Infrared digital number	Gauge	High	date sampled
R	Red digital number	Gauge	High	date sampled
EM38	High density EM38	Gauge	High	Field 2 only
GrainProt	Grain protein (%)	Response	Low	
GrnMoist	Grain moist. content (%)	Response	High	
Yield	Yield (kg ha^{-1})	Response	High	

and spatial resolution. No management variables were recorded in the data set. As a first step, we will construct scatterplot matrices of some of the variables. If we consider endogenous variables to be those variables that contribute directly to grain yield or protein content, either through the harvest index (Donald and Hamblin, 1976; Hay, 1995), or through plant physiological processes, then the only endogenous variables are *CropDens*, *FLN*, and *LeafN*. The exogenous variables are considered to influence grain yield indirectly through their impact on the factors contributing to the harvest index. We use scatterplot matrices to eliminate redundant variables and to suggest relationships that we may wish to explore further. We will start with some of the endogenous and exogenous variables and their relationships with the response variables (we refer to this combination as the "agronomic variables"). The code to construct a scatterplot matrix is as follows.

```
> library(lattice)
> agron.data <- subset(data.Set4.1,
+    select = c(Clay, Silt, SoilP, SoilK, SoilpH, SoilTOC, SoilTN,
+ LeafN, FLN, GrainProt, Yield))
> splom(agron.data, par.settings = list(fontsize=list(text=9),
+   plot.symbol = list(col = "black")), pscales = 0,
+   main = "Agronomic Relationships, Field 4.1") #Fig. 7.25
```

Agronomic Relationships, Field 4.1

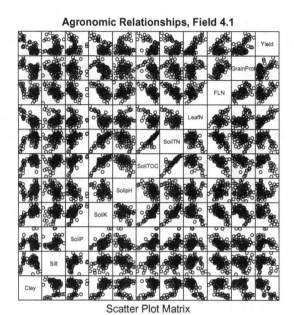

Scatter Plot Matrix

FIGURE 7.25
Scatterplot matrix for "agronomic" data of Field 4.1.

Based on the scatterplot matrix (Figure 7.25), *SoilTOC* and *SoilTN* have a strong positive association. They are sufficiently collinear that one can be removed. *SoilTN*, *SoilK*, and *Clay* are positively associated. *SoilP* appears to have a parabolic relationship with *Clay*. The scatterplots of relationships between *Clay* and some of the other variables appear to indicate that they actually describe two separate relationships. Since *Clay* has a strong north-south trend, it might be suspected that these two separate relationships occur in two separate regions of the field.

Soil phosphorous and potassium are generally regarded as reliable indicators of crop fertilizer demand in California wheat (University of California, 2006). Soil nitrogen level is not, because plant-available forms of nitrogen are highly water soluble and thus very labile, and moreover, a good deal of the nitrogen content of soil is not readily plant-available (Singer and Munns, 1996, p. 215). Fertilization guidelines for nitrogen and potassium in California are based on the level of nitrate-nitrogen and potassium measured in the stem (University of California, 2006). Phosphorous guidelines are based on soil sampling. Data collected in the field were flag leaf nitrogen content, soil phosphorous level, and soil potassium level. In precision agriculture applications, the principal initial questions are whether these data are uniform over the field, and whether any of them are at a level that indicates that an input application is warranted. If an application is warranted, then a further question is whether application at a variable rate is justified, both agronomically and economically, or whether application at a uniform rate are sufficient. The *null hypothesis of precision agriculture* (Whelan and McBratney, 2000) is that precision applications are not economically justified.

The recommended stem nitrate-nitrogen level for wheat in California, measured at the boot stage of plant growth (early in the reproductive phase), is 4,000 ppm, or 0.4% (University of California, 2006). Measurements in Field 4.1 were made between the boot and anthesis stages, so one might expect the level to be a bit lower. The ratio between flag leaf and stem and nitrate concentration at anthesis is about 2.3–2.5 to 1 (Jackson, 1999).

Therefore, one would expect to see a response to fertilizer nitrogen if the flag leaf nitrate-nitrogen level is below about 1 ppm. The soil phosphorous level below which a phosphate application is recommended is 6 ppm (University of California, 2006). Yield response to potassium is rare in California wheat production and is limited to soils with potassium levels less than 60 ppm (University of California, 2006). As an initial check of the measured values relative to these guidelines, we can create stem-and-leaf plots (Tukey, 1977 p. 8; Chambers et al., 1983, p. 26; Tufte, 1990, p. 46) or histograms (Cleveland, 1985, p. 125; Chambers et al., 1983, p. 24); for this application, we choose the former. Applying the function `apply()` to the function `stem()` rapidly performs the calculation of all of the plots as shown here.

```
> apply(with(data.Set4.1, cbind(FLN, SoilP, SoilK)), 2, stem)
```

The results are shown in table 7.6. Based on the stem and leaf plots, one can conclude that nitrogen and potassium levels in the field are adequate, but that phosphorous levels may be low in some areas.

Returning to Figure 7.25, let's explore the relationship between the response variables and some of the quantities that may affect them. *Yield* has a boundary line relation (Webb, 1972) with *Weeds* and, to a lesser extent, with *Disease* and flag leaf nitrogen (*FLN*). Again, this may indicate that in some parts of the field weed level, disease, or nitrogen is the most limiting factor. *Yield*, like *CropDens*, actually has a negative association with *SoilK*, and it has a sort of "sideways parabolic" relationship with *SoilP* and *SoilTN*; that is, if the explanatory variables are plotted on the abscissa and *Yield* on the ordinate, the relationship is parabolic. This is a further indication that difference relationships may occur in different parts of the field, and it indicates the presence of an *interaction*, in which the relationships of yield with these explanatory variables depend on the value of some other explanatory variable, either measured or unmeasured. All of the parabolas appear to have their turning points at a *Yield* value of about 3800 kg/ha.

Table 7.7 shows stem and leaf plots that allow us to examine the distributions of three of the variables. *Clay*, which is representative of soil conditions, is positively skewed and has a bimodal distribution. *Yield*, on the other hand, is negatively skewed. *GrainProt* is fairly symmetric, but is interesting in another context. Wheat at a higher protein content than 13% receives a price premium, which is why this represents a second response variable. About half the field is above this 13% value.

TABLE 7.6

Stem and Leaf Plots for Variables Related to Soil Fertility.

FLN (%)	*SoilP* (ppm)	*SoilK* (ppm)
The decimal point is 1 digit(s) to the left of the |	The decimal point is at the |	The decimal point is 1 digit(s) to the right of the|
	2 | 5815678	
26 | 0	4 | 22333480000233355556689	8 | 869
28 | 21	6 | 0001111244571222357888	10 | 7
30 | 22455779011246788	8 | 112335566779991446788	12 | 360
32 | 013455778900111233445555667	10 | 12236	14 | 045567789
34 | 000133566788900233555688	12 | 014345	16 | 034669023689
36 | 02313478	14 |	18 | 4578899011222445577788
38 | 1137898	16 | 9	20 | 01111367899025669
40 | 1	18 | 1	22 | 0333581144579
		24 | 447069

TABLE 7.7

Stem-and-Leaf Plots for Clay, Grain Protein Percent, and Yield.

Clay	GrainProt	Yield
The decimal point is at the |	The decimal point is at the |	The decimal point is 3 digit(s) to the right of the |
22 | 279		
24 | 68	11 | 77899	1 | 23445566667777777888888999999
26 | 559934578	12 | 0000122344	2 | 0000111111223344555577899999
28 | 0338	12 | 5566777788889	3 | 12334557889
30 | 5	13 | 00001111111122233344444	4 | 0112444888
32 | 2	13 | 555556666777777888899999	5 | 124569
34 | 483456	14 | 0001244	6 | 267
36 | 245670011	14 | 5579	
38 | 1145678935	15 | 2	
40 | 1225567801244555999		
42 | 44456677790355678		
44 | 0135		
46 | 9		

Scatterplots of *Clay* versus several other variables indicate that the boundary *Clay* value between the two groups is about 0.3 (i.e., 30% clay). We can use Thiessen polygons (Section 3.4.4) to visualize the location of sites with a *Clay* value less than 30%. As usual, we need to be careful to ensure that the Thiessen polygon ID variables match with the attribute data ID variables.

```
> all.equal(thsn.spdf$ThPolyID, data.Set4.1$ID)
[1] TRUE
> slot(thsn.spdf, "data")$Clay30 <- factor(data.Set4.1$Clay <= 30,
+    labels = c("Greater than 30%", "Less than 30%"))
> greys <- grey(c(100, 200) / 255)
> spplot(thsn.spdf, zcol = "Clay30", col.regions = greys,
+    main = "Field 4.1 Clay Content Categories") # Fig. 7.26
```

All of these locations are in a contiguous area in the south of the field (Figure 7.26). In Exercise 7.15, you are asked to plot a high spatial resolution version of Figure 7.26 by using linear regression to predict clay content based on December IR value.

A different perspective on the spatial relationships among the attribute values may be obtained using star plots (Chambers et al., 1983, p. 161), sometimes called sun ray plots (Jambu, 1991), in the form of a map. Star plots, like other multivariate graphical devices, come very close to being what Tufte (1983, p. 153) calls a "graphical puzzle," that is, they can be difficult to interpret (Cleveland, 1985, p. 20), but they do provide an overview of general patterns. My own opinion is that star plots with four variables are easiest to interpret; more variables produce a graphical puzzle and fewer variables produce stars with too few "points." The code for a plot of soil nutrients and clay content is as follows.

```
> star.data <- with(data.Set4.1, data.frame(SoilP, Clay,
+    SoilK, SoilTN))
> star.loc <- 2* cbind(data.Set4.1$Column, 13 - data.Set4.1$Row)
> stars(star.data, locations = star.loc,
+    labels = NULL, key.loc = c(18,2),
+    main = "Nutrient Star Plot", cex.main = 2) # Fig. 7.27
```

Field 4.1 Clay Content Categories

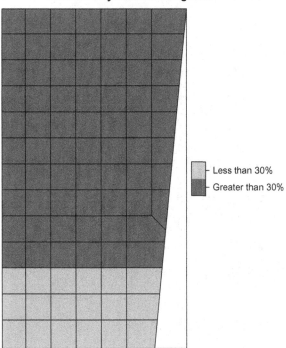

Less than 30%
Greater than 30%

FIGURE 7.26
Thiessen polygon plot of regions of Field 4.1 as defined by clay content. The Thiessen polygons were created in ArcGIS.

The row and column numbers of the attribute data set are manipulated into x and y coordinates. The calculation of these x and y coordinates in line 3 involved a bit of trial and error. The function stars() is called in line 4 with arguments that suppress the plotting of labels (these would be the number of each data record) and that locate the legend (again this involved a bit of trial and error). The star map in Figure 7.27 indicates that all soil nutrient levels are lowest in the southern (high yielding) region and that phosphorous is also low in the northern part of the field.

In Exercise 7.16, you are asked to create a star map for yield in relation to weed and disease level. The highest weed and disease levels occur in the central portion of the field and along the northern and eastern edges. Star plots are useful for identifying interactions. One potential interaction of interest involves soil potassium content and leaf nitrogen content. Potassium deficiency is known to inhibit the utilization of nitrogen (Gething, 1993; Marschner, 1995, p. 299). If such an interaction is present in these data, it should manifest itself in lower values of flag leaf N and yield in low K soils than in high K soils, conditional on soil N levels. This is not indicated in the star plot of Exercise 7.16b. The last phase of the initial data exploration involves recognizing that two quantities, soil phosphorous and grain protein content, are below a meaningful threshold in some parts of the field. To determine these locations precisely, we can apply the function spplot() to the Thiessen polygons file. The results are shown in Figure 7.28.

In summary, the preliminary analyses indicate that the gross trend of declining yield in the north end of the field is probably caused by aeration stress associated with moisture retention by the heavy clay soil. In other parts of the field, weed competition may

Nutrient Star Plot

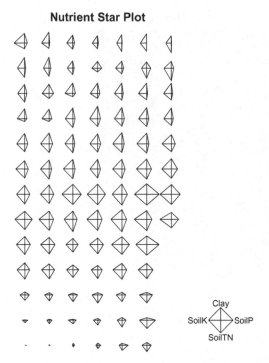

FIGURE 7.27
Star maps of the relationship between soil mineral nutrient levels and clay content.

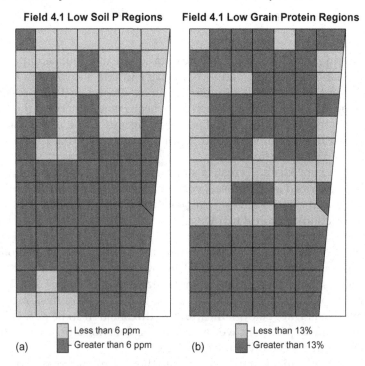

FIGURE 7.28
Relationship to a meaningful threshold for two quantities: (a) soil phosphorous level, and (b) grain protein percent. In each case, the gray lattice cells are those in which the quantity falls below the threshold.

limit yield. The crop appears to have adequate nutrient levels for nitrogen and potassium, but there may be some areas in which phosphorous is deficient. Of the mineral nutrients, neither potassium nor nitrogen is at any sample point below the level at which a yield response to increased fertilization would be expected. However, there is an indication that soil phosphorous levels might be low in the northern part of the field. The "sideways parabolic" relationship of yield with soil phosphorous level is a further indication of this possibility. Yield has a boundary line relationship with weeds, disease, and flag leaf nitrogen level. This might be consistent with a "law of the minimum" (Barbour et al., 1987) interpretation that, for example in the case of nitrogen, in some areas of adequate nitrogen level, some other factor inhibits yield. The area of high flag leaf N level and low yield is in the north of the field. Yield and grain protein content are highly correlated.

7.6 Further Reading

The books by Tufte (1983, 1990) provide an excellent discussion of the use of graphics in data analysis, as do those of Cleveland (1985, 1993). The utility of trellis graphics was first pointed out by Cleveland (1993), who referred to them as "multiway plots." Jambu (1991) provides a discussion of exploratory data analysis using the methods covered in this chapter and others as well. Andrienko and Andrienko (1998) provide a very extensive discussion of map analysis and graphical methods associated with spatial data. Sibley (1988) is also a good early source in this area.

An excellent introduction to mapping with R is provided by the NCEAS Scientific Computing: Solutions Center Use Case: Creating Maps for Publication using R Graphics, located at the link "Create Maps With R Geospatial Classes and Graphics Tools" on the website https://www.nceas.ucsb.edu/scicomp/usecases/CreateMapsWithRGraphics. The free software package GeoDa (https://spatial.uchicago.edu/geoda/) contains many spatial data exploration tools.

Cressie and Wikle (2011) advocate a hierarchical modeling approach that is very similar in concept to the division of variables into categories as represented in Figure 7.1.

Exercises

7.1 By comparing the values ID and obsID, verify that the creation of the object Set1.obs in Section 7.2 correctly links cuckoo observation points and habitat patches. If you have access to a GIS, you should do this by comparing habitat and observation point data files.

7.2 The object Set1.obs, which contains the information about each polygon in Data Set 1 that contains a cuckoo observation point, includes variables *HtClass*, *AgeClass*, and *CoverClass*. When applying the function AreaRatio() to compute the habitat ratios, why is it necessary to use the data in data.Set1 rather than simply using Set1.obs?

7.3 In Section 7.2, a set of size class ratios is computed for use in the Wildlife Habitat Relationships model for the yellow-billed cuckoo expressed in Table 7.1. Carry out a similar construction for the ratio of age classes in this model.

7.4   (a) Compute the suitability score for age class in the CWHR model of Section 7.2. (b) Repeat the computation for the height class suitability score.

7.5   The CWHR model for the yellow-billed cuckoo produces two misclassified patches having a suitability score of zero but where cuckoos are observed. Examine the properties of these locations. Can you conclude anything from them?

7.6   Repeat the contingency table analysis of Section 7.2 with subsets of the explanatory variables and determine which have the most explanatory power.

7.7   Use the function `locator()` to separate out of Data Set 2 that part located in the Klamath range (this is the square shaped northernmost portion of the data). Plot a map of the region and compare it with your maps generated in Exercise 6.1.

7.8   Plot the presence and absence of blue oaks in the Klamath range against elevation.

7.9   Construct scatterplots of precipitation against elevation for the Sierra Nevada and the Coast Range and compare them.

7.10 Use the function `table()` to construct a contingency table of texture class vs. permeability class for Data Set 2. Are the two closely related?

7.11 The application of the function `hexbin()` at the end of Section 7.3 showed that for values of *Precip* less than 800, the distribution of *MAT* is bimodal. Generate the `hexbin()` frequency plot of *MAT* and *Precip* when *Precip* is restricted to this range, Then repeat for *JuMax* and *JaMin*.

7.12 Use the functions of the `spatstat` package to plot the theoretical and observed plots of Ripley's *K* for (a) the random sample pattern of Section 5.3.1; (b) the 32 point grid sample pattern of Section 5.3.2; and (c) the clustered pattern of Section 5.3.5. How do your results (especially those of the clustered pattern) correspond with the interpretation of Ripley's *K*?

7.13 Use the function `xyplot()` to determine whether any of the measured soil quantities in Data Set 3 appears to have a linear association with yield across fields and, if so, which appears to have the highest level of association.

7.14 Repeat Exercise 7.12 plotting yield against management quantities by location and year.

7.15 The C:N ratio, that is, the ratio of total carbon to total nitrogen, is an important ecological quantity in soil science. The soil total organic carbon content is equal to 0.58 times the soil total nitrogen content. Compute the C:N ratio for the 86 sample locations in Field 4.1. If the C:N ratio becomes too high (say, above 30:1), then the demand from soil microorganisms for nitrogen becomes sufficiently high that nitrate will not remain in soil solution and thus will be unavailable for root uptake (Brady, 1974, p. 291).

7.16 Use the December IR image to assist in assessing the distribution of clay content in Field 4.1. (a) Use `over()` to select the pixels values in the December IR image that align with sample points, and compute a regression of IR digital number on clay content. (b) Use this regression relationship to predict clay content at each pixel based on IR values and plot the regions above and below 30% clay.

7.17 (a) Construct a star plot of disease, weeds, and yield for Field 4.1. (b) Construct a star plot of soil total nitrogen, flag leaf nitrogen, potassium, and yield to determine whether there is an interaction between N and K in this field.

7.18 Compare using stem and leaf plots the clay content of Field 4.2 and Field 4.1. Then create a scatterplot showing the relationships between yield and clay content in both fields. Are the relationships similar? Do you expect that clay content will play as significant a role in Field 4.2 as it does in Field 4.1?

7.19 Construct a scatterplot matrix of yield in Field 4.2 with some of the variables that might affect it. Can you identify with this the most likely cause of the large low yield areas on the western side of the field? Check your answer by plotting thematic point maps of yield and this variable.

7.20 Determine whether there are any areas in Field 4.2 that are nitrogen or phosphorous deficient according to University of California standards for wheat in California.

7.21 (a) Construct a star plot of yield, grain moisture, protein content, and leaf nitrogen. (b) Construct a star plot of yield vs. weeds and disease.

8

Data Exploration Using Non-Spatial Methods: The Linear Model

8.1 Introduction

The previous chapter focused on graphical means of data exploration. In the analysis of non-spatial data sets, the methods discussed in this chapter are not necessarily considered exploratory. We classify these methods as exploratory here because in those cases in which a parameter may be estimated or a hypothesis may be tested, spatial data may violate the assumptions underlying these operations, and therefore the estimation or the significance test or both may not be valid. We often do not use the methods to obtain final confirmatory results (parameter estimates and/or significance tests), but rather to gain further insights into the relationships among the data. Confirmatory analysis will be carried out in later chapters in which the analysis incorporates assumptions about the spatial relationships among the data.

All is not bad news, however, when it comes to the impact of spatial autocorrelation on data analysis. Although we must be aware of possible bias in parameter estimates and we may lose the ability to carry out hypothesis tests in the same way that we would with non-spatial data, we have already seen in Chapter 7 that this loss is often more than compensated by our acquisition of the ability to glean further understanding of data relationships through the use of maps and other devices. We will see further examples of this throughout the present and subsequent chapters. In this chapter, Section 8.2 provides an introduction to multiple linear regression and develops the approach to model selection that we will use. Section 8.3 applies this approach to the construction of alternative multiple linear regression models for yield in Field 1 of Data Set 4. Section 8.4 introduces generalized linear models and shows how these are applied to data from Data Sets 1 and 2.

8.2 Multiple Linear Regression

8.2.1 The Many Perils of Model Selection

In a seminar I attended many years ago, a prominent ecologist characterized multiple regression as "the last refuge of scoundrels." He was not actually referring to multiple regression per se, but rather to model selection, the process of using multiple regression to select the "best" variables to explain a particular observed phenomenon. In this context, he had a point. Model selection can be a very useful tool for exploratory purposes, to eliminate some alternatives from a wide variety of competing possible explanations, but it must be used with extreme caution and not abused. In this subsection, we will briefly review some of the traditional automated methods of model selection. The primary purpose of this review is to indicate where they may have problems and to extract any ideas from them that may seem suitable. After that, we discuss two graphical techniques that are often useful in constructing a good multiple regression model: added variable plots and partial residual plots. The third and final subsection reviews the pitfalls associated with model selection and describes the approach that will be used in this book. Readers unfamiliar with the theory of simple linear regression and ordinary least squares (OLS) should consult Appendix A.2.

There are a number of problems in the interpretation of multiple regression results that do not arise with simple linear regression. Two in particular are very important. The first is that when there are more than one explanatory variable it becomes difficult or impossible to visualize the interrelationships among the variables, and to identify influential data records. Second, the explanatory variables themselves may be correlated, a phenomenon called *multicollinearity*. Because of this, it may become difficult or impossible to distinguish between the direct impact a process has on the response variable and the indirect effect it has due to its effect on another explanatory variable. This is why controlled experiments are much more powerful than observational studies. A second negative impact of multicollinearity is its impact on the matrix equation (A.34) in Appendix A defining the computation of the regression coefficients. This computation may become subject to considerable error. The bad effects of multicollinearity, as well as the difficulty of identifying influential data records, are part of what makes the process of selecting explanatory variables to include in a multiple regression model so difficult and dangerous.

In comparing multiple regression methods, one must always keep in mind the objective of the analysis. Generally, the two possibilities are prediction and interpretation. The former objective is much easier to quantify than the latter, and as a result one can, based on the establishment of a precise measure, compare methods and determine which method is optimal according to that measure. If the objective is interpretation, as it generally is in this book, it is not even clear what would define an "optimal" method. The unfortunate fact is that one is using a tool in a way for which it itself is not optimal, and the best one can do is to be careful and try to avoid making serious errors. With these caveats, we can begin discussing the methods.

We will express the multiple linear regression model as (cf. Appendix A.2, Equation A.37)

$$Y_i = \beta_0 + \beta_1 X_{i1} + \beta_2 X_{i2} + ... + \beta_{p-1} X_{i,p-1} + \varepsilon_i, \ i = 1, ..., n, \tag{8.1}$$

where the error terms ε_i are assumed to be normally distributed, of constant variance σ^2, and mutually independent. This last condition, independent errors, is often violated when the data represent measurements at nearby spatial locations. In Chapters 11 through 13 we will discuss the consequences of this failure of the assumptions and the means to deal with them. For the moment, we will simply say that one must consider the results of ordinary least squares linear regression applied to spatial data as strictly preliminary.

The process of selecting which explanatory variables to retain and which to eliminate from a multiple linear regression model can be viewed as a search for balance in the *bias-variance trade-off* (Hastie et al., 2009, p. 37). This concept can be illustrated using an example with artificial data. Our discussion follows that of Kutner et al. (2005, p. 357). Using the notation of Appendix A, let $\hat{Y}_i$ be the value of the response variable predicted by the regression at data location i, that is,

$$\hat{Y}_i = b_0 + b_1 X_{i1} + b_2 X_{i2} + \ldots + b_{p-1} X_{i,p-1}, \ i = 1, \ldots, n. \tag{8.2}$$

Suppose that a linear model with three explanatory variables fits the data exactly, that is, that if we define $\mu_i = E\{Y_i\}$, then

$$\mu_i = \beta_0 + \beta_1 X_{i1} + \beta_2 X_{i2} + \beta_3 X_{i3}, \ i = 1, \ldots, n. \tag{8.3}$$

Suppose, however, that we have measured five explanatory variables, two of which have no association with the response variable, so that $\beta_4 = \beta_5 = 0$. Suppose there are ten measurements of each variable. With five explanatory variables, this means there are only two measurements per variable. This low number is chosen on purpose, as we will see later. We first generate an artificial data set that implements this setup.

```
> set.seed(123)
> X <- matrix(rnorm(50), ncol = 5, byrow = TRUE)
> beta.lm <- matrix(c(1,1,1,0,0), ncol = 1)
> mu <- X %*% beta.lm
```

The matrix X, with five columns of ten rows each, contains the data values X_{ij}. The X_i are normally distributed random variables that are uncorrelated with each other. The matrix beta.lm, with one column, contains the β_j in Equation 8.3, and the last line of code implements the equation $\mu = X\beta$. By the way, "beta" is the name of a function in R, which is why this name is not used.

The total error of the fitted value $\hat{Y}_i$ of the regression is given by $\hat{Y}_i - \mu_i$. Adding and subtracting the expected value $E\{\hat{Y}_i\}$, this error may be written $(\hat{Y}_i - E\{\hat{Y}_i\}) + (E\{\hat{Y}_i\} - \mu_i)$. This divides the error into two components. The *bias component* is given by

$$bias_i = E\{\hat{Y}_i\} - \mu_i, \tag{8.4}$$

and represents the error caused by the model not fitting the data correctly. The *random error component* is given by

$$random \ error_i = \hat{Y}_i - E\{\hat{Y}_i\}, \tag{8.5}$$

which represents the deviation due to random error in the data. The mean square error of the fitted values is given by

$$MSE = \sum_{i=1}^{n} E\{(\hat{Y}_i - \mu_i)\}^2$$

$$= \sum_{i=1}^{n} E\{(E\{\hat{Y}_i\} - \mu_i) + (\hat{Y}_i - E\{\hat{Y}_i\})^2 \tag{8.6}$$

$$= \sum_{i=1}^{n} E\{(bias_i + random \ error_i)^2\}$$

It turns out (Kutner et al., 2005, p. 357) that this equation can be simplified to

$$MSE = \sum_{i=1}^{n}(E\{\hat{Y}_i\} - \mu_i)^2 + \sum_{i=1}^{n}(\hat{Y}_i - E\{\hat{Y}_i\})^2$$

$$= \sum_{i=1}^{n} bias_i^2 + \sum_{i=1}^{n} var\{\hat{Y}_i\}$$

(8.7)

Thus, the total mean square error is the sum of the total squared bias plus the sample variance of the $\hat{Y}_i$. As the complexity of the regression model, measured by the number $p - 1$ of explanatory variables, increases, the bias tends to decrease, because with more explanatory variables in the model the fit will be more precise and therefore the quantity $\Sigma(E\{\hat{Y}_i\} - \mu_i)^2$ will decrease. Conversely, as the complexity increases, the variance $\Sigma(\hat{Y}_i - E\{\hat{Y}_i\})^2$ tends to increase. This decreasing bias and increasing variance as explanatory variables are added to the model represents the bias-variance trade-off.

We can view the bias-variance trade-off from another perspective by using the example begun above. We will fit models with varying numbers of explanatory variables to the data set $\{Y_i\}$ generated by $Y_i = \mu_i + \varepsilon_i$, where μ_i satisfies Equation 8.3 and ε_i is a normally distributed random variable uncorrelated with any of the X_i. We then estimate the squared bias and variance of the fit using Monte Carlo simulation.

First, we need a function `regress()` that computes the regression model.

```
> regress <- function(X, mu, pminus1){
+     Y <- mu + 0.25 * rnorm(10)
+     Yhat <- predict(lm(Y ~ X[,1:pminus1]))
+ }
```

The computation of `Yhat` in this function takes advantage of the fact that the function `lm()` interprets a matrix with $p - 1$ columns on the right-hand side of the formula statement as specifying that the formula consists of the sum of the $p - 1$ explanatory variables making up those columns. Now we are ready to run the Monte Carlo simulation.

```
> b2 <- numeric(5)
> v <- numeric(5)
> for (pminus1 in 1:5){
+     set.seed(123)
+     Yhat <- replicate(1000, regress(X, mu, pminus1))
+     EYhat <- rowMeans(Yhat)
+     b2[pminus1] <- sum((EYhat - mu)^2) / 1000
+     v[pminus1] <- sum((Yhat - EYhat)^2) / 1000
+ }
```

The array `Yhat` generated in the fifth line consists of ten rows and one thousand columns. Each column is the set of predicted values $\hat{Y}_i$, $i = 1,...,10$ of one regression. The values $E\{\hat{Y}_i\}$ are the row means of `Yhat`. The average value of the squared bias and the variance are therefore computed by dividing the sums in Equation 8.7 by the number of replications. The code that created `X` and `mu` is listed at the start of the section.

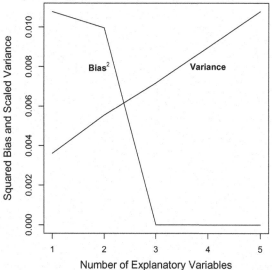

FIGURE 8.1

Plots of squared bias and scaled variance for a regression model with one through five explanatory variables based on artificial data.

Figure 8.1 shows a plot of the squared bias b2 and the variance v, with v scaled to have the same maximum value as b2. The figure shows that as the number of explanatory variables increases, the bias decreases and the variance increases. In this particular example, the model with three or more variables fits the data, so there is no bias.

Pursuing this example a bit further, let's examine a plot of the fits $\hat{Y}_i$ against the observed values Y_i for the models with one, three, and five explanatory variables. (True confession: this example is specifically rigged, in a way shown below, to emphasize the effect and make a point). Once again, we create an artificial data set satisfying Equation 8.3, that is, with three explanatory variables that determine Y and two that are measured but have no effect on Y. First, we fit a regression line to ten data sets, using models with one, three, and five explanatory variables. Figure 8.2a shows the fitted values $\hat{Y}_i$ versus Y_i for the three models. The models with $p - 1 = 3$ and $p - 1 = 5$ both provide a better fit than the model with $p - 1 = 1$; the simplest model *underfits* the data. Next, we generate a new data set and try to fit it with the same three models. Here is where the rigging takes place: to emphasize the effect, the values of X_{i5} are multiplied by ten. Figure 8.2b shows the results. The model with $p - 1 = 5$ does not provide as good a fit as the model with $p - 1 = 3$. The most complex model *overfits* the data. That is, when variables that should not be in the model are included, the effect of these variables is unique to the data set that generates the model and is inappropriate for other data sets. Thus, high bias is associated with underfitting and high variance with overfitting.

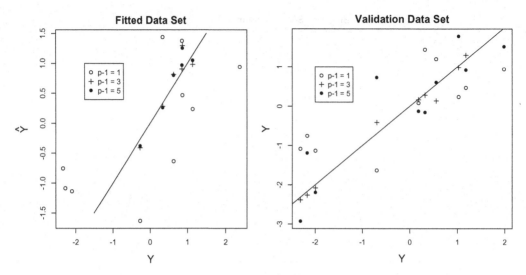

FIGURE 8.2
Regression plots of three linear models constructed with artificial data. The model with $p = 1$ underfits the data; the model with $p = 3$ fits the data; and the model with $p = 5$ overfits the data. (a) Data used to fit the models. (b) Data used to validate the models.

Three statistics are widely used as a means of measuring the trade-off between bias and variance. These are Mallows' C_p (Mallows, 1973), the Akaike information criterion (AIC) (Akaike, 1974), and the Bayesian Information Criterion (BIC) (Schwarz, 1978). These shown in Table 8.1. Each has the form measure = bias penalty + complexity penalty. The value of the *bias penalty* is an error sum of squares *SSE* (equation A.27, Appendix A.2), and the *complexity penalty* is an increasing function of the number of explanatory variables. Mallows' C_p is derived and discussed by Kutner et al. (2005, p. 357), and the AIC and BIC are discussed by Ramsey and Schafer (2002, p. 356). The use of the AIC and BIC is relatively simple: one tries to make them as small (or as negative) as possible. The BIC differs from the AIC only in the complexity penalty. The use of Mallows' C_p is slightly more complex: one searches for the model having the smallest value of C_p that approximately equals the value of p.

In general, automated model selection methods fall into two broad categories: all possible subsets methods and stepwise methods. All possible subsets methods use a statistic such

TABLE 8.1

Statistics that Can be Used in Model Selection

Name	Symbol	Bias Penalty	Complexity Penalty
Mallows' C_p	C_p	$\dfrac{SSE}{MSE_{full}}$	$2p - n$
Akaike info. criterion	AIC	$n \ln SSE$	$2p$
Bayesian info. criterion	BIC	$n \ln SSE$	$p \ln n$

Note: Each of these is the sum of a bias penalty that is computed from the error sum of squares (*SSE*), and a complexity penalty that is computed from the number n of observations and the number p of regression coefficients. In the formula for *Cp*, the term *MSEfull* represents the mean squared error of the full model with all explanatory variables Included.

as Mallows' C_p to compare many or all possible combinations of the explanatory variables. The R package `leaps` (Lumley and Miller, 2017) contains a widely used set of functions for carrying out all possible subsets model selection. Stepwise methods use a statistic such as a p value or the Akaike information criterion to add or remove variables one at a time until the "best" version is obtained. Forward selection starts with no explanatory variables and adds them one at a time, backward selection starts with the full model and removes variables one at a time, and bidirectional selection combines forward and backward selection. If one's goal is to preserve the statistical properties of the full model as well as possible, then backward selection, in which one starts with the full model and removes variables one at a time, may be the best procedure (Snedecor and Cochran, 1989). One problem with stepwise methods is that they may converge to a model that is locally optimal, that is, one that is better than similar models but is not globally optimal over all of the explanatory variables. There are other, more subtle problems with stepwise selection as well (Harrell, 2001, p. 56).

In order to describe the approach to variable selection taken in this book, we need to develop a few more concepts. These include a more detailed look at the effects of multicollinearity as well as at two graphical tools: the added variable plot and the partial residual plot. These are the subjects of the next section.

8.2.2 Multicollinearity, Added Variable Plots, and Partial Residual Plots

Suppose that Y is a response variable and that there are two potential explanatory variables, X_1 and X_2. Consider the regression model

$$Y_i = \beta_0 + \beta_1 X_{i1} + \beta_2 X_{i2} + \varepsilon_i. \tag{8.8}$$

Both X_1 and X_2 are explanatory variables of Y, but if X_1 and X_2 are themselves correlated, that is, if multicollinearity as defined in the previous section exists in the data, then X_1 and X_2 are to some extent providing duplicate information. Before studying an example from a real data set, will first examine two extreme cases using artificial data. Our example is motivated by a similar one in Legendre and Legendre (1998, p. 591). In the first data set (Data Set A), X_1 and X_2 are random variables generated independently from a unit normal distribution (Figure 8.3a).

```
> set.seed(123)
> XA <- cbind(rnorm(100), rnorm(100))
```

Computing the sample variance-covariance matrix shows that X_1 and X_2 are almost uncorrelated.

```
> var(XA)
            [,1]          [,2]
[1,]  0.83323283 -0.04372107
[2,] -0.04372107  0.93506310
```

We construct a response variable Y by adding X_1 and X_2 together with a random error term, and then we compute the coefficients of the linear model. Note that $\beta_0 = 0$, $\beta_1 = \beta_2 = 1$, and the magnitude of the error is roughly one-tenth that of the regression coefficients. Some of the output of the function `summary()` is deleted.

```
> eps <- rnorm(100)
```

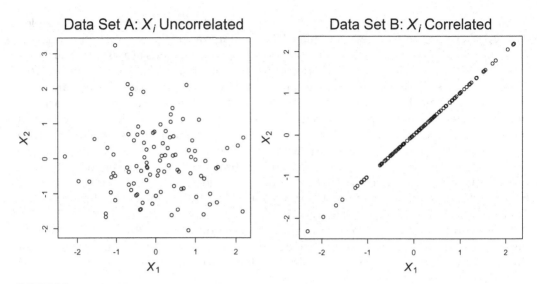

FIGURE 8.3
(a) Scatterplot of Data Set A, two uncorrelated variables X_1 and X_2 generated from a unit normal distribution.
(b) Scatterplot of Data Set B, two highly correlated variables.

```
> YA <- rowSums(XA) + 0.1 * eps
> summary(lm(YA ~ XA))
Coefficients:
            Estimate Std. Error t value Pr(>|t|)
(Intercept) 0.013507   0.009614   1.405    0.163
XA1         0.986683   0.010487  94.087   <2e-16 ***
XA2         1.002381   0.009899 101.256   <2e-16 ***
Multiple R-squared: 0.9947,     Adjusted R-squared: 0.9946
```

Next, we remove X_2 from the model and recompute.

```
> summary(lm(YA ~ XA[,1]))
Coefficients:
            Estimate Std. Error t value Pr(>|t|)
(Intercept) -0.08954    0.09824  -0.911    0.364
XA[, 1]      0.93409    0.10764   8.678 8.89e-14 ***
Multiple R-squared: 0.4345,     Adjusted R-squared: 0.4288
```

Although the R^2 of the model with only X_1 is much lower than that of the model with both X_1 and X_2, the values of the estimated coefficients b_1 of X_1 are similar in both models and are quite close to the true β_1. Because they are independent, neither explanatory variable shares with the other any of the information provided about Y. Therefore, the value of b_1 does not depend on whether or not X_2 is in the model.

In the second case (Data Set B), X_1 is equal to its counterpart in Data Set A. The values of X_2 in Data Set B are a linear combination of X_1 and X_2 from Data Set A, arranged to put almost all the weight on X_1.

```
> XB <- cbind(XA[,1], XA[,1] + 0.0001 * XA[,2])
```

In this case, X_1 and X_2 are very highly correlated (Figure 8.3b).

```
> var(XB)
           [,1]        [,2]
[1,]  0.8332328  0.8332285
[2,]  0.8332285  0.8332241
```

With both variables in the model, the regression coefficients are quite different from the actual values β_i used to compute the response variable. The fit, however, is very good ($R^2 = 0.997$).

```
> YB <- rowSums(XB) + 0.1 * eps
> summary(lm(YB ~ XB))
Coefficients:
              Estimate Std. Error t value Pr(>|t|)
(Intercept)   0.013507   0.009614   1.405    0.163
XB1         -22.824605  98.994169  -0.231    0.818
XB2          24.811288  98.994688   0.251    0.803
Multiple R-squared: 0.9973,     Adjusted R-squared: 0.9973
```

When X_2 is removed from the model, however, unlike the case with Data Set A, the coefficient of X_1 changes dramatically, and the change in R^2 is negligible.

```
> summary(lm(YB ~ XB[,1]))
Coefficients:
              Estimate Std. Error t value Pr(>|t|)
(Intercept) 0.013251   0.009514   1.393    0.167
XB[, 1]     1.986553   0.010424 190.577   <2e-16 ***
Multiple R-squared: 0.9973,     Adjusted R-squared: 0.9973
```

In model B, the explanatory variable X_2 is highly correlated with X_1 (this is the multicollinearity), while in the first model the two explanatory variables are independent. The most obvious effect of multicollinearity is that the value of the regression coefficient of one variable depends on whether the other variable is in the model.

Formally, the value of the regression coefficient b_i represents the marginal effect of the explanatory variable X_i on the response variable Y, assuming that all of the other explanatory variables are held constant (Kutner et al, 2005, p. 216). For example, if the value of X_{i1} changes by 1 unit, and the value of X_{i2} is held constant, the value of Y_i changes by b_1 units. This interpretation is still true mathematically in the presence of multicollinearity, but it no longer has any meaning. For example, suppose that X_1 represents sand content, X_2 represents clay content, and that the silt content is zero. Then $X_2 = 1 - X_1$. It makes no sense to think of holding X_1 constant while varying $1 - X_1$ (Harrell, 2001, p. 97). The best one can say is that, for example, the value of the regression coefficient b_1 represents the marginal effect of the explanatory variable X_{i1} on the response variable Y_i *given that the other variables are in the model*. It makes no sense biophysically, however, to think of holding one variable constant while varying another correlated variable.

A second deleterious effect of multicollinearity can be seen by returning to the output of the `summary()` function for the two artificial data sets displayed above. As we have already mentioned, the R^2 for the full model in Data Set B is about the same as that for the full model in Data Set A. This illustrates that multicollinearity does not substantially affect the fit of the data. Comparing the standard errors of the regression coefficients of Data Set B with those of Data Set A, however, indicates that the standard errors of Data Set B are much higher. Multicollinearity increases the variability of the regression coefficients b_i, and in the presence of extreme multicollinearity these estimated regression coefficients can become so variable that they are almost worthless, as they are in this example. This effect exacerbates the already severe numerical difficulties that are sometimes associated with the spatial regression models discussed in Chapters 12 and 13.

For a physical example of multicollinearity, consider the two explanatory variables *Clay* and *SoilK* in Field 1 of Data Set 4. These variables both have a negative association with yield (Figure 7.25), so that in a regression model with only one explanatory variable, each generates a negative regression coefficient. The data are loaded into R as described in Appendix B.4 and Section 7.5.

```
> print(coef(lm(Yield ~ Clay, data = data.Set4.1)), digits = 4)
(Intercept)          Clay
     9454.9        -176.5
> print(coef(lm(Yield ~ SoilK, data = data.Set4.1)), digits = 3)
(Intercept)         SoilK
     6994.1         -21.5
```

In a model with both of these variables, the regression coefficient of *SoilK* is positive (and much smaller).

```
> print(coef(lm(Yield ~ Clay + SoilK, data = data.Set4.1)), digits = 2)
(Intercept)          Clay         SoilK
    9448.72       -177.09          0.15
```

This indicates that even though the effect of soil potassium by itself has a negative association with yield, the effect of increasing soil potassium given the presence of clay in the model is associated with an increase in yield. It is a bit surprising that the regression coefficient is negative when *SoilK* is the unique explanatory variable in the simple linear regression model for *Yield*. Does this mean that potassium is bad for wheat? Probably not. There are two potential explanations, which are not mutually exclusive. One is the mining effect of non-labile mineral nutrients discussed below in Section 8.3. A more likely explanation in this case, however, is the strong association of soil potassium with soil clay content itself. The area of the field with high clay content is also the area with high potassium content. It is likely that the apparent negative effect of soil potassium on yield occurs because any positive effect of soil potassium is outweighed by the negative effect of the high clay content. The sign of the regression coefficient for *SoilK* changes when *Clay* is added to the model because each variable contains information about the other one. In effect, this "loads" the coefficients of the other variable, that is, part of the effect of an explanatory variable is due to the effect on it of other variables included in the model (Fox, 1997, p. 127). Perhaps one can begin to see the reason for the "last refuge of scoundrels" comment.

The impact of multicollinearity can be assessed by means of *added variable plots*, which are also called *partial regression plots* (Kutner et al., 2005, p. 268; Fox, 1997, p. 281). To avoid confusion with partial residual plots, which are not the same thing and which are described below, we will use the former term. An added variable plot is a way of determining the effect of adding a particular explanatory variable to a model that already contains a given set of other explanatory variables. We will consider the case of two explanatory variables X_1 and X_2; the extension to more variables is straightforward. Consider a multiple regression model $Y_i = \beta_0 + \beta_1 X_{i1} + \varepsilon_i$, having one explanatory variable. We want to understand the effect of adding a potential second explanatory variable X_2 to the model, noting that it may be correlated with X_1. The added variable plot for X_2, given that X_1 is in the model, is constructed as follows. First, we compute the residuals, denoted $e_i(Y \mid X_1)$, of the regression of Y on X_1:

$$\hat{Y}_i = b_0 + b_1 X_{i1}$$

$$e_i(Y \mid X_1) = Y_i - \hat{Y}_i.$$

(8.9)

Next, we compute the residuals of the regression of X_2 on X_1:

$$\hat{X}_{i2} = c_0 + c_1 X_{i1}$$

$$e_i(X_2 \mid X_1) = X_{i2} - \hat{X}_{i2}.$$

(8.10)

Intuitively, the residuals of the regression of Y on X_1 contain the information about the variability of Y not explained by X_1. The residuals of the regression of X_2 on X_1 contain the information about the variability of X_2 not explained by X_1. If X_2 adds information to the model not obtained from X_1, then there will be a pattern in the added variable plot that indicates the effect of adding the explanatory variable X_2. If there is a strong linear relationship between X_2 and X_1, then there will not be much information about the variability of Y remaining in the residuals of the regression of X_2 on X_1, and therefore the added variable plot will be more or less free from pattern.

We can illustrate added variable plots using the two artificial data sets generated above. First, consider Data Set A, in which X_1 and X_2 are independent.

```
> e.YA1 <- residuals(lm(YA ~ XA[,1]))
> e.XA21 <- residuals(lm(XA[,2] ~ XA[,1]))
> plot(e.XA21, e.YA1, #Fig. 8.4a
+     main = "Added Variable Plot of Data Set A", cex.main = 2,
+     xlab = expression(Residuals~of~italic(X)[2]),
+     ylab = expression(Residuals~of~italic(Y)),
+     cex.lab = 1.5)
```

The residuals of the regression of Y on X_1 are very closely associated with the residuals of the regression of X_2 on X_1 (Figure 8.4a). This indicates that X_2 contains a great deal of information about Y that is not contained in X_1. Repeating the same procedure with Data Set B (Figure 8.4b) indicates almost no relationship between the residuals. This indicates that almost all of the information about Y contained in X_2 has been provided by including X_1 in the model.

Another property of added variable plots is illustrated by computing the regression of $e(Y \mid X_1)$ on $e(X_2 \mid X_1)$ (see Equations 8.9 and 8.10). For convenience, the coefficients of the original regressions are also displayed.

```
> coef(lm(YA ~ XA))
(Intercept)            XA1               XA2
0.01350654      0.98668285     1.00238113
> coef(lm(YA ~ XA[,1]))
p(Intercept)       XA[, 1]
  -0.0895413      0.9340863
```

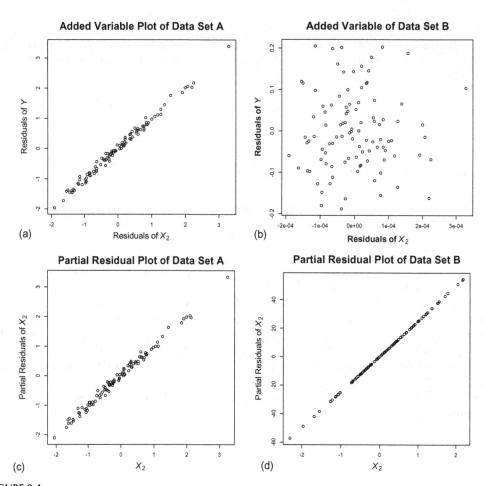

FIGURE 8.4
(a) Added variable plot of Data Set A. (b) Added variable plot of Data Set B. (c) Partial residual plot of Data Set A. (d) Partial residual plot of Data Set B.

```
> coef(lm(e.YA1 ~ e.XA21))
  (Intercept)        e.XA21
-1.334573e-17  1.002381e+00
```

The coefficient of the added variable regression, 1.002381, is equal to the coefficient b_2 of the full regression model. This is of course true for Data Set B as well.

```
> coef(lm(YB ~ XB))
(Intercept)           XB1            XB2
 0.01350654 -22.82460503   24.81128788
> coef(lm(YB ~ XB[,1]))
(Intercept)       XB[, 1]
 0.01325148  1.98655266
> coef(lm(e. YB1 ~ e.XB21))
```

```
(Intercept)          e.XB21
2.167670e-18   2.481129e+01
```

Thus, if we use added variable regression to help build a multiple regression model, we can see the value of the regression coefficient in advance when considering whether or not to add a variable. Added variable plots are also very useful for identifying discordant data records, as we will see in the next section.

Although added variable plots can be constructed using code similar to that given above, it is much easier and more effective to use the function `avPlots()` of the `car` package (Fox and Weisberg, 2011). Not only does this function provide plots similar to (but nicer than) those of Figure 8.4, it also provides the ability to label points on the graph. This is very useful for identifying outliers.

The *partial residual plot* (which is also called the *component residual plot*) provides an alternative to the added variable plot to explore the effects of correlated explanatory variables (Larsen and McCleary, 1972). Again, suppose we wish to consider adding X_2 to the model given that X_1 is already in. Once again, the extension to higher values of p is direct. From Equation 8.9, the vector of residuals of the regression of X_2 on X_1 is denoted $e(X_2 \mid X_1)$. The partial residual $e*(Y \mid X_1, X_2)$ is defined as the sum of the residuals of the full regression plus the X_2 component:

$$\hat{Y}_i = b_0 + b_1 X_{i1} + b_2 X_{i2}$$

$$e_i(Y \mid X_1, X_2) = Y_i - \hat{Y}_i \tag{8.11}$$

$$e_i*(Y \mid X_1, X_2) = e_i(Y \mid X_1, X_2) + b_2 X_{i2}.$$

The partial residual plot is a plot of $e*(Y \mid X_1, X_2)$ against X_2. To see why this works (Fox, 1997, p. 314; Ramsey and Schafer, 2002, p. 323) suppose that the model $Y_i = \beta_0 + \beta_1 X_{i1} + f(X_{i2}) + \varepsilon_i$ is an accurate representation of the data, but we are not sure of the form of the function $f(X_{i2})$. We can write

$$f(X_{i2}) = Y_i - \beta_0 - \beta_1 X_{i1} - \varepsilon_i. \tag{8.12}$$

We would like to plot $f(X_{i2})$ against X_2 to see its form, at least within the error term ε_i but we can't. However, if $f(X_{i2})$ is approximately linear (i.e., $f(X_{i2}) \cong b_2 X_{i2}$), then the regression of Y on X_1 and X_2 will give us estimates of b_0 and b_1 that we can use to approximate the plot. Furthermore, since $e_i(Y \mid X_1, X_2) = Y_i - \hat{Y}$ we have

$$f(X_{i2}) \cong Y_i - b_0 - b_1 X_{i1} - \varepsilon_i \tag{8.13}$$

But

$$
\begin{aligned}
Y_i - b_0 - b_1 X_{i1} &= \hat{Y}_i + e_i(Y \mid X_1, X_2) - b_0 - b_1 X_{i1} \\
&= b_0 + b_1 X_{i1} + b_2 X_{i2} + e_i(Y \mid X_1, X_2) - b_0 - b_1 X_{i1} \\
&= b_2 X_{i2} + e_i(Y \mid X_1, X_2) \\
&= e_i*(Y \mid X_1, X_2)
\end{aligned}
\tag{8.14}
$$

Therefore, a plot of the partial residuals may give an approximation of a plot of the function $f(X_{i2})$.

To illustrate partial residual plots, we again use the example data sets A and B. As with added variable plots, the car package has a function crPlots() (for component residuals) that would actually be used in practice, but we will first do it by hand for illustrative purposes. Here the code for Data Set A.

```
> A.lm <- lm(YA ~ XA)
> e.PRA <- residuals(A.lm)
> Y.PRA <- e.PRA + coef(A.lm)["XA2"] * XA[,2]
> plot(XA[,2], Y.PRA, main = "Partial Residual Plot of Data Set A",
+     cex.main = 2, xlab = expression(italic(X)[2]),
+     ylab = expression(Partial~Residuals~of~italic(X)[2]),
+     cex.lab = 1.5) # Fig. 8.4c
> coef(lm(Y.PRA ~ XA[,2]))
 (Intercept)        XA[, 2]
7.079840e-18 1.002381e+00
```

Your intercept term may be different from mine due to different round-off errors. Figure 8.4c shows the partial residual plot. The code for Data Set B is analogous, but the results are not.

Here are the summaries for the regression of the added variable regression and the partial residual regression.

```
> #Added variable regression
> summary(lm(e.YB1 ~ e.XB21))
Coefficients:
            Estimate Std. Error  t value Pr(>|t|)
(Intercept) 2.168e-18  9.464e-03 2.29e-16    1.000
e.XB21      2.481e+01  9.849e+01    0.252    0.802
Multiple R-squared: 0.0006472,  Adjusted R-squared: -0.00955
F-statistic: 0.06346 on 1 and 98 DF,  p-value: 0.8016
> #Partial residual regression
> summary(lm(Y.PRB ~ XB[,2]))
Coefficients:
            Estimate Std. Error t value Pr(>|t|)
(Intercept) 3.577e-16  9.511e-03 3.76e-14        1
XB[, 2]     2.481e+01  1.042e-02    2381  < 2e-16 ***
Multiple R-squared:   1,      Adjusted R-squared:       1
F-statistic: 5.669e+06 on 1 and 98 DF, p-value: < 2.2e-16
```

The partial residual plot for Data Set B is shown in Figure 8.4d. The two regression coefficients are the same, as the theory predicts, but the plots are somewhat different. To see why, note that due to the instability of this highly multicollinear system, the regression coefficients are very large.

```
> coef(B.lm)
 (Intercept)          XB1          XB2
  0.01350654 -22.82460503   24.81128788
```

The sum $e(Y | X_1, X_2) + b_2 X_2$ is approximately equal to $b_2 X_2$. Therefore, the partial residual regression is approximately a regression of $b_2 X_2$ on X_2, which yields the correct slope b_2 but does not tell us anything explicitly about how much X_2 contributes to the model. This is an example of the numerical problems caused by multicollinearity.

Both the added variable plot and the partial residual plot potentially have other applications besides those described above, and both also have problems that in some cases reduce their effectiveness. Assuming that you want the good news first, let's look at some other potential applications. Both the added variable plot and the partial residual plot can often be used to detect nonlinearity in the contribution of a variable to the model. In addition, both plots can be used to identify the presence of outliers in the data. Partial residuals plots are most effective when the effect of the explanatory variable under consideration is small relative to that of those already in the model (Ramsey and Schafer, 2002, p. 325). Faraway (2002) compares an added variable plot and a partial residual plot for an example data set and concludes that added variable plots are better for detection of discordant data records, and partial residual plots are better for detecting the form of the relation of the response variable to the proposed added variable.

Now for the bad news (Fox. 1997, p. 315; Kutner et al., 2005, p. 389). These plots do not work well if the relation between the explanatory variables is itself nonlinear. Also, the plots may not detect interaction effects. Finally, as shown in the example of Data Set B above, high multicollinearity may invalidate the conclusions drawn from the plot. All this implies that these tools, like all tools, cannot be used blindly. Nevertheless, we shall have occasion to use both added variable and partial residual plots in the analyses of our data sets.

8.2.3 A Cautious Approach Model Selection as an Exploratory Tool

Our objective is to develop a procedure for using model selection as an exploratory tool to help determine potential explanatory variables for ecological processes. Ramsey and Schafer (2002, p. 346) give four good reasons why great caution is necessary when using model selection for this purpose:

1. Inclusion or exclusion of explanatory variables is strongly affected by multicollinearity between them.

2. The interpretation of multiple regression coefficients is difficult in the presence of multicollinearity because their value depends on the other variables in the model.

3. The interpretation of a regression coefficient as the effect of a variable given that all other variables are held constant makes no sense when there is multicollinearity.

4. Causal interpretations are in any case suspect in observational studies when there is no control over the values of explanatory variables.

A fifth problem not included in this list is that several different models often give very similar values for the measures of fit (C_p, AIC, and BIC) used in model selection.

On the other hand, graphical tools such as the scatterplot matrix and the thematic map do not provide conclusive evidence of influence either. The difference is that nobody expects them to. One of the problems with multiple regression, which is perhaps another source of the "last refuge of scoundrels" comment mentioned at the start of this section, is that with its precise numerical values and testable hypotheses, multiple regression has an aura of credibility that graphical methods lack. The trick in using multiple regression as an exploratory tool is not to take the results too seriously (or to present them as anything more than speculation). With that said, let's get on to developing our procedures. Further discussion of methods of model selection (including the very important *shrinkage methods*, which we do not have space to cover) is contained in the references listed in Section 8.7.

We will follow the practice advocated by Henderson and Velleman (1981) and by Nicholls (1989), which is to test the variables for inclusion manually, using ecological knowledge as well as statistical tools. Our procedure follows closely that of Henderson and Velleman (1989), which is to use any available knowledge of the biophysics of the system to guide variable selection. This biophysical knowledge is augmented by tools such as the AIC, added variable plots, partial residual plots, and the general linear test (Appendix A.3) to examine the effects of the variables on the model, and to add or remove variables based on a combination of the computations and knowledge of the ecological processes involved. The procedure will be used to create a set of candidate models that can be further examined, if possible, using other analytical methods. If no further analysis is possible, the candidate models can be used to suggest experiments that can be used to distinguish between the competing models.

Burnham and Anderson (1998, p. 17) refer to the process of iteratively searching for patterns in the data as "data dredging." More recently it has come to be known by the less pejorative term "data mining." By whatever name, even Burnham and Anderson (1998) allow that it is not necessarily a bad thing, but that it must be done openly. They provide a nice analogy of model selection to an auto race (Burnham and Anderson, 1998, p. 54). Before the start of the race there is a qualifying event to determine which drivers can enter. This is analogous to the selection of the candidate models from among all the possible models. The race is run and the winner is determined. Often the winning driver is not necessary the fastest. Random events such as crashes or breakdowns may eliminate other drivers, and indeed in another race with the same set of drivers, a different driver may win. In the same way, the model determined to be "best" for a particular data set may not turn out to be the best for other data sets from the same system. Therefore, one should not read too much into the results of one particular "race" among the models. In this context, it is very important that "data dredging" not lead to "data snooping." This was defined in Section 7.1 as letting the confirmatory process be influenced inappropriately by the exploratory process through an introduced bias. If in this chapter we identify some candidate models through the exploratory process, we must not in a succeeding chapter forget where these models came from and use a hypothesis test inappropriately to assign a significance to a particular model that it does not deserve.

8.3 Building a Multiple Regression Model for Field 4.1

We will carry out our first multiple linear regression analysis on data from Field 4.1. The objective is to develop a set of candidate models for yield as a function of other variables. Our goal is not simply to predict yield but to improve our understanding of how the various explanatory variables interact to influence yield. To summarize the knowledge gained so far about Field 4.1: the field displays the greatest differences in its properties in a north to south direction. Crop yield is highest in the southern part of the field and declines markedly in the north. Dry soil infrared reflectance (Figure 7.23a) indicates that the northernmost part of the field retains water more than the southern part. Clay content is highest in the northern two-thirds of the field. This leads to the speculation that aeration stress may contribute to the low yield in some parts of the north of the field. Of the mineral nutrients, neither soil potassium nor leaf nitrogen is at any sample point below the level at which a yield response to increased fertilization would be expected. The negative association of yield with soil nutrients evident in the scatterplots appears counterintuitive. Yield should increase with increasing nutrient levels. Webster and Oliver (2001, p. 201) point out that this seemingly paradoxical observation of higher nutrient levels with lower yields may be due to increased uptake of these nutrients in the high yielding areas, where they are not limiting. The negative association between yield and both soil nitrogen and potassium occurs in the high yield areas and is consistent with the observation in this data set that these nutrient levels exceed the minimum recommended values. However, there is an indication that soil phosphorous levels might be low in the northern part of the field (Figure 7.28a).

As a first step, we will select from Table 7.6 the explanatory variables that will be tested as candidates for inclusion in the regression model for *Yield*. Since we want to gain insight into the possible ecological processes that might influence yield, we will not consider gauge variables in the sense of Figure 7.1. The variable *SoilTOC* is almost perfectly correlated with *SoilTN* (Figure 7.25), so we will exclude it. The soil texture components sum to 100, and *Sand* has a strong negative association with *Clay* (Figure 7.25), so we will exclude it as well. The scatterplot matrix indicates that all of the endogenous variables (*CropDens*, *LeafN*, and *FLN*) share some degree of association with each of the exogenous variables. For now we will take the endogenous variables out of the model. We can use the term "agronomic" to refer to the variables that remain.

The scatterplot matrix of *Yield* and the agronomic variables (Figure 7.25) displays a "sideways parabola" relationship that indicates that there is an interaction between some of the explanatory variables associated with yield. Some yield values trend in one direction with the explanatory variable, and some are either not associated or trend in the opposite direction. This may indicate that yield has a different relationship with these variables in different parts of the field. In Exercise 8.3, you are asked to create these scatterplots separately for the northern 62 points (nine rows) and the southern 24 points (four rows). These plots indicate that there is a substantial geographic split, with the northern two-thirds of the field displaying a different behavior from the southern third.

The bifunctional relationship between *Yield* and the explanatory variables makes it impossible to construct a meaningful regression model that describes yield response over the entire field without including interaction terms. Once again, we have a decision to make. Since the variables have linear relationships in the northern and southern parts of the field, we could split the field into two separate areas for regression analysis, with 62 data locations in the north and 24 in the south. Unlike the analysis of Data Set 2, however, we will not split the field. The interactions may provide interesting information, and the data set is relatively small, which may lead to a degrees of freedom problem. There are lots of interactions that we could consider, and using the results of previous analyses to pick out some of them looks suspiciously like "data snooping." We will do it nevertheless, because some interactions make neither biophysical nor empirical sense. Based on the analysis in Section 7.5, the almost inescapable conclusion is that the mineral nutrients and pH interact with soil texture. Therefore, our regression model will include interaction terms of *Clay* with *SoilP*, *SoilK*, *SoilTN*, and *SoilpH*. A second possible interaction, justifiable on a biochemical basis, is between *SoilP* and *SoilpH*. In Exercise 8.8, you are asked to test for the possibility of this interaction being significant in the model.

Before we begin the regression analysis, we note that the exogenous variables *Weeds* and *Disease* and the endogenous variable *CropDens* are measured on an ordinal scale. Harrell (2001, p. 34), Kutner et al. (2005, p. 313), and Fox (1997, p. 135) all describe methods

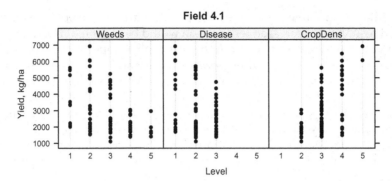

FIGURE 8.5
Dot plots of *Yield* vs. *Weeds*, *Disease*, and *CropDens* in Field 1 of Data Set 4.

of dealing with ordinal variables. These methods may not, however, capture the particular relationships between crop yield and these quantities. Dot plots of these relationships are shown in Figure 8.5. The plots of *Yield* against all three variables look reasonably linear. Mosteller and Tukey (1977, p. 103) describe a method of re-expression suggested for converting ordinal data into interval data. Similar methods may be used to suggest transformations of an interval or ratio scaled explanatory variable, and they may be applied to the response variable as well as the explanatory variables. Henderson and Velleman (1981) present an excellent example in which the response variable is gasoline mileage. They show that transforming miles per gallon to gallons per mile (i.e., inverting the response variable) makes the relationships with the explanatory variables much closer to linear. Returning to the re-expression of ordinal scale variables, it turns out that for our data the re-expression does not provide any improvement over the use of the ordinal variable itself. For this reason, we will stay with the use of *Weeds* and *Disease* in our regression model.

We are now ready to develop a multiple linear regression model for *Yield* in Field 4.1. In doing so, we need to again think about causal relationships and bias, which were touched upon in Section 8.2.3. The following discussion is taken from Fox (1997, p. 128), who in turn borrows from Goldberger (1973). Suppose we have a linear regression model with two explanatory variables:

$$Y_i = \beta_0 + \beta_1 X_{i1} + \beta_2 X_{i2} + \varepsilon_i. \tag{8.15}$$

where β_1 and β_2 are both nonzero. Fox points out that in interpreting Equation 8.15 one must distinguish between an *empirical model* and a *structural model*. In the case of the former, the only objective is to predict the value of Y and there is no assumption of a causal relation between the explanatory variables and the response variable. In the latter case, however (and this is our case), one assumes that the model is a representation of the actual relationships between the variables. Suppose that instead of fitting Equation 8.15, we fit a model with a single explanatory variable,

$$Y_i = \beta_0' + \beta_1' X_{i1} + \varepsilon_i'. \tag{8.16}$$

If X_1 and X_2 are correlated then, since the error term ε' includes the effect of X_2, it is correlated with X_1. This is an error in the specification of the model and results in β_1' being a biased estimate of β_1. Fox (1997, p. 128) points out that the nature of the bias depends on the causal relationship between X_1 and X_2.

To be specific, consider again the model discussed in Section 8.2.2 of Field 4.1 involving *Clay* and *SoilK* as explanatory variables for *Yield*. Suppose first that X_1 represents *Clay* and X_2 represents *SoilK*. Although in the very long-term soil potassium content may influence clay content (Jenny, 1941), for our purposes we may assume that X_1 influences both X_2 and Y, but neither Y nor X_2 influence X_1. In this case, the bias in b_1 simply represents an indirect effect of clay content X_1 on Y by way of soil potassium X_2. Suppose the situation is reversed, however, so that X_2 represents *Clay* and X_1 represents *SoilK*. In this case X_2, which is not in the model, influences both X_1 and Y, so that the bias in b_1 represents a spurious component of the association between X_1 and Y. In our example, it may be that soil potassium X_1 actually does not influence yield at all, but merely appears to do so because of its association with clay content X_2. In order to avoid this sort of situation as much as possible, we will try to keep in mind influence relationships among the variables. In particular, this is why we exclude the endogenous variables *CropDens*, *LeafN*, and *FLN* from the model (and why we worry about distinguishing endogenous and exogenous variables).

The data are loaded as described in Appendix B.4 and Section 7.5. We create a subset of agronomic data as follows.

```
> agron.data <- with(data.Set4.1, data.frame(Clay, Silt,
+    SoilpH, SoilTN, SoilK, SoilP, Disease, Weeds, Yield))
```

There is one more important point to cover. We are constructing a regression model based on estimated values of the response variable *Yield*. Small differences in these estimated values may have dramatic consequences in the variable selection process (see Exercise 8.6). Therefore, your results may be different from those obtained in the book. With that caveat, let's get started. To get a general picture of the various regression relationships, we will first use the function `leaps()` from the package `leaps` to carry out an all possible subsets analysis. This function uses the Mallows C_p criterion (Section 8.2.1) to select the "best" subsets of explanatory variables. If there is no bias in the model, then the expected value of C_p is p. Values of C_p larger than p are taken to indicate bias, and values less than p are taken to indicate random variation (Kutner et al., 2005, p. 357).

The function `leaps()` does not support interaction terms, so we must add these directly to the data set.

```
> agron.data$ClayP <- with(agron.data, Clay*SoilP)
> agron.data$ClayK <- with(agron.data, Clay*SoilK)
> agron.data$ClaypH <- with(agron.data, Clay*SoilpH)
> agron.data$ClayTN <- with(agron.data, Clay*SoilTN)
```

The first argument of `leaps()` is a matrix whose columns are the explanatory variables, and the second argument is a vector representing the response variable.

```
> X <- as.matrix(agron.data[,-which(names(agron.data) == "Yield")])
> Y <- agron.data$Yield
```

We can now apply the function, specifying the names of the columns.

```
> model.Cp <- leaps(X, Y, method = "Cp",
+    names = names(agron.data[,-which(names(agron.data)
+       == "Yield")]))
```

This produces a `leaps` object, `model.Cp`. This is a list, and `model.Cp$which` gives the explanatory variables in each subset, sorted by p and by C_p. A little exploration with the function `str()` indicates that `model.Cp$size` contains the size p of the model (the number of explanatory variables plus one) and `model.Cp$Cp` contains the corresponding C_p values.

Table 8.2 shows some of the output from the following statement.

```
> cbind(model.Cp$Cp, model.Cp$which)
```

Somewhat confusingly, the number in the first column is $p - 1$ rather than p. The table only shows some of the best subsets, and it does not show models that contain interaction terms but not the first order terms. Figure 8.6 shows plots of C_p against $p - 1$ together with the $C_p = p$ line. We are looking for the smallest value of p such that $C_p \leq p$ This occurs at $p = 6$ (i.e., $p - 1 = 5$). We will return to this figure below, and discuss the models shown in the figure. First, we will carry out a stepwise process so we can compare the results.

Since we already have the interaction terms in the object `agron.data`, we will continue for now to use these rather than specifying them directly in the formula argument. To construct candidate models, we will begin by using the Akaike information criterion in the manner suggested by Burnham and Anderson (1998), along with any biophysical knowledge we can apply. We will then see how the resulting models fit in with the `leaps()` results. We start with the full model and use the function `drop1()` to drop variables to arrive at candidates via backward selection. The output of the function `drop1()` is arranged in order of increasing values the AIC would take on if the variable were dropped. In Section 8.4.2, we are going to use `add1()` to construct a regression model for Data Set 1 using forward selection. In any real data analysis project, one should use all three methods, forward, backward, and bidirectional, and draw insight from a comparison of the results.

```
> model.full <- lm(Yield ~ ., data = agron.data)
> d1 <- drop1(model.full)
> d1[order(d1[,4]),]
Single term deletions
Model:
Yield ~ Clay + Silt + SoilpH + SoilTN + SoilK + SoilP + Disease +
    Weeds + ClayP + ClayK + ClaypH + ClayTN
        Df Sum of Sq      RSS     AIC
Clay     1       7779 38814660 1143.7
SoilK    1      10444 38817325 1143.7
ClayK    1      17184 38824065 1143.7
ClaypH   1      31004 38837885 1143.8
SoilpH   1      86604 38893485 1143.9
SoilTN   1     282543 39089424 1144.3
ClayTN   1     305760 39112641 1144.4
SoilP    1     604707 39411588 1145.0
Silt     1     782611 39589492 1145.4
ClayP    1     891388 39698269 1145.7
<none>              38806881 1145.7
Disease  1     991346 39798227 1145.9
Weeds    1    9110692 47917573 1161.8
```

This indicates that the AIC is reduced the most by dropping terms involving *Clay*, *SoilK*, and *ClayK*. Right away we have occasion to use knowledge of the biophysical process,

TABLE 8.2

Output of Function leaps() Applied to Data of Field 4.1, Showing Lowest Mallows' C_p Values for Increasing Number $p-1$ of Explanatory Variables in the Model

```
> print(cbind(model.Cp$Cp, model.Cp$which), digits = 3)
```

$p-1$	C_p	Clay	Silt	SoilpH	SoilTN	SoilK	SoilP	Disease	Weeds	ClayP	ClayK	ClaypH	ClayTN
		*	*	*	*	*	*	*					
3	20.05	1	0	0	0	1	0	0	0	0	1	0	0
3	22.10	0	0	1	0	0	0	0	0	0	0	1	0
3	22.64	1	0	0	0	0	1	0	0	1	0	0	0
		*	*	*DELETED*	*	*	*	*					
4	5.70	1	0	0	0	0	1	1	0	1	0	0	0
4	7.58	1	0	0	0	1	0	0	0	0	1	0	0
4	10.25	1	0	0	1	0	0	0	0	0	0	0	1
		*	*	*DELETED*	*	*	*	*					
5	3.68	0	0	1	0	0	1	1	0	1	0	1	0
5	4.14	1	0	1	0	0	1	1	0	1	0	0	0
		*	*	*DELETED*	*	*	*	*					
6	3.96	0	0	1	0	0	1	1	1	1	0	1	0
6	4.35	1	1	1	0	0	1	1	1	1	0	0	0
6	4.73	1	1	0	0	0	1	1	1	1	0	1	0
		*	*	*DELETED*	*	*	*	*					
7	5.29	0	1	1	0	0	1	1	1	1	0	1	0
7	5.56	0	0	1	0	0	1	1	1	1	1	1	0
7	5.73	0	0	0	1	0	1	1	1	1	0	0	1
		*	*	*DELETED*	*	*	*	*					
8	6.42	0	1	1	1	0	0	0	1	1	0	1	1
8	6.79	0	0	1	0	1	1	1	1	1	1	1	0
8	6.79	0	0	1	1	1	0	1	1	1	0	0	1
		*	*	*DELETED*	*	*	*	*					

Note: For each value of $p-1$, only the models with the lowest C_p are shown. Models that include an interaction without including both first-order terms are also not shown.

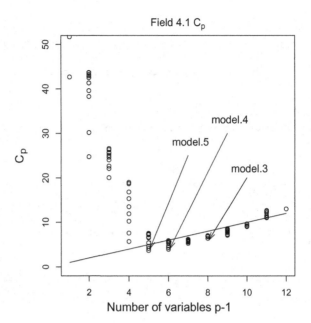

FIGURE 8.6
Plots of Mallows' C_p vs. p for the southern portion of Field 4.1, showing the C_p values of four of the models discussed in the text.

and to see the perils of a blind use of variable elimination based on numerical results. Everything we have seen in our exploration so far indicates that the variable *Clay* is of primary importance, at yet it is ranked as the first to leave. This is probably due to the effects of multicollinearity. Instead, we will drop the interaction terms *SoilK* and *ClayK*, whose resulting model has the same AIC. We will save the full model as model.1, since it includes all the variables.

```
> model.1 <- model.full
> model.test <- update(model.full, Yield ~ Clay + SoilpH + ClaypH +
+     Disease + SoilP + ClayP + Silt + Weeds)
```

Burnham and Anderson (1998) provide several convincing arguments and good examples for the assertion that hypothesis tests should not be used for variable selection. Nevertheless, this application provides a good opportunity to introduce the *general linear test* (Kutner et al., 2005, p. 72; Searle, 1971, p. 110), so we will provisionally ignore their good advice (spoiler alert: in this application it won't make any difference). We want to compare model.test and model.full. The general linear test is described in Appendix A.3. In brief, one can use it to test a null hypothesis of the form $H_0 : \beta = 0$, where β is a regression coefficient, against the alternative hypothesis $H_a : \beta \neq 0$. One regards the model under H_a as the *full* model and the model in which the null hypothesis is satisfied as the *restricted* model, because one or more of its parameters is restricted to a certain value. In our case, model.full is the full model and model.test is the restricted model in which the coefficients of *SoilK* and *ClayK* are restricted to the value zero. The restricted model is said to be *nested* in the full model. If the null hypothesis is true, then a statistic involving the ratio of the sums of squares (Equation A.44 in Appendix A.3) has an F distribution, and thus one can compute a p value based on this distribution. If based on this p value the null hypothesis is not rejected, then we can take this as an indication that the restricted model provides the same functionality as the full model.

In R, the general linear test is carried out using the function anova().

```
> anova(model.test, model.full)
Analysis of Variance Table
Model 1: Yield ~ Clay + Silt + SoilpH + SoilTN + SoilP + Disease +
Weeds + ClayP + ClaypH + ClayTN
Model 2: Yield ~ Clay + Silt + SoilpH + SoilTN + SoilK + SoilP +
Disease + Weeds + ClayP + ClayK + ClaypH + ClayTN
  Res.Df      RSS Df Sum of Sq      F Pr(>F)
1     75 38858288
2     73 38806881     2 51407 0.0484 0.9528
> AIC(model.full)
[1] 1391.757
> AIC(model.test)
[1] 1387.871
```

The p value of 0.95 indicates that we can accept model.test and continue looking for variables to drop. We invoke drop1() again (not shown) and as a result we drop *SoilpH* and *ClaypH*. Continuing in this way, we continue dropping variables until we reach a point where the next dropped variable makes the model worse, either because the AIC is higher or because the null hypothesis is rejected. Note in the list of dropped variables that *Clay* is now at the bottom, indicating the effects of multicollinearity.

```
> model.test <- update(model.full, Yield ~ Clay + Silt + SoilTN +
+      Weeds + ClayTN)
> AIC(model.full)
[1] 1391.757
> AIC(model.test)
[1] 1388.919
> anova(model.test, model.full)
Analysis of Variance Table
Model 1: Yield ~ Clay + Silt + SoilTN + Weeds + ClayTN
Model 2: Yield ~ Clay + Silt + SoilpH + SoilTN + SoilK + SoilP +
Disease + Weeds + ClayP + ClayK + ClaypH + ClayTN
  Res.Df      RSS Df Sum of Sq      F Pr(>F)
1     80 44185294
2     73 38806881  7   5378413 1.4453 0.2006
> d1 <- drop1(model.test)
> d1[order(d1[,4]),]
Single term deletions
Model:
Yield ~ Clay + Silt + SoilTN + Weeds + ClayTN
        Df Sum of Sq      RSS    AIC
<none>               44185294 1142.9
Silt     1   1663033 45848327 1144.0
Weeds    1  11199220 55384514 1160.3
ClayTN   1  11269156 55454450 1160.4
SoilTN   1  11335854 55521148 1160.5
Clay     1  23864449 68049743 1178.0
```

This invocation of `drop1()` indicates that there is no variable that can be dropped that would improve the model's AIC. There is no point in bothering with an application of the general linear test.

We will denote the model that we obtain by this blind use of stepwise regression as `model.2`.

```
> model.2 <- model.test
```

Returning to the Mallows' C_p output in Figure 8.6 and Table 8.2, we see that `model.2` is not included as a leading contender there. We will add three models based on the `leaps()` results.

```
> model.3 <- update(model.full, Yield ~ Clay + SoilpH +
+    SoilP + Weeds + ClayP + ClaypH)
> model.4 <- update(model.full, Yield ~ Clay + SoilpH + SoilP +
+    Weeds + ClayP)
> model.5 <- update(model.full, Yield ~ Clay + SoilP +
+    Weeds + ClayP)
```

Now we need to carry out some checks on the candidate models. We will pick `model.5` as an example, but these tests should also be carried out on the other candidate models as well.

A first question is whether there is substantial multicollinearity. The *variance inflation factor* (VIF) provides a good test for multicollinearity (Kutner et al., 2005, p. 408). This is available using the function `vif()` from the package `car`.

```
> vif(model.5)
     Clay       SoilP      Weeds       ClayP
8.953738   60.727652   1.118789   67.481848
```

Generally, one wants to keep the VIF below ten. The *Clay × SoilP* and *SoilP* terms are considerably above this, but interpretation of the VIF for interactions involving non-centered variables is difficult (Robinson and Schumacker, 2009).

Next, we check for influential variables. The tests for these are described in Appendix A.2.

```
> which(rowSums(influence.measures(model.5)$is.inf) > 0)
48 62 69 80 81 82
```

A call to the function `avPlots()` produces graphics (not shown) that do not support removal of any data.

```
> avPlots(model.5, id.n = 1)
```

This is confirmed by a call to the car function `outlierTest()`.

```
> outlierTest(model.5)
No Studentized residuals with Bonferonni p < 0.05
Largest |rstudent|:
   rstudent unadjusted p-value Bonferonni p
69 2.463534           0.015903           NA
```

Now we construct the partial residuals plot to determine whether either of the variables should have a higher order term. Calls to the function crPlot() (not shown) indicate that linear terms suffice.

The next question is whether there is substantial heteroscedasticity (non-constant variance) among the residuals. One test for this is a plot of the residuals vs. fits (not shown). We will carry out a Breusch-Pagan test (Kutner et al., 2005, p. 118). The function for this is in the lmtest package (Zeileis and Hothorn, 2002).

```
> bptest(model.5)
        studentized Breusch-Pagan test
data:  model.5
BP = 8.614, df = 4, p-value = 0.07151
```

The results indicate a bit of a problem with heteroscedasticity. As of now, however, we will not take any corrective action. The next step is to check for normality of the residuals. The *QQ* plot is a possibility (Kutner et al., 2005, p. 110), and I have included this in the code file. Personally, however, I find these somewhat hard to interpret, so instead I will do a Shapiro-Wilk normality test (Shapiro and Wilk, 1965) on the residuals.

```
> shapiro.test(residuals(model.5))
        Shapiro-Wilk normality test
data:  residuals(model.5)
W = 0.98928, p-value = 0.7065
```

Normality looks good.

The final test we will carry out involves the predicted sum of squares, or *PRESS*, statistic (Allen, 1971; Kutner et al., 2005, p. 360). This statistic provides an indication for all data records of how well a model based on a subset of the data in which the ith data record has been deleted can predict the response variable value Y_i. The formula for the statistic is

$$PRESS = \sum_{i=1}^{n}(Y_i - \hat{Y}_{i(i)})^2, \tag{8.17}$$

where $\hat{Y}_{i(i)}$ is the predicted value of Y_i computed from the model when the ith data record is deleted. It turns out that the statistic can be computed without actually computing each individual regression model (Kutner et al., 2005, p. 360). The *PRESS* statistic has two uses. First, it is a way to perform a final check of the model to ensure against overfitting. One can compute the value of the statistic for the proposed model and for each of the reduced models in which one explanatory variable is omitted, and eliminate any variable leading to a lower value of this statistic. The MPV package (Braun, 2015) contains a function PRESS() that does this computation.

```
> PRESS(model.full)
[1] 56576470
> PRESS(model.5)
[1] 47990460
```

The *PRESS* statistic of model.5 is much smaller than that of the full model.

The second use of the *PRESS* statistics is as one check of model validity. If the regression model is valid, one would expect the value of the *PRESS* statistic to be about the same as

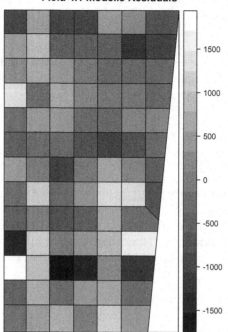

FIGURE 8.7
Thematic map of residuals of linear regression models of the northern and southern portions of Field 4.1, with residuals in the southern portion scaled to those in the north.

the mean square error (Kutner et al., 2005, p. 360). The *MSE* is computed using the R function deviance().

```
> PRESS(model.5) / deviance(model.5)
[1] 1.104985
```

This is a positive indication for the validity of the model.

As a final step in this phase of the analysis, we construct a Thiessen polygon map of the residuals of model.5 (Figure 8.7). We are looking for signs of spatial autocorrelation among the residuals. As is often the case, it is difficult to tell from the figure whether such autocorrelation exists. We will take up this issue in Chapters 12 and 13.

In summary, we have the full model (model.1) plus four selected models to carry forward. These four selected models are:

```
> model.2
Call:
lm(formula = Yield ~ Clay + Silt + SoilTN + Weeds + ClayTN, data =
agron.data)
Coefficients:
(Intercept)        Clay       Silt      SoilTN       Weeds      ClayTN
   19264.13     -487.46      73.43     -157.58     -319.02     4151.94
```

```
> model.3
Call:
lm(formula = Yield ~ Clay + SoilpH + SoilP + Weeds + ClayP +
   ClaypH, data = agron.data)
Coefficients:
(Intercept)       Clay     SoilpH      SoilP      Weeds      ClayP     ClaypH
  -25479.09     544.50    6752.79    -572.53    -380.41      16.66    -140.53
> model.4
Call:
lm(formula = Yield ~ Clay + SoilpH + SoilP + Weeds + ClayP, data =
agron.data)
Coefficients:
(Intercept)       Clay     SoilpH      SoilP      Weeds      ClayP
     6802.9     -314.9     1474.5     -775.7     -369.3       21.9
> model.5
Call:
lm(formula = Yield ~ Clay + SoilP + Weeds + ClayP, data = agron.data)

Coefficients:
(Intercept)       Clay      SoilP      Weeds      ClayP
   16083.29    -324.72    -890.17    -390.78      24.05
```

The same explanatory variables tend to appear in all of them. The coefficients, however, change dramatically. For example, *Clay* has a positive coefficient in model.3, indicating once again the effect of multicollinearity on the values of the coefficients. Nevertheless, they do provide evidence of the importance of soil texture and mineral nutrient concentration. We will continue to accumulate evidence about the effects of these variables as we move forward in the subsequent chapters. The results also show a dramatic difference between those obtained via backwards selection and those obtained by the best subsets method. Again, in a real data analysis project forward and bidirectional selection would also be tested. The function stepAIC() in the package MASS (Venables and Ripley, 2002) provides a good option for this.

8.4 Generalized Linear Models

8.4.1 Introduction to Generalized Linear Models

The response variables in Data Sets 1 and 2 are the indicators of the presence or absence at the sample location of, respectively, yellow-billed cuckoos and blue oaks. These indicator variables take on one of two values: 0, indicating absence, and 1, indicating presence. This is clearly not a normal distribution, and if one were to try to fit a linear regression model of the form of Equation 8.1 to the data, the error terms ε_i would also not be normally distributed, and the model would not be valid. Binary response variables like these are members of a class of data that can be analyzed using *generalized linear models,* or GLM (Nelder and Wedderburn, 1972; Fox, 1997, p. 438; Kutner et al., 2005, p. 555). We will introduce GLM by considering a model with a single explanatory variable. Figure 8.8 shows the distribution

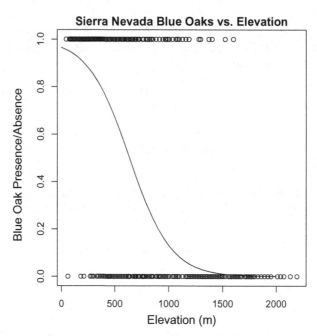

FIGURE 8.8
Plot of presence (= 1) and absence (= 0) of blue oaks as a function of *Elevation*, together with a logistic regression model of this relationship.

of *QUDO* values in the Sierra Nevada subset of Data Set 2 as a function of the single environmental variable *Elevation*. Also shown in the figure is the fit of a *logistic regression* model, which is one form of generalized linear model. The curve symbolizes the expected value of *QUDO* as a function of elevation. We already know that blue oak is almost completely absent above 1300m and that in the Sierra Nevada blue oak presence declines steadily as a function of elevation (Figure 7.11), and this is reflected in the model fit.

Consider the general case of a binary (zero or one) response variable Y_i and a single explanatory variable X_i. Let π_i represent the conditional probability that $Y_i = 1$ given that $X = X_i$, i.e.

$$\pi_i = P\{Y = Y_i \mid X = X_i\}. \tag{8.18}$$

It follows that

$$E\{Y \mid X = X_i\} = \pi_i \times 1 + (1 - \pi_i) \times 0 = \pi_i. \tag{8.19}$$

For example, in the case of the Sierra Nevada blue oak data, the expected value of $Y = QUDO$ should decline with increasing $X = Elevation$ as shown in Figure 8.8. In the case of ordinary linear regression, where the model is $Y_i = \beta_0 + \beta_1 X_i + \varepsilon_i$ and $E\{\varepsilon_i\} = 0$, we have $E\{Y_i\} = \beta_0 + \beta_1 X_i$. With generalized linear models we obtain a curvilinear relationship between $E\{Y_i\}$ and $\beta_0 + \beta_1 X_i$ like that of Figure 8.8 by applying a function like the *logistic function* to the explanatory variable. The logistic function has the form

$$\pi_i = h(\beta_0 + \beta_1 X_i) = \frac{1}{1 + \exp(-[\beta_0 + \beta_1 X_i])}. \tag{8.20}$$

In the case of the blue oak vs. elevation data, β_1 will be negative, so that as X_i becomes a large positive number, $\exp(-[\beta_0 + \beta_1 X_i])$ becomes a very large positive number, and therefore $h(\beta_0 + \beta_1 X_i)$ approaches zero. As X_i becomes a large negative number, $\exp(-[\beta_0 + \beta_1 X_i])$ approaches zero, and therefore $h(\beta_0 + \beta_1 X_i)$ approaches one. Thus, the curve of $h(\beta_0 + \beta_1 X_i)$ resembles that shown in Figure 7.11 and Figure 8.8.

Using a little algebra, one can show (Kutner et al., 2005, p. 561) that the inverse of the logistic function is given by

$$g(\pi_i) = h^{-1}(\pi_i) = \ln\left(\frac{\pi_i}{1 - \pi_i}\right). \tag{8.21}$$

Thus, we can estimate coefficients β_0 and β_1 by fitting the data to the model

$$g(\pi_i) = \ln\left(\frac{\pi_i}{1 - \pi_i}\right) = \beta_0 + \beta_1 X_i. \tag{8.22}$$

The function g is called a *logit transformation*, and when it is used in this way in logistic regression it is called a *link function* (Kutner et al., 2005, p. 623). The quantity $\pi_i / (1 - \pi_i)$ is called the *odds*.

The computation of the regression coefficients β_0 and β_1 is generally carried out using the *method of maximum likelihood*, which is discussed in Appendix A.5. Given that $P\{Y_i = 1\} = \pi_i$ and $P\{Y_i = 0\} = (1 - \pi_i)$, we can represent the probability distribution of the Y_i as

$$f_i(Y_i) = \pi_i^{Y_i}(1 - \pi_i)^{1 - Y_i}, \quad Y_i = 0, 1, \quad i = 1, \ldots n. \tag{8.23}$$

Since the Y_i are assumed independent (this is where we run into problems with spatial data), their joint probability distribution is the product of the individual $f_i(Y_i)$,

$$f(Y_1, \ldots, Y_n) = \prod_{i=1}^{n} \pi_i^{Y_i}(1 - \pi_i)^{1 - Y_i}. \tag{8.24}$$

Taking logarithms yields

$$\ln f(Y_1, \ldots, Y_n) = \ln \prod_{i=1}^{n} \pi_i^{Y_i}(1 - \pi_i)^{1 - Y_i}$$

$$= \sum_{i=1}^{n} [Y_i \ln \pi_i + (1 - Y_i)\ln(1 - \pi_i)] \tag{8.25}$$

$$= \sum_{i=1}^{n} \left[Y_i \ln\left(\frac{\pi_i}{1 - \pi_i}\right)\right] + \sum_{i=1}^{n} \ln(1 - \pi_i).$$

It turns out (based on algebra from earlier equations) that

$$1 - \pi_i = (1 + \exp(\beta_0 + \beta_1 X_i))^{-1}. \tag{8.26}$$

Therefore, substituting from Equations 8.22 and 8.26 into Equation 8.25 yields the log likelihood function

$$\ln L(\beta_0, \beta_1) = l(\beta_0, \beta_1) = \sum_{i=1}^{n} \left[Y_i(\beta_0 + \beta_1 X_i) - \ln(1 + \exp(\beta_0 + \beta_1 X_i)) \right]. \tag{8.27}$$

Given a set of pairs (X_i, Y_i), $i = 1,...,n$, the maximum likelihood estimates of b_0 and b_1 of β_0 and β_1 are obtained by maximizing $l(\beta_0, \beta_1)$ in Equation 8.27. The curve in Figure 8.8 represents the maximum likelihood solution of Equation 8.27 for the Sierra Nevada blue oaks data. The solution to this nonlinear problem must be obtained numerically. The most common method for doing this is called *iteratively reweighted least squares* (McCulloch et al., 2008, p. 144).

The computation of a generalized linear model in R is accomplished with the function glm(), which works just like the function lm() except that one must also specify the family to which the data belong in order to ensure the use of the correct link function. In our case, the data are binomial. With the Sierra Nevada subset of Data Set 2 loaded into the sf object data.Set2S as described in Appendix B.2 and Section 7.3, the code to generate the model is as follows.

```
> glm.demo <- glm(QUDO ~ Elevation, data = data.Set2S,
+     family = binomial)
> summary(glm.demo)
Call:
glm(formula = QUDO ~ Elevation, family = binomial, data = data.Set2S)

Deviance Residuals:
    Min       1Q    Median       3Q       Max
-2.4734   -0.5620  -0.1392   0.6126    3.1737
Coefficients:
               Estimate Std. Error z value  Pr(>|z|)
(Intercept)   3.3291447  0.1722626   19.33  < 2e-16 ***
Elevation    -0.0052236  0.0002481  -21.05  < 2e-16 ***
---
Signif. codes:  0 '***' 0.001 '**' 0.01 '*' 0.05 '.' 0.1 ' ' 1
 (Dispersion parameter for binomial family taken to be 1)
    Null deviance: 2371.3  on 1767  degrees of freedom
Residual deviance: 1366.5  on 1766  degrees of freedom
AIC: 1370.5
Number of Fisher Scoring iterations: 6
```

It is a straightforward matter to extend the logistic regression model to multiple explanatory variables. One simply rewrites the link function in Equation 8.22 as

$$g(\pi_i) = \ln\left(\frac{\pi_i}{1 - \pi_i}\right) = \beta_0 + \beta_1 X_{i1} + \beta_2 X_{i2} + ... + \beta_{p-1} X_{i,p-1}. \tag{8.28}$$

and again obtains the estimates b_i of the β_i by the method of maximum likelihood. The predicted values $\hat{\pi}_i$ are then computed as

$$\hat{\pi}_i = h(b_0 + b_1 X_{i1} + b_2 X_{i2} + \ldots + b_{p-1} X_{i,p-1i}) =$$

$$\frac{1}{1 + \exp(-[b_0 + b_1 X_{i1} + b_2 X_{i2} + \ldots + b_{p-1} X_{i,p-1i}])}. \tag{8.29}$$

Equation 8.27 can be used to develop a statistic analogous to the coefficient of determination R^2 of linear regression. For a given set of regression coefficients having a likelihood function L, the *deviance* G^2 is defined as (Fox, 1997, p. 451)

$$G^2 = -2\ln L. \tag{8.30}$$

Let G_0^2 be the deviance of the model that includes only β_0 present, and let G_1^2 be the deviance of the model being considered. Then one can define

$$R^2 = 1 - \frac{G_1^2}{G_0^2}. \tag{8.31}$$

as a measure of fit analogous to the coefficient of determination of the linear model. In this sense, the deviance can be considered as analogous to the error sum of squares of the linear model.

One cannot simply define the residuals of a generalized linear model as $e_i = Y_i - \hat{Y}_i$ in the manner of the linear model. Such a definition would produce residuals that were not normally distributed and had no meaningful interpretation. Nevertheless, if you apply the R function `residuals()` to a `glm` object, you will get a valid response. As it is with many R functions, polymorphism is implemented in the function `residuals()` by appending the name of the class to the function name. Thus, if you type `?residuals.glm`, you can view the Help file for the calculation of residuals of `glm` objects. As they often do, this help file contains a list of useful references. There are a number of definitions of residuals for the generalized linear model, of which we will focus on two: the *deviance residuals* and the *Pearson residuals*. The deviance residuals are defined as (Kutner et al., 2005, p. 592)

$$e_{Di} = sign(Y_i - \hat{\pi}_i)\sqrt{-2[Y_i \ln \hat{\pi}_i + (1 - Y_i)\ln(1 - \hat{\pi}_i)]}, \tag{8.32}$$

where the $\hat{\pi}_i$ are the fits obtained via Equation 8.29. The Pearson residuals are defined as (Kutner et al., 2005, p. 591)

$$e_{Pi} = \frac{Y_i - \hat{\pi}_i}{\sqrt{\pi_i(1 - \pi_i)}}. \tag{8.33}$$

The deviance residuals focus on the analogy of the deviance to the error sum of squares (Kutner et al., 2005, p. 68), while the Pearson residuals are related to the use of the chi-square statistic in a test of goodness of fit, which is discussed in Chapter 10.

The last step before moving on to the analysis of ecological data is to see how added variable or partial residual plots are applied to generalized linear models. There is considerable literature on the subject. In the case of partial residual plots for the logistic regression model, Landwehr et al. (1984) recommend plotting the *logit partial residuals*, which are similar to the Pearson residuals but involve division by $\hat{\pi}_i(1-\hat{\pi}_i)$ rather than by $\sqrt{\hat{\pi}_i(1-\hat{\pi}_i)}$ and which also include a term involving the explanatory variable in the plot. Nicholls (1989) recommends plotting residuals against variables not in the model. He uses the Pearson residuals. Davison and Snell (1991), on the other hand, advise against using Pearson residuals in general. Kutner et al. (2005, p. 594) compare the various types of residual plots and indicate that, at least in their example, the different types of residuals provide generally similar results. The functions `avPlots()` and `crPlots()` of the `car` package can both be used with `glm` objects. The function `avPlots()` uses a method described by Wang (1985) for generalized linear models, and the function `crPlots()` uses the method of Landwehr et al. (1984) given above, as described by Fox and Weisberg (2011).

To get a better feel for how to use added variable and partial residual plots with generalized linear models we will develop two artificial data sets. The explanatory variables of the data sets are two independent samples drawn from a normal distribution. Each consists of 400 values, in order to make the relationships as clear as possible. The response variable Y of the first data set is a binomial random variable that depends through a link function of the form of Equation 8.22 on the quantity $X_1 + X_2$ (i.e., $\beta_0 = 0$ and $\beta_1 = \beta_2 = 1$). Here is the code for the regression model.

```
> set.seed(123)
> X <- cbind(rnorm(400), rnorm(400))
> p <- 1 / (1 + exp(-rowSums(X)))
> Y <- rbinom(numeric(length(p)), 1, p)
```

Suppose now that we have a regression model with X_1 in as an explanatory variable and we wish to test the effect of including X_2. We can create a partial residual plot with the following statement (note that the reserved word *for* needs to be enclosed in quotes).

```
> model.lin <- glm(Y ~ X[,1] + X[,2], family = binomial)
> library(car)
> crPlots(model.lin, col.lines = c("black", "black"), # Fig. 8.9a
+    main = expression(Partial~Residual~Plots~"for"~italic(X)[2]))
```

The results are shown in Figure 8.9a. There are two things about this plot that we need to discuss. First, note that the points form two clouds. This is a common feature of added variable and partial residual plots for logistic regression models and occurs because the response variable Y can only take on one of two values, zero and one, and thus the residuals are also bimodal. This feature makes the point clouds of added variable and partial residual plots very difficult to interpret, and leads to the inclusion of the second feature that must be explained. The dashed line in Figure 8.9a is a regression line fit to the point clouds, but the more important line for our purposes is the solid line. This is a called a *loess* curve. "Loess" stands for *locally weighted scatterplot smoothing* (Cleveland and Devlin, 1988; Kutner et al., 2005, p. 138; Fox, 1997, p. 417), and for this reason it is sometimes spelled

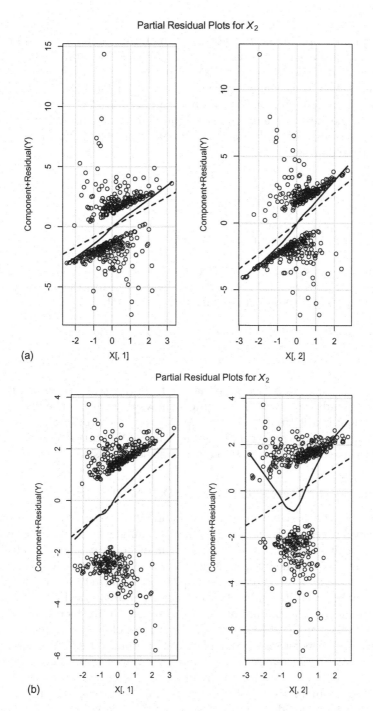

FIGURE 8.9

(a) Partial residual plots for a logistic regression model based on artificial data. (b) Partial regression plots for a model similar to that of part (a), but incorporating a quadratic term in X_2.

"lowess." In any case, unlike the soil type of the same spelling, it is pronounced *"low-ess."* Loess is a form of nonparametric regression, which means that there is no assumption made about the distribution of the regression residuals. In loess fitting, there is no estimation of parameters β_i. Instead, the curve is constructed simply to fit the data. This type of function is also called a *smoothing function*, and we will meet smoothing functions again when we discuss the generalized additive model in Section 9.2.

In Figure 8.9a, the loess curve lies almost right on top of the linear regression fit, which indicates that the contribution of X_2 to the response variable Y is linear. Now consider the following artificial data set.

```
> set.seed(123)
> X <- cbind(rnorm(400), rnorm(400))
> X <- cbind(X, X[,2]^2)
> p <- 1 / (1 + exp(-rowSums(X)))
> Y <- rbinom(numeric(length(p)), 1, p)
```

Now Y also depends on X_2^2, so that the true model would be of the form $Y_i = \beta_1 X_{i1} + \beta_2 X_{i2} + \beta_3 X_{i2}^2 + \varepsilon_i$. Suppose, however, that we again fit the data with a model that excludes the X_2^2 term and construct the partial residual plot.

```
> model.lin <- glm(Y ~ X[,1] + X[,2], family = binomial)
> crPlots(model.lin, col.lines = c("black", "black"), # Fig. 8.9b
+      main = expression(Partial~Residual~Plots~"for"~italic(X)[2]))
```

Now the loess curve shows a distinct parabolic form (Figure 8.9b), which is not visually apparent in the clouds of points. This example shows that we can use the partial residual plots to detect nonlinearity in the relationship of Y with the explanatory variables. We can use the added variable plots to detect discordant values. With this arsenal of graphical tools, we are ready to take up the construction of the logistic regression model for Data Set 2.

8.4.2 Multiple Logistic Regression Model for Data Set 2

We can summarize what we have learned so far about Data Set 2 as follows. Blue oaks are found almost exclusively at elevations less than 1300 m. The fraction of sites without a blue oak declines steadily in the Sierra Nevada but actually increases slightly up to 1100 m in the Coast Range (Figure 7.11). The fraction of sites with a blue oak present in the Sierra Nevada tends to decline steadily with increasing precipitation, while in the Coast Range the decline is more abrupt at between 700 and 900 mm per year (Figure 7.12a). In general, precipitation is more important as a determinant of blue oak presence in the Sierra Nevada than in the Coast Range because the Coast Range is drier over most of its extent. Temperature, and, to a lesser extent, soil properties also may play an important role in blue oak presence. The fraction of blue oak sites increases in both mountain ranges in areas with increasingly fine soil texture (Figure 7.12b). As stated in Chapter 7, this goes against conventional wisdom (McDonald, 1990). Sometimes it is wise to not include in a regression analysis data that violate the conventional wisdom, and indeed our application of the method of Henderson and Velleman (1981) is based on using what might be termed the conventional wisdom. In the present case, however, the conventional wisdom is based primarily on reasoning rather than data and may be incorrect. Finally, there is the "secondary" spatial variable *CoastDist*. In consideration of this quantity, we again have to deal with the fact that we are attempting to use regression for a purpose, explanation, for which it is not really suited. The quantity

CoastDist may have great utility for the purpose for which regression is best suited, namely, prediction. In Section 13.1, we discuss the various ways that spatial variables like this can enter into a model, but for now the important point is that they can affect the bias or lack thereof in the coefficients of the other quantities (in this case, the climate variables) if they are correlated with something that is not included in the model. For example, distance from the coast might be correlated with days per year of fog, which is not in the model. If days of fog has a strong influence, then its effect will be "loaded" onto the other variables, possibly biasing their coefficients. If distance from the coast is highly correlated with days per year of fog, then the former may take most of the load of the latter. It is worth noting that one might expect distance from the coast (i.e., from the ocean) to have a greater influence in the Coast Range than in the Sierra Nevada, both because of the closer proximity and because of the mitigating effects of the Central Valley, which lies between these mountain ranges.

Because of the high level of correlation among the temperature variables (Figure 7.9), we include for now only the summary variable *MAT* (mean annual temperature) among the temperature variables. We represent each parent material type with an indicator variable that simply indicates the presence of that parent material type. The Sierra Nevada subset of Data Set 2 is loaded into the spatial features object `data.Set2S.sf` as described in Appendix B.2 and Section 7.3.

```
> data.Set2S.glm <- with(data.Set2S.sf@data, data.frame(MAT, TempR,
+     Precip, PE, ET, Texture, AWCAvg, Permeab, SolRad6, SolRad12,
+     SolRad, CoastDist, QUDO))
> data.Set2S.glm$PM100 <- as.numeric(data.Set2.sfSierra$PM100 > 0)
```

The last line of code is repeated for each of the other five parent material types.

For illustrative purposes, in this analysis, unlike that of Section 8.3, we will build the model by adding explanatory variables and examining the effect of each addition (i.e., using forward selection). In practice, as was mentioned in that section, one should in an actual analysis program repeat the analysis using both selection methods as well as bidirectional selection and best subsets and compare the results. As a standard of comparison, we first create the global model that includes all of the explanatory variables.

```
> model.glmSfull <- glm(QUDO ~ ., data = data.Set2S.glm,
+     family = binomial)
```

In data sets with a very large number of observations, it is more likely that relatively unimportant explanatory variables will be assigned significance. At the end of our model selection process, we are going to compare models with different numbers of parameters. Since the BIC has a higher complexity penalty for larger data sets (Table 8.1), we will use this to compare these alternative models. According to the Help file for the R function `AIC()`, the BIC is computed using the following second argument.

```
> AIC(model.glmSfull, k = log(nrow(data.Set2Sglm)))
[1] 1134.27
```

To begin, we will create the null model as well as a formula that allows us to use the function `add1()` to examine the effects of adding different explanatory variables.

```
> model.formula <- as.formula("QUDO ~ MAT + TempR + Precip + PE + ET +
+     Texture + AWCAvg + Permeab + SolRad6 + SolRad12 + SolRad +
```

```
+    PM100 + PM200 + PM300 + PM400 + PM500 + PM600 + CoastDist")
> model.glmS0 <- glm(QUDO ~ 1, data = data.Set2S.glm, family =
+    binomial)
```

Now we are ready to start constructing the model. When we use the function add1(), we are comparing models of the same complexity (i.e., the same number of parameters). Therefore, only the bias term is important, and for this comparison we can use the AIC instead of the BIC. Nicholls (1989) recommends a fairly rigid procedure of variable addition based on comparing AIC values. I personally prefer the more flexible approach of Henderson and Velleman (1981), but we will nevertheless be strongly guided by the AIC results. Given all we have seen in Chapter 7, we would expect that the most important single variable would be *Precip*, but let's see what the AIC tells us.

```
> a1 <- add1(model.glmS0, model.formula)
> a1[order(a1[,3]),]
          Df Deviance    AIC
Precip     1    1222.1 1226.1
MAT        1    1404.0 1408.0
PE         1    1593.9 1597.9
CoastDist  1    1610.8 1614.8
   *   *   *    DELETED   *   *   *
SolRad12   1    2363.5 2367.5
<none>          2371.2 2373.2
PM100      1    2370.9 2374.9
```

In this case, the output gives us the increase in AIC that would result from the addition of each variable, so we want to add the term that increases the AIC the least. The variable *Precip* does indeed provide the smallest AIC, as well as making the most sense biophysically, so we bring it into the model.

```
> model.glmS1 <- update(model.glmS0,
+    formula = as.formula("QUDO ~ Precip"))
```

Parent material and *TempR* appear unimportant, so we will remove them.

```
> model.formula2 <- as.formula("QUDO ~ MAT + Precip + PE + ET +
+    Texture + AWCAvg + Permeab + SolRad6 + SolRad12 + SolRad +
+    CoastDist")
> a1 <- add1(model.glmS1, model.formula2)
> a1[order(a1[,3]),]
Single term additions
Model:
QUDO ~ Precip
       Df Deviance    AIC
MAT     1    1158.3 1164.3
ET      1    1174.1 1180.1
PE      1    1188.8 1194.8
   *   *   *   DELETED   *   *   *
```

As expected, *CoastDist* does not have a strong effect in the Sierra Nevada.

The second application of add1() indicates that *MAT* should enter, and this also makes sense based on what we have seen already. The variables *MAT* and *Precip* have a strong linear association, with a negative correlation coefficient.

```
> with(data.Set2S.glm, cor(Precip, MAT))
[1] -0.7488103
```

As elevation increases in the Sierra Nevada, it becomes cooler and wetter. Given this correlation, it is a bit surprising that the coefficient of *Precip* does not change too much when *MAT* is brought into the model.

```
> model.glmS1 <- update(model.glmS0,
+     formula = as.formula("QUDO ~ Precip"))
summary(model.glmS1)
Coefficients:
              Estimate Std. Error z value Pr(>|z|)
(Intercept)  8.0394180  0.3952460   20.34  < 2e-16 ***
Precip      -0.0097772  0.0004698  -20.81  < 2e-16 ***
AIC: 1226.1
> model.glmS2 <- update(model.glmS0,
+     formula = as.formula("QUDO ~ Precip + MAT"))
> summary(model.glmS2)
              Estimate Std. Error z value Pr(>|z|)
(Intercept) -1.3701320  1.2715044  -1.078    0.281
Precip      -0.0071227  0.0005492 -12.970  < 2e-16 ***
MAT          0.5048807  0.0677577   7.451 9.25e-14 ***
AIC: 1164.3
```

Added variable and partial residual plots are not shown, but the latter indicate the possibility that the addition of a second-degree term in *Precip* would improve the model. We will generally save the analysis of higher-order terms and interactions until after the selection of first-order explanatory variables. At this point, however, we can observe something interesting. We first add a *Precip*^{2} term to the model without *MAT*, and a test indicates that its coefficient is significant.

```
> model.glmS1sq <- update(model.glmS0,
+     formula = as.formula("QUDO ~ Precip + I(Precip^2)"))
> summary(model.glmS1sq)
Coefficients:
              Estimate Std. Error z value Pr(>|z|)
(Intercept)  1.064e+01  1.309e+00   8.135 4.13e-16 ***
Precip      -1.599e-02  2.945e-03  -5.431 5.61e-08 ***
I(Precip^2)  3.584e-06  1.633e-06   2.194   0.0282 *
```

The coefficient of *Precip* is negative, as expected, and the coefficient of *Precip*^{2} is positive, indicating that the effect of *Precip* declines as its value increases. As mentioned in Section 3, Burnham and Anderson (1998) provide several convincing arguments and good examples for the assertion that hypothesis tests should not be used for variable selection. Nevertheless, I have found that while this is true for variable selection per se, these tests often can be useful in determining whether or not to include higher order terms or interactions. In this case, there is a significant ($p < 0.05$) difference between the model without and with *Precip*^{2}.

```
> anova(model.glmS1, model.glmS1sq, test = "Chisq")
Analysis of Deviance Table
Model 1: QUDO ~ Precip
Model 2: QUDO ~ Precip + I(Precip^2)
```

```
   Resid. Df Resid. Dev Df Deviance P(>|Chi|)
1      1766     1222.1
2      1765     1218.0  1   4.0801    0.04339 *
```

Let's now add *MAT* to the model that includes *Precip*² and see what happens.

```
> model.glmS2sq <- update(model.glmS0,
+     formula = as.formula("QUDO ~ Precip + I(Precip^2) + MAT"))
> coef(model.glmS2sq)
   (Intercept)          Precip   I(Precip^2)            MAT
  6.942303e-01 -1.167481e-02  2.568349e-06  4.964979e-01
> anova(model.glmS2, model.glmS2sq, test = "Chisq")
Analysis of Deviance Table
Model 1: QUDO ~ Precip + MAT
Model 2: QUDO ~ Precip + I(Precip^2) + MAT
   Resid. Df Resid. Dev Df Deviance P(>|Chi|)
1      1765     1158.3
2      1764     1155.9  1   2.3164    0.1280
```

The variable *MAT* apparently takes some of the "load" from *Precip*². There are a number of possible biophysical interpretations of this. Let's take it as a given that precipitation actually does have an effect on blue oak presence. Since *Precip* and *MAT* are negatively correlated, it is possible that temperature also has an effect, with increased temperatures reducing the probability of blue oak presence. In the alternative, the effect of *Precip* really could decline with increasing value, in which case *MAT* is serving as a surrogate for *Precip*². There may also be the possibility of an interaction. In any case, this is further indication that temperature as well as precipitation may play a role in blue oak presence.

We will for now only include the linear terms. We now have models with different numbers of parameters. Using the BIC to compare model 2, which includes *MAT*, with model 1, which does not, indicates a substantial improvement using model 2.

```
> AIC(model.glmS1, k = log(nrow(data.Set2S.glm)))
[1] 1237.074
> AIC(model.glmS2, k = log(nrow(data.Set2S.glm)))
[1] 1180.691
```

Figure 8.10a shows a thematic map of the predicted values of model.glmS1, which only includes *Precip*. Figure 8.10b shows the effect of adding *MAT* to the model. Both of these can be compared with Figure 7.8a, which shows blue oak presence and absence. Mean annual temperature in the northern Sierra foothills tends to rise as one moves north, due to increased distance from the cooling breezes of the Sacramento–San Joaquin Delta. Thus, the primary effect of adding *MAT* to the model is to slightly increase the presence probabilities in the northern foothills relative to the rest of the region.

Continuing on in this way, we come to candidate models that also include *PE*, *SolRad*, *AWCAvg*, and *Permeab*. Thus, our analysis has identified candidate models that include precipitation, some temperature-related variable or variables, possibly a solar radiation variable, and some variable that has to do with soil water availability.

The last step is to consider *Elevation*. While it is true that many of the explanatory variables in the model are affected by elevation, it is also true that elevation may also affect

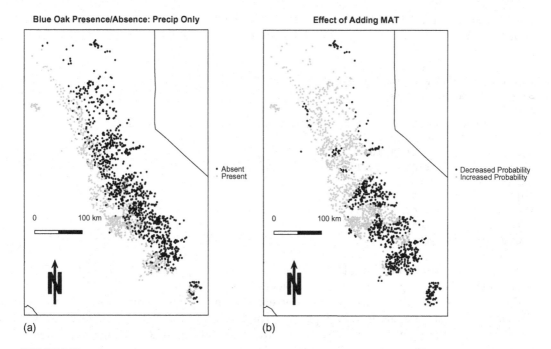

Blue Oak Presence/Absence: Precip Only

• Absent
· Present

Effect of Adding MAT

• Decreased Probability
· Increased Probability

(a)   (b)

FIGURE 8.10
(a) Thematic map of blue oak presence and absence. (b) Thematic map showing the difference in predicted probability of blue oak occurrence between logistic regression models incorporating and not incorporating mean annual temperature.

other quantities not included in the data set that also impact blue oak presence. Therefore, we will test a model that incorporates representatives of all of the variables in our candidate models plus *Elevation*.

```
> data.Set2SglmE <- data.frame(cbind(data.Set2S.glm,
+     Elevation = data.Set2S$Elevation))
> model.glmS5E <- glm(QUDO ~ Precip + MAT + SolRad +
+     AWCAvg + Permeab + Elevation, data = data.Set2SglmE,
+     family = binomial)
> AIC(model.glmS5, k = log(nrow(data.Set2Sglm)))
[1] 1104.036
> AIC(model.glmS5E, k = log(nrow(data.Set2Sglm)))
[1] 1076.657
> summary(model.glmS5E)
Coefficients:
```

	Estimate	Std. Error	z value	Pr(>|z|)	
(Intercept)	-1.070e+00	1.760e+00	-0.608	0.54300	
Precip	-5.531e-03	6.187e-04	-8.939	< 2e-16	***
MAT	2.393e-01	9.251e-02	2.587	0.00968	**
SolRad	7.684e-04	9.652e-05	7.961	1.70e-15	***
AWCAvg	-6.683e-03	1.458e-03	-4.585	4.53e-06	***
Permeab	-3.022e-01	6.642e-02	-4.551	5.35e-06	***
Elevation	-2.511e-03	4.269e-04	-5.880	4.09e-09	***

The model that includes *Elevation* is a substantial improvement over the model that excludes it. We can then use drop1() to determine which variables are removed if *Elevation* is included.

```
> print(d1 <- drop1(model.glmS5E))
Single term deletions
Model:
QUDO ~ Precip + MAT + SolRad + AWCAvg + Permeab + Elevation
          Df Deviance    AIC
<none>           1024.3 1038.3
Precip     1     1129.6 1141.6
MAT        1     1031.3 1043.3
SolRad     1     1096.2 1108.2
AWCAvg     1     1046.1 1058.1
Permeab    1     1045.7 1057.7
Elevation  1     1059.2 1071.2
```

The indication is that none should be dropped, although, surprisingly, *Precip* and *MAT* are the closest. This may, again, be due to the effect of multicollinearity.

The Coast Range data display less correlation among the temperature variables (Figure 7.9c). A similar model development process for the Coast Range (Exercise 8.12) produces the following model. Once again, the high intercorrelation among variables means that this actually represents several candidate models.

```
> summary(model.glmC6)
Call:
glm(formula = QUDO ~ TempR + Permeab + Precip + GS32 + PE + SolRad6,
    family = binomial, data = data.Set2Cglm)
Coefficients:
              Estimate Std. Error z value Pr(>|z|)
(Intercept)  -7.812171   1.525324  -5.122 3.03e-07 ***
TempR         0.295183   0.039035   7.562 3.97e-14 ***
Permeab      -0.737872   0.077277  -9.548  < 2e-16 ***
Precip       -0.001528   0.000334  -4.577 4.72e-06 ***
GS32         -0.021310   0.002750  -7.749 9.23e-15 ***
PE            0.009384   0.001261   7.443 9.83e-14 ***
SolRad6       0.179429   0.040775   4.400 1.08e-05 ***
```

Given the complexity of the models and their preliminary nature, we will postpone any attempts to interpret the results.

The logistic regression model for blue oak presence constructed by Evett (1994) applied to the entire data set; that is, it did not separate the regions into different ranges. The explanatory variables in Evett's model includes *JaMean*, *JaMean*^{2}, *Precip*, *MAT*, *PM100*, *PM200*, *PM300*, *SolRad6*, *SolRad6*^{2}, *SolRad6*^{3}, *Texture*, (*JaMmean* × *MAT*)^2, and *JuMax*. Vayssières et al. (2000) developed a generalized linear model based on 2,085 data records covering all of the ranges. Their explanatory variable set included both exogenous and endogenous variables (i.e., they included *Elevation* and *CoastDist* where we did not). Their model includes *JaMean*, *JaMean*^{2}, *Precip*, *MAT*, *MAT*^{2}, *PM100*, *PM400*, *PM500*, *PM600*, *SolRad6*, *Elevation*, *Elevation*^{2}, *JuMean*, *Texture*, and *ET*. Even given that we have not yet made an effort to include interactions or higher-order terms, there is obviously a considerable range of explanatory variables between the models.

8.4.3 Logistic Regression Model of Count Data for Data Set 1

In Section 7.2, we carried out a preliminary analysis of Data Set 1 as a test of the California Wildlife Habitat Relationships (CWHR) model for habitat suitability for the yellow-billed cuckoo. The model is expressed in terms of *scores* for four explanatory variables given in Table 7.1. The variables *PatchWidth* and *PatchArea* are self-explanatory. The height ratio (*HtRatio*) is the ratio of area with tall trees to total area. The *AgeRatio* is the ratio of low floodplain age (<60 years) to total area. This is a surrogate for the cover fraction of mixed cottonwood/willow. A fifth variable, cover class, is also included in the data set but does not appear to play a major role in determining habitat suitability.

The habitat scores are computed by first computing an individual patch score for each variable according to Table 7.1. The scores in Table 7.1 are non-monotonic (i.e., not strictly increasing or strictly decreasing) in both *HtRatio* and *AgeRatio*, and a regression model for these data will have to take this non-monotonicity into account. The model studied in Section 7.2 was expressed as a contingency table in which the overall habitat score of each patch was computed as the product of the individual values of each of the four variable scores. Any patch with a nonzero habitat score was judged to be suitable. Thus, in this model, any patch is unsuitable if it is unsuitable in any one of the four suitability criteria. The contingency table for this model is as follows (see Section 7.2).

```
> print(cont.table <- matrix(c(length(SP),length(SA),
+     length(UP),length(UA)), nrow = 2, byrow = TRUE,
+     dimnames = list(c("Suit.", "Unsuit."),c("Pres.", "Abs.")))))
        Pres. Abs.
Suit.       5     1
Unsuit.     2    12
```

The results of Exercise 7.6 indicate that the same contingency table can be obtained with a model including only *AreaScore* and *AgeScore*, but that this table is not obtained with any one single explanatory variable.

In this section, we will apply GLM analysis to this data set. We will start with a logistic regression model for presence/absence. Before we begin the regression modeling, it is worthwhile to try to determine whether we really have any chance of distinguishing between the effects of the variable *PatchWidth* and *PatchArea*. We saw in Section 7.2 that these are highly correlated, and the only hope of distinguishing their effects would be if there is some really long, narrow patch that has a low patch width but a high patch area. Figure 8.11 is a plot of scaled values of *PathWidth* vs. *PatchArea*. Except for the smallest areas, which are all unsuitable, there is a virtually linear relationship between *PatchArea* and *PatchWidth*. The two exceptions are Patch 7 and Patch 16. We will use the data frame Set1.obs3 created in Section 7.2. We again delete Patch 191, which has incomplete geographic coverage. Examining the patch habitat scores of the patches further gives us the following.

```
> d.f <- with(Set1.obs3[-which(Set1.obs3$PatchID == 191),],
+     data.frame(obsID, AreaScore, WidthScore, AgeScore, HeightScore,
+     PresAbs))
> d.f[order(d.f$obsID),]
   obsID AreaScore WidthScore AgeScore HeightScore PresAbs
16     2      0.33       0.33     0.00        0.00       0
15     3      0.00       0.33     0.00        0.00       0
```

14	4	0.33	0.33	0.66	1.00	1
13	5	1.00	0.66	0.66	0.66	1
12	6	0.33	0.33	0.00	0.33	0
11	7	0.66	0.66	0.66	0.66	0
10	8	1.00	0.66	0.00	0.00	0
19	9	0.00	0.00	0.00	0.33	0
9	10	0.00	0.33	0.33	0.33	0
8	11	0.66	0.33	0.00	0.00	1
7	12	0.00	0.33	0.00	0.66	0
6	13	0.00	0.33	0.00	0.66	0
20	14	0.66	0.33	0.00	0.66	0
18	15	0.00	0.00	0.33	0.33	0
4	16	0.66	0.66	0.00	0.66	0
5	17	1.00	0.66	0.33	0.66	1
17	18	0.00	0.00	0.33	0.33	1
3	19	0.33	0.33	0.33	0.33	1
2	20	0.00	0.00	0.00	0.00	0
1	21	1.00	0.66	0.33	1.00	1

Patch 16 is unsuitable because of its age score. There are four patches (*obsID* values 3, 10, 12, 13) that have a positive *WidthScore* and a zero *AreaScore*, and none that have the opposite combination. These patches all have about the same width as the patch with *obsID* value 19, which has a slightly larger area (Figure 8.11). The one small area patch (*obsID* = 18) that has a cuckoo present is accounted for in Exercise 7.5. There seems to be no way that we can separate the effects of patch width from those of patch area. Therefore, we will eliminate one of them from the logistic regression. We tentatively choose *PatchWidth* for elimination because, based on Figure 8.11, *PatchArea* seems to be a better predictor of presence vs. absence. Just to be sure, however, we will double check below.

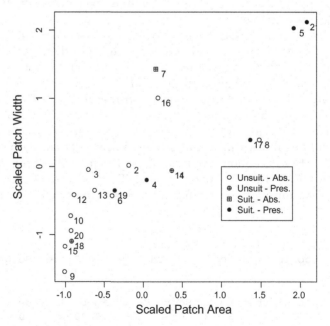

FIGURE 8.11
Scatterplot of patch area vs. patch width for the data of Data Set 1.

To improve the numerical properties of the model, we will center and scale the variables *PatchArea* and *PatchWidth*. The other two variables, *HtRatio* and *AgeRatio*, are fractions whose values range from zero to one, so we will leave these alone. We will not pursue analysis with the variable *CoverRatio*. Here is the code to create the new data set, based on the data frame `Set1.corrected` created in Section 7.2.

```
> Set1.norm1 <- with(Set1.corrected, data.frame(PresAbs = PresAbs,
+       PatchArea = scale(PatchArea), PatchWidth = scale(PatchWidth),
+       HtRatio = HtRatio, AgeRatio = AgeRatio))
```

We will start by generating the null model with nothing but the intercept on the right-hand side.

```
> Set1.logmodel0 <- glm(PresAbs ~ 1, data = Set1.norm1,
+       family = binomial)
```

Next, we will incorporate *PatchArea* and compare this to the null model. Only a portion of the output is shown in the following listings.

```
> Set1.logArea <- update(Set1.logmodel0,
+    formula = as.formula("PresAbs ~ PatchArea"))
> AIC(Set1.logArea)
[1] 25.14631
> anova(Set1.logmodel0, Set1.logArea, test = "Chisq")
Analysis of Deviance Table
Model 1: PresAbs ~ 1
Model 2: PresAbs ~ PatchArea
  Resid. Df Resid. Dev Df Deviance P(>|Chi|)
1        19     25.898
2        18     21.146  1   4.7516    0.02927 *
```

There is indeed a declared significant difference. Just to check that *PatchWidth* does not add to the model, we will add it and compare the two.

```
> Set1.logAreaWidth <- update(Set1.logmodel0,
+    formula = as.formula("PresAbs ~ PatchArea + PatchWidth"))
> AIC(Set1.logAreaWidth)
[1] 26.44410
> anova(Set1.logArea,Set1.logAreaWidth, test = "Chisq")
Analysis of Deviance Table
Model 1: PresAbs ~ PatchArea
Model 2: PresAbs ~ PatchArea + PatchWidth
  Resid. Df Resid. Dev Df Deviance P(>|Chi|)
1        18     21.146
2        17     20.444  1   0.7022    0.4020
```

We cannot use `anova()` to compare a model that only includes *PatchArea* with one that only includes *PatchWidth* because they are not nested. We can, however, use the AIC for such a comparison.

```
> Set1.logWidth <- update(Set1.logmodel0,
+    formula = as.formula("PresAbs ~ PatchWidth"))
> AIC(Set1.logWidth)
[1] 28.03604
```

The AIC of the model including *PatchArea* is substantially lower than that of the model including *PatchWidth*, providing further evidence that we should keep *PatchArea* rather than *PatchWidth*.

The other variable besides *PatchArea* that played a prominent role in the contingency table analysis of Section 7.2 is *AgeRatio*, the ratio of area of the patch floodplain age less than 60 years to total patch area. Figure 7.6 shows a plot of the habitat suitability score as a function of *AgeRatio*. One way to approximate this relationship is with a parabolic function. In Exercise 8.13, you are asked to carry out this analysis. In our analysis, instead of using the raw *AgeRatio* as the second variable in the logistic model, we will use the age suitability score.

The fact that the suitability scores in the contingency table model of Section 7.2 enter into the model as a product rather than a sum is probably best represented in a logistic regression model via an interaction term. This leads us to test a model that includes *PatchArea*, *AgeScore*, and the interaction between the two. We start by introducing *AgeScore* by itself

```
> Set1.logAgeScore <- update(Set1.logmodel0,
+    formula = as.formula("PresAbs ~ AgeScore"))
> summary(Set1.logAgeScore)
Coefficients:
            Estimate Std. Error z value Pr(>|z|)
(Intercept)  -1.8153     0.8123  -2.235   0.0254 *
AgeScore      5.3393     2.5237   2.116   0.0344 *
> anova(Set1.logmodel0, Set1.logAgeScore, test = "Chisq")
Analysis of Deviance Table
Model 1: PresAbs ~ 1
Model 2: PresAbs ~ AgeScore
  Resid. Df Resid. Dev Df Deviance P(>|Chi|)
1        19     25.898
2        18     19.793  1   6.1045   0.01348 *
```

The variable *AgeScore* plays a role analogous to that of *PatchArea*. Now let's try combining the two, in two steps. First, we test *AgeScore* by itself.

```
> Set1.logAgeScore <- update(Set1.logmodel0,
+    formula = as.formula("PresAbs ~ AgeScore"))
> AIC(Set1.logAgeScore)
[1] 23.79338
```

Interestingly, this produces a still lower AIC. Next, we add *PatchArea* but omit the interaction.

```
> Set1.logAreaAge <- update(Set1.logmodel0,
+    formula = as.formula("PresAbs ~ PatchArea + AgeScore"))
> summary(Set1.logAreaAge)
Coefficients:
            Estimate Std. Error z value Pr(>|z|)
(Intercept)  -1.7308     0.8444  -2.050   0.0404 *
PatchArea     0.9876     0.6570   1.503   0.1328
AgeScore      4.8061     2.7064   1.776   0.0758 .
AIC: 23.086
> anova(Set1.logAgeScore, Set1.logAreaAge, test = "Chisq")
Analysis of Deviance Table
Model 1: PresAbs ~ AgeScore
Model 2: PresAbs ~ PatchArea + AgeScore
```

```
   Resid. Df Resid. Dev Df Deviance P(>|Chi|)
1         18      19.793
2         17      17.086  1   2.7069   0.09991 .
```

Both *PatchArea* and *AgeScore* enter in a fairly meaningful way when combined additively, and the model that includes both is somewhat better than the one that only includes *AgeScore*. Remember that the fact that the *p* value of *AgeScore* is slightly lower than that of *PatchArea* should not be interpreted as necessarily meaning that the former is somehow more important than the latter The evidence, however, is beginning to point in that direction. Incorporating the interaction produces the following.

```
> Set1.logAreaAgeInt <- update(Set1.logmodel0,
+    formula = as.formula("PresAbs ~ PatchArea + AgeScore +
+    I(PatchArea * AgeScore)"))
> summary(Set1.logAreaAgeInt)
Coefficients:
                        Estimate Std. Error z value Pr(>|z|)
(Intercept)              -1.7247     0.8553  -2.017   0.0437 *
PatchArea                 0.9555     1.0612   0.900   0.3679
AgeScore                  4.7816     2.7771   1.722   0.0851 .
I(PatchArea * AgeScore)   0.1403     3.6523   0.038   0.9694
AIC: 25.085
> anova(Set1.logAreaAge, Set1.logAreaAgeInt, test = "Chisq")
Model 1: PresAbs ~ PatchArea + AgeScore
Model 2: PresAbs ~ PatchArea + AgeScore + I(PatchArea * AgeScore)
   Resid. Df Resid. Dev Df  Deviance P(>|Chi|)
1         17      17.086
2         16      17.085  1 0.0014860   0.9693
```

This model has a higher AIC than the one without the interaction, and the likelihood ratio test could not be less significant.

In summary, the use of logistic regression as an exploratory tool provides further support for the speculation that patch area and patch age structure, which is a surrogate for species composition, play the most important roles in determining habitat suitability in this stretch of the Sacramento River. In the next sub-section, we use the rather dubious abundance data as a means to introduce the analysis of the zero-inflated Poisson regression model for count data.

8.4.4 Analysis of the Counts of Data Set 1: Zero-Inflated Poisson Data

When the data set of cuckoo responses was created, along with the fact of a bird response to the *kowlp* call, the number of such responses was also recorded. Among our four data sets, this represents the best example of count data, which is often important in ecological studies. It also represents the worst example of reliability in a data set. To recapitulate the problems discussed in Section 7.2, it is impossible to determine whether multiple responses signify multiple observations of the same bird or multiple observations of different birds. Moreover, since the bird count data is an extensive property, it is sensitive to patch area, but patch area is itself one of the explanatory variables so we cannot adjust for it, and moreover it would not make sense to do so. Despite all this, we will press on in this section with an analysis of the abundance data. We do this for two reasons. First, the data provide the opportunity to discuss the analysis of count data. Second, weak as they might be, the data may make some contribution to our understanding of the ecological system.

As in Section 8.4.3, the data are loaded and set up using code from Appendix B.1 and Section 7.2. Before beginning with the analysis, however, we have do decide what to do with the data record that places 135 cuckoos at the southernmost observation point (Figure 7.2d). After pondering this for a bit, I decided on a course of action that might raise a few eyebrows but that seems to be not totally inappropriate. This is to carry out a sort of Winsorization (Section 6.2.1).

If we compare the southernmost habitat patch, with 135 recorded observations, to the northernmost patch, with 16 observations (Figure 7.2a), we can see that they are somewhat similar. From what we know about the cuckoos' perspective so far, both patches seem like pretty good habitat. Both are large and have a reasonable mix of young and old areas. We excluded the northernmost patch because we could not measure the floodplain age over its full extent, but now we will take its response variable value, 16, and substitute it for the value 135 in the southernmost patch.

```
> Set1.norm2 <- with(Set1.corrected, data.frame(Abund = Abund,
+       PatchArea = scale(PatchArea), PatchWidth = scale(PatchWidth),
+       HtRatio = HtRatio, AgeRatio = AgeRatio, CoverRato = CoverRatio,
+       AgeScore = AgeScore))
> mean(Set1.norm2$Abund)
[1] 7.65
> Set1.norm3 <- Set1.norm2
> Set1.norm3$Abund[1] <- 16
> mean(Set1.norm3$Abund)
[1] 1.7
```

The resulting data are contained in the data frame Set1.norm3. This actually produces a reasonable distribution of values. Figure 8.12 is a plot of patch abundance vs. patch area for the data set. Positive vs. zero age ratio suitability scores are indicated by filled vs. open

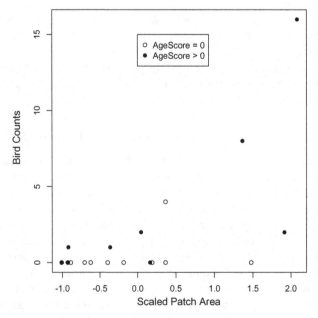

FIGURE 8.12
Scatterplot of abundance vs. scaled patch area of Data Set 1.

circles. The figure leads us to expect that, once again, we will see a regression involving *PatchArea* and *AgeScore*.

The standard method for constructing a regression model for count data is to use Poisson regression (Kutner et al., 2005, p. 618). This is carried out in exactly the same manner as logistic regression, but with a different link function. Instead of the link function given in Equation 8.22 for logistic regression, Poisson regression employs a link function of the form

$$g(Y_i) = \ln Y_i = \beta_0 + \beta_1 X_{i1} + \beta_2 X_{i2} + \ldots + \beta_{p-1} X_{i,p-1}. \tag{8.34}$$

There are, however, two complications. Recall (Larsen and Marx, 1986, p. 187) that the variance of the Poisson distribution equals its mean. The mean of the Winsorized distribution of cuckoo abundances is given above as 1.7.

Figure 8.13 shows a bar plot of the number of observed cuckoo abundance values compared with those predicted for a Poisson distribution with this mean. One feature that strikes the eye is that there are far more counts with zero observations than expected. This is very common with ecological count data. A data set such as this that has more zeroes than expected is said to be *zero-inflated*. A second feature that strikes the eye is that there seems to be a greater spread to the actual data than to the expected count numbers of the Poisson distribution. This is confirmed by a calculation of the sample variance.

```
> var(Set1.norm3$Abund)
[1] 15.16842
```

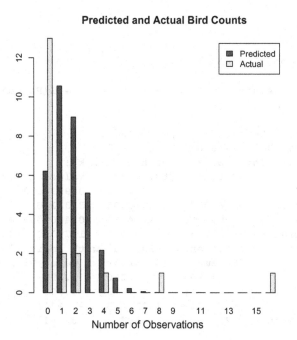

FIGURE 8.13
Bar chart of distribution of bird counts vs. that predicted by a Poisson distribution.

This large variance, called *overdispersion*, is also a common feature of ecological data (Bolker, 2008).

There are a few ways to construct regression models for overdispersed, zero-inflated count data (Cameron and Trivedi, 1998, p. 125). We will only present one: the zero-inflated Poisson, or ZIP method (Lambert, 1992). Briefly, in this method one fits a mixture of two models, a model for the probability that a count will be equal to zero, and a second model for the value of the count when it is greater than zero. This can be expressed as (Gelman and Hill, 2007, p. 127)

$$Y_i = \begin{cases} 0 & if\ S_i = 0 \\ overdispersed\,Poisson(\lambda, \beta) & if\ \ S_i = 1 \end{cases}. \tag{8.35}$$

Here S_i is a random variable that determines whether Y_i is zero or nonzero, λ is the Poisson parameter, and β is an overdispersion parameter that is used to enlarge the variance. The value of S_i can then be modeled using logistic regression.

The R package pscl (Zeileis et al., 2008), among others, contains functions that can be used to construct zero-inflated overdispersed Poisson models in a completely analogous way to that in which the function glm() is used to construct standard generalized linear models. Here is a sequence using the function zeroinfl(), which fits the ZIP model.

```
> library(pscl)
> Set1.zimodel0 <- zeroinfl(Abund ~ 1, data = Set1.norm3)
> AIC(Set1.zimodel0)
[1] 81.99746
> Set1.zimodelArea <- zeroinfl(Abund ~ PatchArea, data = Set1.norm3)
> AIC(Set1.zimodelArea)
[1] 60.31453
> summary(Set1.zimodelAreaAge)
Call:
zeroinfl(formula = Abund ~ PatchArea + AgeScore + I(PatchArea *
AgeScore), data = Set1.norm3)
Pearson residuals:
     Min       1Q    Median       3Q      Max
-1.09855 -0.29880 -0.14634 0.06209  2.00401
Count model coefficients (poisson with log link):
                         Estimate Std. Error z value Pr(>|z|)
(Intercept)                0.6147     0.6411   0.959 0.337586
PatchArea                  2.0875     0.6285   3.321 0.000896 ***
AgeScore                  -0.7066     1.4721  -0.480 0.631261
I(PatchArea * AgeScore)   -2.7942     1.5872  -1.760 0.078332 .
Zero-inflation model coefficients (binomial with logit link):
                         Estimate Std. Error  z value Pr(>|z|)
(Intercept)                 1.803      1.290    1.397    0.162
PatchArea                  -0.476      1.572   -0.303    0.762
AgeScore                 -259.720   1521.219   -0.171    0.864
I(PatchArea * AgeScore)  -267.380   1663.725   -0.161    0.872
```

The coefficient for *PatchArea* is highly significant, that for *AgeScore* is not significant, and that for the interaction is marginally significant. This is pretty much the opposite of the result obtained above using a logistic model for presence and absence. Some insight into this can be gained by constructing plots analogous to that of Figure 8.13 for cuckoo presence/absence, and also by placing *AgeScore* on the abscissa.

Thus far, our initial regression results for Data Set 1 are somewhat equivocal. As with the contingency table analysis of Section 7.2, they are consistent with the interpretation that for this data set, patch area and patch age structure (and hence possibly vegetation composition) play the most important role in determining habitat suitability, and that the interaction between these variables may be important.

8.5 Further Reading

If one wishes to carry out a purely statistical approach to model selection (i.e., one that is based more on the statistical properties of the data than on their scientific interpretation), then one must read a good book on model selection. Some of the best are Fox (1997), Burnham and Anderson (1998), and Harrell (2001). Some authors tend to be a bit zealous in their belief that one or another method is the best. Hastie et al. (2009) provide an even-handed comparison of the methods. Mosteller and Tukey (1977, Ch 13) provide an excellent discussion (appropriately titled "Woes of Regression Coefficients") of the issues associated with interpreting regression coefficients generated from observational data. Venables (2000) provides additional insights into the use of linear models, both in regression and analysis of variance. Belseley et al. (1980) provide a comprehensive discussion of regression diagnostics that is still very relevant, despite its age.

An important class of methods for model fitting and variable selection not discussed here involves a process called *shrinkage*. These are particularly useful when the objective is an accurate prediction of the value of the response variable, as opposed to an improved understanding of process. Shrinkage methods work by accepting a small amount of bias in exchange for a substantial reduction in variance. The best known such method is ridge regression (Kutner et al., 2005, p. 431). Hastie et al. (2009) provide a good discussion. The R package rms (Harrell, 2011) implements several shrinkage methods. This package also contains many other tools for exploratory and confirmatory regression analysis. If you plan to do a lot of regression modeling, it would be a very good idea to learn to use it. Fox (1997) and Fox and Weisberg (2011) are very good sources for added variable and partial regression plots. Wang (1985) and Pregibon (1985) develop somewhat different approaches to added variable plots, and Hines et al. (1993) also present an alternative approach.

Kutner et al. (2005) and Fox (1997) provide excellent introductions to the generalized linear model. The papers of Austin et al. (1983, 1984) are seminal works on use of generalized linear models in vegetation ecology. Nicholls (1989) provides an expository treatment of this subject. Venables and Ripley (2002, Ch. 7) discuss generalized linear model approaches to regression when the response variable is ordinal. Bolker (2008) provides an excellent introduction to zero-inflated models and overdispersion. Cunningham and

Lindenmayer (2005) provide further discussion. It is important to note that just because a data set has too many zeroes does not necessarily mean that it is zero-inflated in the technical sense of the term. Warton (2005) provides a discussion of this issue.

Exercises

8.1   Compute added variable and partial regression plots of Field 4.1 with just *Clay* and *SoilK* and interpret them.

8.2   (a) Use the function `stepAIC()` from the `MASS` package to carry out a stepwise multiple regression (use `direction = "both"`) to select the "best" combination of explanatory variables to construct a linear regression model (use only first-order terms) for yield; (b) find five other linear regression models whose AIC is within 1% of the "best" model; (c) compare the predictions of these models in the context of the variables they contain. (Hint: the function `leaps()` from the `leaps` package is helpful for part b).

8.3   Create scatterplot maps of the agronomic variables for the northern and southern portions of Field 4.1.

8.4   Use the function `leaps()` to carry out a best subsets analysis of the agronomic data for the northern portion of Field 4.1.

8.5   Develop three candidate models for the relationship between *Yield* and the explanatory variables in the northern (first 62 locations) and southern (remaining locations) regions of Field 4.1.

8.6   Use the function `jitter()` to perturb the values of *Yield* by a small amount in the backward selection process of Section 8.3 and observe the effect if you blindly follow the recommendations of the function `drop1()`.

8.7   Let X be a vector of 40 elements, with $X_i = a_i + 0.1\varepsilon_i$, where $a_i = 0$ for $1 \leq i \leq 20$ and $a_i = 2$ for $21 \leq i \leq 40$, and let $Y_i = X_i + 0.1\varepsilon_i'$, with both ε_i and ε_i' being unit normal random variables. (a) Plot Y against X and compute the coefficients of the linear regression of Y on X; (b) Now suppose points 1 through 20 are in the southern half of a field and points 21 through 40 are in the northern half. Compute the regression coefficients of the two halves of the field separately; (c) What relevance does this example have to the analysis of Field 4.1?

8.8   On biophysical grounds one might expect there to be an interaction between *SoilP* and *SoilpH* in a model for *Yield* in Field 4.1. Determine whether such a model is empirically justified.

8.9   In this exercise, we examine the relationship of the endogenous variables to *Yield* and to exogenous variables. (a) Determine which of the two variables *LeafN* and *FLN* is most closely associated with *Yield*; (b) Develop a regression model for the variable that you found in part (a); (c) Repeat part (b) for *CropDens*.

8.10   Develop a regression model for grain protein in terms of the exogenous variables in Field 4.1.

8.11   (a) Construct a scatterplot matrix for Field 4.2; (b) Construct a linear regression model for Field 4.2 in terms of the exogenous data.

8.12 Carry out a regression analysis for the Coast Range subset of Data Set 2 using glm().

8.13 Using Data Set 1, develop a logistic regression model of cuckoo presence/absence using a quadratic term in *AgeRatio* and interpret the results.

8.14 Using Data Set 1, develop logistic regression models of cuckoo presence/absence that includes height class, and determine their significance relative to models that do not include this variable.

8.15 Develop a Poisson regression model for cuckoo abundance in Data Set 1 and compare it with the zero-inflated model developed in Section 8.4.4.

9

Data Exploration Using Non-Spatial Methods: Nonparametric Methods

9.1 Introduction

The previous chapter focused on regression models as a means of data exploration. These are called *parametric* methods because they include assumptions about the form of the error terms, typically that these errors are normally distributed with mean zero and fixed variance σ^2, and part of the analysis involves estimation of the value of the parameter σ^2. The methods of this chapter involve nonparametric statistical analysis, in which these assumptions about the distributional form of the error are not made. As with the methods of Chapter 8, we classify these methods as exploratory since one still typically assumes that the errors are independent, and spatial data may violate this assumption. Confirmatory analysis will be carried out in later chapters in which the analysis incorporates assumptions about the spatial relationships among the data and the errors. This chapter presents a collection of methods, any one of which may be applied to spatial data. I have found generalized additive models (GAMs), recursive partitioning (also known as classification and regression trees or CART), and random forest to be particularly useful in a variety of applications, and therefore these are the focus of this chapter. Section 9.2 deals with GAMs, Section 9.3 deals with recursive partitioning, and Section 9.4 deals with random forest.

Before going on, however, it is worthwhile to point out a famous quote by John Tukey (1986): "The combination of some data and an aching desire for an answer does not ensure that a reasonable answer can be extracted from a given body of data." There is an understandable tendency in all of us to, if our analysis does not initially turn out the way we had hoped it would, move on to a more complicated or sophisticated model in the hope that doing so would reveal patterns that the simpler methods had missed, and sometimes this does happen. That said, the methods described in this chapter are not a panacea and are not guaranteed to produce interesting or meaningful results. They will often produce results, however, and it is up to you to make sure that the results make biophysical sense.

9.2 The Generalized Additive Model

This section contains a brief description of the GAM, introduced by Hastie and Tibshirani (1986). The section is intended to introduce the reader to the GAM, and to do some preliminary exploration of the data. It is *not* intended to provide a thorough introduction to the subject. To do so in a way that would ensure that a GAM could be applied appropriately would require far more space than is available. If, upon reading the material in the section and doing a preliminary exploration such as is presented here, you feel that a GAM would be appropriate for further study, the standard reference, which is an excellent one, is Wood (2006). There are some easily identifiable cases in which the GAM does have certain specific advantages, and for this reason we will look at a simple example to illustrate one of these. This example is based on the plot of blue oak presence vs. elevation in the Coast Range shown in Figure 7.11. Our analysis follows that of Yee and Mitchell (1991).

A good way to introduce the GAM is to compare it to the models we have already seen. Suppose that there are two potential explanatory variables, X_1 and X_2, which do not interact. In this case the linear model is (cf. Equation 8.8).

$$E\{Y_i\} = \beta_0 + \beta_1 X_{i1} + \beta_2 X_{i2}. \tag{9.1}$$

The generalized linear model (GLM) is (cf. Equation 8.28)

$$g(\pi_i) = \beta_0 + \beta_1 X_{i1} + \beta_2 X_{i2}, \tag{9.2}$$

where g is an appropriate link function and $\pi_i = E\{Y_i\}$. In this context, the corresponding generalized additive model would be written

$$g(\pi_i) = \beta_0 + f_1(X_{i1}) + f_2(X_{i2}), \tag{9.3}$$

where again g is an appropriate link function and again $\pi_i = E\{Y_i\}$. The functions f_1 and f_2 are smooth functions, which for our purposes means that they are continuous and have continuous first and second derivatives, so that intuitively they are continuous and have no kinks. To ensure that they are unique, it is required that $f_1(0) = f_2(0) = 0$. Of course, these equations can be extended to a larger number of variables X_i and functions $f_i(X)$. The default method used by ggplot() for the geom geom _ smooth() is a GAM.

In Exercise 8.12, you are asked to construct a generalized linear model analysis for the Coast Range blue oak data. Figure 7.11 indicates that the fraction of sites in which a blue oak occurs reaches a maximum between about 800 m and 1200 m and then declines (leaving aside an apparent anomaly at 1900 m). We will carry out a bit of that analysis here. We begin by fitting generalized linear models to these data (minus the anomaly at 1900 m) using the function glm(). To capture the nonmonotonic relationship between elevation and oak presence our initial model uses the link function

$$g(\pi_i) = \beta_0 + \beta_1 X_i + \beta_2 X_i^2. \tag{9.4}$$

The resulting plot is shown as the solid line in Figure 9.1. It fits the data at the lower elevations reasonably well but does a very poor job at the higher elevations. We can attempt to improve the fit by adding another polynomial degree, that is, using the link function

$$g(\pi_i) = \beta_0 + \beta_1 X_i + \beta_2 X_i^2 + \beta_3 X_i^3, \tag{9.5}$$

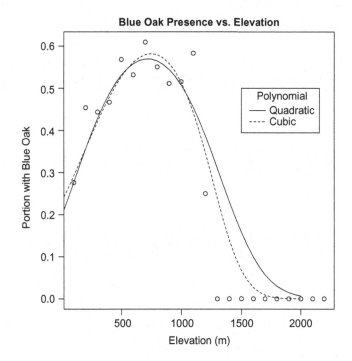

FIGURE 9.1

Plot of generalized linear models of *QUDO* vs. *Elevation* fit to Coast Range subset of the blue oak data from Data Set 2. Also shown are 100 m running averages of fraction of sites occupied by a blue oak.

but, as indicated by the dashed line in the figure, this does not really improve the fit much. Let's see how using a generalized additive model of the form (Equation 9.3) can improve the fit.

The first step is to describe the smooth functions $f_i(X)$. The most commonly used functions, and the only ones we will discuss, are called *splines*. This term comes from a sort of flexible ruler that was used in the days of hand drafting to make irregularly shaped curves. Figure 9.2 shows an example of the most common type of spline (and the one we will use), called a cubic spline, fit to five points. Suppose we have a sequence of points (X_i, Y_i), $i = 1, 2, ..., n$. An *interpolating* spline is sequence of (for us) cubic functions, each of which are fit to two successive points in the sequence, such that the entire sequence of functions fits the sequence of points, as shown in Figure 9.2. For two successive points (X_i, Y_i) and (X_{i+1}, Y_{i+1}), the spline is a sequence of cubic functions $f_i(X)$ that satisfy

$$f_i(X_i) = Y_i, f_i(X_{i+1}) = Y_{i+1},$$

$$f'_{i-1}(X_i) = f'_i(X_i), \tag{9.6}$$

$$f''_{i-1}(X_i) = f''_i(X_i).$$

Here the primes and double primes denote first and second derivatives, respectively. The sequence of points (X_i, Y_i) is called the *knots*, so in words, these spline functions interpolate the points, and their first and second derivatives match at the knots. There is nothing to match at the endpoints, so some other condition is imposed. A common one is $f''_1(X_1) = f''_{n-1}(X_n) = 0$, so that the curvature is zero at the end points. It turns out that the

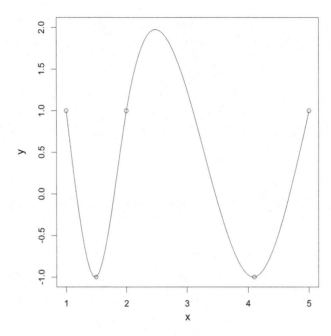

FIGURE 9.2
Plot of a cubic spline interpolation of five data points.

conditions of Equation 9.6 together with the end point conditions are sufficient to uniquely determine the functions (Hastie et al., 2009, Chapter 5).

The interpolating spline function described in the preceding paragraph is not quite what we need for application to the generalized additive model in Equation 9.3. We need something that could fit data, such as that shown in Figure 9.3. This data set was created by using the modulo operator %% to select one-twentieth of the data records of January mean temperature *JaMean* vs June minimum temperatures *JuMin* from the Coast Range data set (see Section 6.3.3 for the use of this operator in this manner). The objective with a data set such as this is not to interpolate it but rather to approximate it with a smooth curve. One way of doing this is by the use of *regression splines.*

We begin our discussion of regression splines by introducing the idea of a *basis.* In thinking about a basis, it is useful to think about the classical vectors *i*, *j*, and *k* that we all encountered in physics as describing coordinates in three-dimensional space. To say that these vectors form a basis for that space means that any vector *v* in this space can be described as a sum $v = i + j + k$. These are not the only possible basis vectors; indeed, any three linearly independent vectors can serve as a basis (cf. Appendix A.1.3).

The idea of a basis in regression splines is analogous. Consider the polynomial function $f(x) = \beta_0 + \beta_1 X + \beta_2 X^2 + \beta_3 X^3$. The four functions 1, X, X^2, and X^3 form a basis for the space of all possible cubic polynomials. As with the basis vectors *i*, *j*, and *k* of physical three-dimensional space, we can make other bases out of linearly independent combinations of these four functions.

Suppose that we wish to construct cubic spline polynomials that we can use on the right-hand side of an equation like Equation 9.3 to create our generalized additive model. Since the spline functions are cubic polynomials, we can express them in terms of basis functions. Thus, the problem is to find the basis functions that satisfy the conditions (Equation 9.6).

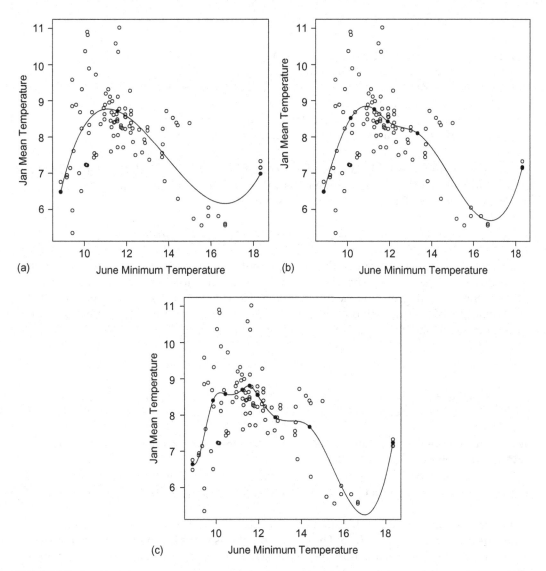

FIGURE 9.3
Cubic spline regression of the variables *JaMean* vs. *JuMin* of the Coast Range subset of the blue oak data from
Data Set 2. (a) 1 interior knot; (b) 5 interior knots; (c) 7 interior knots.

A spline written in terms of these basis functions is called a *b-spline*, where the b stands
for "basis." The actual form of these basis functions is, to quote Simon Wood (2006, p. 124),
"rather opaque." Fortunately, the R package `splines`, which is included with the base R
library, includes a function `bs()` that computes the spline basis for set of regression spline
functions with a specified number of knots. The knots are placed at the percentiles of the
data to ensure that each curve is based on a representative sample.

Because the spline functions are polynomials, a regression based on them is a linear
regression (see the discussion following Equation A.37) in Appendix A if you have problems seeing this). Therefore, the function `lm()` from Chapter 8 can be used to compute the
regression splines themselves. Figure 9.3a shows a regression with one interior knot. The
code to make the data for this plot is as follows.

```
> library(splines)
> ID <- 1:nrow(coast.sf)
> data.plot <- coast.sf[which(ID %% 15 == 0),]
> # Remove a set of anomalous data values
> data.plot <- data.plot[-which((data.plot$JuMin > 11.111) &
+    (data.plot$JuMin < 11.12)),]
> with(data.plot, plot(JuMin, JaMean, xlab = "June Minimum
+    Temperature", ylab = "Jan Mean Temperature")) # Fig. 9.3a
```

The first three lines create the data set by extracting every fifteenth data record. The data set so obtained includes a set of anomalous values, all occurring at *JuMin* = 11.1111, so we remove these values. The code to create the spline is

```
> spl <- lm(JaMean ~ bs(JuMin, df = 4), data = data.plot)
```

The function bs() can accept multiple arguments in many forms. The default for the spline degree is 3 (i.e., cubic). The argument df specifies the number of degrees of freedom, which is the spline degree plus the number of knots. Thus df = 4 specifies one knot. Note that the output of the function bs() is used directly as an argument of lm(). Next, we draw the spline and the knots.

```
> x <- seq(min(data.plot$JuMin), max(data.plot$JuMin), length.out =
+    100)
> lines(x, predict(spl, data.frame(JuMin = x)))
> v1 <- attr(terms(spl), "predvars")[[3]]
> v1
bs(JuMin, degree = 3L, knots = 11.577775, Boundary.knots = c(8.88889,
18.33333), intercept = FALSE)
> x = c(min(data.plot$JuMin),11.5778,max(data.plot$JuMin))
> points(x = x, y = predict(spl1, data.frame(JuMin = x)), pch = 19)
```

The interior knot is placed at the median value of the abscissa values *JuMin*, whose value to four decimal places is 11.577775. The extraction of this value is the subject of Exercise 9.1. As shown in the exercise, it is possible to extract the numerical values of the knots themselves from the function bs().

Figure 9.3b and c show the effect of increasing the number of knots. The fit gets better, but the spline curves get much more "wiggly." If we call the number of knots *k*, then if *k* is too small, then the curve is too smooth, and if *k* is too large, then it is too wiggly. For a simple one-dimensional fit like this it is probably better to select the most appropriate curve by eye, but for more complex, multidimensional fits this may be impossible. Wood (2006, p. 131) recommends using the process of cross validation, which we have already seen in Section 6.3.3, to select the optimal value of *k*. Let a candidate spline function be denoted $\hat{f}(X)$, and let its value at the data point X_i be $\hat{f}(X_i) = \hat{f}_i$. We would like to choose the number of knots *k* to minimize the sum

$$M = \frac{1}{n}\sum_{i=1}^{n}(\hat{f}_i - f(X_i))^2. \qquad (9.7)$$

The problem is that we do not know $f(X_i)$. Therefore, we estimate it by removing the pair (X_i, Y_i) from the data set and computing $\hat{f}_i^{(i)} = f^{(i)}(X_i)$ as the estimated value of Y_i obtained by fitting splines to this reduced data set. The *ordinary cross validation score* v_o is then computed as

$$v_o = \frac{1}{n} \sum_{i=1}^{n} (\hat{f}_i^{(i)} - Y_i)^2. \tag{9.8}$$

This is sometimes called "leave one out" cross validation because the score is based on leaving one data pair out in turn, estimating that value, and then summing the sum of the squared differences of the results.

Wood (2006, p. 132) shows that a process called *generalized cross validation* has for this application both computational and statistical advantages over leave one out cross validation. It turns out, analogous to the computation of the *PRESS* statistic of Equation 8.17, that the value of v_o can be computed without actually going through the process of leaving out one data set at a time. It is equal to

$$v_o = \frac{1}{n} \sum_{i=1}^{n} (\hat{f}_i - Y_i)^2 / (1 - H_{ii})^2, \tag{9.9}$$

where H is the hat matrix of Equation A.36. Wood recommends using a slightly different quantity v_g, given by

$$v_g = \frac{1}{n} \sum_{i=1}^{n} \frac{(\hat{f}_i - Y_i)^2}{(tr(I - H))^2}, \tag{9.10}$$

where the function $tr(A)$ of the matrix A is the *trace*, given by $tr(A) = \Sigma A_{ii}$, that is, the trace is the sum of the diagonal elements. The R core package `stats` contains a function `smooth.spline()` that implements a form of cross validation. This is a bit more sophisticated than the one described here and can return a non-integral value for the number of degrees of freedom. For our purposes, we can just round this value off to the nearest whole number.

```
> spl.sm <- with(data.plot, smooth.spline(JuMin, JaMean))
> round(spl.sm$df,0)
[1] 7
```

The optimal value for `df` is 7, the value shown in Figure 9.3b.

We are now ready to move on to the original purpose of this section, a discussion of the generalized additive model, or GAM, which by analogy with Equation 8.28, we write as

$$g(\pi_i) = \beta_0 + f_1(X_{i1}) + f_2(X_{i2}) + ... + f(X_{i,p-1}) \tag{9.11}$$

The GAM will use cubic splines for the functions $f_i(X)$. We begin with the simple application used to introduce the topic and depicted in Figure 9.1, the fitting of a logistic curve to the Coast Range oak presence/absence vs. elevation data. The figure shows that a cubic polynomial does a better job than a quadratic in fitting the data, although still not a very good one. Since the spline automatically incorporates cubic polynomials, it is not necessary to incorporate them explicitly. Thus for this example we can write

$$g(\pi_i) = \beta_0 + f_1(Elevation_i),$$ (9.12)

There are three primary generalized additive model packages in R: gam (Hastie, 2017), mgcv (Wood, 2006), and gss (Gu, 2014). There are many similarities, and we will focus on the package mgcv. By default the mgcv function gam() uses *thin plate* splines, which are different from cubic splines. However, some experimentation (not shown) indicates that cubic splines work better in our particular case. The code to create the GAM fit with six knots shown in Figure 9.4 is

```
> library(mgcv)
> model.gam1 <- gam(QUDO ~ s(Elevation, bs = "cr", k = 8), data =
+     coast.cor, family = binomial)
```

The function call is very similar to that of glm(), except that the function s() is used to determine the characteristics of the spline function. The package mgcv allows for many options, and we will only scratch the surface. The argument bs specifies the spline basis. The default, as mentioned, is thin plate splines. The argument bs = "cr" specifies cubic splines. The argument k specifies the dimension of the basis, which is two more than the

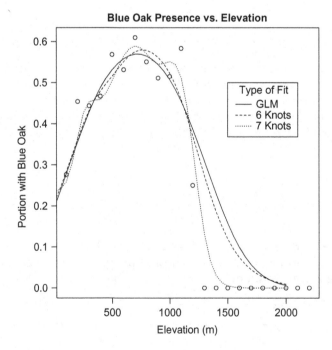

FIGURE 9.4
General linear model (GLM) and GAM fits using cubic splines with 6 and 7 knots, fit to of *QUDO* vs. *Elevation* data from the Coast Range subset of the blue oak data from Data Set 2.

number of knots. The data are difficult to fit because of the steep decline between 1200 and 1300 m. The spline with six knots gives a poor overall fit, while the spline with seven knots give a better overall fit although it is quite wiggly at the lower elevations.

As should be clear from the discussion so far, the objective in developing a GAM is not necessarily to find a better explanatory model, as has been the case with the methods we have discussed earlier. Rather, it is more to find a better fit to the data, that is, a better *predictive* model. In Section 8.4.2, we derived a generalized linear model for the Sierra Nevada data that included *Precip* and *MAT* as explanatory variables. Let's compare this model with a generalized additive model using the same data. We will use the same spline structure as was used for the Coast Range data.

```
> model.glmS2 <- glm(QUDO ~ MAT + Precip, data = sierra.sf, family =
binomial)
> model.gamS2 <- gam(QUDO ~ s(MAT, bs = "cr", k = 9) + s(Precip, bs =
+   "cr", k = 9), data = sierra.sf, family = binomial)
```

How should we do the comparison? The mgcv package includes an anova() function for GAMs, and since a GLM is based on a lower order polynomial model it is in this sense nested in a GAM, so in principle we can compare the two using this function.

```
> # This should be considered as only a rough approximation
> anova(model.glmS2, model.gamS2, test = "Chisq")
Analysis of Deviance Table

Model 1: QUDO ~ MAT + Precip
Model 2: QUDO ~ s(MAT, bs = "cr", k = 9) + s(Precip, bs = "cr", k = 9)
  Resid. Df Resid. Dev     Df Deviance Pr(>Chi)
1    1765.0    1158.3
2    1756.6    1108.8 8.4468   49.437 8.131e-08 ***
---
Signif. codes:  0 '***' 0.001 '**' 0.01 '*' 0.05 '.' 0.1 ' ' 1
```

As it says in the comment, this should be considered as only a rough approximation. The GLM model and the GAM model are computed by two different methods, so basically we are comparing apples and oranges. The p value should therefore not be considered reliable. That said, a value of approximately 10^{-7} is certainly pretty small. We can also compare histograms of the residuals of the two models. These histograms are shown in Figure 9.5 and indicate that the GLM model residuals have a more peaked distribution. A comparison that I find instructive is to plot approximate cross sections of the fit surface. The quantity *Precip* in the Sierra data set ranges from about 325 mm per year to about 2000 mm per year. There are 352 data records whose value is between 900 and 1000 mm. We will take all of these data records, set their value of *Precip* to 1000, and plot the fractional values of *QUDO* at successive intervals of *MAT* between its minimum of 5 and its maximum of 17, and plot the fits of the two models to these data. Figure 9.6 shows the resulting plot. The result is not unexpected. The spline fit of the GAM model is again more wiggly and thus follows the data more closely, for better or worse. Neither the GAM nor the GLM curve is sufficiently influenced by the zero values of *QUDO* at *MAT* values of 16 and 17 to prevent it from taking off into the stratosphere.

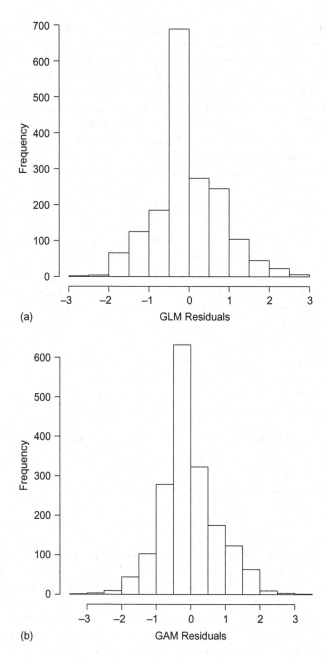

FIGURE 9.5
Histograms of the residuals from (a) GLM and (b) GAM fits of *QUDO* to *Precip* and *MAT* data from the Sierra Nevada subset of the blue oak data from Data Set 2.

In summary, the GAM is clearly better suited for prediction than for explanation, due to its greatly increased flexibility over the GLM. Although most of our interest in this book has been in explanation, there is one application where prediction is the most important, and that is in the estimation of a trend surface. In Exercise 9.3, you are asked to use a GAM to estimate the trend surface computed in Section 3.2.1 using linear regression and median polish. You will be impressed by the result.

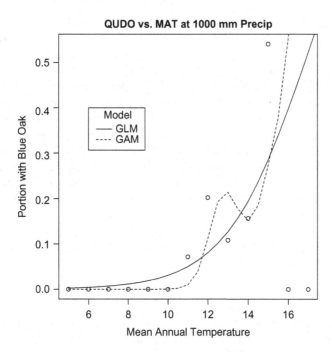

FIGURE 9.6
Cross section at *Precip* = 1000 of the GLM and GAM fits of *QUDO* to *Precip* and *MAT* data from the Sierra Nevada subset of the blue oak data from Data Set 2 (Figure 9.4 Dotplots of the example data in Section 9.2.1).

9.3 Classification and Regression Trees (a.k.a. Recursive Partitioning)

9.3.1 Introduction to the Method

The method described in this section was originally called classification and regression trees, or CART (Breiman et al., 1984), and this is still the better known term. However, the acronym CART was later trademarked and applied to a specific software product. At least in the R community one rarely sees the term "CART." Instead, it is called "recursive partitioning." This name is equally accurate, although unfortunately not as evocative, and we will therefore use it. Recursive partitioning and random forest, which is described in the next section, are two of a group of methods that fall in the category known both as *machine learning* and as *data mining*. These methods are fundamentally different from the other methods in this book in that they focus on an algorithmic approach rather than a data modeling approach to data analysis (Breiman, 2001b). Our application of these methods will not make use of spatial information at all. Instead, we will use our usual tactic, which is to carry out an analysis based solely on the attribute data and then examine the spatial structure of the result as one test of validity.

Recursive partitioning can be illustrated through a simple example. In Section 7.5, a number of factors were identified as potentially being associated with wheat yield in Field 4.1. Two of these were weed infestation level *Weeds*, rated on a scale of 1 through 5, and soil phosphorous content *SoilP*. We will use these not because they are the most important explanatory variables but because they provide a good example of how recursive partitioning works. The response variable (*Yield*) in this example is a ratio scale (i.e., "quantitative") variable (Section 4.2.1). The tree constructed by recursive partitioning in this case is called a *regression tree*. It is also possible to construct trees, called *classification trees*, in which the response variable is nominal (Exercise 9.4).

The data from the 86 sample locations and the interpolated yield data are loaded as described in Appendix B.4. Figure 9.7 shows a bubble plot of the *SoilP–Weeds* data space in which the size of the bubble corresponds to the magnitude of *Yield*. The horizontal and vertical lines are the partitions. The idea of recursive partitioning is to recursively divide the sample space into two parts in such a way that each subdivision is as homogeneous as possible with respect to the response variable. In a regression tree, homogeneity is measured by the sum of the squares of the deviation from the mean over the partition. The data frame data.Set4.1 is created using the code in Appendix B.4. We can apply the function rpart() from the rpart package (Therneau et al., 2011) to the data as follows.

```
> Set4.1.rp1 <- rpart(Yield ~ SoilP + Weeds, data = data.Set4.1,
+    method = "anova")
> plot(Set4.1.rp1) #Fig. 9.8
> text(Set4.1.rp1)
```

The statement in line 1 generates the object Set4.1.rp, and the remaining two lines plot the structure of the object and print the text associated with it. The argument method = "anova" ensures that a regression tree is constructed. If the argument is method = "class" then a classification tree is constructed. If this argument is omitted,

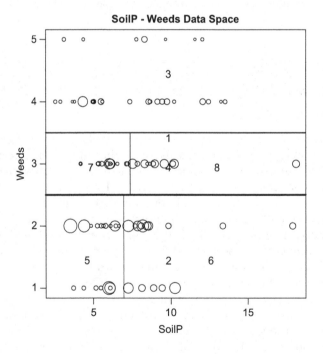

FIGURE 9.7
Plot of the data space of a regression tree for Field 4.1 in which the response variable is *Yield* and the explanatory variables are *SoilP* and *Weeds*. The figure shows the splits that partition the data space into the nodes of the regression tree. The numbers in the figure correspond to the numbers of the splits. The thicker lines correspond to the nodes higher on the tree.

rpart() will try to figure out the appropriate method (classification tree or regression tree) based on the form of the response variable (factor or numerical). The regression tree constructed for these explanatory variables and this response variable is shown in Figure 9.8. The terminal nodes of the regression tree are identified by the mean value of the response variable. The terminal nodes of a classification tree are identified by the value of the classification variable having the most data records in that node.

Using Figures 9.7 and 9.8 as a guide, we can see how recursive partitioning works. The first partition occurs at *Weeds* = 2.5. In tree diagrams generated by rpart(), the left branch always includes data that satisfies the condition listed in the diagram, and the right branch includes data that do not. The program tests every value of each of the explanatory variables (in this case, *SoilP* and *Weeds*) over the indicated subspace of the sample space to determine which value separates that subspace into the two parts whose sums of squares of deviations from the mean of the response variable is minimized. The subsets defined by *Weeds* ≥ 2.5 and *Weeds* < 2.5 are called *nodes*, and the conditions themselves are called *splits* (Breiman et al., 1984, p. 22). Nodes are numbered top to bottom and left to right (Figure 9.8). Thus, node 1 consists of the entire subspace above the thicker of the two horizontal lines, defined by *Weeds* = 2.5, in Figure 9.7, and node 2 is the subspace below that line. This illustrates that splits occur halfway between attribute values. The algorithm recursively searches each node for the split that defines the two most homogeneous subsets. From Figure 9.8 we see that node 3 is the subspace defined by *Weeds* ≥ 3.5 and node 4 is defined by 2.5 < *Weeds* < 3.5. Node 2, defined by *Weeds* < 2.5, is split into the two most homogeneous partitions at the value *SoilP* < 6.94, to form nodes 5 and 6. Finally, node 4 is split at the value *SoilP* < 7.36, forming nodes 7 and 8. The final subsets are called *terminal nodes*. The set of all nodes emanating from a particular *root node* is called a *branch* (Breiman et al., 1984, p. 64).

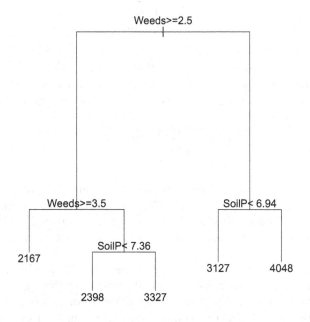

FIGURE 9.8
Regression tree of the recursive partitioning example of Figure 9.7. The numbers at the bottom are the mean *Yield* values for each terminal node.

9.3.2 The Mathematics of Recursive Partitioning

The rpart() function uses the methods described by Breiman et al. (1984), which we will now briefly discuss. It turns out to be easier to discuss them in the context of regression trees; those interested in the parallel discussion for classification trees are referred to the original source. Our discussion simplifies the actual algorithm a bit; for the full complexity, the reader is referred to Therneau and Atkinson (1997).

The algorithm that constructs a regression tree such as that of Figure 9.7 or a classification tree such as that of Exercise 9.4 requires the following three elements (Breiman et al., 1984, p. 229):

1. a way to select a split at every intermediate node
2. a rule for determining when a node is a terminal node
3. a rule for assigning a value of the response variable to every terminal node

Our terminology will be consistent with the rest of this book and different from that of Breiman et al. (1984). What they call a *case* we call a data record, that is, a pair (X, Y), where X is a vector of explanatory variables in a data space, which Breiman et al. (1984) call a *measurement space*, and Y is a response variable.

The third of the required three elements just given, the rule for assigning a response variable to a terminal node, is the easiest to specify, and has already been mentioned informally. Let L be the set of n data records $(X_1, Y_1), ..., (X_n, Y_n)$. Let t denote a terminal node in the regression tree T. In the case in which Y is a categorical variable (the classification tree case), Y is assigned the value of the response variable of the largest number of data records in the node (with some rule for ties). In the case in which Y is a "quantitative" variable (the regression tree case), the rule is defined as follows. For each terminal node t in T we assign the value

$$\bar{Y}(t) = \frac{1}{n(t)} \sum_{X_i \in t} Y_i \tag{9.13}$$

where $n(t)$ is the number of measurements in the subset t.

Having defined the third element of the algorithm, we move on to the first element, the splitting rule. Define the *resubstitution estimate of the relative error* for the tree T to be

$$R(T) = \frac{1}{n} \sum_{t \in T} \sum_{x_i \in t} (Y_i - \bar{Y}(t))^2. \tag{9.14}$$

This same error estimate is used in linear regression, where it is called the mean squared error or *MSE* (Appendix A.2). The notation $R(T)$ emphasizes the dependence on the particular regression tree T. Given this definition, Breiman et al. (1984, p. 231) define the splitting rule (the first element of the three given above) to be the selection of the split that most decreases $R(T)$.

The last remaining element to be determined is the hardest: number 2, the rule to determine when a node is a terminal node. The great advance of the algorithm developed by Breiman et al. (1984) is that, rather than testing at each node to decide whether to stop, recursive partitioning makes this determination by growing the tree beyond the optimal point to a maximal tree T_{max} and then pruning it back to the optimal. In this way, one avoids the problem of stopping too soon based on incomplete information.

After T_{max} has been reached by the splitting process, the determination of the optimal split is made by rewarding a tree for providing a good fit to the data and penalizing the tree for being complex. This concept is related to the *bias-variance tradeoff* that is described more fully in Section 8.2.1 in the regression context. For any subtree T of T_{max}, Breiman et al. (1984, p. 66) define the *complexity* of T, denoted $|\tilde{T}|$, as the number of terminal nodes in T. Let $\alpha \geq 0$ be a real number called the complexity parameter. The *cost complexity measure* $R_\alpha(T)$ is defined as

$$R_\alpha(T) = R(T) + \alpha |\tilde{T}|. \tag{9.15}$$

The cost complexity measure is reduced by trees that reduce the mean square error (the bias), but it is increased by trees that do so by becoming too complex (increasing the variance). In order to find the tree that it declares optimal, the recursive partitioning algorithm begins with the maximal tree and constructs a sequence of trees with increasing cost complexity values, and therefore with decreasing complexity, moving toward the tree that minimizes $R_\alpha(T)$ as defined in Equation 9.15.

The value of $R_\alpha(T)$ is generally estimated by the process of *tenfold cross validation*. We have already seen cross validation once in this chapter, in Section 9.2. Unlike the case with GAM, cross validation is carried out explicitly in recursive partitioning as follows. Let the *sample set* L be defined as the set of all data records, that is, of pairs (X_i, Y_i). Divide the sample set into 10 subsets, each containing (as closely as possible) the same number of cases. Denote these subsets $L_1, L_2, .., L_{10}$. For each k, $k = 1,...,10$, the tenfold cross validation algorithm computes a regression tree $T^{(k)}$ by applying the algorithm to the set $L - L_k$. The cross-validation estimate $R^{CV}(T)$ is

$$R^{CV}(T) = \frac{1}{n} \sum_v \sum_{X_i \in L_v} (Y_i - T^{(k)}(X_i))^2, \tag{9.16}$$

where $T^{(k)}(X_i)$ is the recursive partitioning estimate of Y_i by the tree $T^{(k)}$.

We now have the three elements of the recursive partitioning algorithm in hand. In summary, at each node, the algorithm determines the next split to be the one that minimizes the estimated resubstitution error $R(T)$, given by Equation 9.14. The tree is grown to a maximal tree T_{max} and then pruned back to a size that optimizes the cost-complexity measure given in Equation 9.15, as estimated using the tenfold cross-validation estimate given in Equation 9.16. At that point, the value of the terminal node of the regression tree is the mean value over the node of the response variable. We can now apply the method to our data sets. We first consider Data Set 2.

9.3.3 Exploratory Analysis of Data Set 2 with Regression Trees

The use of the recursive partitioning algorithm to generate classification and regression trees also generates a great deal of auxiliary information that can be very useful in exploring a data set. The function `rpart()` employs two control parameters defined in Section 9.3.2. These are n_{min}, the minimum number of cases allowed in a terminal node, and α, the cost complexity measure defined in Equation 9.15. They are represented in `rpart()` by the parameters `minsplit` and `cp`, which have default values of 20 and 0.05, respectively. The parameter `minsplit` is the smallest node size that will be considered for a split, and thus corresponds exactly to n_{min}. The parameter `cp` (for complexity parameter) is the value of α scaled by $R(T_{root})$, the resubstitution error of the root node T_{root}

of the entire tree. That is, in comparison with Equation 9.15, the resubstitution estimate is given by (Therneau and Atkinson, 1997, p. 22)

$$R_{cp}(T) = R(T) + cp \mid \tilde{T} \mid R(T_{root}). \tag{9.17}$$

The parameters `minsplit` and `cp` govern the complexity of the tree, and therefore the first step in a recursive partitioning analysis is to determine whether the default values for these parameters are appropriate. In particular, it is important to determine whether the parameter `cp`, which controls the complexity of the optimal tree, actually does give us the tree having the most information. The objective is to achieve the correct balance between fit and complexity. Too simple a tree will lead to underfitting, which generates a suboptimal result by not using all of the available data. Too complex a tree leads to overfitting, which uses too much data specific only to this particular data set.

We can obtain information about the sequence of trees generated by `rpart()` in the pruning process by first setting `minsplit` and `cp` to very small values to generate the maximal tree and the using the functions `plotcp()` and `printcp()` to view the properties of the sequence leading to this maximal tree. In our analysis of Data Set 2, we will continue to analyze the Sierra Nevada and Coast Range data separately, focusing on the Sierra Nevada, entered as an `sf` object, in the text and the Coast Range in the exercises. The tree will only be computed for endogenous explanatory variables, and the parent material classes are aggregated.

```
> data.Set2Srp <- with(data.Set2S.sf, data.frame(MAT,
+    Precip, JuMin, JuMax, JuMean, JaMin, JaMax, JaMean, TempR, GS32,
+    GS28, PE, ET, Texture, AWCAvg, Permeab, SolRad6, SolRad12,
+    SolRad, QUDO))
> data.Set2Srp$PM100 <- as.numeric(data.Set2Sierra.sf$PM100 > 0)
> data.Set2Srp$PM200 <- as.numeric(data.Set2Sierra.sf$PM200 > 0)
> data.Set2Srp$PM300 <- as.numeric(data.Set2Sierra.sf$PM300 > 0)
> data.Set2Srp$PM400 <- as.numeric(data.Set2Sierra.sf$PM400 > 0)
> data.Set2Srp$PM500 <- as.numeric(data.Set2Sierra.sf$PM500 > 0)
> data.Set2Srp$PM600 <- as.numeric(data.Set2Sierra.sf$PM600 > 0)
```

We will initially leave `minsplit` at its default value of 20, but reduce the size of `cp` from the default value of 0.05.

```
> library(rpart)
> cont.parms <- rpart.control(minsplit = 20,cp = 0.002)
> Set2S.rp <- rpart(QUDO ~ ., data = data.Set2Srp,
+    control = cont.parms, method = "anova")
```

Now we use `plotcp()` and `printcp()` to determine the appropriate value of `cp`.

```
> plotcp(Set2S.rp) # Fig. 9.9
> printcp(Set2.rp)
Regression tree:
rpart(formula = "QUDO ~ .", data = data.Set2Srp, method = "anova",
    control = cont.parms)
Variables actually used in tree construction:
 [1] AWCAvg    ET        GS28      GS32      JaMax     JaMean    JuMax
```

```
[8] JuMin      MAT       PE          Permeab  PM300    Precip   SolRad
[15] SolRad12 SolRad6   TempR
```

```
Root node error: 422.22/1768 = 0.23881
```

```
n= 1768
```

	CP	nsplit	rel error	xerror	xstd
1	0.4516262	0	1.00000	1.00038	0.010300
2	0.0613849	1	0.54837	0.58686	0.025282
* * *		*DELETED*	* * *		
7	0.0102024	6	0.39782	0.45256	0.022612
8	0.0082386	8	0.37741	0.43808	0.022814
9	0.0077734	9	0.36918	0.43222	0.022589
10	0.0069921	10	0.36140	0.43489	0.022877
11	0.0066791	11	0.35441	0.43432	0.022967
12	0.0062327	12	0.34773	0.43606	0.023056
13	0.0059766	13	0.34150	0.43707	0.023163
14	0.0055890	14	0.33552	0.43782	0.023411
15	0.0055266	15	0.32993	0.43473	0.023438
* * *		*DELETED*	* * *		
31	0.0020000	36	0.25092	0.45350	0.024702

The function `printcp()` gives a table of values of the optimal tree if the corresponding value of cp is used. The table includes the value of cp at which a split occurs; `nsplit`, the number of splits; `rel error`, the resubstitution estimate $R(T)$ given in Equation 9.14 of the relative error; `xerror`, the cross validation estimate $R^{CV}(T)$ given in Equation 9.16 of the error; and `xstd`, the corresponding cross-validation estimate of the standard error. In the example, the first split occurs if a value of cp less than 0.4516262 is specified. One minus the relative error can be used to interpret the "portion of variance explained" by the regression tree, analogous to the R^2 of linear regression. Thus, prior to making any splits (`nsplit` = 0) the relative error is 1, and the regression tree, which consists of a single node, explains none of the variance.

In multiple linear regression, as the number of explanatory variables increases, the coefficient of determination R^2 cannot decrease (Kutner et al., 2005, p. 226). Analogously, as the complexity of the tree increases, the resubstitution estimate $R(T)$ cannot decrease. For this reason, the cross-validation estimate $R^{CV}(T)$ (`xerror` in the table) is considered a better estimate of the "portion of the variance explained by the tree" (Breiman et al., 1984, p. 225). The derivation of the standard error `xstd` estimate is complex (Breiman et al., 1984, p. 304). The greatest usefulness of this statistic is in interpreting the output of the function `plotcp()`.

The function `plotcp()` plots the relative error $R^{CV}(T)$, together with standard error bars, that are obtained in a tree grown using the value of cp on given the abscissa. The values of cp on the abscissa are the geometric means of successive values of cp printed by `printcp()`. Thus, for example in the listing above, the first split occurs at cp = 0.452 (to three decimal places), and the second at cp = 0.061. The first cross-validation computation (which is actually the second point from the left in Figure 9.9) is therefore carried out at cp = 0.166 = $\sqrt{0.452 \times 0.061}$, and successive cross validations are carried out at successive geometric means.

The lowest value of $R^{CV}(T)$ occurs at cp = 0.0055890, at which value the tree has 14 splits. The error bars in the output of `plotcp()` (Figure 9.9) represent one standard error, and the

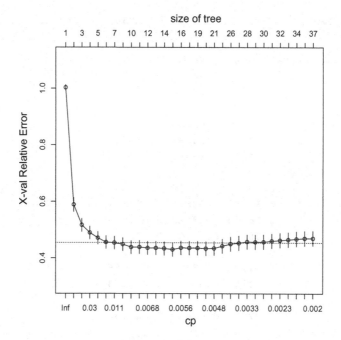

FIGURE 9.9
Output of the function plotcp() for a regression tree grown for Data Set 2 with *QUDO* as the response, showing the relative error for decreasing values of cp.

horizontal dotted line in the figure is drawn at one standard error above the lowest value of the cross validation estimate of the relative error. Therneau and Atkinson (1997, p. 13) recommend using the "1 *SE*" rule to choose the value of cp governing the pruning of the tree. This rule states that a good choice of cp for pruning is often the leftmost value of cp for which the error bar lies entirely below the dotted line. In our case, this is when the "size of the tree" nsplit has the value 9, at which point, from the output of printcp(), cp = 0.0078 to four decimal places.

We can make the regression trees somewhat more attractive (Figure 9.10a) with some arguments to the functions plot() and text().

```
> cont.parms <- rpart.control(minsplit = 20, cp = 0.0078)
> Set2S.rp <- rpart(QUDO ~ ., data = data.Set2Srp,
+    control = cont.parms, method = "anova")
> plot(Set2S.rp,branch = 0.4,uniform = T,margin = 0.1,
+    main = "Data Set 2 Sierra Nevada Regression Tree",
+    cex.main = 2) # Fig. 9.10a
> text(Set2S.rp,use.n = T,all = T, cex = 0.65)
```

The argument uniform = T makes the vertical distance between nodes uniform. The argument branch = 0.4 tilts the branches from vertical. The argument margin = 0.1 creates a small margin around the tree so the bottom values do not get lost. In the text() function, the argument use.n = T causes the value of the number of cases in each node to be printed and the argument all=T puts the values of the response variable and number of cases in all nodes, not just the terminal nodes. As usual, the argument cex controls the size of the text. This value is a compromise between legibility and forcing the text into the tree.

Data Set 2 Sierra Nevada Regression Tree

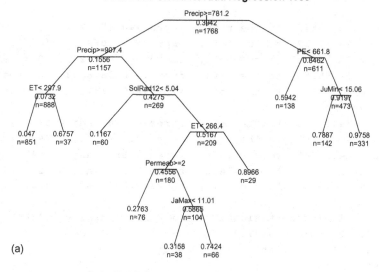

(a)

Data Set 2 Coast Ranges Regression Tree

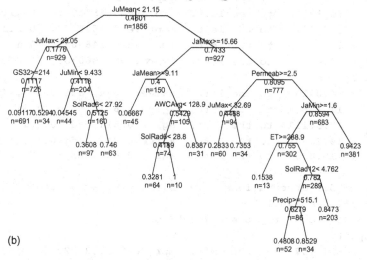

(b)

FIGURE 9.10

(a) Regression tree with QUDO as the response variable for the Sierra Nevada subset of Data Set 2. (b) Same as (a) for the Coast Range subset of Data Set 2.

We can get more information with a call to the function `summary()`, a portion of whose result is shown here.

```
> summary(Set2S.rp)
Call:
rpart(formula = "QUDO ~ .", data = data.Set2Srp, method = "anova",
    control = cont.parms)
  n= 1768
Node number 1: 1768 observations,    complexity param=0.4516262
  mean=0.3942308, MSE=0.2388129
```

```
  left son=2 (1157 obs) right son=3 (611 obs)
  Primary splits:
      Precip  < 781.1516   to the right, improve=0.4516262, (0 missing)
      JuMean  < 24.17917   to the left,  improve=0.4052687, (0 missing)
      JuMax   < 33.70834   to the left,  improve=0.3875277, (0 missing)
      MAT     < 14.85555   to the left,  improve=0.3826863, (0 missing)
      JaMax   < 11.03334   to the left,  improve=0.3647264, (0 missing)
  Surrogate splits:
      JuMax   < 34.90555   to the left,  agree=0.876, adj=0.642, (0 split)
      JaMax   < 11.95      to the left,  agree=0.859, adj=0.592, (0 split)
      JuMean  < 24.51945   to the left,  agree=0.853, adj=0.574, (0 split)
      MAT     < 15.325     to the left,  agree=0.837, adj=0.527, (0 split)
      JaMean  < 6.62083    to the left,  agree=0.836, adj=0.525, (0 split)

Node number 2: 1157 observations,     complexity param=0.06138488
  mean=0.1555748, MSE=0.1313713
  left son=4 (888 obs) right son=5 (269 obs)
  Primary splits:
      Precip  < 907.4277   to the right, improve=0.1705170, (0 missing)
      JuMax   < 33.125     to the left,  improve=0.1572847, (0 missing)
      ET      < 297.942    to the left,  improve=0.1528268, (0 missing)
      JaMax   < 11.01389   to the left,  improve=0.1465055, (0 missing)
      JuMean  < 23.42223   to the left,  improve=0.1360703, (0 missing)
  Surrogate splits:
      JuMax   < 34.09166   to the left,  agree=0.834, adj=0.286, (0 split)
      JaMax   < 11.625     to the left,  agree=0.831, adj=0.275, (0 split)
      JuMean  < 23.56111   to the left,  agree=0.814, adj=0.201, (0 split)
      MAT     < 14.76666   to the left,  agree=0.812, adj=0.190, (0 split)
      PE      < 719.836    to the left,  agree=0.808, adj=0.175, (0 split)

Node number 3: 611 observations,     complexity param=0.02680101
  mean=0.8461538, MSE=0.1301775
  left son=6 (138 obs) right son=7 (473 obs)
  Primary splits:
      PE      < 661.797    to the left,  improve=0.1422702, (0 missing)
      JuMean  < 24.82639   to the left,  improve=0.1357518, (0 missing)
      GS28    < 267.35     to the left,  improve=0.1325340, (0 missing)
      JuMin   < 15.05556   to the left,  improve=0.1260423, (0 missing)
      MAT     < 15.82222   to the left,  improve=0.1239181, (0 missing)
  Surrogate splits:
      JaMin   < 0.76388    to the left,  agree=0.895, adj=0.536, (0 split)
      GS28    < 259.1      to the left,  agree=0.887, adj=0.500, (0 split)
      JuMin   < 14.47778   to the left,  agree=0.881, adj=0.471, (0 split)
      GS32    < 198.4      to the left,  agree=0.866, adj=0.406, (0 split)
      JuMean  < 24.81111   to the left,  agree=0.856, adj=0.362, (0 split)
```

The most important new information is a list of "primary splits" and "surrogate splits." The list of primary splits is easy to explain. Recall that the optimal split is chosen to maximally reduce the estimated resubstitution error $R(T)$. The split at the top of the list of primary splits is the one that satisfies this criterion, and the remainder are those that reduce $R(T)$, but not as much. These other splits are called the *competitors*. Thus, for example the first competitor for Precip in the first node is JuMean. The *surrogate split* is the split that best predicts the optimal split s^* in the sense that it sends cases to the left and right in closest proportions to that of s^* (Breiman et al., 1984, p. 141). Surrogate splits are useful in identifying variables that may be *masked*. This phenomenon is similar to multicollinearity in multiple linear regression and has the same negative consequences. A variable X_k may have almost as much explanatory power at a given split as the variable that produces the optimal split s^*, and may over the entire tree have more explanatory power, but never appear in the tree because locally at each split it has less explanatory power than some other variable. We can informally determine whether a variable is being masked by examining the surrogates at each split and seeing if any variables either appear reasonable from an ecological perspective or appear repeatedly in several splits.

For the Sierra Nevada data in the tree of Figure 9.10a, the output indicates that the main surrogates and competitors of Precip in the first split are JuMean, JuMax, MAT, and JaMax. This is not surprising, given the high degree of correlation among these variables. Interestingly, the variables that determine the splits are almost exclusively climatic, with Permeab entering in one split very far down the tree. Since QUDO is a binary variable ($QUDO = 1$ if blue oak is present, and $QUDO = 0$ if blue oak is absent), the tree could be generated either as a classification tree or a regression tree. The argument method = "anova" in the statement calling the function rpart() specifies that a regression tree will be generated. The regression tree is easier to interpret because it indicates the fraction of the data records in each node for which $QUDO = 1$.

The interpretation of the tree in terms of the response variable $QUDO$ is given by the numbers just below the rule defining the split. This indicates the mean value of QUDO over the node. Thus, for example, in Figure 9.10a the mean value of $QUDO$ at the root node is 0.3942, indicating that 39.42% of the data records in the root node (the whole Sierra Nevada data set) have blue oaks present. In node 2, which contains those data records in which Precip >= 781.2, about 15% of the records have blue oak present, whereas in node 3, where Precip < 781.2, about 84% of the data records have blue oak present. In node 3, the splitting variable is PE (potential evapotranspiration); 92% of the sites where Precip < 781.2 and PE ≥ 15.06 mm contain a blue oak. Evidently precipitation is a very important factor in the Sierra Nevada. Considering that blue oaks are known for their drought tolerance, it is not surprising that a higher fraction of them are found in data locations with low mean annual precipitation and high potential evaporation.

The situation is different in the Coast Ranges (Figure 9.10b, generated in Exercise 9.5). Here, Precip only enters far down the tree, and temperature-related variables seem most important. In the Coast Ranges, precipitation is generally lower (almost all of the records are below 1100 mm), and the data records with low precipitation values span a broader range of elevations. It is possible that the regression tree does not identify Precip as an important splitting variable because virtually all of the sites are located in areas where precipitation is sufficiently low. Recursive partitioning identifies JuMean as the first splitting variable (Figure 9.10b). According to the figure, 75% of the sites where JuMean ≥ 21.5 have blue oaks present, while only 18% of the sites where JuMean < 21.5 have blue oaks present. Both the competitors and the surrogate splits of JuMean are temperature related.

In summary, recursive partitioning has identified precipitation, potential evaporation, and temperature as the most important endogenous variables in the Sierra Nevada, with low precipitation and high potential evaporation especially favoring blue oak presence. In the Coast Range, where precipitation is generally lower, high mean June temperatures and low January maximum temperatures favor blue oak presence. The extremely high correlation values of all of the temperature-related variables, however, render this conclusion highly tentative at best.

9.3.4 Exploratory Analysis of Data Set 3 with Recursive Partitioning

The preliminary analysis of Data Set 3 in Chapter 7 indicated that one of the three regions of rice production, the central region, consistently lagged behind the other two regions in terms of yield. There was, however, considerable variability among the farmers in the northern and central region, and only one farmer was surveyed in the southern region. Climate does not appear to have had an important influence on year-to-year yield variability in the three seasons in which data were collected, and terrain also does not appear to have much influence. The primary characteristic distinguishing soils in the central region was high silt content relative to soils in the northern and southern region. Yields appeared to be more determined by the farmer than by location or year in those cases where different farmers farmed the same field in different years. The farmers in the north applied more fertilizer and appear to have irrigated and managed weeds more effectively (Table 7.4).

The contents of the file *Set3data.csv* are loaded into the data frame `data.Set3`. The data are a mixture of exogenous and management variables. No endogenous plant-based variables were recorded in Data Set 3. Recall that the goal of our analysis is to separate out the management practices from the field conditions in determining the factors associated with high yields. As a step in this direction, we develop a recursive partitioning analysis using only the exogenous variables (Figure 9.11). The procedure is identical to that used in the previous section. *Silt* is the first splitting variable, followed by *pH* in the high silt fields, and then by *Corg* and *Clay*. All of the northern and southern fields have silt content below 44%, and almost all of the central fields have soil silt content above this level. Thus, it may be that the reason silt is identified as the first splitting variable is simply because it separates

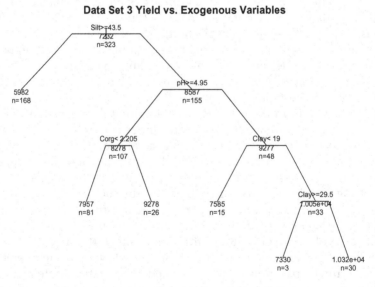

FIGURE 9.11
Regression tree for Yield as a function of the exogenous variables of Data Set 3.

the central region from the other two. What about pH? Soils in the northern region tend to be more acidic than those in the central and southern region. From an agronomic perspective, however, flooding tends to move all soil pH levels towards neutrality (Williams, 2010, p. 10), so that rice in general is less sensitive to soil pH that are terrestrial crops.

Is either silt content or pH a determining factor in yield in these fields? We can address this by first sorting the farmers by mean *Yield* and then by mean *Silt* and mean *pH*.

```
> print(sort(with(data.Set3, tapply(Yield, Farmer, mean))), digits = 4)
    I     E     K     G     H     F     J     L     B     A     D     C
 4982  5112  5245  5559  6011  6331  6958  7948  8219  8738  9542 11456
```

Remember that Farmers *A* through *D* are in the north, Farmer *L* is in the south, and farmers *E* through *K* are in the central region.

```
> print(sort(with(data.Set3, tapply(Silt, Farmer, mean))), digits = 3)
    B     A     D     C     L     E     F     G     K     H     I     J
 29.7  33.3  34.0  36.3  37.9  58.6  59.7  59.8  60.0  61.7  63.2  63.9
```

While it is true that the southern farmers have higher silt content than the northern farmers, the farmer with the highest mean silt content is *J*, the highest yielding central farmer, and the northern farmer with the highest mean silt content is *C*, the highest yielding northern farmer.

```
> print(sort(with(data.Set3, tapply(pH, Farmer, mean))), digits = 3)
    C     D     B     A     E     G     F     H     K     L     I     J
 4.68  4.73  5.14  5.60  5.73  5.77  5.79  5.88  5.89  5.89  5.94  6.02
```

Farmer *J* also has the most basic soil, while Farmer *C* has the most acidic. Evidently, if there is a relation between *Silt*, *pH*, and *Yield*, it is a complex one. In Exercise 9.6, you are asked to plot these means. These plots are interesting as an example of the sort of "barbell"-shaped plots that can produce regression results that are difficult to interpret. Another complicating factor is the issue of spatial scale. Geographically, Data Set 3 has a three level spatial hierarchy: the sample point scale, the field scale, and the region scale. We have seen so far that within each region, the yields at the field scale do not depend on silt content or soil acidity, but between regions the field scale yields do depend on these quantities. We need to try to determine whether this relationship is truly a direct influence or simply an apparent association.

As in Section 9.3.3, we first do a preliminary run with a small value of cp and then use plotcp() and printcp() to select the appropriate value. Figure 9.12 shows the resulting regression tree that includes all of the explanatory variables, both exogenous and management. Exogenous variables play a relatively minor role. The first split is on *N*, the amount of applied nitrogen fertilizer. This split also almost perfectly divides the northern and southern regions from the center: almost all of the former report a fertilizer application rate greater than 60 kg/ha (indeed, all of the northern farmers report the same rate, 61 kg/ha), while all of the central farmers apply at a lower rate (Exercise 9.7). The next splits on each branch are on the variable *Farmer*. We can see this from the output of the function summary().

```
> cont.parms <- rpart.control(minsplit = 20,cp = 0.003)
> Set3.rp2 <- rpart(Yield ~ DPL + Cont + Irrig +
+     N + P + K + Variety + pH + Corg + SoilP + SoilK + Sand +
```

```
+        Silt + Clay + Farmer, data = data.Set3rp, control = cont.parms)
> summary(Set3.rp2)
       *  *  *   DELETED   *  *  *
Node number 1: 323 observations,        complexity param=0.565231
  mean=7232.111, MSE=3271954
  left son=2 (193 obs) right son=3 (130 obs)
  Primary splits:
      N      < 61.3  to the left,      improve=0.5652310, (0 missing)
      Farmer splits as  RRRRLLLLLLLR,  improve=0.5236007, (0 missing)
      Silt   < 43.5  to the right,     improve=0.5176631, (0 missing)
      K      < 11.7  to the left,      improve=0.4108487, (0 missing)
      Cont   < 4.5   to the left,      improve=0.3882669, (0 missing)
  Surrogate   splits:
      K      < 11.7  to the left,      agree=0.923, adj=0.808, (0 split)
      Farmer splits as  RRRRLLLLLLLL,  agree=0.923, adj=0.808, (0 split)
      Silt   < 43.5  to the right,     agree=0.916, adj=0.792, (0 split)
      P      < 67.2  to the left,      agree=0.910, adj=0.777, (0 split)
      Cont   < 4.5   to the left,      agree=0.885, adj=0.715, (0 split)

Node number 2: 193 observations,        complexity param=0.1026634
  mean=6115.995, MSE=1187620
  left son=4 (103 obs) right son=5 (90 obs)
  Primary splits:
      Farmer splits as  ----LRLLLRLR,  improve=0.4733587, (0 missing)
      N      < 44.7  to the left,      improve=0.4578913, (0 missing)
      P      < 67.2  to the right,     improve=0.4578913, (0 missing)
      Irrig  < 2.5   to the left,      improve=0.3517589, (0 missing)
      Cont   < 3.5   to the left,      improve=0.2455773, (0 missing)
  Surrogate   splits:
      N      < 53.5  to the left,  agree=0.720, adj=0.400, (0 split)
      Corg   < 1.52  to the left,  agree=0.674, adj=0.300, (0 split)
      Silt   < 39.5  to the right, agree=0.663, adj=0.278, (0 split)
      Sand   < 30.5  to the left,  agree=0.642, adj=0.233, (0 split)
      Irrig  < 3.5   to the left,  agree=0.637, adj=0.222, (0 split)

Node number 3: 130 observations,     complexity param=0.06751958
  mean=8889.115, MSE=1771316
  left son=6 (120 obs) right son=7 (10 obs)
  Primary splits:
      Farmer  splits as  LLRL-------L,  improve=0.3098846, (0 missing)
      Clay    < 16.5  to the left,      improve=0.2583827, (0 missing)
      Sand    < 50.5  to the right,     improve=0.2272371, (0 missing)
      Irrig   < 4.5   to the left,      improve=0.1897808, (0 missing)
      Variety splits as  RLRR,          improve=0.1585697, (0 missing)
  Surrogate splits:
      SoilP < 1.45 to the right, agree=0.938, adj=0.2, (0 split)
```

Not all farmers are identified in the tree as being in one or the other node, but all are in the tree. For example, in node 3 Farmer *C* is in the right split along with the other

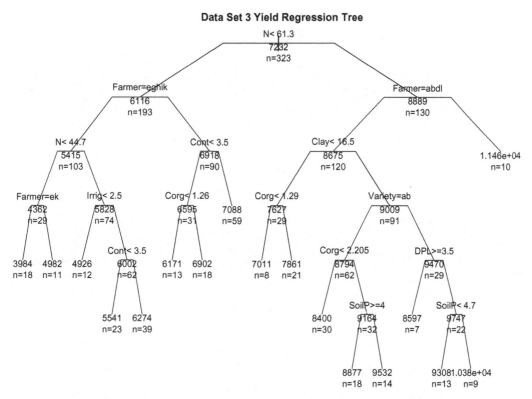

FIGURE 9.12
Regression tree for Yield as a function of both the exogenous and management variables of Data Set 3.

northern and southern farmers. He does not show up in the node label because it identifies the values of splitting variable (*Farmer*) that split to the left, and he splits to the right.

The surrogate and primary splits of the first node indicate that the most important initial splitting factor is the region, with northern separated from central, and southern somewhere in between. Within regions, the most important splitting factor is *Farmer*. We can take advantage of the fact that recursive partitioning works equally well with nominal scale variables as with ratio or interval scale variables to use *Farmer* as a response variable and see what characterizes each individual farmer. Figure 9.13 is a classification tree of exogenous variables with *Farmer* as the response variable. Referring to Figure 7.15, Farmer *C* has the highest median yield among the northern farmers, and *B* has the lowest, while in the central region Farmer *J* has the highest median yield and *I* has the lowest. Farmer *B* does have some sites with very low clay content, but other than that the values of the exogenous variables of Farmers *B* and *C* seem very similar. Farmer *J* farmed fields with a wide variety of properties, some of which are very similar to those of the field farmed by *I*. Let us explore the variables that characterize the management practices and the field environment of each farmer. We have already seen evidence that variety has no substantial effect on yield. In Exercise 9.8, you are asked to examine the regression tree for all management variables including *Variety* and provide further evidence for this lack of effect. Therefore, we remove it from the list of explanatory variables and construct the classification tree (Figure 9.14). The first split, which separates three of

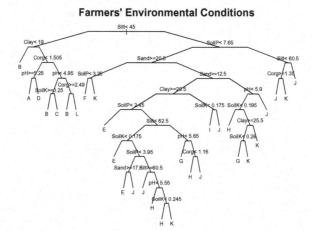

FIGURE 9.13
Classification tree for Data Set 3 with Farmer as the response variable and the exogenous variables as explanatory variables.

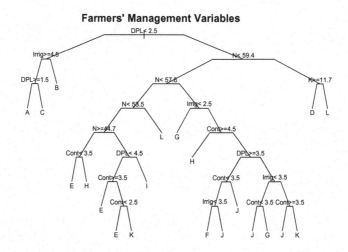

FIGURE 9.14
Classification tree for Data set 3 with Farmer as the response variable and the management variables as explanatory variables.

the northern farmers, is on *DPL*. In the left split, Farmer *C* has the most effective irrigation and the earliest planting date.

Among the farmers with *DPL* > 2.5, the next split, on *N* < 59.4, separates the remaining two non-central farmers. Working our way along the tree, the higher yielding central farmers, *J* and *F*, generally are distinguished by higher values of *Cont* and *Irrig*, while the lower yielding farmers, *E*, *I*, and *K*, are distinguished by lower values of *Cont* and *Irrig*, and, in the case of the lowest yielding farmer, *I*, by a high value of *DPL*. A call to the function summary() (not shown) indicates that the competitors and surrogates at the first split are *N*, *P*, *K*, *Irrig*, and *Cont*.

It appears that the explanatory variables *N*, *P*, *K*, *Cont*, *Irrig*, and *DPL*, play an important role in the trees determining the farmers. In particular, the primary distinguishing characteristics of the northern and southern farmers are an earlier planting date and higher

N fertilization rate. The farmers with better yields tend to irrigate more effectively, plant earlier, and have better weed control. We will conclude the recursive partitioning analysis of Data Set 3 by creating a table that shows the average of these variables as well as *Yield* for each field/farmer/season combination. First, we use the R function ave() to average each variable over all combinations of *Farmer, Season, RiceYear,* and *Field.* Here are the statements for *Yield* and *N;* the statements for *P, K, Cont, Irrig,* and *DPL* are analogous.

```
> Yield <- with(data.Set3, ave(Yield, Farmer, Season, RiceYear, Field))
> N <- with(data.Set3, ave(N, Farmer, Season, RiceYear, Field))
```

Next, we create a data frame with all of these averages except *Farmer,* which we have to treat separately because it is a factor.

```
> df.C <- data.frame(N, P, K, Irrig, Cont, DPL, Var, Yield)
```

This data frame contains many duplicate records because it has an identical record for each of the combinations of *Name, Season, RiceYear,* and *Field.* We can eliminate duplicate records from the data frame using the function remove.dup.rows() from the cwhmisc package (Hoffmann, 2015).

```
> library(cwhmisc)
> Table.C <- remove.dup.rows(df.C)
```

The function remove.dup.rows() doesn't work too well with non-numerical data such as *Farmer.* To include this variable in the table, first we paste together the names of all of the combinations of *Name, Season, Rice,* and *Field* and then eliminate duplicates using the function unique().

```
> Comb <- with(data.Set3, paste(as.character(Farmer),
+    as.character(Season), as.character(RiceYear), as.character(Field)))
> Table.C2 <- data.frame(FrmRySnFld = unique(Comb), Table.C)
```

Finally, we display the table, ordered in terms of yield from highest to lowest.

```
> print(Table.C2[order(Table.C2$Yield, decreasing = TRUE),],
+    digits = 3, right = TRUE)
```

	FrmRySnFld	N	P	K	Irrig	Cont	DPL	Var	Yield
283	C 3 1 2	62.2	72.0	23.4	5.00	5.00	1.00	2.00	11456
323	D 3 1 4	62.2	72.0	23.4	4.60	4.85	3.35	1.00	9542
200	L 2 1 15	62.8	72.8	0.0	4.12	4.64	3.00	2.00	8832
150	A 2 1 1	62.2	72.0	23.4	4.76	4.76	2.00	2.36	8738
175	B 2 1 3	62.2	72.0	23.4	3.64	4.68	2.00	4.00	8473
248	B 3 2 3	62.2	72.0	23.4	3.68	4.64	1.00	4.00	7965
54	J 1 1 5	58.4	55.2	0.0	3.45	3.45	3.00	2.00	7404
52	J 1 1 14	58.4	55.2	0.0	3.83	3.83	3.00	2.00	7153
273	L 3 2 16	58.8	55.5	0.0	3.48	3.56	3.44	1.00	7064
91	J 1 2 12	58.4	55.2	0.0	3.69	3.69	4.00	2.00	6898
221	J 2 2 14	58.4	55.2	0.0	3.83	3.83	3.00	2.00	6742
78	J 1 2 13	58.4	55.2	0.0	4.00	3.14	4.00	2.00	6408

31	F	1	1	7	58.4	55.2	0.0	3.50	3.60	4.00	2.00	6331
303	H	3	1	12	50.2	46.0	0.0	3.15	3.77	3.92	3.85	6023
113	H	1	1	10	58.4	55.2	0.0	3.80	3.80	3.50	2.00	5995
21	K	1	1	6	58.4	55.2	0.0	3.81	3.67	3.00	2.00	5904
290	E	3	1	13	50.2	46.0	0.0	3.00	2.57	3.29	2.57	5698
71	G	1	1	8	58.4	55.2	0.0	2.53	3.59	3.00	2.00	5559
223	I	2	2	5	39.2	72.0	0.0	2.45	2.64	4.45	1.00	4982
125	E	1	1	11	48.8	63.6	0.0	3.42	3.33	3.50	2.00	4771
103	K	1	1	9	39.2	72.0	0.0	2.67	3.00	4.00	1.00	4091

None of the central (or southern) farmers apply any potassium fertilizer. Nitrogen fertilizer shows a definite downward trend as one moves down the table, as does irrigation effectiveness. The northern farmers plant earlier in general, and the latest planting northern farmer gets the lowest yields. The northern farmers have uniformly good weed control. Again, variety does not seem to matter too much (note, however, that we are treating a nominal scale quantity as if it were ratio scale). In Exercise 9.9, you are asked to construct an analogous table using the exogenous variables. The only corresponding pattern that distinguishes the central fields from the others is soil texture.

In summary, we don't yet have any evidence that climatic or environmental factors play as important a role in determining yield as do the farmers' management practices. The data can be analyzed at three levels of spatial hierarchy (the region, the field, and the individual sample point), and the relationships between yield and explanatory variables in this data set appear to change at different levels of this hierarchy. We will explore these differences further in subsequent chapters.

9.3.5 Exploratory Analysis of Field 4.1 with Recursive Partitioning

The recursive partitioning analysis of Field 4.1 proceeds along the same lines as those already carried out in Sections 9.3.2 and 9.3.3. We will include only exogenous variables in the model. Here is the code to specify the rpart() model. In rpart(), when we specify the first argument separately, we use a character string. This differs from the lm() family, where it is a formula object.

```
> Set4.1.model <- "Yield ~ Clay + Silt + Sand + SoilpH + SoilTOC +
+     SoilTN +SoilP + SoilK + Weeds + Disease"
```

We first use the functions printcp() and plotcp() to select values of cp and minsplit, and then apply these values to the recursive partitioning function rpart(). The resulting regression tree is shown in Figure 9.15. The use of the function summary() results in the following list of competitors and surrogates (only a part of the output is shown).

```
> cont.parms <- rpart.control(minsplit = 5, cp = 0.02)
> Set4.1.rp <- rpart(Set4.1.model, data = data.Set4.1,
+     control = cont.parms, method = "anova")
> summary(Set4.1.rp)
     *   *   *   DELETED   *   *   *
Node number 1: 86 observations,    complexity param=0.6553524
   mean=2879.28, MSE=1984255
   left son=2 (62 obs) right son=3 (24 obs)
```

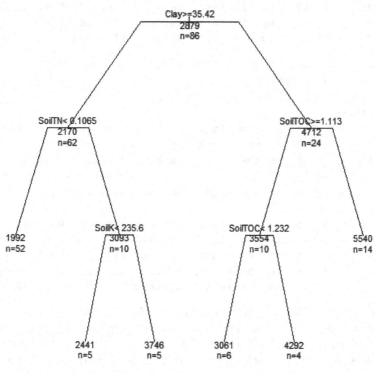

Field 4.1 Yield Regression Tree

FIGURE 9.15
Regression tree for Yield in Field 4.1 with agronomic explanatory variables.

```
    Primary splits:
        Clay    < 35.425   to the right, improve=0.6553524, (0 missing)
        Sand    < 31.245   to the left,  improve=0.5887001, (0 missing)
        SoilK   < 156.85   to the right, improve=0.5454879, (0 missing)
        SoilTOC < 0.8955   to the right, improve=0.3534404, (0 missing)
        SoilTN  < 0.0785   to the right, improve=0.2345896, (0 missing)
    Surrogate splits:
        Sand    < 27.15   to the left,  agree=0.953, adj=0.833, (0 split)
        SoilK   < 156.85  to the right, agree=0.872, adj=0.542, (0 split)
        SoilTOC < 0.8955  to the right, agree=0.802, adj=0.292, (0 split)
        SoilTN  < 0.062   to the right, agree=0.767, adj=0.167, (0 split)
        Silt    < 41.57   to the left,  agree=0.744, adj=0.083, (0 split)

Node number 2: 62 observations,     complexity param=0.0595666
  mean=2169.791, MSE=421854.4
  left son=4 (52 obs) right son=5 (10 obs)
  Primary splits:
        SoilTN  < 0.1065 to the left, improve=0.3886372, (0 missing)
        SoilTOC < 1.375  to the left, improve=0.3193074, (0 missing)
        SoilK   < 252.85 to the left, improve=0.2276935, (0 missing)
        SoilP   < 7.81   to the left, improve=0.2078210, (0 missing)
        Disease < 2.5    to the left, improve=0.1187364, (0 missing)
```

```
Surrogate splits:
    SoilTOC < 1.343    to the left, agree=0.968, adj=0.8, (0 split)
    SoilP   < 15.72    to the left, agree=0.871, adj=0.2, (0 split)
    SoilK   < 230.9    to the left, agree=0.871, adj=0.2, (0 split)

Node number 3: 24 observations,      complexity param=0.134846
  mean=4712.128, MSE=1360739
  left son=6 (10 obs) right son=7 (14 obs)
  Primary splits:
      SoilTOC < 1.113  to the right, improve=0.7046088, (0 missing)
      SoilTN  < 0.0855 to the right, improve=0.6868612, (0 missing)
      SoilK   < 163.8  to the right, improve=0.4792320, (0 missing)
      Sand    < 34.18  to the left,  improve=0.4779311, (0 missing)
      SoilP   < 7.965  to the right, improve=0.4476148, (0 missing)
  Surrogate splits:
      SoilTN  < 0.0855 to the right, agree=0.958, adj=0.9, (0 split)
      SoilP   < 7.965  to the right, agree=0.917, adj=0.8, (0 split)
      SoilK   < 163.8  to the right, agree=0.917, adj=0.8, (0 split)
      Clay    < 30.16  to the right, agree=0.833, adj=0.6, (0 split)
      Sand    < 31.04  to the left,  agree=0.833, adj=0.6, (0 split)
```

The first split subdivides the field into high *Clay* and low *Clay* regions, which is consistent with earlier exploration. The main competitor and surrogate of *Clay* is, unsurprisingly, *Sand*. The second surrogate is *SoilK*. In the high *Clay* region (node 2), the split is on *SoilTN*. The primary competitors and surrogates of *SoilTN* are associated with other mineral nutrients or organic carbon. In the low *Clay* (high *Yield*) region (node 3), the split is on *SoilTOC*, with the main competitors and surrogates again being other mineral nutrient levels.

One of the most useful actions involving a regression tree such as that of Figure 9.15 with spatial data is to create a map of the terminal nodes of the tree. We use a Thiessen polygon map created in ArcGIS for this purpose. Figure 9.16a shows this map. It was created in the same manner as the maps in Chapter 7. The map serves as a reality check on the tree results. The regression tree (Figure 9.15) was created without the use of any spatial information, and displaying the spatial context of the results is an indication of their validity. One expects that the data records in each terminal node will have a spatial structure; a random spatial arrangement would be a clear counter-indication of the validity of the tree. Thus, the map of Figure 9.16a provides both a validation step for the results and a means of interpreting these results spatially. In this context is important to emphasize that the recursive partitioning models of this chapter and the regression models of Chapter 8 do not treat the data in the same way. A regression model incorporates the contribution of each explanatory variable to the response variable over the whole data space. The recursive partitioning model, on the other hand, recursively splits the data space, and at each split separately incorporates the contributions of the response variables only over that subset of the data space. It is of particular interest to determine whether these splits in the data space correspond to splits in the geographic space. The corresponding yield map of Figure 9.16b provides a spatial representation of the response variable of the tree. One counterintuitive result is that in the high-yielding southern end of the field, *Yield* is higher in the presence of low *SoilTOC* than high *SoilTOC*. This likely indicates that in this part of the variable *SoilTOC* is serving as a surrogate for some other variable, and the most likely candidates are *SoilTN*, *SoilP*, and *SoilK*.

We conclude this section by demonstrating a property of regression trees that can cause very serious problems. We would expect that the results of our analysis should not be

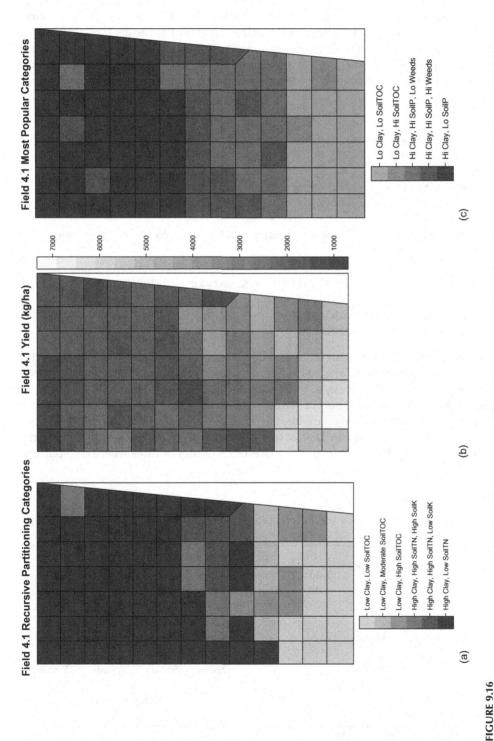

FIGURE 9.16

(a) Thematic map of the regression tree of Figure 9.15. (b) Thematic map of yield values at sample points in Field 4.1. (c) Thematic map of the ninth perturbation in Table 9.1 as a representative of the most "popular" regression tree.

sensitive to small perturbations of the data. We will test this property by employing a form of sensitivity test commonly used in engineering: one applies small random perturbations to the data and observes the effect (Clements, 1989, p. 129; Boyd, 2001; Ravindran et al., 2006, p. 590). Although this process could be automated, we will simply run the following code sequence (except the first two lines) repeatedly and enter the results into a table. The function `rnorm()` in the sixth line generates a random matrix, and the operation in the eighth row multiplies each element of the data frame by one plus a small random number. Note that in R the matrix operation A * B creates a matrix whose elements are $a_{ij} \times b_{ij}$.

```
>set.seed(123)
> n <- 0
> data.Perturb <- with(data.Set4.1, data.frame(Yield, Clay, Silt, Sand,
+     SoilpH, SoilTOC, SoilTN, SoilP, SoilK, CropDens, Weeds, Disease))
> epsilon <- 0.05
> mm <- matrix(rnorm(nrow(data.Perturb) * ncol(data.Perturb)),
+    nrow = nrow(data.Perturb))
> df2 <- data.Perturb * (1 + epsilon * mm)
> Perturb.rp <- rpart(Set4.1.model, data = df2, control = cont.parms,
+     method = "anova")
> n <- n + 1
> plot(Perturb.rp,branch = 0.4,uniform = T,margin = 0.1,
+     main = paste("Field 4-1 Yield Regression Tree ",
+     as.character(n)), cex.main = 2)
> text(Perturb.rp,use.n = T,all = F)
```

The R function `jitter()` could also be used, but in this case the explicit generation of the matrix seems more transparent. The ordinal scale variables *Weeds*, *Disease*, and *CropDens* are treated as if they were ratio scale, so that this analysis is very rough and ready. Table 9.1 shows the results of compiling the output of nine regression trees. Although nodes 1, 2, and 5 generally contain the same splitting variable, the other nodes do not. This illustrates the fact that regression trees can be *unstable*. That is, trees grown with small changes in the parameters or their values can change dramatically. This often occurs in cases where the splitting

TABLE 9.1

Splitting Variables at Each Node of a Succession of Regression Trees for Field 4.1, Each Generated with a Small Random Perturbation of the Data

Split	1	2	3	4	5	6	7	8	9
1	Clay	Clay	Clay	Sand	Clay	Clay	Clay	Clay	Clay
2	SoilTN	SoilP	SoilP	SoilP	Silt	SoilP	SoilP	SoilK	SoilK
3	SoilP	SoilTOC	SoilTOC		SoilTOC	SoilTOC	SoilK	SoilTOC	SoilTOC
4	SoilP				SoilP				
5		Weeds	Weeds	Weeds		Weeds	Weeds	SoilTN	SoilTN
6	SoilpH								
7						Disease	Weeds	Weeds	Weeds
8									
9	Weeds	SoilP			Weeds			SoilTOC	SoilTOC
>9									

Note: Nodes left blank were not generated.

variable that defines a particular node changes, because this change tends to propagate and amplify as one moves through the branch emanating from that node. A second indication of Table 9.1 is that some trees are more "popular" than others. Figure 9.16c shows a thematic map of Tree 3 in Table 9.1 as a representative of the "popular" trees. Breiman (2001a) recognized these properties, instability and popularity, and both he and others have developed procedures to reduce the impact of tree instability by focusing on popular trees. In the next section, we will briefly describe *random forest*, one of the most useful of these procedures.

9.4 Random Forest

9.4.1 Introduction to Random Forest

In the previous section's sensitivity analysis, we used a method common to engineering, the perturbation of a data set by applying a small random variable to its values. The outcome of the sensitivity analysis, shown in Table 9.1, is that the regression tree is unstable, that is, it is highly sensitive to these perturbations. The *random forest* method developed by Breiman (2001a) also generates large numbers of trees, each of which is built based on a small perturbation of the original data set. The perturbations, however, are of a different form from those in Section 9.3.5. Rather than adding a small random number to each of the n data values, a random sample of the data records is drawn by sampling from these records *with replacement*. This procedure, repeatedly sampling with replacement from the existing data set and then carrying out a statistical analysis of the sample of samples, is called *bootstrapping* and is discussed in Chapter 10. The application of bootstrapping to an algorithm like recursive partitioning is called *bagging* (for "bootstrap aggregation") (Breiman, 1996). One practical advantage of bootstrapping over the sensitivity analysis of Section 9.3.5 is immediately apparent. As we saw in that section, adding a small random number to the data is an inappropriate modification for nominal or ordinal scale data such as the *Weeds*, *CropDens*, and *Disease* data fields in Data Set 4. The bootstrap procedure just described does not affect the form of the data themselves (see Exercise 9.12).

The random forest algorithm works with bootstrap samples as follows. Suppose n_{boot} bootstrap samples (i.e., samples with replacement from the original data set) are drawn. For each bootstrap sample, a classification or regression tree is generated using a modification of the recursive partitioning algorithm described in Section 9.3.2. The modification differs from the recursive partitioning algorithm discussed in that section in two ways. First, the maximal tree is grown and not pruned back. Second, somewhat surprisingly, the random forest algorithm is found to work better if not every possible explanatory variable is tested at each node to determine the next split. Therefore, at each node only a subset of size n_{test} of the variables is tested, and the splitting variable and value are selected from this subset.

The action of drawing n_{boot} samples with replacement and using them to grow a maximal tree on a randomly selected subset n_{test} of the variables is repeated N_T times. In this way, a "random forest" of N_T trees is grown. Of course, due to the instability of recursive partitioning, each of these trees may have different splitting variables at each node. The splitting variable or value ultimately selected at each node is determined by having the trees "vote." For classification trees, the predicted variable is that of the majority of the trees and for regression trees the predicted value is the average value of the majority variable.

One of the advantages of the random forest algorithm is that it provides a built-in data set that can be used for validation. Each of the n_{boot} bootstrap samples omit some data records. It turns out that on average about 36% of the data records are not included in each bootstrap sample (Breiman, 2001a). Since these data records are not used in constructing that particular tree, they can be used as a validation set by predicting the value of the response variable for each of them and comparing the predicted and actual values. For reasons explained by Breiman (2001a), the data records not included in the bootstrap sample are called *out of bag* (or OOB) data records.

The nodal structure of the random forest is not nearly as important as other information that the algorithm provides. To describe this, we need to have an example, and we will use the data from Field 4.1. After loading the data into the data frame data.Set4.1, we enter the following code.

```
> library(randomForest)
> data.rf <- with(data.Set4.1, data.frame(Yield, Clay, Silt, Sand,
+    SoilpH, SoilTOC, SoilTN, SoilP, SoilK, CropDens, Weeds, Disease))
> Set4.1.rf <- randomForest(Yield ~ ., data = data.rf,
+    importance = TRUE, proximity = TRUE)
```

The variable *Sand* is excluded on agronomic grounds that it is simply the reflection of *Clay*. Two points must be noted, one subtle and one obvious. The obvious point is that there are two arguments, importance and proximity, that we have not discussed yet. The more subtle point is that when we used the function rpart() to do recursive partitioning we first adjusted the tuning parameters minsplit and cp; here we did not adjust anything. There are two tuning parameters for the random forest algorithm, N_T and n_{test} defined above, but the algorithm is not very sensitive to them and we won't bother to adjust them from their default values.

The arguments importance and proximity instruct the random forest algorithm to compute the *variable importance* and the *proximity matrix*. The variable importance can be defined in two different ways, which are described in the ?importance file of the function importance(). We will discuss this only for regression trees; the concept is analogous for classification trees. The first way to compute the importance is to rank the *increase in node purity*. For regression trees, this is the reduction in the residual sums of squares over the nodes. The second way is to rank the *percent increase in the mean square error*. Without going into detail, this involves computing the error based on the OOB data records. The function varImpPlot() generates a dotplot of the variable importance.

```
> varImpPlot(Set4.1.rf,
      main = "Data Set 4.1 Variable Importance") # Fig. 9.17
```

Figure 9.17 supports the notion that soil texture is the most important variable (or, perhaps better said, is the most important variable over the largest area of the field), and that mineral nutrients are also important. The variable *Weeds* is not listed as very important, which shows that with spatial data a variable that is very important in one region may not be judged important because it is not important over the whole area.

The proximity matrix, computed when the argument proximity is set at TRUE, is defined as follows (Liaw and Weiner, 2002). The (i, j) element of the proximity matrix is the fraction of the trees in which data records i and j fall in the same terminal node. The elements of the proximity matrix thus give a measure of the attribute proximity of two data records. The proximity matrix for Field 4.1 is an 86 by 86 matrix, so obviously it is not

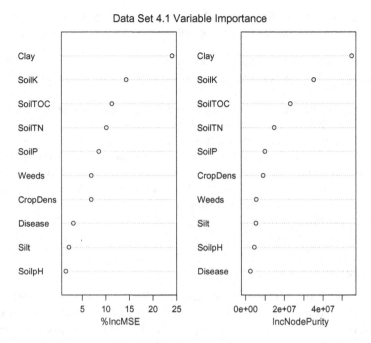

FIGURE 9.17
Variable importance values of the agronomic variables of Field 4.1.

very useful to just display it in its raw form. Hastie et al. (2009, p. 595) describe a graphical device called the proximity plot that can in principle be used to visualize the proximity matrix, but they point out that these plots are often of limited utility. With spatial data, however, one can obtain some information from this matrix. For example, for each data record we can identify the data record with the highest proximity. This provides an indication of the spatial structure. Unlike the spatial weights matrix discussed in Chapter 4, the proximity matrix generated by `randomForest()` has the number 1 on the diagonal, so we first need to eliminate these values. The function `closest()` defined here eliminates the diagonal elements and then identifies for a given row or column of the proximity matrix the largest off-diagonal element.

```
closest <- function(x){
    x[which(x > 0.999)] <- 0
    which(x == max(x))
}
```

We can use the function `apply()` to apply this function row-wise (or column-wise) to the proximity matrix.

```
> apply(Set4.1.rf$proximity, 1, closest)
  1  2  3  4  5  6  7  8  9 10 11 12 13 14 15 16 17 18 19 20 21 22 23 24
 18 17 10  7  4  5  4  9  8  3 56 10 60 52 39 22  2 17 20 21 20 23 22 29

 25 26 27 28 29 30 31 32 33 34 35 36 37 38 39 40 41 42 43 44 45 46 47 48
 27 28 25 26 30 29 29 27 27 40 29 37 38 37 22 34 40 44 63 42 49 53 45 45
```

```
49 50 51 52 53 54 55 56 57 58 59 60 61 62 63 64 65 66 67 68 69 70 71 72
45 62 57 54 54 53 53 61 51 61 65 64 56 50 67 59 59 72 68 67 86 71 79 74

73 74 75 76 77 78 79 80 81 82 83 84 85 86
74 80 76 84 76 70 71 74 82 83 82 76 86 71
```

Blank lines have been inserted in the output to make it easier to read. In each pair of rows, the first row is the data record number and the second row is the corresponding closest data record. Although the proximity matrix is symmetric, the closest proximity relation is not reflexive. That is, for example, although data record 18 is closest to data record 1, meaning that data record 18 has the most nodes in common with data record 1, the data record having the most nodes in common with 18 is data record 17. One can use this proximity matrix together with the record number map of Figure 4.5a to obtain a sense of the spatial structure of these attribute value relationships.

9.4.2 Application to Data Set 2

Turning to Data Set 2, we will again consider the Sierra Nevada and Coast Ranges separately. The `SpatialPointsDataFrame data.Set2S` created in Section 7.3 for the Sierra Nevada is loaded as described in Appendix B.2. Since the presence-absence variable *QUDO* is binary, we will convert it to a factor so that `randomForest()` develops a forest of classification trees rather than regression trees. Because there are so many data records, we will not develop the proximity matrix for this case. Here is the code to produce Figure 9.18a.

```
> library(randomForest)
> attr.S <- slot(data.Set2S, "data")
> data.rfS <- with(attr.S, data.frame(QUDO = factor(QUDO),
+    MAT, Precip, SolRad6, SolRad12, Texture, AWCAvg,
+    Permeab, PM100, PM200, PM300, PM400, PM500, PM600))
> set.seed(123)
> Set2.rf <- randomForest(QUDO ~ ., data = data.rfS,
+    importance = TRUE)
> varImpPlot(Set2.rf, # Fig. 9.18a
+    main = "Data Set 2 Sierra Variable Importance")
```

The soil-related explanatory variables are identified as the least important. Sometimes random forests can be used to indicate unimportant variables that may be candidates for elimination from the model (Liaw and Wiener, 2002). We can determine the effect of eliminating the soil variables as follows. The `randomForest` object itself contains the relative error, which is determined from attempting to predict the out-of-bag data records.

```
> Set2.rf
        OOB estimate of error rate: 12.39%
Confusion matrix:
    0   1 class.error
0 969 102   0.0952381
1 117 580   0.1678623
```

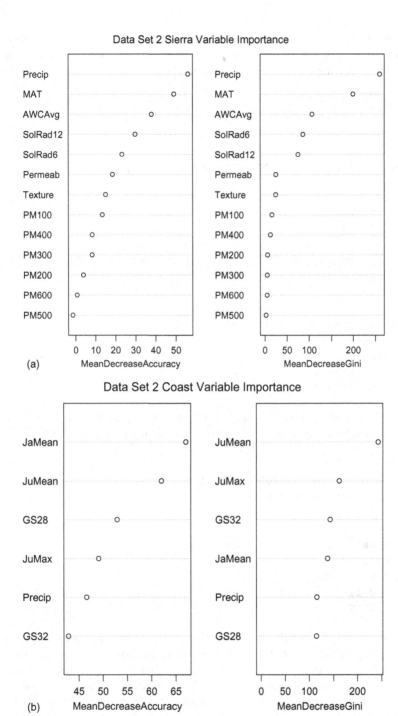

FIGURE 9.18
(a) Variable importance values of the Sierra Nevada subset of Data Set 2. (b) Same as (a) for the Coast Range subset of Data Set 2.

It turns out that if we eliminate all the soil variables from the model, then the prediction accuracy declines, but leaving one in retains a reasonable accuracy. By a process of trial and error, we find that the following model has almost as much predictive accuracy as the full model.

```
> set.seed(123)
> Set2.rf4 <- randomForest(QUDO ~ MAT + Precip + SolRad6 +
+   AWCAvg + Permeab, data = data.rfS, importance = TRUE)
> Set2.rf4
        OOB estimate of error rate: 12.9%
Confusion matrix:
    0    1 class.error
0 963 108    0.1008403
1 120 577    0.1721664
```

Carrying out the same exercise with the Coast Ranges data set (Exercise 9.14) produces Figure 9.18b and the following output.

```
> Set2.rf
        OOB estimate of error rate: 14.33%
Confusion matrix:
    0    1 class.error
0 874 128    0.1277445
1 138 716    0.1615925
```

Again, by using a process of trial and error to eliminate explanatory variables, we find that the following model has about as much predictive accuracy as the full model.

```
> set.seed(123)
> Set2.rf2 <- randomForest(QUDO ~ JuMax + JuMean + JaMean + GS32 +
+     GS28 + JaMean + Precip, data = data.rfC,
+   importance = TRUE)
> Set2.rf2
        OOB estimate of error rate: 15.95%
Confusion matrix:
    0    1 class.error
0 855 147    0.1467066
1 149 705    0.1744731
```

It is important to emphasize that this exercise in variable selection does not guarantee that the surviving variables are the most important explanatory variables for prediction of habitat suitability for blue oak. There is evidence that the importance measures show a bias toward correlated explanatory variables. We can draw the tentative conclusion from the recursive partitioning and random forest analyses that precipitation is important for explanation of blue oak presence in both the Sierra Nevada and the Coast Range, although precipitation is generally lower throughout most of the Coast Range. Some combination of temperature variables may also be important, particularly in the Coast Range, and soil properties seem less important.

9.5 Further Reading

As mentioned in Section 9.2, Wood (2006) is an excellent source of information on the generalized additive model. The standard source for recursive partitioning is Breiman et al. (1984), although discussions of the method are beginning to make it into standard regression texts (e.g., Kutner et al., 2005, p. 453). Two excellent discussions of machine learning techniques are Hastie et al. (2009) and James et al. (2013). In particular, Hastie et al. present in Chapter 5 a nice discussion of basis splines. For examples of the application of recursive partitioning to ecological problems, see for example Plant et al. (1999), De'ath and Fabricus (2000), De'ath (2002), Roel and Plant (2004a,b), and Lobell et al. (2005).

Classification and regression trees provide a means for the detection of boundaries, delimiting what ecologists call "patches" and agronomists call "management zones." For a more extensive treatment of this topic and the use of other methods to identify patches, see Fortin and Dale (2005). It is also worth noting that recursive partitioning has a natural application in conjunction with boundary line analysis (Webb, 1972; Lark, 1997). This is briefly discussed in Section 17.3.

Exercises

9.1 Determine the value of the knot displayed in Figure 9.3a directly from the function bs(). What is the class of the resulting quantity? Note that it is possible for elements of this class to have names.

9.2 Repeat Exercise 9.2 for the knots of Figure 9.3b and determine the difference between each of them.

9.3 Generalized additive models can be a very effective way to compute trend surfaces. In Section 3.2.1, we computed trend surfaces for a set of data shown in Figure 3.2a. Compute a trend surface using a GAM, plot the result as a perspective plot, and compare the result with the actual trend data shown in Figure 3.2b.

9.4 This exercise involves the construction of a classification tree. As mentioned in Chapter 2, R contains a number of built-in data sets, which are invoked through the function data(). One of these is Fisher's iris data, which we obtain by typing data(iris). This is a famous data set employed by R.A. Fisher (1936) in a publication in a journal whose name reveals a dark side of the thinking of some prominent early twentieth-century statisticians, including Fisher, Galton, and others. The data set consists of 50 samples from each of three species of iris (*I. setosa*, *I. virginica*, and *I. versicolor*). Four features were measured from each sample: the length and width of the sepal (the sepal is the green part of the flower that separates the petals) and the length and width of the petal. Construct a classification tree using only the variables *Sepal width* and *Sepal length* and construct a plot of the data space showing the partitions.

9.5   Carry out a recursive partitioning analysis of the factors influencing the presence or absence of blue oak in the Coast Range of Data Set 2. Construct the regression tree shown in Figure 9.10b.

9.6   With the data of Data Set 3, display sorted means by farmer of yield, soil pH, and silt content.

9.7   With the data of Data Set 3, plot yield vs. nitrogen rate N with separate symbols for the northern, central, and southern regions.

9.8   With the data of Data Set 3, construct the regression tree for all management variables.

9.9   With the data of Data Set 3, construct a table in which the rows are mean values of the exogenous variables for each field and the columns are arranged in order of decreasing mean yield over the field.

9.10 Carry out a recursive partitioning with *Yield* as the response variable for Field 4.2.

9.11 Instead of checking the stability of a regression tree by perturbing the data by adding a small value to each attribute value, one can perturb them by replacing the set of n data records with a set of n records randomly selected from the original n with replacement. Use the function sample() to carry out this method to check for stability of the regression tree generated with the Field 4.1 data in Section 9.3.5 and compare the results with those obtained in that section.

9.12 Using a perturbation method similar to that employed in Exercise 9.11, check the Data Set 2 trees for stability. What does this imply about how the size of the data set influences the stability of the recursive partitioning?

9.13 Using *Yield* as a response variable, carry out a random forest analysis of Field 2 of Data Set 4.

9.14 Carry of a random forest analysis of the data from the Coast Range subset of Data Set 2.

10

Variance Estimation, the Effective Sample Size, and the Bootstrap

10.1 Introduction

An important statistical consequence of positive spatial autocorrelation is that it inflates the variance of the test statistic, and this may in turn result in an inflated Type I error rate in significance tests (Section 3.3). Let us review and summarize the discussion in that section. Suppose we have obtained a sample $\{Y_1, Y_2, ..., Y_n\}$ from a normally distributed, spatially autocorrelated population, and we are testing the null hypothesis $H_0 : \mu = 0$ against the alternative $H_a : \mu \neq 0$ (Equation 3.6). Recall from Equation 3.9 that the variance σ^2 of the population from which the Y_i are drawn can be estimated by the sample variance s^2, which satisfies

$$s^2 = \frac{1}{n-1} \sum_{i=1}^{n} (Y_i - \bar{Y})^2. \tag{10.1}$$

We use the t statistic (Equation 3.11), $t = \bar{Y} / s\{\bar{Y}\}$, where $s^2\{\bar{Y}\} = s^2 / n$ and $s\{\bar{Y}\} = \sqrt{s^2\{\bar{Y}\}}$, to carry out the test. Reproducing Equation 3.17,

$$\text{var}\{\bar{Y}\} = \frac{\sigma^2}{n} + \frac{2}{n} \sum_{i \neq j} \text{cov}\{Y_i, Y_j\}, \tag{10.2}$$

which indicates that if $\text{cov}\{Y_i, Y_j\} > 0$ (i.e., if the values are autocorrelated), then $s^2\{\bar{Y}\}$ underestimates the true variance of the mean (see Section 3.5.2 for a more complete discussion). Thus, the denominator of the t statistic is smaller than it should be, so that the value of t is artificially inflated, resulting in an increased probability of rejecting the null hypothesis (i.e., an increased Type I error rate).

Cressie (1991, p. 14) provides a very simple example of this effect that illustrates some important points. Suppose the Y_i represent the values of an autocorrelated random process that takes place along a line, with the n sites arranged in numerical order. Suppose the process is such that the covariance between the values of any two sites is given by

$$\text{cov}\{Y_i, Y_j\} = \sigma^2 \rho^{|i-j|}, \quad i = 1, 2, ..., n, \tag{10.3}$$

where $0 < \rho < 1$. In this case, the variance of the mean can be worked out exactly. It turns out (Cressie, 1991, p. 14) that it satisfies

$$\text{var}\{\bar{Y}\} = \frac{\sigma^2}{n}\left[1 + \left(\frac{2\rho}{1-\rho}\right)\left(1 - \frac{1}{n}\right) - \frac{2}{n}\left(\frac{\rho}{1-\rho}\right)^2\left(1 - \rho^{n-1}\right)\right]. \tag{10.4}$$

Figure 10.1a shows a plot of $\text{var}\{\bar{Y}\}$ in Equation 10.4 as a function of ρ for $n = 25$ and $\sigma^2 = 1$. As expected, $\text{var}\{\bar{Y}\}$ increases with increasing ρ, slowly at first but dramatically for values of ρ near 1.

There is a second way to view the effect of positive autocorrelation on the variance of the mean. Recall that an intuitive explanation of the inflated Type I error rate in a significance test involving positively autocorrelated data is that each data value contains information about the values of surrounding data, and therefore cannot be counted as a single independent sample. Thus, the *effective sample size* is less than the sample size n. For given values of the population variance σ^2 of the Y and variance of the mean var$\{\bar{Y}\}$, we can make this notion precise by defining the effective sample size n_e to be the value that satisfies the equation

$$\text{var}\{\bar{Y}\} = \frac{\sigma^2}{n_e}. \tag{10.5}$$

In the simple example of Equation 10.4, the effective sample size can be computed exactly. It has the form (Cressie, 1991, p. 15)

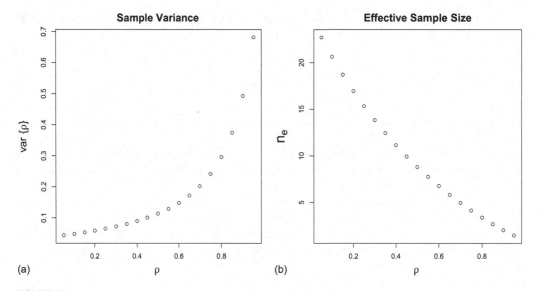

FIGURE 10.1
(a) Variance of the mean defined by Equation 10.3 for $n = 25$ and $\sigma^2 = 1$; (b) Effective sample size for the same values.

$$n_e = \frac{\sigma^2}{\text{var}\{\bar{Y}\}}$$

$$= n\left[1 + \left(\frac{2\rho}{1-\rho}\right)\left(1 - \frac{1}{n}\right) - \frac{2}{n}\left(\frac{\rho}{1-\rho}\right)^2\left(1 - \rho^{n-1}\right)\right]^{-1}. \tag{10.6}$$

The effective sample size for this example is shown in Figure 10.1b. It declines from a value of 25 for $\rho = 0$ to a value of approximately one for ρ near one.

Now let us consider the general case, in which we are given a sample $\{Y_1, Y_2, ..., Y_n\}$ of spatially autocorrelated data. If there is substantial spatial autocorrelation, then we know that $s^2\{\bar{Y}\}$ may underestimate the true variance of the mean. Suppose, however, that we could obtain a second, independent estimate $\text{var}_{est}\{\bar{Y}\}$ of the variance of the mean. If this second estimate is more accurate than $s^2\{\bar{Y}\}$, then we can insert $\text{var}_{est}\{\bar{Y}\}$ into Equation 10.5 to obtain an estimate $\hat{n}_e$ of the effective sample size n_e. We define $\hat{n}_e$ to be the value satisfying

$$\text{var}_{est}\{\bar{Y}\} = \frac{\sum_{i=1}^{n}(Y_i - \bar{Y})^2}{\hat{n}_e}. \tag{10.7}$$

In other words, $\hat{n}_e$ is our estimate of the sample size that would yield a variance $s^2\{\bar{Y}\}$ if the data were independent rather than spatially autocorrelated. Solving for $\hat{n}_e$ yields

$$\hat{n}_e = \frac{\sum_{i=1}^{n}(Y_i - \bar{Y})^2}{\text{var}_{est}\{\bar{Y}\}}. \tag{10.8}$$

As an example of the effect of spatial autocorrelation on the results of a hypothesis test involving real data, consider EM38 electrical conductivity measurements taken in the furrows and beds of Field 1 of Data Set 4 on two separate dates, April 25, 1995, and May 20, 1995. As usual, we create a data frame `data.Set4.1` from the contents of the file *Set4.196sample. csv* as described in Appendix B.4. Some measurements were not made on April 25, so these data records will be eliminated.

```
> EM425B <- data.Set4.1$EM38B425[!is.na(data.Set4.1$EM38B425)]
> EM520F <- data.Set4.1$EM38F520[!is.na(data.Set4.1$EM38B425)]
```

It happens that the April 25 bed readings are very similar to the furrow readings made on May 20. Here is the code to plot histograms of these two data sets.

```
> h1 <- hist(EM425B, breaks = seq(50,100,5), plot = FALSE)
> h2 <- hist(EM520F, breaks = seq(50,100,5), plot = FALSE)
> plot(h1$mids,h1$density, xlim = c(50,100), ylim = c(0,0.06),
+     type = "o", cex.main = 2, cex.lab = 1.5, # Fig. 10.2
+     xlab = "EC (mS/M)", ylab = "Frequency",
+     main = "Histograms of EM38 Readings")
> lines(h2$mids,h2$density, xlim = c(50,100),
+     type = "o", lty = 2)
> legend(80, 0.05, c("4/25 Bed", "5/20 Furrow"), lty = c(1,2))
```

Histograms of EM38 Readings

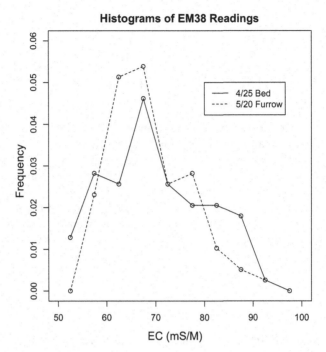

FIGURE 10.2
Histograms of the EM38 samples from Field 4.1.

The result is shown in Figure 10.2. The means of the April 25 bed and May 20 furrow EM38 readings are 70.38 and 69.24 mS/M, respectively. Based on this small difference and on the histograms in Figure 10.2 one might expect that a test of the null hypothesis that the two means are equal would not be rejected. Here are the results of a paired t test of the null hypothesis of equality.

```
> t.test(EM425B,EM520F, paired = TRUE, alternative = "two.sided")
        Paired t-test
data: EM425B and EM520F
t = 1.9629, df = 77, p-value = 0.05327
```

The test returns a suspiciously low p value, considering the variability displayed in Figure 10.2, and we might suspect that spatial autocorrelation has reduced the effective sample size.

We can easily compute the value of s^2 in Equation 10.1: for the paired t test it is the variance of the difference between the two data vectors, which is $s^2 = 26.37$. This is the variance estimate assuming that the EM38 readings are spatially independent. However, in order to obtain the value of n_e we need the second estimate $\text{var}_{est}\{\overline{Y}\}$ of the variance of the mean. This chapter discusses a simple and often remarkably effective resampling method for doing this called the *bootstrap*. The name comes from the phrase "pulling oneself up by one's bootstraps," because this is what the bootstrap method effectively does. The method was developed by Bradley Efron (1979), who showed that by resampling a data set many times, one could generate an approximation of the population distribution of that data set that is sufficiently accurate to permit an estimate of the variance of the mean (as well as other statistics). The basic bootstrap is described in the next section. A complication arises when dealing with autocorrelated data in that the basic bootstrap does not preserve the correlation structure of the data. Section 10.3 introduces the bootstrap for autocorrelated data by describing two methods for dealing with time series data, and Section 10.4 extends the ideas developed in Section 10.3 to spatially autocorrelated data. This chapter is a bit of a detour in that none of the analyses contribute directly to the resolution of any of the research problems posed for the four data sets. Rather, the discussion introduces a powerful method that we will use to good effect in subsequent chapters. Section 10.5 provides a resolution to the EM38 comparison problem that we use as an illustration throughout this chapter.

10.2 Bootstrap Estimation of the Standard Error

The bootstrap method (Efron, 1979) is described by Efron and Tibshirani (1993). It is a remarkably robust method of estimating variance statistics from the sample data itself. The reasoning behind the method is that the sample provides the best available information about the population itself. Therefore, to estimate statistics associated with the variability of the population, one draws a sample from the sample, that is, a *resample*, and computes the statistics associated with the variability of this resample. Bootstrapping can be used to estimate a variety of statistics associated with distributions; in this chapter, we will use it to estimate the standard error.

In order to demonstrate the bootstrap, we will begin with a finite population whose parameters are known exactly, and whose standard error $\sigma/\sqrt{n}$ we can estimate as $s/\sqrt{n}$. The objective will be to show that we can generate a standard error estimate as good as $s/\sqrt{n}$ directly from the data. Our population has a size of 1,000 and is generated from a lognormal distribution. The lognormal distribution is chosen because it is very highly skewed, and thus far from normal, as can be seen in the histogram of Figure 10.3a.

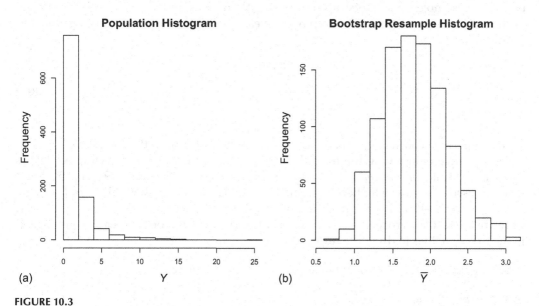

FIGURE 10.3

(a) Histogram of the finite population of size 1,000 used to demonstrate the bootstrap; (b) Histogram of the means of the bootstrap resamples.

```
> set.seed(123)
> pop.Y <- rlnorm(1000)
> print(Y.bar <- mean(pop.Y))
[1] 1.671687
> print(pop.var <- sum((pop.Y - Y.bar)^2) / 1000)
[1] 4.44692
```

The parameters of the population are $\mu = 1.671687$ and $\sigma^2 = 4.44692$.

We draw a sample of size 20 (without replacement) from the population. The variability parameter we will be interested in is the standard error $\sigma / \sqrt{n}$, which is estimated by the sample standard error $se = sd\{\bar{Y}\} = s / \sqrt{n}$.

```
> set.seed(123)
> print(sample.Y <- sample(pop.Y, size = 20), digits = 3)
 [1] 5.160 0.141 0.583 1.611 0.503 0.325 3.048 0.354 1.896 1.545
[11] 7.439 0.128 2.747 1.106 1.293 1.576 4.464 0.812 0.262 0.914
> print(mean(sample.Y), digits = 3)
[1] 1.80
> print(sd(sample.Y) / sqrt(20), digits = 3)
[1] 0.431
> print(sqrt(pop.var / 20), digits = 3)
[1] 0.472
```

This sample contains the best information available about the population. The sample mean is $\bar{Y} = 1.80$ and the sample standard error is $s\{\bar{Y}\} = 0.431$. The true value of $\sigma / \sqrt{n}$ is 0.472.

To apply the bootstrap method, we repeatedly, a total of B times, sample *with replacement* from the sample. Each "sample of the sample," or resample, has the same size as the original sample (in our example, this size is 20). The sampling must be done with replacement, because otherwise each resample would simply consist of a reordering of the 20 original values. Let the samples be denoted $S_i(Y)$, $i = 1,...,B$. Let $\bar{S} = \Sigma S_i(Y) / B$ be the mean of the bootstrap resamples. The bootstrap estimate $\hat{s}_{boot}\{\bar{Y}\}$ of the standard error is then obtained as the standard error of the bootstrap resamples, and is given by the formula (Efron and Tibshirani, 1993, p. 13).

$$\hat{s}_{boot}\{\bar{Y}\} = \sqrt{\frac{\sum_{i=1}^{B} (S_i(Y) - \bar{S})^2}{B-1}}. \tag{10.9}$$

If this seems a bit confusing, read on to the end of the section and then come back.

The principle of applying the formula for the standard error to the resampled data is known as the *plug-in principle* (Efron and Tibshirani, 1993, p. 35; Sprent, 1998, p. 35). There is some discussion over whether the number in the denominator should be B or $B-1$, but, as Efron and Tibshirani (1993, p. 43) point out, it usually doesn't make much difference. Here are the results of three successive resamples. The function boot.sample() is created in the first two lines to draw the resample.

```
> boot.sample <- function(x) sample(x, size = length(x),
+      replace = TRUE)
> # One bootstrap resample
> print(b <- boot.sample(sample.Y), digits = 3)
 [1] 0.812 1.106 2.747 0.914 1.106 1.293 7.439 0.128 0.325 0.583
```

```
[11] 0.914 0.262 1.106 1.576 5.160 1.545 1.576 0.503 3.048 0.503
> print(mean(b), digits = 3)
[1] 1.63
> # A second bootstrap resample
> print(b <- boot.sample(sample.Y), digits = 3)
 [1] 0.583 1.896 1.896 0.354 1.611 0.583 0.503 1.545 0.325 0.812
[11] 5.160 1.896 1.576 0.583 0.128 0.503 0.583 1.576 0.812 0.354
> print(mean(b), digits = 3)
[1] 1.16
> # A third bootstrap resample
> print(b <- boot.sample(sample.Y), digits = 3)
 [1] 1.106 0.141 0.354 0.325 4.464 1.896 4.464 4.464 1.576 1.896
[11] 1.576 2.747 1.293 5.160 1.545 0.503 0.354 2.747 0.354 0.583
> print(mean(b), digits = 3)
[1] 1.88
```

You can see that each mean of the resamples is different, and so we can generate a lot of them and compute the variance. The square root of this variance is the bootstrap estimate of the standard error.

Now we are ready to draw a large number of resamples. As with the Monte Carlo simulations done in earlier chapters, we can accomplish the repeated resampling in R by creating a function and then replicating that function. Unlike Monte Carlo simulation, we don't usually need a really large number of resamples; Efron and Tibshirani (1993, p. 54) state that 200 is in almost all cases sufficient to achieve an accurate estimate of the standard error. We will start with 1000 to be on the safe side. The replicated resampling is shown here.

```
> sample.mean <- function(x){
+     Y <- sample(x, size = length(x), replace = TRUE)
+     mean(Y)
+ }
> set.seed(123)
> boot.dist <- replicate(1000, sample.mean(sample.Y))
> print(mean(boot.dist), digits = 3)
[1] 1.79
> print(sd(boot.dist), digits = 3)
[1] 0.419
```

Figure 10.3b shows the histogram of the distribution of means of the resamples. As expected based on the central limit theorem (Larsen and Marx, 1986, p. 322), the resampling distribution is approximately normal, even though the population distribution is lognormal. Also as expected from the central limit theorem, the mean of the means of the resamples is not a very good estimate of the population mean, but it is a good estimate of the sample mean. The standard error of the resample of means, however, does provide an estimate of the population standard error of 0.472 that is almost as good as the sample standard error of 0.431. This illustrates the remarkable fact that simply sampling with replacement many times from an original sample (i.e., by resampling), one can generate a surprisingly accurate estimate of the standard error, and therefore of the variance, of the population.

The bootstrap doesn't always work, but it is often very effective. The accuracy of the result is constrained by the extent to which the sample represents the population. The advantage of the method might not be evident from this example, since the standard error can easily be calculated directly from the data and is a good estimate of the standard deviation of the mean, but such an easy calculation is not always possible. The great advantage of the bootstrap is that it can be applied in cases where there is no easy way to estimate the standard deviation of the mean directly.

It is not necessary to write one's own function for bootstrapping: R has the function `boot()` in the package `boot` (Canty and Riplyy, 2017) that carries out a bootstrap estimate of the standard error.

```
> library(boot)
> set.seed(123)
> mean.Y <- function(Y,i) mean(Y[i])
> boot.Y <- boot(sample.Y, mean.Y, R = 1000)
> print(sd(boot.Y$t), digits = 3)
[1] 0.421
```

The function `boot()` typically employs a user defined function as shown. There is one feature that must be explained. In the third line, the function `mean.Y()` is defined so it can be passed as an argument of the function `boot()` in the fourth line. In addition to the first argument `Y` of the function `mean.Y()`, which identifies an array consisting of the data to be resampled, the definition of the function *must include* a second argument that is an index, in this case `i`, of this array. The index argument must be represented in the function definition as shown in line 3.

There are two sources of randomness in the bootstrap, the original random sample from the population and the randomness of the resampling. In the example above, the first source of randomness led to the large difference between the sample mean and the population mean. The difference between the estimated standard errors of the homemade bootstrap and the function `boot()` illustrates the second effect. When a reasonably large number of resamples are computed, the main source of variability is the first of the two, which depends on the sample size. It is not necessarily the case that increasing the number of resamples increases the accuracy of the bootstrap estimate (Exercise 10.2). This is particularly true for small sample sizes (i.e., small values of n), because a small sample may not be very representative of the population.

Finally, recall that there is another statistical method, the permutation method, in which the sample is repeatedly resampled (Section 3.3). Let us compare the two methods so that we can eliminate confusion between them. In Section 3.3, the permutation test was introduced as a way to test the hypothesis that two samples come from the same population. In Section 4.3, the method was used by the function `joincount.mc()` to test the null hypothesis that high weed levels are not spatially autocorrelated against the alternative hypothesis that they are positively autocorrelated, and in Section 4.4 it was used by the function `moran.mc()` to carry out the same test on detrended sand content data. The permutation test involves sampling the data values *without replacement* in order to rearrange their configuration. One computes the statistic for each rearrangement, and then tabulates the number of rearrangements for which the computed value is more extreme than that of the observed data. If the observed statistic is more extreme than all but a few of the rearrangements, then we reject the null hypothesis. If it is more of a "typical" value, then we cannot reject. Thus, the permutation method involves repeated sampling of the data without replacement in order to rearrange them and in so doing carry out a statistical test in which the null hypothesis is that the data are randomly arranged. The bootstrap method involves repeated sampling of the data *with* replacement to estimate the variability associated with a statistic. For a more in-depth comparison of the bootstrap and permutation methods, see Efron and Tibshirani (1993, p. 202).

10.3 Bootstrapping Time Series Data

10.3.1 The Problem with Correlated Data

In Section 3.4.1, we studied a model of the form

$$Y_i = \mu + \eta_i,$$

$$\eta_i = \lambda \eta_{i-1} + \varepsilon_i, \ i = 2,3,\ldots,$$

(10.10)

where Y_1 and η_1 are given, the $\varepsilon_i \sim N(0,\sigma^2)$ are independent, and $-1 < \lambda < 1$. When $\lambda > 0$, a test of the null hypothesis that the mean μ of the distribution of the Y_i is zero had an inflated Type I error rate. The variance of the Y_i, and therefore the standard error, is inflated because the Y_i are autocorrelated, and therefore the t statistic based on the assumption of independent Y_i is artificially high, so that a t test using this statistic will have an increased probability of rejecting the null hypothesis. Based on the discussion in the previous section it seems like it might be possible to construct a bootstrap estimate of the standard error and thus obtain a test with an improved Type I error rate. This idea is explored in the present section. We first verify that the bootstrap produces an adequate t statistic with uncorrelated data. Here is a t test of a random sample of size 100.

```
> set.seed(123)
> Y <- rnorm(100)
> t.test(Y, alternative = "two.sided")$p.value
[1] 0.3243898
```

Now we carry out the test on the same data set using a bootstrap estimate of the standard error, computing the t statistic by hand and using the function pt() to compute the p value.

```
> library(boot)
> mean.Y <- function(Y,i) mean(Y[i])
> boot.Y <- boot(Y, mean.Y, R = 1000)
> t.boot <- mean(Y) / sd(boot.Y$t)
> print(p <- 2 * (1 - pt(q = abs(t.boot),
+     df = length(Y) - 1)))
[1] 0.3302481
```

The results are similar.

We now try using the bootstrap estimate of the standard error on a t test involving autocorrelated data. Recall that in Section 3.3 we carried out a Monte Carlo simulation of a test of the null hypothesis $\mu = 0$ in Equation 10.10 in which the value of μ had been set at zero. When λ had the value 0.4, the Type I error rate in 10,000 simulations was 0.1936, well above the α value of 0.05. As we did in Chapter 3, in the present simulation study, we first carry out the simulation with uncorrelated data.

```
> set.seed(123)
> ttest <- function(){
+   Y <- rnorm(100)
+   t.ttest <- t.test(Y,alternative = "two.sided")
+   TypeI <- as.numeric(t.ttest$p.value < 0.05)
+ }
```

```
> U <- replicate(10000, ttest())
> mean(U)
[1] 0.0488
```

Next, we run the same simulation except that we use autocorrelated data. As we did in Section 3.5.1, we drop the first 10 values to avoid the effect of an initial transient. Also, since $\mu = 0$ implies that $Y_i = \eta_i$, we express the model in terms of Y rather than η, writing Equations 10.10 as

$$Y_i = \lambda Y_{i-1} + \varepsilon_i, \ i = 2, 3, \dots,. \tag{10.11}$$

The value of λ is set to 0.4.

```
> lambda <- 0.4 #Autocorrelation term
> set.seed(123)
> ttest <- function(lambda){
+   Y <- numeric(110)
+   for (i in 2:110) Y[i] <- lambda * Y[i - 1] + rnorm(1, sd = 1)
+   Ysamp <- Y[11:110]
+   t.ttest <- t.test(Ysamp,alternative = "two.sided")
+   TypeI <- as.numeric(t.ttest$p.value < 0.05)
+ }
> U <- replicate(10000, ttest(lambda))
> mean(U)
[1] 0.1944
```

As expected, the Type I error rate is inflated. Now run the same simulation but compute the standard deviation of the t statistic using the bootstrap method. First, we create a function t.boot() to carry out a t test using the bootstrap estimate of the standard error. To save time, the bootstrap only uses 200 resamples, but based on the results of Exercise 10.2, this should not cause a serious problem.

```
> t.boot <- function(Y){
+     mean.Y <- function(Y,i) mean(Y[i])
# Bootstrap estimate of the variance
+     boot.Y <- boot(Y, mean.Y, R = 200)
+     se <- sd(boot.Y$t)
+     t.stat <- mean(Y) / se
# t test based on the bootstrap s.e. estimate
+     p <- 2 * (1 - pt(q = abs(t.stat),df = length(Y) - 1))
+     return(c(p, se))
+ }
```

Next, we modify our function ttest() to use the bootstrap standard error. We will have the function return the two standard error estimates as well.

```
> ttest <- function(lambda){
+     Y <- numeric(110)
+     for (i in 2:110) Y[i] <- lambda * Y[i - 1] + rnorm(1, sd = 1)
+     Ysamp <- Y[11:110]
+     t.ttest <- t.boot(Ysamp)
+     TypeI <- as.numeric(t.ttest[1] < 0.05)
+     Yse <- sd(Ysamp) / 10
+     Yse.boot <- t.ttest[2]
+     return(c(TypeI, Yse, Yse.boot))
+ }
```

TABLE 10.1

Mean Standard Error and Error Rate for Three Monte Carlo simulations of the Test of the Null Hypothesis of Zero Mean for Equation 10.1, $n = 100$

ρ		0.0	0.2	0.4	0.6	0.8
Uncorrected	mean *se*	0.099	0.101	0.107	0.122	0.157
	E.R.	0.051	0.114	0.206	0.337	0.527
Block bootstrap	mean *se*	0.093	0.113	0.146	0.206	0.346
	E.R.	0.093	0.099	0.109	0.126	0.182
Parametric bootstrap	mean *se*	0.098	0.122	0.161	0.236	0.441
	E.R.	0.059	0.062	0.0680	0.080	0.108

Note: The standard error or independent samples would be $s\{\bar{Y}\} = 1/\sqrt{100} = 0.1$. The symbol *se* represents the bootstrap estimate of the standard error, denoted $\hat{s}_{boot}\{\bar{Y}\}$ in the text (e.g., Equation 10.9). Statistics are based on 10,000 simulations.

Now we are ready to carry out the Monte Carlo simulation.

```
> set.seed(123)
> U <- replicate(10000, ttest(lambda))
> mean(U[1,]) # Error rate
[1] 0.2062
```

The result is disappointing: the Type I error rate produced by the bootstrap standard error estimate is no better than that produced by the ordinary sample standard error estimate. Indeed, as indicated in the first two rows of Table 10.1, the bootstrap standard error does not change with increasing values of the autocorrelation parameter λ, and therefore the error rate increases with λ.

What happened? The problem is that the inflated standard error is due to the autocorrelation structure of the Y_i. By randomly resampling the sequence of Y_i, the bootstrap resamples break up this autocorrelation structure so that the bootstrap resamples are not autocorrelated and therefore reproduce the incorrect estimate of the standard error. We can see this by printing out the average sample standard error and the average of the bootstrap estimate of the standard error.

```
> mean(U[2,]) # Avg std error
[1] 0.1079641
> mean(U[3,]) # Avg bootstrap s.e.
[1] 0.1073135
```

To effectively use the bootstrap method to estimate the standard error for autocorrelated data, some way has to be found to generate bootstrap resamples that preserve the autocorrelation structure of the data. There are two commonly used approaches: the block bootstrap and the parametric bootstrap. We will describe each of these in turn in the next two subsections.

10.3.2 The Block Bootstrap

The problem in the previous subsection is that the ordinary bootstrap disrupts the correlation structure of autocorrelated data. One way to partially preserve this structure is, instead of resampling the sample data for individual values, to resample the data in blocks, each of some specified length *l* (Efron and Tibshirani, 1993, p. 99). More complex methods involve moving the block along the time series in a continuous manner (Efron and Tibshirani, 1993, p. 101), but we will not be concerned with these. The advantage of block resampling is that serial autocorrelation is preserved within each block and is only disrupted at the boundaries between the blocks. At these block boundaries, the data will in general be uncorrelated.

To create a block bootstrap system for serially autocorrelated data, the first step is to create the blocks. We will do this by placing the data into a matrix in which each block is a column of the matrix. By default, the function `matrix()` places a vector of values into a matrix by columns. We will take our autocorrelated sample of 100 values from the previous subsection and split it into 10 blocks of 10 values each. The first 10 values of *Y*, which are the initial transient, are ignored.

```
> lambda <- 0.4
> set.seed(123)
> Y <- numeric(110)
> for (i in 2:110) Y[i] <- lambda * Y[i - 1] + rnorm(1, sd = 1)
> Ysamp <- Y[11:110]
> n.blocks <- 10
> blocks <- matrix(Ysamp, ncol = n.blocks)
```

Next, we create a function `block.samp()` to randomly sample the blocks with replacement and covert them into a vector.

```
> block.samp <- function(n.blocks, blocks){
+     cols <- sample(1:ncol(blocks), n.blocks, replace = TRUE)
+     as.vector(blocks[,cols])
+ }
```

Here is a simple example of the application of `block.samp()`.

```
> set.seed(123)
> print(A <- matrix(1:9,3))
     [,1] [,2] [,3]
[1,]   1    4    7
[2,]   2    5    8
[3,]   3    6    9
> print(B <- block.samp(6, A))
 [1] 1 2 3 7 8 9 4 5 6 7 8 9 7 8 9 1 2 3
```

Next, we create a function `mean.b.boot()` that calls the function `block.samp()` and computes the mean of the resulting sample.

```
> mean.b.boot <- function(n.blocks, blocks){
+     mean(block.samp(n.blocks, blocks))
+ }
```

For example, if we apply `mean.b.boot()` to the simple matrix A, we get the following result.

```
> set.seed(123)
> mean.b.boot(6, A)
[1] 5.5
> mean(B)
[1] 5.5
```

Recall that the matrix `blocks` contains the same data as the sample `Ysamp` and therefore has the same mean.

```
> mean(Ysamp)
[1] 0.05088889
> mean(blocks)
[1] 0.05088889
```

Therefore, we can conduct a block bootstrap t test by generating a set of B resamples by applying the function `mean.b.boot()` B times and then, using the plug-in principle, estimating the standard error by the standard deviation of the vector of resampled block bootstrap means. We set B at 200.

```
> set.seed(123)
> u <- replicate(200, mean.b.boot(10, blocks))
> t.stat <- mean(blocks) / sd(u)
> print(p <- 2 * (1 - pt(q = abs(t.stat),
+     df = nrow(blocks)*ncol(blocks) - 1)))
[1] 0.772062
```

Now we are ready to construct a Monte Carlo simulation of a t test of autocorrelated data using the block bootstrap. First, we create a function `t.b.boot()` that carries out a single t test using the block bootstrap.

```
> t.b.boot <- function(lambda){
+ # Generate the sample, eliminating the initial transient
+     Y <- numeric(110)
+     for (i in 2:110) Y[i] <- lambda * Y[i - 1] + rnorm(1, sd = 1)
+     Ysamp <- Y[11:110]
+ # Break into blocks
+     n.blocks <- 10
+     blocks <- matrix(Ysamp, ncol = n.blocks)
+ # Compute the bootstrap resample
+     U <- replicate(200, mean.b.boot(n.blocks, blocks))
+ # Carry out the t-test
+     t.stat <- mean(blocks) / sd(U)
+     p <- 2 * (1 - pt(q = abs(t.stat),
+       df = nrow(blocks)*ncol(blocks) - 1))
+     return(c(as.numeric(p < 0.05), sd(U)))
+ }
```

Now we carry out the Monte Carlo simulation.

```
> set.seed(123)
> lambda <- 0.4
> U <- replicate(10000, t.b.boot(lambda))
> mean(U[1,]) # Error rate
[1] 0.1089
> mean(U[2,]) # Mean est. std. error
[1] 0.1459619
```

The results of several such simulations are given in the second row of Table 10.1. As shown in the table, the block bootstrap actually has about the same or even a higher error rate than the uncorrected t test for uncorrelated or weakly autocorrelated data. For values of λ greater than 0.4, the block bootstrap outperforms the uncorrected t test, although the error rate is still inflated by a factor of two to four. This relatively poor result is due in part to the interruption of the correlation structure on the boundaries of the blocks, and in part to the fact that a relatively small number of independent entities (the blocks) are involved in the bootstrap simulation. There is a trade-off: the smaller the number of boundaries, the smaller the number of independent entities. The parametric bootstrap, discussed in the next section, avoids this problem but does bring another problem of its own.

10.3.3 The Parametric Bootstrap

Efron and Tibshirani (1993, p. 92) describe a second method for structured data such as time series that has come to be called the *parametric bootstrap* (Davison and Hinckley, 1997, p. 389). The reason for this term is that, unlike the block bootstrap, the parametric bootstrap assumes a model for the random process, estimates the parameter or parameters of the model, and uses these estimated parameters to construct the bootstrap resamples. Suppose we have a data sequence $\{Y_1, Y_2, ..., Y_n\}$ that we assume has been generated by a first-order autoregressive process of the form of Equation 10.11, and we wish to test the null hypothesis $\mu = 0$. We can fit the Y_i with the first-order autoregressive model (Equation 10.11) and compute the least squares estimate $\hat{\lambda}$ for λ. Let $\{\hat{\varepsilon}_i\}$ be the set of residuals of this estimate. The $\hat{\varepsilon}_i$ are given by the equation

$$\hat{\varepsilon}_i = Y_i - \hat{\lambda} Y_{i-1}, \; i = 1, 2, ... \tag{10.12}$$

The errors ε_i in Equation 10.11 are independent. Even, if we assume (and this is a very important assumption) that the model (Equation 10.11) provides an accurate description of the data set $\{Y_i\}$, the error estimates $\hat{\varepsilon}_i$ in Equation 10.12 will not be independent because they sum to zero. However, the correlation structure associated with Equation 10.11 will be removed. Therefore, one can generate a bootstrap sequence with an error structure that approximates the error structure of the solution of Equation 10.12 by setting Y_1 equal to $\hat{Y}_1$ and then computing the remaining Y_i iteratively according to the equation

$$\hat{Y}_i = \hat{\lambda} \hat{Y}_{i-1} + \tilde{\varepsilon}_i \; i = 2, 3, ..., n \tag{10.13}$$

where at each step $\tilde{\varepsilon}_i$ is selected randomly with replacement from the set of random variables $\{\hat{\varepsilon}_i\}$, $i = 1, ..., n$. In other words, the bootstrap errors are selected randomly in the same way as the original bootstrap values are selected, by sampling from the original data set with replacement, and then they are fitted into the parametric model of the random process.

R contains built-in facilities for estimating the parameters of time series such as Equation 10.10. These are described by Venables and Ripley (2002, Ch. 14), and the reader is referred to that reference for further discussion. We first use the R function `ts()` to convert the vector `Ysamp` generated in the previous section into an R time series. Then we use the function `arima()` to construct a first-order autoregressive integrated moving average model based on the time series data.

```
> Ysamp.ts <- ts(Ysamp)
> Ysamp.ar <- arima(Ysamp.ts, c(1,0,0))
```

Next, we obtain our estimate $\hat{\lambda}$ and our set of residuals $\hat{\varepsilon}_i$.

```
> print(lambda.hat <- coef(Ysamp.ar)[1], digits = 3)
  ar1
0.401
> e.hat <- residuals(Ysamp.ar)
```

Now we construct a function to generate the simulated time series based on Equation 10.13.

```
> para.samp <- function(Ysamp, e.hat, lambda.hat){
+     Y.sim <- numeric(length(Ysamp))
+     Y.sim[1] <- Ysamp[1]
+     for (i in 2:length(Ysamp)){
+        ei.boot <- sample(e.hat, 1)
+        Y.sim[i] <- lambda.hat * Y.sim[i - 1] + ei.boot
+     }
+   return(mean(Y.sim))
+ }
```

In theory, the series `Y.sim` has the same autocorrelation structure as `Ysamp`, so that it can function as a bootstrap resample. Thus, we can use the function `replicate()` to carry out our bootstrap resampling.

```
> set.seed(123)
> U <- replicate(200,para.samp(Ysamp, e.hat, lambda.hat))
```

Using the plug-in principle, we can obtain an estimate of the standard error as the standard deviation of the mean of the bootstrap resamples.

```
> print(sd(U), digits = 3)
[1] 0.151
```

By comparison, the estimate $s\{\overline{Y}\} = s/\sqrt{n}$, where $n = 100$, gives us the estimate we would expect from an independent sample.

```
> print(sd(Ysamp) / 10, digits = 3)
[1] 0.0986
```

The Monte Carlo simulation of the t test using the parametric bootstrap is set up in the same way as that of the block bootstrap, and the code is not shown (if you try to run this, you will find that it takes a *very* long time). The results are shown in the last two rows of Table 10.1. The parametric bootstrap outperforms the block bootstrap at all values of λ and has a reasonable error rate for all but the highest value.

The success of the parametric bootstrap method may be highly dependent on the adequacy of the parametric model in simulating the data. There has been relatively little study of the effects of misspecification of the model on the accuracy of the parametric bootstrap. It is important to emphasize that the parametric model used in the test whose results are reported in Table 10.1 is the actual model itself, so there is no misspecification error. That is, we are estimating the parameters of the model of a data set that is actually the same as the model used to generate the data set in the first place. The existence of misspecification error in a real application could seriously impact the performance of the method.

10.4 Bootstrapping Spatial Data

10.4.1 The Spatial Block Bootstrap

The preceding section introduced the bootstrap estimate of the standard error of data sets with temporally autocorrelated errors. As with other aspects of the analysis of spatial data, bootstrapping data with spatially autocorrelated errors proceeds in a completely analogous way. Where the time series data is specified by Equation 10.10, the spatial error model is specified by Equation 3.25, which is repeated here as

$$Y = \mu + \eta$$
$$\eta = \lambda W \eta + \varepsilon. \tag{10.14}$$

Assuming that the spatial weights matrix is row normalized (Equation 3.22), the values of λ range between -1 and 1. Once again, we will test the null hypothesis $\mu = 0$ against the alternative $\mu \neq 0$. A spatial block bootstrap can be developed that is precisely analogous to the time series block bootstrap of Section 10.3.

We will proceed in the same manner as we did with time series data in Section 10.3. We will run Monte Carlo simulations of a test of the null hypothesis $\mu = 0$ on an artificial data set satisfying Equation 10.14 in which the value of μ is zero. The data set is generated as described in Section 3.5.3, by first generating a vector ε of independent pseudorandom numbers from a standard normal distribution and then plugging this into Equation 3.26, $\eta = (I - \lambda W)^{-1} \varepsilon$. Under the null hypothesis $\mu = 0$, and therefore the first of Equations 10.14 becomes $Y = \eta$. Therefore, we have

$$Y = (I - \lambda W)^{-1} \varepsilon. \tag{10.15}$$

The transformation of Equation 10.15 is applied to a 24 by 24 lattice and then a two-layer border is trimmed to avoid edge effects. The Monte Carlo simulation is therefore carried out on a 20 by 20 cell square lattice.

We can use the array feature of R to develop our sampling code. We briefly mentioned in Chapter 2 that R is capable of constructing arrays of more than two dimensions. To create the block resampling system, we employ a three-dimensional array of blocks, so that the blocks are stacked on top of each other like a raster brick or a Rubik's cube. We will then sample these blocks with replacement.

We illustrate the method using a 6 by 6 lattice with 2 by 2 = 4 blocks, each of size 3 by 3. The cell numbering is set up as follows.

```
> n.row <- 6
> n.block <- 2
> print(A <- matrix(1:n.row^2, nrow = n.row, byrow = TRUE))
     [,1] [,2] [,3] [,4] [,5] [,6]
[1,]    1    2    3    4    5    6
[2,]    7    8    9   10   11   12
[3,]   13   14   15   16   17   18
[4,]   19   20   21   22   23   24
[5,]   25   26   27   28   29   30
[6,]   31   32   33   34   35   36
```

This cell numbering scheme mimics the one generally used with data on a lattice, in which the index value starts in the northwest corner and moves across the rows and down the columns. In R, the natural array numbering is by columns, so we will work with the transpose of A, which we define explicitly to increase the transparency of the code.

```
> At <- t(A)
```

The size of each block is 6 / 2 = 3. We define a three-dimensional array B consisting of four 3 by 3 matrices.

```
> blk.size <- n.row / n.block
> B <- array(0, c(blk.size, blk.size, n.block^2))
```

Next, we fill the elements of B with the corresponding elements of the transpose of A. The first 3 by 3 element of B (the upper left corner of the transpose At) is displayed.

```
> k <- 1
> for(i in 1:n.block){
+     for(j in 1:n.block){
+         B[,,k] <- At[(1+(i-1)*blk.size):(i*blk.size),
+             (1+(j-1)*blk.size):(j*blk.size)]
+         k <- k+1
+ }}
> B[,,1]
     [,1] [,2] [,3]
[1,]    1    7   13
[2,]    2    8   14
[3,]    3    9   15
```

Next, we create a sample with replacement of the block number, which is the third index of B.

```
> set.seed(123)
> print(i.samp <- sample(1:n.block^2, replace = TRUE))
[1] 2 4 2 4
```

Next, we create another three-dimensional array of the same size as B and fill it with the resampled blocks of B.

```
> Ct <- matrix(0, nrow = n.row, ncol = n.row)
> k <- 1
> for(i in 1:n.block){
+     for(j in 1:n.block){
+         Ct[(1+(i-1)*blk.size):(i*blk.size),
+             (1+(j-1)*blk.size):(j*blk.size)] <- B[,,i.samp[k]]
+         k <- k+1
+ }}
```

Finally, we transpose back to get our block resampled version of the original lattice A.

```
> print(C <- t(Ct))

     [,1] [,2] [,3] [,4] [,5] [,6]
[1,]   19   20   21   19   20   21
```

```
[2,]   25    26    27    25    26    27
[3,]   31    32    33    31    32    33
[4,]   22    23    24    22    23    24
[5,]   28    29    30    28    29    30
[6,]   34    35    36    34    35    36
```

The blocks are sampled columnwise, which does not affect the bootstrap distribution.

The block bootstrap is implemented by modification of the uncorrected bootstrap analogous to that of the previous section on time series. Code similar to that just displayed is used to construct a three-dimensional array based on the transpose of the data lattice, and the third dimension of the array is then sampled with replacement. A weakness of the block bootstrap, at least as implemented here, is that it is limited to rectangular data lattices.

The Monte Carlo simulation is accomplished by developing a function t.b.boot() that is called by the function replicate().

```
> t.b.boot <- function(lambda){
+ # Generate the sample, eliminating the boundary
+    Y <- IrWinv %*% rnorm(24^2)
+    Ysamp <- matrix(Y, nrow = 24, byrow = TRUE)[3:22,3:22]
+ # Break into blocks
+    Yt <- t(Ysamp)
+    n.block <- 4
+    blk.size <- 20 / n.block
+    blocks <- array(0, c(blk.size,blk.size,n.block^2))
+    k <- 1
+    for(i in 1:n.block){
+        for(j in 1:n.block){
+            blocks[,,k] <- Yt[(1+(i-1)*blk.size):(i*blk.size),
+                (1+(j-1)*blk.size):(j*blk.size)]
+            k <- k+1
+    }}
+ #Compute the bootstrap resample
+    U <- replicate(200, mean.b.boot(n.block, blocks))
+ # Carry out the t-test
+    t.stat <- mean(blocks) / sd(U)
+    p <- 2 * (1 - pt(q = abs(t.stat),
+      df = nrow(Ysamp)*ncol(Ysamp) - 1))
+    return(c(as.numeric(p < 0.05), sd(U)))
+ }
```

The function creates the random field on a 24 by 24 and carries out the block resampling on a 20 by 20 sub-grid. It calls the function mean.b.boot(), which in turn calls block.samp() to generate the block resample and then computes the mean.

```
> mean.b.boot <- function(n.block, blocks){
+    mean(block.samp(n.block, blocks))
+ }
```

The function block.samp() uses the matrix code shown above to generate the block resample.

```
> # Block sample function
> block.samp <- function(n.block, blocks){
```

```
+     i.samp <- sample(1:n.block^2, replace = TRUE)
+     Ct <- matrix(0, nrow = 20, ncol = 20)
+     k <- 1
+     blk.size <- 20 / n.block
+     for(i in 1:n.block){
+         for(j in 1:n.block){
+             Ct[(1+(i-1)*blk.size):(i*blk.size),
+                 (1+(j-1)*blk.size):(j*blk.size)] <- blocks[,,i.samp[k]]
+             k <- k+1
+     }}
+     return(t(Ct))
+ }
```

Here is a Monte Carlo simulation for $\lambda = 0.4$.

```
> set.seed(123)
> lambda <- 0.4
> nlist <- cell2nb(24, 24)
> IrWinv <- invIrM(nlist, lambda)
> U <- replicate(10000, t.b.boot(lambda))
> mean(U[1,]) # Error rate
[1] 0.1008
> mean(U[2,]) # Mean est. std. error
[1] 0.07199279
```

The results of a set of simulations for increasing values of λ are shown in Table 10.2. Similar to the case with time series data (Table 10.1), the error rate of the block bootstrap corrected test is actually higher than that of the uncorrected bootstrap test for small values of λ. As λ increases the error rate of the block bootstrap corrected test increases, but not as dramatically as that of the uncorrected test. This poor performance of the block bootstrap was also observed by Davison and Hinkley (1997, p. 399). Efron and Tibshirani (1993, p. 101) note that the block size can have a major impact on the performance of the block bootstrap. In addition, Davison and Hinkley (1997, p. 426) point out that one might expect the performance of the spatial block bootstrap to be worse than that of

TABLE 10.2

Mean Standard Error and Error Rate for Three Monte Carlo simulations of the Test of the Null Hypothesis of Zero Mean for Equation 10.1 for a 20 by 20 Spatial Lattice

ρ		0.0	0.2	0.4	0.6	0.8
Uncorrected	mean(se)	0.049	0.050	0.052	0.057	0.070
	E.R.	0.049	0.109	0.198	0.338	0.531
Block bootstrap	mean(se)	0.048	0.057	0.072	0.100	0.171
	E.R.	0.076	0.085	0.100	0.123	0.173
Parametric bootstrap	mean(se)	0.050	0.061	0.080	0.116	0.220
	E.R.	0.053	0.055	0.058	0.064	0.073

Note: The standard error for independent samples would be $se = 1/\sqrt{400} = 0.05$. The symbol se represents the bootstrap estimate of the standard error, denoted $\hat{s}_{boot}\{\bar{Y}\}$ in the text (e.g., Equation 10.9). Statistics are based on 10,000 simulations. The block bootstrap uses 25 blocks, each of size 4 by 4.

the time series block bootstrap for data sets of the same size. Comparing Tables 10.1 and 10.2 indicates a slightly poorer performance of the spatial block bootstrap. The reason for this is related to a phenomenon known as the "curse of dimensionality" (Bellman, 1957), discussed in this context by Hastie et al. (2017, p. 23). The problem is that as the dimension of the space in which the sampling takes place increases, a greater portion of the sample region near a block boundary also increases. Intuition for this can be gained by printing out a vector representing an individual block of the time series and a matrix representing a block of the spatial block bootstrap. We will use a total of 400 data records and 16 blocks. Here is sequence of random numbers that simulates a time series of length 400 / 16 = 25.

```
> print(rnorm(25), digits = 2)
 [1]    0.450   2.092  0.035  0.733 0.118   0.772  1.837 1.376  -0.544
[10]   -1.069  -0.430 -0.111 -0.034 0.629  -0.298 -0.271 0.988  -0.216
[19]    0.755  -0.683 -0.729 -1.018 0.249  -0.721  0.649
```

Here is matrix of random numbers that simulates a single simulated spatial block.

```
> print(matrix(rnorm(25), nrow = 5), digits = 2)
       [,1]    [,2]    [,3]    [,4]     [,5]
[1,] -1.54  -0.031  2.752  -1.59  -0.0094
[2,]  1.22  -0.574 -0.127  -0.15   0.6734
[3,] -3.28  -0.941 -1.663   0.35  -1.2237
[4,] -0.15  -1.053 -0.092  -1.96  -2.2888
[5,] -2.96  -0.122 -0.434   0.46   0.8126
```

Recall that the autocorrelation structure of the data is disrupted at the block boundaries. Of the 25 points in the simulated time series block, only 2 (the first and last in the sequence) are on the boundary. The simulated spatial data block has 16 boundary points and only 9 interior points. Therefore, a larger fraction of the data records is associated with disrupted autocorrelation structure in the case of the spatial block bootstrap than in that of the time series block bootstrap.

10.4.2 The Parametric Spatial Bootstrap

One can construct a parametric bootstrap for spatial data, analogous to that constructed for time series data in Section 10.3.3. As in the time series case, one fits a parametric model to the data and then samples with replacement from the residuals of the model to generate the bootstrap resamples. The parametric bootstrap procedure for a spatial error model (Section 3.5.3; these will be discussed in Chapter 13) is to fit the data to a model of the form of Equation 10.14 and compute the residuals $\hat{\varepsilon}_i$. One then resamples these residuals to generate bootstrap resamples $\tilde{\varepsilon}_i$. Under the conditions of the null hypothesis, $Y = (I - \lambda W)^{-1}\varepsilon$, so we may apply the plug-in principle to obtain

$$\tilde{Y} = (I - \lambda W)^{-1}\tilde{\varepsilon}. \tag{10.16}$$

One then computes the sample mean of the $\tilde{Y}$. This procedure is repeated many times, and the variance of these means becomes the bootstrap variance estimate for use in computing $\hat{n}_e$ via Equation 10.8.

Similar to previous analyses, to analyze the parametric bootstrap we will create a 24 by 24 grid and excise the outer two grid cells around the boundary. Once again, we will use

Monte Carlo simulation. First, we create a function `t.para.boot()` that will be called by the function `replicate()`. The function returns the outcome of the t test and the standard error.

```
> # Monte Carlo simulation of parametric bootstrap
> t.para.boot <- function(IrWinv, nlist.small, W){
+ # Generate the sample, eliminating the boundary
+    Y <- IrWinv %*% rnorm(24^2)
+    Y.samp <- matrix(Y, nrow = 24, byrow = TRUE)[3:22,3:22]
+ # Fit a spatial error model to the data
+    Y.mod <- errorsarlm(as.vector(Y.samp) ~ 1, listw = W)
+    e.hat <- residuals(Y.mod)
+    lambda.hat <- Y.mod$lambda
+    IrWinv.mod <- invIrM(nlist.small, lambda.hat)
+ # Bootstrap resampling
+    u <- replicate(200,para.samp(e.hat, IrWinv.mod))
+ # Carry out the t-test
+    t.stat <- mean(Y.samp) / sd(u)
+    p <- 2 * (1 - pt(q = abs(t.stat),
+     df = nrow(Y.samp)*ncol(Y.samp) - 1))
+    return(c(as.numeric(p < 0.05), sd(u)))
+ }
```

The function has three arguments: `IrWinv`, `nlist.small`, and `W`. The first is the matrix $(I - \lambda W)^{-1}$ on the 24 by 24 lattice, the second is the neighbor list of the 20 by 20 lattice, and the third is the `listw` object of the 20 by 20 lattice (see Section 3.5.2). The sp function `errorsarlm()`, discussed in Chapter 13, is used to estimate the parameter λ. This estimate is used to construct a matrix of the form $(I - \lambda W)^{-1}$ on the 20 by 20 lattice. The residuals are then resampled, and the resampled errors are used to generate via Equation 10.16 a set of values Y_i, whose mean is computed. This process is bootstrapped B times (in this case, $B = 200$) to generate the bootstrap distribution.

The function `t.para.boot()` calls a function `para.samp()` to generate the resampled data using Equation 10.16. Here is the code for that function.

```
> para.samp <- function(e.hat, IrWinv.mod){
+    e.samp <- sample(e.hat, length(e.hat), replace = TRUE)
+ +    Y.boot <- IrWinv.mod %*% e.samp
+    return(mean(Y.boot))
+ }
```

The Monte Carlo simulation is carried out for $\lambda = 0.4$ as follows (as with the code in Section 10.3.3, this takes a *really* long time to run!).

```
> lambda <- 0.4
> nlist <- cell2nb(24, 24)
> IrWinv <- invIrM(nlist, lambda)
> nlist.small <- cell2nb(20, 20)
> W <- nb2listw(nlist.small)
> U <- replicate(10000, t.para.boot(IrWinv, nlist.small, W))
> mean(U[1,]) # Error rate
[1] 0.0576
> mean(U[2,]) # Mean est. std. error
[1] 0.07996112
```

TABLE 10.3

Power of the Test of the Methods for Several
Values of δ, with $\lambda = 0.6$

δ	0.125	0.25	0.375	0.5
Uncorrected	0.554	0.862	0.983	0.999
Block bootstrap	0.295	0.667	0.920	0.994
Parametric bootstrap	0.203	0.569	0.875	0.986

Note: Based on 10,000 Monte Carlo simulations.

As was the case with time series data, the parametric bootstrap outperforms the block bootstrap at all values of λ (Table 10.2), and, as with the case of time series data, one must interpret these results with caution, recognizing that in the test of the parametric bootstrap there is no misspecification error, and therefore the results of this test may give an unrealistically optimistic picture of the method.

10.4.3 Power of the Tests

The block bootstrap and especially the parametric bootstrap have been shown to be reasonably effective at reducing the Type I error rate for spatially correlated data. One might ask whether these methods have adequate power. Recall (Section 3.3) that the power of a test is defined as $1 - \Pr\{$Type II error$\}$, where a Type II error is the failure to reject the null hypothesis when it is false. Monte Carlo experiments were carried out in which a constant δ was added to the spatially autocorrelated data of Equation 10.16 with a value of λ equal to 0.6. Simulations were conducted of t tests using no correction, the block bootstrap, and the parametric bootstrap. The results are shown in Table 10.3. The methods lack the power of the simple t test but do have reasonable power. The parametric bootstrap is actually slightly less powerful than the block bootstrap in this particular simulation.

10.5 Application to the EM38 Data

Both the block bootstrap and the parametric bootstrap were tested on the data shown in Figure 10.2 (Section 10.1) in a paired t test of the null hypothesis that the mean of the difference between the distributions of the EM38 samples is zero. The block bootstrap works best when applied to a regularly spaced rectangular region, and this particular data set fits that criterion. For our test, the bottom row and eastern column of the sample were removed to create a rectangular lattice of size 6 by 12 cells. No trend removal was carried out, since it was assumed that the data followed the same trend on the two dates, meaning that the difference between them has no trend.

Here is the code for a block bootstrap resample. First, we modify the function `block. samp()`, which constructs and samples the blocks, from that of Section 10.4.1 to accept rectangular lattices.

```
> block.samp <- function(n.blockx, n.blocky, blk.size, blocks){
+ # Resample the block index
+     i.samp <- sample(1:(n.blockx*n.blocky), replace = TRUE)
+ # Construct the 3D array of blocks
+     Ct <- matrix(0, nrow = n.blocky * blk.size,
+         ncol = n.blockx * blk.size)
+     k <- 1
+ # Place resampled blocks in the array
+     for(i in 1:n.blocky){
+         for(j in 1:n.blockx){
+             Ct[(1+(i-1)*blk.size):(i*blk.size),
+                 (1+(j-1)*blk.size):(j*blk.size)] <- blocks[,,i.samp[k]]
+             k <- k+1
+     }}
+ # Since we are just computing the mean, we wouldn't realy need
+ # to compute the transpose here
+     return(t(Ct))
+ }
```

The function `mean.b.boot()` is unchanged, beyond passing the extra arguments describing the lattice.

```
> mean.b.boot <- function(n.blockx, n.blocky, blk.size, blocks){
+     mean(block.samp(n.blockx, n.blocky, blk.size, blocks))
+ }
```

After loading the data as described in Appendix B.4, we construct the Y matrix, deleting column 7 and row 13 from the data.

```
> EM38.rect <- data.Set4.1[which(data.Set4.1$Column <= 6 &
+     data.Set4.1$Row <= 12),]
> EM38.rect$EMDiff <- with(EM38.rect, EM38B425 - EM38F520)
> Ysamp <- matrix(EM38.rect$EMDiff, nrow = 12, byrow = TRUE)
> Yt <- t(Ysamp)
```

We will use a 3 by 3 block, meaning that there are two blocks in the east–west direction and four blocks in the north–south direction.

```
> n.blockx <- 4
> n.blocky <- 2
> blk.size <- 3
```

Next, we create the array of blocks.

```
> blocks <- array(0, c(blk.size,blk.size,n.blockx * n.blocky))
> k <- 1
> for(i in 1:n.blocky){
+     for(j in 1:n.blockx){
+         blocks[,,k] <- Yt[(1+(i-1)*blk.size):(i*blk.size),
+             (1+(j-1)*blk.size):(j*blk.size)]
+         k <- k+1
+ }}
>
```

Now we are ready to carry out the test.

```
> set.seed(123)
> u <- replicate(200, mean.b.boot(n.blockx, n.blocky, blk.size, blocks))
> # Carry out the t-test
> Y.bar <- mean(Ysamp)
> se.boot <- sd(u)
> t.stat <- Y.bar / se.boot
> print(p <- 2 * (1 - pt(q = abs(t.stat),
+     df = nrow(Ysamp)*ncol(Ysamp) - 1)))
[1] 0.2262729
```

We can compute the effective sample size using Equation 10.8.

```
> print(ne.hat.bb <- var(EM38.rect$EMDiff) / se.boot^2)
[1] 15.68952
```

The effective sample size as estimated by the block bootstrap is about $\hat{n}_e = 16$, compared with the sample size of $n = 72$.

Turning to the parametric bootstrap, the function para.samp(), which uses the objects e.hat and IrWinv.mod, is unchanged from Section 10.4.2.

```
> para.samp <- function(e.hat, IrWinv.mod){
+     e.samp <- sample(e.hat, length(e.hat), replace = TRUE)
+     Y.boot <- IrWinv.mod %*% e.samp
+     return(mean(Y.boot))
+ }
```

To carry out the parametric bootstrap, we first construct the objects describing the geometry of the lattice.

```
> library(spdep)
> set.seed(123)
> coordinates(EM38.rect) <- c("Easting", "Northing")
> nlist.w <- dnearneigh(EM38.rect, d1 = 0, d2 = 61)
> W <- nb2listw(nlist.w, style = "W")
```

Next, we construct the spatial error model of the data and compute its residuals and parameter estimate.

```
> Y.mod <- errorsarlm(EMDiff ~ 1, data = EM38.rect, listw = W)
> e.hat <- residuals(Y.mod)
> print(lambda.hat <- Y.mod$lambda)
   lambda
0.533286
> IrWinv.mod <- invIrM(nlist.w, lambda.hat)
```

Now we carry out the bootstrap resample.

```
> U <- replicate(200,para.samp(e.hat, IrWinv.mod))
```

Finally, we are ready to carry out the *t* test and estimate the effective sample size.

```
> se.para <- sd(U)
> t.stat <- mean(EM38.rect$EMDiff) / se.para
> print(p <- 2 * (1 - pt(q = abs(t.stat),
+       df = length(EM38.rect$EMDiff) - 1)))
[1] 0.1489184
> print(ne.hat.para <- var(EM38.rect$EMDiff) / se.para^2)
[1] 22.42287
```

The block bootstrap produces a *p* value of 0.22, and the parametric bootstrap produces a *p* value of 0.14. Recall that in Section 10.1 the naïve *t* test gave us a *p* value of 0.053.

In summary, the results indicate that there is no valid evidence to indicate that a significant difference exists between the means of the two EM38 data sets.

There is absolutely no agronomic importance associated with the issue of whether or not the mean of the EM38 values sampled from the bed on April 25, 1996 does or does not happen to equal the mean of the EM38 values sampled from the furrow on May 20, 1996 (aside possibly from the question of consistency of the EM38 instrument). This was just a convenient question to ask to illustrate the methods. A great deal of caution must be exercised, however, in interpreting the results of similar tests in the case that these methods are applied to questions that do have some importance. The methods are relatively new and have not been sufficiently tested for a substantial body of literature to have accumulated. Using this example as an illustration, the strongest inferences that can be made are that the data cannot be assumed to satisfy the independence of errors assumption associated with classical statistical analysis, and that the effective sample size is considerably reduced by spatial autocorrelation.

10.6 Further Reading

This chapter makes use of the bootstrap only to estimate the standard error. This is the simplest use of this tool, but there are many other uses. Efron and Tibshirani (1993), Manly (1997), and Sprent (1998) provide good entry level discussions. At a slightly higher level are Efron (1979) and Davison and Hinkley (1997). Both Efron and Tibshirani (1993) and Davison and Hinkley (1997) discuss the block bootstrap and the parametric bootstrap. Davison and Hinckley (1997) provide an extensive discussion of the spatial block bootstrap, and Zhu and Morgan (2004) have used it to carry out a statistical analysis of the distribution of nematodes in a field in Wisconsin. Lahiri and Zhu (2006) apply the block bootstrap to another agricultural problem. Kutner et al. (2005, p. 458) and Fox (1997, p. 493) discuss the application of bootstrapping to regression analysis to generate confidence intervals and carry out hypothesis tests involving the regression coefficients. Rizzo (2008) provides R code for the bootstrap method.

Exercises

10.1 Generate a population of 10,000 values from a standard normal distribution. Draw a sample of size 10 from this population and compute the standard error. Use the function boot() to compute the bootstrap standard error based on 50 resamples, and compare the two standard errors.

10.2 Compare accuracy of the bootstrap in problem 10.1 with 20, 50, 1000, and 10,000 resamples.

10.3 Enclose the work you did in Exercise 10.1 in a function and use the function replicate() to replicate the bootstrap process 1000 times. Compute the average difference between the bootstrap standard error estimate and the population standard error, and plot a histogram of these differences.

10.4 Repeat Problem 10.2 for a population of 10,000 values generated from a lognormal distribution. What can you say about the effect of asymmetry of the population distribution on the outcome of the bootstrapping?

10.5 Using the data from Field 4.1, test the null hypothesis that the expected value of the EM38 data measured in the bed (EM38425B) is the same as that of the furrow (EM38425F) using a parametric bootstrap correction based on a spatial lag model. How does your result compare with that obtained in the text using the spatial error model? (This exercise may require reading of Chapter 13).

10.6 (a) Carry out a Monte Carlo simulation of the test of the null hypothesis $\mu = 0$ against the alternative $\mu \neq 0$ for a spatially autocorrelated data set similar to that of Section 10.4.2 in which you repeatedly generate a set of pseudorandom numbers that satisfy this parametric model, and use the plug-in principle to estimate the standard error. (b) Compare the results with those obtained in the text.

11

Measures of Bivariate Association between Two Spatial Variables

11.1 Introduction

The calculation of measures of bivariate association among attribute values is often one of the first steps taken following the initial exploratory analysis of a data set. In this sense, measures of bivariate association form a sort of bridge between the exploratory and confirmatory stages of the analysis. When dealing with continuously valued data, two of the most commonly used measures of bivariate association are the Pearson product moment correlation coefficient r and the Spearman rank correlation coefficient r_s. These statistics measure *linear* association between two quantities. The exploratory phase of the data analysis should, however, have established a general working hypothesis about the form of the relationships between variables, and the computation of the correlation coefficient should be carried out if the results of this exploratory analysis indicate that a linear association is possible.

Consider the case of the wheat data from Field 1 of Data Set 4. The scatterplot matrix of Figure 11.1a indicates that there appears to be a more or less linear relationship between the variables *SPAD* and *LeafN*. *SPAD* refers to the reading on a Minolta SPAD® meter, which is an optical device that estimates the chlorophyll content of a leaf based on the relationship between its transparency to electromagnetic radiation in the near-infrared and in the red regions (Markwell et al., 1995). *LeafN* is the total leaf nitrogen concentration. The SPAD reading has been proposed as a measure of *LeafN* content (Bullock and Anderson, 1998), so one would expect that a relationship between these two quantities would exist. One might expect in turn that leaf nitrogen content would be related to grain protein content, which is a measure of grain quality. Figure 11.1b appears to indicate a relationship between grain protein content and SPAD meter reading, although there is a lot of variability. If this relationship were demonstrated to be significant and consistent, it would provide a relatively low-cost method to estimate the spatial variability of grain quality, which is an economically important quantity. Thus, it is of interest to study the relationship between SPAD reading and grain protein content.

When dealing with categorical data, the most common descriptor of association is the contingency table (Larsen and Marx, 1986, p. 442; Agresti, 1996, 2002). As a very simple example, we can consider the distribution of oak species in Data Set 2 as a function of elevation. According to Pavlik et al. (1991), the blue oak (*Quercus douglasii*) is found primarily at elevations less than about 1100 m. Using a similar approach to that taken in Section 7.3, we can construct a contingency table of the relationship between the presence or absence of blue oaks at a site and the relationship of the elevation of the site to 1100 m.

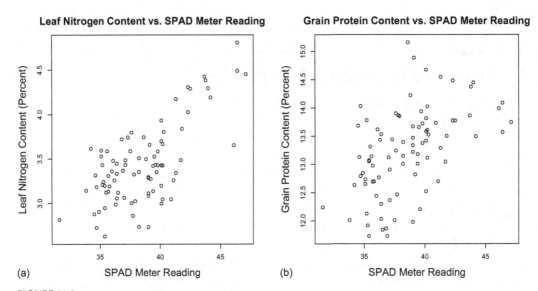

FIGURE 11.1
(a) Scatterplot of leaf nitrogen concentration vs. SPAD ® reading for the data of Field 4.1; (b) Scatterplot of SPAD ® reading vs. grain protein content.

```
> with(data.Set2,
+     matrix(c(sum(Elevation <= 1100 & QUDO == 1),
+     sum(Elevation <= 1100 & QUDO == 0),
+     sum(Elevation > 1100 & QUDO == 1),
+     sum(Elevation > 1100 & QUDO == 0)), nrow = 2, byrow = TRUE,
+     dimnames = list(c("Low", "High"), c("Pres", "Abs")))))
      Pres  Abs
Low   1609 1776
High    37  679
```

It is evident that blue oaks do indeed prefer elevations below 1100 m, but they are not ubiquitous at these elevations. There are about the same number of sites below 1100 m with and without blue oaks (recall that all sites in Data Set 2 contain some species of oak).

One does not need much of a statistical analysis to conclude that the fraction of sites at which a blue oak is present is higher below 1100 m than it is above that elevation. With some contingency tables, however, one may need to test a null hypothesis of randomness in order to be able to conclude that there is a pattern in the data. One way to accomplish this is to use a chi squared test, which is asymptotically valid as the sample size increases. A second method, asymptotically equivalent to the chi squared test, involves the maximum likelihood estimator. A third test, the Fisher test, is used in cases where the sample size is not large enough for the chi squared or likelihood tests to be approximately valid.

When one evaluates a contingency table or a statistic such as the correlation coefficient for spatial data, one faces the same sort of issue that arises with the hypothesis tests that were discussed in Chapter 10: the existence of spatial autocorrelation may reduce the effective sample size, inflating the apparent significance level determined by the test. The first two sections of this chapter discuss methods for addressing this problem. Section 11.2 addresses continuously varying data, and Section 11.3 addresses categorical data. There are other issues that arise in the evaluation of measures of bivariate association, however. One of these can be understood by considering again the relationship between SPAD reading and grain protein level. In trying to understand this relationship, one could simply compute the correlation coefficient between the former and the latter. It is reasonable to suppose, however, that SPAD reading is more directly related to leaf nitrogen level, and leaf nitrogen level to grain protein level, than SPAD is directly to grain protein. Nevertheless, it is grain protein and not leaf nitrogen that provides the farmer with economic yield, so this is the relationship we would like to understand. There may be other factors that strongly influence this relationship. The partial Mantel test, which is described in Section 11.4, provides one way to determine whether a third factor influences the behavior of the associated quantities.

The data describing the relationship between SPAD meter reading and leaf nitrogen level come from 86 sample points arranged 61 m apart in a grid in the field. In dealing with the relationship between, for example, NDVI and yield as measured by a yield monitor, there are thousands of measurements that are either contiguous or very close to each other. These data must be aggregated before they can be compared. The *scale*, as defined in Chapter 6, of this aggregation is not naturally imposed, but rather must be selected. It turns out that the scale at which the data are represented may play a major role in the statistical properties of these data, particularly those that describe the relation between two quantities. This phenomenon is called the *modifiable areal unit problem* (MAUP). Furthermore, the relationship between two quantities when they are aggregated may be different from that between these same quantities when compared as individuals. Using aggregated data to form a conclusion about a relationship between individuals is called the *ecological fallacy*. The MAUP and the ecological fallacy are very closely related, and indeed Cressie (1997) considers them to be manifestations of the same phenomenon, the "change of support problem" discussed in Section 6.4.3. Conceptually they are different, however, in the following sense. The MAUP deals with ecological systems for which there may be no meaningful geographical subdivision, such as an agricultural field. The ecological fallacy concerns systems where the concepts of an individual and a collection of individuals make sense. We will therefore treat the problems as distinct phenomena. They are discussed in Section 11.5.

Finally, but perhaps most importantly, there is one problem that does not necessarily have a statistical solution. This is the temptation to confuse association with causality. Although this problem arises in dealing with any data set, it is particularly pervasive in dealing with spatial data, where similar patterns in thematic maps can be particularly compelling. There is a well-known and easily conceptualized analogy in time series, where the term *nonsense correlation* was used by Yule (1926) to describe the phenomenon. A good example is the high correlation observed between the stork population and the human birthrate in the German city of Oldenberg (Box et al., 1978, p. 8). The stork population and human birthrate both increased due a growth in the human population and an associated increase in the construction of buildings. In the same way, in data arising from an observational study, spatial variables can exhibit a spurious correlation due to a common response to a third factor that has a geographic trend. Perhaps the best way to avoid this pitfall is to constantly be on the lookout for it.

11.2 Estimating and Testing the Correlation Coefficient

11.2.1 The Correlation Coefficient

The most common statistic measuring the linear association between two variables is the Pearson product moment correlation coefficient r. Let the two variables be denoted Y_1 and Y_2. The population correlation coefficient ρ is defined as (Larsen and Marx, 1986, p. 435)

$$\rho = \frac{\text{cov}\{Y_1, Y_2\}}{\sigma_1 \sigma_2}, \tag{11.1}$$

where σ_1^2 is the population variance of Y_1 and $\sigma_1 = \sqrt{\sigma_1^2}$ is the standard deviation, and similarly for σ_2. An estimator of ρ is Pearson product moment correlation coefficient r, defined as (Larsen and Marx, 1986, p. 453; Wackerly et al., 2002, p. 568)

$$r = \frac{\sum_{i=1}^{n}(Y_{i1} - \bar{Y}_1)(Y_{i2} - \bar{Y}_2)}{\sqrt{\sum_{i=1}^{n}(Y_{i1} - \bar{Y}_1)^2 \sum_{i=1}^{n}(Y_{i2} - \bar{Y}_2)^2}} \tag{11.2}$$

$$= \frac{\text{cov}\{Y_1, Y_2\}}{s_1 s_2}.$$

where s_1 and s_1 are defined implicitly by the second equation as $s_1 = \sqrt{\Sigma(Y_{i1} - \bar{Y}_1)^2}$ and similarly for s_2. In this section, we use the symbols Y_1 and Y_2 rather than X and Y to represent the variables to emphasize the fact that we are only testing association. There is no implication of a direction of influence. A computational formula for r that will come in handy in Section 11.4 is (Larsen and Marx, 1986, p. 437)

$$r = \frac{n \sum_{i}^{n} Y_{i1} Y_{i2} - n^2 \bar{Y}_1 \bar{Y}_2}{s_1 s_2}. \tag{11.3}$$

A value of r can be computed for any pair of random variables, but statistical tests on r are only valid if the probability distribution of the random pair (Y_1, Y_2) is *bivariate normal*. The bivariate normal distribution has five parameters, μ_1, μ_2, σ_1^2, σ_2^2, and ρ.

We will work with the three data fields from Field 1 of Data Set 4 discussed in Section 11.1: SPAD meter reading, leaf nitrogen content, and grain protein content. These data can be used to address two questions, namely whether SPAD meter reading is linearly related to *LeafN* content, and whether meter reading is linearly related to grain protein content. As stated in Section 11.1, scatterplots of the relationship between these variables appear to indicate a positive linear association (Figure 11.1), but, particularly in the case of the relationship between grain protein content and SPAD meter reading, the significance of this relationship requires testing.

To formulate the problem in a general way, let $Y_1(x_i, y_i)$ and $Y_2(x_i, y_i)$ be two quantities measured at a set of locations with coordinates (x_i, y_i), $i = 1, \ldots, n$. If, conditional on Y_1, the values of Y_2 are independent and identically distributed, then the variance of the sampling distribution of the random variable r under the null hypothesis $r = 0$ is

$$\sigma_r^2 = 1/(n-1) \tag{11.4}$$

(Kendall and Stuart, 1979). If Y_1 and Y_2 are normally distributed, the null hypothesis $r = 0$ can be tested using the statistic

$$t = \frac{r\sqrt{n-2}}{\sqrt{1-r^2}}, \tag{11.5}$$

which has a t distribution with $n-2$ degrees of freedom under the null hypothesis (Wackerly et al., 2002, p. 568). A preliminary visual inspection of Figure 11.1a casts some doubt on the issue of whether the quantities in the scatterplot are normally distributed (they appear skewed low). In case they are not, the question of a linear relationship between them can be addressed using the Spearman rank correlation coefficient r_s, which is computed by replacing the values for Y_1 and Y_2 in Equation 11.1 with their respective ranks. The statistic r_s does not follow a t distribution under the null hypothesis $r_s = 0$, but corresponding acceptance regions can be either computed directly or tested via a permutation test (Section 3.3).

The first question we address is whether the effect of positive spatial autocorrelation on a significance test of the correlation coefficient is to make the test more "anticonservative," that is, to artificially inflate the Type I error rate. This question can be addressed through Monte Carlo simulation in the same manner as was done in Section 3.4. To set up the simulation, we first run a single test. We create a 14 by 14 square grid of point pairs. Unfortunately, the same symbol ρ is traditionally used for both the correlation coefficient and the autocorrelation term, and this tradition is preserved in the R functions we will be using in this section, so we will follow it. The meaning will generally be clear from the context. Here we set the autocorrelation term ρ to 0.6 and generate the data.

```
> library(spdep)
> rho <- 0.6
> nlist <- cell2nb(14, 14)
> IrWinv <- invIrM(nlist, rho)
> set.seed(123)
> Y1 <- IrWinv %*% rnorm(14^2)
> Y2 <- IrWinv %*% rnorm(14^2)
```

Next, we eliminate the two outer cell layers to reduce edge effects.

```
> Y1samp <- matrix(Y1, nrow = 14, byrow = TRUE)[3:12,3:12]
> Y2samp <- matrix(Y2, nrow = 14, byrow = TRUE)[3:12,3:12]
```

Now we use the R function cor.test() to test the null hypothesis of zero correlation between Y_1 and Y_2, which we have constructed to be uncorrelated.

```
> cortest <- cor.test(Y1samp, Y2samp,
+     alternative = "two.sided", method = "pearson")
> print(r <- cortest$estimate, digits = 3)
  cor
0.205
```

```
> print(t.stat <- cortest$statistic, digits = 3)
   t
2.07
> print(p <- cortest$p.value, digits = 3)
[1] 0.0407
```

Setting the argument method in cor.test() to "pearson" causes the test to be carried out using the t statistic given in Equation 11.5. Assuming a significance level of $\alpha = 0.05$, a Type I error (rejecting the null hypothesis when it is true) occurs in this particular test. Table 11.1 contains the experimental Type I error rate generated via Monte Carlo simulation as the autocorrelation parameter ρ is increased (the table also contains similar results for two methods of correcting for spatial autocorrelation that are discussed below). As with the t test of Chapter 3, the Type I error rate of the test increases with increasing ρ, although it is moderate for small values of ρ. The next section describes a widely used method of correcting for spatial autocorrelation.

11.2.2 The Clifford et al. (1989) Correction

Clifford and Richardson (1985) and Clifford et al. (1989) have developed a correction method that reduces the effect of positive spatial autocorrelation on the outcome of the significance test. Recall (Section 3.4) that the intuitive explanation for the increased Type I error rate is that if the values of $Y_1(x, y)$ and $Y_2(x, y)$ are spatially positively autocorrelated, then each value provides some information about the values of its spatial neighbors, and therefore the actual number of degrees of freedom provided by the set of observations is less than the sample size n. Equation 10.8 is a formula for the estimated effective sample size $\hat{n}_e$ for the t test of the null hypothesis of the equality of two means. By analogy, an estimate $\hat{n}_e$ of the effective sample size n_e for the correlation coefficient could be obtained if one had an independent estimate $\hat{\sigma}_r^2$ of the variance of the sampling distribution of r. This estimate would be obtained by inverting Equation 11.4 to obtain

$$\hat{n}_e = 1 + \frac{1}{\hat{\sigma}_r^2}. \tag{11.6}$$

When $Y_1(x, y)$ and $Y_2(x, y)$ are positively autocorrelated the sampling distribution $\hat{\sigma}_r^2$ is generally larger than predicted under zero autocorrelation, so therefore $\hat{n}_e$ will generally be less than n in the case of positive spatial autocorrelation.

TABLE 11.1

Values of the Type I Error Rate for a Test of the Null Hypothesis $r = 0$ for the Uncorrected Test, the Correction of Clifford et al. (1989), and the Parametric Bootstrap Correction

ρ	0.0	0.2	0.4	0.6	0.8
Uncorrected	0.051	0.053	0.068	0.113	0.230
Clifford	0.052	0.052	0.057	0.073	0.111
Bootstrap	0.050	0.52	0.053	0.057	0.065

We can rewrite Equation 11.2 as

$$r = \frac{s_{12}}{s_1 s_2},$$

$$s_{12} = \frac{1}{n} \sum_{i=1}^{n} (Y_{1i} - \bar{Y}_1)(Y_{2i} - \bar{Y}_2). \tag{11.7}$$

Using an ingenious Taylor series argument, Clifford et al. (1989) showed that under fairly general conditions the value of $\hat{\sigma}_r^2$ can be approximated by

$$\hat{\sigma}_r^2 \cong \frac{\text{var}\{s_{12}\}}{E\{s_1^2\}E\{s_2^2\}}. \tag{11.8}$$

They further showed that the numerator of this equation can be written as

$$\text{var}\{s_{12}\} = n^{-2} \sum_{i-1}^{n} \sum_{j=1}^{n} (Y_1(x_i, y_i) - \bar{Y}_1)(Y_2(x_j, y_j) - \bar{Y}_2)\text{cov}\{Y_1, Y_2\}. \tag{11.9}$$

Clifford et al. (1985, 1989) suggest using the covariogram (Section 4.6.2) to approximate the covariance term in this equation. Recall from that section (Equation 4.27) that the experimental covariogram of lag group k is given by

$$\hat{C}(k) = n_k^{-1} \sum_{i,j \in H(k)} (Y(x_i, y_i) - \bar{Y})(Y(x_j, y_j) - \bar{Y}), \tag{11.10}$$

where $H(k)$ is the kth lag group and n_k is the cardinality (i.e., the number of pairs of data records) in $H(k)$.

Suppose there are a total of H_T lag groups, and that the covariance is isotropic. Of these H_T lag groups, the first H are used in the estimate of $\text{var}\{s_{12}\}$ in Equation 11.8. This implicitly assumes that the covariogram declines to zero as the lag distance k increases. Under these assumptions, Clifford et al. (1989) show that the variance $\text{var}\{s_{12}\}$ may be approximated by

$$\text{var}\{s_{12}\} \cong n^{-2} \sum_{h=1}^{H} n_k \hat{C}_1(k) \hat{C}_2(k). \tag{11.11}$$

where

$$\hat{C}_m(k) = \sum_{i,j \in H(k)_k} n_k^{-1}(Y_m(x_i, y_i) - \bar{Y}_m)(Y_m(x_j, y_j) - \bar{Y}_m) \tag{11.12}$$

is the experimental covariogram for Y_m, $m = 1, 2$. Clifford et al. (1985, 1989) suggest estimating the values of $E\{s_1^2\}$ and $E\{s_2^2\}$ in Equation 11.8 by the sample variances s_1^2 and s_2^2. Since the experimental correlogram $\hat{r}_1(k)$ of Y of the kth lag group $H(k)$ is given by $\hat{r}_1(k) = \hat{C}_1(k) / s_1^2$, and similarly for $\hat{r}_2(k)$ (see Section 4.6.2), we may substitute Equation 11.11 into the numerator of Equation 11.8. Based on the definitions in Section 4.6.2 of the relationship between the covariogram and the correlogram, we can express Equation 11.8 in terms of the correlogram as

$$\hat{\sigma}_r^2 \cong n^{-2} \sum_{k=1}^{H} n_k \hat{r}_1(k) \hat{r}_2(k). \tag{11.13}$$

Since Equation 11.13 involves the product of the experimental correlograms for Y_1 and Y_2, if either Y_1 or Y_2 is spatially uncorrelated, then its correlogram at a positive lag is zero, and therefore the spatial autocorrelation of the other variable has no effect on the hypothesis test. Haining (1990, 1991) has extended the work of Clifford et al. (1989), testing the approximation under various combinations of values H_T and H of lag groups and also showing that the approximation may be used for Spearman's rank correlation coefficient as well as Pearson's r.

We can apply the correction of Clifford et al. (1989) to the hypothesis test carried out on the artificial data set in Section 11.2.1. First, we calculate the experimental correlograms $\hat{r}_1(k)$ and $\hat{r}_2(k)$ in Equation 11.13 using the function `correlogram()` of the R package `spatial` (Venables and Ripley, 2002). There are a number of packages that calculate the correlogram, but the function `correlogram()` of the `spatial` package is particularly useful for this application because it also returns the value of n_k in Equation 11.13. The calculation of the correlogram for Y_1 is accomplished by first using the function `surf.ls()` to express the data as a surface, and then passing this surface to the function `correlogram()`.

```
> library(spatial)
> coords.xy <- expand.grid(1:10,10:1)
> Y1.vec <- as.vector(t(Y1samp))
> Y1.surf <- surf.ls(0,coords.xy[,1], coords.xy[,2], Y1.vec)
> r1 <- correlogram(Y1.surf,20)
> Y2.vec <- as.vector(t(Y2samp))
> Y2.surf <- surf.ls(0,coords.xy[,1], coords.xy[,2], Y2.vec)
> r2 <- correlogram(Y2.surf,20)
```

The second argument of the function `correlogram()` specifies the number of lag groups, denoted H_T above. The object r1 contains the correlogram values $\hat{r}_1(k)$ in Equation 11.13. The object r2, containing the values of $\hat{r}_2(k)$, is computed similarly. Next, we implement Equation 11.13 with $H = 10$ to compute $\hat{\sigma}_r^2$.

```
> nr1r2 <- r1$cnt * r1$y * r2$y
> print(sr.hat <- sum(nr1r2[1:10]) / 10^4, digits = 3)
[1] 0.0116
```

Finally, we compute $\hat{n}_e$ from Equation 11.6, substitute this for n in Equation 11.5 to compute a corrected t statistic t_{corr}, and then compute a p value, again using $\hat{n}_e$ to represent the number of degrees of freedom.

```
> print(ne.hat <- 1 + 1 / sr.hat, digits = 3)
[1] 87.2
> print(t.corr <- r * sqrt(ne.hat - 2) / sqrt(1 - r^2), digits = 3)
  cor
1.93
```

```
> print(p.corr <- 2 * (1 - pt(q = abs(t.corr),
+       df = ne.hat - 2)), digits = 3)
   cor
0.0564
```

In this particular case, the p value is increased from 0.040 to 0.056, and effective sample size is reduced from 100 to 87. The second row of Table 11.1 shows the results of a Monte Carlo simulation for increasing values of ρ.

11.2.3 The Bootstrap Variance Estimate

A possible alternative method for obtaining an estimate of σ_r^2 for use in Equation 11.6 is to compute a bootstrap estimate. Efron and Tibshirani (1993, p. 49) discuss the bootstrap estimate of the variance of the correlation coefficient for independent pairs of data values. This method can in principle be extended to autocorrelated data using the block bootstrap or parametric bootstrap in the same way that these methods were used to estimate the variance of the mean in Chapter 10. Here we will only discuss the parametric bootstrap. As with the test of the null hypothesis that the difference between two means is zero carried out in Chapter 10, one must fit a parametric model to the data. In comparing the means of two random variables $Y_1(x, y)$ and $Y_2(x, y)$ in Chapter 10, we set $Y = Y_1 - Y_2$ and applied the model

$$Y = \mu + \eta$$
$$\eta = \lambda W \eta + \varepsilon. \tag{11.14}$$

with the null hypothesis $\mu = 0$. In the present application, we must fit Y_1 and Y_2 separately. The use of the model (Equation 11.14) for data such as those in Section 11.1, where we are determining the associations between SPAD meter reading, leaf nitrogen level, and grain protein content, would require the assumption that the deterministic component of each variable is constant over the entire region of measurement. This is generally not appropriate. One possibility to correct for this lack of stationarity would be to insert a trend $T(x, y)$ into the model, and a second would be to insert explanatory variables, as will be done in Chapter 13. A third, simpler possibility is to replace the spatial error model (Equation 11.14) with a different model that may provide a better empirical description of the behavior of the measured quantities. This is the approach we will take. Before going on, however *a word of caution is in order*. So far as I know, using the parametric bootstrap in this way has not been validated theoretically, and the conditions, if any, under which its use is justified have not been delimited. Until it receives further scrutiny, use it at your own risk.

The model we will use is the spatial lag model (Anselin, 1992). This model is discussed in greater detail in Chapter 13, but for the present we will write it as

$$Y = \mu + \rho W Y + \varepsilon. \tag{11.15}$$

where ρ is an autocorrelation parameter and again ε is an independent normally distributed random variable with mean zero and variance σ^2. The reason for the name "spatial lag model" should be evident: the lag term ρW is applied directly to the measured variable Y as opposed to applying it to the error η as in the spatial error model of Equation 11.14. We will not make any attempt to justify the use of the spatial lag model on biophysical grounds at this point (indeed, such a justification may be impossible), but rather simply use it as an empirical description of the observed process.

With this model, we can implement the parametric bootstrap in the usual manner. To apply the model to the artificial data analyzed in Section 11.2.2, we first estimate ρ in Equation 11.15 and generate the matrix $(I - \rho W)^{-1}$ based on this estimate. In the case that the estimated value for ρ is negative, we replace it with zero. The following code uses the same artificial data set as Section 11.2.2:

```
> nlist.10 <- cell2nb(10,10)
> W <- nb2listw(nlist.10) > Y1.vec <- as.vector(t(Y1samp))
> Y1.mod <- lagsarlm(Y1.vec ~ 1, data = data.frame(Y1.vec), W)
> IrWinv.1 <- invIrM(nlist.10, max(Y1.mod$rho, 0), style = "W")
> Y2.vec <- as.vector(t(Y2samp))
> Y2.mod <- lagsarlm(Y2.vec ~ 1, data = data.frame(Y2.vec), W)
> IrWinv.2 <- invIrM(nlist.10, max(Y2.mod$rho, 0), style = "W")
```

Next, we write a function to generate the bootstrap resamples.

```
> boot.samp <- function(Y1.mod,Y2.mod,IrWinv.1,IrWinv.2){
+    e1 <- sample(Y1.mod$residuals, replace = TRUE)
+    Y1 <- IrWinv.1 %*% e1
+    e2 <- sample(Y2.mod$residuals, replace = TRUE)
+    Y2 <- IrWinv.2 %*% e2
+    r.boot <- cor(Y1,Y2)
> }
> U <- replicate(200, boot.samp(Y1.mod,Y2.mod,IrWinv.1,IrWinv.2))
> sigmar.hat <- var(U)
```

Finally, we estimate the value of $\hat{\sigma}_r^2$ and use it to compute the corrected t statistic and corresponding p value.

```
> print(ne.hat <- 1 + 1 / sigmar.hat, digits = 3)
[1] 77
> print(t.corr <- sqrt(ne.hat - 2) * r / sqrt(1 - r^2), digits = 3)
 cor
1.82
> print(p.corr <- 2 * (1 - pt(q = abs(t.corr),
+    df = ne.hat - 2)), digits = 3)
   cor
0.0735
```

The parametric bootstrap is more conservative than the Clifford et al. (1989) correction, with an effective sample size of 77 and a p value of 0.07. As usual, we can test this in a Monte Carlo simulation.

Table 11.1 gives a comparison of the three methods (uncorrected, Clifford correction, and parametric bootstrap) for increasing values of ρ. The method of Clifford et al. (1989) provides a considerable improvement over the uncorrected data except at the highest ρ value, while for this particular example the parametric bootstrap method provides a relatively accurate Type I error rate over the range of ρ values. Once again, it must be emphasized that the parametric bootstrap method in this example has an unfair advantage in that it is being applied to data that are known to satisfy the parametric model. The application of this method to data for which the model is incorrectly specified has not been studied, and how the method would fare under these circumstances is an open question. Indeed, the method is, for the present, best used in conjunction with a Clifford correction as a check on this method.

11.2.4 Application to the Example Problem

The example discussed in Section 11.1 involves the correlation of Minolta SPAD meter reading with leaf nitrogen content and with grain protein content. The scatterplot of SPAD reading with *LeafN* content (Figure 11.1) suggests that the distribution of SPAD readings and/or *LeafN* might be skewed. Although we are carrying out a correlation test at this point, ultimately we will be interested in using the SPAD meter to predict leaf and grain N levels. Regression is more sensitive to data heteroscedasticity than to non-normality (Kutner et al., 2005, p. 110), so we will test for a need to stabilize the variance via a transformation. We will try the original data set and log transformed data.

One way to test the variance stabilizing effect of a log transformation is to carry out a linear regression of both the transformed and untransformed data and then conduct a homogeneity of variance test on the residuals of the linear regression (Kutner et al., 2005, p. 116). Here is the code for a Breusch-Pagan test (Section 9.3) of the two sets of residuals.

```
> spadn.lin <- lm(LeafN ~ SPAD, data = data.Set4.1)
> library(lmtest)
> bptest(spadn.lin)
        studentized Breusch-Pagan test
data:  spadn.lin
BP = 4.4284, df = 1, p-value = 0.03534
> spadn.log <- lm(log(LeafN) ~ log(SPAD), data = data.Set4.1)
> bptest(spadn.log)
        studentized Breusch-Pagan test
data:  spadn.log
BP = 1.0646, df = 1, p-value = 0.3022
```

The results of both the Breusch-Pagan test and a Levene's test (not shown) indicate that homogeneity of variance is considerably improved with the log transformed data. Therefore, these are what we will use. Note, by the way, that the R function for the natural logarithm is log(), not ln().

```
> log.SPAD <- log(data.Set4.1$SPAD)
> log.LeafN <- log(data.Set4.1$LeafN)
> cor.test(log.SPAD, log.LeafN, alternative = "two.sided",
+     method = "pearson")
        Pearson's product-moment correlation
data:  log.SPAD and log.LeafN
t = 8.1567, df = 84, p-value = 2.945e-12
alternative hypothesis: true correlation is not equal to 0
95 percent confidence interval:
 0.5271963 0.7684462
sample estimates:
     cor
0.6648121
```

The p value of the correlation coefficient, uncorrected for spatial autocorrelation, is $p = 2.945 \times 10^{-12}$.

We are now ready to carry out the correction of Clifford et al. The function `correlogram()` of library `spatial` is again used to compute the correlograms of the ln(*SPAD*) and ln(*LeafN*) data. The code for ln(*SPAD*) is the following:

```
> library(spatial)
> spad.surf <- surf.ls(0, data.Set4.1$Easting,
+     data.Set4.1$Northing,
+     log.SPAD)
> spad.cgr <- correlogram(spad.surf,10, cex.main = 2, # Fig. 11.2a
+     main = "Log SPAD Correlogram", cex.lab = 1.5,
+     xlab = "Lag Group", ylab = "Experimental Correlogram")
```

The code for ln(*LeafN*) is similar. The correlograms are shown in Figure 11.2. The failure of the correlograms to approach zero is an indication of the nonstationarity of the data (Section 4.6.1). Clifford et al. (1989) indicate that their method should still be effective with data such as these that display a global trend. Based on the figures, only the first three correlogram values are used to compute the estimated covariance. Here is the code to generate the value of $\hat{\sigma}_r$ from Equation 11.13.

```
> n <- nrow(data.Set4.1)
> cgr.prod <- spad.cgr$y * LeafN.cgr$y * LeafN.cgr$cnt
> sigr.hat <- sum(cgr.prod[1:3]) / n^2
```

Next, we generate the value of $\hat{n}_e$ from Equation 11.6 and plug this into Equation 11.5 to generate the t statistic.

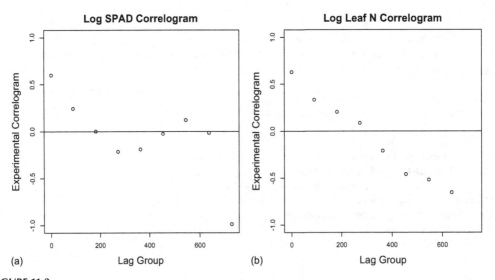

(a)   (b)

FIGURE 11.2
(a) Correlograms of log *SPAD*; (b) log *LeafN* for the data of Field 4.1.

```
> print(ne.hat <- 1 + 1 / sigr.hat, digits = 3)
[1] 43.3
> print(r <- cor(log.SPAD,log.LeafN), digits = 3)
[1] 0.665
> print(t.uncorr <- sqrt(n - 2) * r / sqrt(1 - r^2), digits = 3)
[1] 8.16
> print(t.corr <- sqrt(ne.hat - 2) * r / sqrt(1 - r^2), digits = 3)
[1] 5.72
> print(p.uncorr <- 2 * (1 - pt(q = t.uncorr, df = n - 2)), digits = 3)
[1] 2.94e-12
> print(p.corr <- 2 * (1 - pt(q = t.corr, df = ne.hat - 2)),digits = 3)
[1] 1.05e-06
```

Not surprisingly, the correlation is still highly significant. Perhaps the most interesting result is that the estimated effective sample size is 43.

Next, we will address the same problem using the parametric bootstrap. The first step is to use the spatial lag model of Equation 11.15 to generate the models for ln(*SPAD*) and ln(*LeafN*).

```
> library(spdep)
> coordinates(data.Set4.1) <- c("Easting", "Northing")
> nlist.w <- dnearneigh(data.Set4.1, d1 = 0, d2 = 61)
> W <- nb2listw(nlist.w)
> Y1.mod <- lagsarlm(log.SPAD ~ 1, data = data.Set4.1, W)
> Y1.mod$rho
      rho
0.7659298
> IrWinv.1 <- invIrM(nlist.w, max(Y1.mod$rho, 0), style = "W")
> Y2.mod <- lagsarlm(log.LeafN ~ 1, data = data.Set4.1, W)
> Y2.mod$rho
      rho
0.7497912
> IrWinv.2 <- invIrM(nlist.w, max(Y2.mod$rho, 0), style = "W")
```

Next, the bootstrap resampling function is constructed. The model is then executed to generate the bootstrap resample.

```
> boot.samp <- function(Y1.mod,Y2.mod,IrWinv.1,IrWinv.2){
+    e <- sample(Y1.mod$residuals, replace = TRUE)
+    Y1 <- IrWinv.1 %*% e
+    e <- sample(Y2.mod$residuals, replace = TRUE)
+    Y2 <- IrWinv.2 %*% e
+    r.boot <- cor(Y1,Y2)
+ }
```

Finally, the bootstrap sample is generated and used to create the results.

```
> set.seed(123)
> U <- replicate(200, boot.samp(Y1.mod,Y2.mod,IrWinv.1,IrWinv.2))
> print(sigmar.boot <- var(U), digits = 3)
[1] 0.0308
> print(r <- cor(log.SPAD, log.LeafN), digits = 3)
[1] 0.665
```

```
> print(ne.hat <- 1 + 1 / sigmar.boot, digits = 3)
[1] 33.5
> print(t.corr <- sqrt(ne.hat - 2) * r / sqrt(1 - r^2), digits = 3)
[1] 5
> print(p.corr <- 2*(1-pt(q=abs(t.corr), df = ne.hat - 2)),digits = 3)
[1] 2.09e-05
```

The value of $\hat{n}_e$ generated by the parametric bootstrap is in this particular example smaller than that generated by the Clifford correction. Repeating the parametric bootstrap for *GrainProt* yields an effective sample size estimate of 39 and a corrected p value of 0.0008 for a correlation coefficient of $r = 0.52$ (Exercise 11.1).

In summary, after correction for spatial autocorrelation of the two variables, the correlation between the logarithms of SPAD reading and leaf nitrogen level is still highly significant. Both the Clifford et al. (1989) and the parametric bootstrap correction yield estimated effective sample sizes considerably less than the nominal value of 86. In order to more fully understand the processes underlying the formation of grain protein, however, it is of interest to determine whether there are any other quantities associated with grain protein. This issue is taken up in Section 11.4.

11.3 Contingency Tables

11.3.1 Large Sample Size Contingency Tables

There are three commonly used methods of dealing with contingency table data: the chi-square test, the likelihood test, and Fisher's exact test (Larsen and Marx, 1986, p. 422; Agresti, 1996, p. 39). The chi-square test and the likelihood tests are more appropriate for large data sets and Fisher's exact test is more appropriate for small data sets (the terms *small* and *large* will be made more precise below). We will discuss the first two tests first. Our discussion is restricted to the 2×2 contingency table case, as in Table 11.2. The extension to the more general case can be found in the references.

Let X and Y represent two categorical variables. We use X and Y in this section because the notation is much simpler, with fewer subscripts to deal with. Also, it is often (although not always) the case that X is considered an explanatory variable for Y. Let $p_1 = \Pr\{X = X_1\}$, and $p_2 = \Pr\{X = X_2\} = 1 - p_1$, and let $q_1 = \Pr\{Y = Y_1\}$, and $q_2 = \Pr\{Y = Y_2\} = 1 - q_1$. Let n_{ij} be the number of observations in which $X = X_i$ and $Y = Y_j$, $i, j = 1, 2$. Assume that there are a total of n observations, that is, that $n_{11} + n_{12} + n_{21} + n_{22} = n$. Assume further that the observations

TABLE 11.2

A two by two Contingency Table

	Y_1	Y_2		
X_1	n_{11}	n_{12}	n_{1m}	$\hat{p}_1 = n_{1m} / n$
X_2	n_{21}	n_{22}	n_{2m}	$\hat{p}_2 = n_{2m} / n$
	n_{m1}	n_{m2}	n	
	$\hat{q}_1 = n_{m1} / n$	$\hat{q}_2 = n_{m2} / n$	1	

are mutually independent (this is the assumption we will drop when we take up the analysis of spatial data). Let the marginal sums n_{im} and n_{mj} be defined as $n_{im} = n_{i1} + n_{i2}$ and $n_{mj} = n_{1j} + n_{2j}$ (Table 11.2).

In the case of the blue oak presence/absence vs. elevation data discussed in Section 11.1, the column variable X represents the elevation of the site with $X_1 = \{$less than or equal to 1100 m$\}$ and $X_2 = \{$greater than 1100 m$\}$ (Table 11.2), and the row variable Y represents the presence (Y_1) or absence (Y_2) of oaks. In this case, one can think of X as an explanatory variable and Y as a response variable (Agresti, 1997, p. 17). The question of interest to us is whether X and Y are independent. If they are, then the probability that $X = X_i$ and $Y = Y_j$ is just the product $p_i q_j$ of the individual probabilities, and X has no explanatory value concerning Y.

If we let H_0 be the null hypothesis that X and Y are independent and H_a be the alternative that they are not, then under H_0 the statistic

$$C^2 = \sum_{i=1}^{2} \sum_{j=1}^{2} \frac{(n_{ij} - np_i q_j)^2}{np_i q_j} \tag{11.16}$$

has asymptotically as $n \to \infty$ a χ^2 distribution with one degree of freedom (Larsen and Marx, 1986, p. 423; Agresti, 1996, p. 30). The maximum likelihood estimate for p_1 is $\hat{p}_1 = n_{1m} / n$, and that for p_2 is $\check{p}_2 = n_{2m} / n$, and similarly for $\hat{q}_1$ and $\hat{q}_2$. Therefore, one way to test the null hypothesis is to compute the estimate $\hat{C}^2$ of the statistic C^2 in Equation 11.16 and compare this with the percentiles of the χ_1^2 distribution. Because Equation 11.16 holds only asymptotically as $n \to \infty$, it cannot be reliably applied to data sets with small sample sizes. Snedecor and Cochran (1989, p. 127) and Agresti (1997, p. 28) suggest the rule of thumb that this test can be used if $n > 20$ and if the smallest expected value $n\hat{p}_i \hat{q}_j$ is larger than 5.

The blue oak presence vs. elevation contingency table computed in Section 11.1 is not particularly interesting to analyze, since blue oak presence is clearly dependent on elevation. We will focus on another aspect of Data Set 2 that is more interesting and more helpful to the process of gaining an understanding of the data set. In Section 9.4.2, we constructed logistic regression models of the response of blue oaks to environmental variables. These environmental variables can be roughly divided into three classes: those related to temperature and solar radiation, those related to soil, and those related to precipitation. There is a high level of correlation between variables in these classes, particularly among temperature-related variables (Figure 7.9).

In the exploratory analysis of Section 7.3, we found that in the Coast Range in a multivariate frequency plot with *Precip*, mean annual temperature, or *MAT*, has a bimodal distribution for values of *Precip* less than 800 mm (Figure 7.13). The mean value of *QUDO* tends for a given value of *Precip* to be lower in the Coast Range (Figure 7.12a), so it makes sense to ask whether this could be due to a temperature effect. As a preliminary means of investigating this issue, we will consider the data set consisting of all sites in the Coast Range where *Precip* ≤ 800, that is, where *MAT* is roughly bimodal. Within this data set, we will subdivide the data records into those with values of *MAT* less than 15°C (LoMAT = 1), and those with values greater than or equal 15°C (LoMAT = 0). Let X_1 represent the property of a site having a value of LoMAT equal one, and let X_2 represent the property of a site having value of LoMAT equal zero. Again, let $Y = Y_1$ represent the presence of blue oak at the site and $Y = Y_2$ represent the absence.

First, we set up the data to determine the values of *X* and *Y* from the sf object `coast.sf`. These are reflected in the R variables LoMAT and QUDO

```
> coast.800 <- coast.sf[which(coast.sf$Precip <= 800),]
> coast.800$LoMAT <- 0
> coast.800$LoMAT[which(coast.800$MAT <= 15)] <- 1
```

Next, we compute the contingency table. With a two by two table, it is easy to do this by hand. With a larger table, we would use the function `table()`.

```
> print(n.mat <- with(coast.800, matrix(c(
+     sum(LoMAT == 1 & QUDO == 1),
+     sum(LoMAT == 1 & QUDO == 0),
+     sum(LoMAT == 0 & QUDO == 1),
+     sum(LoMAT == 0 & QUDO == 0)),
+     nrow = 2, byrow = TRUE,
+     dimnames = list(c("LoMAT", "HiMAT"), c("Pres", "Abs")))))
        Pres Abs
LoMAT   508 523
HiMAT   293 182
> print(n.tot <- sum(n.mat))
[1] 1506
```

A total of 1,506 of the Coast Range sites have values of *Precip* less than or equal 800. The estimated probabilities for use in Equation 11.16 are computed as follows:

```
> print(p1 <- (n.mat[1,1]+n.mat[1,2])/n.tot, digits = 2)
[1] 0.68
> print(p2 <- (n.mat[2,1]+n.mat[2,2])/n.tot, digits = 2)
[1] 0.32
> print(q1 <- (n.mat[1,1]+n.mat[2,1])/n.tot, digits = 2)
[1] 0.53
> print(q2 <- (n.mat[1,2]+n.mat[2,2])/n.tot, digits = 2)
[1] 0.47
```

Plugging these into Equation 11.16 leads to the following:

```
> print(C2.hat <- (
+     (n.mat[1,1]-n.tot*p1*q1)^2/(n.tot*p1*q1) +
+     (n.mat[1,2]-n.tot*p1*q2)^2/(n.tot*p1*q2) +
+     (n.mat[2,1]-n.tot*p2*q1)^2/(n.tot*p2*q1) +
+     (n.mat[2,2]-n.tot*p2*q2)^2/(n.tot*p2*q2)))
[1] 20.11941
```

Before one tests the null hypothesis of independence of *X* and *Y* using the normal approximation, one must make one last computation, which is called the Yates' correction for continuity (Snedecor and Cochran, 1989, p. 126). The continuity correction results from the fact that we are applying a normal distribution to data that are only defined for integer values. We will not go into the details here; they are provided by Snedecor and Cochran (1989), and the correction is computed automatically in R. The R function `chisq.test()`

provides two methods of carrying out a significance test, one based on the normal approximation and the other on a permutation test. The continuity correction is applied only to the normal approximation test.

```
> chisq.test(n.mat, simulate.p.value = FALSE)
        Pearson's Chi-squared test with Yates' continuity correction
data:   n.mat
X-squared = 19.624, df = 1, p-value = 9.428e-06
> chisq.test(n.mat, simulate.p.value = TRUE)
        Pearson's Chi-squared test with simulated p-value (based on 2000
replicates)
data:   n.mat
X-squared = 20.1194, df = NA, p-value = 0.0004998
```

Both tests result in a very low p value, although the simulated p value is by far the higher of the two despite the fact that the computed chi-squared statistics are very close.

Now we incorporate spatial autocorrelation on the data analysis. Based on what we have seen in earlier chapters, we would expect that spatial autocorrelation in the data would have the effect of reducing the effective sample size and thus increasing the p value of the significance test. In order to carry out this analysis, however, we need to first describe the second, asymptotically equivalent method for testing the null hypothesis of independence of X and Y. This is to use a maximum likelihood estimator (Agresti, 1996, 2002). The explanation is easiest to understand by using the concept of the *odds ratio* (Agresti, 1996, p. 23). We will use for intuition the interpretation of X as the temperature type in Coast Range (*LoMAT* = 0 vs. *LoMAT* = 1) and Y as the presence vs. absence of blue oaks. We denote by $p_{i,j}$ the *joint probability* $\Pr\{Y = Y_i, X = X_j\}$, that is, the probability that $Y = Y_i$ and that $X = X_j$. We then denote by $p_{i|j}$ the *conditional probability*, that is, the probability that $Y = Y_i$ *given* that $X = X_j$. Recall (e.g., Larsen and Marx, 1986, p. 42; Kutner et al., 2005, p. 1300) that the conditional probability $\Pr\{Y = Y_i \mid X = X_j\}$ satisfies $\Pr\{Y = Y_i \mid X = X_j\} = \Pr\{Y = Y_i, X = X_j\} / \Pr\{X = X_j\}$.

Define the *odds* of the outcome of Y in row j as the ratio of the two conditional probabilities: $odds_j = \Pr\{Y = Y_1 \mid X = X_j\} / \Pr\{Y = Y_2 \mid X = X_j\}$. In our example, $\Pr\{Y = Y_1 \mid X = X_1\}$ represents the probability that a site will contain a blue oak, given that the site is in a low *MAT* environment, and $\Pr\{Y = Y_2 \mid X = X_1\}$ represents the probability that a site will not contain a blue oak given that the site is in a low *MAT* environment. The quantity $odds_1$ represents the odds (ratio of probabilities) that a blue oak will be present in a low *MAT* environment, and $odds_2$ represents the odds that a blue oak will be present in a high *MAT* environment. Intuitively, if the presence or absence of blue oaks is independent of climate type, then these odds should be the same.

To continue, we define the *odds ratio* θ by $\theta = odds_1 / odds_2$. Then

$$\theta = \frac{\Pr\{Y = Y_1 \mid X = X_1\} / \Pr\{Y = Y_2 \mid X = X_1\}}{\Pr\{Y = Y_1 \mid X = X_2\} / \Pr\{Y = Y_2 \mid X = X_2\}}. \tag{11.17}$$

We can estimate θ as follows: The estimator for the joint probability $\Pr\{Y = Y_i, X = X_j\}$ is $\hat{p}_{ij} = n_{ij} / n$. The estimator for p_j is $\breve{p}_j = (n_{1j} + n_{2j}) / n$. Therefore, the estimator for the conditional probability $\Pr\{Y = Y_i \mid X = X_j\}$ is $\hat{p}_{i|j} = n_{ij} / (n_{1j} + n_{2j})$, where we make use of the fact that the n values cancel. Using these relationships, we define the estimator $\hat{\theta}$ as

$$\hat{\theta} = \frac{\dfrac{n_{11}/(n_{11}+n_{12})}{n_{12}/(n_{11}+n_{12})}}{\dfrac{n_{21}/(n_{21}+n_{22})}{n_{22}/(n_{21}+n_{22})}} \cdot$$

$$= \frac{n_{11}n_{22}}{n_{12}n_{21}}$$

(11.18)

Now, suppose X and Y are independent. In this case, we have $\Pr\{Y = Y_i \mid X = X_j\} = \Pr\{Y = Y_i\}\Pr\{X = X_j\}$. Therefore, under this condition Equation 11.17 becomes

$$\theta = \frac{\Pr\{Y = Y_1\}\Pr\{X = X_1\}/\Pr\{Y = Y_2\}\Pr\{X = X_1\}}{\Pr\{Y = Y_1\}\Pr\{X = X_2\}/\Pr\{Y = Y_2\}\Pr\{X = X_2\}}$$

$$= \frac{\Pr\{Y = Y_1\}\Pr\{Y = Y_2\}\Pr\{X = X_1\}\Pr\{X = X_2\}}{\Pr\{Y = Y_1\}\Pr\{Y = Y_2\}\Pr\{Y = Y_2\}\Pr\{X = X_1\}\Pr\{X = X_2\}} \cdot$$

$$= 1$$

(11.19)

Therefore, we can test for independence by testing the hypothesis $\theta = 1$ against the alternative $\theta \neq 1$ using the estimator $\hat{\theta}$ given in Equation 11.18. Agresti (1996, p. 24; 2002, p. 425) shows that if we define the statistic W as

$$W = \frac{n(\ln\hat{\theta})^2}{\hat{\sigma}^2},$$

(11.20)

where

$$\hat{\sigma}^2 = \frac{n}{n_{11}} + \frac{n}{n_{21}} + \frac{n}{n_{12}} + \frac{n}{n_{22}}$$

(11.21)

then W is the maximum likelihood estimator for the standard error of $\ln\hat{\theta}$. Agresti further shows that (under the assumption that the X_i and Y_i are not spatially autocorrelated) W is asymptotically distributed as χ_1^2 as $n \to \infty$. Thus, the statistic W could be used in a chi-squared test instead of the statistic $\hat{C}^2$ defined earlier, provided that there is no spatial autocorrelation in the data.

The problem, of course, is that for our data the X_i and Y_i *are* autocorrelated, or at least they may be. Motivated by the work of Clifford et al. (1989) discussed in the previous section, Cerioli (1997) developed a means of estimating the effect of spatial autocorrelation using the covariogram. In a slight change of notation from Equation 4.27, we express the covariogram for X as

$$\hat{C}_X(k) = n_k^{-1} \sum_{i,j \in H(k)} (X(x_i, y_i) - \bar{X})(X(x_j, y_j) - \bar{X}).$$

(11.22)

and similarly for $\hat{C}_Y(k)$. Recall that both X and Y are binary variables. With these definitions, Cerioli (1997) showed the following: Define a quantity λ by

$$\lambda = \frac{2}{np_1q_1p_0q_0} \sum_{h=1}^{K} n_k\hat{C}_X(k)\hat{C}_Y(k). \tag{11.23}$$

Then the adjusted statistic W_{adj} defined by

$$W_{adj} = \frac{W}{1+\lambda} \tag{11.24}$$

is asymptotically distributed as a χ_1^2 random variable under the null hypothesis that X and Y are independent. In other words, W_{adj} is adjusted to take into account the effects of spatial autocorrelation. Cerioli (1997) also derives a second adjustment statistic based on the Moran correlogram. We will not use this second adjustment, but rather base our analysis on Equation 11.24.

As with the correction of Clifford et al. (1989) (Equation 11.13), Equation 11.24 involves the product of the experimental covariograms for X and Y, and therefore if either X or Y is spatially uncorrelated, then its covariogram is zero and the autocorrelation of the other variable has no effect on the significance test. Moreover, if both X and Y are positively autocorrelated, then since λ must be greater than zero, the value of W_{adj} as calculated in Equation 11.24 will be less than the value of W, so that the adjustment for autocorrelation reduces the value of the test statistic.

The computation of W based on Equations 11.20, 11.23, and 11.24 proceeds as follows. First, we compute W in Equation 11.20.

```
> print(theta.hat <-
+     (n.mat[1,1]*n.mat[2,2]) / (n.mat[1,2]*n.mat[2,1]), digits = 3)
[1] 0.603
> print(sig2.hat <- sum(1/(n.mat/n.tot)), digits = 3)
[1] 19.3
> print(W <- n.tot * log(theta.hat)^2 / sig2.hat, digits = 4)
[1] 19.96
```

The value of W is quite close to, and in between, those of the chi-square statistics computed above. To implement Equation 11.23, we must compute the covariograms of X and Y. The covariogram at lag k can be estimated from the variogram through the use of Equation 4.26, $\gamma(h) = C(0) - C(h)$, which, when written in terms of lag groups instead of lags, implies that $\hat{C}(k) = \hat{C}(0) - \hat{\gamma}(k)$. Computation of the nugget estimator $\hat{C}(0)$ ordinarily requires that we fit a model to the variogram. For this application, however, it is simpler, and probably sufficiently accurate, to assume that at large values of k the variogram has reached its sill and therefore $C(k_{large}) \cong 0$. This implies that

$$\hat{C}(0) \cong \hat{\gamma}(k_{large}) \tag{11.25}$$

is an adequate estimator, and we will use it. As in the case of the Clifford correction of the previous section, we choose the function `variogram()` of the `spatial` package for our variogram estimator. Here is the code.

```
> library(spatial)
> qudo.surf <- with(coast.800, surf.ls(2, Longitude, Latitude, QUDO))
> qudo.vgr <- variogram(qudo.surf,200)
```

The raw experimental variograms of the *QUDO* and *LoMAT* data are fairly ugly, as might be expected for a binary variable defined over a long, narrow domain. We are only interested in the variogram at the smallest lags, however, and fortunately these are relatively well behaved (Figure 11.3). We will sum over the first 20 terms in the experimental variogram and use the maximum over these terms as an estimate of the sill. Applying Equations 11.23, 11.24, and 11.25, we also check to make sure that the numbers of pairs in the lag groups of X and Y are the same.

```
> CX.hat <- max(qudo.vgr$y[1:20]) - qudo.vgr$y[1:20]
> nkX <- qudo.vgr$cnt[1:20]
> CY.hat <- max(lomat.vgr$y[1:20]) - lomat.vgr$y[1:20]
> nkY <- lomat.vgr$cnt[1:20]
> all.equal(nkX,nkY)
[1] TRUE
> print(lambda <-
+     (2 * n.tot* lambda.sum)/(n.mat[1,1]*n.mat[2,2]), digits = 3)
[1] 6.72
> print(W.adj <- W / (1 + lambda), digits = 3)
[1] 2.59
```

Since W_{adj} is asymptotically chi square, we can test its significance as follows:

```
> print(p <- 1 - pchisq(W.adj, df = 1), digits = 3)
[1] 0.108
```

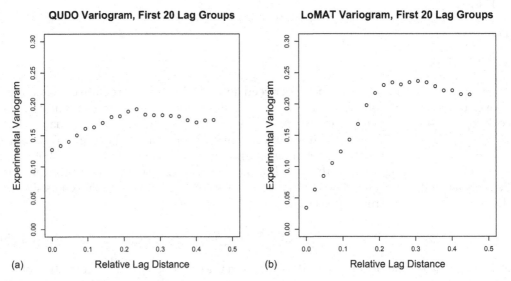

(a) Relative Lag Distance (b) Relative Lag Distance

FIGURE 11.3
Experimental variograms of (a) *QUDO* and *LoMAT* from Data Set 2; (b) Only the first 20 lag groups are plotted.

The statistic is not significant, which calls into question whether there is a significant effect of mean annual temperature on the presence or absence of blue oaks in the Coast Range. We must remember, however, not to take the p value of these significance tests too seriously, since they are an approximation of a quantity that is somewhat questionable in the first place. The main conclusion is that we need to be cautious about interpretation of our results. Interestingly, when this analysis is repeated using *JuMax*, the maximum July temperature, the resulting contingency table is significantly different from random even when spatial autocorrelation is incorporated (Exercise 11.2). This may be meaningful, or it may just be data snooping, but we will pursue it.

11.3.2 Small Sample Size Contingency Tables

In addition to the contingency table developed in Chapter 7 for the distribution of blue oaks with elevation, a second contingency table was developed in Section 7.2, involving Data Set 1. The contingency table has as entries the relationship between X, representing suitability of habitat patches for the yellow-billed cuckoo as defined by the California Wildlife Habitat Relationships (CWHR) model, and Y, representing the observation or failure to observe at least one cuckoo in the patch. The contingency table has values $n_{11} = 5$, $n_{12} = 1$, $n_{21} = 2$, and $n_{22} = 12$. The number of observations in this contingency table does not fall into the range in which the asymptotic chi-square approximation is valid. One means of overcoming this problem is to use a Monte Carlo simulation. When it is called with the argument `simulate.p.value = TRUE`, the R function `chisq.test()` carries out a Monte Carlo simulation in which contingency tables are generated whose entries n_{ij} are sets of random numbers such that their values add up to the marginal values of the contingency table being tested and the fraction of the tables whose chi-square statistic is more extreme is recorded (Hope, 1968).

An alternative method, Fisher's exact test (Fisher, 1922), relies on the fact that, under the assumption of independence of X and Y and for fixed marginal values, the elements of the contingency table are distributed according to a hypergeometric probability distribution (Larsen and Marx, 1986, p. 91). For a given set of marginal values n_{im} and n_{mj}, the process of specifying the value of n_{11} in a two by two contingency table also specifies the others by the equations $n_{12} = n_{1m} - n_{11}$, $n_{21} = n_{m2} - n_{11}$ and $n_{22} = n - n_{11} - n_{12} - n_{21}$. The probability distribution for n_{11} under the independence assumption has the form (Agresti, 1995, p. 39)

$$\Pr\{n_{11} = k\} = \binom{n_{m1}}{k}\binom{n_{m2}}{n_{m1} - k}\bigg/\binom{n}{n_{m1}}, \tag{11.26}$$

which defines a hypergeometric distribution. If n is relatively small, the total number of possible enumerations (i.e., of values of k in Equation 11.26) will be small as well, and therefore it is relatively easy to compute the number of enumerations that are "more extreme" than the observed one, and thereby obtain a p value for the test of the null hypothesis of independence of X and Y. Agresti (2002, p. 93) points out that for small sample sizes n, the p values resulting from Fisher's exact test can only take on a relatively small number of discrete values, and for this reason using the test in a hypothesis test context with a fixed α value of, say, 0.05 will result in the test being conservative. Agresti (2002, p. 94), therefore, recommends simply reporting the p value. There are some other technical issues with the test, which are discussed by Campbell (2007).

In Section 7.2, we generated a contingency table for presence vs. absence of the yellow-billed cuckoo based on suitable vs. unsuitable habitat as defined by the model. The contingency table was based on an array `PresAbs` of bird observations (1 = observed, 0 = not observed), and a second array `HSIPred` (1 = suitable, 0 = not suitable). The data are loaded as described in Appendix B.1 and Section 7.2.

```
> Set1.corrected$HSIPred
 [1] 1 0 1 0 1 0 0 0 0 0 1 0 1 1 0 0 0 0 0 0
> Set1.corrected$PresAbs
 [1] 1 0 1 0 1 0 0 1 0 0 0 0 1 1 0 0 1 0 0 0
> UA <- with(Set1.corrected, which(HSIPred == 0 & PresAbs == 0))
> UP <- with(Set1.corrected, which(HSIPred == 0 & PresAbs == 1))
> SA <- with(Set1.corrected, which(HSIPred == 1 & PresAbs == 0))
> SP <- with(Set1.corrected, which(HSIPred == 1 & PresAbs == 1))
> print(cont.table <- matrix(c(length(SP),length(SA),
+     length(UP),length(UA)), nrow = 2, byrow = TRUE,
+     dimnames = list(c("Suit.", "Unsuit."),c("Pres.", "Abs."))))
        Pres. Abs.
Suit.       5    1
Unsuit.     2   12
```

Here are the results of a Monte Carlo based chi-square test, an asymptotic chi-square test, and a Fisher test for the cuckoo data. The R function `fisher.test()` returns the estimated value $\hat{\theta}$ of the odds ratio (Equation 11.17) as well as the p value.

```
> chisq.test(cont.table, simulate.p.value = TRUE)
X-squared = 8.8017, df = NA, p-value = 0.009995
> chisq.test(cont.table, simulate.p.value = FALSE)
X-squared = 6.0283, df = 1, p-value = 0.01408
Warning message:
In chisq.test(cont.table, simulate.p.value = FALSE) :
  Chi-squared approximation may be incorrect
> fisher.test(cont.table)
p-value = 0.007224
sample estimates:
odds ratio
  22.96080
```

The declared significance value of the table in both the permutation test and the Fisher test is about 0.01. Once again however, this data set, like Data Set 2, contains spatially autocorrelated data, which may invalidate the results of the test. With only 20 values, there are not enough data to compute a correlogram as would be required by the method of Cerioli used in the previous section.

Since an adjusted test based on the asymptotic properties of the distribution is not possible, it is natural to consider employing a permutation test. There is, however, an important difference between the application of a permutation test to this problem and the applications discussed in Chapter 4. As was the case with the bootstrap employed in Chapters 9 and 10, the process of randomizing spatially autocorrelated data breaks up the spatial pattern, which defeats the purpose of employing the randomization. Analogous to the bootstrap case, permutation tests carried out on spatially autocorrelated data must involve a restricted randomization that preserves the spatial structure of the data.

Fortin and Jacquez (2000) discuss restricted randomization for the purpose of carrying out permutation tests on spatial data. They list three approaches. One approach, generally the most effective when it can be used, is the application of conditional simulation (Cressie, 1991, p. 207; Burrough and McDonnell, 1998, p. 152) to spatially autocorrelated data. Conditional simulation is a form of restricted randomization that uses the kriged interpolation estimate of the quantity being simulated to generate random data arrangements having the same spatial autocorrelation structure as the original data set. Unfortunately, our data set has only 20 values, which is not sufficient to generate the variogram necessary to compute a kriged estimate.

A second approach discussed by Fortin and Jacquez (2000) is the *toroidal shift* (Upton and Fingleton, 1985, p. 253). In this approach the map is converted into a *torus*, that is, a donut-shaped object, by first connecting the east and west edges to form a cylinder and then looping the cylinder so that the north and south edges connect. Coordinates are then repeated shifted a randomly selected distance on the torus. Fortin and Dale (2005, p. 240) give an example of this method. However, this method works best when the data are continuously defined on a rectangular surface. Data Set 1 does not fit this criterion.

The third approach discussed by Fortin and Jacquez (2000) is to restrict the randomization to geographically defined subsets. This is considered the least powerful of the three methods, but it can be applied to our present data set. Its use is described by Manly (1997, p. 188). The data are first divided into geographic blocks (Figure 11.4). Randomization only takes place within these blocks. That is, one variable is randomly rearranged (we will use the presence-absence data) within each block and then the resulting rearrangement is compared with the original arrangement of the other variable, in this case, the habitat suitability index. We will run a permutation test using this restricted randomization according to blocks in the same manner as the permutation tests described in Chapter 4. We use the function `fisher.test()`

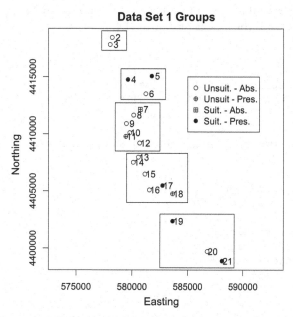

FIGURE 11.4
Subdivision into 5 blocks of the 20 observation points making up the contingency table for the data of Data Set 1.

to compute an odds ratio statistic, and determine where the observed odds ratio statistic lies along the spectrum of statistics computed from permuted data.

This method bears a superficial resemblance to the block bootstrap discussed in Section 10.4.1. There is, however, an important difference. In the case of the block bootstrap, the spatial arrangement of data is preserved within the blocks, and the randomization is accomplished by rearranging the blocks. In the case of the permutation test described in this section, the arrangement of the blocks is preserved and the randomization is accomplished by rearranging the data with each block. The reason for the difference is the following. The objective of the block bootstrap is to estimate the sample variance. Spatial autocorrelation disrupts the estimate because of the correlation of values of with their neighbors. This effect is indicated by Equation 3.28, which is reproduced here:

$$\text{var}\{\bar{Y}\} = \frac{\sigma^2}{n} + \frac{2}{n^2} \sum_{i \neq j} \text{cov}\{Y_i, Y_j\}. \tag{11.27}$$

The arrangement of the data values within blocks is preserved in the block bootstrap in order to preserve the covariance term on the right-hand side, which involves values of Y_i in relation to those at nearby locations. In the case of the permutation test discussed in this chapter, the effect of autocorrelation is indicated by Equation 11.23,

$$\lambda = \frac{2}{n p_1 q_1 p_0 q_0} \sum_{h=1}^{K} n_k \hat{C}_X(k) \hat{C}_Y(k). \tag{11.28}$$

The effect of spatial autocorrelation in this case depends on the spatial correlation structure of both X and Y. By rearranging the Y_i only within blocks, we do our best to preserve the correlation structure of the Y_i and its relation to that of the X_i.

We begin the implementation of block-based randomization by creating five blocks of observation data to match the blocks of Figure 11.4.

```
> obs.block1 <- Set1.corrected$PresAbs[1:2]
> obs.block2 <- Set1.corrected$PresAbs[3:5]
> obs.block3 <- Set1.corrected$PresAbs[6:11]
> obs.block4 <- Set1.corrected$PresAbs[12:17]
> obs.block5 <- Set1.corrected$PresAbs[18:20]
```

Next, we create a function `sample.blocks()` to sample without replacement within each block.

```
> sample.blocks<- function(){
+     s1 <- sample(obs.block1)
+     s2 <- sample(obs.block2)
+     s3 <- sample(obs.block3)
+     s4 <- sample(obs.block4)
+     s5 <- sample(obs.block5)
+     c(s1,s2,s3,s4,s5)
+ }
```

Next, we create a function called `calc.fisher()` that computes the estimate of the odds ratio statistic for the block permutation of sites.

```
> calc.fisher <- function(){
+     PA <- sample.blocks()
+     UA <- sum(Set1.corrected$HSIPred == 0 & PA == 0)
+     UP <- sum(Set1.corrected$HSIPred == 0 & PA == 1)
+     SA <- sum(Set1.corrected$HSIPred == 1 & PA == 0)
+     SP <- sum(Set1.corrected$HSIPred == 1 & PA == 1)
+     n <- matrix(c(SP, SA, UP, UA), nrow = 2, byrow = TRUE)
+     odds.ratio <- fisher.test(n)$estimate
+ }
```

To carry out the permutation test, we first recompute the observed odds ratio statistic for the data in their observed arrangement.

```
> obs.test <- fisher.test(cont.table)
> print(obs.stat <- obs.test$estimate)
odds ratio
  22.96080
```

Next, we sample 1,999 block permutations, letting the observed data serve as the two thousandth.

```
> set.seed(123)
> U <- replicate(1999,calc.fisher())
```

We now determine where the observed statistic lies in the ranks of all of the permutations.

```
> print(p <- sum(U >= obs.stat - 0.001) / 2000)
[1] 0.0725
```

If you are wondering what the 0.001 is doing in there, see Exercise 11.7.

The value $p = 0.073$ generated by the restricted permutation test is considerably larger than the value $p = 0.0072$ from the original Fisher test. Here we must be careful not to fall into the $p = 0.05$ trap, that is, we must not take these p values too seriously. The fact that we get a p value of 0.07 rather than, say, 0.04, really doesn't matter much. To get some perspective, the original value $p = 0.0072$ is roughly the probability of tossing a fair coin and getting heads seven times in a row. This is fairly unlikely. The value $p = 0.073$ is close to the probability of tossing the same coin and getting heads four times in a row. This is not all that unlikely. At this stage, all we can say that there may be a relationship between the habitat suitability index, as computed by multiplying the quantities in the model, and the presence or absence of cuckoos in a patch, but we cannot be too sure.

The significance test of independence of X and Y carried out in this section does not really satisfy the original objective laid out in Chapter 1, which was to test the CWHR model for habitat suitability. A simple test of independence provides a fairly weak test of the model. It simply shows that the model is at least as good as a random assignment of classes to observations. We will next see Data Set 1 in Section 17.2, where we draw our final conclusions using a different, and arguably more appropriate, statistic.

11.4 The Mantel and Partial Mantel Statistics

11.4.1 The Mantel Statistic

The autocorrelation statistics discussed in Sections 4.3 and 4.4, namely the join–count statistics, Moran's I, and Geary's c, are all special cases of the Mantel statistic Γ defined in Equation 4.1. The theory of these statistics was first studied by Mantel (1967) and Mantel and Valand (1970) in the more general context of comparing two matrices. Let the $n \times n$ matrices A and B have elements a_{ij} and b_{ij}, where $1 \leq i, j \leq n$. By analogy with the Pearson's product moment correlation coefficient (Equation 11.3), Mantel (1967) defined the correlation coefficient between these matrices as (Manly, 1997, p. 174)

$$r = \frac{\Sigma_{ij} a_{ij} b_{ij} - \Sigma_{ij} a_{ij} \Sigma_{ij} b_{ij} / m}{\sqrt{(\Sigma_{ij} a_{ij}^2 - \left[\Sigma_{ij} a_{ij}\right]^2 / m)(\Sigma_{ij} b_{ij}^2 - \left[\Sigma_{ij} b_{ij}\right]^2 / m)^2}} \qquad (11.29)$$

where the symbol Σ_{ij} indicates summation over both the indices i and j and $m = n(n-1)$ is the number of off-diagonal elements. If A and B are symmetric, then the sums can be taken over only the lower triangular elements, setting $m = n(n-1)/2$. Mantel (1967) showed that the distribution of r is asymptotically normal. Although this is true, nevertheless, in practice the significance of r is generally tested using a permutation test. As described in Section 4.2, this involves permuting the ordinal relationship between the a_{ij} and the b_{ij} and computing the statistic for each permutation. The only term in Equation 11.27 that is affected by the order of the a_{ij} and b_{ij} is the term

$$Z = \Sigma_{ij} a_{ij} b_{ij}, \qquad (11.30)$$

where again the sum is over both i and j, so this quantity may be used in place of r in a permutation test of the significance of the relation between A and B.

We will illustrate the permutation test process by working through a simple example. We examine the relationship between the logarithms of the leaf nitrogen level *LeafN* and SPAD reading *SPAD*, respectively, in the four points in the two by two square in the northwest corner of Field 4.1. We will use the package vegan (Oksanen et al., 2017). Let A be a matrix measuring the distance between ln(*LeafN*) values, where the distance between two values Y_i and Y_j is defined as $d_{ij} = |Y_i - Y_j|$, and let B be a distance matrix measuring the distance between corresponding ln(*SPAD*) values. By their definition, A and B are symmetric, and therefore we may restrict the computation of Z in Equation 11.30 to the lower triangular parts of A and B. The lower triangular part of A is given by

$$A = \begin{bmatrix} a_{12} & & \\ a_{13} & a_{23} & \\ a_{14} & a_{24} & a_{34} \end{bmatrix}, \qquad (11.31)$$

and the lower triangular part of B is defined similarly.

Here is the computation of A using the function vegdist() from the vegan package. As usual, the data frame data.Set4.1 holds the contents of the file *Set4.196Sample.csv* and is loaded using the statements in Appendix B.4.

```
> library(vegan)
> data.2by2 <- with(data.Set4.1,data.Set4.1[which((Row <= 2) &
+     (Column <= 2)),])
> print(log.LeafN <- log(data.2by2$LeafN), digits = 3)
[1] 1.28 1.14 1.14 1.23
> print(A <- vegdist(log.LeafN, method = "euclidean"), digits = 3)
        1      2      3
2 0.1335
3 0.1415 0.0080
4 0.0454 0.0881 0.0961
```

The matrix B of distances of log *SPAD* is computed similarly.

```
> log.SPAD <- log(data.2by2$SPAD)
> B <- vegdist(log.SPAD, method = "euclidean")
```

The function `mantel()` of the vegan package carries out a permutation test.

```
> mantel(A, B)
'nperm' >= set of all permutations: complete enumeration.
Set of permutations < 'minperm'. Generating entire set.
Mantel statistic based on Pearson's product-moment correlation
Call:
mantel(xdis = A, ydis = B)
Mantel statistic r: 0.305
     Significance: 0.25
Upper quantiles of permutations (null model):
  90%    95% 97.5%    99%
0.329 0.608 0.652 0.662
Permutation: free
Number of permutations: 23
```

Although the Mantel statistic can be used in this manner as a measure of association between two quantities, that is what the Pearson product moment correlation does, so the Mantel statistic is not usually used in this way. The distinguishing feature of the Mantel statistic is that it can serve as a means of measuring the association between two bivariate quantities, or between a univariate and a bivariate quantity. Thus, one can measure the association between a quantity Y_i measured at location (x_i, y_i) and the same quantity Y_j measured at location (x_j, y_j). Here Y is the univariate quantity and (x, y) is the bivariate quantity. In this way, the Mantel statistic becomes a measure of spatial autocorrelation. Indeed, as mentioned in Chapter 4, both the Moran's I and the Geary's c can be considered as special cases of a Mantel statistic.

One of the primary uses of the Mantel statistic Z as defined in Equation 11.30 is in the *partial Mantel* test (Smouse et al., 1986). This test is used primarily to determine the relationship between two quantities after taking into account their spatial relationship. In order to describe the partial Mantel test, we first briefly review the concept of partial regression, which was introduced in Section 8.2.2 the context of partial regression plots, which in that section are called added variable plots.

Recall from that section that if we have a set of measurements of a response variable Y and corresponding sets of measurement of two explanatory variables, X_1 and X_2, then we can use partial regression to help determine whether the explanatory variable X_2 provides a substantial amount of information about Y independently of the information provided by X_1.

Specifically, partial regression measures the effect of adding X_2 to the linear regression model given that X_1 is already in the model. To carry out a partial regression analysis (Kutner et al., 2005, p. 268) one computes the residuals of a regression of the response variable Y on X_1, which we denote $Y \mid X_1$ and the residuals of X_2 on X_1, which we denote $X_2 \mid X_1$. One then carries out a linear regression of $Y \mid X_1$ on $X_2 \mid X_1$. Intuitively, the residuals $Y \mid X_1$ contain the information about the variability of Y not explained by X_1. The residuals $X_2 \mid X_1$ contain the information about the variability of X_2 not explained by X_1. If X_2 contains substantial information about Y that is not provided by X_1, then the coefficient of determination R^2 of the regression of $Y \mid X_1$ on $X_2 \mid X_1$ will be high. On the other hand, if most of the information that X_2 provides about Y is also provided by X_1, then the coefficient of determination R^2 of the regression of $Y \mid X_1$ on $X_2 \mid X_1$ will be low. In simple linear regression, there is a close relationship between the correlation coefficient and regression. This relationship was exploited by Smouse et al. (1986). They viewed the Mantel statistic as a measure of a linear regression and applied the partial regression concept to develop the partial Mantel test. In order to explain their work, we must first cover one further aspect of the theory of simple linear regression.

In Appendix A.2, mention is made of two types of simple linear regression. In the first type, which is discussed in the appendix, the explanatory variable X is a *mathematical variable*, that is, it is a deterministic quantity. A simple example of this is an experiment involving wheat yield vs. applied fertilizer, in which the experimenter applies fertilizer at fixed rates X of, say, 0, 50, 100, 150, and 200 kg ha^{-1} and measures the yield Y. In the second type of regression, both X and Y are random variables. An example of this would be an experiment carried out at many locations in which the explanatory variable X is total seasonal rainfall and the response variable Y is again wheat yield. Suppose X and Y are normally distributed random variables with means μ_X and μ_Y and variances σ_X^2 and σ_Y^2. Changing the notation slightly from Equation 11.1, let

$$\rho = \frac{\text{cov}\{Y, X\}}{\sigma_Y \sigma_X}, \tag{11.32}$$

where $\text{cov}\{Y, X\}$ is the covariance between Y and X, σ_Y is the standard deviation of Y, and σ_X is the standard deviation of X. Then the correlation coefficient ρ is related to simple linear regression of Y on X through the equation (Larsen and Marx, 1986, p. 446)

$$E\{Y \mid X\} = \mu_Y + \frac{\rho \sigma_Y}{\sigma_X}(X - \mu_X). \tag{11.33}$$

That is, if the regression equation is written as $Y_i = \alpha + \beta X_i + \varepsilon_i$, then the regression coefficient β satisfies the equation $\beta = \rho \sigma_Y / \sigma_X$. The estimator b of β obtained by least squares satisfies a similar relationship,

$$b = r s_Y / s_X, \tag{11.34}$$

One further regression property simplifies our analysis. The regression of the residuals of Y on X_1 against the residuals of X_2 on X_1 is a simple linear regression, and therefore the correlation coefficient between these sets of residuals is the square root of the coefficient of determination (Kutner et al., 2005, p. 78). For this reason, we do not need to actually carry out the regression in order to compute the coefficient of determination. Instead, we can simply compute the correlation coefficient between the sets of residuals $Y \mid X_1$ and $X_2 \mid X_1$. This is called the *partial correlation coefficient* (Kutner et al., 2005, p. 270).

11.4.2 The Partial Mantel Test

Smouse et al. (1986) exploited the close relationship between correlation and regression statistics in their development of the partial Mantel test. They showed that the Mantel statistic of Equation 11.29 is related to a simple linear regression in the same way that r in Equation 11.2 is related to the regression model of Equation A.22. The most powerful aspect of their work, and the most widely used, is the partial Mantel test. Suppose one has three distance matrices as defined in Section 11.4.1, denoted A, B, and C. One can compute the residuals of the simple linear regression of the matrix A on C and of B on C, and then compute the Mantel statistic of these residuals. Since the Mantel statistic is a matrix correlation, this will provide information about the contribution of distance matrix B to the regression model for A given that matrix C is already included in the model. Smouse et al. (1986) refer to this as a *partial Mantel test*. Since the terms of the distance matrices are not independent, the Student t statistic cannot be employed to test significance. Therefore, Smouse et al. (1986) recommend testing for significance using a permutation test.

Although the partial Mantel test can be applied to any distance matrix, when dealing with spatial data one often carries out the test using a matrix C that represents actual geographic distances. Then a partial Mantel test of A and B on C tests whether or not there is spatial information not in the model that significantly contributes to explaining the variation in A. For example, in the relationship between $\ln(SPAD)$ and $\ln(LeafN)$ in Section 11.2.1, let A be the distance matrix of $\ln(LeafN)$, let B be the distance matrix for $\ln(SPAD)$, and let C be the Euclidean distance matrix between locations i and j. Then a significant value of the partial Mantel statistic indicates that there is spatial information besides that provided by $\ln(SPAD)$ that contributes to explaining the variation in $\ln(LeafN)$.

In our derivation, we will first consider only the "attribute distance" matrices A and B, and after we have worked through these, we will incorporate the geographic distance matrix C. For purposes of generality Smouse et al. (1986), consider matrices A and B that include upper as well as lower diagonal elements. In the case of symmetric distance matrices such as those in our example, this simply means including the upper as well as the lower triangular part. Let $\bar{A} = A/(n[n-1])$, the denominator being the total number of off-diagonal elements, and let $\bar{B}$ be defined similarly. Regarding the elements a_{ij} and b_{ij} as samples of random variables, the quantities $\bar{A}$ and $\bar{B}$ are the sample means. Smouse et al. (1986) define the *corrected sum of products SP(A, B)* as

$$SP(A,B) = Z - n(n-1)\bar{A}\bar{B},$$ (11.35)

where Z is the Mantel statistic defined in Equation 11.30, and they define the *corrected sums of squares SS(A)* and *SS(B)* as

$$SS(A) = \Sigma_{ij}(A - \bar{A})^2,$$
$$SS(B) = \Sigma_{ij}(B - \bar{B})^2.$$ (11.36)

Smouse et al. (1986) point out that in a permutation test, $SP(A, B)$ is affected by a reordering of the values of the a_{ij} and b_{ij} in exactly the same way as is Z (since $n(n-1)\bar{A}\bar{B}$ is not affected by permutations of a_{ij} and b_{ij}), and $SS(A)$ and $SS(B)$ are not affected by permutations. Therefore, the results of a permutation test involving a function of these statistics will be the same as a result of a permutation test on the Mantel statistic Z.

Now let

$$b_{AB} = SP(A,B)/SS(A),\tag{11.37}$$

and consider the equation

$$b_{ij} - \mu_b = \beta(a_{ij} - \mu_a) + \varepsilon_{ij},\tag{11.38}$$

where μ_a and μ_b are the population means of the distributions from which the a_{ij} and the b_{ij} are drawn, respectively, and the ε_{ij} are independent, identically distributed normal random variables with mean zero and variance σ^2 Taking an expectation over B conditional on A yields (recall that $E\{\varepsilon\} = 0$)

$$E\{B \mid A\} = \mu_b + \beta E\{A - \mu_a\}.\tag{11.39}$$

Comparing Equation 11.39 with Equation 11.33 indicates that they have the same form. Define the quantity r_{AB} by

$$r_{AB} = SP(A,B)/[SS(A)SS(B)]^{1/2}.\tag{11.40}$$

This has the same form as that of r in Equation 11.32. Based on Equation 11.37, we can write

$$b_{AB} = r_{AB}\sqrt{SS(B)}/\sqrt{SS(A)},\tag{11.41}$$

which has the same form as Equation 11.34. Therefore, computing a Mantel statistic is equivalent to computing a simple linear regression of B on A.

Now, consider the case where there is a third distance matrix C. By analogy with the discussion of the partial regression coefficients in Section 11.4.1, we can compute the partial correlation coefficient between the residuals of the regression of A on C and the residuals of the regression of B on C. Assume the elements of C are the geographic distances between the (x,y) coordinates of the locations of measurement of A and B, that is, $c_{ij} = \sqrt{(x_i - x_j)^2 + (y_i - y_j)^2}$. The residuals of the regression of A on C contain all the information about A not predicted by a linear regression of A on location, and the residuals of B on C contain similar information about B. If the correlation coefficient between these two sets of residuals is approximately zero, then this is an indication that B provides little information about A beyond that which is due to how A and B vary together with location. If the correlation coefficient is large, then this is an indication that B provides more information about A than simply that due to their shared location. Finally, as Smouse et al. (1986) point out, because the relationship between these sets of residuals is one of simple linear regression, in which the correlation coefficient equals the square root of the coefficient of determination, it is not necessary to make any regression-related assumptions about the relationship between predictors and response variables. One can equally consider the partial correlation coefficient as measuring the association between the three distance matrices.

When C is a matrix containing geographic distances, it is often recommended that inverse distances be used in order to put greater weight on small distances rather than large distances (Manly, 1997, p. 177, cf. the discussion on the spatial weights matrix in Section 3.2.2). We will consider the cases where C represents a Euclidean distance matrix and an inverse Euclidean distance matrix between the sampling points at which the data for matrices A

and *B* are measured. Smouse et al. (1986) discuss the more general case where the matrices *A*, *B*, and *C* are arbitrary distance matrices. In the context of the example of the relationship between ln(*LeafN*) content and ln(*SPAD*), we are addressing the issue of whether these two quantities really are associated or whether they are both responding to some environmental variable that is associated with the measurement locations. We use functions from the vegan package. First, we use the function `vegdist()` to compute the distance matrices *A* for ln(*LeafN*) and *B* for ln(*SPAD*).

```
> LeafN.dist <- vegdist(log(data.Set4.1$LeafN), method = "euclidean")
> SPAD.dist <- vegdist(log(data.Set4.1$SPAD), method = "euclidean")
```

Next, we use the function `dist()` to compute the matrix *C* representing the Euclidean distance between sample points. Since *C* is symmetric, we compute only the lower triangular part to reduce the amount of computational complexity. Here is the code to compute the distance matrix and the matrix of inverse distances.

```
> dist.mat <- with(data.Set4.1,
+    dist(cbind(Easting, Northing), upper = FALSE))
> invdist.mat <- 1 / dist.mat
```

Next, we use the vegan function `mantel.partial()` to compute the partial Mantel statistic.

```
> mantel.partial(SPAD.dist, LeafN.dist, dist.mat)
Partial Mantel statistic based on Pearson's product-moment correlation
Call:
mantel.partial(xdis = SPAD.dist, ydis = LeafN.dist, zdis = dist.mat)
Mantel statistic r: 0.4556
      Significance: 0.001
Upper quantiles of permutations (null model):
   90%    95%  97.5%     99%
0.0708 0.0920 0.1141 0.1404
Permutation: free
Number of permutations: 999
```

The results of using the distance matrix are shown; results using the inverse distance (not shown) are very similar. The value of the Mantel statistic is highly significant. This indicates that the linear relationship between ln(*SPAD*) and ln(*LeafN*) is not based simply on their common reaction to a location effect. As with the partial regression coefficient described in Section 9.2.2, the partial Mantel statistic can be very difficult to interpret (Legendre and Legendre, 1998, p. 559). The value and significance of the statistic depends on the relationship between *A* and *B*, between *A* and *C*, and between *B* and *C*. One of the original goals of this chapter was to determine whether the relation between SPAD meter reading and grain protein content is based strictly on geographic location. You are asked to make this determination in Exercise 11.9. A number of other three-way distance matrix tests besides the partial Mantel test have been proposed (Manly, 1997, p. 181). Oden and Sokal (1992) compared several of these and concluded that all of the methods had problems dealing with spatially autocorrelated data, but that the method of Smouse et al. (1986) is conservative, that is, it is unlikely to falsely reject the null hypothesis of zero correlation.

11.5 The Modifiable Areal Unit Problem and the Ecological Fallacy

11.5.1 The Modifiable Areal Unit Problem

Spatial analysis often involves aggregating data that have been collected at different spatial scales, as discussed in Section 6.4. In addition, data collected at one spatial scale are often used to make inferences about processes at a different spatial scale. Sometimes this mixing of scales results in incorrect or misleading results. The main purpose of this section is to describe two ways this can happen, each of which is so prevalent that it has a recognized name. The first is the *modifiable areal unit problem* (MAUP), which has to do with the effect of scale on conclusions about relationships between quantities. The second is the *ecological fallacy*, which is the use of aggregated data to draw conclusions about the behavior of individuals.

Data Set 4 includes measurements from a yield monitor that are separated by a distance of about 1 m in the direction of travel of the harvesting equipment and about 7 m in a direction perpendicular to the direction of travel. The data set also includes aerial images made up of pixels, each of which represents an area of about 9 m^2. We will use these data to demonstrate the MAUP. This term itself was first used by Openshaw and Taylor (1979), who note that it "is in reality two separate but interrelated problems" (Openshaw and Taylor, 1979, p. 128). The first problem, which they call the *scale problem* was recognized as early as 1934 by Gehlke and Biehl (1934). Cressie (1997) calls it the "change of support problem." The scale problem is, again quoting Openshaw and Taylor (1979, p. 128), "the variation in results that may be obtained when the same areal data are combined into sets of increasingly larger areal units of analysis." The most commonly observed manifestation of the scale problem involves the Pearson product moment correlation coefficient (Equation 11.2). Long (1996) demonstrates the scale problem with field crop data. He shows that the correlation coefficient between aggregated NDVI data and aggregated yield data depends on the scale over which the data are aggregated. We can demonstrate this same effect with data from Field 4.1. Figure 11.5a shows the boundary of the field, together with a rectangular area measuring 700 m north to south and 350 m east to west over which the NDVI values and the yield data are to be compared. The figure shows a set of fifty square cells over which the data are aggregated. It also shows a subset of the yield monitor data, which are housed in a `SpatialPointsDataFrame` called `data.Yield4.1`. The square cells are established by creating Thiessen polygons in the `SpatialPolygons` object `thsn.sp` (Section 3.5), after first using the function `expand.grid()` to create a set of points to use to define the polygons. The function `over()` of package `sp` is then used to aggregate the data by computing the mean over each polygon (see Section 6.4.2 for a discussion). We could also aggregate the data by using block kriging over each Thiessen polygon (Isaaks and Srivastava, 1989, p. 323), but simply computing the mean is simple, effective, and, works even when each block contains a relatively small number of data records. Here the object `img.rect` is a `SpatialPointsDataFrame` created from the May 1996 aerial image. The attribute data field `img.rect$NDVI` contains the NDVI computed according to Equation 7.1. Here is the computation of the mean over each polygon.

```
> data.ovrl <- over(thsn.sp, data.Yield4.1, fn = mean)
> image.ovrl <- over(thsn.sp, img.rect, fn = mean)
```

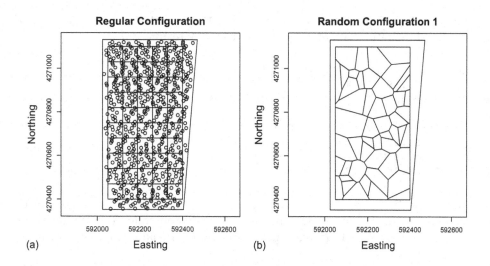

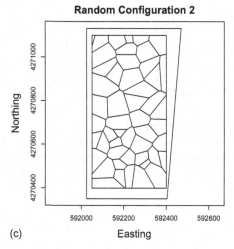

FIGURE 11.5
Illustrations of the modifiable areal unit problem: (a) aggregation of data into a regular array of square cells, illustrating the scale problem; (b) and (c) aggregation of data into two arrays of randomly arranged cells, illustrating the zoning problem. In Figure 11.5a, every 25th yield value is shown.

Finally, the correlation between the aggregated data is computed.

```
> print(cor(data.ovrl$Yield, image.ovrl$NDVI), digits = 3)
[1] 0.665
```

This code sequence is placed into a function `yield.vs.NDVI(cell.size)` to carry out the computations in sequence. The argument `cell.size` is expressed in terms of the width of the rectangle divided by the number of cells running from west to east. In addition to the correlation coefficient *r*, the function yield.vs.NDVI() returns the numerator and the denominator of the ratio that determines *r*. Here are the results.

```
> print(yield.vs.NDVI(350/20), digits = 2)
[1]    0.51   76.78 150.44
> print(yield.vs.NDVI(350/10), digits = 2)
[1]    0.59   79.71 135.95
> print(yield.vs.NDVI(350/5), digits = 2)
[1]    0.67   81.62 122.66
> print(yield.vs.NDVI(350/2), digits = 2)
[1]    0.86   85.96  99.71
```

The results indicate that as the cell size increases from 350/20 = 17.5 m to 350/2 = 175 m, the value of the correlation coefficient increases. This demonstrates a commonly observed phenomenon: as two data sets are aggregated over larger and larger areas, their correlation coefficient increases.

Wong (1995) gives a lucid explanation for the scale problem as it relates to the correlation coefficient. Referring to Equation 11.2, and examining the output just presented, we see that the covariance, which is the numerator of *r*, is relatively stable, increasing by about 11% over the range of areas. The denominator (the product of the standard deviations), however, decreases by about 50% over this range. Wong (1995) points out that it is often the case that as the aggregation area increases, a smoothing effect occurs in which the standard deviation decreases while the covariance remains relatively constant.

The second of Openshaw and Taylor's (1979) modifiable areal unit problems is called the *zoning problem*. This can also be demonstrated with our Field 4.1 data. Figure 11.5b and c show two Thiessen polygon sets generated for randomly placed locations in the data rectangle. The first polygon set was generated using the spatstat function runifpoint() following code as follows:

```
> E <- 592400
> W <- 592050
> S <- 4270400
> N <- 4271100
library(spatstat)
> ran1.ppp <- runifpoint(nrow(cell.ctrs), win = owin(c(W, E), c(S, N)))
> thsn.pp <- dirichlet(ran1.ppp)
> thsn.sp <- as(thsn.pp, "SpatialPolygons")
```

Carrying out the computation of the correlation coefficient, covariance, and product of the standard deviations yields the following:

```
> print(c(cor(data.ovrl$Yield, image.ovrl$NDVI),
+     cov(data.ovrl$Yield, image.ovrl$NDVI),
+     sd(data.ovrl$Yield) * sd(image.ovrl$NDVI)), digits = 2)
[[1]    0.76  97.82 128.42
```

Now we compute a second randomization.

```
> ran2.ppp <- runifpoint(nrow(cell.ctrs), win = owin(c(W, E), c(S, N)))
```

Repeating the computation yields the polygons of Figure 11.5c and the following coefficients:

```
> print(c(cor(data.ovrl$Yield, image.ovrl$NDVI),
+     cov(data.ovrl$Yield, image.ovrl$NDVI),
+     sd(data.ovrl$Yield) * sd(image.ovrl$NDVI)), digits = 2)
[1]    0.59  69.41 117.77
```

The correlation coefficients of the two configurations are 0.76 and 0.59. The covariance values are 97.82 and 69.41, and the denominator values are 128.42 and 117.77. Both the numerator and the denominator of Equation 11.2 are sensitive to the configuration of the polygons.

Now that the effects of the MAUP are apparent, the question arises as to what can be done about it. One possibility is to ignore it. Openshaw and Taylor (1981) consider three hypothetical justifications for this. The first is that it is insoluble, the second is that it is of trivial importance, and the third is that "to acknowledge its existence would cast doubts on the applicability of nearly all applications of quantitative techniques to zonal data" (Openshaw and Taylor, 1981, p. 67). They assert that the first two are incorrect, but that the third is true.

Wong (1995) discusses several approaches to a resolution of the MAUP. He divides these into three groups. The first approach, which is advocated by Moellering and Tobler (1972), involves computing the deviation from the mean of data aggregated at several scales. In the example of NDVI vs. yield in Field 4.1, these would be the regular Thiessen polygons. One then computes the sums of squares of these errors at each scale and assumes that the total of these sums of squares represents in some sense a total sum of squares. One then carries out significance tests for these sums of squares and considers the optimal scale to be the one which is most significant. Wong (1995) points out that the most serious problem with this method is that it does not rest on a very firm theoretical foundation. A second approach discussed by Wong (1995) is to attempt to construct a model, for example some form of parametric model that is immune to the effect of scale on the data relationships. The difficulty with this approach is that such a parametric model may not exist. The third approach is to construct a model that explicitly accounts for the spatial relationships among the errors. Once again, however, this approach may not be feasible.

Wong (1995) and Waller and Gotway (2004, p. 107) offer advice for dealing with the MAUP that we will summarize and interpret using the NDVI vs. yield example. Somewhat obviously, if appropriate scale invariant techniques are available to carry out the analysis, then these should be used. If such techniques are not available, then one must determine the objectives of the analysis and use statistics and a spatial scale that best fit these objectives. If the goal is to infer relationships at a fine scale, one should report the results of data analysis using fine scales. In our example, we would report the correlation coefficient between NDVI and yield as 0.51, the value at the smallest scale considered. This also has the virtue of being the most conservative. If the goal is to use NDVI to predict yield at the sub-field scale, it is appropriate to use a linear regression model over polygons of a size needed to generate a useful model. One should report the effects on the results of aggregating the data. In our example, this would involve reporting the values of the correlation coefficients at each of the scales. In dealing with the zoning effect, it would be reasonable, given a data set like that of Field 4.1, to analyze a large number of random arrangements of a fixed number of sample Thiessen polygon aggregation points similar to those in Figure 11.5b and c and report the mean and standard deviation of

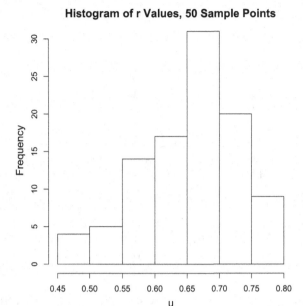

FIGURE 11.6
Histogram of values of Pearson's *r* between *Yield* and *NDVI* for 50 random arrangements of 50 polygons of the type shown in Figure 11.5b and c.

the correlation coefficient. Figure 11.6 shows a histogram of 100 random arrangements of 50 sample points for Field 4.1. The mean value of the correlation coefficient is 0.66, and the standard deviation is 0.075.

11.5.2 The Ecological Fallacy

The *ecological fallacy* is the fallacy that one may infer relationships at the individual level based on aggregated data collected about many individuals. Initial recognition of the ecological fallacy is attributed to Robinson (1950), although he did not use that specific term in the paper. Instead, he used the term *ecological correlation* to refer to a statistic describing a relationship that is defined for individuals but computed using aggregated data. An example from that paper can be used to illustrate the concept. Robinson (1950) presents statistics in the form of a contingency table in which the X variable is the number of foreign born vs. native born individuals in a sample of 97,272 US residents in 1930, and the Y variable is then number of literate vs. illiterate individuals in the sample. He shows that at the individual level, being illiterate is positively (albeit weakly) correlated with being foreign born. He then presents the same data, aggregated by region over nine regions in the US, and shows that when percent foreign born is related to percent illiterate, the correlation is negative. In the interest of full disclosure, it should also be pointed out that the relationship as presented in Figure 3 of Robinson (1950) is also highly nonlinear, so that the use of a correlation coefficient is probably not appropriate.

In any case, there are many instances where the process of ecological inference, which Haining (2003, p. 138) describes as the use of grouped data to infer individual relationships, breaks down. We have already seen an example of this phenomenon in Section 7.4 in the exploration the relationship between rice yield and silt content in Data Set 3 (Figure 7.20). This could be considered an example of the MAUP, but the distinction (possibly only a technical one) is that here there are identifiable spatial entities, the individual fields, whose data are aggregated. The ecological correlation between yield and silt content, that is, the correlation coefficient computed over the entire data set, has the value −0.61. Displaying the range of values of the individual field correlations provides the opportunity to demonstrate the use of the function by().

```
> Data <- with(data.Set3, data.frame(Yield, Silt))
> Field <- data.Set3$Field
> print(sort(by(Data, Field, function(x) cor(x)[1,2])[1:16]),
+ digits = 2)
Field
      8       7      15       2       1      14       9       4      13
-0.709  -0.630  -0.250  -0.246  -0.238  -0.183  -0.142   0.081   0.106
      5      12       6       3      16      11      10
 0.113   0.201   0.208   0.331   0.440   0.684   0.722
>
```

There is no evident pattern to the relationships between yield and silt content at the field scale.

The third line of the code sequence provides a good opportunity to review some R programming concepts. The first argument of the function by() is Data, a data frame whose columns are the quantities we wish to compare. The second argument, Field, is the index defining the groups of data. The function cor(), applied to these data, produces a two by two correlation matrix whose diagonal elements are unity and whose off-diagonal elements are the correlation coefficients. The third argument, function(x) cor(x)[1,2], is a function whose argument is a two by two matrix and whose return value is one of the off-diagonal elements. Thus, the function by() returns an array of correlation coefficients of the data by field. The function sort() sorts this array from most negative to most positive.

The term *ecological bias* is also used to describe the difference between a statistical result based on data observed at one geographic scale and that based on the same data at a different scale (Greenland and Morgenstern, 1989). One reason that data may display an ecological bias is that individuals, or data measured at a small scale, may encompass less variability than the same data measured at a larger scale, and the variability at the larger scale may be necessary for it to manifest the relationship in question. For instance, Hill et al. (1992) describe a strong relationship in rice between grain moisture content at harvest and percent fractured grains. This observation is based on aggregated data over many fields. At the individual field scale, however, such variation is not necessarily seen (Marchesi, 2009). Thompson and Mutters (2006) have found that the relationship between grain moisture content at harvest and percent fractured grains in rice depends in a complex way on meteorological conditions during the period leading up to harvest. The difference in observations between the regional and field scale is probably due to the fact that the regional observations encompass a variety of meteorological conditions, whereas these conditions are more or less fixed for the harvest of an individual field. This is related to a statistical phenomenon called Simpson's paradox (Cressie, 1997).

11.6 Further Reading

Almost all statistics texts discuss the Pearson correlation coefficient, but the discussion by Snedecor and Cochran (1989) is particularly comprehensive and easy to understand. Haining (2003) provides an excellent discussion of the method of Clifford et al. (1989). Alternative methods have also been proposed (Dutilleul, 1993).

The books by Agresti (1996, 2002) are the standard references for categorical data analysis. Freeman (1987) also provides a good basic introduction to the subject. Thompson (2009) provides an excellent discussion of the use of R in the analysis of problems involving categorical data.

Cressie (1997) discusses the MAUP in geostatistical terms and provides an excellent discussion of its relationship to other statistical phenomena. Gotway and Young (2002, 2007) give an extensive discussion with many references of the MAUP and the ecological fallacy, and of approaches to its resolution. Openshaw (1984) provides an extensive discussion of the ecological fallacy and provides many references. Haining (2003) provides an excellent discussion of the analysis of aggregated data. For further discussion of the ecological fallacy, see Duncan et al. (1960), Alker (1969), and Langbein and Lichtman (1978).

For more on the Theory of the Stork, see Höfer et al. (2004).

Exercises

11.1   Use the Clifford correction and the parametric bootstrap to adjust the correlation coefficient between *LeafN* and *GrainProt* in Data Set 4, Field 1.

11.2   For the data of Data Set 2, using the definitions of Section 11.3.1, repeat the analysis carried out in that section, but instead of using $\{MAT \leq 15\}$ as the X variable, use $\{JuMax \leq 30\}$.

11.3   An alternative question relating to temperature to that posed in Section 11.3.1 is whether especially cool winters might impede blue oak establishment due to reduced seedling survival. As a preliminary means of investigating this issue, consider for the Coast Range only those sites that are on opposite sides of the median in their values of *JaMean*, the mean January temperature and *JuMean*, the mean July temperature. Let X_1 represent the property of a site having a value of *JaMean* below the median of all values of *JaMean* and a value of *JuMean* above the median of all values of *JuMean* (cool winters and hot summers), and let X_2 represent the property of a site having a value of *JaMean* above the median of all values of *JaMean* and a value of *JuMean* below the median of all values of *JuMean* (warm winters and warm summers). Again, let $Y = Y_1$ represent the presence of blue oak at the site and $Y = Y_2$ represent the absence. For the data of Data Set 2, using the definitions of Section 11.3.1, compute the contingency tables of WW-WS vs. CW-HS for the Coast Range and carry out a significance test both with and without accounting for spatial autocorrelation.

11.4   Construct scatterplot matrices of *Precip*, *JaMean*, *JuMean*, and *Elevation* for the Sierra Nevada and Coast Range subsets of Data Set 2. Do you see any striking differences?

11.5 Create a binomial variable in Data Set 2 that is 1 when *Precip* > median{*Precip*} and 0 otherwise. Create a contingency table for this variable and CW-HS. What does this tell you about the results of Section 11.3.1?

11.6 With the Sierra Nevada and Coast Range subsets of Data Set 2, construct scatterplots of *Precip* vs. *Elevation*, *MAT* vs. *Elevation*, and *Precip* vs. *MAT*, using different symbols for $QUDO = 1$ and $QUDO = 0$. The plots are easiest to visualize if you only plot every fourth point.

11.7 One might wonder why in Section 11.3.2 the factor 0.001 is subtracted from the computed value obs.chisq in computing the p value based on repeated evaluation of the observed chi-square statistics. Display the computed value of these statistics in a permutation test and use the function unique() to find all the possible values that the array u of permutation values takes on. Can you use this information to answer the question?

11.8 Repeat the block restricted randomization for the permutation test for the subsets of Data Set 1 determined in Exercise 7.4 to be important. What can you conclude about the effect of spatial autocorrelation on the conclusions drawn from the analysis?

11.9 Compute the partial Mantel statistic for the relationship between log SPAD meter reading vs. log Grain N. What can you say in an ecological context about the comparison between these results and those obtained for the relationship between log SPAD meter reading and log Leaf N?

12

The Mixed Model

12.1 Introduction

The exploratory analysis of Data Set 3, the Uruguayan rice fields, in Chapters 7 and 9 indicated that irrigation effectiveness plays an important role in distinguishing high-yielding from low-yielding fields. Irrigation effectiveness is represented by the ordinal variable *Irrig*, which is the expert agronomist's rating of irrigation effectiveness at each measured site. Can the association between *Irrig* and *Yield* be scaled down? That is, is irrigation effectiveness consistently associated with yield at the single-field scale, or, alternatively, is this an example of the ecological fallacy discussed in Section 11.5.2, in which an observation that holds at the scale of many fields does not "scale down" to the individual field? Figure 12.1 shows a trellis plot, made with the `lattice` package (Sarkar, 2008) function `xyplot()`, of yield versus irrigation effectiveness for each field. Yield generally seems to increase with irrigation effectiveness, although there is a lot of variability, and some fields actually seem to show a *decreasing* yield with increasing effectiveness. The individual plots generally look as though they would be appropriate for linear regression. The variable *Irrig*, however, is an ordinal scale variable, and as such the operations of multiplication and addition, which are used to compute the regression coefficients, are not meaningful when applied to its values. In principle, therefore, it is not appropriate to interpret a regression as indicating that yield increases by a certain percent for each unit increase in irrigation effectiveness. For the present application, however, our interest is not in predicting a value of yield based on irrigation effectiveness but rather on computing the regression coefficient in order to test the null hypothesis that this coefficient is equal to zero. We will consider this an acceptable breach of the rules.

We can use the function `lmList()` of the `nlme` package (Pinhiero et al., 2011) to examine the coefficients of the fits of a simple linear regression to each of the fields. The data frame `data.Set3` is loaded using the code in Appendix B.3.

```
> library(nlme)
> data.lis <- lmList(Yield ~ Irrig | Field, data = data.Set3)
> print(coef(data.lis), digits = 3)
  (Intercept)    Irrig
1        5680   642.50
2       11456       NA
3        4645   976.43
4        3664  1277.83
5        2042  1404.97
6        5919    -3.93
7        4361   562.78
```

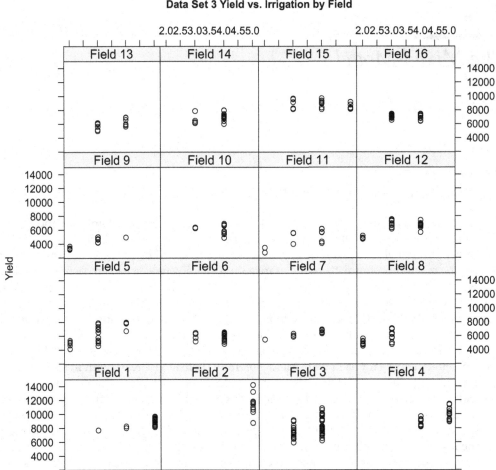

FIGURE 12.1
Yield versus irrigation effectiveness for each of the fields in Data Set 3.

8	2681	1138.03
9	1491	975.14
10	7414	-373.62
11	1828	861.34
12	4108	687.39
13	3568	710.00
14	5648	339.05
15	9846	-246.03
16	7042	6.51

This code is an example of the use of *grouped data*. Grouped data, which we have seen before in working with trellis graphics, are data in which there exists a *grouping factor*, that is, an index that divides the data into meaningful groups (Pinhiero and Bates, 2000, p. 99).

In the current example, the grouping factor is the field identification number. The vertical line in the formula `Yield ~ Irrig | Field` indicates that the variable *Field* is the grouping factor, and the function `lmList()` calculates a separate linear regression for each group, indexed by this factor. Grouped data are used extensively in the mixed-model analyses that are the subject of this chapter.

The regression on Field 2 above returns a value of `NA` because all of the data records associated with this field have a value of *Irrig* equal to 5 (Figure 12.1). The field is uniformly well irrigated (and has the highest mean yield). The fact that all of the data records in Field 2 have a value of *Irrig* equal to 5 can be verified by applying the functions `tapply()` and the argument `FUN` set to unique to apply the function `unique()`.

```
> tapply(data.Set3$Irrig, data.Set3$Field, unique)
$`1`
[1] 5 3 4

$`2`
[1] 5

$`3`
[1] 3 4
      *    *    *    DELETED    *    *    *
```

This field is therefore extracted from the data set.

```
> data.Set3a <- data.Set3[-which(data.Set3$Field == 2),]
```

If in fact the regression relationship between yield and irrigation effectiveness were consistent across fields, then one could use the analysis of covariance (ANCOVA) (Kutner et al., 2005, p. 314; Searle, 1971, p. 348; Sokal and Rohlf, 1981, p. 509) to construct and analyze a model of the relationship between field, irrigation effectiveness, and yield. The appropriate ANCOVA model is

$$Y_{ij} = \mu + \alpha_i + \beta X_{ij} + \varepsilon_{ij}, \ j = 1,...,n_i, \ i = 1,...,m, \tag{12.1}$$

where Y_{ij} is value of the response variable (*Yield*) at the *jth* location of the *ith* group (*Field*), μ is the population grand mean, α_i is the effect of Field i (initially assumed fixed), the quantity βX_{ij} is the effect explained by the variate X_{ij} (*Irrig*), and the $\varepsilon_{ij} \sim N(0,\sigma^2)$ are independent and identically distributed random variables for all i and j. As a first step, we carry out a traditional fixed effects analysis of covariance. This is accomplished by leaving `Irrig` as a `numeric` object rather than converting it to a factor in the arguments to the function `aov()`.

```
> Yld.aov <- aov(Yield ~ factor(Field)+ Irrig , data = data.Set3a)
> summary(Yld.aov)
               Df   Sum Sq   Mean Sq  F value     Pr(>F)
factor(Field)  14 630893068 45063791  74.899  < 2.2e-16 ***
Irrig           1  42761520 42761520  71.072  1.51e-15 ***
Residuals     297 178693447   601661
---
Signif. codes:  0 '***' 0.001 '**' 0.01 '*' 0.05 '.' 0.1 ' ' 1
```

The result, not surprisingly, is a highly significant effect of both field and irrigation effectiveness.

There are several very serious problems with this analysis. The first is that by assuming the regression coefficient β in Equation 12.1 to be fixed, the analysis implicitly assumes an affirmative answer to one of the questions being asked, namely, whether the response of individual fields to irrigation effectiveness can be considered consistent across fields. As such, the result does not really provide assistance in answering the other question, whether there is a significant field scale yield response to irrigation effectiveness. These are not the only problems, however. The field data, because they are spatially autocorrelated, may violate the assumptions of the model, which include independence of errors. In addition, the assumption that the α_i are fixed means that the results of the analysis should be interpreted only for these specific 16 fields. We are not, however, interested in just these fields, but rather in using these fields as a representative of a larger population of rice fields in the region.

These defects in the model can be addressed by analyzing the model of Equation 12.1 as a *mixed effects model*, which is generally called simply a *mixed model* in the modern literature. The traditional interpretation of a linear model such as the ANCOVA model of Equation 12.1 treats the effects (field effect α_i and irrigation effect β) as *fixed effects*, that is, as nonrandom, fixed quantities. The results of the analysis are then valid only for these values of the predictors, which means in the present example that the results are only valid for these specific 16 fields. An alternative is to consider some or all of the effects as *random effects*, that is, as random variables whose values are considered to be drawn from a hypothetical larger pool of effects. In our case, this is interpreted as meaning that the fields are randomly drawn from a population of fields, that the irrigation effectiveness of each field is itself a random variable drawn from a population of irrigation effectiveness values, or both. A mixed effects model is one in which some of the effects are random and some are fixed.

In a mixed-model analysis, both the objectives and the procedure are different from those of a fixed effects study. Suppose, for example, that in the ANCOVA model (Equation 12.1), the field effect α_i is random. The procedure is different in that instead of selecting certain specific fields, one hypothetically draws the fields at random from the entire population of fields. These fields represent the variability of the total population. The study forgoes the opportunity to meaningfully compare the effect on yield response of the specific fields, and instead focuses on the question of variability of the population of fields.

Of course, as is common with field studies, reality and theory do not match very well. The fields in Data set 3 were not chosen at random; they were selected haphazardly from a set of volunteers (cf. Section 5.1 and Cochran, 1977, p. 16). On the other hand, there is no specific interest in these fields, or in these particular farmers. The researchers really do intend these fields to be representative of rice fields throughout the region. The region itself is not precisely specified, but is assumed to consist of similarly managed rice fields throughout east-central Uruguay. The question that must be answered in deciding whether an effect is fixed or random is whether the effect level can reasonably be assumed to represent a probability distribution (Henderson, 1982; Littell et al., 2002, p. 92; Searle et al., 1992, p. 16). It is important to emphasize that it is not the *factor* but rather the *effect* that is assumed random. For example, suppose a measurement is made in successive years. The years themselves are certainly not randomly chosen, but if there is no trend in the data, and there is no interest in the specific years in which the measurements were made, then it may be that the year effects can be taken to be random (although they may be autocorrelated). This permits us to relax slightly the assumption that the fields are randomly drawn; we must instead assume that, however the fields are drawn, their effects are a random sample of the population of field effects.

Mixed model analyses are based on a grouping factor. The grouping factor "groups" the data in the sense that data values within the same group are presumed to be more related than data records from different groups. In our example, the grouping factor is the field. This will be treated as a random effect. Although it is not always evident from the computer output, mixed model analysis is based on the accurate estimation of the variance of the grouping factor. It is necessary to have more than a few factor levels to get good variance estimates and, hence, results from the mixed-model analysis. For this reason, it is generally not recommended that a mixed-model analysis be carried out if there are fewer than about six to eight levels of the grouping factor (Littell et al., 2002, p. 168). If there are insufficient levels, the analysis should proceed using a fixed-effects model, with no pretense of applying the results to a wider range of cases.

The basic mixed model is discussed in Section 12.2, and this model is then applied to Data Set 3 in Section 12.3. The application of mixed models to spatially autocorrelated data is discussed in Section 12.4. The remaining two sections are devoted to a pair of other methods that are similar to the mixed model. Section 12.5 discusses generalized least squares, which is applied to data for which there is no grouping variable. Section 12.6 discusses the quasibinomial model, which is an iterative method used to deal with autocorrelated data.

12.2 Basic Properties of the Mixed Model

Consider again the ANCOVA model of Equation 12.1. The grand mean μ is a fixed effect, and we initially continue with the assumption that the irrigation effectiveness β is a fixed effect as well. We replace the fixed effect assumption on the field effects α_i with the assumption that the α_i are random variables with mean zero and variance σ_α^2. In addition, the α_i and the ε_{ij} are assumed to be mutually independent. The quantities σ^2 and σ_α^2 are called the *variance components* of the model (Searle, 1971, p. 376). As described in the previous section, the random variable α whose levels are the α_i may be considered as a grouping factor of the data.

One crucial difference between the fixed-effects model and the mixed-effects model concerns the covariance relationship between the response variables. In the case of the fixed-effects model, the only random variables are the independent errors, and therefore the response variables Y_{ij} are also independent with the same variance σ^2 as the error terms ε_{ij}. In the case of the mixed model, this is no longer true. The quantities α_i and ε_{ij} are random variables and both contribute to the variance and covariance of Y_{ij}. We would expect the Y_{ij} values from the same group (i.e., same Field i in our example) to be more related, and this is indeed the case. It turns out that the covariance between the Y_{ij} is (Stanish and Taylor, 1983)

$$\text{cov}\{Y_{ij}, Y_{i'j}\} = \frac{\sigma_a^2}{\sigma_a^2 + \sigma^2}, \ i = i'$$

$$\text{cov}\{Y_{ij}, Y_{i'j'}\} = 0, \ i \neq i'.$$

(12.2)

To carry out a hypothesis test on the coefficient β of the model (Equation 12.1), we use the general linear test, which is described in Appendix A.3 and discussed in Section 8.3.

To recapitulate, if, for example, one wishes to test the null hypothesis $H_0 : \beta = 0$ against the alternative hypothesis $H_a : \beta \neq 0$, then one regards model (Equation 12.1) as the *full* model, denoted with the subscript F, and develops a model called the *restricted* model, denoted with the subscript R, in which the null hypothesis is satisfied. In our case, the restricted model is

$$Y_{ij} = \mu + \alpha_i + \varepsilon_{ijk} \tag{12.3}$$

The restricted model of Equation 12.3 is nested in the full model of Equation 12.1 because the model of Equation 12.3 can be derived from that of Equation 12.1 by restricting one or more of the parameters of the model (Equation 12.1) to certain values. In this particular example, the value of β is restricted to $\beta = 0$. If the null hypothesis is true, then a statistic involving the ratio of the sums of squares (Equation A.44 in Appendix A.3) has an F distribution, and thus one can compute a p value based on this distribution. However, the statistic G of Equation A.44 has an F distribution only if the Y_{ij} are independent (Kutner et al., 2005, p. 699). A consequence of the nonzero covariance structure of Equation 12.2 is, therefore, that the general linear test on the sums of squares can no longer be used to test H_0. The maximum likelihood method, however, remains valid when the response variables are not independent. This method is described in Appendix A.5.

We continue the discussion considering the case in which the full model is given by Equation 12.1, $Y_{ij} = \mu + \alpha_i + \beta X_{ij} + \varepsilon_{ij}$ and the restricted model is given by restricting $\beta = 0$ to obtain Equation 12.3, $Y_{ij} = \mu + \alpha_i + \varepsilon_{ij}$. According to the discussion in Appendix A.5, if there are k_F parameters in the full model and k_R in the restricted model, with $k_F > k_R$, and if L_F is the maximum likelihood estimate of the full model and L_r that of the restricted model, then asymptotically as n approaches infinity the quantity

$$G = -2\log(L_F / L_R) = 2(l_F - l_R) \tag{12.4}$$

has a chi-square distribution with $(k_F - k_R)$ degrees of freedom (Kutner et al., 2005, p. 580; Theil, 1971, p. 396). In our case, $l_F = l(\mu, \alpha, \beta, \sigma^2 \mid Y)$, $l_R = l(\mu, \alpha, \sigma^2 \mid Y)$ and $k_F - k_R = 1$.

The likelihood ratio test is implemented by computing the maxima of the log likelihood statistics l_F and l_R and subtracting them. Finding these maxima is a nonlinear optimization problem. There are two methods used to carry out this optimization, the *restricted maximum likelihood* (REML) method and the ordinary maximum likelihood (ML) method (Appendix A.5). The REML provides better variance estimates than the ML. However, it cannot be used in a likelihood ratio test of a fixed-effect term. Therefore, we cannot use the REML to test the null hypothesis $\beta = 0$ and must instead use the full maximum likelihood method.

There is one further advantage to using the maximum likelihood method, or the restricted maximum likelihood method if it were available, in our ANCOVA problem of testing the effect of irrigation effectiveness on yield. An examination of the data set reveals that different numbers of measurements were made in different fields. This is expressed in Equation 12.1, which describes the model, in the condition $j = 1,...,n_i$, $i = 1,...,m$, that is, in the fact that the value of n_i, the number of data records in Field i, depends on i. Such an experimental design is said to be *unbalanced*. The maximum likelihood method can be adapted to deal with unbalanced data in simple ANCOVA models (Henderson, 1982). McCulloch et al. (2008, p. 95) provide the mathematics for the adaptation for unbalanced data of the model of Equation 12.1 in which the term α_i is random.

12.3 Application to Data Set 3

We now return to the problem of determining the field scale relationship between irrigation effectiveness and yield. To summarize, the initial full model is given by Equation 12.1, $Y_{ij} = \mu + \alpha_i + \beta X_{ij} + \varepsilon_{ij}$, $j = 1, ..., n_i$, $i = 1, ..., m$. The ultimate goal is to determine whether irrigation effectiveness has a consistent, positive effect on yield at the field scale. We will carry out the analysis in two stages. First, we will let the field effect α be a random variable, and then we will include a random component in the irrigation effect. With β as a fixed effect, we test the null hypothesis $H_0 : \beta = 0$ against the alternative $H_a : \beta \neq 0$. Since β is a fixed effect, we must use the ML rather than REML method. The restricted model, obtained by setting $\beta = 0$ in Equation 12.1 is given by Equation 12.3.

We will test this and other hypotheses involving mixed models using functions from the nlme package. The use of the functions in this package is extensively described by Pinhiero and Bates (2000). The package has the major (for us) advantage that it includes the capacity to model spatially autocorrelated data. The fundamental mixed linear modeling function in this package is lme(). The primary arguments of this function are (1) the *fixed* part of the model, (2) the data source, (3) the *random* part of the model, and (4) the optional method specification. The function anova(), when applied to an object created by lme(), carries out a likelihood ratio test as described in the previous section.

The first call to lme() generates the restricted model.

```
> Yld.lmeR1 <- lme(Yield ~ 1, data = data.Set3a,
+       random = ~ 1 | Field, method = "ML")
```

The function call and the model are interpreted together as follows:

$$Y_{ij} = \mu \, \| + \alpha_i + \varepsilon_{ij} \tag{12.5}$$

```
lme(Yield ~ 1,...,‖ random = ~ 1 | Field)
```

The vertical double line in the model (which is included for clarification and is not part of the R code) equation separates the fixed part on the left from the random part on the right. The corresponding symbol in the line of R code represents the same separation of fixed and random effects. The only fixed component is the mean μ, which is represented by the number 1. As usual, the random error term ε_{ij} is implicitly included. The random part is the field effect α_i. This is represented by the statement random = ~ 1 | Field. The quantity on the right of the vertical bar is the grouping factor, which in this case is the field. The quantity on the left of the bar is called the *primary covariate* and represents the factor (if any) that is grouped by the grouping factor. In this case, the only fixed effect is the grand mean, and this is indicated by the 1 to the left of the bar, so there is no primary covariate. The second call to lme() generates the full model.

```
> Yld.lmeF1 <- lme(Yield ~ Irrig, data = data.Set3a,
+       random = ~ 1 | Field, method = "ML")
```

The function call and the model are interpreted together as follows:

$$Y_{ij} = \mu + \beta X_{ij} \, \| + \alpha_i + \varepsilon_{ij} \tag{12.6}$$

```
lme(Yield ~ Irrig,‖ random = ~ 1 | Field)
```

Since there is another fixed effect besides the grand mean, the mean itself can be made implicit. The variable representing irrigation effectiveness is incorporated into the fixed effects, and the random effect is unchanged. Both of these models could be examined using the function summary(). We will put this off for now. The likelihood ratio test is carried out using the function anova().

```
> anova(Yld.lmeR1, Yld.lmeF1)
          Model df      AIC      BIC    logLik   Test  L.Ratio p-value
Yld.lmeR1     1  3 5186.099 5197.337 -2590.049
Yld.lmeF1     2  4 5117.571 5132.556 -2554.785 1 vs 2 70.52764  <.0001
```

The effect of irrigation effectiveness is highly significant. There is a complication involved in interpreting the results of this hypothesis test of a fixed effect that will be discussed below.

The next step in the analysis is to determine whether the effect of irrigation effectiveness on yield can be considered as constant over fields. We treat irrigation effectiveness as the sum of a fixed effect β that represents the mean irrigation effectiveness effect and a random effect γ that represents the individual field's irrigation effectiveness effect (Searle, 1971, p. 355). The former full model of Equation 12.1 becomes the new restricted model and the new full model, along with its representation in lme(), is

$$Y_{ij} = \mu + \beta X_{ij} \, \| + \alpha_i + \gamma_i X_{ij} + \varepsilon_{ij} \tag{12.7}$$

```
lme(Yield ~ Irrig, ‖ random = ~ Irrig | Field)
```

where γ_{ij} represents the random component of the total irrigation effect. We can update the model in R using the function update(). We can use the restricted maximum likelihood method in this case because our null hypothesis involves a random effect, so we drop the argument method = "ML".

```
> Yld.lmeR2 <- lme(Yield ~ Irrig, data = data.Set3a,
+      random = ~ 1 | Field)
> Yld.lmeF2 <- update(Yld.lmeR2, random = ~ Irrig | Field)
> anova(Yld.lmeR2, Yld.lmeF2)
          Model df      AIC      BIC    logLik   Test L.Ratio p-value
Yld.lmeR2     1  4 5093.738 5108.697 -2542.869
Yld.lmeF2     2  6 5077.556 5099.994 -2532.778 1 vs 2 20.1826  <.0001
```

The results of the call to anova() indicate that we cannot drop the random component of the irrigation effectiveness from the model, that is, that the effect of irrigation cannot be considered constant across fields. Instead, it is the sum of a fixed effect, which is the mean irrigation effect over all the fields, and a random effect, which is the effect of that individual field.

Returning to the test carried out of the null hypothesis that the fixed effect β equals zero, the function anova() applied to the comparison of the models (Equation 12.5) and (Equation 12.6) implements a likelihood ratio test and gives a p value less than 0.0001, which we would like to interpret as indicating the fixed effect of irrigation effectiveness is significant in the model. As was mentioned earlier, there is a complication that must be addressed.

Recall that the log likelihood ratio asymptotically follows a χ_k^2 distribution, where k is the difference in the number of degrees of freedom between the full and restricted model. The function logLik(), when applied to a model, returns the log likelihood and the number of degrees of freedom, so this can be used to determine k. Pinhiero and Bates (2000, p. 89) indicate that the likelihood ratio test applied to the fixed effects of a model may be "anticonservative," that is, may have an inflated Type I error rate. Since we have rejected the null hypothesis in this case at a p value less than 0.0001, we should not be too concerned about making a Type I error. However, we can double check the result using the function simulate.lme() from the nlme package. This function generates a Monte Carlo simulation of the model to produce an empirical distribution of p values. These can be plotted, and if the curve of nominal p values falls below the 45° line, then this indicates "anticonservative" behavior. Here is the code to carry out this test.

```
> logLik(Yld.lmeR1)
'log Lik.' -2590.049 (df=3)
> logLik(Yld.lmeF1)
'log Lik.' -2554.785 (df=4)
> sim.lme <- simulate.lme(Yld.lmeR1,Yld.lmeF1,
+    nsim = 1000, seed = 123)> plot(sim.lme, df = 1)
```

Figure 12.2 shows a plot of the values for the models Yld.lmeR1 and Yld.lmeF1. For $df = 2$, the p values lie along the 45° line, which indicates that there should be little problem with accepting the failure to reject the null hypothesis in this case. For more details, see Pinhiero and Bates (2000, p. 89).

In the test of the model of Equation 12.7, we are not actually testing the hypothesis $\gamma_i = 0$. The γ_i are random variables with mean zero, and the null hypothesis is $H_0 : \sigma_\gamma^2 = 0$ against the alternative $H_a : \sigma_\gamma^2 > 0$, where σ_γ^2 is the variance of γ. Of course, if γ has mean zero and variance zero, then it is identically zero, so in effect the test is the same. There is, however, a complication with the p values of this test. Pinhiero and Bates (2000, p. 86), citing the results of Stram and Lee (1994), point out that the value of σ_γ^2 that satisfies the null hypothesis is "on the boundary of the parameter space," (i.e., the boundary of the range of possible values of σ_γ^2). This may make the likelihood ratio test conservative, that is, it may cause an inflated Type II error rate, failing to reject the null hypothesis when it is false. Pinhiero and Bates (2000, p. 86) suggest simulating the test at one degree of freedom less than the nominal difference between the restricted and full as well as computing a range of p values at a mixture of these two degrees of freedom values. Here is the code to carry out this simulation.

```
> logLik(Yld.lmeR2)
'log Lik.' -2542.869 (df=4)
> logLik(Yld.lmeF2)
'log Lik.' -2532.778 (df=6)
> sim.lme <- simulate.lme(Yld.lmeR2,Yld.lmeF2, nsim = 1000, seed = 123)
> plot(sim.lme, df = c(1,2))
```

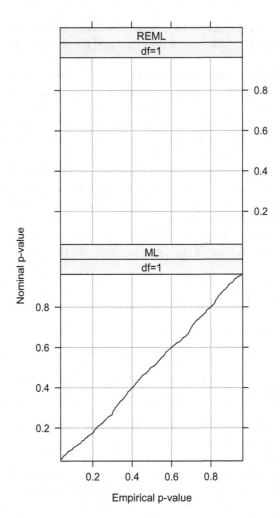

FIGURE 12.2
Plot of the nominal p value of the test of the hypothesis $\beta = 0$ against the empirical value generated by a Monte Carlo simulation. The approximately straight line of equal values for maximum likelihood indicates that the nominal p value can be accepted.

The result, shown in Figure 12.3, indicates that the nominal p values tend to be higher than the empirical ones; in other words, the test tends to be conservative. Since the test rejects the null hypothesis ($p < 0.0001$), we can conclude that the rejection of the null hypothesis is appropriate and that the effect of irrigation effectiveness is not constant across fields.

12.4 Incorporating Spatial Autocorrelation

Until now, the mixed model analysis has included the assumption that the residuals ε_{ij} are independent, normally distributed random variables with fixed variance σ^2. We now incorporate the effects of spatial autocorrelation into the model. One of the reasons for

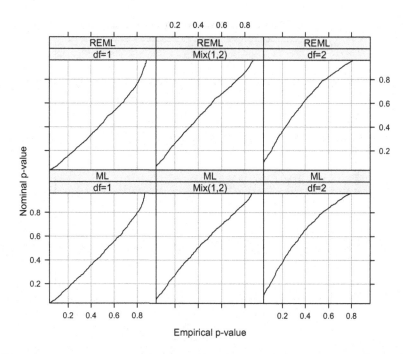

FIGURE 12.3
Plots of nominal versus empirical p values for the REML and ML methods for the test of $H_0: \sigma_\gamma^2 = 0$ against the alternative $H_a: \sigma_g^2 > 0$.

using mixed model analysis for the testing of hypotheses involving spatially autocorrelated data is that the maximum likelihood method does not require the assumption of independent errors in a model such as that of Equation 12.7. The function lme() has the ability to incorporate spatial autocorrelation by calculating a variogram and incorporating a variogram model (or other correlation models) into the correlation structure. The available variogram models include exponential, Gaussian, linear, quadratic, and spherical.

The nlme package contains the function Variogram() (capital V), which computes an empirical variogram (Section 4.6) for the residuals of the mixed model. This variogram can be plotted and inspected and, if appropriate, a variogram model can then be used to define the error correlation structure of the data. This correlation structure is then incorporated into the analysis. The function Variogram() takes two primary arguments. The first argument is an lme object that is used to obtain the regression residuals. Pinhiero and Bates (2000, p. 245) provide examples of the use of this function. Taking up from where we left off in Section 12.2, here is how we initially apply it to compute and plot the variogram for the residuals of the full model Yld.lmeF2 for our problem.

```
> var.lmeF <- Variogram(Yld.lmeF2, form = ~ Easting + Northing)
```

The variogram in Figure 12.4 was produced by a call to the function plot(). A word of caution is advisable here about this function. It is the basic workhorse plotting function that produces many of the graphs displayed throughout this book. These graphs are not, however, all produced by the same function, and not even all by the same type of function. This is an example of the use of polymorphism, described in Sections 2.5 and 2.6.2. In brief, the specific form of the function depends on the arguments. In the traditional graphics version of plot(), the first argument is a vector of x coordinates. In other applications of

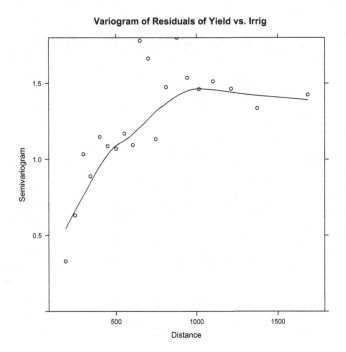

FIGURE 12.4
Variogram of the residuals of the full model of Equation 12.9 incorporating spatial autocorrelation.

`plot()`, the first argument is an R object that determines the form of the function `plot()` that will be used. In the creation of Figure 12.3, the first coordinate is a `simulate.lme` object. In the creation of Figure 12.4, the first argument is a `Variogram` object. Why does this matter? Because it crucially affects how we exercise special control over the graphics. The plot in Figure 12.4 looks like an ordinary traditional graphics plot. In its unmodified form, the plotted material is blue, but for best reproduction we would like it to be black. Following the normal procedure with the traditional graphics version of `plot()`, we would enter the following.

```
> plot(var.lmeF, col = "black") # This doesn't work
```

This doesn't work (try it!). Typing `?plot.Variogram` brings up the Help screen, which indicates that, although it creates what looks like a traditional graphics plot, `plot. Variogram()` is actually a trellis graphics function based on the `lattice` package. Therefore, in order to obtain a black-and-white graph, we must use a trellis graphics statement such as the following.

```
> trellis.device(color = FALSE)
```

Figure 12.4 shows that the variogram of the residuals of the model of Equation 12.7 has the shape associated with a spherical correlation structure of Equation 4.22 (Isaaks and Srivastava, p. 374; Pinhiero and Bates, 2000, p. 233), and so this correlation structure will be used in the model.

We are now almost ready to carry out a likelihood ratio test for the model with spatial autocorrelation against the model without it. The correlation structure is introduced into the residuals through the use of an argument in `lme()` of the form `correlation = corSpher(form = ~ X + Y, nugget = TRUE)`. This describes the correlation structure of the data. The symbols X and Y represent the data fields that identify the x and y coordinates of the data record. In our example, these are `Easting` and `Northing`. Let's try plugging these in.

```
> # This produces an error
> Yld.lmeF3 <- update(Yld.lmeF2, correlation
    + = corSpher(form = ~ Easting + Northing, nugget = TRUE))
Error in getCovariate.corSpatial(object, data = data) :
  Cannot have zero distances in "corSpatial"
```

The error occurs because some of the data records contain data that are measured in successive years at the same location, and therefore the distance between some of the data records is zero. We can get around this by using the function `jitter()` to add a very small random quantity to every x and y coordinates, not enough to affect the result but enough to avoid the zero distance problem. As usual, we set a random number seed. It turns out that our usual seed value of 123 produces a particularly extreme result (try it), so we use a different one.

```
> set.seed(456)
> data.Set3a$EAST2 <- jitter(data.Set3a$Easting)
> data.Set3a$NORTH2 <- jitter(data.Set3a$Northing)
```

Now we revise the model to include the jittered coordinates.

```
> # Redo the model with the new data.Set3a
> Yld.lmeF2a <- update(Yld.lmeF2, data = data.Set3a)
> var.lmeF2a <- Variogram(Yld.lmeF2a, form = ~ EAST2 + NORTH2)
> # Check that the variogram hasn't changed visibly
> plot(var.lmeF2a)
```

Now we are ready to carry out the hypothesis test involving spatial autocorrelation. Formally, the null hypothesis is $H_0: cor\{\varepsilon_i, \varepsilon_j\} = 0$, $i \neq j$, and the alternative is that the correlation is not zero for some i and j. We must be sure that the restricted model is nested in the full model.

The full model is given by

$$Y_{ij} = \mu + \beta X_{ij} + \alpha_i + \gamma_i X_{ij} + \varepsilon_{ij},$$
$$(\alpha_i, \gamma_i)' \sim N(0, \Psi), \ \varepsilon_i \sim N(0, \sigma^2 \Lambda_i),$$

(12.8)

where Λ_i is a variance-covariance matrix whose off-diagonal terms are not necessarily zero. This is analogous to the general form of the mixed model given by Pinhiero and Bates (2000, p. 202), which is known as the *Laird-Ware* form. The expression $(\alpha_i, \gamma_i)' \sim N(0, \Psi)$ indicates that the random effects may be correlated. This correlation was also generally present in the analysis of the mixed model with independent errors, as indicated in Equation 12.2.

The expression $\varepsilon_i \sim N(0, \sigma^2 \Lambda_i)$ indicates that the magnitude of the variance of the error terms is represented by the parameter σ^2, and that error terms from the same grouping factor (same *i*) may be correlated. One way (the way we will use) to represent the correlation is by using a correlogram (Section 4.2.6). We can write (Pinhiero and Bates, 2000, p. 205)

$$\Lambda_i = V_i R_i V_i, \tag{12.9}$$

where V_i is a diagonal matrix and R_i is a correlation matrix. Since we are modeling this correlation structure by fitting a spherical variogram model to the residuals, lme() uses the relationship $\rho(h) = (\gamma(h) - C(0))/C(0)$ obtained by combining Equations 4.25 and 4.26. It models the correlation structure in Equation 12.9 by incorporating the range of the variogram into the matrix R_i and the square root of the sill into the V_i.

The restricted model is given by Equation 12.7, which we repeat here, including explicitly the conditions on the random effects and errors:

$$Y_{ij} = \mu + \beta X_{ij} + \alpha_i + \gamma_i X_{ij} + \varepsilon_{ij},$$

$$(\alpha_i, \gamma_i)' \sim N(0, \Psi), \ \varepsilon_i \sim N(0, \sigma^2 I). \tag{12.10}$$

As just mentioned, the expression $\varepsilon_i \sim N(0, \sigma^2 I)$ indicates that the error terms are independent and with constant variance σ^2. Therefore, the restricted model is indeed a special case of the full model in which $\text{cov}\{\varepsilon_i, \varepsilon_j\} = 0$ for $i \neq j$, and so the restricted model is nested in the full model. Therefore, we are justified in using the likelihood ratio test.

Here is an implementation in R.

```
> Yld.lmeF3 <- update(Yld.lmeF2a, correlation
+ = corSpher(form = ~ EAST2 + NORTH2, nugget = TRUE))
> anova(Yld.lmeF2, Yld.lmeF3)
          Model df     AIC      BIC    logLik   Test L.Ratio p-value
Yld.lmeF2     1  6 5077.556 5099.994 -2532.778
Yld.lmeF3     2  8 5078.361 5108.279 -2531.180 1 vs 2 3.195141  0.2024
```

Based on the results of the test, we can conclude that the simpler model without autocorrelation is adequate. We conclude that the most appropriate model is that of Equation 12.9, represented by the R object Yld.lmeF2, in which both the parameter representing the field and that representing irrigation effectiveness are random effects, and the errors are spatially uncorrelated.

Turning to the interpretation of the results, the function summary() as usual provides us with a lot of information.

```
> summary(Yld.lmeF2)
Linear mixed-effects model fit by REML
 Data: data.Set3a
       AIC      BIC    logLik
  5077.556 5099.994 -2532.778
Random effects:
 Formula: ~Irrig | Field
 Structure: General positive-definite, Log-Cholesky parametrization
             StdDev    Corr
(Intercept) 2092.8975 (Intr)
Irrig        451.3202 -0.83
```

```
Residual      730.4511        Fixed effects: Yield ~ Irrig
              Value Std.Error  DF   t-value p-value
(Intercept) 4410.58 615.9446 297 7.160676       0
Irrig        660.28 141.5584 297 4.664363       0
Correlation:
      (Intr)
Irrig -0.867
Standardized Within-Group Residuals:
        Min           Q1         Med          Q3          Max
-3.21519988 -0.60943706 0.03328884  0.52766754 3.11924207

Number of Observations: 313
Number of Groups: 15
```

The estimated variance components are given as standard deviations in the information about random effects. They are $\sigma = 730.4511$, $\sigma_\gamma = 451.3202$, $\sigma_\alpha = 2092.8975$. The value of β, the mean effect of irrigation effectiveness, is given in the fixed effects as 660.28. Since irrigation effectiveness is an ordinal variable, we cannot assign too much meaning to this beyond the statement that on average, irrigation effectiveness has a significantly positive effect on yield. The correlation between the intercept term a and the irrigation effect term γ is −0.867. The intercept term α represents the random effect of the field on yield and can be interpreted as the effect of everything about the field but irrigation effectiveness. The random effect γ represents the variation with field of the rate of change of yield with increasing irrigation effectiveness. The negative correlation indicates that the association between irrigation effectiveness is reduced as the intercept of the plot field's yield is increased. It is important to note, however, that the intercept is *not* the mean yield of the field.

Because irrigation effectiveness is an ordinal variable, we cannot with any sort of theoretical justification use these results to predict yield. In other cases, however, the concomitant variable may be interval or ratio scale, in which case such a prediction would be justified. To see what we could do under those circumstances, we will pretend for the moment that irrigation effectiveness is of an appropriate scale to justify this sort of prediction. To get an idea of the regression relationships of each field we can use the nlme function augPred() (for "augmented prediction"). This is most effectively used with the function plot() as follows.

```
> plot(augPred(Yld.lmeF2, as.formula("~Irrig"))) # Fig. 12.5
```

Figure 12.5 shows the results of this statement, which generates a regression line for each field. The second argument in the function augPred() specifies the primary covariate of the grouped data as defined above. These plots demonstrate that because the regression predictor variables X_i data are not *centered*, that is, because the model in Equation 12.9 is written as it is and not as $Y_{ij} = \mu + \alpha_i + (\beta + \gamma_i)(X_{ij} - \bar{X}_i) + \varepsilon_{ij}$, the intercept of the regression line is not the mean yield for the field but rather is related both to the slope and the mean yield. The figure indicates that the negative covariance between the intercept and the slope of the regression line is due partly (or maybe mostly) to the fact that regression lines with a steep slope tend to have a lower intercept. Inspection of the figure indicates that there is not much, if any, relationship between regression slope and mean yield.

Finally, we can compare the values of the coefficients generated by the individual models using lmList() with those generated using the model Yld.lmeF2. They are displayed side by side for ease of comparison.

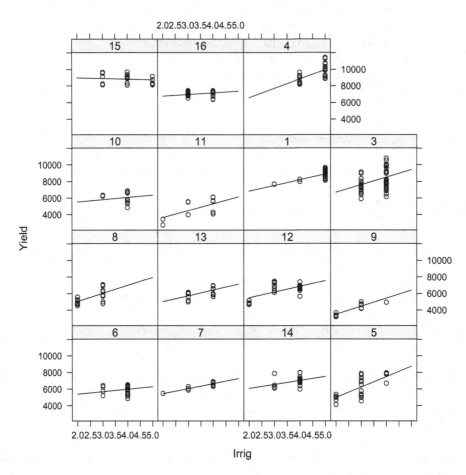

FIGURE 12.5
Regression of yield versus irrigation effectiveness by field, as generated by the model in Equation 12.9.

	> coef(data.lis)			> coef(Yld.lmeF2)	
	(Intercept)	Irrig		(Intercept)	Irrig
1	5679.500	642.500000	1	5467.691	683.22006
3	4644.995	976.433155	3	4881.098	908.86211
4	3664.417	1277.833333	4	4321.939	1127.59501
5	2041.734	1404.966805	5	2487.202	1252.04669
6	5918.529	-3.926471	6	4795.256	295.92892
7	4360.978	562.777778	7	4224.195	606.28036
8	2680.694	1138.027778	8	3122.792	964.31254
9	1490.786	975.142857	9	1564.360	967.92928
10	7414.375	-373.625000	10	4994.641	272.53620
11	1828.181	861.337349	11	1966.939	837.58022
12	4107.708	687.388740	12	4118.344	685.14777
13	3567.714	710.000000	13	3609.932	703.03172
14	5647.850	339.050000	14	5105.962	480.40071
15	9845.544	-246.025316	15	9121.889	-77.50652
16	7041.615	6.512821	16	6376.453	196.83426

Statistically, these two outputs of the function `coef()` represent two completely different things. The left column, which was generated by the function `lmList()`, represents a set of best linear unbiased estimates (BLUE) of the fixed effect β for a set of 15 independent regressions of yield on irrigation effectiveness. The column on the right represents a set of best linear unbiased predictors (BLUP) of the random effect γ_i added to the estimate of the fixed effect β for a single regression of yield on irrigation effectiveness. Each coefficient in the left column is generated independently with no influence from the data of other fields, while the coefficients in the right column are based on the data from all the fields. Generally the two sets of coefficients tend to be similar, but the extreme cases in the left-hand column are not as extreme in the right-hand column.

12.5 Generalized Least Squares

We now consider applying the mixed model concept to the data of Field 1 of Data Set 4. In Section 9.3, we constructed a multiple linear regression model for the field. We would like to be able to apply the mixed-model method to regression models such as this one to try to account for spatial autocorrelation of the errors, but there are no variables in these regression models that can reasonably be considered to be random effects. In other words, there is no grouping variable. In this case, Equation 12.8 becomes, in matrix notation,

$$Y = \mu + \beta X + \varepsilon,$$

$$\varepsilon \sim N(0, \sigma^2 \Lambda).$$

(12.11)

Such a model, with no grouping variable, is referred to as a *generalized least squares* model (Pinhiero and Bates, 2000, p. 204), and can be solved using the same techniques as those described for the solution of Equation 12.8. Before discussing this, it might be worthwhile to review some of the other models with similar names. The term *generalized* has appeared several times, so let's make sure we have them all straight. The first use was the generalized linear model (GLM) in Section 8.4. This has a form such as $g(\pi_i) = \beta_0 + \beta_1 X_i$ and is used, for example, when the response variable is binomial. The next use was in Section 9.2 with the generalized additive model (GAM), which has the form $g(\pi_i) = \beta_0 + f(X_i)$ where $f(X)$ is a smooth function such as a spline. This is used to provide a better fit to the data. The generalized least squares model of Equation 12.11 is used as one form of solution of a linear regression model where the error structure can be modeled using a variogram.

The nlme package contains a function `gls()` that can be applied to generalized least squares regression in the same way that `lm()` is applied to ordinary linear regression. We will apply this function to the analysis of the data from Field 4.1 following an example of similar analysis carried out by Pinhiero and Bates (2000, p. 260) on data from wheat variety trials. We developed five candidate models, but we will continue to focus on `model.5.` as an example. Our objective is to see whether the incorporation of spatial autocorrelation in the errors affects the result.

```
> model.5lm <- lm(Yield ~ Clay + SoilP + I(Clay*SoilP) + Weeds,
+ data = data.Set4.1)
> summary(model.5lm)
              Estimate Std. Error t value Pr(>|t|)
(Intercept)   16083.290 1523.702 10.555 < 2e-16 ***
```

```
Clay                     -324.716 37.852 -8.578 5.42e-13 ***
SoilP                    -890.167 205.394 -4.334 4.17e-05 ***
I(Clay * SoilP)   24.051 5.277 4.558 1.81e-05 ***
Weeds                    -390.776 71.283 -5.482 4.64e-07 ***
```

This forms the null model for the analysis in which the errors may be correlated. We can re-express it in terms of gls() as follows.

```
> library(nlme)
> model.5gls1 <- gls(Yield ~ Clay + SoilP + I(Clay*SoilP) +
+     Weeds, data = data.Set4.1)
```

The summary() function, shown below, indicates that the coefficients are all exactly the same. We now develop the variogram model that can be used to characterize the spatial structure of the residuals. To exclude lag groups with very few members, the variogram is calculated for lag values less than or equal 300 m.

```
> plot(Variogram(model.5gls1, form = ~ Easting + Northing,
+     maxDist = 300), xlim = c(0,300), # Fig. 12.6
+     main = "Variogram of Residuals, model.5, Field 4.1")
```

The shape of the variogram indicates that the best model is probably spherical, although the variogram looks like it is practically pure nugget (Figure 12.6).

```
> model.5gls2 <- update(model.5gls1,
+     corr = corSpher(form = ~ Easting + Northing, nugget = TRUE))
```

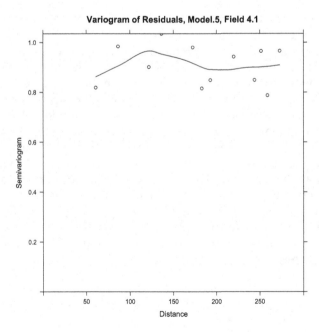

FIGURE 12.6
Variogram of the residuals of the generalized least squares model for the northern portion of Field 4.1.

Since the models are nested, we can compare them using the likelihood ratio test.

```
> anova(model.5gls1, model.5gls2)
            Model df      AIC       BIC     logLik    Test  L.Ratio p-value
model.5gls1     1  6 1344.236 1358.603 -666.1179
model.5gls2     2  8 1336.940 1356.096 -660.4700 1 vs 2 11.29591  0.0035
```

There is a significant difference, indicating that the effect of spatial autocorrelation of the errors is sufficient to make the generalized least square model preferable to the ordinary least squares model. The effect of including spatial autocorrelation on the Akaike information criterion (AIC) is relatively minor, as is shown by these excerpts from the summary() function.

```
> summary(model.5gls1)
       AIC      BIC     logLik
  1344.236 1358.603 -666.1179
Coefficients:
                      Value Std.Error    t-value p-value
(Intercept)       16083.290 1523.7018 10.555405         0
Clay               -324.716   37.8523 -8.578485   ,     0
SoilP              -890.167  205.3942 -4.333945         0
I(Clay * SoilP)      24.051    5.2771  4.557588         0
Weeds              -390.776   71.2825 -5.482070         0
> summary(model.gls2)
       AIC      BIC     logLik
  1344.236 1358.603 -666.1179
Coefficients:
                      Value Std.Error    t-value p-value
(Intercept)       11815.244  55202.73  0.214034  0.8311
Clay               -201.002     51.46 -3.905613  0.0002
SoilP              -890.766    228.12 -3.904832  0.0002
I(Clay * SoilP)      21.850      5.79  3.777035  0.0003
Weeds              -255.535     77.38 -3.302258  0.0014
```

All of the coefficients except *SoilP* are reduced in magnitude in the correlated errors model. Autocorrelation of the residuals does not bias the coefficients by itself, but there are several possible things that may be going on. In Chapter 13, we will pursue this issue much more deeply.

12.6 Spatial Logistic Regression

12.6.1 Upscaling Data Set 2 in the Coast Range

This section describes a method called the *quasibinomial model* that allows the incorporation of spatial autocorrelation into regression models for binomial variables. The quasibinomial model is related to the generalized linear model discussed in Section 8.4. In Exercise 8.10, the blue oak presence-absence data of the Coast Range subset of Data Set 2 was analyzed using a multiple logistic regression model. One of the candidate models for blue oak presence versus absence in the Coast Range (as presented in the solution given in Section 8.3) includes the explanatory variables *TempR, Permeab, Precip, GS32, PE,* and *SolRad6*. Mention was made in Section 8.3 of the complexity of earlier logistic regression models (Evett, 1994; Vayssieres et al., 2000) for this data set. It is possible that this complexity is in part due to the spatial scale of the data. Data Set 2 has over 4000 individual records. Relative to the extent of the data set, the support (Section 6.4.1) of the raw data in Data Set 2 is much smaller than that of the other data sets.

The method that we will present in Section 12.6.2 for incorporating spatial autocorrelation effects into regression models for binomial variables involves data that have been aggregated over a set of subregions. For this reason, we examine in greater detail the effect of data aggregation on analysis of the blue oak presence-absence data in the Coast Range subset of Data Set 2. There are no natural boundaries that we can use for aggregation, so the first order of business is to create a set of regions over which to aggregate. We would like to have regions that are geographically close and, since elevation is a major predictor of blue oak presence, we would like the regions to also be close in elevation. Therefore, we will search for clusters of similar values of longitude, latitude, and elevation. We will do this using *k*-means cluster analysis (Jain and Dubes, 1984). In this algorithm, *k* points in the data space are initially selected as cluster *seeds*. Clusters are formed by assigning all other points in the data space to the seed having the closest value. The means of each cluster are then selected as the new set of *k* seeds, and a set of clusters are formed around this set of seeds. This process is repeated iteratively. In theory, the process may be iterated to convergence, but in practice it is stopped after a predefined maximum set of iterations. In the present case, the data space is comprised of the triples of longitude, latitude, and elevation.

Clustering can be carried out in R using the function kmeans(). In this function the initial set of seeds, if it is not specified in the argument of the function, is selected at random. In our implementation of the method, we will search for *k* = 50 clusters. The sf object data.Set2C contains the Coast Range subset of Data Set 2, created in Section 7.3. We create a new version of this data set, Set2.kmeans, in which the quantities Longitude, Latitude, and Elevation are centered and scaled, and then Elevation is multiplied by 0.5 in order to put more emphasis on spatial contiguity.

```
> Set2.kmeans <- with(data.Set2C, data.frame(scale(Longitude),
+    scale(Latitude), 0.5 * scale(Elevation)))
> cluster.k <- 50
> set.seed(123)
> cl <- kmeans(Set2.kmeans, cluster.k, iter.max = 100)
> data.Set2C$clusID <- cl$cluster
```

Figure 12.7 shows the resulting clusters at the southern end of the Coast Range (a plot of the full Coast Range would be impossibly busy). Cluster membership of each point is represented by the point's symbol, and the cluster membership symbols are superimposed on the contours of a digital elevation model created by interpolating the elevation data. The heavy line is the California coast, and the map illustrates two of the problems associated with map overlay and the issues of scale discussed in Section 6.4. The first problem is that the contour map is an extrapolation in the sense of Bierkens et al. (2000, cf. Figure 6.10). That is, it extends the elevation values beyond the geographic region of the data, and therefore its extrapolated elevation values include regions where the predicted elevation of the ocean is over 200 m. The second problem is that the map of California on which the coastline is based was digitized at the extent of the entire United States, and therefore its accuracy does not match that of the site location data. This results in some tree locations

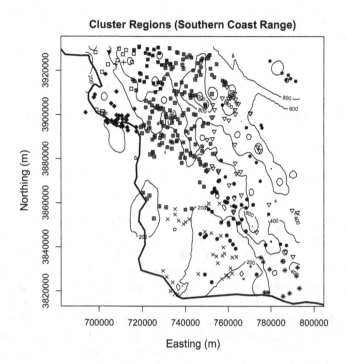

FIGURE 12.7
Cluster regions at the southern end of the Coast Range, shown superimposed over a digital elevation model of the region.

apparently being in the ocean. Since the map is used for purposes of visualization only, these problems will not affect our analysis.

A data frame called data.Set2Cpts is created to hold the attribute values on with the regression model will be based. We use the suffix "pts" in this case to emphasize that these are pointwise data, as opposed to clusterwise data set data.Set2Cclus that will be created next.

```
> data.Set2Cpts <- with(data.Set2C, data.frame(MAT, TempR, GS32,
+     Precip, PE, ET, Texture, AWCAvg, Permeab, SolRad6, SolRad12,
+     SolRad, QUDO, clusID))
> data.Set2Cpts$PM100 <- as.numeric(data.Set2C$PM100 > 0)
```

There are similar statements to the last one for each class of parent material.

Now we will create a data frame data.Set2Cclus to hold the aggregated data in the spatial clusters. We use the function tapply() to compute the averages of each data field over each cluster.

```
> data.Set2C.mat <- matrix(nrow = length(unique(data.Set2Cpts$clusID)),
+     ncol = ncol(data.Set2Cpts))
> for (i in 1:ncol(data.Set2Cpts)){
+    data.Set2C.mat[,i] <- tapply(data.Set2Cpts@data[,i],
+       data.Set2Cpts$clusID, mean)
+ }
> data.Set2Cclus <- data.frame(data.Set2C.mat)
> names(data.Set2Cclus) <- names(data.Set2Cpts)
```

Next, we convert the resulting matrix into a data frame and assign the appropriate name to each data field.

```
> data.Set2Cclus <- data.frame(data.Set2C.mat)
> names(data.Set2Cclus) <- names(data.Set2Cpts)
```

To summarize, let's look at the number of data records in each data frame.

```
> nrow(data.Set2Cpts)
[1] 1852
> nrow(data.Set2Cclus)
[1] 50
```

Your answer may vary slightly depending on how you constructed the Coast Range data set in Section 7.3. The data frame `data.Set2Cpts` contains the pointwise data, and the data frame `data.Set2Cclus` contains the data averaged over each cluster.

Now we are ready to construct the quasibinomial model. The response variable QUDO in `data.Set2Cclus` represents the fraction of sites in the cluster that have a blue oak present. The model for this type of data is also sometimes called a *logistic-binomial* model (Gelman and Hill, 2007, p. 116). Recall that in the case of the binomial model the response variable Y_i took on the value 1 with probability π_i and zero with probability $1 - \pi_i$, and that the explanatory variables were related to the response variable through the link function (Equation 8.28)

$$g(\pi_i) = \ln\left(\frac{\pi_i}{1 - \pi_i}\right) = \beta_0 + \beta_1 X_{i1} + \beta_2 X_{i2} + \ldots + \beta_{p-1} X_{i,p-1}. \tag{12.12}$$

The quasibinomial model uses the same link function, but now the response variable Y_i takes on any value between zero and n_i and is distributed as *binomial*(n_i, π_i). In our implementation, we have divided the number of sites with blue oaks by the total number of sites in the region, so that $n_i = 1$ for all i, and Y_i is allowed to take on fractional values. This avoids the problem that different clusters have different numbers of sites (i.e., it makes Y an *intensive* variable in the sense of Section 7.2).

One of the key differences between the binomial and quasibinomial models is that the latter frequently will have a variance different from the variance of a binomial distribution. This is an example of the *overdispersion* phenomenon discussed in Section 8.4.4. The function `glm()` can be used to fit quasibinomial models by specifying the family argument as `family = quasibinomial`. This introduces a variable dispersion parameter.

Rather than using a likelihood function, the function `glm()` when fitting a quasibinomial model uses a *quasi-likelihood* function (McCulloch et al., 2008, p. 152). A major problem with this approach is that the AIC cannot be computed for quasi-likelihood functions. Burnham and Anderson (1998, p. 52) suggest a quasi-AIC (QAIC) statistic, but this is not implemented in `glm()`, and we will instead break the rules a little and use hypothesis testing to compare models.

First, we compute the coefficients of the candidate model for both the pointwise data and the aggregated data.

```
> model.glmC6clus <- glm(QUDO ~ TempR + Permeab + Precip +
+    GS32 + PE + SolRad6, data = data.Set2Cclus,
+    family = quasibinomial)
```

```
> print(coef(model.glmC6clus), digits = 2)
(Intercept)      TempR     Permeab       Precip         GS32
   -8.9e+00    1.6e-01    -1.6e+00      4.8e-05     -1.6e-02
         PE     SolRad6
    8.9e-03    2.6e-01
> model.glmC6pts <- glm(QUDO ~ TempR + Permeab + Precip +
+     GS32 + PE + SolRad6, data = data.Set2Cpts,
+     family = binomial)
> print(coef(model.glmC6pts), digits = 2)
(Intercept)      TempR     Permeab       Precip         GS32
    -7.7864     0.2966     -0.7391      -0.0015      -0.0213
         PE     SolRad6
     0.0093     0.1783
```

The coefficient of *Precip* changed sign. Perhaps more importantly from a model selection perspective (since we have agreed to use hypothesis testing for model selection purposes) is the effect on significance. Unlike the pointwise case, there is no significant difference (at the $\alpha = 0.05$ level) between aggregated data models with three and six explanatory variables.

```
> model.glmC3clus <- update(model.glmC6clus,
+     formula = as.formula("QUDO ~ Permeab +
+     GS32 + PE"))
> model.glmC3pts <- update(model.glmC6pts,
+     formula = as.formula("QUDO ~ Permeab +
+     GS32 + PE"))
> anova(model.glmC3clus, model.glmC6clus, test = "Chisq")
Analysis of Deviance Table
Model 1: QUDO ~ Permeab + GS32 + PE
Model 2: QUDO ~ TempR + Permeab + Precip + GS32 + PE + SolRad6
  Resid. Df Resid. Dev Df Deviance P(>|Chi|)
1        46     6.8125
2        43     5.9361  3  0.87647    0.06527 .
> anova(model.glmC3pts, model.glmC6pts, test = "Chisq")
Model 1: QUDO ~ Permeab + GS32 + PE
Model 2: QUDO ~ TempR + Permeab + Precip + GS32 + PE + SolRad6
  Resid. Df Resid. Dev Df Deviance P(>|Chi|)
1      1852     1660.5
2      1849     1554.7  3   105.76 < 2.2e-16 ***
```

Further reduction in the number of variables does produce a significant difference (the code is not shown). Thus, it appears that we can make do with an aggregated data model containing only *Permeab*, *GS32*, and *PE*. Alternative models are certainly possible as well.

The next question is how the models compare in their predictive ability. The large dark circles in Figure 12.8a represent a comparison between the predictions of the cluster model, with three explanatory variables, and the means over the clusters of the pointwise model, with six explanatory variables. The difference is not large. The smaller circles show the values of each predicted probability of the pointwise model, indicating the considerable variability within each cluster in the predicted value of blue oak presence probability. Figure 12.8b shows observed versus predicted fractions of the sites in each cluster with

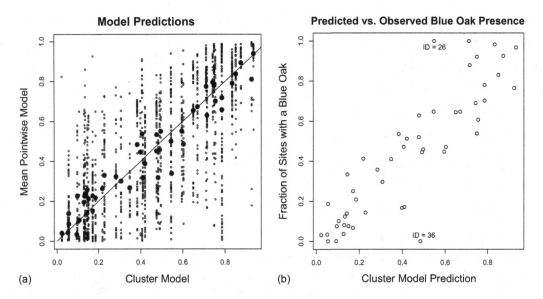

FIGURE 12.8
(a) Plot of the predicted values of the pointwise data generalized linear model of *QUDO* versus *Elevation* for the Coast Range subset of Data Set 2 versus the clustered data model. The large filled circles are based on the mean over each cluster region of the pointwise data model, and the small circles are based on the pointwise data; (b) Observed versus predicted values of the portion of sites with a blue oak for the clustered data model. Two discordant values are identified by cluster ID number.

a blue oak. By and large the fit is good. There is a pair of seeming outliers, whose cluster numbers are identified. These may be of interest if some corresponding biophysical anomaly can be identified.

In summary, the explanatory variables selected by the pointwise and aggregated data models are quite different, although their predictions are, as far as comparisons are possible, similar. The questions addressed by the two models are, however, completely different. The pointwise model predicts the probability that an individual site will contain a blue oak, and the aggregated model predicts the fraction of sites in each cluster that will contain a blue oak. Each model provides information about the ecology of blue oaks, but they cannot simply be interchanged with each other. We now move on to the incorporation of spatial autocorrelation in the cluster model.

12.6.2 The Incorporation of Spatial Autocorrelation

For the quasibinomial model used in Section 12.6.1, if Y_i represents the portion of sites in the region having a blue oak, and π_i is the probability that a site contains a blue oak, so that $E\{Y_i\} = \mu_i = \pi_i$, and if there are $p-1$ explanatory variables X_i, then by combining Equations 8.19 and 8.28, we can express the model as

$$g(E\{Y \mid X_i\}) = g(\pi_i) = \ln\left(\frac{\pi_i}{1-\pi_i}\right) = \beta_0 + \beta_1 X_{1i} + \ldots + \beta_{p-1} X_{p-1,i}, \tag{12.13}$$

where $\pi_i = \Pr\{Y_i = 1\}$ and $g(\pi_i)$ is the logit link function. The response variable Y_i is binomially distributed and normalized, so that if the measurements X_i are independent, then its variance is

$$Var\{Y_i\} = \pi_i(1-\pi_i). \tag{12.14}$$

In other words, unlike the case with the normal distribution, the variance of the binomial distribution depends on the mean, and hence indirectly on the explanatory variables. When Y_i approaches its limits of zero or one, the variance declines to zero. When the errors are spatially autocorrelated we expect that $Var\{Y_i\}$ will be larger than predicted by Equation 12.14. This equation is sometimes modified to

$$Var\{Y_i\} = a\pi_i(1-\pi_i), \tag{12.15}$$

where a is called the *scale* or *overdispersion* parameter.

In Section 12.5, we discussed the incorporation of spatial autocorrelation into linear regression equation via generalized least squares. In this context, in Equation 12.11 above, the equation $\Lambda_i = V_i R_i V_i$ describes the modification of the variance equation for the linear regression model to consider spatial autocorrelation. The analogous equation for the variance of the GLM, taking into account spatial autocorrelation, is

$$Var\{Y\} = \Lambda = aV(\mu)RV(\mu), \tag{12.16}$$

where a is again a variance scaling term, R is, as in Equation 12.11, a correlation matrix that depends on the variogram model, μ is the vector of means, and V is a diagonal matrix whose elements are $v_{ii} = \sqrt{\mu_i(1-\mu_i)}$. In the mixed model of Equation 12.10 there are two indices: the index i refers to the *group*, in which measurements are indexed by j, $j = 1,...,n_i$. In the quasibinomial model of Equation 12.13, there is only one index: i indexes the measurements of the regions, which are not grouped. In this sense, the analysis of this section is roughly analogous to that of the generalized least square model discussed in Section 12.5.

The objects data.Set2Cpts and data.Set2Cclus are data frames. We will be creating variograms, so we will also need a SpatialPointsDataFrame that contains the same data. Because we are working with distances, the data will be projected in UTM Zone 10 (all of the sites are in this zone), so we will need the Easting and Northing. We therefore use spTransform() to transform the data.

```
> coordinates(data.Set2Cpts) <- c("Longitude", "Latitude")
> proj4string(data.Set2Cpts) <- CRS("+proj=longlat +datum=WGS84")
> data.Set2Cutm <- spTransform(data.Set2Cpts,
+    dCRS("+proj=utm +zone=10 +ellps=WGS84"))
> data.Set2Cpts$Easting <- coordinates(data.Set2Cutm)[,1]
> data.Set2Cpts$Northing <- coordinates(data.Set2Cutm)[,2]
```

The mean values of the Easting and Northing of the points in each cluster region will be used to identify the coordinates of that region in data.Set2Cclus.

```
> data.Set2C.mat <- matrix(nrow = length(unique(data.Set2Cpts$clusID)),
+    ncol = ncol(data.Set2Cpts))
> for (i in 1:ncol(data.Set2Cpts)){
+    data.Set2C.mat[,i] <- tapply(data.Set2Cpts@data[,i],
+       data.Set2Cpts$clusID, mean)
+ }
> data.Set2Cclus <- data.frame(data.Set2C.mat)
> names(data.Set2Cclus) <- names(data.Set2Cpts)
```

The analysis follows closely the example presented by Gotway and Stroup (1997) as Example 1 (see also Waller and Gotway, 2004, p. 385). Gotway and Stroup (1997) refer to this as a *marginal* formulation of the problem, because the quantity $E\{Y \mid X_i\}$ in Equation 12.13 is a marginal mean. Gotway and Stroup (1997) also discuss an alternative method called the *conditional* formulation. Shabenberger and Pierce (2002, p. 684) give an example of this approach, and Gotway and Wolfinger (2003) discuss a different type of marginal formulation analysis.

The marginal analysis procedure of Gotway and Stroup (1997) is called *iteratively reweighted generalized least squares* (IRWGLS) (see also Waller and Gotway, 2004, p. 337, 388) and is a generalization of the iteratively reweighted least squares algorithm mentioned in Section 8.4.1. It consists of four steps that are repeated iteratively in order to obtain estimates of the regression coefficients β_i in Equation 12.13.

Step 1. The algorithm is initiated by obtaining starting estimates $b_{i,0}$ for the regression coefficients. This is most easily done by simply solving the generalized linear model with independent errors using the function glm(). In our example, they can be carried over from the solution obtained in Section 12.6.1, which is repeated here.

```
> oaks.irwgls <- oaks.glm <- glm(QUDO ~ Permeab + GS32 + PE,
+    data = data.Set2Cclus, family = quasibinomial)
> print(b.irwgls <- coef(oaks.irwgls), digits = 3)
(Intercept)     Permeab        GS32          PE
     0.8958     -1.7213     -0.0265      0.0123
```

Now we must compute the predicted values of π_i using Equation 8.20. To do this, we use the function predict() to obtain the predictions p_i. Note the second argument (use ?predict.glm for an explanation).

```
> data.Set2Cclus$p.irwgls <- p.irwgls <- predict(oaks.irwgls,
+ type = "response")
```

Step 2. The second step is to compute a variogram of the residuals of the model in order to estimate their spatial autocorrelation. The gstat package (Pebesma, 2004) function variogram() can be made to compute the generalized least square residuals of a generalized linear model by specifying the link function in the variogram formula (Pebesma, 2004).

```
> data.Set2Cgeo <- data.Set2Cclus
> coordinates(data.Set2Cgeo) <- c("Easting", "Northing")
> oaks.vgm <- variogram(log(p.irwgls / (1 - p.irwgls)) ~
+    Easting + Northing, data = data.Set2Cgeo)
```

Step 3. Based on the experimental variogram of the residuals, estimate the parameters of a variogram model. If the residuals do not display autocorrelation (i.e., if the variogram is a pure nugget), then the process can stop. The experimental variogram of the residuals is shown in Figure 12.9. Because of the discordant values of $\hat{\gamma}(h)$ at the largest values of the lag h, models were fit by eye. Based on the figure, the exponential model shown in the figure was selected.

```
> b.exp <- 1.0
> alpha.exp <- 20000
```

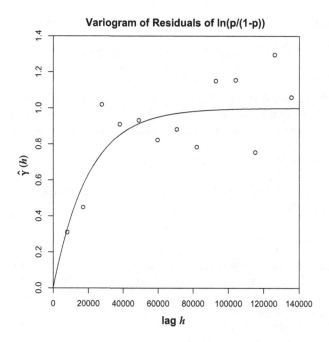

FIGURE 12.9
Variogram of the residuals of the generalized linear model of probability of a blue oak as a function of elevation. Fit by an exponential variogram model is also shown.

Step 4. This step involves the implementation of the *generalized estimating equations* to obtain an updated estimate of the regression coefficients. In order to facilitate cross-referencing with their paper, we will employ the same notation as Gotway and Stroup (1997), with one exception: where they use the symbol Z to represent their response variable, we will retain our practice of using Y. To further ease cross-referencing, we will implement the R code following their equations exactly, although it results in extremely inefficient code.

Similar to Equation 12.16, the variance of Y is written (Equation 2.13 of Gotway and Stroup, 1997)

$$Var\{Y\} = v^{1/2}(\mu)R(\alpha)v^{1/2}(\mu), \tag{12.17}$$

where $R(\alpha)$ is the correlation matrix, which depends on a parameter or parameters α, and $v^{1/2}(\mu)$ is the diagonal matrix whose diagonal elements are the square roots of products of the expected values of the Y_i with the scale factors a_i. Since Y is a binomially distributed random variable, $\mu_i = E\{Y_i\} = \pi_i$, $var\{Y_i\} = \pi_i(1-\pi_i)$, and $a_i = 1/n_i$ (Gotway and Stroup, 1997). Therefore, the matrix $v^{1/2}(\mu)$ is written

$$v^{1/2}(\mu) = \begin{bmatrix} \sqrt{\pi_1(1-\pi_1)/m_1} & 0 & \cdots & 0 \\ 0 & \sqrt{\pi_2(1-\pi_2)/m_2} & \cdots & 0 \\ \cdots & \cdots & \cdots & \cdots \\ 0 & 0 & \cdots & \sqrt{\pi_n(1-\pi_n)/m_n} \end{bmatrix} \tag{12.18}$$

where m_i is the number of data sites in polygon i. This is implemented as

```
> n.sites <- as.numeric(table(data.Set2C$clusID))
> v.half <- diag(as.vector(p.irwgls * (1 - p.irwgls) / n.sites))
```

In the first line above, the function `table()` computes a contingency table consisting of counts of the number of occurrences of each value of `data.Set2C$clusID`.

Because we are using an exponential variogram model, the correlation matrix $R(\alpha)$ in Equation 12.17 is given by

$$
R(\alpha) = \begin{bmatrix}
1 & \exp(-\alpha h_{12}) & \ldots & \exp(-\alpha h_{1n}) \\
\exp(-\alpha h_{12}) & 1 & \ldots & \exp(-\alpha h_{2n}) \\
\ldots & \ldots & \ldots & \ldots \\
\exp(-\alpha h_{1n}) & 0 & \ldots & 1
\end{bmatrix}
\tag{12.19}
$$

where based on the results of Step 3, the parameter α of the variogram model is set at $\alpha = 20,000$ and h_{ij} is the distance from polygon i to polygon j. Here is the implementation

```
> x <- data.Set2Cclus$Easting
> y <- data.Set2Cclus$Northing
> lag.mat <- as.matrix(dist(cbind(x, y)))
> R <- exp(-lag.mat / alpha.exp)
```

The estimated variance-covariance matrix V is then computed from Equation 12.17.

```
> V <- sqrt(v.half) %*% R %*% sqrt(v.half)
```

Now we are ready to compute the updated estimates for the regression coefficients. In Appendix A, it is shown that if we write the linear regression equation in matrix form as $Y = X\beta + \varepsilon$ (Equation A.33), then normal equations defining the regression coefficients may be written as $X'Xb = X'Y$. This allows the computation of the regression coefficients via the formula $b = (X'X)^{-1}X'Y$ (Equation A.35). Regression software such as R does not actually compute this inverse directly, because the computation can be done more efficiently, but in principle the regression coefficients could be computed this way.

Gotway and Stroup (1997), citing Nelder and Wedderburn (1972), Wedderburn (1974), McCullagh (1983), Liang and Zeger (1986), and McCullagh and Nelder (1989), develop the iterative approximation to the regression coefficients via IRWGLS as follows: Write the inverse link function h of equation (8.20) as $\pi = \mu = h(\eta)$, $\eta = X\beta$. Then let D be the diagonal matrix whose diagonal elements are $\partial \pi_i / \eta_i$. For the binomial case, $\pi_i = h(\eta_i) = 1 / (1 + \exp(-\eta_i))$ and therefore $\partial \pi_i / \partial \eta_i = \pi_i(1 - \pi_i)$. This is implemented as follows:

```
> D <- diag(as.vector(p.irwgls * (1 - p.irwgls)))
```

We then define the matrix W by $W = D'V^{-1}D$. This is implemented (very inefficiently) as follows:

```
> V.inv <- solve(V)
> W <- t(D) %*% V.inv %*% D
```

Now, let z^* be defined by $z^* = \eta + D^{-1}(z - \mu)$. We can estimate the μ_i using the values $h(\Sigma b_{0,j}X_{ij})$, where the $b_{0,i}$ are obtained in Step 1 and the X_{ij} are the mean values of the explanatory variables of cluster region i. We can estimate η as $g(p_{i,0}) = \ln(p_{i,0} / (1 - p_{i,0}))$, and substitute for the z_i the values p_i representing for each polygon the fraction of sites occupied by a blue oak, and $z*$.

```
> eta <- log(p.irwgls / (1 - p.irwgls))
> mu <- as.vector(p.irwgls)
> z <- p
> D.inv <- solve(D)
> z.star <- eta + D.inv %*% (z - mu)
```

The last step is to use an estimating equation given as Equation 3.1 by Gotway and Stroup (1997). This corresponds to Equation A.34 for linear regression and is written as

$$X'WX\beta = X'Wz*. \tag{12.20}$$

Premultiplying by $(X'WX)^{-1}$ allows us to compute the values $b_{i,1}$ as

$$b_{i,1} = (X'WX)^{-1}X'Wz*. \tag{12.21}$$

Here is the implementation.

```
> X <- with(data.Set2Cclus, cbind(rep(1,length(p.irwgls)),
+     Permeab, GS32, PE))
> XpWX.inv <- solve(t(X) %*% W %*% X)
> b.old <- b.irwgls
> t(b.irwgls)
              Permeab          GS32         PE
[1,]  0.4304769  -1.522903  -0.02922488  0.01358672
```

Now we can compare the updated value $b_{j,1}$ of the regression coefficients with the original values.

```
> t(b.old)
      (Intercept)      Permeab          GS32         PE
[1,]   0.8957865   -1.721327   -0.0264501  0.01225389
> t(abs((b.irwgls - b.old) / b.old))
              Permeab          GS32         PE
[1,]   0.5194426   0.1152737   0.1049063  0.1087677
```

Step 5. Since there is a fairly large change in the coefficients, we use these values to generate a second iteration by plugging $b_{j,1}$ into Step 2, working again through Steps 2, 3, 4, and 5, and continuing this iteration until reaching a sufficiently small change in the estimates for the β_i.

```
> data.Set2Cclus$p.irwgls <- p.irwgls <-
+     1 / (1 + exp(-X %*% b.irwgls))
```

In this case, the second iteration produces the following result:

```
> t(b.irwgls)
                       Permeab              GS32               PE
[1,]      0.4333026    -1.519584    -0.02928080    0.01360120
> t(b.old)
                       Permeab              GS32               PE
[1,]      0.4304769    -1.522903    -0.02922488    0.01358672
> t(abs((b.irwgls - b.old) / b.old))
                       Permeab              GS32               PE
[1,]  0.006564155 0.002179365     0.001913198    0.00106632
```

A third iteration produces this:

```
> t(b.irwgls)
                       Permeab              GS32               PE
[1,]      0.4345514    -1.519843    -0.02928375    0.01360108
> t(b.old)
                       Permeab              GS32               PE
[1,]      0.4333026    -1.519584    -0.02928080    0.01360120
> t(abs((b.irwgls - b.old) / b.old))
                       Permeab              GS32               PE
[1,]  0.002882207 0.0001701886     0.0001008558 8.808101e-06
```

We will accept this answer as the estimates for β_j. Here is the comparison with the solution obtained in Section 11.6 by ignoring spatial autocorrelation.

```
> print(t(b.glm <- coef(oaks.glm)), digits = 3)
     (Intercept) Permeab     GS32       PE
[1,]       0.896    -1.72  -0.0265   0.0123
> print(t(b.irwgls), digits = 3)
          Permeab     GS32       PE
[1,] 0.435    -1.52  -0.0293   0.0136
> print(t((b.irwgls - b.glm) / b.glm), digits = 3)
          Permeab  GS32     PE
[1,] -0.515  -0.117  0.107  0.11
```

Since this has been a long analysis, it is worthwhile to look back and recall our objective. This was to determine the effect of the incorporation of spatial autocorrelation on the coefficients of a model for blue oak distribution. Rather than using pointwise data, in which each point takes on a value of 0 or 1, we used aggregated data in which for each region the value represents the fraction of sites in a that region in which an oak is present. Our conclusion is that there is about a 10% difference in each of the coefficients between the models incorporating and not incorporating spatial autocorrelation.

Shabenberger and Pierce (2002, p. 686) point out an obvious problem with the IRWGLS method. Even though the method could be made much more efficient numerically by not following exactly the equations of Gotway and Stroup (1997) as we have done here, nevertheless, it remains necessary in Equation 12.21 to invert an $n \times n$ matrix, where n may be quite large. Shabenberger and Pierce (p. 691) compare several methods, including IRWGLS, other marginal methods, and a conditional method, and find that in many cases they provide similar answers.

Finally, we note in passing that the term *logistic regression* has another different but (possibly confusingly) similar application. This is in the fitting of a logistic function to a set of data that are not probabilities but whose values follow a logistic shaped (also known as sigmoid) curve. An example is a plant growth curve (Hunt, 1982), in which an annual plant's growth rate, measured, for example, in biomass or leaf area index, starts out slowly, increases through a "grand growth phase," and eventually declines to zero as the plant matures. Such data may be fit with a logistic equation, but the asymptotic values are typically not 0 and 1, and the assumptions associated with the error terms are different from those of the probabilistic model discussed in this section. Pinhiero and Bates (2000) describe software for the fitting of this form of data using nonlinear regression (see Exercise 12.5).

12.7 Further Reading

The precise definitions of the terms *fixed effect* and *random effect* vary considerably in the literature, despite efforts to constrain them (Littell et al., 2002, p. 92). Gelman and Hill (2007, p. 245) identify five different uses of the terms, and someone who was really interested in such things could no doubt find some more.

Pinhiero and Bates (2000) is the single most important book for all the material in this chapter except Section 12.6. For the numerical method used by lme() to solve a mixed model such as Equation 12.7, see Pinhiero and Bates (2000, p. 202). Other books on the mixed model that provide a useful supplement are Searle (1971), Searle et al. (1992), and McCulloch et al. (2008). In particular, Searle (1971, p. 458) provides a good introduction to the application of the maximum likelihood method to mixed models. McCulloch et al. (2008) provide a good discussion of the differences between ML and REML. Diggle et al. (2002) discuss the use of the mixed model in the analysis of data models with autocorrelated errors. The specific case of unbalanced data such as that encountered in Data Set 3 is well covered by Searle (1987). Rutherford (2001) gives an excellent introduction to the analysis of covariance using least squares.

Zuur et al. (2009) and Bolker et al. (2009) provide very well organized discussions of the application of mixed models to ecology. Waller and Gotway (1994) and Schabenberger and Pierce (2002) provide good discussions of the generalized linear model with autocorrelated errors. Carl and Kuhn (2007) discuss the use of generalized estimating equations in species distribution models.

The process of variogram estimation for the generalized linear model as used in gstat is described by Christensen (1991) and ver Hoef and Cressie (1993). In addition to Shabenberger and Pierce (2002), McCulloch et al. (2008) discuss the conditional versus marginal formulation. Venables and Ripley (2002) also discuss the generalized linear mixed model and in particular the function glmmPQL(), which addresses overdispersion in binomial and Poisson models. Hadfield (2010) provides in the MCMCglmm package a very nice implementation of the Markov Chain Monte Carlo (MCMC) method, discussed in Chapter 14, for the solution of generalized linear mixed models.

Exercises

12.1 Create a scatterplot of the regression coefficients of the intercepts computed in Section 12.4 using the mixed model and the mean yields of each of the fields.

12.2 Repeat the analysis using `gls()` of Section 12.5 for a model describing the data in Field 4.2.

12.3 Load the `debug` package (Bravington, 2013). Use the function `debug()` to monitor the progression of code to predict values of π_i for the model `glm.demo` of Section 9.4.1.

12.4 Repeat the IRWGLS analysis of Section 12.6.2 for a model of the data in the Sierra Nevada subset of Data Set 2.

12.5 Read about the function `nls()` in Pinhiero and Bates (2000). Use this function to fit the Coast Range clusters data in Section 12.6.2. Compare this fit with the one obtained using the quasibinomial model. What happens to the `nls()` model as the explanatory variable approaches ∞ and $-\infty$?

13

Regression Models for Spatially Autocorrelated Data

13.1 Introduction

In this chapter, we discuss regression models specifically developed for processes described by spatially autocorrelated random variables. The effect of spatial autocorrelation, or apparent spatial autocorrelation, on regression models depends on how these spatial effects influence the data (Cliff and Ord, 1981, p. 141). Apparent autocorrelation may or may not be the result of real autocorrelation. Miron (1984) discusses three sources of real or apparent spatial autocorrelation: interaction, reaction, and misspecification. These are not mutually exclusive and may exist in any combination. We will discuss these in the context of a population of plants (for specificity, let us say they are oak trees in an oak woodland like that of Data Set 2) growing in a particular region. Suppose Y_i represents a measure of plant productivity such as tree height or population density at location (x_i, y_i), and that the population is sufficiently dense relative to the spatial scale that the productivity measure may be modeled as varying continuously with location. For purposes of discussion, suppose that we model tree productivity via a linear regression of the form

$$Y_i = \beta_0 + \beta_1 X_{i1} + \beta_2 X_{i2} + \varepsilon_i, \tag{13.1}$$

where X_{i1} represents the amount of light available at location i, X_{i2} represents the amount of available nutrients. In matrix notation, the ordinary least squares model is

$$Y = \beta X + \varepsilon. \tag{13.2}$$

We will set up models using artificial data to demonstrate the various effects of spatial autocorrelation.

The first source of autocorrelation we will consider is *interaction*. Negative autocorrelation may occur if trees in close proximity compete with each other for light and nutrients, so that relatively productive tree populations tend to inhibit the growth of other trees. Positive autocorrelation would occur if existing trees produce acorns that don't disperse very far, which in turn results in more trees in the vicinity. We will only consider positive autocorrelation, which seems to be much more common in ecology.

First, we generate a set of data that satisfies the model.

```
> set.seed(123)
> X1 <- rnorm(100)
> X2 <- rnorm(100)
> eps <- 0.01 * rnorm(100)
> b <- c(0.5,0.3)
> Y <- b[1]*X1 + b[2]*X2 + eps
> print(coef(lm(Y ~ X1 + X2)), digits = 2)
(Intercept)          X1          X2
     0.0014      0.4987      0.3002
```

If Y is positively autocorrelated, then in place of Equation 13.2, the true model for its dependence on X is

$$Y = \beta X + \rho WY + \varepsilon. \tag{13.3}$$

Here the term ρWY is the *spatial lag* (Section 3.5.2). We can generate a set of artificial data satisfying this model by employing the method of Haining (1990, p. 116) described in Section 3.5.3. Subtracting ρWY from both sides of Equation 13.3 and premultiplying the result by $(I - \rho W)^{-1}$ yields

$$Y = (I - \rho W)^{-1}(\beta X + \varepsilon). \tag{13.4}$$

Here is the code.

```
> rho <- 0.6
> library(spdep)
> nlist <- cell2nb(10, 10)
> IrWinv <- invIrM(nlist, rho)
> Y <- IrWinv %*% (b[1]*X1 + b[2]*X2 + eps)
```

Now let's try the regression again.

```
> print(coef(lm(Y ~ X1 + X2)), digits = 2)
(Intercept)          X1          X2
      0.021       0.542       0.324
```

The regression coefficients have all moved a bit away from their true values of 0, 0.5, and 0.3. The difference is not very great because Y is not very large. Nevertheless, this example does illustrate the effect of positive interaction among the Y. The reason that omitting interactive spatial autocorrelation affects the regression coefficients can be seen by comparing Equations 13.2 and 13.3. Because Equation 13.2 does not include the lag term, when the data are fit to this model, some of the variability that would be assigned to this term if it were present is instead assigned to the regression coefficients. For example, the true marginal effect of nutrient level X_2, which is measured by β_2, will be incorrectly estimated because it will include some of the effect of the lag WY.

The second source of autocorrelation is called *reaction*. This would manifest itself in a plant population if nearby plants were reacting to the values of some external factor, such as soil nutrient content, that varies in space. If the primary influence on the process is one of reaction to the general ambient conditions, the explicit inclusion of spatial autocorrelation terms may not be necessary or even appropriate. Instead, a simple regression model including the external factor, or inclusion of a trend model, may be appropriate. Miron (1984) discusses the potential effects of an omitted spatial influence on the parameters of a regression model. This effect can be visualized by supposing that the oak woodland we are modeling is a riparian area, and that the trees are affected by their distance to the river, either due to the soil properties or the availability of water or both. Replicating the discussion of Miron (1984) in an ecological context, the failure to include this effect means that there is a missing variable that we can denote X_{i3}, the distance from the river at location i. Instead of Equation 13.1, the model should be

$$Y_i = \beta_0 + \beta_1 X_{i1} + \beta_2 X_{i2} + \beta_3 X_{i3} + \varepsilon_i. \tag{13.5}$$

Suppose, in addition, that X_3 is correlated with one of the other explanatory variables, say, with nutrient level X_2. In that case, the exclusion of distance from the river X_3 from the model may bias the coefficient β_2. The reason is again that the effect of X_3 has been "loaded" onto the explanatory variable X_2. We saw an example of this in Section 8.4.2 with the spatial variable *CoastDist*. In the analysis of the Sierra Nevada data carried out in the text, *CoastDist* did not have a strong effect, but if you did Exercise 8.12, you had the opportunity to show that in the Coast Range excluding *CoastDist* did impose a small bias on the other coefficients.

We will use the simulation to show the effects of excluding X_3 from the model when the process depends on this variable. In the model, the equation defining X_3 is a pure trend in the x direction. Nutrient level X_2 also depends on distance from the river so that X_2 and X_3 are highly correlated, although there is not necessarily any causal relationship. We imbed the models in a function that can be used for Monte Carlo simulation.

```
> set.seed(123)
> coords <- expand.grid(1:10,10:1)
> b <- c(0.5,0.3,0.8)
> reg.mc <- function(coords, b, incl.X3){
+    X1 <- rnorm(100)
+    X2 <- coords[,1]+rnorm(100)
+    X3 <- coords[,1]+rnorm(100)
+    Y <- b[1]*X1 + b[2]*X2 + b[3]*X3 + rnorm(100)
+    {if (incl.X3 == T)
+       Y.lm <- lm(Y ~ X1 + X2 + X3)
+    else
+       Y.lm <- lm(Y ~ X1 + X2)}
+    return(coef(Y.lm))
+ }
```

If the third argument of the function reg.mc(coords, b, incl.X3) is TRUE, then X_3 is included in the model; otherwise, it is excluded. We first simulate the case in which X_3 is in the model.

```
> U <- replicate(1000,reg.mc(coords, b, T))
> print(rowMeans(U), digits = 2)
(Intercept)        X1            X2            X3
      0.014       0.499         0.303         0.796
```

The coefficients β_1 and β_2 are estimated without apparent bias.

In the second simulation, X_3 is not included in the model.

```
> U <- replicate(1000,reg.mc(coords, b, F))
> print(rowMeans(U), digits = 2)
(Intercept)            X1             X2
      0.46          0.49           1.02
```

Your numbers may be slightly different depending on the pseudorandom number generator. The estimate of β_2 is biased while that of β_1 is not. In this example, although X_3 is interpreted as a "spatial" variable, its role in the model is identical to that which it would play if it were interpreted as purely an attribute value, with no spatial connotation.

A second potential reaction effect of a spatial variable not included in the model is as follows. Suppose X_3 is not correlated with any of the other explanatory variables in the model. If X_3 is spatially autocorrelated and is not included in the model, then because its effect is loaded entirely into the error term, the result will be that the errors are autocorrelated. As we have seen in Section 3.3.2, in the presence of autocorrelated errors the estimates of the coefficients remain unbiased, but the error variance is inflated, so that the estimates of the coefficients become inefficient (i.e., needing more observations to achieve a conclusion). We can again demonstrate this effect through Monte Carlo simulation by modifying the data so that X_1, X_2, and X_3 are mutually uncorrelated and comparing simulations including and excluding X_3 in the model.

```
> library(spdep)
> set.seed(123)
> coords <- expand.grid(1:10,10:1)
> b <- c(0.5,0.3,0.8)
> lambda <- 0.6
> nlist <- cell2nb(10, 10)
> IrWinv <- invIrM(nlist, lambda)
> reg.mc <- function(coords, b, incl.X3){
+     X1 <- rnorm(100)
+     X2 <- rnorm(100)
+     X3 <- IrWinv %*% rnorm(100)
+     Y <- b[1]*X1 + b[2]*X2 + b[3]*X3 + rnorm(100)
+     {if (incl.X3 == T)
+         Y.lm <- lm(Y ~ X1 + X2 + X3)
+     else
+         Y.lm <- lm(Y ~ X1 + X2)}
+     return(coef(Y.lm))
+ }
```

First, we run a simulation in which X_3 is included in the model.

```
> U <- replicate(1000,reg.mc(coords, b, T))
> print(rowMeans(U), digits = 2)
(Intercept)            X1             X2             X3
     0.0088        0.4981         0.3029         0.7954
> print(apply(U, 1, var), digits = 2)
(Intercept)            X1             X2             X3
     0.0117        0.0105         0.0114         0.0084
```

The variable X_3 is autocorrelated. When it is included in the model, the variances of the coefficients are all about 0.01. Next, we exclude X_3 from the model and repeat the simulation.

```
> U <- replicate(1000,reg.mc(coords, b, F))
> print(rowMeans(U), digits = 2)
(Intercept)        X1            X2
  -0.00084      0.49357       0.29666
> print(apply(U, 1, var), digits = 2)
(Intercept)        X1            X2
     0.051        0.020         0.020
```

When X_3 is not included, the errors in the model become autocorrelated due to the autocorrelation of X_3. Unlike the previous example, because X_2 is independent of X_3 the estimate of β_2 is unaffected by removing X_3 from the model. The variance of the estimates of β_1 and β_2, however, are roughly doubled.

The third problem discussed by Miron (1984) is the effect of *model misspecification*. In this case, the measured autocorrelation is not due to interaction or reaction but to the incorrect form of the model. In particular, specification of homoscedastic errors when in fact the errors are heteroscedastic can lead to both biased estimates of the regression coefficients and indication of spatial autocorrelation when none really exists. To use Miron's term, this can "warp" the error terms of the model, creating a spatially heteroscedastic error, that is, non-constant variance (see Section 8.3). Suppose, for example, that in the tree growth model, nutrient level X_2 is spatially autocorrelated and that the error variance (i.e., the randomness associated with X_2) is an increasing function of X_2, but is not itself spatially autocorrelated (remember that it is spatial autocorrelation in the errors, not the in explanatory variables, normally introduces problems into regression results). This heteroscedasticity can induce apparent spatial dependency into the errors. We will demonstrate this with a simulation arranged so that X_2 has a spatial pattern similar to that displayed in Figure 3.9, with $\lambda = 0.8$. The error terms are uncorrelated, but because the error variance is a function of X_2 and high values of X_2 tend to be near other high values of X_2, a test for spatial autocorrelation of the residuals has a high Type I error rate. We use the function lm.morantest(), discussed in Section 13.2, to test for spatial autocorrelation of the residuals.

In the code below X_2 is spatially autocorrelated. The model for Y_1 is correctly specified, with constant variance errors, while the model for Y_2 is incorrectly specified, with heteroscedastic but not directly correlated errors.

```
> set.seed(123)
> coords <- expand.grid(1:10,10:1)
> b <- c(0.5,0.3)
> nlist <- cell2nb(10,10)
> IrWinv <- invIrM(nlist, 0.8)
> W <- nb2listw(nlist)
> reg.mc <- function(coords, b, W){
+    X1 <- rnorm(100)
+    X2 <- IrWinv %*% rnorm(100)
+    Y1 <- b[1]*X1 + b[2]*X2 + rnorm(100)
+    Y2 <- b[1]*X1 + b[2]*X2 + exp(1 + 2 * X2) * rnorm(100)
+ # Model correctly specified
```

```
+    Y1.lm <- lm(Y1 ~ X1 + X2)
+    t1 <- lm.morantest(Y1.lm, W)
+    C1 <- as.numeric(t1$p.value < 0.05)
+ # Model mis-specified, heteroscedastic errors
+    Y2.lm <- lm(Y2 ~ X1 + X2)
+    t2 <- lm.morantest(Y2.lm, W)
+    C2 <- as.numeric(t2$p.value < 0.05)
+    return(c(C1, C2))
+ }
> U <- replicate(1000,reg.mc(coords, b, W))
> rowMeans(U)
[1] 0.045 0.112
```

Significant spatial autocorrelation of the residuals is detected at the nominal rate of 5% in the correctly specified models, but it is detected in about 10% of the incorrectly specified ones.

Anselin and Griffith (1988) carry out a systematic investigation of the impacts of spatial effects on least squares regression models. In particular, they study the impact of autocorrelated errors on the Mallows C_p (Section 9.1). Anselin and Griffith find that positively autocorrelated errors can artificially inflate the size of C_p. Anselin and Griffith also find that spatial autocorrelation of the errors can affect the outcome of tests for homogeneity of the residuals such as the Breusch-Pagan test (Kutner et al., 2005, p. 118). They find that autocorrelation of sufficient magnitude can increase the Type I error rates of certain tests, although the effect is not consistent across all types of tests.

In summary, models can be impacted by failure to account for spatial effects in regression model, either through the exclusion of a reaction to a spatially autocorrelated explanatory variable or by the presence of interactive spatial autocorrelation in the response variable or the error terms or both. They can also be impacted by misspecification of the model. Depending on the manner in which the effects of spatial effects enter into the system, these impacts can include biased estimates of the regression coefficients, biased estimates of the error variance, inefficient estimates of the regression coefficients, or spurious spatial autocorrelation.

Spatial autocorrelation often manifests itself in autocorrelation of the components of the error vector ε in Equation 13.1. Section 13.2 contains a discussion of the detection of this type of autocorrelation. The field of spatial econometrics has generated a particularly useful approach to spatial regression, and much of the material in this chapter is motivated by that approach. Spatial autocorrelation is frequently incorporated into linear regression through one of two models, the spatial lag model or the spatial error model. Section 13.3 discusses these, and Section 13.4 describes a method for determining which is most appropriate for a particular data set. Section 13.5 describes the procedure for fitting the model. Section 13.6 describes an alternative model called the conditional autoregressive model, and Section 13.7 demonstrates the application of these models to field data.

13.2 Detecting Spatial Autocorrelation in a Regression Model

Consider the ordinary linear regression model, written in matrix notation (Appendix A.2) as

$$Y = X\beta + \varepsilon, \tag{13.6}$$

where Y is a vector of n response variables, β is a vector of $p-1$ parameters, X is an $n \times p$ design matrix, and ε is a vector of n error terms. The Gauss-Markov theorem states (Kutner et al., 2005, p. 18) that if the data conform to the model (Equation 13.6) and the components ε_i of the error vector are independent, identically distributed random variables with mean zero and variance σ^2, then the method of least squares provides the best linear unbiased estimate of the parameters β. Spatial autocorrelation can disrupt least squares regression if the error terms ε_i are spatially autocorrelated. One of the first questions to be addressed is, therefore, whether spatial autocorrelation exists among the error terms.

A number of tests for autocorrelation, prominent among them the use of Moran's I, are discussed in Chapter 4. Since the regression residuals are generally used as a model for the error terms, one might think that spatial autocorrelation could be detected by carrying out a Moran's I test on these residuals. Unfortunately, things are not that simple (Cliff and Ord, 1981, p. 200; Miron, 1984). Suppose a least squares regression is carried out on a data set $Y = \begin{bmatrix} Y_1 & Y_2 & \dots & Y_n \end{bmatrix}$ resulting in a regression model

$$Y = Xb + e. \tag{13.7}$$

The regression coefficients b satisfy (Kutner et al., 2005, p. 200)

$$b = (X'X)^{-1}X'Y, \tag{13.8}$$

and therefore the regression residuals satisfy

$$e = Y - Xb$$

$$= (I - X(X'X)^{-1}X')Y \tag{13.9}$$

$$= MY$$

where $M = I - X(X'X)^{-1}X'$. The formula for the variance–covariance matrix of e is then $\text{var}\{e\} = \text{var}\{MY\}$ (Kutner et al., 2005, p. 196). Therefore,

$$\text{var}\{e\} = \text{var}\{MY\}$$

$$= M \, \text{var}\{Y\} M' \tag{13.10}$$

$$= M \sigma^2 I M'$$

$$= \sigma^2 MM'$$

It turns out (Kutner et al., 2005, p. 204) that the matrix M is *idempotent*, that is, it satisfies $MM' = M$. We get the final formula

$$\text{var}\{e\} = \sigma^2 M. \tag{13.11}$$

Since M has nonzero off-diagonal terms, it follows that $\text{cov}\{e_i, e_j\} \neq 0$ in general. Therefore, the regression residuals are correlated, and a test using Moran's I on these residuals may indicate autocorrelation even if the error terms ε_i are themselves uncorrelated. It is worth noting that one could have determined that the residuals e_i are correlated without actually computing Equation 13.11. Indeed, they must satisfy (Kutner et al., 2005, p. 23) $\Sigma_i e_i = 0$, and given any $n-1$ of them, one can compute the last. Therefore, they must be correlated with each other. To say that the residuals are correlated does not mean that they are *spatially* autocorrelated, but the possibility exists that they are. Cliff and Ord (1981, p. 200) have derived formulas for the mean and variance of Moran's *I* for regression residuals. Any test of autocorrelation of ordinary least squares residuals must be carried out using these formulas rather than those given in Chapter 4.

The spdep package includes a function lm.morantest() that takes into account the correlation of residuals in carrying out the test for autocorrelation. This test was used in Section 13.1 to detect apparent spatial autocorrelation in a misspecified model.

As an illustration of the test of a real model for spatially autocorrelated residuals, we use the linear model model.5 for Field 4.1 developed in Section 9.3.

```
> model.5 <- lm(Yield ~ Clay + SoilP + I(Clay*SoilP) + Weeds,
+      data = data.Set4.1)
> library(spdep)
> coordinates(data.Set4.1) <- c("Easting", "Northing")
> nlist <- dnearneigh(data.Set4.1, d1 = 0, d2 = 61)
> W <- nb2listw(nlist, style = "W")
> lm.morantest(model.5, W)
        Global Moran's I for regression residuals
data :
model: lm(formula = Yield ~ Clay + SoilP + I(Clay * SoilP) + Weeds, data
= data.Set4.1)
weights: W
Moran I statistic standard deviate = 1.6864, p-value = 0.04586
alternative hypothesis: greater
sample estimates:
Observed Moran's I       Expectation          Variance
      0.091175267      -0.040897372        0.006133274
```

The resulting *p* value supports the conclusion that the errors in this model are spatially autocorrelated. Therefore, to safely use this model we must determine the most appropriate form in which to include the effects of spatial autocorrelation. This is the subject of the next section.

13.3 Models for Spatial Processes

13.3.1 The Spatial Lag Model

In Chapter 3, it was pointed out that a close relationship exists between spatial processes and time series. Two spatial models were briefly discussed, the *spatial lag model* and the *spatial error model*. In this section, we consider these and other models in greater detail. We begin our discussion with the spatial lag model. In time series, a fundamental model is the so-called *first-order autoregressive model*, (Box and Jenkins, 1976, p. 51), which has the form

$$Y_i = \rho Y_{i-1} + \varepsilon_i, \, i = 1, 2, \dots \tag{13.12}$$

The equation in matrix notation for the analogous spatial model, which describes a spatially lagged random variable with zero mean and no explanatory variables, has the form

$$Y = \rho WY + \varepsilon. \tag{13.13}$$

As discussed in Chapter 3, the quantity WY is called the *spatial lag* by analogy with the time lag Y_{i-1} in a time series model. The interpretation of model (Equation 13.13) is that the value of the process Y at one location is directly associated with the values of the process at nearby locations. In the example discussed in Section 13.1, Y represents a measure of oak tree productivity, and high productivity at one location is associated with high productivity at other nearby locations. It is important to remember that nothing in the model says that high productivity at one location actually *causes* high productivity at nearby locations.

The spatial lag model, which is also sometimes called the *simultaneous autoregressive model*, was initially studied by Whittle (1954) in a paper that played a pivotal role in initiating the study of spatial processes. Whittle considered (in the terminology of Chapter 3) a binary rook's case spatial relationship. Anselin (1992, p. 34) describes higher-order contiguity models (cf. the discussion of the correlogram in Section 4.6), but in this text we will only consider first-order contiguity. Inverting Equation 13.13 leads to

$$Y = (I - \rho W)^{-1}\varepsilon, \tag{13.14}$$

Equations of this form have been used repeatedly throughout the book to generate spatially autocorrelated random variables.

In the analysis of the mixed model in Chapter 12, we made an assumption about the variance of Y, and therefore of the residuals, that enabled a model of the autocorrelation structure to be constructed (Equations 12.8 and 12.9). In the analysis of the autoregressive models of this chapter, the form of the variance of Y is enforced by the model. From Equation 13.14 we have

$$\begin{aligned} \text{var}\{Y\} &= E\{YY'\} \\ &= E\{(I - \rho W)^{-1}\varepsilon\varepsilon'[(I - \rho W)^{-1}]'\} \\ &= (I - \rho W)^{-1}E\{\varepsilon\varepsilon^T\}(I - \rho W')^{-1} \\ &= (I - \rho W)^{-1}\sigma^2 I(I - \rho W')^{-1}, \end{aligned} \tag{13.15}$$

where in going from the second to the third equation we have used the fact that the transpose of the inverse equals the inverse of the transpose (Appendix A.1). As with the mixed model, the variance of the autoregressive model is made up of a matrix product of terms that describe the spatial structure with a term that describes the magnitude of the variance. However, the roles of the terms in the last of Equations 13.15 is the reverse of that of the corresponding matrices in Equation 12.9.

It is evident that if we have a set of explanatory variables to incorporate into a spatial regression model, then this can be accomplished by augmenting the spatial lag model of Equation 13.13. Adding a term $X\beta$ to the spatial lag model of Equation 13.13 yields

$$Y = \rho WY + X\beta + \varepsilon. \tag{13.16}$$

This is sometimes called the mixed regressive–spatial auto regressive model (Anselin, 1988, p. 35), or sometimes simply the *spatial autoregressive model*. We will still use the term *spatial lag model* (Anselin, 1992, p. 48). In other words, we regard the model of Equation 13.13 as simply a special case of Equation 13.16 in which $\beta = 0$.

The model of Equation 13.16 can be interpreted in three different ways (Anselin, 1992, p. 36). In the first interpretation, one is interested in the model of the spatial process for its own sake. A common (non-ecological) example of this is the study of the spread of technological innovations. For example, Anselin (1992, p. 57) studies the state-by-state spread of the adoption of home freezers in the United States. In this case, the proper specification of the spatial weights matrix W and the estimation of the parameter ρ become important in their own right as indicators of the nature and strength of spatial interaction.

A second interpretation is obtained by bringing the first term on the right-hand side of Equation 13.16 to the left side, resulting in

$$(I - \rho W)Y = X\beta + \varepsilon. \tag{13.17}$$

In this formulation, one is primarily interested in the influence of the predictor variables X on the response variable after controlling for the effect of spatial autocorrelation. In this context, the quantity $(I - \rho W)$ is sometimes called the *spatial filter*.

A third interpretation is obtained by multiplying both sides of Equation 13.16 by $(I - \rho W)^{-1}$ and taking the expectation of both sides. This yields

$$E\{Y\} = E\{(I - \rho W)^{-1}X\beta\} + E\{(I - \rho W)^{-1}\varepsilon\}$$
$$= E\{(I - \rho W)^{-1}X\beta\}. \tag{13.18}$$

since $E\{\varepsilon\} = 0$. In this interpretation, one focuses on the nonlinear effect of the spatial auto-correlation on the expected values of the components of Y.

13.3.2 The Spatial Error Model

In Chapter 3, we introduced the autoregressive time series model (Equation 3.13)

$$Y_i - \mu = \lambda(Y_{i-1} - \mu) + \varepsilon_i, \ i = 1, 2, \dots \tag{13.19}$$

which (see Exercise 13.1) can be written as (Equation 3.12)

$$Y_i = \mu + \eta_i$$
$$\eta_i = \lambda \eta_{i-1} + \varepsilon_i, \ i = 1, 2, \dots \tag{13.20}$$

Note that in this case the lagged quantity is the *error* rather than the response variable. The spatial version of this model, including explanatory variables, is

$$Y = X\beta + \eta$$
$$\eta = \lambda W\eta + \varepsilon. \tag{13.21}$$

This formulation is called the *spatial error model*. By tradition, the symbol ρ is used with the spatial lag model, and λ is used with the spatial error model. The first equation of Equations 13.21 is that of an ordinary linear regression model, but because of the second equation, the errors have a correlated structure. Spatial autocorrelation of the form (Equation 13.21) is generally considered a nuisance (Anselin, 1992, p. 38), because the primary interest is in the relationship between the explanatory variables X and the response variable Y, and the effect of spatial autocorrelation is to prevent the data from satisfying the assumptions of ordinary least squares regression.

There is an algebraic relationship between the spatial error model and the spatial lag model (see also Exercise 13.1 for the time-series analogy). Indeed, subtracting λWY from both sides of the first of Equations 13.21 yields

$$Y - \lambda WY = X\beta - \lambda WY + \eta. \tag{13.22}$$

Substituting $X\beta + \eta$ for Y and $-\lambda W\eta + \eta$ for ε from Equations 13.21 gives

$$
\begin{aligned}
Y - \lambda WY &= X\beta - \lambda WY + \eta \\
&= X\beta - \lambda W(X\beta + \eta) + \eta \\
&= X\beta - \lambda WX\beta + (-\lambda W\eta + \eta) \\
&= X\beta - \lambda WX\beta + \varepsilon.
\end{aligned}
\tag{13.23}
$$

Thus, the pair of equations (Equation 13.21) defining the spatial error model is equivalent to the single autoregressive model

$$Y - \lambda WY = X\beta - \lambda WX\beta + \varepsilon, \tag{13.24}$$

in which the spatial lag operator is applied to the explanatory variables as well as the response variable. Since the fixed term $X\beta$ in Equation 13.24 does not contribute to the variance of Y, we can set it equal to zero and see that the expression for the variance of the spatial error model is the same as that of the spatial lag model except that λ is substituted for ρ.

It follows that if one knew the value of λ one could define variables

$$
\begin{aligned}
\tilde{Y} &= Y - \lambda WY \\
\tilde{X} &= X - \lambda WX,
\end{aligned}
\tag{13.25}
$$

and carry out a regression of $\tilde{Y}$ on $\tilde{X}$. This process is implemented in generalized least squares, which is discussed in Section 12.5 (Anselin, 1992, p. 44; Pinhiero and Bates, 2000, p. 203). Since the value of λ is not known, it must be estimated by a method such as maximum likelihood. One can write a spatial autoregressive moving average model as (Anselin, 1992, p. 35)

$$Y = \rho WY + X\beta + \lambda W\varepsilon + \varepsilon, \tag{13.26}$$

and this can be generalized to higher orders as well. However, the theory of these models is not as well developed as is the theory for the spatial error and the spatial lag model, and we will not consider the possibility of fitting data to a model of the form (Equation 13.26). See LeSage and Pace (2009) for a discussion of the model.

13.4 Determining the Appropriate Regression Model

13.4.1 Formulation of the Problem

Suppose spatial autocorrelation has been detected in the residuals of a linear regression model, as it was, for example, in Section 13.2 for the model of yield in Field 1 of Data Set 4. The first step in augmenting the model to take this autocorrelation into account is generally to determine whether the spatial lag model (Equation 13.16) or the spatial error model (Equation 13.21) provides a better fit to the data. Anselin et al. (1996) formulate the general problem of distinguishing between models (Equation 13.16) and (Equation 13.21) by combining them into a single model of the form

$$Y = \rho W_1 Y + X\beta + \eta$$
$$\eta = \lambda W_2 \eta + \varepsilon. \tag{13.27}$$

where $\varepsilon \sim N(0, \sigma^2 I)$, and W_1 and W_2 are row-normalized spatial weights matrices. In order for the model to be identifiable, either W_1 cannot equal W_2 or the matrix X cannot be the trivial matrix consisting of a vector of ones (Anselin et al., 1996, p. 81). We will assume that X does indeed contain at least one true explanatory variable. The problem of distinguishing between the two models then becomes one of testing two separate hypotheses. The first is a test of $H_0 : \rho = 0$ against $H_a : \rho \neq 0$, and the second is a test of $H_0 : \lambda = 0$ against $H_a : \lambda \neq 0$. Anselin et al. (1996) carry out the test using an alternative to the likelihood ratio test called the *Lagrange multiplier test*. The next subsection contains a description of this test.

13.4.2 The Lagrange Multiplier Test

Suppose we have a statistical model with a parameter θ and we wish to test the null hypothesis $H_0 : \theta = \theta_0$ against the alternative $H_a : \theta \neq \theta_0$ using the method of maximum likelihood. As discussed in Section 12.2, one way to do this is to use the likelihood ratio test. If l_F is the value of the log likelihood function $l(\theta \mid z)$ at maximum likelihood estimate when θ is not restricted to the value $\theta = \theta_0$ (the full model) and l_R is the value of the log likelihood function when θ is restricted to $\theta = \theta_0$ (the restricted model), then the likelihood ratio test involves computing the statistic $G = -2\log(L_F / L_R) = 2(l_F - l_R)$ (Equation 12.4) and comparing its value with that of a chi-square distribution. For complex likelihood functions like those that arise from Equation 13.27, however, the likelihood function can be prohibitively difficult to compute, and therefore one would like to find a test involving a simpler function. One such test is the Lagrange multiplier test, also known as the score test.

The method of Lagrange multipliers is described in Appendix A.4. It is applied to the maximum likelihood test described in the first paragraph of this section as follows (Theil, 1971, p. 527; Fox, 1997, p. 570). Let $l(\theta \mid Y)$ be the log likelihood function, and let θ_{ML} be the maximum likelihood estimator of θ. Consider the constrained maximization problem of maximizing $l(\theta \mid Y)$ subject to the constraint $\theta = \theta_{ML}$. Formulating this as a Lagrange multiplier problem implies finding the maximum of $l(\theta \mid Y) + \gamma(\theta_{ML} - \theta)$. Applying the Lagrange multiplier equations (Equation A.46) yields

$$\frac{\partial l}{\partial Y} - \gamma\theta = 0$$
$$\frac{\partial l}{\partial \gamma} = \theta_{ML} - \theta = 0. \tag{13.28}$$

Thus, the value of the Lagrange multiplier γ is the slope $\partial l / \partial Y$ of the log likelihood function. The larger this slope at a particular value of θ, the further this value of θ is from θ_{ML}.

The slope of the log likelihood function at the value θ is called the *score* and is denoted $S(\theta)$. This does not appear to have simplified the problem, since we still must compute the derivative of the log likelihood. It turns out, however, that our effort does lead to a simplification. Anselin (1988a) developed a way to test hypotheses involving the model of Equation 13.27 using the Lagrange multiplier test, and Anselin et al. (1996) developed a simpler and more robust form of this test that functions equally well. Anselin and Rey (1991) compared the Lagrange multiplier test applied to model (Equation 13.27) with a hypothesis test based on the Moran's I statistic and concluded that the Lagrange multiplier test is more appropriate, and this conclusion was further supported by the work of Anselin et al. (1996).

Both the original Lagrange multiplier test of Anselin (1988a) and the robust version of Anselin et al. (1996) are available in the function lm.LMtests() in the package spdep. The tests are carried out on the model

$$Y = \rho WY + X\beta + \eta$$

$$\eta = \lambda W\eta + \varepsilon. \tag{13.29}$$

This is the same as the model of equations (Equation 13.27) except that $W_1 = W_2 = W$. This is acceptable since we have specified that X is nontrivial. The tests are successively that $\lambda = 0$ and that $\rho = 0$. If either of these tests returns a nonsignificant p value, then we cannot reject the null hypothesis that the corresponding autocorrelation coefficient is zero.

To implement the test, we first generate the spatial lag (Equation 13.16) and spatial error (Equation 13.21) models for the regression problem. Then we call the function lm.LMtests() with the argument tests = "all", which runs all of the tests on the two models.

```
> model.5 <- lm(Yield ~ Clay + SoilP + I(Clay*SoilP) + Weeds,
>     data = data.Set4.1)
> library(spdep)
> coordinates(data.Set4.1) <- c("Easting", "Northing")
> nlist <- dnearneigh(data.Set4.1, d1 = 0, d2 = 61)
> lm.LMtests(model.5, W, test = "all")
        Lagrange multiplier diagnostics for spatial dependence
LMerr = 1.2303, df = 1, p-value = 0.2673
LMlag = 10.9906, df = 1, p-value = 0.0009157
RLMerr = 8.1514, df = 1, p-value = 0.004303
RLMlag = 17.9116, df = 1, p-value = 2.314e-05
SARMA = 19.142, df = 2, p-value = 6.972e-05
```

Most of the output of the function is suppressed; only the relevant p values are shown. The p values indicate a preference for the spatial lag model (Equation 13.16). The ecological interpretation of this result is that the crop yield values are directly associated with each other, as opposed to being associated with unmeasured, spatially autocorrelated processes that are loaded into the error. This may seem a bit counterintuitive and warrants further discussion, which we give at the end of Section 13.7. The SARMA, or spatially autoregressive moving average, model is also not rejected. This model is not covered here. In the next section, we will develop the methods necessary to fit both a spatial lag model and a spatial error model to these data.

13.5 Fitting the Spatial Lag and Spatial Error Models

In this section we discuss the problem of estimating the coefficients of the spatial lag model (Equation 13.16) and the spatial error model (Equation 13.21). The fitting of both models involves a similar procedure, and so we only discuss the spatial error model. The mathematics of this section is fairly complex and need not be totally mastered on first reading.

To begin, we review the solution by maximum likelihood of the simple linear regression problem

$$Y = X\beta + \varepsilon, \tag{13.30}$$

where ε is a vector with elements $\varepsilon_i \sim N(0,\sigma^2)$ and the ε_i are mutually independent (i.e., there is no spatial autocorrelation). Assume that the explanatory variables X_i are mathematical variables as defined in Appendix A.2. The case in which the X_i are random variables requires a longer argument to reach the same conclusion.

For notational simplicity, we will consider the case in which $p = 2$. The problem we must resolve is that when we compute the likelihood function $L(\beta_0,\beta_1,\sigma^2 \mid Y)$, we are computing the likelihood in which the vector Y, with components consisting of the random variables Y_i, is the parameter, but the distribution we know is that of the random variables ε_i. Specifically, the ε_i satisfy $\varepsilon_i \sim N(0,\sigma^2)$. Since the ε_i are the only random variables in Equation 13.30, the Y_i have the same distribution as the ε_i, and therefore the Y_i are also independent and normally distributed with variance σ^2. The fact that the Y_i are independent permits us to write the likelihood as a product of normal density functions, but if they were not independent, then in order to write the likelihood as a product we would have to transform the variables from the Y_i into the ε_i. In reality, we are also doing a transformation when the errors ε_i are independent, but this transformation is just the identity $Y = I\varepsilon = \varepsilon$, so we can ignore it. When we deal with autocorrelated errors, however, the transformation will not be the identity and we will not be able to ignore it.

When the ε_i are independent the likelihood function for Equation 13.30 is (Kutner et al., 2005, p. 27)

$$L(\beta_0,\beta_1,\sigma^2 \mid Y) = \prod \frac{1}{(2\pi\sigma^2)^{1/2}} \exp\left[-\frac{1}{2}\left(\frac{Y_i - \beta_0 - \beta_1 X_i}{\sigma}\right)^2\right]$$

$$= \frac{1}{(2\pi\sigma^2)^{n/2}} \exp\left(-\frac{1}{2\sigma^2}\sum_{i=1}^n (Y_i - \beta_0 - \beta_1 X_i)^2\right). \tag{13.31}$$

Computing $l(\beta_0,\beta_1,\sigma^2 \mid Y) = \log L(\beta_0,\beta_1,\sigma^2 \mid Y)$ and differentiating yields

$$\frac{\partial l}{\beta_0} = \frac{1}{\sigma^2}\sum_{i=1}^n (Y_i - \beta_0 - \beta_1 X_i)$$

$$\frac{\partial l}{\partial \beta_1} = \frac{1}{\sigma^2}\sum_{i=1}^n X_i(Y_i - \beta_0 - \beta_1 X_i) \tag{13.32}$$

$$\frac{\partial l}{\partial \sigma^2} = -\frac{n}{2\sigma^2} + \frac{1}{2\sigma^4}\sum_{i=1}^2 (Y_i - \beta_0 - \beta_1 X_i)^2.$$

Solving these yields

$$\sum_{i=1}^{n}(Y_i - \hat{\beta}_0 - \hat{\beta}_1 X_i) = 0$$

$$\sum_{i=1}^{n}X_i(Y_i - \hat{\beta}_0 - \hat{\beta}_1 X_i) = 0 \tag{13.33}$$

$$\frac{\sum_{i=1}^{n}(Y_i - \hat{\beta}_0 - \hat{\beta}_1 X_i)^2}{n} = \hat{\sigma}^2.$$

The first two equations are identical to the normal equations of ordinary least squares regression (Kutner et al., 2005, p. 17), and the third equation is a biased estimator of σ^2.

Now consider the autoregressive case, specifically, the spatial error model given by Equation 13.29. Analogous to Equations 13.30 and 13.31, we must compute a likelihood function $L(\beta, \sigma^2, \lambda \mid Y)$, where β is the vector of regression parameters. However, Y now has the same probability distribution as η rather than ε, and whereas the ε are independent, the η are not. Moreover, in order to determine the probability distribution of the η we need to know λ, but λ is one of the parameters we are trying to estimate. Therefore, in order to compute the likelihood function, we must use the relationship $\eta = \lambda W\eta + \varepsilon$ to re-express in terms of η the probability density that we currently have in terms of ε. In doing so, we have to be careful to preserve properties of the likelihood as a probability density function (Appendix A.5.1).

In Appendix A.6, it is shown that the equation for the likelihood L, expressed in terms of η, is

$$L(\beta, \lambda, \sigma^2 \mid \eta) = |\det A| \frac{1}{(2\pi\sigma^2)^{n/2}} \exp\left(-\frac{1}{2\sigma^2}(A\eta)'A\eta\right). \tag{13.34}$$

where the matrix A is defined by

$$\varepsilon = (I - \lambda W)\eta \equiv A\eta. \tag{13.35}$$

Taking logarithms and expressing the equation $\eta = Y - X\beta$ yields

$$l(\beta, \lambda, \sigma^2 \mid Y) = \log|\det A| - \frac{n}{2}\log\sigma^2 - $$
$$\frac{1}{2\sigma^2}\left(Y'A'AY - 2\beta'X'A'AY + \beta'X'A'AX\beta\right) + C \tag{13.36}$$

where C is a constant incorporating all the terms that do not depend on the parameters of the model. In order to obtain maximum likelihood estimates of the parameters, one must differentiate Equation 13.36 with respect to these parameters and set these partial derivatives equal to zero.

There are a number of different ways to maximize the log likelihood (Equation 13.36). The method described by Upton and Fingleton (1985, p. 285) provides a good example. If λ were known, it would be a straightforward if somewhat messy task to differentiate (Equation 13.36) and solve the resulting equations. Indeed, differentiating l with respect to β and σ^2 yields

$$\frac{\partial l}{\partial \beta} = -\frac{1}{2\sigma^2}(-2X'A'AY + 2X'A'AX\beta)$$

$$\frac{\partial l}{\partial \sigma^2} = -\frac{n}{\sigma^2} + \frac{1}{2\sigma^4}(Y'A'AY - 2\beta'X'A'AY + \beta'X'A'AX\beta)$$

(13.37)

Setting these equal to zero and solving yields

$$\hat{\beta} = (X'A'AX)^{-1}X'A'AY$$

$$\hat{\sigma}^2 = \frac{1}{n}(Y'A'AY - 2\beta'X'A'AY + \beta'X'A'AX\beta)$$

(13.38)

The matrix depends on λ, which we do not know. If we did, we could compute A and substitute the second of Equations 13.37 into Equation 13.36 to get

$$l(\beta, \lambda, \sigma^2 \mid Y) = \log|\det A| - \frac{n}{2}\log\hat{\sigma}^2 + C$$

(13.39)

We do not know λ, however, so we must also estimate its value. Since $A = I - \lambda W$, $\det A$ is an nth degree polynomial in λ, which considerably complicates things (Cliff and Ord, 1981, p. 155). Usually one assumes a value of λ and then uses a numerical scheme analogous to the Newton-Raphson method described in Appendix A.5.2 to find the zeroes of the derivatives in Equations 13.37, and then one iterates this with solutions of Equation 13.39 until a maximum of l is reached. Most software packages make use of some variant of this approach. We will postpone the application of the theory to data until after the next section, in which we discuss an alternative model for spatial autoregression.

13.6 The Conditional Autoregressive Model

The *conditional autoregressive*, or CAR, model is based on an alternative description of the probabilistic structure of the data. The models discussed in Section 13.3 describe the *joint* probability density functions of the Y_i. Although the terminology is not completely consistent, many authors refer to models such as these as *simultaneous autoregressive* or SAR models. The CAR model is constructed by specifying the *conditional* probability density of one of the Y_i given all of the others. To introduce the CAR formalism, we return to the simple time series model $Y_i - \mu = \rho(Y_{i-1} - \mu) + \varepsilon_i$, $i = 1, 2, \ldots$ given in Equation 13.12 for the special case $\mu = 0$. This model could also be expressed as (Cliff and Ord, 1981, p. 146)

$$E\{Y_i \mid Y_{i-1}, Y_{i-2}, \ldots\} = \mu + \rho(Y_{i-1} - \mu),$$

$$\text{var}\{Y_i \mid Y_{i-1}, Y_{i-2}, \ldots\} = \sigma^2.$$

(13.40)

Equations 13.40 convey the idea that the properties of Y_i are conditional on those of the preceding values in the time series. In the case of a time series, the formulations, Equation 13.40 and Equation 13.12, are equivalent, but in the case of spatial models they are not (Cliff and Ord, 1981, p. 146). Nevertheless, the CAR model can be developed by analogy to the conditional time series formulation (Equation 13.40). Specifically, let Y_{-i} denote the set of all members of the vector Y *except* Y_i. Then we can write the CAR formulation as (Waller and Gotway, 2004, p. 371)

$$E\{Y_i \mid Y_{-i}\} = X\beta + \sum_{j \neq i} c_{ij}(Y_j - X\beta),$$

$$\text{var}\{Y_i \mid Y_{-i}\} = \sigma^2,$$

(13.41)

where $c_{ij} = c_{ji}$ and the conditional expectations are assumed to be normally distributed. Waller and Gotway (2004, p. 371) provide a nice analogy of this process in a classroom setting. Imagine a classroom in which the desks are arranged in a set of rows and columns. The instructor hands each student a slip of paper that only he or she can see and that contains the value of some quantity, say, temperature, measured at the location of that student's desk. The instructor then tells the students to raise their hands to a level that indicates the value of that quantity. The students are aware that the temperature of one desk would be close to the temperature at a nearby desk, so they surreptitiously peek around and see how high their neighbors are raising their hands, and each student adjusts his or her own hand accordingly. The students are thus conditioning their response on the values of those of their neighbors.

Equations 13.41 can easily be generalized to the case $\text{var}\{Y_i \mid Y_{-i}\} = \sigma_i^2$, that is, to the case in which the variances are not identical. We will not make this generalization here; the interested reader is referred to Waller and Gotway (2004, p. 371). There (see also Cliff and Ord, 1981, p. 180) it is shown that for fixed σ^2 the joint distribution of the Y_i is multivariate normal with mean $X\beta$ and variance $\Sigma_Y = \sigma^2(I - C)^{-1}$, where C is the symmetric matrix whose elements are the c_{ij}.

In order to ensure a well-specified probability structure, a set of conditions contained in a theorem known as the Hammersley-Clifford theorem must be satisfied. Haining (1990, p. 87) provides a discussion of these conditions, but for our purposes it is sufficient to say that the conditions are satisfied by any practical example that we will consider (Waller and Gotway, 2004, p. 372).

It turns out that any SAR model of the form (Equation 13.21) may be expressed as a CAR model where $C = W + W' - WW'$. However, CAR models have different properties and applications from SAR models and should be considered differently (Ripley, 1981, p. 88). One advantage of CAR models is that they may be solved using ordinary least squares (Haining, 1990, p. 130). A disadvantage of the CAR model is that it may happen that in order to ensure a valid least-squares estimate, part of the data may have to be left unused in the estimation (Upton and Fingleton, 1985, p. 363; Haining, 1990, p. 133). This latter problem, however, should not affect models constructed with a first-degree nearest-neighbor spatial weights matrix. From the perspective of this book, perhaps the most significant aspect of the CAR formalism is that it fits well into a Bayesian analysis, which will be covered in Chapter 14.

13.7 Application of Simultaneous Autoregressive and Conditional Autoregressive Models to Field Data

13.7.1 Fitting the Data

We will continue to work with model.5 for the data of Field 1 of Data Set 4. As a reminder, the model is

$$Yield_i = \beta_0 + \beta_1 Clay_i + \beta_2 SoilP_i + \beta_3 Clay_i \times SoilP_i + \beta_4 Weeds_i + \varepsilon_i \qquad (13.42)$$

The results in Section 13.4.2 lead us to believe that the errors ε_i in this model may display significant spatial autocorrelation.

Our analysis follows those of Bivand et al. (2013b, Ch. 10), Waller and Gotway (2004, Ch. 9), and Schabenberger and Gotway (2005, Ch. 6). Before we begin the analysis, we must make one very important modification. Fitting the data to a spatial lag or spatial error model involves computing the determinant of the matrix A in Equation 13.39, which is a *Jacobian* matrix (Appendix A.6). It turns out that if the ranges of the variables in the model are very different, then the numerical methods involved in the computation may fail (Bivand et al., 2013b, p. 292). For this reason, before beginning the analysis we will center and scale the variable *Yield*, which will reduce the size of the regression coefficients and make them more uniform. As usual, the data are loaded into R using the code in Appendix B.4. The data frame data.Set4.1 holds the contents of *set4.196sample.csv*, and the data frame data.Yield4.1idw holds the yield monitor values interpolated to the 86 sample points.

```
> data.Set4.1$Yield <- scale(data.Yield4.1idw$Yield)
```

Here is a portion of the output of the function summary() for model.5 with the scaled yield.

```
> formula.5 <- as.formula("Yield ~ Clay + SoilP +
+     I(Clay*SoilP) + Weeds")
> model.5 <- lm(formula.5, data = data.Set4.1)
> summary(model.5)
Coefficients:
                  Estimate Std. Error t value Pr(>|t|)
(Intercept)       9.318959   1.075379   8.666 3.64e-13 ***
Clay             -0.229174   0.026715  -8.578 5.42e-13 ***
SoilP            -0.628251   0.144961  -4.334 4.17e-05 ***
I(Clay * SoilP)   0.016974   0.003724   4.558 1.81e-05 ***
Weeds            -0.275797   0.050309  -5.482 4.64e-07 ***
```

Since heteroscedasticity of errors can lead to problems with spatial data, we will check for it using the Levene test and the Breusch-Pagan test (Kutner et al., 2005, p. 116). The former is available in the car package (Fox and Weisberg, 2011), and the latter in the lmtest package (Zeileis and Hothorn, 2002).

```
> library(car)
> library(lmtest)
> bptest(model.5)
        studentized Breusch-Pagan test
data:   model.5
BP = 8.614, df = 4, p-value = 0.0715
```

```
> fits.med <- median(fitted(model.5))
> fits.groups <- fitted(model.5) <= fits.med
> leveneTest(residuals(model.5),factor(fits.groups))
Levene's Test for Homogeneity of Variance (center = median)
      Df F value Pr(>F)
group  1  8.4328 0.004707 **
```

The tests give very different results in this case, but there is an indication of heteroscedasticity. If this is indeed a problem, the best option to address it is often to use a weighted regression (Kutner et al., 2005, p. 421). This is especially true if the error variance is a function of one of the explanatory variables. A plot of the residuals against each of the explanatory variables (not shown), however, did not reveal any such relationship. Therefore, we will not attempt any weighting.

We saw in Section 13.5 that, based on the output of the function lm.LMtests(), the spatial lag model appeared to fit the data better than the spatial error model. We will now fit both of these models and compare them. First, we fit a spatial lag model to the data.

```
> model.5.lag <- lagsarlm(formula.5,
+ data = data.Set4.1, listw = W)
> summary(model.5.lag)
Type: lag
Coefficients: (asymptotic standard errors)
                  Estimate Std. Error z value Pr(>|z|)
(Intercept)      5.9514012  1.1511819  5.1698 2.343e-07
Clay            -0.1431834  0.0286813 -4.9922 5.969e-07
SoilP           -0.4732151  0.1295233 -3.6535 0.0002587
I(Clay * SoilP)  0.0124313  0.0033396  3.7224 0.0001974
Weeds           -0.1866198  0.0460086 -4.0562 4.988e-05

Rho: 0.47975, LR test value: 15.146, p-value: 9.9522e-05
Asymptotic standard error: 0.098315
    z-value: 4.8798, p-value: 1.0621e-06
Wald statistic: 23.812, p-value: 1.0621e-06
Log likelihood: -55.11086 for lag model
ML residual variance (sigma squared): 0.19638, (sigma: 0.44314)
Number of observations: 86
Number of parameters estimated: 7
AIC: 124.22, (AIC for lm: 137.37)
LM test for residual autocorrelation
test value: 2.8897, p-value: 0.089145
```

Next, we fit the spatial error model.

```
> model.5.err <- errorsarlm(formula.5,
+    data = data. Set4.1, listw = W)
> summary(model.5.err)
Type: error
Coefficients: (asymptotic standard errors)
                  Estimate Std. Error z value  Pr(>|z|)
(Intercept)      8.8116785  1.1684911  7.5411 4.663e-14
Clay            -0.2153939  0.0291188 -7.3971 1.392e-13
SoilP           -0.5946674  0.1540439 -3.8604 0.0001132
I(Clay * SoilP)  0.0156392  0.0039488  3.9605 7.478e-05
Weeds           -0.2389056  0.0528680 -4.5189 6.216e-06
```

```
Lambda: 0.28077, LR test value: 2.121, p-value: 0.14529
Asymptotic standard error: 0.13267
    z-value: 2.1163, p-value: 0.034318
Wald statistic: 4.4788, p-value: 0.034318
Log likelihood: -61.62322 for error model
ML residual variance (sigma squared): 0.23978, (sigma: 0.48968)
Number of observations: 86
Number of parameters estimated: 7
AIC: 137.25, (AIC for lm: 137.37)
```

The spatial error model and the spatial lag model are not nested, so we cannot compare them directly. The Akaike information criterion is lower for the spatial lag model, so this would lead us to tend to prefer that model, which supports the conclusion reached in Section 13.4 based on the Lagrange Multiplier test. The coefficients of the two models are somewhat different, but not substantially so.

Before moving on, we demonstrate the problem of working with an unscaled *Yield* variable.

```
> data. Set4.1$YieldUS <- yield.pts$Yield
> lagsarlm(YieldUS ~ Clay + SoilP +
+    I(Clay*SoilP) + Weeds, data = data. Set4.1, listw = W)
Error in solve.default(inf, tol = tol.solve):
  system is computationally singular: reciprocal condition number =
1.22513e-13
```

If you get this message with your own data and the simple expedient of scaling does not make it go away, don't despair! There are several alternative things that you can try, which are listed in the Help pages for the functions.

In addition to the spatial error model and the spatial lag model, the spdep package also has the capacity to compute a solution to a CAR model (Bivand et al., 2013b, p. 282). This is done using the function spautolm(). This function has an argument family, which if unspecified, will by default lead to the function fitting a spatial error model. A CAR model solution is obtained by specifying family = "CAR".

```
> W.B <- nb2listw(nlist, style = "B")
> model.5.car <- spautolm(formula.5,
+    data = data. Set4.1, family = "CAR", listw = W.B)
> summary(model.5.car)
Coefficients:
                  Estimate Std. Error z value  Pr(>|z|)
(Intercept)      8.7688570  1.1577547  7.5740  3.619e-14
Clay            -0.2136640  0.0289191 -7.3883  1.488e-13
SoilP           -0.5934295  0.1528116 -3.8834  0.000103
I(Clay * SoilP)  0.0155192  0.0039264  3.9525  7.733e-05
Weeds           -0.2418481  0.0532106 -4.5451  5.491e-06
Lambda: 0.16705 LR test value: 3.0812 p-value: 0.0792
Log likelihood: -61.14311
ML residual variance (sigma squared): 0.22943, (sigma: 0.47899)
Number of observations: 86
Number of parameters estimated: 7
AIC: 136.29
```

Comparing all of the coefficients yields the following.

```
> print(coef(model.5), digits = 2)
   (Intercept)          Clay          SoilP I(Clay * SoilP)          Weeds
         9.319        -0.229         -0.628           0.017         -0.276
> print(coef(model.5.lag), digits = 2)
      rho    (Intercept)    Clay      SoilP    I(Clay * SoilP)          Weeds
    0.480          5.951  -0.143     -0.473              0.012         -0.187
> print(coef(model.5.err), digits = 2)
   lambda    (Intercept)    Clay      SoilP    I(Clay * SoilP)          Weeds
    0.281          8.812  -0.215     -0.595              0.016         -0.239
> print(coef(model.5.car), digits = 2)
   (Intercept)     Clay      SoilP    I(Clay * SoilP)          Weeds        lambda
         8.769   -0.214     -0.593              0.016         -0.242         0.167
```

In general, the coefficients are similar between the models.

In summary, we have found that the "best" linear regression model as determined by Mallow's C_p indicates spatially autocorrelated errors, and that the indicated error structure is best represented by a spatial lag model of the form (Equation 13.16). It seems counterintuitive that the spatial lag model fits the data better than the spatial error model, since it is unlikely that wheat plants would influence each other at a distance of 61 m. It is very important in this context to remember that correlation does not imply causality. Each of the explanatory variables *SoilP*, *Weeds*, and *Clay* is likely to be autocorrelated. The values of *SoilP* and *Clay* are highly influenced by soil-forming processes, and weed level is likely influenced by the weed seed bank, which likely displays interactive autocorrelation. However, yield response to these autocorrelated explanatory variables, as well as to autocorrelated variables not included in the model, may cause it to display a level of positive autocorrelation sufficient for the lag model to provide the best fit to the data. This is especially true if the uncorrelated errors ε tend to be larger in magnitude than the correlated errors η.

13.7.2 Comparison of the Mixed Model and Spatial Autoregression

Lambert et al. (2004) compared the generalized least squares (GLS) approach of Section 12.5 and the spatial autoregression (SAR) approach of this section with each other and with OLS, the trend surface (TS) method (Tamura et al., 1988) and the nearest-neighbor (NN) method (Papadakis, 1937), in the analysis of data from a precision agriculture experiment. The latter two methods are discussed in Chapter 16. They found that the GLS and SAR methods were able to detect the interaction of topography with nitrogen in the explanation of yield, while the other three methods were not. The fit of the GLS method was generally considered more accurate than that of the ordinary least squares (OLS) methods but somewhat less accurate than that of the SAR method. Lambert et al. (2004) also found that the standard error of the regression coefficients was considerably less with the GLS and SAR methods than with the other methods. They point out that the GLS method needs more data than the SAR method because it must estimate a variogram. Wall (2004) has criticized the use of the SAR and CAR models for data on an irregular mosaic of polygons. Using an example from the continental United States, she shows that, partly because of the issue of topological invariance discussed in Section 3.5.2 and illustrated in Figure 3.8, these models can lead to counterintuitive results. An alternative for this situation is the mixed model, but unfortunately, the calculation of the experimental variogram necessary to implement this approach is also fraught with difficulties with data on a small, irregular mosaic. As always, the best approach is to try as many methods as possible and attempt to understand the reason behind any difference that appear in the results.

13.8 Further Reading

Anselin (1992) provides an excellent, nontechnical introduction to spatial regression. Other good sources are Upton and Fingleton (1985, 1989), Griffith and Layne (1999), Waller and Gotway (2004), and Schnabenberger and Pierce (2001). LeSage and Pace (2009) provide a thorough discussion at a somewhat higher level. Bivand et al. (2013b) contains an extensive discussion and comparison of functions for spatial autoregression. Dormann et al. (2007) presents a comparative study of several different models for dealing with spatially autocorrelated data, and in a supplement several of the authors provide examples of R code to implement these models. A method called the autologistic model has been proposed for the analysis of binary data (Besag, 1974; Augustin et al., 1996). Dormann et al. show that his method has several problems that make it generally less desirable than other methods for the analysis of this type of data.

Lichstein et al. (2002) present a very thorough and extensive comparison of OLS and CAR models, with and without trend terms, for habitat suitability models for southern Appalachian songbirds. Diniz-Filho et al. (2003) and Hawkins et al. (2007) questioned the superiority of spatial regression methods to ordinary least squares in ecological applications. These papers prompted a lively exchange in the literature (Beale et al., 2007; Diniz-Filho et al., 2007). Waller and Gotway (2004) provide a thorough discussion of the issue of heteroscedastic residuals in spatial regression models. Faraway (2002, p. 62) discusses the implementation of weighted regression in R.

For a basic discussion of the Lagrange multiplier test, see Fox (1997) and Theil (1971). Further information is provided by Buse (1982) and Engle (1984).

Exercises

13.1 Show that the autoregressive time series model $Y_i - \mu = \lambda(Y_i - \mu) + \varepsilon_i$ can be written as a spatial error model.

13.2 The test for spatial autocorrelation in Equation 13.2 was carried out on the model model.5, which has the smallest number of explanatory variables of any of the candidate models. Test other models with more explanatory variables for autocorrelated errors. Why might you observe what you do?

13.3 The most complex leaps() model developed in Section for Field 4.1 is model.3. Test this model to distinguish between the spatial lag model and the spatial error model. Is there a difference between model.3 and model.5?

13.4 Using the spatial lag model, compare the effect of the inclusion of spatial autocorrelation on the coefficients of each of the six candidate models for yield in Field 4.1.

13.5 In this book we always generate autocorrelated data using the method of Haining (1990). However, one can also generate a *Gaussian random field* (Wood and Chan, 1994). The geoR package (Ribeiro and Diggle, 2005) contains the function grf(), which develops a Gaussian random fields simulation of autocorrelated data. (a) Read about this function and then use it to generate a variable X distributed as a Gaussian random field on a 20 by 20 grid; (b) Use the geoR function variog() to display the experimental variogram of X.

13.6 Use the Gaussian random field data from Exercise 13.4 to generate a binomially distributed random variable Y based on a logistic probability on X. Then use the GLM model to fit these data and compute the bias in the fit.

14

Bayesian Analysis of Spatially Autocorrelated Data

14.1 Introduction

Bayesian statistics is based on *Bayes' rule*, which is named after the Rev. Thomas Bayes, who discussed it in a paper published posthumously in 1763. We can introduce the concept as follows. Let A and B be two events. For example, suppose we are sampling leaves in a wheat field to determine whether the mean nitrogen level of the plants in the field exceeds the recommended minimum for an adequate supply of nitrogen. Let A be the event that the mean N level of a particular leaf sample exceeds the recommended minimum, and let B be the event that mean N level over all of the plants in the field exceeds the recommended minimum. The *joint probability*, denoted $P\{A, B\}$ is the probability that both events A and B occur. The *conditional probability* that A occurs given that B occurs is denoted $P\{A \mid B\}$, and one can similarly define the conditional probability $P\{B \mid A\}$ that B occurs given that A occurs. The laws of probability state that (Larsen and Marx, 86, p. 42)

$$P\{A, B\} = P\{B \mid A\}P\{A\} = P\{A \mid B\}P\{B\}, \tag{14.1}$$

and these can be use to formally define the conditional probability. Bayes' rule is obtained by dividing the second equation in (Equation 14.1) by $P\{A\}$ (Koop, 2003, p. 1):

$$P\{B \mid A\} = \frac{P\{A \mid B\}P\{B\}}{P\{A\}}. \tag{14.2}$$

If we have observed the event A, that the mean N level of the sample exceeds the minimum, then Bayes' rule gives us a means to compute the probability that B occurs, that is, that the field mean N level exceeds the minimum.

Let us now see how Bayesian statistics is applied to data rather than events. Instead of expressing the model in terms of events A and B as in the previous paragraph, we will employ as data a vector (which may have only one component) Y of observed values of a random variable or variables, and a vector θ of parameters describing the distribution of these random variables. Rather than considering the probability of events, we consider probability *densities*, which we denote with a lower case p. For example, let Y again be mean N level of a sample of plants in the field (in this case the vector Y does have only one component), and let θ be the mean plant leaf nitrogen content for the entire population of plants. Then Bayes' rule becomes

$$p(\theta \mid Y) = \frac{p(Y \mid \theta)p(\theta)}{p(Y)}. \tag{14.3}$$

The key distinction between Bayesian statistics and classical statistics is in how the parameters contained in the parameter vector θ are treated. In the classical, or *frequentist*, interpretation of statistics, the vector θ is not a random variable but rather a parameter or set of parameters fixed by nature, and thus it makes no sense to speak of its probability distribution. In the *Bayesian* interpretation of statistics, the vector θ *is* a random variable, and its distribution represents the state of our certainty about its value (in this case, the mean plant N content in the field). This interpretation is called *subjective probability*. The density $p(\theta)$ is called the *prior probability density*, or just the *prior*. It represents the state of our knowledge about the value of θ before making an observation. The density $p(Y \mid \theta)$ is called the *likelihood*, and indeed it has the same form as the likelihood described in Appendix A.5 and in our earlier discussions about maximum likelihood, although its interpretation is different in a Bayesian context. In the context of Appendix A.5, the likelihood is considered as a function $L(\theta \mid Y)$, and the objective is to determine a value of Y to maximize θ. In the Bayesian context, $p(Y \mid \theta)$ describes the probability density of observations Y for each value of θ, or, roughly speaking, the probability of observing values of Y given values of θ. The density $p(Y)$ is called the *marginal density* and in principle can be computed as

$$p(Y) = \int p(Y \mid \theta)p(\theta)d\theta. \tag{14.4}$$

The density $p(\theta \mid Y)$ in Equation 14.3 is called the *posterior density* and represents the updated belief about the value of θ taking into account the observation Y. Some measure of central tendency such as the mean or median of these densities may be taken to be a good point estimate of θ given the available knowledge, and a measure of spread such as the variance or quantile may be taken as representative of the degree of uncertainty about that value.

The following illustration of how Bayes' rule works in this context is taken from Box and Tiao (1973, p. 15). Suppose there is a parameter θ of a variable Y that is to be estimated. We continue with the example that Y is the concentration of nitrogen in the tissue of a sample of wheat plants in a field, and θ is the mean value over the whole field. Suppose further that one scientist, called Scientist A, is fairly sure that the value of θ lies between 880 and 920. We interpret "fairly sure" to mean that one standard deviation of Scientist A's belief has the value 20. Therefore, we can represent this belief using a normal probability density as

$$p_A(\theta) \sim N(900, 20^2). \tag{14.5}$$

This is the subjective probability of Scientist A prior to collecting data. Suppose that another scientist, Scientist B, is a little less sure of the value of θ but thinks it is lower, specifically between 720 and 880. Scientist B's prior can be represented as

$$p_B(\theta) \sim N(800, 80^2). \tag{14.6}$$

The curves in Figure 14.1 show the prior distributions $p_A(\theta)$ and $p_B(\theta)$. Suppose now that a single observation Y_1 of the variable Y is made, with mean $Y_1 = 850$. Let σ_1 be the standard deviation of the measurement, representing the magnitude of uncertainty in the

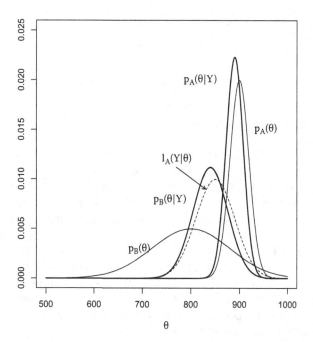

FIGURE 14.1
Prior, likelihood, and posterior densities of the beliefs of Scientist A and Scientist B regarding the value of the parameter θ.

measurement, and suppose $\sigma_1 = 40$. This measurement is interpreted as the likelihood function, so that we can write

$$l(Y \mid \theta) = l(850 \mid \theta) \sim N(850, 40^2). \tag{14.7}$$

With this information, we can compute the posterior densities $p_A(\theta \mid Y)$ and $p_B(\theta \mid Y)$. The formula for $p_A(\theta \mid Y)$ is

$$
\begin{aligned}
p_A(\theta \mid Y) &= \frac{l(Y \mid \theta)p_A(\theta)}{\displaystyle\int_{-\infty}^{\infty} l(Y \mid \theta)p_A(\theta)d\theta} \\[2mm]
&= \frac{f_N(\theta; 850, 40^2)f_N(\theta, 900, 20^2)}{\displaystyle\int_{-\infty}^{\infty} f_N(\theta; 850, 40^2)f_N(\theta, 900, 20^2)d\theta},
\end{aligned} \tag{14.8}
$$

where $f_N(\theta, \mu, \sigma^2)$ represents the normal density function. It turns out that this computation can be done analytically. The usual symbols for the parameters of the prior and posterior distribution involve bars above and below the character. However, since we use the bars to denote sample means, we will use the very nonstandard notation $\hat{\theta}$ and $\hat{\sigma}^2$ to denote parameters of the prior distribution and $\breve{\theta}$ and $\breve{\sigma}^2$ to denote parameters of the posterior distribution. Given a normal prior $p(\theta) \sim N(\hat{\theta}, \hat{\sigma}^2)$ and a likelihood $l(\theta \mid Y) \sim N(Y, \sigma_1^2)$, Box and Tiao (1973, p. 74) show that $p(\theta \mid Y) \sim N(\breve{\theta}, \breve{\sigma}^2)$ where

$$\breve{\theta} = \frac{(\hat{\tau}\hat{\theta} + \tau_1 Y)}{\hat{\tau} + \tau_1}, \quad \frac{1}{\breve{\sigma}^2} = (\hat{\tau} + \tau_1)$$

$$\hat{\tau} = \frac{1}{\hat{\sigma}^2}, \quad \tau_1 = \frac{1}{\sigma_1^2}.$$

(14.9)

The quantities $\hat{\tau}$ and τ_1, which are the inverse of the variances $\hat{\sigma}^2$ and σ_1^2, are called the *precision* of the prior and the data, respectively, and are commonly used instead of the variance in formulas in Bayesian statistics. Plugging in the numbers for Scientist A, we have $\hat{\theta} = 900$, $\hat{\sigma}^2 = 20^2 = 400$, $Y_1 = 850$, and $\sigma_1^2 = 40^2 = 1600$. The result is that $p_A(\theta \mid Y)$ is a normal density with mean 890 and standard deviation 17.9. Similarly, $p_B(\theta \mid Y)$ is a normal density with mean 840 and standard deviation 35.7. Figure 14.1 shows these results. The measurement $Y_1 = 850$ is less than the mean value of Scientist A's prior and greater than that of Scientist B, so Scientist A's posterior is reduced a bit and Scientist B's posterior is increased. The standard deviations of both scientists' posteriors are reduced, reflecting greater certainty as a result of new information. The effect both on the mean and the standard deviation is greater for Scientist B, who was less certain in the first place.

This simple example illustrates two very important aspects of Bayesian analysis. The first is that in order to compute the posterior density one must compute the marginal density by evaluating an integral. In the case of the normal density, this can be done analytically, but for more general densities it would have to be done numerically. The second aspect is that in this example the posterior density has the same form as the prior; both are normal. This is a very special and convenient property of the normal density. It means that if a second measurement is made, the posterior just computed can serve as the prior for a second calculation. For other densities besides the normal, the likelihood must be carefully chosen, if it exists at all, in order for the posterior density to have the same form as the prior. A prior that, given the appropriate likelihood function, produces a posterior whose density has the same form as the prior is called a *conjugate prior*. These two factors, the need to evaluate an integral and the need for a conjugate prior and corresponding likelihood function, were the primary reasons why, before the advent of reasonably powerful computers, Bayesian calculations had to be carried out using a very restricted set of prior and likelihood functions (e.g., Plant and Wilson, 1985). The great advance in Bayesian statistics in recent years has been the development of powerful numerical techniques that remove this restriction.

When, as in this example, the prior and posterior form a conjugate pair, the posterior $p(\theta \mid Y)$ can be used as the prior for a second measurement, and the process may be repeated. To continue the example of Box and Tiao (1973, p. 18), suppose that n independent observations of Y are drawn from a normal population with mean θ and variance σ^2. Then the likelihood is normally distributed with mean $\bar{Y}$ variance σ^2/n. Therefore, from Equation 14.9, the posterior is normally distributed with mean and variance (or precision) given by

$$\breve{\theta} = \frac{(\hat{\tau}\hat{\theta} + \tau_1 \bar{Y})}{\hat{\tau} + \tau_n}, \quad \frac{1}{\breve{\sigma}^2} = (\hat{\tau} + \tau_n)$$

$$\hat{\tau} = \frac{1}{\hat{\sigma}^2}, \quad \tau_n = \frac{n}{\sigma_1^2}.$$

(14.10)

In our example, with 100 observations drawn from a population with a mean of 870 and a variance of 40^2, the posterior has a distribution $p_A(\theta \mid Y) \sim N(871.2, 3.9^2)$.

An important feature of Bayes' rule as formulated in Equation 14.3 is that the denominator does not depend on θ, and thus for a given set of data Y, the denominator can be regarded as a constant. Thus we can write

$$p(\theta \mid Y) \propto p(Y \mid \theta)p(\theta), \tag{14.11}$$

where the symbol $\propto$ means "is proportional to." The verbal statement of Expression 14.11, "the posterior is proportional to the likelihood times the prior," is something of a Bayesian mantra. It means that, although the exact value of the posterior density $p(\theta \mid Y)$ cannot be determined, one can determine its shape from Expression 14.11 without having to compute the integral in Equation 14.4.

The last concept in Bayesian statistics to discuss in this section is the *noninformative prior* (Box and Tiao, 1973, p. 32; Koop, 2003, p. 6). Informally speaking, a prior density is noninformative if the posterior density $p(\theta \mid Y)$ is unaffected, or almost unaffected, by the parameters of the prior. To continue the example above, if the prior has mean 890 and variance 10^6 and again the likelihood has mean 850 and variance 40^2, then the posterior has mean 850.06 and variance 1597.44. Thus, the high value of the prior variance causes the prior to provide little information about the values of the posterior.

The availability of inexpensive, powerful computers beginning in the 1980s motivated the development of numerical methods that eliminate the need to restrict Bayesian analysis to likelihood functions that are associated with conjugate priors. In the next section, we discuss the most commonly used of these methods, and their application to linear and generalized linear regression models.

14.2 Markov Chain Monte Carlo Methods

Following Koop (2003), the first application we consider from a Bayesian perspective is simple linear regression. To further simplify the discussion, we initially assume that the explanatory variable X and the response variable Y have been centered, so that there is no intercept term. The model is then

$$Y_i = \beta X_i + \varepsilon_i, \ i = 1,\ldots,n, \tag{14.12}$$

where the $\varepsilon_i \sim N(0, 1/\tau)$ (from now on we will express the normal distribution in terms of τ rather than σ^2) are independent and identically distributed and the X_i are either a set of values of a mathematical variable or a set of values of a random variable independent of the ε_i. For now we will treat the values of the explanatory variable X as fixed (the mathematical variable case). The *gamma distribution* plays a major role in Bayesian linear regression analysis. This distribution is characterized by two parameters that we will denote μ and v. The formula for the gamma density function can be written in several ways; the formula we will use can be specified in R as

$$f_G(Y \mid \mu, v) = \begin{cases} \dfrac{v^{-\mu} Y^{\mu-1} e^{-xv}}{\Gamma(\mu)}, & \text{if } 0 < Y < \infty \\ 0 & \text{otherwise}, \end{cases} \tag{14.13}$$

where $\Gamma(\mu)$ is the gamma function (Abramowitz and Stegun, 1964). The gamma density has mean μ/v and variance μ/v^2. The parameter μ is called the *shape* parameter, and v is called the *rate* parameter. We will use the notation $G(\mu,v)$ to denote a gamma density with parameters μ and v as in Equation 14.13.

The unknown parameters in the linear regression model (Equation 14.12) are the regression coefficient β and the error precision $\tau = 1/\sigma^2$. Remember that in the Bayesian formulation, these are random variables, as was the case in Section 14.1. We begin the analysis with a prior probability density for the vector $\theta = (\beta,\tau)'$ that represents our subjective belief in the values of the parameters. The subjective formulation requires a prior probability density $p(\beta,\tau)$ that represents our belief prior to data collection, and a likelihood $p(Y \mid \beta,\tau)$ that represents the probability of observing the data that we collect given the values of β and τ. If we don't know anything about these values, we use a noninformative prior.

The conjugate prior for the normal regression model is a probability density called the *normal-gamma*. The mathematics of this density are discussed by Koop (2003, Ch 2–3). The advantage of using a conjugate prior, as we did in Section 14.1, is that we can obtain an analytical expression for the posterior. The problem with using the normal-gamma density as a prior is that one cannot independently specify prior probabilities for the regression coefficient β and the error precision τ. Instead, one must first specify the prior for the error precision τ and then specify the prior for β conditional on τ. This is often inconvenient and motivates the consideration of numerical methods for dealing with non-conjugate priors. This is the approach we will take.

It is very common, and convenient, in the numerical Bayesian analysis of linear regression models to specify a normal prior for β and a gamma prior for the precision τ. We will write the priors as

$$p(\beta) \sim N(\hat{\beta},\hat{\tau}_\beta), \;\; p(\tau) \sim G(\hat{\mu}_\tau,\hat{v}_\tau), \tag{14.14}$$

to indicate that the prior for β is a normal density with mean $\hat{\beta}$ and precision $\hat{\tau}_\beta$, and the prior for τ is a gamma density with shape and rate parameters $\hat{\mu}_\tau$ and $\hat{v}_\tau$. Remember that the hats are used to denote the parameter values of the prior.

Since β and τ are independent, the joint prior $p(\beta,\tau)$ is the product of the two priors in Equation 14.14. The likelihood is the same as the likelihood function developed for linear regression in Appendix A.5 and given in Equation A.50, except of course that β_0 is not present in Equation 14.12 because in the present case it has the value 0. From Equation A.50, the likelihood is given by

$$L(\beta,\sigma^2 \mid Y) = \frac{\tau}{(2\pi)^{n/2}} \exp(-\tau \sum_{i=1}^{n} (Y_i - \beta X_i)^2 / 2). \tag{14.15}$$

Using the relationship in Expression 14.11, that the posterior is proportional to the likelihood times the prior, gives the following

$$p(\beta,\tau \mid Y) \propto p(Y \mid \beta,\tau)p(\beta)p(\tau). \tag{14.16}$$

Here we have used the fact that β and τ are independent, so that $p(\beta,\tau) = p(\beta)p(\tau)$, where $p(\beta)$ is the normal prior of β and $p(\tau)$ is the gamma prior of the precision τ. Plugging in the thee densities (likelihood plus two priors) and substituting gives

$$p(\beta,\tau \mid Y) \propto \tau^{n/2} \exp\left(-\frac{\tau}{2}\left[\sum_{i=1}^{n}(Y_i - \beta X_i)^2\right]\right) \exp\left(-\frac{\hat{\tau}_\beta}{2}(\beta - \hat{\beta})^2\right) \times$$

$$\tau^{\hat{\mu}-1} \exp(-\tau \hat{v}).$$

(14.17)

Note that the first part of the right-hand side (the first exponential function), which comes from the likelihood, contains the parameters τ and β, while the second part, which comes from the prior, contains their values in the prior $\hat{\beta}$ and $\hat{\tau}_\beta$. Koop (2003, p. 61) shows that given these densities, the posterior is proportional to the product of a normal density and a gamma density,

$$p(\beta,\tau \mid Y) \propto N(\breve{\beta}, \breve{\sigma}_\beta^2) G(\breve{\mu}_\tau, \breve{v}_\tau),$$

(14.18)

where the symbols N and G again represent normal and gamma densities, respectively. Because of Expression 14.18, if τ is fixed, then the density of β conditional on τ is a normal density, and if β is fixed, then the density of τ conditional on β is a gamma density. The parameters in Expression 14.18 are, in our notation,

$$\breve{\tau}_\beta = \hat{\tau}_\beta + \tau \sum_{i=1}^{n} X_i^2$$

$$\breve{\beta} = \frac{1}{\breve{\tau}_\beta}\left(\frac{\hat{\beta}}{\hat{\sigma}_\beta^2} + \tau \sum_{i=1}^{n} X_i Y_i\right)$$

(14.19)

$$\breve{\mu}_\tau = n + \hat{\mu}_\tau$$

$$\breve{v}_\tau = \frac{\hat{v}_\tau + \sum_{i=1}^{n}(Y_i - \beta X_i)^2}{2}$$

These equations are a bit complicated, so let's sort through them before moving on. Before we do, however, note that the first two equations define the parameters of the conditional probability density $p(\beta \mid \tau, Y)$, and the second two equations define the parameters of the gamma density $G(\mu, v)$. Keep this in mind, because it will be important when we discuss the Gibbs sampler below.

The quantities we are trying to estimate are the regression coefficient β and the precision τ, which is the inverse of the error variance. These appear themselves in the right-hand sides of three of the four equations (Equation 14.19) because the first two equations define the parameters of the posterior density of β conditional on τ, and the second two equations define the parameters of the posterior density of τ conditional on β. The parameters of the prior densities, with the hats (i.e., $\hat{\beta}$ and so forth), also appear on the right-hand sides, and the posterior parameter $\breve{\tau}_\beta$ also appears on the right-hand side of the second equation, because the value of β is conditional on the value of τ. On the left-hand sides are all of the parameters of the posterior density, with the inverted hats. This means, however, that the equations that we use to estimate β and τ have β and τ themselves on their right-hand sides. Since we need β and τ to estimate β and τ, we are going to have to use some sort of iterative procedure.

The posterior density defined in Expression 14.18 does not have a convenient form that would allow us to express its mean and variance analytically, the way we can with a normal or a gamma density. Therefore, to compute these posterior parameters, we must use a numerical algorithm. The algorithm we will describe can be used to calculate a wide range of parameters, but we will focus on $\breve{\beta}$, the posterior mean, the estimated value of the regression coefficient β, and $\breve{\tau}$, the estimated precision (inverse of the variance) of the error, both after taking into account the observed values of the data.

To simplify things, let us return for the moment to the notation of Section 14.1, where we were dealing with only a single parameter θ. Consider the numerical calculation of the posterior mean,

$$\breve{\theta} = E\{\theta \mid Y\} = \int_{-\infty}^{\infty} \theta p(\theta \mid Y)d\theta. \tag{14.20}$$

A theorem of probability theory called the *weak law of large numbers* states (Larsen and Marx, 1986, p. 191) that if we draw a sequence of random numbers $\{\theta^{(0)}, \theta^{(1)}, ..., \theta^{(n)}\}$ from the probability distribution $p(\theta \mid Y)$, then as n increases the sample mean $\Sigma\theta^{(i)}/n$ converges to the expected value $\breve{\theta} = E\{\theta \mid Y\}$ of this distribution. This use of a sequence of random numbers to compute an expected value is called *Monte Carlo integration*. We mention in passing that this can also be extended to state that if we draw a sequence $\{g(\theta_1), g(\theta_2), ..., g(\theta_n)\}$ from the distribution, then the sample mean of this sequence converges to $E\{g(\theta)\} = \int_{-\infty}^{\infty} g(\theta)p(\theta \mid Y)d\theta$. However, for now we will stick to the case $g(\theta) = \theta$.

Thus, we would like to draw a sequence of random values $\theta^{(i)}$ from the posterior distribution $p\{\theta \mid Y\}$, because this sequence would enable us to compute a quantity that converges to $\breve{\theta}$. Although we don't know $p\{\theta \mid Y\}$ (if we did, we could compute $E\{\theta \mid Y\}$ directly), we do know its shape from Expression 14.18 ("the posterior is proportional to the likelihood times the prior"), and we can use this in the numerical calculation of the parameters in Equation 14.19 by employing Monte Carlo methods.

One very common algorithm for the numerical calculations described in the previous paragraph is called the *Gibbs sampler* (German and German, 1984). In order to minimize the notational complexity, we will describe the Gibbs sampler for the specific problem of Equation 14.12, $Y_i = \beta X_i + \varepsilon_i$, $i = 1, ..., n$, $\varepsilon_i \sim N(0, 1/\tau)$, in which we are interested in two quantities, the regression coefficient β and the precision τ.

From Expression 14.18, the posterior density $p(\beta, \tau \mid Y)$ is proportional to the product of a normal density and a gamma density. The conditional probability density $p(\beta \mid \tau, Y)$ is obtained by holding τ to a fixed value, and therefore it must be the case that this conditional density is normal, that is, $\beta \mid \tau, Y \sim N(\hat{\beta}, \hat{\tau}_\beta)$. Similarly, $p(\tau \mid \beta, Y)$ must be a gamma density, that is, $\tau \mid \beta, Y \sim G(\hat{\mu}, \hat{v})$. Consider the case of calculating $\breve{\beta}$. We would like to use Monte Carlo integration by drawing random numbers from the posterior density $p(\beta, \tau \mid Y)$ so that we could apply the weak law of large numbers to obtain a sequence $\beta^{(0)}, \beta^{(1)}, \beta^{(2)}, ...$ that converges to the expected value in Equation 14.20, which in this case is $\breve{\beta}$. We can't draw from $p(\beta, \tau \mid Y)$ because we don't know it. We can, however, draw from the conditional densities $p(\beta \mid \tau, Y)$ and $p(\tau \mid \beta, Y)$ since they are a normal and a gamma, respectively. The Gibbs sampler is based on drawing from these densities.

The Gibbs sampler works like this. Suppose we could find an initial value $\tau^{(0)}$ drawn from $p(\tau \mid \beta, Y)$. We can't, because we don't know the parameters of the distribution, but suppose we could. Step 1 of the Gibbs sampler algorithm is to substitute this value for τ into the first two equations of Equation 14.19, which define the parameters of the conditional density $p(\beta \mid \tau, Y)$. Step 2 is to draw a value of $\beta^{(0)}$ from this density and substitute it for β in the second two equations of Equation 14.19 to produce a gamma density. From this density, we draw second value $\tau^{(1)}$, which we then plug into the first two equations to produce a density from which to draw $\beta^{(1)}$. We continue alternating in this way and generate random sequences of $\beta^{(i)}$ and $\tau^{(i)}$ that we can use in Monte Carlo integration.

The difficulty with this algorithm is that we do not have a density from which to draw the $\tau^{(0)}$ to start with. However, it turns out that the starting values of the sequences often do not matter, and that ultimately the alternating $\beta^{(i)}$ and $\tau^{(i)}$ do come from the correct densities even if the initial values of the sequence do not. Since we compute the expected value of a parameter by averaging all of the values of the sequence, we don't want to include the initial values that do not come from the correct density. Therefore, the usual procedure in employing the Gibbs sampler is to initialize the process by first generating sequences of length B of values $\beta^{(0)}, ..., \beta^{(B-1)}$ and $\tau^{(0)}, ..., \tau^{(B-1)}$, called the *burn-in values*, and then to discard these. One continues with the sequence and uses the values $\beta^{(B)}, ..., \beta^{(n)}$ and $\tau^{(B)}, ..., \tau^{(n)}$ in Monte Carlo integration to estimate $\bar{\beta}$ and $\bar{\tau}$.

The Gibbs sampler is one of a class of methods called *Markov Chain Monte Carlo* (or *MCMC*) methods. We have already seen enough Monte Carlo methods in earlier chapters to know where this term comes from, but what about the Markov Chain part? A *Markov Chain* is a stochastic sequence in which the value of each member of the sequence depends explicitly only on the value in the sequence that immediately precedes it (as well as random noise). We have already seen an example of a Markov Chain: the autoregressive time series $Y_i - \mu = \lambda(Y_{i-1} - \mu) + \varepsilon_i$, $i = 1, 2, ...$ of Equation 3.13. Note that each value Y_i in this sequence depends explicitly only on Y_{i-1} and not on earlier values of Y. This characterizes a Markov Chain. In the same way, the values $\{\beta^{(i)}, \tau^{(i)}\}$ of the Gibbs sampler depend explicitly only on $\{\beta^{(i-1)}, \tau^{(i-1)}\}$, and therefore this sequence forms a Markov Chain. For this reason, the sequence is often referred to as a *chain*.

We will illustrate the Gibbs sampler using artificial data. The code is based on a similar example given by Ntzoufras (2009).

```
> beta.true <- 1
> set.seed(123)
> n <- 20
> print(X <- rnorm(n))
 [1] -0.56047565 -0.23017749  1.55870831  0.07050839  0.12928774
 [6]  1.71506499  0.46091621 -1.26506123 -0.68685285 -0.44566197
[11]  1.22408180  0.35981383  0.40077145  0.11068272 -0.55584113
[16]  1.78691314  0.49785048 -1.96661716  0.70135590 -0.47279141
> print(Y <- beta.true * X + rnorm(n))
 [1] -1.62829935 -0.44815240  0.53270387 -0.65838284 -0.49575153
 [6]  0.02837168  1.29870325 -1.11168812 -1.82498979  0.80815295
[11]  1.65054602  0.06474234  1.29589711  0.98881620  0.26573995
[16]  2.47555339  1.05176813 -2.02852887  0.39539324 -0.85326241
> summary(lm(Y ~ X))
Call:
lm(formula = Y ~ X)
Coefficients:
            Estimate Std. Error t value Pr(>|t|)
(Intercept) -0.04017    0.19197  -0.209 0.836588
X            0.92174    0.20027   4.603 0.000221 ***
```

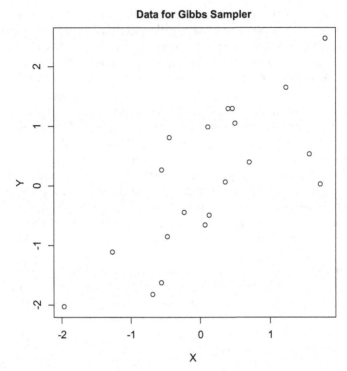

FIGURE 14.2
Scatterplot of Y versus X for the artificial data set used to illustrate the Gibbs sampler.

Figure 14.2 shows a scatterplot of X and Y. Next, we choose parameters $\hat{\beta}$ and $\hat{\tau}_\beta = 1/\hat{\sigma}_\beta^2$ for the prior $p(\beta)$ and $\hat{\mu}_\tau$ and $\hat{v}_\tau$ for the prior $p(\tau)$. Recall that a noninformative prior is one that has a large variance, reflecting little or no knowledge of the values. We will use noninformative priors, with the priors for both β and τ having a variance of 100. The mean of $p(\beta)$ is set at 0 and the mean of $p(\tau)$ at 1. Recall that the gamma density has mean μ/v and variance μ/v^2. Therefore, if $\hat{\mu}_\tau$ and $\hat{v}_\tau$ are each assigned the value 0.01, then the prior density $p(\tau)$ has mean 1 and variance 100, which means that it is basically noninformative.

```
> beta.pri <- 0
> s2.pri <- 100
> mu.pri <- 0.01
> nu.pri <- 0.01
```

The next step is to pick initial value $\tau^{(0)}$. We want it to be small to reflect a low precision.

```
> cur.tau <- 0.01
```

Now we generate an array to hold our chain of $\beta^{(i)}$ values and carry out the Gibbs sampler algorithm. We generate 5000 iterations using a for loop. This could also be done using the replicate() function, but we use a for loop in order to compare this code with WinBUGS code in the next section.

```
> MCMC.data <- numeric(5000)
> for (i in 1:5000){
+ # Use  current tau to generate conditional posterior of beta
+     sb2.post <- 1/(1/s2.pri + cur.tau*sum(X^2))
+     beta.post <- sb2.post*(beta.pri/s2.pri+ cur.tau*sum(X*Y))
+ # Draw beta from the current conditional posterior
+     cur.beta <- rnorm(1, beta.post, sqrt(sb2.post))
+ # Use current beta to generate conditional posterior of tau
+     mu.post <- n + mu.pri
+     nu.post <- (nu.pri + sum((Y-cur.beta*X)^2)) / 2
+ # Draw tau from the current conditional posterior
+     cur.tau <- rgamma(1, shape = mu.post, rate = nu.post)
+     MCMC.data[i] <- c(cur.beta)
+ }
```

Now we get rid of the first 1000 iterations as the burn-in values and compute the mean of the remaining ones. It is reasonably close to the value generated by least squares regression above.

```
> burn.in <- 1000
> beta.MCMC <- MCMC.data[-(1:burn.in)]
> mean(beta.MCMC)
[1] 0.9153175
```

Figure 14.3 shows the values of $\beta^{(i)}$ for the first 200 iterations of the burn-in sequence. The values of $\beta^{(i)}$ appear to have in this example already settled into a steady sequence of random movement about the mean.

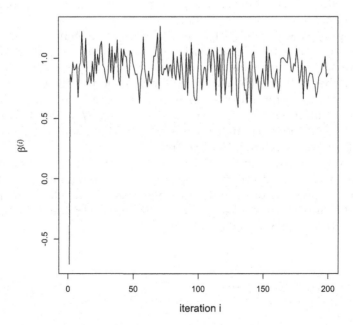

FIGURE 14.3
Plot of the first 200 iterations of the Gibbs sampler applied to the artificial data set of Figure 14.2.

Markov Chain Monte Carlo methods are not always so well behaved, and the most important phase of the analysis when using them is the diagnostic phase. This is a circumstance where most people who practice Bayesian analysis turn to a program specialized for this process. The most commonly used programs are descendents of the BUGS (for Bayesian inference Using Gibbs Sampling) program developed by statisticians at the Medical Research Council in Cambridge, England. Currently the two most powerful versions are WinBUGS (Lunn et al., 2000), which, as its name suggests, works under Microsoft Windows, and OpenBUGS (http://www.openbugs.net/w/FrontPage), developed at the University of Helsinki by a group led by Andrew Thomas. We will use the former. The next section introduces WinBUGS and describes how to access it from R.

14.3 Introduction to WinBUGS

14.3.1 WinBUGS Basics

As of the writing of this book, the most recently developed version of WinBUGS is Version 1.4.3. This will probably remain the most recent version for the foreseeable future, since, according to the website, future development efforts will all be devoted to OpenBUGS. Both programs are associated with R packages that can generate scripts from within R. WinBUGS is associated with R2WinBUGS (Sturtz et al., 2005) and OpenBUGS is associated with BRugs (Thomas et al., 2006), although OpenBUGS can also be run from R2WinBUGS. At present, WinBUGS is more widely used and arguably more stable, and so it is the one we will choose. WinBUGS Version 1.4.3, along with a user manual and associated data and scripts, can be downloaded from the BUGS website (https://www.mrc-bsu.cam.ac.uk/software/bugs/the-bugs-project-winbugs/). The program is free to use, and a permanent license key is provided.

Although graphical methods exist for developing WinBUGS models, we will focus on the WinBUGS programming language, which is quite similar in many respects to the R language. We will not make any attempt to provide a comprehensive introduction to WinBUGS but rather will provide just enough information to allow the reader to create, execute, and carry out diagnostic analysis of models developed in this chapter. Several sources of further information are given in Section 14.7. We begin by reanalyzing the simple linear regression model of the previous section. The initial description of the use of WinBUGS is taken from McCarthy (2007, p. 249).

In this section we will work in WinBUGS itself, and in the next section we will return to R. Once you have installed WinBUGS, open it in the usual manner by clicking on the WinBUGS icon. You will be presented with an application window. The first step is to open a new code window using *File->New*. Copy the following code from the file 14.3.1.txt in the R code for this chapter into the window.

```
model
{
beta ~ dnorm(0, 0.01)
tau ~ dgamma(0.01, 0.01)
for (i in 1:20)
  {
  Y.hat[i] <- beta * X[i]
```

```
  Y[i] ~ dnorm(Y.hat[i], tau)
  }
}
```

This code specifies the Bayesian regression model. The first two lines after the initial curly bracket specify that the parameters of the model are β and τ, that the prior density for β is normal with mean 0 and precision 0.01 (i.e., variance 100), and that the prior for τ is a gamma density with mean 1 and variance 100 (i.e., scale parameter 0.01 and rate parameter 0.01). Note that in WinBUGS one specifies in dnorm() the precision (the inverse of the variance) of the normal distribution, and in R one specifies in rnorm() the standard deviation (the square root of the variance). The statement Y[i] ~ dnorm(Y. hat[i], tau) says that the variable Y is normally distributed, but this statement does not actually generate variables, as would an R statement of the form Y <- rnorm(Y. hat, sigma2). The rest of the model statement specifies the likelihood, which is the part that describes the regression model. This is specified using a for loop, where the first line specifies the deterministic part of the model, βX_i and the second line specifies that Y_i is drawn from a normal distribution with mean βX_i and precision τ.

Next, we input the data, which are the values of the X_i and corresponding Y_i specified in the previous section. This is done with a list statement.

```
list(
X=c(-0.56047565, -0.23017749, 1.55870831, 0.07050839, 0.12928774,
1.71506499, 0.46091621, -1.26506123, -0.68685285,
 -0.44566197, 1.22408180, 0.35981383, 0.40077145, 0.11068272,
-0.55584113, 1.78691314, 0.49785048, -1.96661716,
0.70135590, -0.47279141),
Y=c(-1.62829935, -0.44815240, 0.53270387, -0.65838284, -0.49575153,
0.02837168, 1.29870325, -1.11168812, -1.82498979, 0.80815295,
1.65054602, 0.06474234, 1.29589711, 0.98881620, 0.26573995,
2.47555339, 1.05176813, -2.02852887, 0.39539324,
 -0.85326241))
)
```

Notice that the model statement uses curly brackets and the list statement uses parentheses. The concatenate function c() is used in the same manner as in R, and the data were just copied in from the R output in Section 14.2. Finally, we must specify the values of $\beta^{(0)}$ and $\tau^{(0)}$.

```
list(beta=0, tau = 0.01)
```

This completes the WinBUGS program. There are many similarities between the WinBUGS and R programming languages, and these may serve to make the subtle differences more confusing. Let's go through the two code sequences and compare them. The first two lines in the model statement of the WinBUGS code are

```
beta ~ dnorm(0,0.01)
tau ~ dgamma(0.01, 0.01)
```

These establish the prior densities of β and τ. These statements involve the use of a tilde. In R, the tilde is used, for example, in functions such as lm() to express the form of a linear model for example with a statement like lm(Y ~ X, data = field.data). In WinBUGS, the tilde is used to establish the distribution to which a random variable belongs (cf. expressions like [Equation 14.5]), so that for example the statement beta ~ dnorm(0,0.01) indicates that the prior distribution of β is normal with mean $\hat{\beta} = 0$ and precision $\hat{\tau} = 1/\hat{\sigma}^2 = 0.01$. In the R code of Section 14.2, we did not explicitly specify the form of the priors; rather we simply set their parameter values.

```
> s2.pri <- 100
> beta.pri <- 0
> mu.pri <- 0.01
> nu.pri <- 0.01
```

The next set of code in the WinBUGS model statement above is a for loop.

```
for (i in 1:20)
  {
  Y.hat[i] <- beta * X[i]
  Y[i] ~ dnorm(Y.hat[i], tau)
  }
```

This establishes the relationship between X and Y. WinBUGS does not have the facility to vectorize assignment statements. In R, which does have this ability, the first assignment would be done with a simple statement like Y.hat <- beta * X. Note that the second statement in the WinBUGS code establishes for each Y_i the regression relationship $Y \sim N(\beta X_i, 1/\tau)$, which is equivalent to $Y_i = \beta X_i + \varepsilon_i$, $\varepsilon_i \sim N(0, 1/\tau)$. Note also that in the WinBUGS code the values of the X[i] have not yet been set. In R, this would generate an error, since R executes code line by line. WinBUGS code is compiled, however, so it is only executed when all statements have been loaded. That completes the WinBUGS model statement. In R we had to explicitly specify the form of the likelihood and the Gibbs sampler algorithm, but in WinBUGS this is handled for us.

To execute the model, do the following.

1. Open the Specification window by selecting *Model->Specification*.
2. Highlight the word model in the model description window (the window in which you did the typing) by double clicking on it.
3. Click on *check model* in the Specification window. The phrase "model is syntactically correct" should appear in the lower left corner of the WinBUGS widow, and the *load data* and *compile* buttons in the Specification window should no longer be grayed out.
4. Highlight the first word list and click on the *load data* button. The phrase *data loaded* should appear in the lower left corner of the WinBUGS window.
5. Click on the *compile* button in the Model Specification window. The phrase *model compiled* should appear in the lower left corner of the WinBUGS window and the two *inits* buttons should no longer be grayed out.
6. Highlight the second word list and click on the *load inits* button. The phrase *model initialized* should appear in the lower left corner of the WinBUGS window.

Now we are ready to run the model.

7. Select *Model->Update* and *Inference->Samples*. These two windows will pop up right on top of the Model Specification window, so drag them out of each other's way.

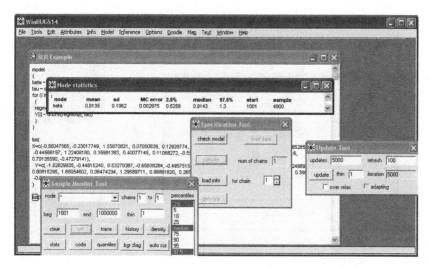

FIGURE 14.4
WinBUGS window after completion of the fitting of the regression model to the artificial data set.

8. In the Update window, change *updates* from 1000 to 5000. In the Sample Monitor window, change *beg* from 1 to 1001. This establishes the first 1000 iterations as the burn-in.

9. In the *node* box of the Sample Monitor window, type *beta* and click on *set*.

10. In the Update window, click on *update*. This runs the model.

11. In the *node* box of the Sample Monitor window, type a "*" and click on *stats*. Your screen should now resemble Figure 14.4. The estimated value of β is 0.9135, about the same as the values we have obtained previously.

In this case things look pretty good, but unfortunately in Bayesian inference things can sometimes look pretty good and not be so good at all. The most important phase of a Bayesian analysis is the diagnostic phase, which is the topic of the next subsection.

14.3.2 WinBUGS Diagnostics

Our principal use of WinBUGS diagnostics will be to determine that the MCMC method has produced a convergent sequence like that of Figure 14.3 so that the integral in Equation 14.20 is accurately estimated. The initial checks are graphical and are accomplished with the *Sample Monitor* window. Assuming that the WinBUGS run from the previous section is still available, from this window click on the *history, density, quantiles*, and *auto cor* buttons. You should see something like Figure 14.5, in which the other windows have been closed to make room (don't close your windows yet). The history window, labeled *Time series*, shows the values of $\beta^{(i)}$ from the end of the burn-in to the end of the chain. The sequence should look like patternless noise around a constant mean, as it does in Figure 14.5. In particular, there should be no trend and no pattern in the variation. The density window, labeled *Kernel density*, shows a histogram of the $\beta^{(0)}$ values. It should look smooth and should resemble the appropriate density, if the form of this is known. In the present case, it looks like a normal density (more or less). The quantiles window, labeled *Running quantiles*, shows an approximation of the moving quantiles of the sequence. The quantiles should be fairly constant. The *auto cor* window, labeled *Autocorrelation function*, shows the

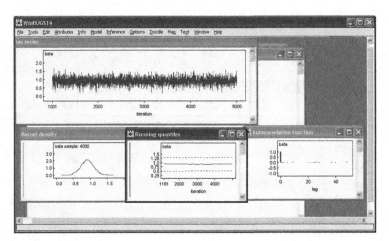

FIGURE 14.5
WinBUGS diagnostic windows.

autocorrelation value as a function of the lag. Since the sequence $\beta^{(i)}$ is an autoregressive time series, values of $\beta^{(i)}$ are correlated with values of $\beta^{(i-1)}$, $\beta^{(i-2)}$, and so forth. In the case of the model shown in Figure 14.5, this autocorrelation dies off quickly, but sometimes it does not. It is desirable to use an uncorrelated sequence of values in calculations such as Equation 14.20. If the *auto cor* window indicates a substantial correlation at a particular lag, the appropriate procedure is to run the model with the *thinning* parameter, which represents the number of values of the time series to skip in assembling the sequence used in the calculations, to a value just greater than the autocorrelation lag. This is set using the *thin* button in the *update* window. While the chain is still in memory, click the *coda* button. This will open two windows, an *index* window and a *chain* window. Copy the data from these windows into a text file and save the two files. We will use them shortly.

The button denoted *brg diag* in the Sample Monitor window allows the user to carry out a convergence check using the Gelman-Rubin diagnostic (Gelman and Rudin, 1992). This basically carries out an analysis of variance on the data (Ntzoufras, 2009, p. 143), and in order to employ it one must generate several chains simultaneously. Start WinBUGS and again go through the steps 1 through 11 given above, with the following changes.

1. Prior to clicking on *check model* in the Specification window, change the number in the *num of chains* window to 5.
2. Click on the *gen inits* button. This generates a random set of initial values for each of the five chains.

After the run, click the *brg diag* button in the Sample Monitor window. A plot similar to that in the upper left corner of Figure 14.6 should appear. This plot shows the variance ratio and convergence statistics for the run, which should both converge to 1. The actual numerical values displayed graphically can be obtained as follows. First double click on the graph window, and then hold down the *Ctrl* key and left click again. This brings up the window shown on the right in Figure 14.6. There are other diagnostics that we can run, but we will defer a discussion of these until the next section. This completes our introduction of WinBUGS. We now turn to a discussion of the package R2WinBUGS, which permits the user to execute WinBUGS from within R.

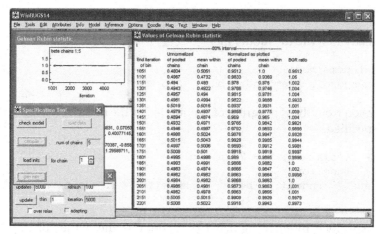

FIGURE 14.6
WinBUGS convergence statistics.

14.3.3 Introduction to R2WinBUGS

The `R2WinBUGS` package does not eliminate the need to know how to construct a WinBUGS model, but it does make it very easy to execute the model from within R. We will demonstrate the use of the package by continuing with our simple example. The first part of the program is identical to the R program of Section 14.2.

```
> library(R2WinBUGS)
> beta.true <- 1
> set.seed(123)
> n <- 20
> X <- rnorm(n)
> Y <- beta.true * X + rnorm(n)
```

The next step is to specify the model. This is done by writing an R function that contains within it the WinBUGS code.

```
> demo.model <- function(){
+ beta ~ dnorm(0, 0.01)
+ tau ~ dgamma(0.01, 0.01)
+ for (i in 1:n)
+   {
+   Y.hat[i] <- beta * X[i]
+   Y[i] ~ dnorm(Y.hat[i], tau)
+   }
+ }
```

Notice that this is the same code (literally) as is in the WinBUGS program of Section 14.3.2. It is *not* R code. Next, we use the function `write.model()` to write this model to disk. We will use the character string `mybugsdir` to contain the directory where this file is written. You need to establish this directory on your disk. Suppose you have set up the folders (i.e., directories), as described in Section 2.1.2, and you wanted to use a subfolder of your folder *C:\rdata\SDA2* called *WinBUGS* to house your WinBUGS files. First, you would use the Windows explorer to create the folder, and then you would enter the following statement.

```
> mybugsdir <- "c:\\rdata\\SDA2\\WinBUGS\\"
```

Now we are ready to write the model file.

```
> write.model(demo.model,
+     paste(mybugsdir,"demomodel.bug", sep = ""))
```

The first diversion from the R program of Section 14.2 is the manner in which the data and the inits are specified.

```
> XY.data <- list(X = X, Y = Y, n = n)
> XY.inits <- function(){
+     list(beta0 = rnorm(1), beta1 = rnorm(1), tau = exp(rnorm(1)))
+}
```

The data are specified by creating a list that contains the names of the data vectors in the model on the left side of the equals sign and the source of the data on the right side. The inits are specified by a function that tells how to generate the initial values. We want the values of tau to be positive, so we make its distribution lognormal. The use of a function to generate the inits corresponds roughly to the use of the *gen inits* button in WinBUGS as discussed in Section 14.3.2, and is necessary if there is to be more than one chain. There is some discussion in the literature of the merits of one long chain versus several short ones, but there are some important advantages to the use of more than one chain (the advantages to be described below), and we will always use more than one.

Now we call the function bugs(), which actually runs the model. The first time we run it, we do so as a debugging test by setting the argument debug = TRUE and by specifying a very small number of iterations with the value of the argument n.iter. The argument debug has the default value FALSE. When debug is FALSE, if there is a problem in WinBUGS, the error messages will flash by on the screen too quickly for you to read, and then the program will close. When debug is set to TRUE, WinBUGS does not close, and therefore you can read the error messages. Just make sure that once the program is running correctly, you remove the assignment demo = TRUE. Otherwise, when the program is finished, you and the computer will patiently wait for each other to make the next move, and the computer is much more patient than you are.

```
> demo.test <- bugs(data = XY.data, inits = XY.inits,
+     model.file = paste(mybugsdir,"demomodel.bug", sep = ""),
+     parameters = c("beta0", "beta1", "tau"), n.chains = 1,
+     n.iter = 10, n.burnin = 2, debug = TRUE,
+     bugs.directory=mybugsdir)
```

Assuming this runs in WinBUGS, hit the ESC key while in the RStudio Console to return to R.

The next step is a full run including diagnostics. Diagnostic evaluation in R is carried out using the coda (Plummer et al., 2011) or BOA (Smith, 2007) package. We will use coda. Use the assignment codaPkg = TRUE to be able to view the coda diagnostics. We can remove the debug = TRUE assignment and specify nchains = 5, n.iter = 5000, and n.burnin = 1000.

```
> library(coda)
> demo.coda <- bugs(data = XY.data, inits = XY.inits,
+     model.file = paste(mybugsdir,"demomodel.bug", sep = ""),
+     parameters = c("beta0", "beta1", "tau"), n.chains = 1,
```

```
+     n.iter = 5000, n.burnin = 1000, codaPkg = TRUE,
+     bugs.directory=mybugsdir)
```

This time you won't see anything in the WinBUGS screen because it is written to a file. Use read.bugs() to read the coda file and then type codamenu(). Your screen should look like this.

```
> library(coda)
> demo.mcmc <- read.bugs(coda.file)
> codamenu()
CODA startup menu

1: Read BUGS output files
2: Use an mcmc object
3: Quit

Selection:
```

The coda package is menu driven. The first step is to load the output from the WinBUGS files. Type 2. Then, when prompted for the name of the saved object, type it in as shown.

```
Enter name of saved object (or type "exit" to quit)
1:demo.mcmc
Checking effective sample size...OK
CODA Main Menu
1: Output Analysis
2: Diagnostics
3: List/Change Options
4: Quit
Selection:
```

Type 1 to access the output analysis.

```
CODA Output Analysis menu
1: Plots
2: Statistics
3: List/Change Options
4: Return to Main Menu
Selection: 1
```

To access the plots, type 1. This produces the plots shown in Figure 14.7. Now type 4 to return to the main menu, and select choice 2 from the menu. This leads to the following.

```
CODA Diagnostics Menu

1: Geweke
2: Gelman and Rubin
3: Raftery and Lewis
4: Heidelberger and Welch
5: Autocorrelations
6: Cross-Correlations
7: List/Change Options
8: Return to Main Menu

Selection:
```

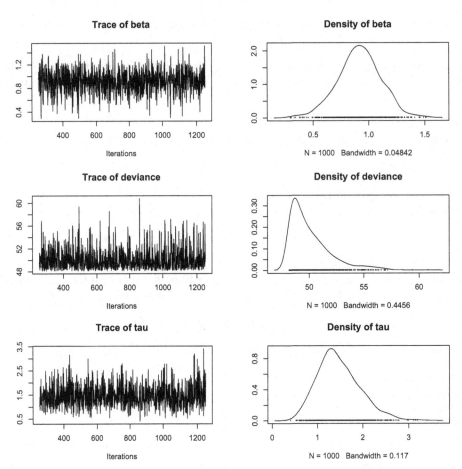

FIGURE 14.7
Diagnostic plots generated by coda for the artificial data set.

We have already discussed the Gelman-Rubin diagnostic as implemented in WinBUGS, so here we will briefly describe the diagnostics of Geweke (1992), Raftery and Lewis (1992), and Heidelberger and Welch (1992). The year 1992 was obviously a very good one for MCMC diagnostics. The Gewecke test generates a set of z scores based on subsamples of the elements of the chain; these z scores should fall between −2 and 2. The Raftery and Lewis diagnostic generates a statistic denoted I that measures the relative effect on the effective sample size due to autocorrelation. The value of I may be taken as an estimate of the appropriate thinning interval. The diagnostic test of Heidelberger and Welch checks subsets of the MCMC chain for stationarity. These various tests each focus on a different aspect of the chain, and a failure of any one of them should be taken as an indication of non-convergence. Further information can be found in Ntzoufras (2009) as well as other sources listed in Section 14.7.

The chains all appear stable, and the diagnostics are within the acceptable range, so we can exit the diagnostics. Next, run the simulation by removing the assignment codaPkg = TRUE. First, we will run the model with one chain.

```
> demo.sim <- bugs(data = XY.data, inits = XY.inits,
+    model.file = paste(mybugsdir,"demomodel.bug", sep = ""),
+    parameters = c("beta0", "beta1", "tau"), n.chains = 1,
```

```
+      n.iter = 5000, n.burnin = 1000,
+      bugs.directory = mybugsdir)
```

This runs the model, as you will see on your screen. The results can be displayed by the functions `print()` and `plot()`. We will only display the former.

```
> print(demo.sim, digits = 3)
          *        *        *
n.sims = 1000 iterations saved
            mean     sd    2.5%     25%     50%     75%   97.5%
beta       0.915 0.192   0.528   0.796   0.920   1.039   1.263
tau        1.469 0.464   0.698   1.156   1.407   1.745   2.524
deviance 50.102 1.902  48.200  48.690  49.530  50.932  55.453

DIC info (using the rule, pD = Dbar-Dhat)
pD = 1.9 and DIC = 52.0
DIC is an estimate of expected predictive error (lower deviance is
better).
```

For a discussion of the DIC (deviance information criterion), see Gelman and Hill (2007, p. 525). Now we will run the model with five chains.

```
> demo.sim <- bugs(data = XY.data, inits = XY.inits,
+ model.file = paste(mybugsdir,"demomodel.bug", sep = ""),
+ parameters = c("beta", "tau"), n.chains = 5,
+ n.iter = 5000, n.burnin = 1000,
+ bugs.directory=mybugsdir)
> print(demo.sim, digits = 3)
          *        *        *
            mean     sd    2.5%     25%     50%     75%   97.5%  Rhat n.eff
beta       0.920 0.198   0.520   0.789   0.920   1.043   1.305 1.002  1000
tau        1.450 0.466   0.696   1.107   1.391   1.741   2.497 1.002  1000
deviance 50.177 1.994  48.210  48.720  49.645  51.045  55.346 1.006   440

For each parameter, n.eff is a crude measure of effective sample size,
and Rhat is the potential scale reduction factor (at convergence,
Rhat=1).
DIC info (using the rule, pD = Dbar-Dhat)
pD = 2.0 and DIC = 52.2
DIC is an estimate of expected predictive error (lower deviance is
better).
```

Note that the output from the second run includes two diagnostic quantities, `Rhat` and `n.eff`. These can only be computed when there is more than one chain. The former is a convergence diagnostic and is approximately equal to the square root of the variance of the mixture of all of the chains, divided by the average within-chain variance (Gelman and Hill, 2007, p. 358). As it says in the output, the fact that `Rhat` is approximately 1 indicates convergence. The second diagnostic quantity, `n.eff`, is the "effective number of simulation draws." This represents the extent to which autocorrelation in the Markov Chain has affected the assumption of independent draws. Gelman and Hill (2007, p. 358) recommend that `n.eff` be at least 100.

The extension of WinBUGS analysis to multiple linear regression is straightforward. We will demonstrate it using a simplified linear regression model for yield as a function of only soil clay content and soil potassium content for Field 1 of Data Set 4. First, we normalize all of the data. For example, we use a normalized yield *YieldN* = *Yield* / max(*Yield*), and both *Clay* and *SoilK* are similarly normalized. There are two reasons for this. The first is to keep numerical problems to a minimum by avoiding very large or very small numbers. The second is to avoid inadvertently specifying priors that are not noninformative (Gelman and Hill, 2007, p. 355). Since the values of *Yield* are measured in the thousands and the values of *Clay* and *SoilK* are in one or two digits, a regression model involving *Yield*, *Clay*, and *SoilK* could have coefficients in the tens or hundreds. Other models might have coefficients with even larger magnitude. It would be easy to specify a prior with a mean of 0 and an insufficiently large variance and inadvertently (and inappropriately) pull the value of the coefficients toward 0.

```
> data.Set4.1$YieldN <- with(yield.pts, Yield / max(Yield))
> data.Set4.1$ClayN <- with(data.Set4.1, Clay / max(Clay))
> data.Set4.1$SoilKN <- with(data.Set4.1, SoilK / max(SoilK))
```

Debugging a WinBUGS model can be a maddening experience, so it is always best to move in the smallest possible steps from a model that works to the model that one is trying to construct. In this case, I started with the simple demo.model from this section. The first step was to delete the line

```
> n <- 20
```

and change the line

```
> XY.data <- list(X = X, Y = Y, n = n)
```

to

```
> XY.data <- list(X = X, Y = Y, n = nrow(data.Set4.1))
```

The next step was to change the line

```
> X <- rnorm(n)
> Y <- X + rnorm(n)
```

to

```
> X <- data.Set4.1$ClayN
> Y <- data.Set4.1$YieldN
```

In this way, one baby step at a time, I first created a model of *YieldN* regressed only on *ClayN* and then expanded it to include *SoilKN*. Here is the full resulting WinBUGS program.

```
> Set4.1.model <- function(){
+ beta0 ~ dnorm(0, 0.01)
+ beta1 ~ dnorm(0, 0.01)
+ beta2 ~ dnorm(0, 0.01)
+ tau ~ dgamma(0.01, 0.01)
+ for (i in 1:n)
+   {
+   Y.hat[i] <- beta0 + beta1 * Clay[i] + beta2 * SoilK[i]
+   Yield[i] ~ dnorm(Y.hat[i], tau)
+   }
```

```
+ }
>
> write.model(Set4.1.model,
+     paste(mybugsdir,"\\Set4.1model.bug", sep = ""))
>
> XY.data <- list(Clay = data.Set4.1$ClayN, SoilK = data.Set4.1$SoilKN,
+     Yield = data.Set4.1$YieldN, n = nrow(data.Set4.1))
> XY.inits <- function(){
+     list(beta0 = rnorm(1), beta1 = rnorm(1), beta2 = rnorm(1),
+     tau = exp(rnorm(1)))
+ }
```

Next, I went through the test and coda steps. Reading the coda file gives us something we have not seen before.

```
> codamenu()
CODA startup menu
1: Read BUGS output files
2: Use an mcmc object
3: Quit
Selection: 2
Enter name of saved object (or type "exit" to quit)
1:Set4.1.mcmc
Checking effective sample size...
*********************************************
WARNING !!!
Some variables have an effective sample size of less than 200 in at least
one chain. This is too small, and may cause errors in the diagnostic
tests. HINT: Look at plots first to identify variables with slow mixing.
(Choose menu Output Analysis then Plots). Re-run your chain with a larger
sample size and thinning interval. If possible, reparameterize your model
to improve mixing
*********************************************
CODA Main Menu
```

When it loads the file, the function codamenu() runs a test of the Gelman and Rubin statistic described in the previous section and flags values below an acceptable threshold. The term *mixing* refers to the selection of random samples from the full posterior distribution by the MCMC algorithm (Congdon, 2010, p. 21). The warning suggests that we increase the sample size and the thinning interval, so we will try that. After the increase, we still get the same message, but the plots of the chains look pretty good. Let's try the simulation steps. Here is the output of the simulation.

```
> Set4.1.sim <- bugs(data = XY.data, inits = XY.inits,
+     model.file = paste(mybugsdir,"\\Set4.1model.bug", sep = ""),
+     parameters = c("beta0", "beta1", "beta2", "tau"), n.chains = 5,
+     n.iter = 10000, n.burnin = 2000,
+     bugs.directory = mybugsdir)
> print(Set4.1.sim, digits = 3)
Inference for Bugs model at "c:\aardata\book\Winbugs\\Set4.1model.bug",
fit using WinBUGS,
 5 chains, each with 10000 iterations (first 1000 discarded), n.thin = 5
 n.sims = 9000 iterations saved
```

```
        mean     sd    2.5%    25%     50%     75%   97.5%  Rhat  n.eff
beta0   1.362  0.092   1.181  1.300   1.362   1.422  1.543 1.002   3700
beta1  -1.216  0.157  -1.533 -1.319  -1.215  -1.115 -0.905 1.006    670
beta2   0.036  0.143  -0.241 -0.060   0.036   0.131  0.318 1.004    910
tau    59.022  8.959  42.610 52.850  58.610  64.890 77.591 1.001   9000
```

Your values may be slightly different. The values of the diagnostics Rhat and n.eff look pretty good. We can compare the output with the model computed using lm().

```
> Set4.1.lm <- lm(YieldN ~ ClayN + SoilKN, data = data.Set4.1)
> coef(Set4.1.lm)
(Intercept)         ClayN       SoilKN
 1.364794826 -1.199936005  0.005542428
>
> t(Set4.1.sim$mean)
     beta0     beta1      beta2        tau    deviance
[1,] 1.364603 -1.19636 0.002188471 59.15018 -107.0198
```

Remember that tau is the inverse of the error variance. Here is a computation of the inverse of the error variance from lm().

```
> 1 / (sum(residuals(Set4.1.lm)^2) / (nrow(data.Set4.1) - 3))
[1] 60.03328
```

All of the values are reasonably close. My own experience indicates that, although it is useful, the warning message produced by codamenu() often seems quite conservative. I generally put more weight in the chain plots and the statistics Rhat and n.eff and, when possible, a comparison of the results with another method. So far, however, we have not seen any particular advantage that the Bayesian approach provides over the classical approach. The greatest advantages occur in the analysis of more complex models, so we will not pursue ordinary linear regression any further. In the next section, we continue to build our Bayesian toolbox by considering generalized linear models.

14.3.4 Generalized Linear Models in WinBUGS

Once you know how to analyze a multiple linear regression model using WinBUGS, the extension to generalized linear models is in most cases fairly straightforward. We will describe it by continuing our analysis of the logistic regression model for presence/absence of blue oaks (Data Set 2). Recall that the logistic regression model is specified by (Section 8.3.1)

$$E\{Y \mid X_i\} = \pi_i,$$

$$\ln\left(\frac{\pi_i}{1-\pi_i}\right) = \beta_0 + \sum_{j=1}^{p-1} \beta_j X_{ij}, \tag{14.21}$$

where Y is a binomially distributed random variable taking on values 0 and 1.

In Section 8.4, we developed a logistic regression model for the relationship between blue oak presence/absence and the explanatory variables in the Sierra Nevada subset of Data Set 2. We will use the sf object data.Set2S created in that section and develop a Bayesian model for the data. As in the analysis of Section 14.3.2, we will normalize the variable *Elevation*. The data are loaded into R as spatial features objects and can be used in this form.

```
> data.Set2S$ElevN <- data.Set2S$Elevation / max(data.Set2S$Elevation)
```

The logistic regression model and the coefficients are as follows:

```
> glm.demo <- glm(QUDO ~ ElevN,
+    data = data.Set2S, family = binomial)
> coef(glm.demo)
(Intercept)        ElevN
   3.329145  -11.463550
```

Now let's set up the model to pass to WinBUGS. The logistic regression form of the generalized linear model uses a normal density prior for the values of the coefficients β_i (Ntzoufras, 2009, p. 255). With binomial values of Y_i (i.e., $Y_i = 1$ implies presence and $Y_i = 0$ implies absence), the variance of the Y_i is the binomial variance, which is equal to $np(1-p)$, so there is no precision term τ as there is in the linear model. The code for the model statement is the following:

```
> library(R2WinBUGS)
> demo.model <- function(){
+ beta0 ~ dnorm(0, 0.01)
+ beta1 ~ dnorm(0, 0.01)
+ for (i in 1:n)
+   {
+   logit(p.QUDO[i]) <- beta0 + beta1 * ElevN[i]
+   QUDO[i] ~ dbin(p.bound[i], 1)
+   p.bound[i] <- max(0, min(1, p.QUDO[i]))
+   }
+ }
```

The first line inside the `for` loop sets up the model using the WinBUGS function `logit()` to create a logistic regression model. The last two lines express the fact that QUDO is a binomially distributed quantity. The variable p.bound is used instead of p.QUDO in order to avoid numerical problems in case a computed value of p.QUDO is less than 0 or greater than 1. Remember that the second line in the `for` loop is not an assignment statement but rather a statement of the relationship between the variable p.bound in the model and the data values in the vector QUDO.

The next step is to provide WinBUGS with the data values and to initialize the model.

```
> XY.data <- list(QUDO = data.Set2S$QUDO,
+    ElevN = data.Set2S$ElevN, n = nrow(data.Set2S))
> XY.inits <- list(list(beta0 = 0, beta1 = 0))
```

Next, we write a file providing information to WinBUGS.

```
> write.model(demo.model,
+    paste(mybugsdir,"\\demoglm.bug", sep = ""))
```

Before we continue, we must change the XY.inits. Since there will be five chains, we must have five initial sets of values. This is similar to the change made in the second run of WinBUGS in Section 14.3.1.

```
> XY.inits <- function(){
+    list(beta0 = rnorm(1), beta1 = rnorm(1))
+ }
```

We then go through the usual debug and coda steps, with codamenu() giving us the same warning as in the last section. Next, then we execute the model.

```
> demo.sim <- bugs(data = XY.data, inits = XY.inits,
+    model.file = paste(mybugsdir,"\\demoglm.bug", sep = ""),
+    parameters = c("beta0", "beta1"), n.chains = 5,
+    n.iter = 5000, n.burnin = 1000,
+    bugs.directory = mybugsdir)
```

Let's compare the Bayesian results with those of the function glm().

```
> t(demo.sim$mean)
        beta0     beta1     deviance
[1,]  3.321083 -11.44447  1368.549
```

The values are quite close to those displayed above. In Exercise 14.4, you are asked to develop a WinBUGS version of the model selected in Section 8.4.2 to represent the candidates for the Sierra Nevada subset. My results of the glm() model and the WinBUGS model are the following (yours may vary slightly).

```
> coef(glm. Set2S)
  (Intercept)          Precip            MAT           SolRad
-6.3122558455  -0.0063894184   0.6026153179   0.0006361065
       AWCAvg         Permeab
-0.0066932919  -0.3157297346
> t(Set2S.sim$mean)
        beta0      betaPre        betaMAT    betaSol
[1,]  -4.029703  -0.006347872   0.537417   0.0004254458
      betaAWC       betaPerm      deviance
[1,]  -0.006252348 -0.3136346   1071.546
```

Again, the values of the β_i are reasonably close, with the greatest difference from the glm() output being between the coefficients for *MAT* and *SolRad*.

We are finally ready to tackle the models that take greatest advantage of what the Bayesian approach has to offer. These are hierarchical models, which are described in the next section.

14.4 Hierarchical Models

The mixed-model theory discussed in Chapter 12 deals with *grouped data*. These are data in which each data record is associated with a grouping factor, and data records associated with the same grouping factor are considered more closely related to each other than to data records associated with a different grouping factor. In the case of spatial data, the grouping factor is generally defined by geography. For example, in the case of Data Set 3 the grouping factor is *Field*. The grouping factor is considered to be a random effect. The mixed model contains a mixture of fixed and random effects, where a fixed effect is one in which only the values of the effect actually present in the data are considered, and there is no attempt to extend the results of the analysis to other values. A random effect is...well, that is a problem. Statistics is a discipline that insists on precise definitions, but the definition of a random effect is remarkably vague and inconsistent in the literature (Gelman and Hill, 2007, p. 245). Some authors focus on the idea of a random effect as being defined by a probability distribution. This is, however, often not very realistic. Other authors emphasize the idea that a random effect is one for which the study is interested in values beyond those included in the observed data. The difficulty is that this implies that if the investigators change their interest, then the nature of the variable could change. It is not clear that this is desirable.

Another problem is that the one thing upon which almost all authors agree is that the values of a random effect should be randomly selected. In the case of ecological or agronomic data, however, this is usually not even remotely close to the truth. For example, consider the fields in Data Set 3. These should have been selected at random from the population of fields in northeastern Uruguay. In fact, as is true of virtually every such study, the selection of the fields was based on a combination of personal relationships, convenience, and availability. Researchers tend to work with cooperating growers with whom they have a relationship, the fields must be accessible and situated in a way that permits all of them to be visited in the allotted time, and the grower, on whose cooperation the study depends, may have some reason to prefer the use of one field over another. As with all such nonrandom selection, this opens up the possibility of bias, but more generally it calls into question the appropriateness of the assumption of randomness.

In Bayesian theory, the parameters as well as the data are considered random variables, and the probability distribution represents subjective belief about their value rather than some distribution occurring in nature. This allows one to avoid many of the definitional problems associated with random effects. For this reason, Bayesian theory does not generally use the term *mixed model*; instead, the term *hierarchical model* is used. We will introduce the concept of the hierarchical model using the same example as was used in Chapter 12 to introduce the mixed model: the relationship between irrigation effectiveness *Irrig* and *Yield* in Data Set 3.

We will start with the pooled data from all the fields. We found in Chapter 12 that all of the data records in Field 2 have a value of *Irrig* equal to 5 (the highest level), so that a regression of *Yield* on *Irrig* involving only this field would not work. We therefore remove Field 2 from the data set (however, see Exercise 14.5).

```
> data.Set3a <- data.Set3[-which(data.Set3$Field == 2),]
```

Once again, in order to avoid potential numerical problems we normalize *Yield*.

```
> data.Set3a$YieldN <- data.Set3a$Yield / max(data.Set3a$Yield)
```

Since *Irrig* ranges from one to five anyway, we will not normalize it. Here is the pooled model WinBUGS code for use with R2WinBUGS. It is derived by modification of the code in Section 14.3.4.

```
> Set3.model <- function(){
+ beta0 ~ dnorm(0, 0.01)
+ beta1 ~ dnorm(0, 0.01)
+ tau ~ dgamma(0.01, 0.01)
+ for (i in 1:n)
+   {
+   Y.hat[i] <- beta0 + beta1*Irrig[i]
```

```
+    Yield[i] ~ dnorm(Y.hat[i], tau)
+    }
+ }
> write.model(Set3.model,
+    paste(mybugsdir,"\\Set3model.bug", sep = ""))
> XY.data <- list(Irrig = data.Set3a$Irrig, Yield = data.Set3a$YieldN,
+    n = nrow(data.Set3a))
> XY.inits <- function(){
+    list(beta0 = rnorm(1), beta1 = rnorm(1), tau = exp(rnorm(1)))}
```

We now run through the usual debug, coda, simulation sequence. The coda analysis gives us the same warning message is in the last section, which we will ignore. Here are the results of the simulation.

```
> Set3.sim <- bugs(data = XY.data, inits = XY.inits,
+    model.file = paste(mybugsdir,"\\Set3model.bug", sep = ""),
+    parameters = c("beta0", "beta1", "tau"), n.chains = 5,
+    n.iter = 5000, n.burnin = 500,
+    bugs.directory = mybugsdir)
> print(Set3.sim, digits = 2)
Inference for Bugs model at "c:\aardata\book\Winbugs\\Set3model.bug", fit
using WinBUGS,
 5 chains, each with 5000 iterations (first 500 discarded), n.thin = 22
 n.sims = 1025 iterations saved
             mean     sd    2.5%      25%      50%      75%    97.5% Rhat n.eff
beta0       0.203  0.030   0.145    0.183    0.203    0.221    0.263 1.002  980
beta1       0.114  0.008   0.098    0.109    0.114    0.119    0.130 1.001 1000
tau        79.680  6.288  67.898   75.410   79.240   83.900   91.572 0.999 1000
deviance -482.2   2.40  -484.90  -484.00  -482.90  -481.30  -475.86 0.999 1000
```

As a check, we will print the coefficients of the ordinary least square (OLS) model.

```
> print(coef(lm(YieldN ~ Irrig, data = data.Set3a)),
+    digits = 3)
(Intercept)         Irrig
      0.204         0.114
```

Everything looks reasonably good: the model converges and the number of effective simulation draws is good (sometimes by random chance I get results that are not quite this good). We will therefore construct our first hierarchical model. The process is so effortless (except for the debugging) that you may find yourself asking whether this is indeed all there is to it. The answer is basically yes, at least at the operational level. At the level of interpretation, the issues are a bit subtler and will be discussed in a small way in Section 14.6.

In Section 12.3, we began by introducing a term α_i that represented the random overall effect of the field. We will do the same thing here. There is one little complication that must be addressed. We eliminated Field 2 earlier, so we now have values of the variable *Field* consisting of 1 and 3 through 16. In Chapter 12, when we were dealing with mixed models, the variable *Field* was considered a factor (i.e., a nominal quantity), so the missing value of 2 was not important. In the WinBUGS model, the easiest way to proceed is to use *Field* as an index quantity, but we will get an error message if we ignore the missing value of 2 (try it!). Therefore, we change the *Field* value of Field 16 from 16 to 2.

```
> data.Set3a$Field[which(data.Set3a$Field == 16)] <- 2
```

Now we are ready to proceed. To create a hierarchical model analogous to the mixed model of Equation 12.6 in which the intercept is allowed to depend on the group, we simply specify that the intercept β_0 be allowed to depend on the group (i.e., the field). We introduce a `for` loop defining the prior distribution of the β_{0i} and introduce two new parameters μ_0 and τ_0 that characterize their priors. In a hierarchical model, the parameters without subscripts, which are defined over all groups, are called *hyperparameters* (Gelman and Hill, 2007, p. 258), although this term is also sometimes used to characterize all the parameters associated with the distributions.

```
> Set3.model2 <- function(){
# For loop defining the beta0
+ for (i in 1:n.F){
+     beta0[i] ~ dnorm(mu.0, tau.0)
+ }
# Parameters for the priors
+ mu.0 ~ dnorm(0, 0.001)
+ tau.0 ~ dgamma(0.01, 0.01)
+ beta1 ~ dnorm(0, 0.001)
+ tau ~ dgamma(0.01, 0.01)
+ for (i in 1:n)
+     {
+     Y.hat[i] <- beta0[Field[i]] + beta1*Irrig[i]
+     Yield[i] ~ dnorm(Y.hat[i], tau)
+     }
+ }
```

The `write.model()` statement has the same form as before and is not shown. The modifications to the data and inits statements are straightforward.

```
> XY.data2 <- list(Irrig = data.Set3a$Irrig, Yield = data.Set3a$YieldN,
+     n = nrow(data.Set3a), n.F = length(unique(data.Set3a$Field)),
+     Field = data.Set3a$Field)
> n.F <- length(unique(data.Set3a$Field))
> XY.inits2 <- function(){
+     list(beta0 = rnorm(n.F), beta1 = rnorm(1), tau = exp(rnorm(1)),
+     mu.0 = rnorm(1), tau.0 = exp(rnorm(1)))}
```

We again go through the debug, coda, and simulate steps. Here are the results of the simulation step.

```
> Set3.sim2 <- bugs(data = XY.data2, inits = XY.inits2,
+     model.file = paste(mybugsdir,"\\Set3model.bug", sep = ""),
+     parameters = c("beta0", "beta1", "tau", "mu.0", "tau.0"),
+     n.chains = 5,
+     n.iter = 10000, n.burnin = 2000, n.thin = 20,
+     bugs.directory = mybugsdir)
> print(Set3.sim, digits = 2)
Inference for Bugs model at "c:\aardata\book\Winbugs\\Set3model.bug", fit
using WinBUGS,
 5 chains, each with 10000 iterations (first 2000 discarded), n.thin = 10
 n.sims = 4000 iterations saved
           mean   sd    2.5%    25%    50%    75%   97.5% Rhat n.eff
beta0[1]   0.48  0.03   0.41   0.46   0.48   0.51   0.55 1.01   430
```

```
beta0[2]    0.41  0.03    0.36    0.39    0.41    0.43    0.46 1.01    560
beta0[3]    0.50  0.03    0.45    0.48    0.50    0.52    0.55 1.01    360
    *    *    *   DELETED
beta0[15]   0.53  0.03    0.47    0.51    0.53    0.55    0.59 1.01    420
beta1       0.06  0.01    0.05    0.05    0.06    0.06    0.07 1.01    380
tau       213.97 17.52  181.50  201.90  213.50  225.60  250.10 1.00   2800
mu.0        0.38  0.04    0.30    0.35    0.37    0.40    0.45 1.00   2000
tau.0      78.26 30.56   31.23   55.83   74.17   95.75  147.50 1.00   4000
```

Now we move on to a model with both β_0 and β_1 varying with field, analogous to Equation 12.7.

```
> Set3.model3 <- function(){
+ for (i in 1:n.F){
+    beta0[i] ~ dnorm(mu.0, tau.0)
+    beta1[i] ~ dnorm(mu.1, tau.1)
+ }
+ mu.0 ~ dnorm(0, 0.001)
+ tau.0 ~ dgamma(0.01, 0.01)
+ mu.1 ~ dnorm(0, 0.01)
+ tau.1 ~ dgamma(0.01, 0.001)
+ tau ~ dgamma(0.01, 0.01)
+ for (i in 1:n)
+    {
+    Y.hat[i] <- beta0[Field[i]] + beta1[Field[i]]*Irrig[i]
+    Yield[i] ~ dnorm(Y.hat[i], tau)
+    }
+ }
```

The rest of the setup and execution is analogous to the model with only β_0 dependent on group. Here are the results of the simulation step.

```
> Set3.sim3 <- bugs(data = XY.data3, inits = XY.inits3,
+    model.file = paste(mybugsdir,"\\Set3model.bug", sep = ""),
+    parameters = c("beta0", "beta1", "tau", "mu.0", "tau.0", "mu.1",
"tau.1"), n.chains = 5,
+    n.iter = 10000, n.burnin = 2000, n.thin = 20,
+    bugs.directory = mybugsdir)
> print(Set3.sim, digits = 2)
Inference for Bugs model at "c:\aardata\book\Winbugs\\Set3model.bug", fit
using WinBUGS,
5 chains, each with 10000 iterations (first 2000 discarded), n.thin = 20
n.sims = 2000 iterations saved
5 chains, each with 10000 iterations (first 2000 discarded), n.thin = 20
n.sims = 2000 iterations saved
            mean    sd    2.5%    25%    50%    75%   97.5% Rhat
n.eff
beta0[1]    0.47  0.08    0.32    0.42    0.47    0.51    0.61 1.01    540
beta0[2]    0.51  0.07    0.38    0.46    0.51    0.55    0.65 1.00    730
beta0[3]    0.43  0.06    0.31    0.39    0.43    0.47    0.54 1.00   1100
    *    *    *   DELETED    *    *    *
beta1[13]   0.05  0.02    0.01    0.04    0.05    0.07    0.10 1.00   2000
beta1[14]   0.05  0.02    0.01    0.03    0.05    0.06    0.09 1.00   1200
beta1[15]   0.01  0.02   -0.02    0.00    0.01    0.03    0.05 1.01    230
```

tau	236.19	20.06	198.19	222.50	235.60	249.52	276.11	1.00	2000
mu.0	0.39	0.05	0.30	0.35	0.39	0.42	0.49	1.00	1400
tau.0	49.33	23.38	16.15	32.37	45.78	60.76	104.91	1.00	1000
mu.1	0.06	0.01	0.03	0.05	0.06	0.06	0.08	1.00	2000
tau.1	1081.64	601.61	319.97	652.62	956.20	1342.25	2608.22	1.01	480
devian	-825.72	9.25	-842.10	-832.30	-826.20	-819.70	-806.30	1.00	1000

We can use this result to get some insight into the question of the importance of the field versus the farmer in influencing yield. The fields in the central region tend to have a higher silt content than those in the other regions. Soil texture has a strong influence on soil water-holding capacity and drainage, and it is possible that improvements in irrigation effectiveness would not improve yield as much in the central region as in the other two regions. Let us compare the relationship between irrigation effectiveness and yield in the central region with that in the combined northern and southern regions. That is, we will compare the slopes b_1 of the regression lines. This analysis could also be carried out using the mixed model results of Chapter 12.

First, we compute the mean values of b_1 for the two regions (noting that Fields 1 through 4 and 15 are in the north or south of data.Set3a).

```
> param.est <- t(unlist(Set3.sim3$mean))
> b1 <- param.est[16:30]
> print(b1NS <- b1[c(1:4, 15)], digits = 2)
[1] 0.062 0.031 0.080 0.090 0.015
> print(b1C <- b1[5:14], digits = 2)
 [1] 0.100 0.031 0.052 0.074 0.060 0.032 0.058 0.058 0.054 0.049
> mean(b1NS)
[1] 0.05563988
> mean(b1C)
[1] 0.05674888
> mean(b1NS) - mean(b1C)
[1] -0.001108998
```

The mean value of b_1 is actually higher in the central region, which would indicate that on average, an increase in irrigation effectiveness would have a greater effect in this region. There is a lot of variability, however. Let's carry out a permutation test (Section 4.2.3) to determine whether the difference is significant. Recall that we concatenate the two vectors of b_1, randomly rearrange their elements many times, and determine where our observed difference sits among the rearrangements.

```
> b <- c(b1NS, b1C)
> print(d0 <- mean(b[1:5] - b[6:length(b)]))
[1] -0.001108998
> u <- numeric(1000)
> sample.d <- function(b){
+      b.samp <- sample(b)
+      mean(b.samp[1:5] - b.samp[6:length(b)])
+ }
> set.seed(123)
> u <- replicate(999, sample.d(b))
> u[1000] <- d0
> print(p <- length(which(u > d0)) / length(u))
[1] 0.537
```

The observed difference is not statistically significant. The results of this little analysis support the conclusion that improvements in irrigation effectiveness would have the same average impact on yield in the central region as they would in the northern or southern region.

The hierarchical analysis carried out in this section made implicit use of the spatial relationships among the data records through the grouping structure of the fields. None of the analyses completed so far have made explicit use of the spatial relationships of the data, that is of the geographical coordinates of the data records. We will take up explicit spatial analyses in the next section.

14.5 Incorporation of Spatial Effects

14.5.1 Spatial Effects in the Linear Model

The Bayesian approach to regression modeling of spatial data was largely initiated in a seminal paper by Besag (1974) in the context of the conditional autoregressive (CAR) model. The CAR model was introduced in Section 13.6. The equation for a linear CAR model with constant variance, given as Equation 13.42, is

$$E\{Y_i \mid Y_{-i}\} = X\beta + \sum_{j \neq i} c_{ij}(Y_j - X\beta)$$

$$\text{(14.22)}$$

$$\text{var}\{Y_i \mid Y_{-i}\} = \sigma^2,$$

where Y_{-i} is the vector of all of the Y values except Y_i. As discussed in Section 13.6, the weights matrix C is not the same as the spatial weights matrix W of the spatial autoregressive (SAR) model, although they are related. Moreover, in the model of Equation 14.22, we must have $c_{ij} = c_{ji}$. Our discussion of the issues associated with the CAR model is simplified a bit. For example, we assume that the variance term σ^2 is constant. Readers who wish to see the more general discussion may find it in the references, beginning with Waller and Gotway (2004, p. 270) and continuing with other sources listed in Section 14.7.

One of the issues associated with the conditional formulation is that in many cases the adoption of a standard form for the spatial relationships suitable for the CAR model results in a density function for the joint distribution of the Y values that cannot be integrated (Besag et al., 1991; Banerjee et al., 2004). Such a formulation can be used only to describe the prior densities, and not the model itself. Such a set of densities is called an *improper prior*. It is possible to formulate proper densities, but these have disadvantages that, for us, outweigh their advantages. Therefore, we will use improper priors in our analysis.

We will introduce the Bayesian approach to spatial modeling through the analysis of a simple model using artificial data, similar to the models used to introduce the spatial autoregression in Chapter 13. We will generate artificial data relating Y to X on a ten by ten square grid. First, we set up the model.

```
> lambda <- 0.4
> nlist <- cell2nb(10,10)
> IrWinv <- invIrM(nlist, lambda)
> set.seed(123)
> X <- rnorm(100)
> eps <- as.numeric(IrWinv %*% rnorm(100))
```

```
> sig2 <- 0.1
> Y <- X + sig2*eps
```

Now we will display SAR and CAR analyses using the function `spautolm()` described in Chapter 13. To keep the output simple, we will only display the regression coefficients.

```
> W <- nb2listw(nlist, style = "B")
> Y.SAR <- spautolm(Y ~ X, listw = W)
> print(coef(Y.SAR), digits = 3)
(Intercept)          X        lambda
   -0.0204       0.9975       0.1161
> Y.CAR <- spautolm(Y ~ X, listw = W, family = "CAR")
> print(coef(Y.CAR), digits = 3)
(Intercept)          X        lambda
   -0.0208       0.9967       0.1845
```

The estimates of β_0 and β_1 are close to their true values of 0 and 1, but the estimates of λ differ substantially from the value used to create the data set. This may be due to random variation, or it may be partly because no effort was made to take boundary effects into account.

Now we will use `R2WinBUGS` to analyze the CAR model. We will go through this first example in some detail. First, we will set up the independent errors WinBUGS model, and then we will modify the necessary lines to analyze the full CAR model. We will also anticipate the use of this code as the template for a subsequent model involving real data. Here is the initial WinBUGS formulation.

```
> library(R2WinBUGS)
> # WinBUGS formulation
> demo.model <- function(){
+ beta0 ~ dnorm(0, 0.01)
+ beta1 ~ dnorm(0, 0.01)
+ tau ~ dgamma(0.01, 0.01)
+ for (i in 1:100)
+   {
+   Y.hat[i] <- beta0 + beta1*X[i]
+   Y[i] ~ dnorm(Y.hat[i], tau)
+   }
+ }
> write.model(demo.model,
+     paste(mybugsdir,"\\demomodel.bug", sep = ""))
> data.XY <- data.frame(X = X, Y = Y)
> XY.data <- list(X = data.XY$X, Y = data.XY$Y)
> XY.inits <- function(){
+     list(beta0 = rnorm(1), beta1 = rnorm(1), tau = exp(rnorm(1)))}
```

The data frame `data.XY` is created to facilitate later transfer to the real data model. We go through the debug, coda, and simulate sequence to produce this WinBUGS simulation.

```
> demo.sim <- bugs(data = XY.data, inits = XY.inits,
+     model.file = paste(mybugsdir,"\\demomodel.bug", sep = ""),
+     parameters = c("beta0", "beta1", "tau"), n.chains = 5,
+     n.iter = 10000, n.burnin = 1000, n.thin = 10,
+     bugs.directory = mybugsdir)
```

```
> print(demo.sim, digits = 2)
Inference for Bugs model at "c:\aardata\book\Winbugs\\demomodel.bug",
fit using WinBUGS,
 5 chains, each with 10000 iterations (first 1000 discarded), n.thin = 10
 n.sims = 4500 iterations saved
          mean    sd   2.5%    25%    50%    75%  97.5% Rhat n.eff
beta0    -0.02  0.01  -0.04  -0.02  -0.02  -0.01   0.00    1  4500
beta1     0.99  0.01   0.97   0.99   0.99   1.00   1.02    1  4500
tau      90.12 12.87  66.60  80.97  89.52  98.44 117.10    1  4500
```

Now we will modify the code for the CAR analysis. The program WinBUGS contains a built-in program called GeoBUGS (Thomas et al., 2004) that facilitates the construction of CAR models. In particular, GeoBUGS provides a function called car.normal() that can be used to specify an improper prior Gaussian distribution. This function takes four arguments: adj, num, weights, and tau. The vector adj contains the adjacency values, that is, the data records spatially adjacent to data rerecord *i*. The vector weights contains the values of the spatial weights matrix. The vector num contains the number of neighbors for each data record. The scalar tau contains the inverse variance of the normal distribution.

The spdep package contains a function nb2WB() that creates the arguments for the function car.normal() from a neighbor list.

```
> W.WB <- nb2WB(nlist)
```

The CAR model is implemented by first adding a term v[i] in the model statement as shown below. The first modified line establishes the vector v[i] as describing the autocorrelation effect, and the second modified line inserts the vector in the linear model.

```
> CARlm.model <- function(){
+ beta0 ~ dnorm(0, 0.01)
+ beta1 ~ dnorm(0, 0.01)
+ tau ~ dgamma(0.01, 0.01)
+ # This line is modified.
+ v[1:100] ~ car.normal(adj[], weights[], num[], tau)
+   for (i in 1:100)
+   {
+ # This line is modified.
+   Y.hat[i] <- beta0 + beta1*X[i] + v[i]
+   Y[i] ~ dnorm(Y.hat[i], tau)
+   }
+ }
```

The write.model() statement is unchanged and is not shown. The xy.data() and xy.inits() statements incorporate the objects created by the function nb2WB().

```
> XY.data <- list(X = data.XY$X, Y = data.XY$Y,
+   adj = W.WB$adj, weights = W.WB$weights,
+   num = W.WB$num, n = nrow(data.XY))
> XY.inits <- list(list(beta0 = 0, beta1 = 0, tau = 0.01,
+   v = rep(0.01,100)))
```

Once again, we run the debug, coda, and simulate sequence. Here is the result of the WinBUGS simulation.

```
> demo.sim <- bugs(data = XY.data, inits = XY.inits,
+   model.file = paste(mybugsdir,"\\demomodel.bug", sep = ""),
```

```
+    parameters = c("beta0", "beta1", "tau", "v"), n.chains = 5,
+    n.iter = 10000, n.burnin = 1000, n.thin = 10,
+    bugs.directory = mybugsdir)
> print(demo.sim, digits = 2)
Inference for Bugs model at "c:\aardata\book\Winbugs\\demomodel.bug", fit
using WinBUGS,
 5 chains, each with 10000 iterations (first 1000 discarded), n.thin = 10
 n.sims = 4500 iterations saved
          mean    sd    2.5%     25%     50%     75%   97.5% Rhat n.eff
beta0    -0.02  0.01   -0.03   -0.02   -0.02   -0.01    0.00    1  4500
beta1     0.99  0.01    0.97    0.99    0.99    1.00    1.02    1  4500
tau     132.81 19.14   98.38  119.30  131.80  145.22  172.40    1  4500
v[1]     -0.03  0.06   -0.14   -0.07   -0.03    0.01    0.08    1  4500
v[2]     -0.01  0.05   -0.11   -0.04   -0.01    0.02    0.09    1  3700
   *    *    *   DELETED    *    *    *
v[100]   -0.06  0.06   -0.18   -0.10   -0.06   -0.03    0.05    1  4500
```

The results indicate that the process has converged and that the effective number of simulation draws is acceptable. The coefficients are comparable with the results from the other methods. As a check on the CAR model, you are asked in Exercise 14.6 to run this model with lambda set at zero and compare the results of the CAR model with those of the model that does not include spatial autocorrelation.

It all looks very easy, right? Well, maybe not always.

14.5.2 Application to Data Set 3

As a demonstration of the application of the Bayesian approach to a real data set, we will return to the relationship between *Yield* and *Irrig* analyzed in Section 14.4. We will start with the pooled data model and then move to the hierarchical model. Once again, the easiest way to proceed in coding the model is to modify an existing model that works. In this case, I will use the simple demonstration model from Section 14.5.1. To emphasize the benefits of moving one small step at a time, I will change the code of Section 14.5.1 as little as possible to get a working model for the pooled data of Data Set 3. The model function does not need to be changed at all. The statement defining the spatial weights matrix W.WB is changed.

```
> nlist <- dnearneigh(data.Set3a, d1 = 0, d2 = 600)
> W.WB <- nb2WB(nlist)
```

Next, the assignment statement defining data.XY is changed, and the names of the data records Irrig and Yield are substituted for X and Y in the statement defining the data.

```
> data.XY <- data.Set3a
> XY.data <- list(X = data.XY$Irrig, Y = data.XY$YieldN,
+    adj = W.WB$adj, weights = W.WB$weights, num = W.WB$num,
+    n = nrow(data.XY))
```

Everything else remains the same. The coda analysis looks reasonable, so we run the simulation. Here is the result.

```
> Set3.sim <- bugs(data = XY.data, inits = XY.inits,
+    model.file = paste(mybugsdir,"\\Set3model.bug", sep = ""),
+    parameters = c("beta0", "beta1", "tau", "v"), n.chains = 5,
```

```
+      n.iter = 5000, n.burnin = 1000, n.thin = 10,
+      bugs.directory = mybugsdir)
>
> print(Set3.sim, digits = 2)
Inference for Bugs model at "c:\aardata\book\Winbugs\demomodel.bug", fit
using WinBUGS,
 5 chains, each with 5000 iterations (first 1000 discarded), n.thin = 10
 n.sims = 2000 iterations saved
              mean    sd   2.5%      25%     50%     75%   97.5%   Rhat n.eff
beta0          0.4   0.0    0.3      0.4     0.4     0.4     0.4   1.00   600
beta1          0.1   0.0    0.1      0.1     0.1     0.1     0.1   1.00   540
tau          223.2  18.1  187.3    211.0   223.1   234.8   260.3  1.00  2000
v[1]          -0.1   0.0   -0.2     -0.1    -0.1    -0.1    -0.1   1.00  2000
v[2]          -0.1   0.0   -0.2     -0.1    -0.1    -0.1    -0.1   1.00  1600
   *     *     *    DELETED   *      *       *
v[313]         0.15  0.02   0.11     0.1     0.1     0.2     0.2   1.00  2000
deviance    821.75 13.66 -847.61  -830.8  -822.1  -812.4  -795.2  1.00  2000
```

Let's compare the results of four different models.

```
> Y.lm <- lm(YieldN ~ Irrig, data = data.Set3a)
> W.B <- nb2listw(nlist, style = "B")
> Y.SAR <- spautolm(YieldN ~ Irrig,
+      data = data.Set3a, listw = W.B)
> Y.CAR <- spautolm(YieldN ~ Irrig,
+      data = data.Set3a, family = "CAR", listw = W.B)
> print(coef(Y.lm), digits = 3)
(Intercept)        Irrig
      0.204        0.114
> print(coef(Y.SAR), digits = 3)
(Intercept)        Irrig       lambda
     0.3112       0.0868       0.0522
> print(coef(Y.CAR), digits = 3)
(Intercept)        Irrig       lambda
     0.2781       0.0961       0.0528
> print(t(Set3.sim$mean[1:3]), digits = 3)
     beta0 beta1   tau
[1,] 0.389 0.0633  223
```

The results of the Bayesian analysis are a bit different from the others, which is disconcerting. We will keep this in mind but move on to the multi-intercept hierarchical model.

Once again, we build the more complex model starting from a simpler one, this time the pooled data model that we have just discussed. The symbols are adjusted for consistency with the model in Section 14.4. Here is the code.

```
> data.Set3a$Field[which(data.Set3a$Field == 16)] <- 2
>
> Set3.model <- function(){
+ # Priors for beta0
+ for (i in 1:n.F){
+      beta0[i] ~ dnorm(mu.0, tau.0)
+ }
+ # Parameters for the priors
+ mu.0 ~ dnorm(0, 0.001)
```

```
+ tau.0 ~ dgamma(0.01, 0.01)
+ beta1 ~ dnorm(0, 0.001)
+ tau ~ dgamma(0.01, 0.01)
+ # This line is modified.
+ v[1:n] ~ car.normal(adj[], weights[], num[], tau)
+ for (i in 1:n)
+   {
+   Y.hat[i] <- beta0[Field[i]] + beta1*Irrig[i] + v[i]
+   Yield[i] ~ dnorm(Y.hat[i], tau)
+   }
+ }
```

The XY.data and XY.inits statements are set appropriately.

```
> XY.data <- list(Irrig = data.Set3a$Irrig, Yield = data.Set3a$YieldN,
+    n = nrow(data.Set3a), n.F = length(unique(data.Set3a$Field)),
+    adj = W.WB$adj, weights = W.WB$weights,
+    num = W.WB$num, n = nrow(data.Set3a), Field = data.Set3a$Field)
> n.F <- length(unique(data.Set3a$Field))
> XY.inits <- function(){
+    list(beta0 = rnorm(n.F), beta1 = rnorm(1), tau = exp(rnorm(1)),
+    mu.0 = rnorm(1), tau.0 = exp(rnorm(1)),
+    v = rep(0.01,nrow(data.Set3a)))}
```

As usual, we run the test and coda steps. Here is the code for the coda step.

```
> Set3.coda <- bugs(data = XY.data, inits = XY.inits,
+    model.file = paste(mybugsdir,"\\Set3model.bug", sep = ""),
+    parameters = c("beta0", "beta1", "tau", "mu.0", "tau.0", "v"),
+    n.chains = 5, n.iter = 10000, n.burnin = 2000, n.thin = 10,
+    codaPkg = TRUE, bugs.directory = mybugsdir)
```

The resulting graphs provide a good example of what chains should not look like (Figure 14.8). Rather than looking like noise, several of the chains have a distinct pattern, indicating that they have not converged. Unfortunately, this occurrence seems to happen more often than not in models of more than the smallest level of complexity. Gelman and Hill (2007) devote a full chapter to dealing with failed WinBUGS programs, but for our particular case, probably the best (certainly the easiest) thing to do, assuming we have the patience to wait for the computer, is to increase the number of iterations, burn-in interval, and thinning interval.

```
> Set3.sim <- bugs(data = XY.data, inits = XY.inits,
+    model.file = paste(mybugsdir,"\\Set3model.bug", sep = ""),
+    parameters = c("beta0", "beta1", "tau", "mu.0", "tau.0", "v"),
+    n.chains = 5, n.iter = 30000, n.burnin = 20000, n.thin = 25,
+    bugs.directory = mybugsdir)
> print(Set3.sim)
Inference for Bugs model at "c:\aardata\book\Winbugs\Set3model.bug", fit
using WinBUGS,
5 chains, each with 30000 iterations (first 20000 discarded), n.thin = 25
 n.sims = 2000 iterations saved
            mean    sd   2.5%    25%    50%    75%  97.5% Rhat  n.eff
beta0[1]     0.4   0.1    0.2    0.3    0.4    0.5    0.6  1.0    200
beta0[2]     0.4   0.1    0.2    0.3    0.4    0.5    0.6  1.0    260
```

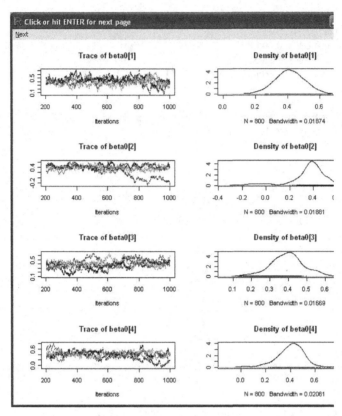

FIGURE 14.8
Diagnostic plots generated by coda for the model of *Yield* versus *Irrig* for Data Set 3. Note that several of the chains appear to wander, unlike those of Figure 14.7.

```
beta0[3]    0.4    0.1    0.3    0.4    0.4    0.5    0.6    1.1    29
beta0[4]    0.4    0.1    0.2    0.4    0.4    0.5    0.6    1.1    41
beta0[5]    0.4    0.1    0.3    0.4    0.4    0.5    0.5    1.0   150
              *      *      *
```

Even at very large values of n.iter, n.burnin, and n.thin, the results are discouraging. A comparison of the first five values of the intercept β_0 with those obtained in the mixed model analysis of Section 12.3 is even more discouraging.

```
> library(nlme)
> print(t(coef(lme(YieldN ~ Irrig, data = data.Set3a,
+      random = ~ 1 | Field)))[1,1:5], digits = 3)
    1     2     3     4     5
0.480 0.411 0.501 0.558 0.366
> print(t(unlist(Set3.sim$mean))[1:5], digits = 3)
[1] 0.398 0.399 0.413 0.404 0.416
```

The purpose of this section has been to show that the incorporation of spatial autocorrelation into a Bayesian model may not work, at least not without a lot of effort. In such a case, to carry out a Bayesian analysis one would have to fall back on the model that does not include spatial autocorrelation. In the present case, for example, the mixed model analysis of Chapter 12 indicated that spatial autocorrelation does not have a strong effect on the model.

14.5.3 The spBayes Package

Bayesian analysis in general and Bayesian analysis of spatial data in particular is one of the most active areas of research and of the development of new R packages. One of the most popular and useful is spBayes (Finley et al. 2015) and we will use this as an example of the capabilities of these packages. The package spBayes implements a variety of models using the MCMC methods that vary in two important ways from those described previously in this chapter. The first difference is the probability model itself. We will illustrate this difference by reexamining the Sierra Nevada oak model that we studied in Section 14.3.4 using the WinBUGS. Recall from Equation 14.21 in that section that the model is specified by the expected value of the binary distributed random variable Y, $E\{Y \mid X_i\} = \pi_i$ where π_i is specified via the link function as $\ln(\pi_i / (1 - \pi_i)) = \beta_0 + \Sigma\beta_j X_j$. This same formulation is used in the CAR model described in Equation 14.22. In spBayes the error is treated as a geostatistical random variable that is described by a variogram. For this reason the model is not expressed in terms of the expected value. Rather it is given as

$$Y_i \sim Binomial(\pi_i),$$

$$\ln\left(\frac{\pi_i}{1 - \pi_i}\right) = \beta_0 + \sum_{i=1}^{n} \beta_i X_i + w_i. \tag{14.23}$$

Here w_i describes the site-specific Gaussian random field (cf Sections 3.1 and 3.2). If we let W be the vector whose components are the w_i, then W is a multivariate normal vector with mean 0 and variance-covariance matrix Σ. The value of the components σ_{ij} depends via a variogram model on the distance h_{ij} that separates them (cf. Section 4.6.1). We will use the exponential model, so that

$$\sigma_{ij} = \sigma^2 \exp(-\varphi h_{ij}). \tag{14.24}$$

The second way in which the algorithms in spBayes are different is that, although they use the MCMC method, they do not use the Gibbs sampler described in Section 14.2. Instead, they use a more sophisticated method called the Metropolis-Hastings algorithm (Metropolis et al., 1953; Hastings, 1970). With this algorithm, the prior distribution of the parameters is specified as before, but instead of iterating directly on this distribution a second probability density called the *proposal distribution* is used. Very roughly speaking, at each iteration, instead of drawing from the existing distribution to generate the updated one, as does the Gibbs sampler, distributional parameter values are drawn from the proposal distribution. If the resulting probability density represents an improvement over the previous iteration, it is accepted. If not, it is rejected. The parameters of the proposal distribution are called the *tuning parameters.*

Turning to the Sierra Nevada oak versus elevation generalized linear model (GLM) from Section 14.3.4, after reading in the data, we first normalize the elevation data and scale the coordinates to avoid numerical problems.

```
> data.Set2S.sf$ElevN <-
+    data.Set2S.sf$Elevation / max(data.Set2S.sf$Elevation)
> data.Set2S.sf$x <- scale(data.Set2S.sf$Longitude)
> data.Set2S.sf$y <- scale(data.Set2S.sf$Latitude)
> glm.demo <- glm(QUDO ~ ElevN, data = data.Set2S.sf, family =
```

```
+      binomial)
> coef(glm.demo)
(Intercept)        ElevN
   3.329145  -11.463550
```

The full Sierra Nevada data set has almost 2000 records. To simplify things for this demonstration, we will use the modulo operator to reduce that number by a factor of ten.

```
> ID <- 1:nrow(data.Set2S.sf)
> mod.fac <- 10
> data.Set2Sr.sf <- data.Set2S.sf[which(ID %% mod.fac == 0),]
> nrow(data.Set2Sr.sf)
[1] 176
```

Recall that data.Set2Sr.sf is a spatial features object. Just to make sure this data reduction has not affected the analysis too much, we can rerun the GLM.

```
> glm.small <- glm(QUDO ~ ElevN, data = data.Set2Sr.sf, family =
+      binomial)
> coef(glm.small)
(Intercept)        ElevN
   4.213447  -13.457494
```

Again, to reduce the possibility of numerical problems, we rescale the coordinate values.

```
> Set2Sr.coords <- as.matrix(data.Set2Sr.sf[, c("x","y")])[,1:2]
```

Next, we need to set the values of the starting and tuning parameters β_i.

```
> beta.start <- as.numeric(coef(glm.small))
> beta.tuning <- t(chol(vcov(glm.small)))
```

The starting values are those of the GLM. The tuning parameters require some explanation. The spGLM documentation recommends using for the tuning parameters the lower triangular Cholesky decomposition of the variance-covariance matrix of the desired parameters of the proposal distribution. The function vcov() computes the variance-covariance matrix of this model. The function chol() computes the *Cholesky decomposition* of the variance covariance matrix. As used by spBayes, the Cholesky decomposition of a matrix A, sometimes called the "square root" of A (Press et al., 1986, p. 311), is a lower triangular matrix (i.e., one with zeroes above the diagonal) denoted L such that $LL' = A$, (where the prime denotes transpose). The function chol() returns it in upper triangular form so the transpose function t() converts this to lower triangular form.

The function call to the spBayes function spGLM() is a long one, so we will give the code first and then explain it.

```
> n.samples <- 100000
> model.glm.bayes <- spGLM(QUDO ~ ElevN, data = data.Set2Sr.sf, family +
+         = "binomial", coords = Set2Sr.coords,
+    starting = list("beta" = beta.start, "phi"=0.05,"sigma.sq"=0.1,"w"
+         = 1),
```

```
+        tuning = list("beta" = beta.tuning, "phi"=0.05, "sigma.sq"=0.1,
+          "w"=0.01),
+        priors=list("beta.Flat", "phi.Unif"=c(0.01, 0.50), "sigma.sq.IG"
+          =c (2, 1)),
+        cov.model = "exponential", verbose=FALSE, n.samples = n.samples)
```

The number of MCMC samples is set at 100,000, which should give us a good result if one is possible. The first four arguments are straightforward. The argument `starting` is a list of starting values for each parameter with each tag corresponding to a parameter name. These correspond to the parameters in Equations 14.23 and 14.24. Because only one value is given for the vector W in Equation 14.23, that value is used for all of its components. The same thing applies to the values of the argument `tuning` of the proposal density and the parameters of the prior density. The last line specifies that we use the exponential model as in Equation 14.24, that intermediate output is suppressed, and gives the number of samples.

The function `spGLM()` returns a lot of information. We can see what we get by first looking at the names of all of its elements.

```
> names(model.glm.bayes)
 [1] "p.beta.theta.samples" "acceptance"      "p.w.samples"
 [4] "weights"              "family"          "Y"
 [7] "X"                    "coords"          "is.pp"
[10] "cov.model"            "run.time"
```

According to the `spGLM()` documentation, the variable `p.beta.theta.samples` is "a coda object of posterior samples for the defined parameters." Since it is a coda object we can plot it as we did with WinBUGS. We will set a burn in period of the first half of the samples, and use the R function `window()` to extract the second half. First, we plot the results.

```
> burn.in <- 0.5 * n.samples
> plot(window(model.glm.bayes$p.beta.theta.samples, start = burn.in))
```

Figure 14.9 shows the results. These don't look as pretty as the ones in the textbooks, but mine never seem to. Next, we plot the results.

```
> summary(window(model.glm.bayes$p.beta.theta.samples, start =
+    burn.in))

Iterations = 50000:1e+05
Thinning interval = 1
Number of chains = 1
Sample size per chain = 50001

1. Empirical mean and standard deviation for each variable,
    plus standard error of the mean:

              Mean       SD   Naive SE Time-series SE
(Intercept)  3.6827 0.74469 0.0033303        0.11844
ElevN      -14.8351 2.30977 0.0103295        0.37997
sigma.sq     1.8162 0.57684 0.0025797        0.15159
phi          0.3851 0.08042 0.0003596        0.04585
```

2. Quantiles for each variable:

	2.5%	25%	50%	75%	97.5%
(Intercept)	2.3027	3.0314	3.7452	4.1623	5.4000
ElevN	-20.4809	-16.1254	-15.0191	-13.0516	-10.7612
sigma.sq	1.2118	1.3616	1.6269	2.1382	3.2950
phi	0.2127	0.3447	0.4207	0.4519	0.4784

Because some of the distributions are far from normal, the median values are probably more appropriate to use as the reported output. The values are reasonably close to those generated by glm(), which is a good sign. In the next section, we will summarize the results of the various analyses we have carried out in this chapter and how they apply to the issued of frequentist versus subjective probability in spatial data analysis.

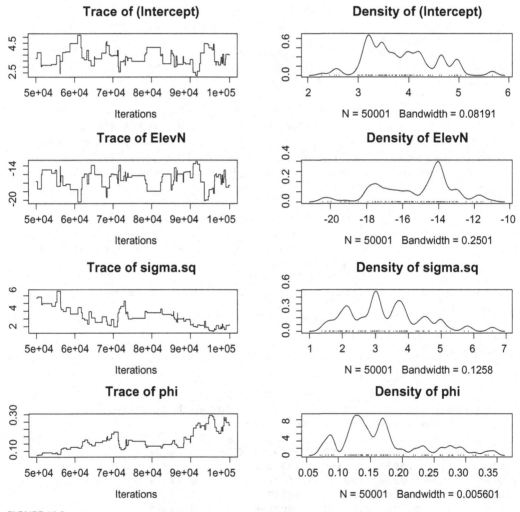

FIGURE 14.9
Diagnostic information from spBayes() model of *QUDO* versus *ElevN*.

14.6 Comparison of the Methods

First of all, it is important to emphasize that Bayes' theorem is exactly that, a mathematical theorem, and the results of a correct application of that theorem are mathematically correct. Bayes' theorem has been used to derive many highly useful probabilistic results and tools, such as the Kalman filter (Gelb, 1974), that do not necessarily invoke subjective probability in the manner that it is presented in this chapter. Moreover, some operations, such as search and rescue, that do involve using the repeated incorporation of new data to improve an estimate, fit very naturally into a subjective probability framework. Indeed, we will have occasion to apply these methods when we discuss spatiotemporal data in Chapter 15. Finally, the power of the MCMC method is increasingly being used to attack so-called "Big Data" problems, where the size of the data set precludes analysis by more traditional methods (Miroshnikov and Conlon, 2014). In this context, it is probably unfortunate that the term *Bayesian analysis* has come to be the standard term to characterize this approach, as opposed to "subjective probability," which more appropriately emphasizes the distinction between it and the other, more "traditional" methods that are generally called the "frequentist" approach.

At a theoretical level, subjective probability does remove some of the imprecision in the interpretation of parameters and the probabilities associated with them. At a more practical level, it is often easier to formulate multilevel models in a Bayesian rather than a mixed-model form. Often the precise statement of a complex mixed effects model involving several interacting variables will leave even the mixed-model experts tied in knots. Both mixed models and Bayesian models require numerical solution using methods that are often subject to instability. The numerical solutions associated with Bayesian computation, however, often seem even more intractable than those associated with the mixed model. The WinBUGS manual (Spiegelhalter et al., 2003) has the statement on the first page, in red letters, "Beware: MCMC sampling can be dangerous!" For the practitioner who does not want to devote a career to working out statistical intricacies, this is not very reassuring.

At a very practical level, both WinBUGS and spBayes programs can be excruciatingly difficult to debug (it is probably for this reason that Gelman and Hill (2007) devote an entire chapter to the subject, one that should be read by anyone venturing into this territory). Using a terminology that is somewhat dated, we can divide the computer programs we construct in this book into two classes, interpreted and compiled. R programs are interpreted; that is, they are executed one line at a time. WinBUGS programs are compiled; that is, the entire program is translated into a machine readable form and then this is executed. Interpreted programs are much easier to construct and debug because you can see the results of each statement directly. Indeed, most modern compilers include a debugging module that allows the programmer to step through the code line by line on the screen as it executes. In this context, the spBayes package may appear to offer some advantage since it runs directly in R, but this advantage is sometimes illusory because actually most of the action takes place with a function such as spGLM() itself, behind the locked doors of the compiled C++ program. It is a poor carpenter who blames his or her tools, but nevertheless the efficiency of use of one's time is something that one must take into account in carrying out a data analysis project. To repeat something I have said already, in my own experience the most effective way to construct a new Bayesian program is to start with an existing program that works and move toward the new one in a series of baby steps.

There are, however, a number of examples of the successful application of Bayesian analysis to difficult problems. Several examples are given in the references in Section 14.7.

Anyone who is even remotely aware of the subject is also aware of the endless amounts of ink that have been spilled on one side or another of the debate between the "frequentist" and the "subjective" approaches. There is no point in adding to this, and as a practitioner I am not qualified to do so anyway. My own feeling is that the best approach is that presented by Bolker et al. (2009). One should be knowledgeable about both approaches and able to use both effectively and appropriately.

I want to conclude with one point that I consider especially important. In Section 12.1 it was stated that in a mixed-model analysis involving grouped variables, at least six or so levels of the grouping factor are needed to do a valid analysis. This is because a reasonably sized sample of values is needed to get a good estimate of the range of variation of the effect. Thus, for example, we can use the fields in Data Set 3 as a grouping factor because there are 16 of them, but we cannot use the fields in Data Set 4 in this way because there are only two. Some authors claim that Bayesian analysis eliminates this requirement for a minimum number of levels of a grouping factor because if its use of subjective probability. In my opinion, this is absolutely false. Critics of the subjective probability approach used to say that Bayesian analysis made it possible to carry out data analysis without actually having any data. One cannot make something out of nothing, however, and ultimately no method will ever permit you to draw justifiable conclusions without having the data to back them up. The amount of data needed to draw a valid conclusion is no different whether the method used is Bayesian analysis or mixed model analysis.

14.7 Further Reading

If you don't know anything about Bayesian analysis, McCarthy (2007) provides a nice introduction. Korner-Nievergelt et al. (2015) provide an excellent introduction to the use of Bayesian methods in the analysis of ecological data. Koop (2003) is an excellent introduction to Bayesian linear regression analysis and Markov Chain Monte Carlo methods. Gelman and Hill (2007) provide very complete coverage of both classical and Bayesian regression that is comprehensive, highly readable, and oriented to the practitioner. At a more intermediate level, Ntzoufras (2009) provides many good examples. Congdon (2003) and Gelman et al. (2014) provide a complete description of modern Bayesian methodology at a more advanced level. Cressie and Wilkes (2011) develop a sort of middle ground between the "frequentist" and the "subjective" approaches that focuses on conditional probability. This may gain traction in the future.

The term *Gibbs sampler* was coined by Geman and Geman (1984), who developed the method. They used this term because their data followed a Gibbs distribution, named for the physicist J. Willard Gibbs (1839–1903), who is often considered the founder of modern statistical thermodynamics. The distributions we work with are generally not Gibbs distributions, but the Gibbs sampler has since been shown to have far more general applicability. Clark and Gelfand (2006) and Gilks et al. (2010) contain collections of papers on Bayesian modeling. Jiang et al. (2009) carry out a Bayesian analysis of spatial data in an agricultural context. There are a number of R packages devoted to Bayesian analysis and the Markov Chain Monte Carlo method. For example, Hadfield (2010) provides in the MCMCglmm package a very nice implementation of the MCMC method for the solution of generalized linear mixed models.

Finally, those contemplating a Bayesian analysis may find it interesting to read Efron (1986) and the discussant papers that follow it.

Exercises

14.1 Develop and run in R2WinBUGS a model using artificial data for the regression relation $Y_i = \beta_0 + \beta_1 X_i + \varepsilon_i$. In other words, the same model as in Section 14.3.3, but including an intercept.

14.2 In Section 9.3.2 we analyzed two linear models, model A, in which the two explanatory variables are independent, and model B, in which the two explanatory variables are highly collinear. (a) Develop a Bayesian model analysis of model A. (b) Using the same values for n.iter and n.burnin as in part (a), compute the solution for model B. What do you observe?

14.3 Carry out a Bayesian analysis using R2WinBUGS of model.5 from Section 8.3 and compare the results with those obtained in that section.

14.4 In Section 8.4.2 a GLM was developed for the Sierra Nevada subset that includes *Precip, MAT, SolRad, AWCAvg,* and *Permeab.* Analyze this model using R2WinBUGS.

14.5 In our mixed model and hierarchical analyses of *Yield* versus *Irrig,* we have always removed Field 2 because all of the values of *Irrig* are the same in that field. However, although this makes it impossible to carry out a regression analysis on data from this field in isolation, a mixed model or hierarchical analysis can still be carried out that incorporates the data. Repeat the analysis of Section 14.4 with data from Field 2 included, and compare the results with those obtained in that section.

14.6 The CAR model discussed in Section 14.5.1 should give the similar results to the ordinary WinBUGS model when $\lambda = 0$. Check to see whether it does.

14.7 (a) Read about the function spLM() in the package spBayes. Load the data for Field 4.1 and construct a model using spLM() for the linear regression of *Yield* on *Clay.* (b) Read about the values returned by spLM() and display a plot of the Bayesian parameters. Then read about the function spRecover() and use it to display the quantiles of the Bayesian estimates of β_0 and β_1. (c) Plot *Yield* versus *Clay* and plot the curve fits of both the linear regression computed using lm() and that computed using spLM().

15

Analysis of Spatiotemporal Data

15.1 Introduction

Many ecological data sets include not only a spatial but also a temporal component. Data may be collected in an agricultural field trial over a period of years in the same field. An uncultivated ecosystem may be monitored over an extended period of time. Both of these processes lead to spatiotemporal data sets. In this chapter, we discuss methods for dealing with data that contain both spatial and temporal components. The theory for spatiotemporal data is much less well developed than that for dealing with purely spatial or purely temporal data, and the contents of this chapter are somewhat *ad hoc* in nature. Perhaps unsurprisingly, this is also one of the most active areas of research, and one of the most dynamic, both in the development of the theory and in the development of analytical methods and software. The development of the theory took a major step forward with the publication of the book by Cressie and Wikle (2011). As Gräler et al. (2016) point out, however, the implementation of the theory has lagged behind. For this reason, the chapter consists of a diverse selection of methods that attempt to put some of the material discussed by Cressie and Wikle (2011) into the context of the data used in this book. Section 15.2 deals with spatiotemporal variograms and kriging, that is, the interpolation of spatial quantities in both space and time. Section 15.3 introduces the concept of spatiotemporal process models derived from physicochemical models for diffusion-reaction processes. Section 15.4 introduces discrete time approximations to spatiotemporal processes, sometimes called "state and transition models." Finally, Section 15.5 describes a Bayesian approach to the analysis of spatiotemporal processes.

Because spatiotemporal data analysis is such a dynamic area, both in development of the theory and in development of the R code, it is not unlikely that some aspects of R for analysis of spatiotemporal data may have changed by the time you are reading this. If you run into problems, use the resources discussed in Section 2.7 to help you solve them.

15.2 Spatiotemporal Data Interpolation

15.2.1 Representing Spatiotemporal Data

Figure 15.1 shows a set of spatiotemporal data that we will be analyzing in this section. The data set is four years of yield monitor data from Field 4.1. The crops grown in the four successive years were wheat, tomato, bean, and sunflower. The open circles

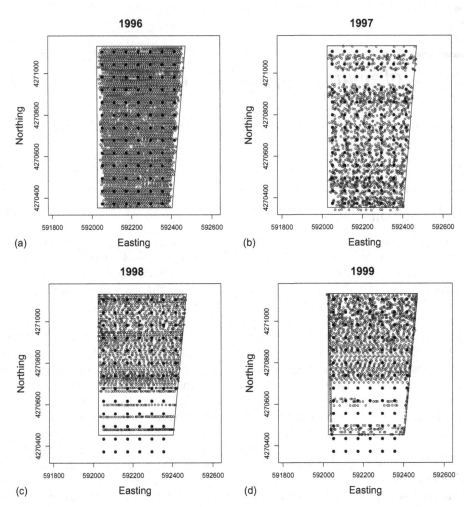

FIGURE 15.1
Yield monitor data from four years of harvests in Field 1 of Data Set 4. The location of every tenth data record is shown. The black dots indicate the hand-sampling locations. Yields from (a) 1996; (b) 1997; (c) 1998; (d) 1999.

represent yield data measured in each year, and the dark circles represent locations where manually collected data were recorded. Although the dark circles are shown in each figure, these data were only recorded in the first year. So far we have only analyzed the data from the first year, 1996. A glance at the figure shows some of the issues that arise with spatiotemporal data sets. The figure also illustrates some of the difficulties of field research (difficulties that are sometimes not fully appreciated by our laboratory-based colleagues). The first problem revealed in the figure is that the southern portion of the field was removed from the experiment after two years. The reason is this. One of the principles of precision agriculture is that spatially precise information should be useful to the farmer in increasing profitability by permitting a greater efficiency in resource use. The cooperator who farms the fields in Data Set 4 is very astute and a very good farmer, and he put this principle into practice in our study. Although he knew that the south end of Field 4.1 had better soil than the rest of the field, he did not know exactly where this good soil was located; nor did he know how good it was relative to the rest of the field. After seeing the results of our first two years of the experiment, he made a decision that was good for his profitability but bad for the experiment: he took the southern part of the field out of field crop production and put it into a high-value crop (seed onions). Therefore, the analysis of the four years of data must exclude the southernmost 12 data locations.

The second problem is that yield monitor coverage was incomplete in each of the years after the first. This was partly due to equipment problems and partly due to the need of the cooperator to harvest the material in a timely manner. In 1998, when not all the field could be surveyed, yield monitor passes were made along a line that passed close to the sample points, as shown in Figure 15.1c. In 1997 and 1999, however, there are sample points with no yield data near to them. The third apparent problem really isn't one. The GPS located on the yield monitor in 1997 appears to have been in error in the southern part of the field, since some of the recorded data locations are apparently outside the boundary of the field (Figure 15.1b). This is actually not a problem. The boundary polygon was adjusted in this book for expository purposes and is actually slightly smaller than the field itself.

In this section we are going to put the data into objects of the class spacetime, from the R package of the same name (Pebesma, 2011, 2012). This has rapidly become the standard format for spatiotemporal data, particularly for purposes of interpolation. To prepare the data for analysis, we first follow the procedure used in Section 6.3.1 to interpolate the 1997, 1998, and 1999 yield data to the 1996 sample points using inverse distance weighted interpolation. Since we did not eliminate the last two rows of the sample points from 1996, the 1998 and 1999 data include records for data that do not really exist. We therefore eliminate these data records from each year. To keep the code concise, we use much shorter names for the variables than we have in the previous chapters.

```
> Y96 <- data.Set4.1.96[(1:74),]
> Y97 <- data.Set4.1.97[(1:74),]
> Y98 <- data.Set4.1.98[(1:74),]
> Y99 <- data.Set4.1.99[(1:74),]
```

We are going to put the data into an object of class STFDF. This extends the sp classes for points, lines, and polygons to include a time component. The class is one of those developed in the spacetime package. First, we create the attribute data values for the STFDF object. Because the absolute magnitudes of the yields for the four different crops are very different, we normalize each year's yield to a scale of 0 to one 100, creating a of normalized values for each year.

```
> Y96.norm <- 100 * Y96[,2] / max(Y96[,2])
> Y97.norm <- 100 * Y97[,2] / max(Y97[,2])
> Y98.norm <- 100 * Y98[,2] / max(Y98[,2])
> Y99.norm <- 100 * Y99[,2] / max(Y99[,2])
```

The attribute values are stored as a data frame with a single data field, so we concatenate them to create the data set we will use.

```
> Yield.norm <- data.frame("Yield" = c(Y96.norm,
+ Y97.norm, Y98.norm, Y99.norm))
> nrow(Yield.norm)
[1] 296
```

The spatial portion of a `spacetime` object is stored as an `sp` object called `Yield.sites`, and this is created using the function `coordinates()` in the usual way.

```
> Yield.sites <- data.frame(Easting = Y96$Easting, Northing =
+    Y96$Northing)
> coordinates(Yield.sites) <- c("Easting", "Northing")
> proj4string(Yield.sites) <- CRS("+proj=utm +zone=10 +ellps=WGS84")
```

There are 74 position coordinates and a data frame four times that long of attribute values, so the `spacetime` object will automatically recycle through the position coordinates to obtain the correct location for each attribute value.

We have not yet had to represent time sequences in this book. R treats temporal reference systems in a way analogous to how it treats spatial coordinate reference systems, which are briefly discussed in Section 2.4.3. Specifically, with spatial coordinates there are a variety of packages and functions such as `st_crs()` and `proj4string()` that have been created by contributors to deal with standard coordinate reference systems such as longitude-latitude and UTM. Similarly, there are base and contributed packages in R that allow the user to deal with standard temporal reference systems. One of these is `ISOdate()`, which is included in the base package. We will use this to create the time data.

```
> Yield.year <- c(1996, 1997, 1998, 1999)
> Yield.month <- c(5, 9, 9, 9)
> Yield.day <- rep(15, length(Yield.year))
> print(Yield.time <- ISOdate(year = Yield.year, month =
+    Yield.month, day = Yield.day))
[1] "1996-05-15 12:00:00 GMT" "1997-09-15 12:00:00 GMT"
[3] "1998-09-15 12:00:00 GMT" "1999-09-15 12:00:00 GMT"
```

The values are represented in the International Standards Organization (ISO) format for year, month, day, hour, minute, and second. In our case, we just approximate the harvest dates, which were in the spring for the 1996 wheat crop and in the fall for the other crops. The `spacetime` package makes extensive use of the contributed R package zoo, described by Zeileis and Grothendieck (2005), to handle temporal data, and this reference provides a good discussion of the representation of temporal data.

We are now ready to create an `STFDF` object. In order to be able to plot it nicely, we will create one with polygon data. Here we use the very convenient property of `sp` objects that they can for many operations be treated as data frames. This allows us to simply drop the lowest two rows of polygons from the data frame.

```
> thsn.sf <- st_read("auxiliary\\set4.1thiessen.shp")
> thsn.sp <- as(thsn.sf, "Spatial")
> proj4string(thsn.sp) <- CRS("+proj=utm +zone=10 +ellps=WGS84")
> thsn9899.sp <- thsn.sp[1:74,]
```

Now we create the polygon object. We can use the fact that the `spacetime` plotting function `stplot()` works through the `sp` function `spplot()`, which in turn works through the `lattice` package (Section 2.6.3). Figure 15.2 shows the result.

```
> YieldPolys.stfdf <- STFDF(thsn9899.sp, Yield.time,
+    data = Yield.norm)
> library(lattice)
```

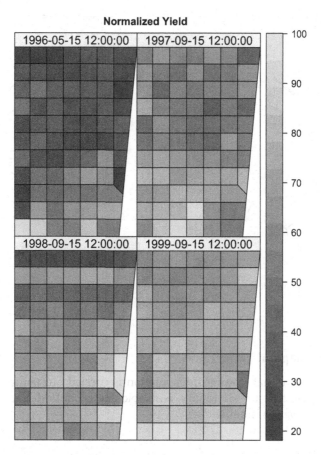

FIGURE 15.2
Plot of the data space of normalized yields in Field 4.1 in 1996 and 1997 showing the four clusters obtained via *k*-means clustering with $k = 4$.

```
> greys <- grey(seq(5, 19) / 22)
> trellis.device(color = FALSE)
> stplot(YieldPolys.stfdf, col.regions = greys, main = "Normalized
+ Yield")
```

Recall that the crop sequence was wheat, tomato, bean, and sunflower. Wheat is the only winter crop, and clearly suffered more from the heavy soil in the north.

Now that we have assembled the spatiotemporal data, in the next two sections we will discuss one important application: the spatiotemporal variogram and its application to spatiotemporal kriging. For purposes of comparison in this section, we fit four individual variograms to the sequence of four years of normalized yield data. Figure 15.3a shows a perspective plot created using the function persp() of the four empirical variograms. Figure 15.3b shows the same plot, but with the dist variable replaced by cutoff – dist, where cutoff is the cutoff distance of the variogram, 400 m. This figure is shown to provide ease of comparison the figures that will be shown in the next section. These are created using the lattice function wireframe(), which arranges the axes differently from that of persp(). After 1996, the individual variograms tend to be mostly nugget, indicating little spatial autocorrelation and even less temporal autocorrelation.

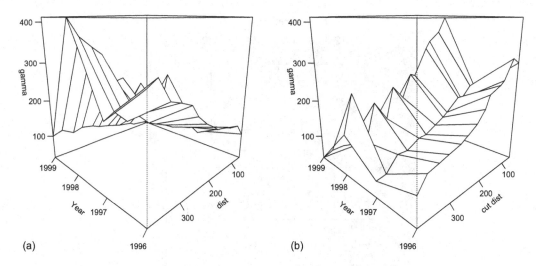

FIGURE 15.3
(a) Perspective plot of four empirical variograms of yield in Field 4.1. (b) Same variogram rotated 180°.

15.2.2 The Spatiotemporal Variogram

In Section 6.3.2, we saw that kriging interpolation is carried out using the covariogram. Recall (Equation 4.20) that for stationary, isotropic spatial data defined on a domain D (Section 1.2.1) the *variogram* is defined as

$$\gamma(h) = \frac{1}{2} \text{var}\{Y(x+h) - Y(x)\}, \tag{15.1}$$

where Y is the measured quantity and x is a position vector with coordinates (x, y). The quantity h is the spatial lag. In Section 4.6.2, we noted that for a purely spatial process if we define the covariogram as (Equation 4.24)

$$C(h) = \text{cov}\{Y(x), Y(x+h)\}, \tag{15.2}$$

then from Equation 4.26 the covariogram is related to the variogram by

$$C(h) = C(0) - \gamma(h), \tag{15.3}$$

where $C(0)$ is the sill of the variogram. Consider now a spatiotemporal domain, which we will denote $D \times T$ with elements (x_i, t_j). Let u represent a temporal lag analogous to the spatial lag h. Then we can write the spatiotemporal variogram as (Cressie and Wikle, 2011, p. 315)

$$\gamma(h, u) = \frac{1}{2} \text{var}\{Y(x+h, t+u) - Y(x, t)\}. \tag{15.4}$$

Similarly, if we define the spatiotemporal covariogram as

$$C(h,u) = \text{cov}\{Y(x+h,t+u) - Y(x,t)\}, \tag{15.5}$$

then

$$C(h,u) = C(0,0) - \gamma(h,u). \tag{15.6}$$

This leads to a complication in estimating the variogram. The spatial dimension is assumed stationary and isotropic (Section 3.2.2), which permits to express the covariogram in the simple form of Equation 15.2. We cannot, however, extend this spatial isotropy assumption to the covariance function $C(h, u)$ defined in Equation 15.5 because although the process is stationary in time, it is not isotropic between the time dimension and the spatial dimension.

In computing a purely spatial empirical variogram in Section 4.6.2, we used a method of moments estimator (the formula is given by Equation 4.21). In the case of the spatiotemporal variogram, it is still possible to compute a simple estimator of this form. This simple estimator may not, however, be consistent with the spatiotemporal covariance of the data. It is, therefore, often desirable to make a particular assumption about the relationship between the variogram in the spatial dimension and that in the temporal dimension, that is, in the form of the covariogram function $C(h, u)$ in Equation 15.6. We will present a few of the options discussed by Gräler et al. (2016). Our discussion is based on code written by Gräler (2016) and by Veronesi (2015).

We start with the same four yield maps used in the previous section. If we are going to compute a variogram, however, we are going to need more than four points in the time direction. Webster and Oliver (1992) suggest at least 100 points for a spatial variogram. We will, therefore, create some fake data that gives us at each of the 74 spatial coordinates a sequence of 100 temporal values. To create the 100 temporal values, we will for each of the 74 locations first fit the 4 normalized year values with a cubic spline (see Section 9.2) denoted s and then generate a sequence of 100 values Y_i according to the equation

$$Y_i = s_i + \eta_i,$$
$$\eta_i = \lambda \eta_{i-1} + \varepsilon_i. \tag{15.7}$$

This is a first-order autoregressive time series of the type discussed in Section 3.5.1, and provides some degree of temporal autocorrelation to the data. Here η_i are the autocorrelated random variables, λ is a constant whose value is between 0 and 1, and the ε_i are independent, normally distributed random variables with mean 0 and variance σ^2. The code is very straightforward and is not shown. Figure 15.4 shows for the first spatial data location the four yield values and the full temporal data set, with the four "real" data values shown as black dots and the 96 artificial data values shown as open circles. The dashed line shows the spline fit, and the solid line connects the 100 data values. To remove any connection with yield and years for this artificial data set we simply call the data values $Y(x, t)$. The temporal autocorrelation is apparent. The data are stored in a data frame.

```
> Y.data <- data.frame(Y = Y)
> nrow(Y.data)
[1] 7400
```

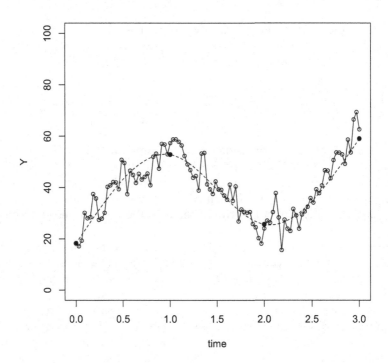

FIGURE 15.4
Artificial data set created by fitting a first-order autoregressive model to the time series of four years of yield data in Field 4.1. The yield values at the first data location are shown; the other 73 set of values are similar.

The location data object is created in the same way as in Section 15.2.1 except that in this case we create point rather than polygon data.

```
> Y.sites <- data.frame(Easting = Y96$Easting, Northing = Y96$Northing)
> coordinates(Y.sites) <- c("Easting", "Northing")
> proj4string(Y.sites) <- CRS("+proj=utm +zone=10 +ellps=WGS84")
```

In creating the temporal data, we need to be a bit careful. The functions that we will be using to create the spatiotemporal variogram work better if the time values are evenly spaced (this property is called being *strictly regular*). If we use the four successive years 1996 through 1999 as the time values, then because 1996 is a leap year the data will not be strictly regular. Since the time variable, like the attribute data, no longer has any physical meaning, we will simply use the function as.date() to create a sequence of 100 days as members of the date class. This class numbers days sequentially beginning with January 1, 1960 as day 0.

```
> as.date(0:3)
[1] 1Jan60 2Jan60 3Jan60 4Jan60
> class(as.date(0:3))
[1] "date"
```

Therefore, we set a time of the first 100 days in 1960. R has two basic time classes, POSIXct and POSXlt. The class POSIXct represents calendar days beginning with January 1, 1970 as day 0. We can use a coercion function to covert our date values to POSIXct values.

```
> as. POSIXct (as.date(0:3))
[1] "1959-12-31 16:00:00 PST" "1960-01-01 16:00:00 PST"
[3] "1960-01-02 16:00:00 PST" "1960-01-03 16:00:00 PST"
```

Note that because I am writing this in California and my computer is therefore set to California time, as.POSITct() automatically converts GMT to Pacific Standard Time. This doesn't matter in our case since it is all fake data anyway. We can now create the temporal sequence and the STFDF object.

```
> Y.time <- as.POSIXct (as.date(0:(nt - 1)))
> Y.stfdf <- STFDF(Y.sites, Y.time, data = Y.data)
```

We are now ready to begin trying various forms of spatiotemporal variogram estimation. We begin with the simple empirical variogram, making no assumptions about the form of the covariogram $C(h, u)$. Here is the code.

```
> Y.empvgm <- variogramST(Y ~ 1, data = Y.stfdf, tlags = 0:7,
+    cutoff = 400, width = 100)
```

The first two arguments are obvious. The argument tlags specifies the number of time lags at with the variogram will be estimated, cutoff specifies the spatial separation distance up to which point pairs are included in semivariance estimates, and width specifies the width of subsequent distance intervals into which data point pairs are grouped for semivariance estimates.

Figure 15.5 shows a wireframe plot of the empirical variogram. Comparing this figure with Figure 15.3b (note that the time axes in the figures move in opposite directions), we see that the empirical spatiotemporal variogram is smoother than the corresponding set of independent variograms, presumably reflecting the temporal autocorrelation of the former. Once again, however, the variograms after the initial time are mostly nugget.

Now we turn to constructing variograms reflecting models for the structure of the covariogram $C(h, u)$. Gräler et al. (2016) describe five models of increasing complexity. We will discuss two of them. The simplest is the *separable* model, which assumes a relationship of the form

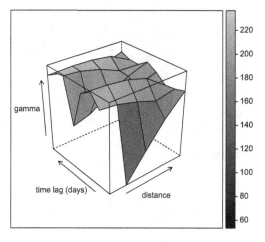

FIGURE 15.5
Wireframe plot of the spatiotemporal variogram of the artificial data set.

$$C(h,u) = C_s(h)C_t(u). \tag{15.8}$$

The empirical variogram assuming a variogram model is computed using the gstat function vgmST() and then fit using the function fit.Stvariogram(). Both of these functions have very many arguments, and I am not going to show the code, which can be viewed in the accompanying R code file. Veronesi (2015) discusses these arguments to some extent. They are mostly control parameters and, as he points out, they are mostly selected by trial and error. In the examples of this section, I did not find the results to be strongly dependent on the values of the control parameters.

Figure 15.6 shows a wireframe plot of the fitted surface. The model is practically pure nugget after the first year. The process of selecting the best of the candidate models in an actual fitting process would ordinarily involve minimizing the mean square error. This is provided as an attribute of the StVariogramModel object.

```
> attr(Y.sep.fit, "MSE")
[1] 324.2339
```

Veronesi (2015), who analyzed a real data set, found that his separable model also yielded very poor fit.

The sum-metric model assumes a relationship of the form

$$C(h,u) = C_s(h) + C_t(u) + C_j(\sqrt{h^2 + \kappa u^2}), \tag{15.9}$$

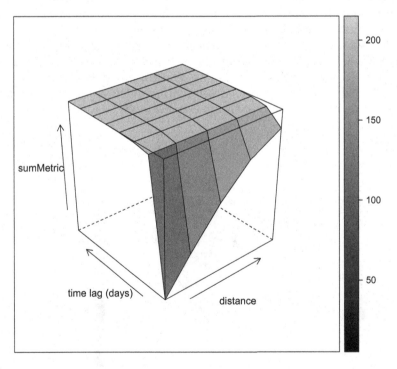

FIGURE 15.6
Wireframe plot of the spatiotemporal variogram of the artificial data set fitter using a separable model given by Equation 15.8.

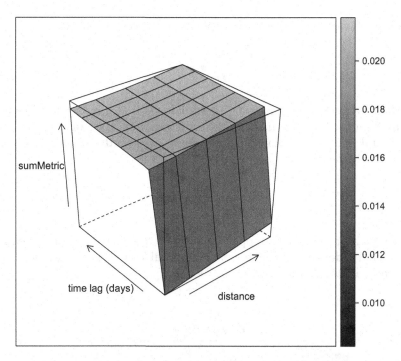

FIGURE 15.7
Wireframe plot of the spatiotemporal variogram of the artificial data set fitter using a sum-metric model given by Equation 15.9.

where κ is a spatial anisotropy parameter whose value can be obtained using the function stAni(). Figure 15.7 shows the resulting wireframe plot. It again almost pure nugget after the first time step. We can see that the sum-metric model provides a slightly better fit.

```
> attr(Y.smm.fit, "MSE")
[1] 310.3402
```

One application of the spatiotemporal variograms, as with the purely spatial variogram, is the visualization of spatial relationships. In that context, the empirical spatiotemporal variogram provides an indication that after the first year the spatiotemporal variogram is mostly pure nugget, meaning that the data are not highly spatially autocorrelated. As with the pure variogram, probably the most important use of the spatiotemporal variogram model is in its application to kriging interpolation. This is discussed in the next section.

15.2.3 Interpolating Spatiotemporal Data

Recall (Equation 6.2) that an interpolator $\hat{Y}(x)$ is called a *linear interpolator* if for some set of coefficients ϕ_i, $i = 1, 2, ..., n$,

$$\hat{Y}(x) = \sum_{i=1}^{n} \phi_i Y(x_i). \qquad (15.10)$$

An interpolator $\hat{Y}(x)$ is *unbiased* if the interpolation functions ϕ_i satisfy

$$\sum_{i=1}^{n} \phi_i = 1. \tag{15.11}$$

The important property of the kriging interpolator is that it is the best linear unbiased estimator in the sense that is minimizes the error variance. The same conditions apply for a spatiotemporal linear interpolator. The equations are exactly the same, except that now Y and $\hat{Y}$ are functions of x and t, so we have

$$\hat{Y}(x,t) = \sum_{i=1}^{n} \varphi_i Y(x_i, t_j). \tag{15.12}$$

The package gstat provides a wrapper function krigeST() that, once a spatiotemporal variogram model is developed, makes spatiotemporal kriging very easy. We will continue our discussion with the example begun in Section 15.2.2.

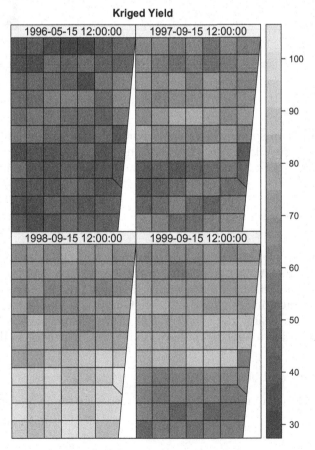

FIGURE 15.8
Thiessen polygons showing kriged spatiotemporal "yield" values.

We will use the sum-metric variogram model computed in the previous section in our interpolation. This variogram is in the object Y.fit.smm. In order to compare the interpolated values with the original in as simple a manner as possible, we will plot the kriged values at the times corresponding to the four original harvests.

```
> Y.predtime <- as. POSIXct (as.date (seq (24.5,nt - 0.5, 25)))
```

We then use the code from Section 15.2.1 to enter the year values 1996 through 1999 as the temporal data. The kriging is carried out using a call to the function krigeST().

```
> Y.pred <- krigeST(Y ~ 1, data = Y.stfdf, modelList = Y.fit.smm,
+     newdata = STF(Y.sites, Y.predtime))
```

Figure 15.8 shows the results, plotted as Thiessen polygons. As with many interpolations, the interpolated values are quite different from the original values in Figure 15.2. In particular, the lower yields in the northern end in 1996 and 1997 have been shifted toward the southern end in 1997 and 1999. The results of this section do not lead to any further biophysical insights about the factors influencing crop yield in Field 4.1, and should only be considered as an introduction to how spatiotemporal kriging is carried out.

15.3 Spatiotemporal Process Models

15.3.1 Models for Dispersing Populations

Cressie and Wikle (2011, Chapter 6) recommend the use of process models to characterize spatiotemporal processes. The model that they find generally most appropriate is the diffusion-reaction model, and that is the subject of this section. We begin by discussing a model for diffusion. Consider a petri dish with a very shallow layer of water, and suppose a drop ink is released into the water. If $Y(x, y, t)$ describes the concentration of ink at location (x, y) at time t, then Y can be modeled using the *diffusion equation*. This equation is written

$$\frac{\partial Y}{\partial t} = D\left(\frac{\partial^2 Y}{\partial x^2} + \frac{\partial^2 Y}{\partial y^2}\right), \qquad (15.13)$$

together with appropriate boundary conditions (Dettman, 1969, p. 145). We will assume that the equation is defined on an infinite plane with zero flux in the limit. Here D is the diffusion coefficient and measures the rate at which the ink disperses into the water. The diffusion equation also has a long history modeling the movement of species in the environment (Okubo, 1980). If $Y(x,y)$ represents a population that is entirely concentrated at the origin $(x,y) = (0,0)$ at time $t = 0$ with a population Y_0, and if there is no mortality, then for any $t > 0$ the population $Y(x,y,t)$ will be described by the solution of Equation 15.13 on an infinite plane. This solution is a normal density (Plant and Cunningham, 1991),

$$Y(x,y,t) = \frac{Y_0}{4\pi Dt}\exp\left(\frac{-(x^2 + y^2)}{4Dt}\right). \qquad (15.14)$$

If mortality is included in the population model, then Equation 15.13 is modified to

$$\frac{\partial Y}{\partial t} = D\left(\frac{\partial^2 Y}{\partial x^2} + \frac{\partial^2 Y}{\partial y^2}\right) - \mu Y, \tag{15.15}$$

where μ is the mortality rate (assumed fixed), and Equation 15.14 becomes

$$Y(x,y,t) = \frac{Y_0 e^{-\mu t}}{4\pi Dt}\exp\left(\frac{-(x^2+y^2)}{4Dt}\right). \tag{15.16}$$

This model assumes that the organisms behave as passively diffusing entities, with no directed behavior, and that the entire population is concentrated at a single point at time $t = 0$. More complex initial conditions are relatively easy to take care of, but the issue of behavior may not be.

Plant and Cunningham (1991) developed a model of the form of Equation 15.15 to describe the results of an experiment in which marked sterile Mediterranean fruit flies (medflies, *Ceratitis capitata* W.) were released from a single location in a macadamia orchard on the island of Hawaii. Traps using a pheromone bait were arranged in a regular grid throughout the orchard. Because medflies are fairly weak fliers, the attractive properties of the bait were not thought to seriously disrupt the spatial distribution of the population.

This biophysical system is one that should be amenable to a diffusion model if any such system is. Insects, particularly weak fliers like the medfly, may not display substantial nonrandom dispersal behavior, and the orchard provided a relatively uniform environment. Plant and Cunningham found that it was necessary to modify the model of Equation 15.15 to include a convection term that accounted for a prevailing wind. In addition, to obtain a good fit it was necessary to divide the population into a mobile and a "settled" subpopulation, with the latter no longer changing position. In the model, flies moved from the mobile to the settled population at a fixed rate. The model as modified was fit to the data using a nonlinear least squares algorithm (see Exercise 12.5), and found to provide a reasonable fit.

Returning to the analogy of ink dispersing in a petri dish, the "mortality" term μY may be thought of as modeling a chemical reaction in which the ink is gradually transformed into some other chemical, so that its concentration declines. From this interpretation, equations such as Equation 15.15 are often called *diffusion-reaction equations* (Winfree, 1977). It is also possible to include more complex "reaction" functions $f(Y)$.

Of course the annual crop yields that are the subject of Section 15.2 do not in any sense "diffuse" or "react." Many spatiotemporal processes in ecology have do not have biophysical properties from which a diffusion equation could be derived from first principles. The interesting question is whether a diffusion-reaction model that incorporates an error term can provide a purely empirical description of a spatiotemporal process, in the same way that a linear regression model does not have to imply any causal relationship between the explanatory and response variables but only an empirically observed relationship. In the next section, we will explore the application of this approach to our yield data.

15.3.2 A Process Model for the Yield Data

The yield process represented in Figure 15.1 is continuous in space (approximately) and discrete in time. The data, however, are discrete in both space and time. Although we will represent the "diffusion-reaction" model for this process abstractly using a continuous spatiotemporal model such as Equation 15.15 we will implement this model using discrete space and time. Our approach is motivated by the methods used in the solution of ordinary and partial differential equations. We begin by considering an ordinary differential equation of the form

$$\frac{dY}{dt} = f(Y),$$

$$Y(0) = Y_0.$$

(15.17)

The simplest method to compute a numerical solution to this equation is called *Euler's method* (Press et al., 1986, p. 543). We approximate the derivative on the left-hand side of the first equation by a forward difference operator

$$\delta_t \cong \frac{Y(t + \Delta t) - Y(t)}{\Delta t}.$$

(15.18)

This allows us to write symbolically

$$\delta_t Y = f(Y),$$

$$Y(0) = Y_0$$

(15.19)

and to implement this numerically as

$$Y(t + \Delta t) \cong Y(t) + \Delta t Y(t).$$

(15.20)

Starting with $Y(0) = Y_0$, we can march forward in time. There are of course much more sophisticated numerical methods for the solution of ordinary differential equations, but we will base our representation of the temporal part of the discrete "diffusion-reaction" process on Equation 15.20.

The spatial part of the model is based on the so-called second centered difference. We write

$$\frac{\partial^2 Y}{\partial x^2} \cong \frac{\frac{\partial Y}{\partial X}(x + \Delta x, y, t) - \frac{\partial Y}{\partial X}(x - \Delta x, y, t)}{2\Delta x}.$$

(15.21)

The first partial derivative in the numerator of Equation 15.21 may be approximated as a forward difference

$$\frac{\partial Y}{\partial X}(Y(x + \Delta x, y, t) \cong \frac{Y(x + \Delta x, y, t) - Y(x, y, t)}{\Delta x},$$

(15.22)

and the second partial derivative may be approximated by the corresponding backward difference. Plugging these into Equation 15.21 allows us to write

$$\frac{\partial^2 Y}{\partial x^2} \cong \frac{Y(x + \Delta x, y, t) - 2Y(x, y, t) + Y(x - \Delta x, y, t)}{2\Delta x^2},$$

(15.23)

and we define the second centered difference δ_{xx}^2 to be the quantity on the right-hand side. The second centered difference in the y direction is defined similarly. This allows us to write a discrete spatiotemporal analog to Equation 15.15 as

$$\delta_t Y = D\left(\delta_{xx}^2 Y + \delta_{yy}^2 Y\right) + \mu Y + \varepsilon_{xyt}. \tag{15.24}$$

where an error term ε_{xyt} is explicitly incorporated. In the numerical solution of partial differential equations, Equation 15.24, without the error term, is a very crude method called an explicit method (Ames, 1977, p. 62), but we are interested in testing it as the basis for an empirical model of a spatiotemporal process. We have changed the sign of the coefficient μ because this term no longer represents mortality or any other biophysical quantity; Equation 15.24 is purely an empirical fit to data.

Expanding δ_t, δ_{xx}, and δ_{yy} yields after a little algebra

$$Y(t+\Delta t,x,y) = Y(t,x,y) + D\left(\frac{Y(t,x+\Delta x,y) - 2Y(t,x,y) + Y(t,x-\Delta x,y)}{2\Delta x}\right) + $$

$$D\left(\frac{Y(t,x,y+\Delta y) - 2Y(t,x,y) + Y(t,x,y-\Delta y)}{2\Delta y}\right) + \mu Y(t,x,y) + \varepsilon_{xyt} \tag{15.25}$$

This can be simplified by letting $\Delta t = 1$ and noting that $\Delta x = \Delta y$ so that we can express D in units of one cell width. This yields

$$Y(t+1,x,y) = Y(t,x,y) + \frac{D}{2}\left[Y(x+\Delta x,y,t) - 2Y(x,y,t) + Y(x-\Delta x,y,t)\right] + $$

$$\frac{D}{2}\left[Y(x,y+\Delta y,t) - 2Y(x,y,t) + Y(x,y-\Delta y,t)\right] + \mu Y(t,x,y) + \varepsilon_{xyt} \tag{15.26}$$

Equation 15.26 cannot be solved without boundary conditions. If it represented a biophysical process, these would ordinary be a fixed set of values for either Y or its partial derivatives on the boundary. However, for our application we simply set the differentials δ_{xx} and δ_{yy} equal to zero outside the region on which Y is defined.

It is important to recognize that under ordinary circumstances, Equation 15.26 would be treated as an approximation to a parabolic partial differential equation and solved for $Y(x,y,t+1)$, the quantity of the left-hand side of Equation 15.26, by plugging. In our application, Equation 15.26 represents an approximation to the spatiotemporal autocorrelation structure and is solved for D and μ by plugging in the appropriate values of Y and minimizing the mean square error ε.

The set of normalized yield values in the code are represented by a 74 by 4 matrix called *Y.norm*. The code to carry out the solution is as follows.

```
> MSE <- function(P){
+    Yest <- numeric(74)
+    for (i in 1:74) Yest[i] <- Y1[i] +
+        P[1] * (dxx[i] + dyy[i]) + P[2] * Y1[i]
+    E <- sum((Yest[i] - Yp1[i])^2) / length(Y1)
+    return(E)
+}
>
```

```
> D <- numeric(3)
> mu <- numeric(3)
>
> for (n in 1:3){
+     Yp1 <- Y.norm[, n+1]
+     Y1 <- Y.norm[, n]
+     soln <- nlm(MSE, c(0,0))
+     D[n] <- soln[[2]][1]
+     mu[n] <- soln[[2]][2]
+ }
```

Here the function `nlm()` is a nonlinear minimization function (Press et al. 1986, p. 521) that computes the minimum value of the first argument using the second argument as the vector of initial values. Here are the results.

```
> D
[1] -0.3034430 -0.2006647 -0.3416246
> mu
[1] 0.4827022 0.3203651 0.7631263
```

Interestingly, the "diffusion coefficient" D takes on negative values. This is of course impossible in a physical system since it would contradict the second law of thermodynamics, but it is possible in this purely empirical curve fit. The values of D are fairly consistent for each of the three time periods. It is unclear whether this is anything more than a coincidence. Indeed, the results are sensitive to the method used to scale the Y values (Exercise 15.2).

Again, the results of this section do not add anything to our understanding of the processes underlying crop growth and development in Field 4.1. They serve strictly as an illustration of how an empirical model could be constructed for a spatiotemporal process. Cressie and Wrikle (2011, p. 332) encourage ecological modelers to build their models from first principles where possible. Nothing could be further from the truth regarding the model constructed in this section. The code written for the section, however, will work for a model that is developed from first principles with one very important exception: the boundary conditions. The model of Plant and Cunningham (1991) described in Section 15.3.1 used boundary conditions that specify zero flux as the spatial coordinates tend to infinity. The correct specification of boundary conditions is perhaps somewhat less crucial for a parabolic equation such as Equation 15.14 than it is for elliptic or hyperbolic partial differential equations, but it cannot be completely ignored. Those wishing to apply models developed from first principles to spatiotemporal processes in ecology are advised to consult a good text on the subject. My favorite is Street (1973).

15.4 Finite State and Time Models

15.4.1 Determining Finite State and Time Models Using Clustering

As an alternative to modeling the spatiotemporal structure of the data, one might consider a discrete time model in which each spatial location can move between finite states. The state and transition model (Westoby et al., 1989) is one such approach. It provides a framework for the modeling of successional dynamics in ecosystems. In the model, ecosystems can exist in a set of discrete *states*, and *transitions* can occur between these states in finite time. Roel and Plant (2004a and 2004b) developed an exploratory method for organizing spatiotemporal data by organizing the data into clusters in space and time. In this section, we discuss a method used by for generating and analyzing such clusters. We apply this approach to the four years of normalized yield data in Field 4.1. The spatiotemporal method discussed in this section works by first forming clusters of the response variable in the data space of each year's normalized yield data. We demonstrate the method with the first two years of data, since these can be visualized, and then apply the method to the data from all four years. In Exercise 15.3, you are asked to analyze centered and scaled yield data, which provide a slightly different perspective.

The analysis uses the k-means clustering algorithm introduced in Section 12.6.1. It is important to note that no spatial information is used in the algorithm; the only values used are attribute values. This permits us to use the spatial data to assess the validity of the model. Clustering is carried out in R using the function kmeans(). Our two-year demonstration uses $k = 4$ clusters. First, we read in the data and set it up for the cluster analysis.

```
> Yield96 <- read.csv("created\\set4.1yld96ptsidw.csv", header = TRUE)
> Yield97 <- read.csv("created\\set4.1yld97ptsidw.csv", header = TRUE)
> Yield96.norm <- 100 * Yield96$Yield / max(Yield96$Yield)
> Yield97.norm <- 100 * Yield97$Yield / max(Yield97$Yield)
> Yield.2Yr <- data.frame(Yield96 = Yield96.norm,
+     Yield97 = Yield97.norm)
```

Generating the clusters just requires a simple call to kmeans().

```
> set.seed(123)
> cl <- kmeans(Yield.2Yr, 4)
```

One way to visualize the clusters is in the data space, as shown in Figure 15.9. A second way is in geographic space, as shown in Figure 15.10. The relationship in the data space of Figure 15.9 appears roughly parabolic. In the geographic space of Figure 15.10, the clusters are spatially organized; that is, they appear to form a meaningful spatial pattern. This is not guaranteed to occur, and its occurrence may be taken as an indication that the clusters are biophysically meaningful. This will be discussed more fully below.

Cluster 1 contains points from the large northern area of the field that are low in both years. Cluster 2 is in the south and is the highest yielding region in 1996 but yields lower in 1997. Cluster 3 is located mostly in the center of the field and is low in 1996 and moderate in 1997. Cluster 4 is in the south-center and is the highest yielding region in 1997.

Now that we have an idea of the process, let's move on to the full four-year data set, in which we remove the southernmost two rows of sample points from the interpolated yield data. The data are loaded and organized into a data frame as in the previous sections. To begin, we must recognize that, since the clustering algorithm starts with a random set of seeds, the process will not always result in the same set of clusters. Let's see what happens when we generate nine replications of the clustering algorithm applied to the four-year yield data set. We will generate these replications of clusters with $k = 3$ and collect them in a cluster matrix called CM for display. Your results may be different from those displayed here due to slight differences in your interpolated yield values.

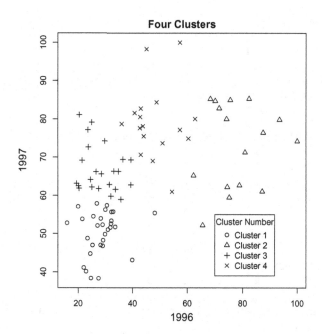

FIGURE 15.9
Plot of the data space of normalized yields in Field 4.1 in 1996 and 1997 showing the four clusters obtained via k-means clustering with $k = 4$.

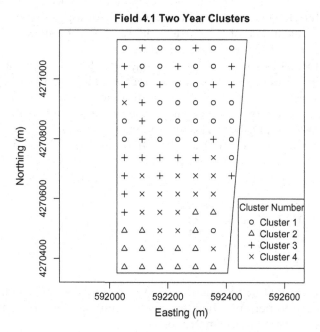

FIGURE 15.10
Thematic map showing the locations of the four clusters obtained via k-means clustering with $k = 4$.

```
> cluster.k <- 3
> n.reps <- 9
> n.sites <- nrow(Yield.4Yr)
> CM <- matrix(nrow = n.sites, ncol = n.reps)
> set.seed(123)
> for (i in 1:n.reps) CM[, i] <-
+      kmeans(Yield.4Yr, cluster.k)$cluster
> CM
       [,1] [,2] [,3] [,4] [,5] [,6] [,7] [,8] [,9]
 [1,]    1    3    1    3    3    3    1    2    3
 [2,]    1    3    1    3    3    3    1    2    3
 [3,]    1    3    1    3    3    3    1    2    3
 [4,]    1    3    1    3    3    3    1    2    3
 [5,]    1    3    1    3    3    3    1    2    3
 [6,]    1    3    1    3    3    3    1    2    3
 [7,]    1    3    1    3    3    3    1    2    3
 [8,]    3    1    2    1    2    2    3    1    1
   *    *    *    DELETED    *    *    *
[74,]    2    2    3    2    1    1    2    3    2
```

Since the initial cluster seeds are assigned randomly, the clusters are not identified by the same number in each replication. In addition, comparison of rows 7 and 8 of the matrix CM indicates that the k-means algorithm does not generate the same set of clusters in each replication. In dealing with spatial data, this property can be used to advantage. The objective in using cluster analysis with spatiotemporal data is to develop clusters that can be used to identify factors underlying the observed pattern in space and time of the response variable. That is, the clusters must be "biophysically meaningful." One criterion for being biophysically meaningful is that the cluster pattern should be *stable*, that is, it should be attainable starting from different sets of initial cluster seeds.

In order to test for stability it is necessary to align the cluster identification numbers. We will do this by minimizing the number of different ID values between them. For example, let's determine the permutation of cluster identification numbers in replication 2 (i.e., column 2 of the matrix CM above) that produces the best alignment with the identification numbers in each other replication. The first step is to generate a matrix called perms of all the possible permutations of the k identification numbers. This can be done using the function permutations() of the e1071 package (Dimitriadou et al., 2017).

```
> library(e1071)
> print(perms <- t(permutations(cluster.k)))
      [,1] [,2] [,3] [,4] [,5] [,6]
[1,]    1    2    2    1    3    3
[2,]    2    1    3    3    1    2
[3,]    3    3    1    2    2    1
```

Next, we create an array test to contain the permuted identification numbers. For example, suppose we want to permute the identification numbers in the second column of CM using the third permutation in the matrix perm.

```
> test <- numeric(nrow(CM))
> for (m in 1:nrow(perms)) test[which(CM[,2] == m)] <- perms[m,3]
> CM[,2]
> CM[,2]
```

```
[1] 3 3 3 3 3 3 3 1 1 1 1 1 1 1 3 3 3 3 3 3 3 3 2 1 1 3 1 3 1 1 1 1 1
[33] 1 1 1 1 1 1 1 1 1 1 1 1 1 1 1 2 1 1 2 1 2 2 2 1 3 2 2 2 2 2 1 2
[65] 2 2 2 2 2 2 2 2 2 2
> test
[1] 1 1 1 1 1 1 1 2 2 2 2 2 2 2 2 1 1 1 1 1 1 1 3 2 2 1 2 1 2 2 2 2 2
[33] 2 2 2 2 2 2 2 2 2 2 2 2 2 2 2 2 3 2 2 3 2 3 3 3 3 2 1 3 3 3 3 3 2 3
[65] 3 3 3 3 3 3 3 3 3 3
```

The 1's have been replaced by 2's, the 2's by 3's, and the 3's by 1's.

Suppose we want to determine the permutation of column 2 of the matrix CM that produces the best match with column 4. We will create a vector diff.sum to hold the number of elements of each permutation that differ from the corresponding element of column 4 of CM.

```
> diff.sum <- numeric(ncol(perms))
> for (n in 1:length(diff.sum)){
+    for (m in 1:nrow(perms)) test[which(CM[,2] == m)] <- perms[m, n]
+    diff.sum[n] <- sum((test != CM[,4]))
+}
> diff.sum
[1]  0 57 74 39 74 52
```

For my data set, Permutation 1 has the smallest sum of differences. Your results may vary.

We now embed this code in a function min.sum() that for each *i* and *j* returns the minimum sum of differences and the permutation that produces it.

```
> min.sum <- function(i, j, perms, CM){
+      test <- numeric(nrow(CM))
+      diff.sum <- numeric(ncol(perms))
+      for (n in 1:length(diff.sum)){
+ # test replaces the IDs in rep i with those in column n of perms
+      for (m in 1:nrow(perms)) test[which(CM[, j] == m)] <- perms[m, n]
+      diff.sum[n] <- sum((test != CM[, i]))
+      }
+ # which.min automatically selects the first in a tie
+      n.min <- which.min(diff.sum)
+      opt.id <- diff.sum[n.min]
+      return(c(n.min, opt.id))
+}
```

Now we are ready to compute the matrix of minimum ID differences and of optimal permutations for each pair of replications.

```
> dist.mat <- matrix(nrow = n.reps, ncol = n.reps)
> min.id <- matrix(nrow = n.reps, ncol = n.reps)
> for (i in 1:n.reps){
+    for (j in 1:n.reps){
+         x <- min.sum(i, j,perms, CM)
+         min.id[i, j] <- x[1]
+         dist.mat[i, j] <- x[2]
+    }
+}
> dist.mat
```

```
        [,1]  [,2]  [,3]  [,4]  [,5]  [,6]  [,7]  [,8]  [,9]
[1,]     0     0     2     0     0     0     0     0     2
[2,]     0     0     2     0     0     0     0     0     2
[3,]     2     2     0     2     2     2     2     2     0
[4,]     0     0     2     0     0     0     0     0     2
[5,]     0     0     2     0     0     0     0     0     2
[6,]     0     0     2     0     0     0     0     0     2
[7,]     0     0     2     0     0     0     0     0     2
[8,]     0     0     2     0     0     0     0     0     2
[9,]     2     2     0     2     2     2     2     2     0
> min.id
        [,1]  [,2]  [,3]  [,4]  [,5]  [,6]  [,7]  [,8]  [,9]
[1,]     1     6     4     6     3     3     1     5     6
[2,]     6     1     5     1     2     2     6     4     1
[3,]     4     3     1     3     6     6     4     2     3
[4,]     6     1     5     1     2     2     6     4     1
[5,]     5     2     6     2     1     1     5     3     2
[6,]     5     2     6     2     1     1     5     3     2
[7,]     1     6     4     6     3     3     1     5     6
[8,]     3     4     2     4     5     5     3     1     4
[9,]     6     1     5     1     2     2     6     4     1
```

Your matrix may be different from mine. The matrix of minimal distances is not always zero. Based on the matrix dist.mat, it appears that the replications divide into two groups. The alignment procedure is as follows. We first pick one replication to serve as the reference, and then we align all the other replications by minimizing the distance to this reference. We will pick replication 1.

```
> ref.rep <- 1
> best <- min.id[ref.rep,]
> CM.align <- matrix(nrow = n.sites, ncol = n.reps)
> for (i in 1:n.reps){
+     for (j in 1:cluster.k){
+         CM.align[which(CM[, i] == j),i] <- perms[j, best[i]]
+     }
+}
> CM.align
        [,1]  [,2]  [,3]  [,4]  [,5]  [,6]  [,7]  [,8]  [,9]
[1,]     1     1     1     1     1     1     1     1     1
[2,]     1     1     1     1     1     1     1     1     1
[3,]     1     1     1     1     1     1     1     1     1
   *  *  * DELETED * * *
[73,]    2     2     2     2     2     2     2     2     2
[74,]    2     2     2     2     2     2     2     2     2
```

Figure 15.11 shows the results. The clusters form into two primary groups. The first group in turn splits the northern, low-yielding area into two subgroups, corresponding to numbers 1 and 3 in the figure. The second group covers most of the southern section (number 2). The stability and spatial contiguity evident in the figure provides one indication that the clusters are biophysically meaningful.

A second check that the clusters are biophysically meaningful is that they must be spatially organized (i.e., arranged in a spatially meaningful pattern). To proceed further, and not have to rely on visualization, we will use a statistic that measures internal consistency

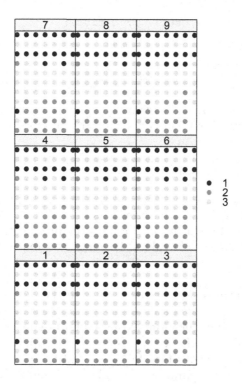

FIGURE 15.11
Thematic maps showing the locations of the clusters obtained via k-means clustering of four years of yield data with $k = 3$.

(Roel and Plant, 2004b). Suppose there are n data locations, and that the set under consideration has m replications (i.e., m of the random reorderings of initial seeds result in a member of this set). Let an indicator c_i, $i = 1,...,n$, be defined for each location i by

$$c_i = \begin{cases} 1 \text{ if in all members of the set location } i \text{ belongs to the same cluster} \\ 0 \text{ if at least one location } i \text{ belongs to a different cluster.} \end{cases}$$

Then the measure of consistency γ of the set is defined by

$$\gamma = \frac{\sum_{i=1}^{n} c_i - \dfrac{n}{m^k}}{n - \dfrac{n}{m^k}}. \tag{15.27}$$

The motivation for this definition of γ is similar to that for Cohen's κ (Cohen, 1960). The quantity n/m^k is the expected number of cells that all belong to the same cluster by chance. If all of the members of the set are identical then γ has the value 1, and if each cell of each member of the set is randomly assigned one of the k possible values, then $\gamma = 0$. The code to compute γ in Equation 15.27 is as follows:

```
> c.clus <- function(x) as.numeric(sum(abs(x[2:length(x)] - x[1]))
+      == 0)
> gamma.clus <- function(CM, set, k){
```

```
+    sum.c <- sum(apply(CM[, set], 1, c.clus))
+    m <- length(set)
+    n <- nrow(CM)
+    return((sum.c - n/m^k) / (n - n/m^k))
+}
```

With this as background, we are now ready to begin the analysis. The computations are identical to those just carried out, but now we set the number of clusters to $k = 2$ and the number of replications to $m = 25$.

```
> gamma.clus(CM.align, 1:ncol(CM.align), 2)
[1] 1
```

The γ statistic for $k = 2$ equals 1, indicating that all 25 replications are in alignment. Figure 15.12a indicates that the clusters are highly structured spatially. Figure 15.13a shows a plot of the mean normalized yield values. This indicates that average over the locations in cluster 2 in the southern part of the region is in every year higher than the regions in cluster 1 in the north.

Besides consistency from several initial cluster seeds and a spatial pattern, a third indication of whether the clusters are biophysically meaningful is whether they are hierarchically arranged. This question involves a bit of subtlety. Consider the case of an increase in the number of clusters from k to $k+1$. Suppose, for example, $k = 2$, so that we go from two to three clusters. Suppose that there is a single factor, say, soil clay content, underlying the separation of yield values into the two clusters. When the elements are partitioned into three clusters, the physical property underlying yield variability may continue to be the single factor clay content, it may switch to one or two completely new factors, or it may be comprised of clay content from the original two-cluster partition and some other factor, say, organic matter, within one of the two original clusters. It is only in the latter case that one would expect the clusters to be

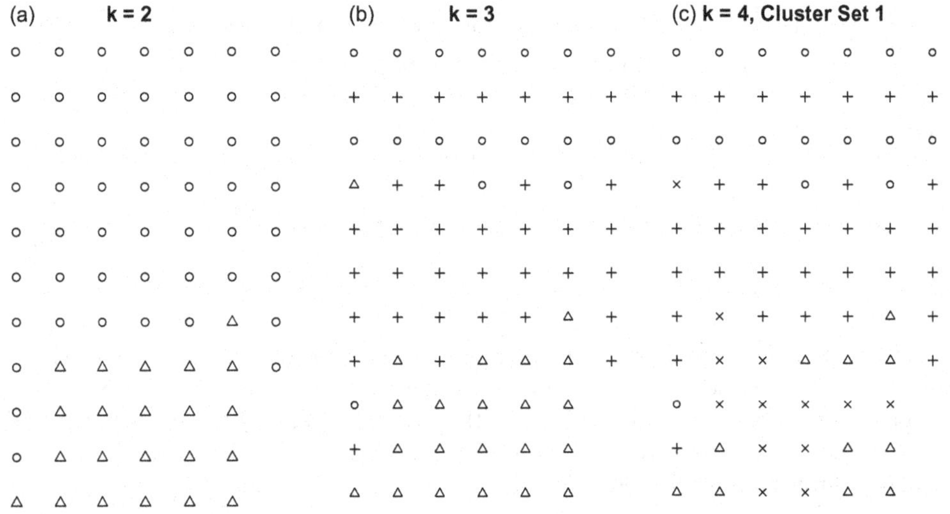

FIGURE 15.12
Locations of the clusters obtained using different values of k for the four years of yield data in Field 4.1. (a) $k = 2$; (b) $k = 3$; (c) $k = 4$.

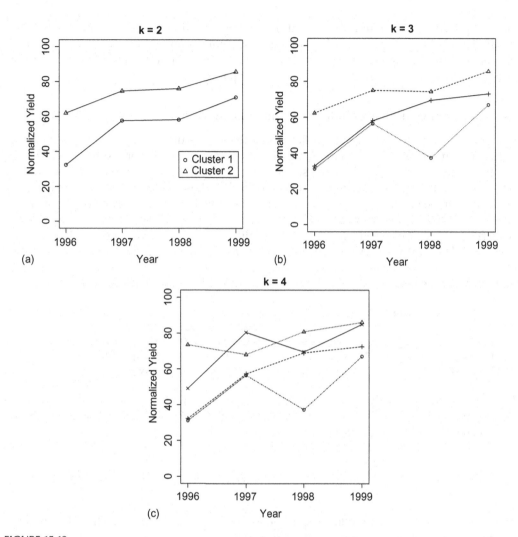

FIGURE 15.13
Plots of cluster mean yields over the four years for each of the cluster arrangements shown in Figure 15.5:
(a) $k = 2$; (b) $k = 3$; (c) $k = 4$.

hierarchically arranged. Therefore, we will not test statistically for the presence or absence of a hierarchical arrangement, but rather simply examine the pattern of mean yields as k is increased. Let's increase k to 3 and see what happens.

```
> gamma.clus(CM.align, 1:ncol(CM.align), 3)
[1] 0.9729712
```

For $k = 3$ the γ statistic has the value 0.97, indicating that the replications again divide almost perfectly into two groups (your results may be different). The lower yielding northern cluster splits into two, one in the north and one in the center (Figures 15.12b and 15.13b). For $k = 4$ the γ value for the entire set is 0.66, indicating a loss of consistency. It turns out that the clusters can be divided into two groups that are internally consistent, and that these two groups are very similar. We only show the results for the first group. The higher yielding southern cluster splits into two (Figures 15.12c and 15.13c). The patterns of cluster splitting do seem to indicate that the clusters are hierarchical. The number of data locations in each cluster for $k = 4$ has become sufficiently small that we will not carry out a cluster analysis for higher values of k. Therefore, we move on to the problem of determining the factors that underlie the observed cluster patterns.

15.4.2 Factors Underlying Finite State and Time Models

Our primary tools for attempting to determine the factors underlying the observed cluster patterns from Section 15.4.1 will be classification trees and random forest (Sections 9.3 and 9.4). Before carrying out this analysis, however, it is useful to examine climatic variables to determine what, if any, influence they may have had on these patterns. We have already seen evidence that the northern portion of the field suffered yield loss probably caused by aeration stress in the 1995–1996 season due to heavy rainfall. The file *dailyweatherdata.csv* was developed from data downloaded from the CIMIS (California Irrigation Management Information System) website, http://wwwcimis.water.ca.gov/cimis/welcome.jsp. The data were collected at that CIMIS weather station located on the UC Davis campus, which is located about 15 kilometers east of the fields in Data Set 4. Some values are missing from the data set.

```
> names(weather.data)
 [1] "Date"          "JulianDay"   "CIMIS_Eto_mm" "Precip_mm"
 [5] "SolRad_W.sq.m" "AvgVap_kPa"  "MaxAirTemp_C" "MinAirTemp_C"
 [9] "AvgAirTemp_C"  "MaxRelHum"   "MinRelHum"    "AvgRelHum"
[13] "DewPt_C"       "AvgwSpd_m.s" "WndRun_Km"    "AvgSoilTemp_C"
```

In order to compare the three summer seasons (remember that the wheat was grown in the winter), we will define the season of each year to begin on April 1 and end on September 1. This corresponds to the following.

```
> season.2 <- 548:701
> season.3 <- 913:1066
> season.4 <- 1278:1431
```

The quantity most likely to reflect the influence of climatic variability on the yield pattern is reference evapotranspiration CIMIS _ Eto _ mm.

```
> which(is.na(weather.data$CIMIS_Eto_mm))
integer(0)
> ETo1997 <- sum(weather.data$CIMIS_Eto_mm[season.2])
> ETo1998 <- sum(weather.data$CIMIS_Eto_mm[season.3])
> ETo1999 <- sum(weather.data$CIMIS_Eto_mm[season.4])
> c(ETo1997, ETo1998, ETo1999)
[1] 983.04 816.66 948.03
```

The third season, the summer of 1998, had a reference ET about 150 mm lower than the other two summer seasons. This indicates that the summer of 1998 may have been cooler than the other two. The usual way to measure seasonal temperature is with accumulated degree-days (Zalom et al., 1983). However, four days in the 1997 season and three in the 1998 season are missing daily minimum temperature data.

```
> which(is.na(weather.data[,8]))
[1] 353 648 649 650 651 652 653 1032 1349 1410
```

Before plotting the degree days, we will impute the missing values by linearly interpolating between the temperature values just before and just after each sequence of missing data. This interpolation is left for the Exercises (Exercise 15.5). After imputation, degreedays are computed using simple averaging with a lower threshold of 10°C.

```
> DD.per.day <- function(max.temp, min.temp) (max.temp+min.temp)/2 - 10
```

The function `cumsum()` is then used to generate the data to plot. Consistent with the evapotranspiration data, the highest degree-day accumulation occurs in 1997 (Figure 15.14a). Although the ultimate accumulation is about the same in 1998 as in 1999, the midseason is warmer in 1999. Precipitation is not really a factor in any season (Figure 15.14b, Exercise 15.6).

Now that we have developed the climatic data, we will go through successive analyses of the clusters generated with three successive values of k, beginning with $k = 2$. The code for the analysis is analogous to that used in Section 9.3. The output is heavily redacted to show only the first competitor and the first two surrogates at nodes where splits occur.

```
Node number 1: 74 observations, complexity param=0.5909091
   Primary splits:
       Clay < 35.52 to the right, improve=15.574660, (0 missing)
       SoilP < 7.21 to the left, improve=12.392600, (0 missing)
   Surrogate splits:
       Sand < 26.895 to the left, agree=0.932, adj=0.615, (0 split)
       SoilK < 156.45 to the right, agree=0.851, adj=0.154, (0 split)
Node number 2: 61 observations, complexity param=0.1818182
   Primary splits:
       SoilTN < 0.1065 to the left, improve=7.301125, (0 missing)
       SoilK < 230.9 to the left, improve=5.674194, (0 missing)
   Surrogate splits:
```

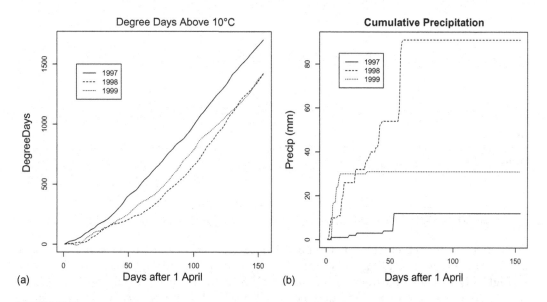

(a)   (b)

FIGURE 15.14
(a) Accumulated degree-days over 10°C in Fields 4.1 and 4.2 for the 1997, 1998, and 1999 seasons. (b) Cumulative precipitation in each season.

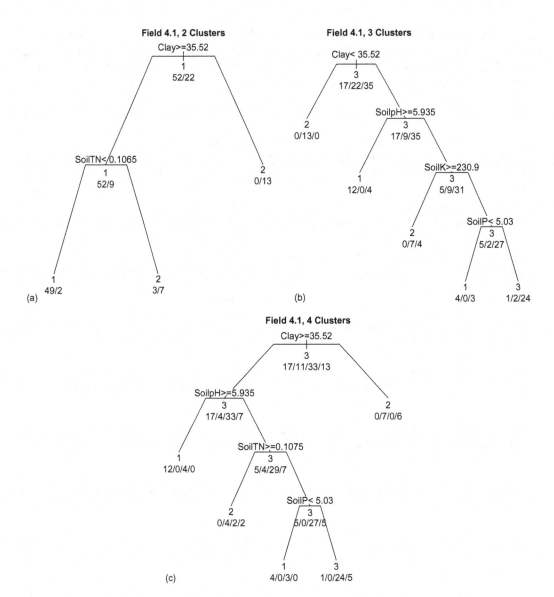

FIGURE 15.15
Classification trees with three different cluster values as the response variable for the clusters shown in Figure 15.12. (a) 2 clusters, (b) 3 clusters, (c) 4 clusters.

```
SoilP < 15.72 to the left, agree=0.869, adj=0.2, (0 split)
SoilK < 230.9 to the left, agree=0.869, adj=0.2, (0 split)
```

In the two-cluster set, cluster 1 is the northern, lower mean yield cluster and cluster 2 is the southern, higher yielding one (Figure 15.12a). The first splitting variable is *Clay*, with all high clay locations partitioned into a node identified as cluster 1 (Figure 15.15a). The second split, acting on the node defined by *Clay* ≥ 35.52, is on *SoilTN*, with low values going into cluster 1. For $k = 3$ the higher yielding north splits on *SoilpH*, then on *SoilK*, and then on *SoilP* (Figure 15.15b). Based on the instability discussed in Section 9.3.5, we would

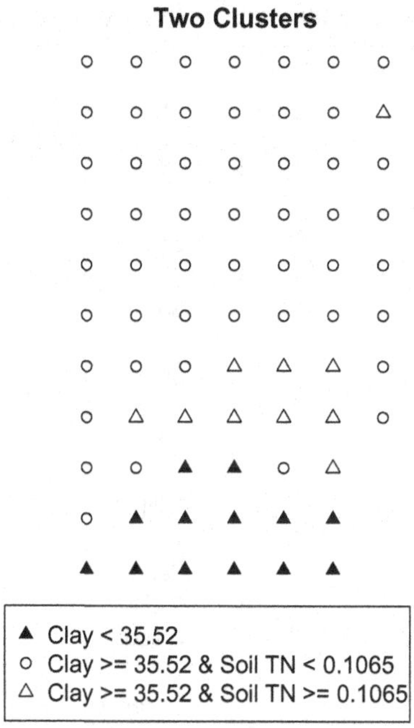

Two Clusters

```
▲  Clay < 35.52
○  Clay >= 35.52 & Soil TN < 0.1065
△  Clay >= 35.52 & Soil TN >= 0.1065
```

FIGURE 15.16
Plot of the nodes of the classification tree shown in Figure 15.15a. The white symbols are in the Cluster 1, the low mean yield cluster, and the black symbols are in Cluster 2.

not expect the lower splits to be very predictive. For $k = 4$ the same splits take place in the lower yielding region and no splits are identified in the south (Figure 15.15c). Figure 15.16 shows a map of the nodes of the $k = 2$ tree.

Figure 15.17 shows the results of a random forest analysis of the three-cluster data sets. Somewhat surprisingly, *Weeds* are rated fairly low. This may be because weeds only affect yield at high levels, primarily at *Weeds* = 5, and this only occurs in a relatively small number of locations. Thus, when these locations are not part of the random sample, weed level appears unimportant.

In Section 9.4.2, we saw that classification and regression trees may be unstable, that is, small changes in the variable values may have large effects on the tree. An alternative way of dealing with tree instability to those discussed in that section is to perturb the response variable slightly and test the effect of these perturbations on the tree structure. We will try this by raising or lowering the value of roughly 10% of the cluster numbers by one. We will also ensure that the cluster ID numbers are all between 0 and 4. We won't worry about the fact that some of the ID numbers will be both raised and lowered, with no net effect. Investigation of the rpart object using our old friend str() indicates that the first column of the data field frame contains the name of the splitting variable. As usual, see the R file for this section for the full code to set up the tree. Here is a quick simulation, showing only the first six nodes.

```
> n.clus <- 3
> data.Perturb <- data.Set4.1clus
> p <- 0.1
>set.seed(123)
>for (i in 1:10){
```

Cluster Variable Importance

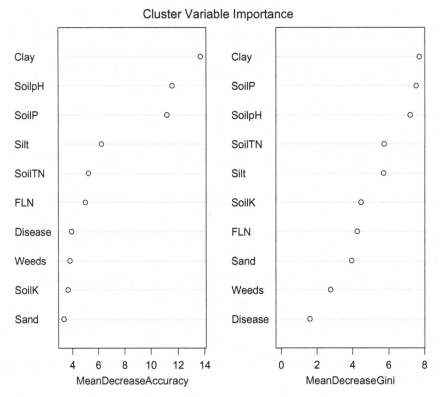

FIGURE 15.17
Variable importance plots from a random forest model of the $k = 2$ cluster data.

```
+    data.Perturb$Cluster <- clusters. Set4.1$x +
+        rbinom(74, 1, p) - rbinom(74, 1, p)
+    data.Perturb$Cluster[which(data.Perturb$Cluster < 0)] <- 0
+    data.Perturb$Cluster[which(data.Perturb$Cluster > 4)] <- 1
+    data.Perturb$Cluster
+    Perturb.rp <- rpart(model.1, data = data.Perturb,
+        method = "class")
+    print(as.character(Perturb.rp$frame[1:6,1]))
+}
[1] "Clay"    "<leaf>"  "SoilpH"  "<leaf>"  "SoilK"   "Weeds"
[1] "Clay"    "<leaf>"  "SoilP"   "<leaf>"  "SoilpH"  "Sand"
[1] "Clay"    "<leaf>"  "SoilP"   "<leaf>"  "SoilTN"  "<leaf>"
[1] "SoilpH"  "<leaf>"  "Sand"    "<leaf>"  "SoilP"   "<leaf>"
[1] "Clay"    "<leaf>"  "SoilTN"  "<leaf>"  "Silt"    "SoilK"
[1] "Clay"    "<leaf>"  "SoilP"   "<leaf>"  "SoilpH"  "<leaf>"
[1] "Clay"    "<leaf>"  "SoilpH"  "SoilTN"  "<leaf>"  "<leaf>"
[1] "Clay"    "<leaf>"  "SoilpH"  "<leaf>"  "SoilTN"  "<leaf>"
[1] "Clay"    "<leaf>"  "SoilTN"  "SoilpH"  "<leaf>"  "<leaf>"
[1] "Clay"    "<leaf>"  "SoilP"   "<leaf>"  "SoilpH"  "<leaf>"
```

The printout shows the splitting variable at each node. The symbol "<leaf>" indicates that the node is a terminal node. *Clay* and soil nutrients are consistently identified as important.

Soil properties such as texture and organic matter content, and the content of non-labile nutrients such as P and K, should not change much over years. In Exercise 15.4, you are asked to construct a scatterplot matrix of these quantities against yield of all four years. We will postpone an agronomic interpretation of these results until Chapter 17.

15.5 Bayesian Spatiotemporal Analysis

15.5.1 Introduction to Bayesian Updating

Bayesian statistical theory is based on the concept of subjective probability and on the combination of data with prior subjective probability to obtain posterior subjective probability. The concept of updating is central to the application of Bayesian theory to temporal data analysis. The Kalman filter (Gelb, 1974), which is used in optimal control of the trajectory of a dynamic system, and statistical decision theory (Raiffa and Schlaifer, 1961; Melsa and Cohn, 1978), which is concerned with the optimal use of observational data in the decision-making process, both make effective use of Bayesian updating.

To see how the updating process works, let's return to the example of Scientist A and Scientist B that we used in Section 14.1 to introduce subjective probability. Recall that Scientist A and Scientist B are both interested in determining the value of a parameter that we will call θ. Scientist A is about 95% certain that the value of θ lies between 860 and 940. This says that Scientist A's prior subjective probability distribution (assumed normal) has a mean $\hat{\theta} = 900$ and a standard deviation $\hat{\sigma} = 20$. A measurement of θ is made that results in a value plus or minus one standard deviation is $Y_1 = 850 \pm 40$. Equation 14.9 specifies the mean $\breve{\theta}$ and variance $\breve{\sigma}^2$ of the posterior $p(\theta \mid Y_1)$ as

$$\breve{\theta} = \frac{(\hat{\tau}\hat{\theta} + \tau_1 Y_1)}{\hat{\tau} + \tau_1}, \quad \breve{\tau} = (\hat{\tau} + \tau_1)$$

$$\hat{\tau} = \frac{1}{\hat{\sigma}^2}, \quad \tau_1 = \frac{1}{\sigma_1^2}. \tag{15.28}$$

For our data, Scientist A's updated parameters are $\breve{\theta} = 890$ and $\breve{\sigma} = 17.9$. The reduction in the standard deviation from $\hat{\sigma} = 20$ to $\breve{\sigma} = 17.9$ represents the increase in the certainty of Scientist A's subjective belief about the value of θ taking into account the data. We could now make a second observation Y_2 and repeat the process.

Before we go on, we must add a caveat. Equations 15.28 indicate that the only quantities affecting the change in standard deviation of Scientist A's subjective probability are the precision $\hat{\tau} = 1/\hat{\sigma}^2$ of the prior and the precision $\tau_1 = 1/\sigma_1^2$ of the data. While they have the nice, intuitively reasonable property that they predict an increased certainty of belief as more data are accumulated, they do not necessarily model the way a scientist would really think in all cases. For example, suppose that instead of the values given above, the observation is $Y_1 = 100 \pm 40$. According to Equations 15.28, Scientist A would dutifully modify his or her belief in the value of θ to $\breve{\theta} = 740$ with the same value for $\breve{\tau}$. In reality, however, Scientist A would probably consider the data value to be discordant and consider applying one of the options discussed in Section 6.2.1.

There are two fundamental issues at play here. The first is whether the Bayesian formulas represent an accurate model for how people modify their beliefs in the presence of new evidence, and the second is whether people, when confronted with new evidence, modify their beliefs in an appropriate way. These are both complex and controversial issues. Plant and Stone (1991, Ch. 3) provide an introductory discussion, and further references are provided in Section 15.6. Since this section is about the use of Bayesian inference, we will accept the assumptions of Bayesian updating of subjective probability and see where they lead.

We first consider a simple, schematic example of how Bayesian updating might be used to incorporate new data into a regression model. We begin with the same artificial data set introduced in Section 14.2 and shown in Figure 14.2. Since we are going to be basing a second regression on this regression, we will add the subscript 1 to the variables.

```
> library(R2WinBUGS)
> set.seed(123)
> n <- 20
> X1 <- rnorm(n)
> Y1 <- X1 + rnorm(n)
> print(coef(lm(Y1 ~ X1)), digits = 3)
(Intercept)        X1
    -0.0402      0.9217
```

Here are the results of the same Markov Chain Monte Carlo run as was done in Section 14.2.

```
> print(t(cbind(demo.sim1$mean, demo.sim1$sd)), digits = 3)
       beta0 beta1 tau   deviance
[1,] -0.0502 0.917 1.41  51.2
[2,]  0.199  0.206 0.467 2.62
```

The means and the inverse of the squares of the standard deviation of beta0 and beta1 can be used directly as the prior mean and precision in a second simulation, with one cautionary comment. Recall that the joint prior density $p(\beta, \tau)$ is the product of a normal prior for β and a gamma prior for τ. This is not a conjugate prior, however, so the posterior density $p(\beta, \tau \mid Y)$ does not have this form. Therefore, we do not actually use the posterior from one observation as the prior for the subsequent one. Instead, we formulate a new prior based on the statistical properties of the preceding posterior.

The random variable τ is distributed according to a gamma distribution with parameters μ and v, where the mean is equal to μ / v and the variance is equal to μ / v^2. Let us denote the mean and standard deviation of τ by m_τ and sd_τ. Then we have $m_\tau = \mu / v$ and $sd_\tau^2 = \mu / v^2$. Therefore, $v = m_\tau / sd_\tau^2$ and $\mu = m_\tau v$. The values of these parameters are computed as follows:

```
> print(mu.0 <- demo.sim1$mean$beta0, digits = 3)
[1] -0.0502
> print(tau.0 <- 1/demo.sim1$sd$beta0^2, digits = 3)
[1] 25.3
> print(mu.1 <- demo.sim1$mean$beta1, digits = 3)
[1] 0.917
> print(tau.1 <- 1/demo.sim1$sd$beta1^2, digits = 3)
[1] 23.5
> print(nu.t <- demo.sim1$mean$tau / demo.sim1$sd$tau^2, digits = 3)
[1] 6.45
```

```
> print(mu.t <- demo.sim1$mean$tau * nu.t, digits = 4)
[1] 9.06
```

We will use these values in the prior distributions of a second data set, which in our example has a slightly different relationship between X and Y.

```
> set.seed(456)
> X2 <- rnorm(n)
> Y2 <- 0.7 * X2 + rnorm(n)
```

Here is the code that will be passed to WinBUGS. Only those parts that are changed from
the first run are shown.

```
> demo.model <- function() {
+ beta0 ~ dnorm(mu.0, tau.0)
+ beta1 ~ dnorm(mu.1, tau.1)
+ tau ~ dgamma(mu.t, nu.t)
+ for (i in 1:n)
+   {
+   Y.hat[i] <- beta0 + beta1 * X[i]
+   Y[i] ~ dnorm(Y.hat[i], tau)
+   }
+}
> XY.data <- list(X = X2, Y = Y2, n = n,
+   mu.0 = mu.0, tau.0 = tau.0, mu.1 = mu.1, tau.1 = tau.1,
+   mu.t = mu.t, nu.t = nu.t)
```

Here are the results of the second simulation.

```
> print(t(cbind(demo.sim1.2$mean, demo.sim1.2$sd)), digits = 3)
      beta0  beta1 tau   deviance
[1,] -0.155 0.762 1.2    56.5
[2,]  0.144 0.134 0.29   1.78
```

The estimated slope decreases from 0.92 to 0.76 as a consequence of the new data, and the
standard deviation of the posterior has declined from 0.21 to 0.13. In Exercise 15.9, you are
asked to verify that this same result is obtained (approximately) if all of the data are used
at once in a simulation with a noninformative prior.

Let us compare this result with that obtained by using ordinary least squares regression.

```
> print(coef(lm(Y1 ~ X1)), digits = 3)
(Intercept)        X1
    -0.0402      0.9217
> Y12 <- c(Y1,Y2)
> X12 <- c(X1,X2)
> print(coef(lm.12 <- lm(Y12 ~ X12)), digits = 3)
(Intercept)        X12
     -0.146      0.769
```

The results are almost identical. This is comforting, but you may be wondering what the
point is of going to the trouble to use the Bayesian approach if we can get the same result
using ordinary least squares (OLS). We will address this question by moving on to an
example with real data.

15.5.2 Application of Bayesian Updating to Data Set 3

We have tentatively identified planting date *DPL*, irrigation effectiveness *Irrig* and nitrogen fertilizer rate *N* as the most important distinguishing factors in obtaining a high yield in east central Uruguay rice production (Data Set 3). Let us consider irrigation effectiveness. Improving irrigation effectiveness generally requires reworking the land to obtain a more uniform water depth given the topography of the field (see Figure 1.7). This requires a financial investment, and therefore a farmer will be interested in predicting the improvement in yield that would occur if the improvement was made. Probably not too many Uruguayan rice farmers (or any other farmers, for that matter) would use Bayesian updating to address this question, but let's suppose they did. In particular, consider the case of a farmer in the third year of the study who wants to use the results of the first two years to help predict the effect of an improvement in irrigation on the yield in his particular field. Our approach will follow that of Levy and Crawford (2009).

To study the procedure of Bayesian updating using Data Set 3, we will break up the data by season. As usual we will work with normalized yield.

```
> data.Set3$YieldN <- data.Set3$Yield / max(data.Set3$Yield)
> data.Set3S12 <- data.Set3[-which(data.Set3$Season == 3),]
> data.Set3S1 <- data.Set3[which(data.Set3$Season == 1),]
> data.Set3S2 <- data.Set3[which(data.Set3$Season == 2),]
> data.Set3S3 <- data.Set3[which(data.Set3$Season == 3),]
> data.Set3F3S2 <- data.Set3S2[which(data.Set3S2$Field == 3),]
> data.Set3F3S3 <- data.Set3S3[which(data.Set3S3$Field == 3),]
```

We will consider the specific case of Field 3. This field was farmed in both the second and third seasons by Farmer *B*. The irrigation effectiveness in the field was not uniformly good (Exercise 15.10), and the mean yield was lower in the regions whose irrigation effectiveness was lower.

```
> with(data.Set3F3S2, tapply(Yield, Irrig, mean))
       3        4
8046.667 8712.562
```

Will investing in improved irrigation increase yield? We will use Bayesian analysis to incorporate the experience of the other farmers to address this question. Our treatment will of course be very superficial. Readers who are interested in addressing this issue in greater depth are referred to the sources in Section 15.6.

The key idea is to employ Bayes' theorem to quantify the experiences of Farmer *B* as well as the other farmers and to use this experience to formulate a subjective probability distribution for the effect of irrigation effectiveness on rice yield. This requires that we treat the ordinal scale variable *Irrig* as if it were measured on a ratio scale, and that we assume that *YieldN* is linearly related to *Irrig*, at least in the range of values of *Irrig* that characterizes our data set. Considering the difficulties encountered in Section 14.5.2, and the results of Section 12.3, we will not incorporate spatial autocorrelation into the model.

Let's start by examining the relationship between *YieldN* and *Irrig* in Season 2 using OLS regression.

```
> model.lm <- lm(YieldN ~ Irrig, data = data.Set3F3S2)
> summary(model.lm)
Coefficients:
```

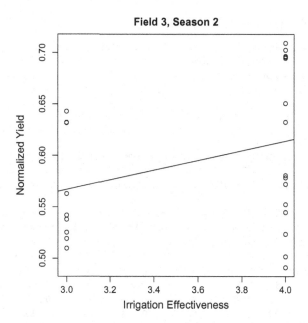

FIGURE 15.18
Plot of *Yield* vs. *Irrig* for Field 3 in Season 2 of the data of Data Set 3, showing the OLS regression line.

```
             Estimate Std. Error t value Pr(>|t|)
(Intercept)  0.42622    0.10908    3.908 0.000708 ***
Irrig        0.04692    0.02971    1.579 0.127919
---
Signif. codes: 0 '***' 0.001 '**' 0.01 '*' 0.05 '.' 0.1 ' ' 1

Residual standard error: 0.0713 on 23 degrees of freedom
Multiple R-squared: 0.09784,      Adjusted R-squared: 0.05861
F-statistic: 2.494 on 1 and 23 DF, p-value: 0.1279
```

The regression coefficient b_1 is not significantly different from zero. The low R^2 indicates a high variability in the data, which is confirmed in Figure 15.18. The figure also indicates, however, that there are two influence points at *Irrig* = 3.

Now we will run a Bayesian analysis of the data using a noninformative prior. First, we set up the noninformative prior model.

```
> Set3model.ni <- function(){
+ beta0 ~ dnorm(0, 0.001)
+ beta1 ~ dnorm(0, 0.001)
+ tau ~ dgamma(0.01, 0.01)
+ for (i in 1:n)
+   {
+   Y.hat[i] <- beta0 + beta1*Irrig[i]
+   Yield[i] ~ dnorm(Y.hat[i], tau)
+   }
+}
```

Next, we set the data and inits.

```
> XY.data.F3S2 <- list(Irrig = data.Set3F3S2$Irrig,
+    Yield = data.Set3F3S2$YieldN, n = nrow(data.Set3F3S2))
> XY.inits <- function(){
+    list(beta0 = rnorm(1), beta1 = rnorm(1), tau = exp(rnorm(1)))}
```

Now we run the model. The test and coda steps are not shown.

```
> Set3.sim. F3S2ni <- bugs(data = XY.data.F3S2, inits = XY.inits,
+    model.file = paste(mybugsdir,"\\Set3model.ni.bug", sep = ""),
+    parameters = c("beta0", "beta1", "tau"), n.chains = 5,
+    n.iter = 10000, n.burnin = 1000, n.thin = 10,
+    bugs.directory = mybugsdir)
> print(Set3.sim. F3S2ni, digits = 2)
Inference for Bugs model at "c:\aardata\book\Winbugs\Set3model.ni.bug",
fit using WinBUGS,
5 chains, each with 10000 iterations (first 1000 discarded), n.thin = 10
n.sims = 4500 iterations saved
```

	mean	sd	2.5%	25%	50%	75%	97.5%	Rhat	n.eff
beta0	0.43	0.12	0.18	0.36	0.44	0.52	0.67	1.01	310
beta1	0.04	0.03	-0.02	0.02	0.04	0.07	0.11	1.01	300
tau	169.13	50.01	86.95	133.10	164.25	198.90	279.20	1.00	2600
deviance	-59.46	2.93	-62.88	-61.61	-60.20	-58.09	-51.95	1.00	1200

The values of the regression coefficients are roughly the same as those of the OLS regression. The credibility interval includes zero, so Farmer B's subjective probability could not exclude a lack of improvement of yield following an increase in irrigation effectiveness.

Now suppose that Farmer B wants to incorporate the experience of other farmers into his subjective probability prior to making a decision at the start of Season 2. First, we must compute the parameters of the posterior from Season 1 to use in the prior for Season 2.

```
> XY.data.S1 <- list(Irrig = data.Set3S1$Irrig,
+ Yield = data.Set3S1$YieldN, n = nrow(data.Set3S1))
```

Now we run the model with a noninformative prior.

```
> Set3.sim. S1ni <- bugs(data = XY.data.S1, inits = XY.inits,
+ model.file = "c:\\aardata\\book\\Winbugs\\Set3model.ni.bug",
+ parameters = c("beta0", "beta1", "tau"), n.chains = 5,
+ n.iter = 10000, n.burnin = 1000, n.thin = 10,
+ bugs.directory = mybugsdir)
> print(Set3.sim. S1ni, digits = 2)
Inference for Bugs model at "c:\aardata\book\Winbugs\Set3model.ni.bug",
it using WinBUGS,
5 chains, each with 10000 iterations (first 1000 discarded), n.thin = 10
n.sims = 4500 iterations saved
```

	mean	sd	2.5%	25%	50%	75%	97.5%	Rhat	n.eff
beta0	0.22	0.03	0.16	0.20	0.22	0.24	0.29	1.01	700
beta1	0.06	0.01	0.04	0.05	0.06	0.06	0.08	1.01	590
tau	192.72	24.65	146.89	175.70	191.60	208.92	242.45	1.00	2300
deviance	-305.71	2.67	-308.70	-307.70	-306.40	-304.50	-298.90	1.01	930

The mean value of $\hat{\beta}_1$ over all of the fields in the first year is 0.06 and the credibility interval, which is much smaller because there are more data, does not include 0.

Next, we compute the parameter estimates of the posterior distributions for β_0, β_1, and τ.

```
> print(mu.0 <- Set3.sim. S1ni$mean$beta0, digits = 3)
[1] 0.222
> print(tau.0 <- 1/Set3.sim. S1ni$sd$beta0^2, digits = 3)
[1] 973
> print(mu.1 <- Set3.sim. S1ni$mean$beta1, digits = 3)
[1] 0.0585
> print(tau.1 <- 1/Set3.sim. S1ni$sd$beta1^2, digits = 3)
[1] 11818
> print(nu.t <- Set3.sim. S1ni$mean$tau / Set3.sim. S1ni$sd$tau^2,
+    digits = 3)
[1] 0.316
> print(mu.t <- Set3.sim. S1ni$mean$tau * nu.t, digits = 3)
[1] 60.9
```

Now we will set up a model that incorporates the parameters of this posterior into the prior for Season 3.

```
> Set3.model.ip <- function(){
+ beta0 ~ dnorm(mu.0, tau.0)
+ beta1 ~ dnorm(mu.1, tau.1)
+ tau ~ dgamma(mu.t, nu.t)
+ for (i in 1:n)
+   {
+   Y.hat[i] <- beta0 + beta1 * Irrig[i]
+   Yield[i] ~ dnorm(Y.hat[i], tau)
+   }
+}
```

Next, we set the data to include the prior parameters.

```
XY.data.F3S2ip <- list(Irrig = data.Set3F3S2$Irrig,
  Yield = data.Set3F3S2$YieldN, n = nrow(data.Set3F3S2),
  mu.0 = mu.0, tau.0 = tau.0, mu.1 = mu.1, tau.1 = tau.1,
  mu.t = mu.t, nu.t = nu.t)
```

Now we are ready to run the model.

```
> Set3.sim. F3S2ip <- bugs(data = XY.data.F3S2ip, inits = XY.inits,
+     model.file = "c:\\aardata\\book\\Winbugs\\Set3model.ip.bug",
+     parameters = c("beta0", "beta1", "tau"), n.chains = 5,
+     n.iter = 10000, n.burnin = 1000, n.thin = 10,
+     bugs.directory = mybugsdir)
> print(Set3.sim. F3S2ip, digits = 2)
Inference for Bugs model at "c:\aardata\book\Winbugs\Set3model.ip.bug",
it using WinBUGS,
5 chains, each with 10000 iterations (first 1000 discarded), n.thin = 10
n.sims = 4500 iterations saved
           mean     sd    2.5%     25%     50%     75%   97.5%  Rhat  n.eff
beta0      0.30   0.02    0.25    0.28    0.30    0.32    0.35     1   3400
beta1      0.08   0.01    0.06    0.07    0.08    0.08    0.09     1   2200
tau      191.98  22.73  149.75  176.30  191.40  206.72  239.00     1   4500
deviance -59.84   2.26  -62.25  -61.34  -60.51  -59.05  -53.52     1   4500
```

The estimated mean goes from $\hat{\beta}_1 = 0.04$ to $\hat{\beta}_1 = 0.08$, and the credibility interval no longer includes zero.

Suppose Farmer B does not make any irrigation improvements in Season 2 but decides to explore the possibility again in Season 3 (i.e., the subsequent season in which the field is planted to rice). He has to decide which farmers to include in his data set. One possibility would be to only incorporate data from Season 2 from his own field. This, however, would throw away information from the other farms, and one of the advantages of the Bayesian approach is precisely this capacity to incorporate such information (Hilborn and Mangel, 1997, p. 203). Farmer B could incorporate only the data from other northern farmers, under the assumption that they are more like him than the others (Exercise 15.11). Alternatively, he could incorporate data from all the farms. Another alternative would be to somehow weight the data, or simply to make up his own parameter values. Since he is describing his own subjective probability, there are no rules that he must follow, and that is a difficulty. When one assigns subjective probabilities, if the assignments are made in an arbitrary manner, then it is easy to slip, possibly subconsciously, into a practice of assigning these probabilities in a way that confirms the desired outcome, that is, of data snooping (Walters, 1986, p. 166). If this methodology is to be applied sensibly, rules must be established in advance and strictly followed.

15.6 Further Reading

From the practitioner's point of view, the literature on spatiotemporal problems is still fairly sparse. Cressie and Wikle (2011) is very comprehensive, but the mathematical level may be a challenge for some practitioners. Hengl et al. (2008) provide case studies of spatiotemporal modeling. Fortin and Dale (2005) devote a chapter to the topic, including some material on spatiotemporal clustering and animal movement modeling. LeSage and Pace (2009) also include a chapter on spatiotemporal models. In addition to spacetime, the R package Stem (Cameletti, 2011) deals with spatiotemporal modeling, including kriging and bootstrapping.

Okubo (1980) is the standard source for dispersal. See for example, Dennis et al. (1995) or Cantrell and Cosner (2003) for a discussion of diffusion-reaction models in ecology. For another example of a state and transition model, see Whalley (1994). George et al. (1992) and Huntsinger and Bartolome (1992) developed state and transition models for oak woodlands of the type described by Data Set 1. These models were studied in a spatial context by Plant et al. (1999). As an alternative, Alonso (2008) developed a state and transition simulation system using Bayesian networks (Jensen, 2001). Wiles and Brodahl (2004) and Ferraro et al. (2012) apply recursive partitioning to clusters in a similar way to that done in Section 15.4.2.

Although it is now a bit dated, Walters (1987) is still an excellent source for Bayesian updating. Hilborn and Mangel (1997) also has some very good information. Duncan et al. (1995) provide an example of Bayesian updating in a social science setting. Pearl (1986) provides a good discussion of Bayesian inference. The Dempster-Schafer theory of evidence (Dempster, 1990) is a major competitor of Bayesian updating. See also Cohen (1985). Banerjee et al. (2004) devote a chapter to Bayesian analysis of spatiotemporal data. They focus on the use of a matrix transformation called the singular value decomposition to reduce the number of variables in the analysis.

Exercises

15.1 The analyses in this chapter used adjusted values of *Yield* because of the large variation in mean and variance between years. Many of the results of this chapter are sensitive to how the data are adjusted to account for this variation. In the text the data were normalized, and in the exercises the same data will be analyzed using data that have been centered and scaled. Repeat the construction of the spatiotemporal semivariograms carried out in Section 15.2.2. with data that have been centered and scaled using the function `scale()`.

15.2 Repeat the process analysis of Section 15.3.1 using centered and scaled values. Note that the resulting estimates of the diffusion coefficient D are no longer aligned.

15.3 Re-run the cluster analysis of two clusters using centered and scaled values of *Yield*. (a) Create a plot of the cluster means similar to Figure 15.12a. What new information does this provide? and (b) Create a plot showing the scaled yield values at each location, with pch values chosen so you can distinguish the clusters.

15.4 Soil texture, organic carbon, and the mineral nutrients K and P should remain fairly stable over a four-year period in Field 4.1. Create a scatterplot matrix of *Yield* in each year against these quantities.

15.5 Some of the weather data from the CIMIS file http://wwwcimis.water.ca.gov/cimis/welcome.jsp are missing. Use linear regression of nearby data to impute these values.

15.6 Create the plot of cumulative precipitation in the fields in Data Set 4 shown in Figure 15.14b.

15.7 Plot individual regression trees for *Yield* in Field 4.1 in each year against the same explanatory variables as you used in Exercise 15.2. Do you see any patterns emerging?

15.8 Create a plot of the means by year of normalized yield over each of the terminal nodes of the classification tree of the two-cluster model (Figure 15.12a). What can you infer from the plot?

15.9 In Section 15.5.1, a Bayesian regression model was computed for a set of artificial data, and then the model was updated using a second set of artificial data. Using the same data, verify that approximately the same results are obtained if all of the data is used at once with a noninformative prior.

15.10 Create a map showing the irrigation effectiveness of Field 3 in Season 2. What does this say about the economics of Farmer B's decision as to whether to improve the irrigation effectiveness in this field?

15.11 Suppose Farmer B does not make any irrigation improvements in Field 3 in Season 2. (a) Estimate the regression coefficient for *Irrig* based on a noninformative prior. (b) Estimate the regression coefficient for *Irrig* based on an informative prior using only data from Field 3 in Season 2. (c) Estimate the regression coefficient for *Irrig* based on an informative prior using data from all of the northern fields in Season 2.

16

Analysis of Data from Controlled Experiments

16.1 Introduction

At the landscape scale, controlled experiments are often logistically or financially difficult, and precision landscape measurement tools such as remotely sensed images and yield monitors may provide observational data that supports scientific interpretation. Nevertheless, if it is possible to carry out a controlled, replicated experiment under the same conditions as an observational study, then the controlled experiment provides more powerful results. Often the results of observational studies and controlled experiments can be used in a complementary manner to address a question from different perspectives. The controlled experiment in this ideal scenario provides precise, readily interpretable answers to specific questions, and the observational study permits those answers to be extended to more general conditions than those under which the controlled experiment is conducted. One must, however, confront the same issues in the presence of spatial autocorrelation in data from controlled experiments that one confronts with observational data. The purpose of this chapter is to discuss measures that can be taken in dealing with these issues.

The three fundamental concepts of the design of a controlled experiment, as put forward by Fisher (1935, pp. 17, 21, 48), are randomization, replication, and blocking. In discussing these concepts, we will refer to the experimental unit, that is, the component that is assigned the treatment and whose response is measured, as the *plot*, and we refer to the environment that defines the response of the plot to the treatment as the *substrate*. The notion of *randomness* is itself difficult to define (Kempthorne, 1952, p. 121), and we will adopt the practical definition given by Kempthorne in the just-cited reference that a process is random if it is based on a set of random numbers. This definition works for us because the use we need to make of randomness is to randomly assign treatments to experimental plots. In this context, the treatments are assigned at random if at the start of the assignment process each plot has the same probability of being assigned every treatment (Kutner et al., 2005, p. 653). This occurs if the treatments are assigned based on a set of random numbers. The primary purpose of randomization is to avoid bias. Even in the presence of autocorrelated variability of the substrate properties, the comparison of treatment effects is unbiased if the assignment of treatments to plots is random (Kempthorne, 1952, p. 140). The effect of modifications to the experimental design to take into account heterogeneity of the substrate of a randomized experiment is not to remove bias but rather to increase precision (Brownie et al., 1993).

The primary purpose of replication, that is, of assigning each treatment to more than one plot, is to decrease the error in treatment comparisons (Kempthorne, 1952, p. 177). That said, it is necessary to have at least two replications of a treatment to be able to estimate the experimental variance at all. A second, almost equally important effect of replication is to provide what Hurlbert (1984) calls *interspersion* of the plots. This means that the plots are scattered among each other on the substrate, thus reducing the chance that a group of plots that each receive the same treatment will be located near to each other.

The incorporation of blocking into the design of an experiment involving spatial data represents an explicit recognition that spatial autocorrelation is present in the substrate. The purpose of blocking is to reduce the effect of spatial heterogeneity on the measurement of the treatment effect by making the individual blocks as homogeneous as possible (Fisher, 1935, p. 56; Kempthorne, 1952, p. 210; Kutner et al., 2005, p. 661). Having defined these three fundamental concepts of experimental design, we now turn to a discussion of the fundamentals of the analysis of variance.

16.2 Classical Analysis of Variance

Both linear regression and analysis of variance (ANOVA) are particular cases of the general linear model $Y = X\beta + \varepsilon$. In linear regression, the explanatory variables are assumed to be interval or ratio scale, or at least to function as if they had one of these measurement scales, and the model assumes a relation between the response and the explanatory variables (e.g., $Y_i = \beta_0 + \beta_1 X_{i1} + \varepsilon_i$). In ANOVA the explanatory variables, or *treatments*, may be nominal, and in any case the model does not assume a particular relationship between the treatments and the response. For example, suppose we are conducting an experiment on the effects of a type of fertilizer on crop yield, and we are going to analyze the data using the linear model. If the explanatory variable is the amount of fertilizer applied, we would probably choose (at least initially) to analyze these data using linear regression. If the explanatory variables are different types of fertilizer, then ANOVA would probably be a more appropriate choice. We will assume that the reader has been exposed to ANOVA already and simply provide a brief review to set up the notation. References are provided in Section 16.5.

We will adopt a notation based on that of Brownie et al. (1993). We restrict ourselves to rectangular plot arrangements such as that shown in Figure 16.1. The rows of plots are indexed by i, $i = 1,...,l$ and the columns are indexed by j, $j = 1,...,m$. For example, in Figure 16.1 $l = 12$ and $m = 6$. The treatment applied to plot ij is denoted $\tau_{k(ij)}$, signifying that plot ij receives the kth treatment, $k = 1,...,r$. Assume that there are n replications per plot (we will not deal with unbalanced designs), so that $lm = rn$. The simplest ANOVA model is the *completely randomized design*. It is expressed in the so-called *effects formulation* as

$$Y_{ij} = \mu + \tau_{k(ij)} + \varepsilon_{ij}, \tag{16.1}$$

where ε_{ij} is the error term. We assume that the ε_{ij} are normal random variables with mean 0 and variance σ^2.

The analysis of variance was developed by R.A. Fisher (1935) as a result of his experience with field plot research at the Rothamsted Experiment Station in Harpenden, England. Fisher recognized the first axiom of geography (Tobler, 1970, See Section 3.1), that nearby experimental plots would generally be more similar in their properties than more distant plots, and that this effect could reduce the statistical power of field experiments. It is important to reemphasize that if the assignment of treatments to plots is randomized (i.e., if each plot has at the start of the assignment of treatments an equal probability of

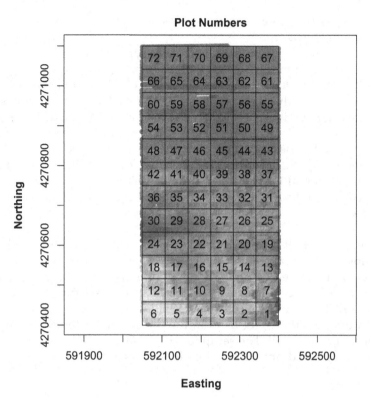

FIGURE 16.1
Plot numbers for a uniformity trial held on a rectangular subsection of Field 4.1.

being assigned any treatment), then the greater similarity of properties of nearby plots does not invalidate the results, but only reduces their power (Kempthorne, 1952, p. 135; Brownie et al., 1993). In order to alleviate this reduction in power, Fisher proposed the idea of *blocking*. The most common blocked design is the randomized complete block (RCB), which is represented as

$$Y_{ij,b} = \mu + \tau_{k(ij)} + \rho_b + \varepsilon_{ij}, \tag{16.2}$$

where all of the symbols are as before, and ρ_b is the effect of the *b*th block, to which plot *ij* belongs. In laying out an RCB experiment, each block should be as uniform as possible in the blocking factor, and encompass as much variability as possible of any other factors that might enter into the experiment, to avoid confounding these factors with the treatment.

To clarify these concepts, we will introduce the data set that will be used throughout this chapter to compare methods. It is the same data set that was used to discuss the modifiable areal unit problem in Section 11.5.1. The data are yield monitor output from Field 4.1 restricted to a 700 m by 350 m rectangular region. This region is divided into a checkerboard of 72 square plots (Figure 16.1). Our "experiment" is a *uniformity trial*, that is, an experiment in which there are plots but no applied treatments (Cochran, 1977, p. 41), or, perhaps better said, different plots are labeled as having different "treatments," but there is nothing purposefully done to distinguish these plots. Since the "treatment" effect is therefore zero, the difference between treatment means should ideally be approximately zero.

The checkerboard layout is not representative of a typical field experiment, but it is common in uniformity trials. One of the most famous agricultural data sets, and certainly one of the data sets most analyzed by statisticians, is that of Mercer and Hall (1911), which is a uniformity trial. A major advantage of the uniformity trial in terms of comparing alternative methods for dealing with spatial autocorrelation is that, because the treatments have zero effect, alternative randomization schemes can be applied in a Monte Carlo simulation study. This approach has been used by a number of authors, including Wilkinson et al. (1983), Besag and Kempton (1986), and Zimmerman and Harville (1991), in comparative analyses of methods for addressing spatial autocorrelation in experimental data. Indeed, Brownie et al. (1993) point out that if there are actual treatment effects then, because these effects are not known, it is impossible to make a completely valid comparison of the different methods.

In our example, we will have 18 treatments $(k = 1,...18)$ and four blocks $(b = 1,...,4)$. Figure 16.2 shows a grayscale plot of the yield values with a set of treatment numbers for a completely randomized design (CRD) experiment overlaid. Figure 16.3a shows the block layout of an RCB experiment, and Figure 16.3b shows a grayscale plot of the yield values with the RCB treatment numbers overlaid. Prior to carrying out a comparative analysis of different methods, let us examine the results of the RCB experiment and the CRD experiment in detail. The setup for this analysis follows what should by now be a familiar pattern, and not all of the code is shown. It is important to emphasize,

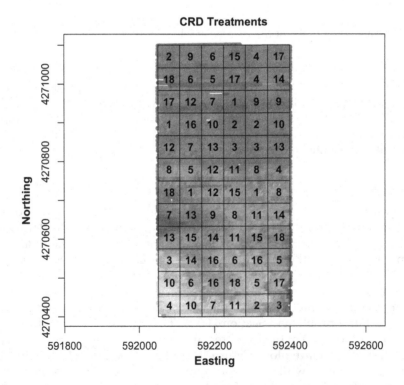

FIGURE 16.2
Treatment numbers for a completely randomized design (CRD) experiment carried out in the plots of Figure 4.1.

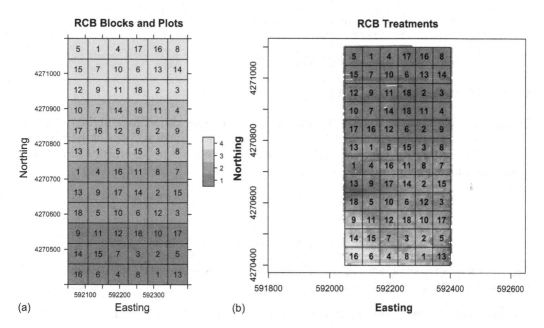

FIGURE 16.3
Layout of a randomized complete block (RCB) experiment carried out on the plots of Figure 4.1. (a) Gray scale showing blocks. (b) Gray scale showing yield.

however, that in order to obtain correct results, factors and not numbers must be used to identify the plots and blocks.

```
> set.seed(123)
> plots.spdf@data$trtmt.CRD <- factor(sample(rep((1:18), 4)))
> plots.spdf@data$block <- factor(sort(rep(1:4, 18)))
> plots.spdf@data$trtmt.RCB <- factor(unlist(tapply(rep(1:18,4),
+     sort(rep(1:4,18)), sample)))
```

As an exercise, work through the code for the assignment of plot numbers for the CRD and RCB cases and verify that they produce the plot layouts shown in Figures 16.2 and 16.3. We obtain the plot yields by applying the function over().

```
> plots.spdf@data$Yield <-
+     over(Yield.pts, plots.spdf, fn = mean)$Yield
```

The blocks in Figure 16.3a are laid out appropriately since most of the soil variability is in the north to south direction. Stroup (2002) points out that an incomplete block design is generally superior to an RCB in cases such as the present where there are many treatments. The most common such design is the lattice (Gomez and Gomez, 1984, p. 39), but this requires that the number of treatments be a square. Since our main interest is in comparison of methods for dealing with spatial autocorrelation, we will not pursue incomplete block designs.

Zimmerman and Harville (1991) used the RCB data as the standard of comparison, and we will do the same. We first tabulate the mean yields by treatment, from lowest mean response to highest.

```
> print(sort(tapply(plots.spdf@data$Yield,
+    plots.spdf@data$trtmt. RCB, mean)), digits = 4)
  13   17   18   11    3    5    9    7   12   10    4
2469 2571 2584 2757 2761 2790 2799 2801 2802 2934 2962
   2    1   15    8    6   14   16
2973 3005 3108 3158 3288 3390 3421
```

Examination of Figure 16.3b indicates that, by the luck of the draw, three of the treatment 13 plots and three of the treatment 17 plots fall in the lower yielding areas of their blocks. The analysis of variance reveals a significant block effect but no significant differences among treatments.

```
> RCB.aov <- aov(Yield ~ block + trtmt.RCB,
+    data = plots.spdf@data)
> summary(RCB.aov)
            Df   Sum Sq  Mean Sq F value Pr(>F)
block        3 68033791 22677930 72.9854 <2e-16 ***
trtmt.RCB   17  5046710   296865  0.9554 0.5189
Residuals   51 15846650   310719
---
Signif. codes:  0 '***' 0.001 '**' 0.01 '*' 0.05 '.' 0.1 ' ' 1
```

An aligned dot plot shows that the yield by treatment trends upward gradually from lowest yielding to highest (Figure 16.4). With this preparation we can now move on to the comparison of methods.

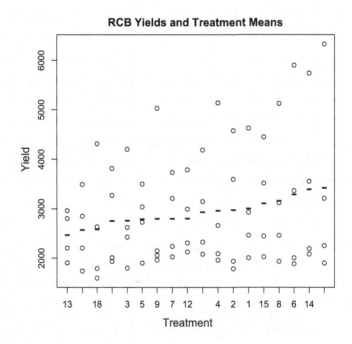

FIGURE 16.4
Dot plot of treatment responses and treatment means for the RCB experiment.

16.3 The Comparison of Methods

16.3.1 The Comparison Statistics

Zimmerman and Harville (1991) use comparison statistics that are based on a comment of Besag (1983). They are defined as follows:

$$EMP = \frac{2}{r(r-1)} \sum_{i=1}^{r} \sum_{j=i+1}^{r} (\hat{\tau}_i - \hat{\tau}_j)^2$$

$$PRE = \frac{2}{r(r-1)} \sum_{i=1}^{r} \sum_{j=i+1}^{r} est.var(\hat{\tau}_i - \hat{\tau}_j),$$

(16.3)

where $\hat{\tau}_i$ is the estimated effect of treatment i, and $est.var(\hat{\tau}_i - \hat{\tau}_j)$ is the estimated variance of the difference in treatment effects. *EMP* stands for "empirical," and *PRE* stands for "predicted." Because the treatment effects are zero, the statistic *EMP* represents the empirical estimate of the variance of $(\tau_i - \tau_j)$, and therefore a small value of *EMP* indicates high accuracy. Put another way, since all of the τ_i should be the same, the quantities $(\hat{\tau}_i - \hat{\tau}_j)^2$ should be close to zero. *PRE* is the mean squared error of the difference, as predicted by the analysis, and should be approximately equal to *EMP* (Besag, 1983). To keep the discussion as simple as possible, we will focus on *EMP*.

Under the CRD model, the values of $\hat{\tau}_i$ are just the treatment means, computed in the last section. Let's compute the values $\hat{\tau}_i - \hat{\tau}_1$ for this model.

```
> trt.mean <- tapply(plots.spdf@data$Yield,
+    plots.spdf@data$trtmt.RCB, mean)
> print((trt.mean - trt.mean[1])[2:18], digits = 4)
      2        3        4        5        6        7        8        9       10
 -31.29 -243.50  -42.89 -214.38  283.10 -203.77  153.91 -205.44  -70.84
     11       12       13       14       15       16       17       18
-248.06 -202.35 -535.74  385.84  103.51  416.79 -433.74 -420.28
```

We can compute the value of *EMP* for the randomized complete block design directly from these treatment means. In order to facilitate the computation of the value of *EMP* for the other methods, however, we will obtain the values $\hat{\tau}_i - \hat{\tau}_j$ in a different way.

We will generate the same RCB analysis of variance model in the previous section, but using the function lm() instead of aov().

```
> RCB.lm <- lm(Yield ~ trtmt.RCB + block,
+    data = plots.spdf@data)
> print(RCB.sum <- summary(RCB.lm))

Call:
lm(formula = Yield ~ trtmt. RCB + block, data = plots.spdf@data)

Residuals:
     Min       1Q   Median       3Q      Max
-1229.30  -315.90    22.78   219.38  1341.03
```

```
Coefficients:
              Estimate Std. Error t value Pr(>|t|)
(Intercept)   4568.41     301.04  15.175  < 2e-16 ***
trtmt.RCB2     -31.29     394.16  -0.079    0.937
trtmt.RCB3    -243.50     394.16  -0.618    0.539
   *    *    *   DELETED    *    *    *
trtmt.RCB18   -420.28     394.16  -1.066    0.291
block2       -1524.73     185.81  -8.206 6.93e-11 ***
block3       -2210.73     185.81 -11.898 2.49e-16 ***
block4       -2519.91     185.81 -13.562  < 2e-16 ***
---
Signif. codes:  0 '***' 0.001 '**' 0.01 '*' 0.05 '.' 0.1 ' ' 1

Residual standard error: 566.2 on 51 degrees of freedom
Multiple R-squared: 0.8168,      Adjusted R-squared: 0.745
F-statistic: 11.37 on 20 and 51 DF,  p-value: 1.709e-12
```

Notice that the values of the Estimates column of `trtmt.RCB2` through `trtmt.RCB18` are the same as the values $\hat{\tau}_i - \hat{\tau}_1$ printed in the previous output. These quantities are called the *treatment contrasts*. Without going into detail (See Kutner et al., 2005, p. 741, for information about contrasts in general and Venables and Ripley, 2002, p. 146 for information specific to R), the second contrast represents the difference $\hat{\tau}_2 - \hat{\tau}_1$ between the sample mean of treatment 2 and treatment 1, and the *ith* contrast represents the difference $\hat{\tau}_i - \hat{\tau}_1$ between the sample mean of treatment *i* and treatment 1. After the contrasts for treatments come the contrasts for blocks. One must be careful to place `trtmt.RCB` before `block` in the model specification to obtain this set of contrasts. We can then obtain the values $\hat{\tau}_i - \hat{\tau}_1$ as

```
> tau.dif1 <- coef(RCB.sum)[2:18,1]
```

We don't just need $\hat{\tau}_i - \hat{\tau}_1$, we need $\hat{\tau}_i - \hat{\tau}_j$ for all values of *j*. However, as far as the differences are concerned, the value of $\hat{\tau}_1$ doesn't matter. Each difference can be computed as $(\hat{\tau}_i - \hat{\tau}_1) - (\hat{\tau}_j - \hat{\tau}_1) = \hat{\tau}_i - \hat{\tau}_j$. Therefore, we can compute the quantities $\hat{\tau}_i - \hat{\tau}_j$ by just setting $\hat{\tau}_1 = 0$ and using the treatment contrasts directly.

```
> tau.hat <- c(0, tau.dif1)
```

We can use the function `vegdist()` from the vegan package (Oksanen et al., 2017) to compute the distances between the $\hat{\tau}_i$. Here are the distances for the first four values of $\hat{\tau}_i$.

```
> print(vegdist(tau.hat[1:4], method = "euclidean"), digits = 3)
             trtmt.RCB2 trtmt.RCB3
trtmt.RCB2   31.3
trtmt.RCB3  243.5          212.2
trtmt.RCB4   42.9           11.6        200.6
```

Now we compute the full distance matrix.

```
> tau.dist <- vegdist(tau.hat, method = "euclidean")
```

The value of *EMP* in Equation 16.3 is computed from the sum of the squares of these values.

```
> print(EMP.RCB <- 2 * sum(tau.dist^2) / (18*17))
[1] 148432.6
```

As usual, the Monte Carlo simulation is carried out using the function `replicate()`. We first define the function `emp.calc()`.

```
> EMP.calc <- function(trt.data, form.lm){
+      trt.data$trtmt <- factor(unlist(tapply(rep(1:18,4),
+         sort(rep(1:4,18)), sample)))
+      trt.sum <- summary(lm(form.lm, data = trt.data))
+      tau.dif1 <- coef(trt.sum)[2:18,1]
+      tau.hat <-  c(0, tau.dif1)
+      tau.dist <- vegdist(tau.hat, method = "euclidean")
+      EMP <- 2 * sum(tau.dist^2) / (18*17)
+      return(EMP)
+ }
```

Now we carry out the simulation.

```
> set.seed(123)
> RCB.form <- as.formula(Yield ~ trtmt.RCB + block)
> U <- replicate(100, EMP.calc(plots.spdf@data, RCB.form))
> print(EMP.RCB <- mean(U))
[1] 148432.6
```

Following Zimmerman and Harville (1991), we will express all values of *EMP* as percentages of the RCB values. In Exercise 16.1, you are asked to repeat this exercise for the completely randomized design for this field. The result of this exercise indicates that the *EMP* value of the CRD is over three times as large as that of the RCB design.

In the next three subsections, we discuss the R implementation of three commonly used methods for dealing with spatial autocorrelation: the Papadakis nearest-neighbor method, the trend method, and the correlated errors method. Following this we review the results of published comparisons of these and other methods for dealing with a spatially autocorrelated substrate in field plot experiments.

16.3.2 The Papadakis Nearest-Neighbor Method

Next to the randomized complete block method, probably the oldest method still used for dealing with spatial autocorrelation is the *Papadakis method* (Papadakis, 1937). There are several variations on this method; we will discuss the one described by Brownie et al. (1993). Let e_{ij} denote the residual $e_{ij} = Y_{ij} - \overline{Y}_{k(ij)}$, where $\overline{Y}_{k(ij)}$ is the treatment mean of the treatment applied to cell ij. Then the Papadakis nearest-neighbor method is to fit the analysis of covariance (see Equation 12.1) model

$$Y_{ij} = \mu + \tau_{k(ij)} + \beta X_{ij} + \varepsilon_{ij}, \tag{16.4}$$

where, for interior plots,

$$X_{ij} = \frac{1}{4}(e_{i,j-1} + e_{i,j+1} + e_{i-1,j} + e_{i+1,j}). \tag{16.5}$$

For border plots, X_{ij} is the mean of the two or three neighboring plots. The Papadakis method does not include a blocking variable in the model, because the covariate term βX_{ij} replaces it.

Equation 16.5 can be implemented using a row normalized, rook's case spatial weights matrix as defined in Equations 3.22 and 3.23 of Section 3.4.2. Wilkinson et al. (1983) criticized the Papadakis nearest-neighbor method as being statistically inefficient, and a number of more complex methods, many of them involving iteration, have been developed. Brownie et al. (1993) and Stroup (2002) describe several of these. The improvement provided by these methods is often not very great, so we will not discuss them in detail.

In order to implement the Papadakis method, we must first compute the error terms $e_{ij} = Y_{ij} - \bar{Y}_{k(ij)}$. The treatment means are computed using the function `tapply()`, which we have seen before. The differences are computed using `sapply()`, which returns the value of the treatment mean of the appropriate treatment. The R code in this and the other subsections of this section require the results of Section 16.2 to be in memory.

```
> trt.means <- tapply(plots.spdf@data$Yield,
+    plots.spdf@data$trtmt.CRD, mean)
> Yield.e <- plots.spdf@data$Yield -
+       sapply(plots.spdf@data$trtmt.RCB, function(x) trt.means[x])
```

The X_{ij} in Equation 16.5 are computed by generating a `listw` object and using the function `listw2mat()` to convert the `listw` object into a matrix with which to multiply the error terms.

```
> library(spdep)
> nb6x12 <- cell2nb(6,12)
> W.papa <- nb2listw(nb6x12)
> X <- listw2mat(W.papa) %*% Yield.e
```

We can now apply the function `replicate()` to compute the value of *EMP*.

```
> set.seed(123)
> RCB.form <- as.formula(Yield ~ trtmt + X)
> U <- replicate(100, EMP.calc(papa.RCB, RCB.form))
> print(EMP.papa <- mean(U))
[1] 104232.1
> print(EMPbar.papa <- 100 * EMP.papa / EMP.RCB)
[1] 70.2218
```

For this data set, the Papadakis method presents, by this measure, a considerable improvement over the RCB design, with a value of *EMP* roughly 70% of that of the RCB.

16.3.3 The Trend Method

The trend method was proposed by Federer and Schlottfeld (1954) and analyzed by Tamura et al. (1988). It involves the use of a model of the form

$$Y_{ij} = \mu + \tau_{k(ij)} + T_{ij} + \varepsilon_{ij}, \tag{16.6}$$

where T_{ij} represents a trend (Section 3.2.1). The point of trend analysis is to improve on blocking by recognizing that the properties of the substrate are likely to vary smoothly rather than discretely at boundaries of the blocks (Brownie et al., 1993). Therefore, as with the Papadakis method, a block term is not added to the model. Although, as discussed in Sections 3.2.1 and 9.2, a model of T_{ij} may be developed by a method such as median polish or a generalized additive model, almost all published investigations of trend analysis have used a polynomial model, most commonly of the form

$$T_{ij} = \beta_0 + \beta_1 x_{ij} + \beta_2 y_{ij} + \beta_3 x_{ij}^2 + \beta_4 y_{ij}^2 + \beta_5 x_{ij} y_{ij} + \varepsilon_{ij}, \tag{16.7}$$

where x_{ij} and y_{ij} are x and y coordinates taken as the location of plot ij. To implement a trend analysis using Equation 16.7, we first compute the trend, normalizing the geographic coordinates to avoid computing with very large numbers.

```
> x <- (cell.ctrs[,1] - W) / (E - W)
> y <- (cell.ctrs[,2] - S) / (N - S)
> Yield.T <- lm(Yield ~ x + y + I(x^2) +
+    I(y^2) + I(x*y), data = plots.spdf@data)
> trend.RCB <- with(plots.spdf@data, data.frame(Yield = Yield,
+    trtmt = trtmt.RCB, Trend = predict(Yield.T)))
```

Next, we replicate the model and compute the relative value of *EMP*.

```
> trend.form <- as.formula(Yield ~ trtmt + Trend)
> set.seed(123)
> U <- replicate(100, EMP.calc(trend. RCB, trend.form))
> print(EMP.trend <- mean(U))
[1] 120175.6
> print(EMPbar.trend <- 100 * EMP.trend / EMP.RCB)
[1] 80.96304
```

Although it represents a considerable improvement over the RCB, in this particular case the trend method does not fare quite as well as the Papadakis method.

16.3.4 The "Correlated Errors" Method

The work of Zimmerman and Harville (1991) introduced the concept of using a generalized least squares model of the type described in Section 12.5 to the analysis of agricultural field plot data. This work in turn motivated further comparative studies by Brownie et al. (1993) and Stroup et al. (1994). Brownie et al. (1993) used the term "correlated errors model" to describe this approach. The model may be written

$$Y_{ij} = \mu + \tau_{(k)ij} + T_{ij} + \varepsilon_{ij},$$
$$\varepsilon_{ij} \sim N(0, \sigma^2 \Lambda). \tag{16.8}$$

where Λ is a variance-covariance matrix. The model of Equation 16.8 has the form, in the terminology of Brownie et al. (1993), of a "correlated errors plus trend" model. A purely correlated errors model would be obtained by eliminating the trend term T_{ij}. Brownie et al. (1993) found that the performance of the pure correlated errors was inferior to that of the correlated errors with trend model. It would be easy to develop the former by modifying the code given here, but we will only analyze the model in which a trend is included.

Following the methods of Section 12.5, it is straightforward to implement this model using the function `gls()` of the `nlme` package (Pinhiero et al., 2011). We first compute the generalized least squares (GLS) model with independent errors.

```
> library(nlme)
> trend.RCB$x <- x
> trend.RCB$y <- y
> model.gls <- gls(Yield ~ trtmt + Trend, data = trend.RCB)
```

Next, we use the function `Variogram()` to construct the variogram of the residuals (Figure 16.5) and by inspection, set the range and nugget to 0.35 and 0.5, respectively.

```
> plot(Variogram(model.gls, form = ~ x + y,
+      maxDist = 1), xlim = c(0,1),
+      main = "Variogram of Residuals")
```

We modify the function `emp.calc()` of Section 16.3.3 to invoke the function `gls()` rather than `lm()`, and to take into account that for a `gls` object the function `coef()` returns a vector containing only the treatment contrasts. We have to use the same variogram parameters for all of the models, but the values of the model coefficients are not very sensitive to these parameters (Exercise 16.2).

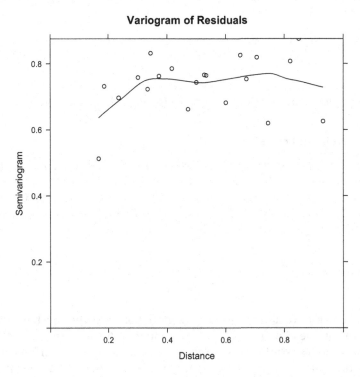

FIGURE 16.5
Experimental variogram of the residuals of the generalized least squares model for the example data set.

```
> EMP.gls <- function(trt.data) {
+     trt.data$trtmt <- factor(unlist(tapply(rep(1:18,4),
+         sort(rep(1:4,18)), sample)))
+     trt.sum <- summary(gls(Yield ~ trtmt + Trend,
+         corr = corSpher(value = c(0.35, 0.5),
+         form = ~ x + y, nugget = TRUE), data = trt.data))
+     contr <- coef(trt.sum)
+     tau.hat <- c(0, contr[2:18])
+     tau.dist <- vegdist(tau.hat, method = "euclidean")
+     EMP <- 2 * sum(tau.dist^2) / (18*17)
+     return(EMP)
+ }
```

Finally, we run the simulation.

```
> set.seed(123)
> U <- replicate(100, EMP.gls(trend. RCB))
> print(EMP.gls <- mean(U))
[1] 80315.73
> print(EMPbar.gls <- 100 * EMP.gls / EMP.RCB)
[1] 54.10921
```

The results indicate that the correlated errors plus trend model outperforms both the Papadakis and the trend methods, with a value of *EMP* roughly half of that of the RCB.

16.3.5 Published Comparisons of the Methods

Since the work of Zimmerman and Harville (1991) was the first to include a generalized least square method, we begin our discussion with their results. In their comparison tests, Zimmerman and Harville (1991) compared a variety of versions of the Papadakis nearest-neighbor method, a pair of GLS (i.e., "correlated errors") models with the assumption of aniso-tropic covariance, a correlated errors model with the assumption of isotropic covariance, and a few methods not discussed here. Several different blocking schemes were tested on three data sets. The correlated errors methods consistently performed the best, with the version of the Papadakis method described above doing rather less well but still better than the simple RCB.

Following on the work of Zimmerman and Harville (1991), Brownie et al. (1993), and Stroup et al. (1994) conducted further comparisons of the various methods. Both of these latter papers used data from actual field trials rather than uniformity trials, so that they could not use Monte Carlo simulation of the *EMP* and *PRE* statistics of Besag (1973) in the comparisons. Brownie et al. (1993) used the RCB as a baseline and compared the Papadakis nearest-neighbor method, the trend method, the "correlated errors" method without incorporation of a trend, and the "correlated errors" plus trend method. They used what is essentially the square root of the *PRE* statistic plus an analysis of plots of the residuals versus yield values as their primary standard of comparison. Brownie et al. (1993) compared the performance of the methods applied to several data sets. Although the results varied somewhat, they found that the trend plus correlated errors model provided the best performance, with the trend model generally second and the Papadakis method third. All of the methods outperformed the RCB. Thus, the relative ranking of the trend and Papadakis method in the trials of Brownie et al. (1993) was the opposite of that of our simple test, but there was agreement in the superiority of the correlated errors plus trend model.

Stroup et al. (1994) compared the RCB method with two versions of the Papadakis method and also conducted a limited comparison of the correlated errors model. Their comparison was based on the magnitude of the experimental error, the coefficient of variation, and the effect of the test on the relative ranking of the treatments. Like Brownie et al. (1993), Stroup et al. (1994) found that the correlated errors model performed the best, and that there was little difference between the versions of the Papadakis method. Based on these results, it seems that the generalized least squares (or "correlated errors") model incorporating a trend term is the superior method for dealing with spatial autocorrelation in replicated experiments. In addition to performing better, it is less dependent on the shape of the experimental region than the Papadakis method, and in a sense it is unlikely to do worse than the model with a trend but independent errors. Brownie et al. (1993) point out that the success of the trend and correlated errors models may depend on the accuracy of the trend model and the model for the correlation structure of the errors. We saw in Exercise 9.3 that a generalized additive model (GAM) model can provide an accurate prediction of a trend surface when the trend surface is known. In Exercise 16.4, you are asked to repeat the analysis of this section using a GAM model to predict the trend surface.

16.4 Pseudoreplicated Data and the Effective Sample Size

16.4.1 Pseudoreplicated Comparisons

In carrying out a large-scale, landscape-level experiment, it may happen that for logistical or economic reasons it is impossible to lay out the experiment using a traditional repli-cated plot design. For example, Lee et al. (2006) report on an experiment to compare the response of carbon sequestration rate to tillage method (conventional vs minimum till) in a commercial corn field. The experiment involves the measurement of carbon sequestra-tion rate using the eddy covariance method. Without going into detail about this method, the fetch (i.e., the ground area measured) of the measurement instruments is sufficiently large that no more than two could be placed in a single field. The cost of the equipment precluded more than one trial at this initial stage of the program, but it was nevertheless desirable to obtain some statistical measure of the difference between various quantities, such as yield and normalized difference vegetation index (NDVI), in the conventional and minimum till plots of the field.

The characteristic of an experiment such as this is that a single large area is divided into two halves. Half the area receives treatment 1 and the other half treatment 2. The response variable is measured at n locations within each half. Under the usual ANOVA procedure, these data would be modeled using a nested design (Kutner et al., 2005, p. 662). In this case, however, the data are autocorrelated, but not perfectly so. Depending on the level of autocorrelation, it may not be appropriate to model the experiment using either a nested design or an independent errors model. In calculating degrees of freedom in a comparison of factor level means, the n pairs of measurements provide an effective sample size of n_e pairs, where $1 \leq n_e \leq n$ (see Section 10.1). The next subsection describes a method for the estimation of n_e, followed by the application to an unreplicated comparison.

Prior to describing the estimation of n_e we must take this opportunity to emphasize the hazards of carrying out an unreplicated experiment. As was mentioned earlier, although the nominal reason for replicating treatments in an experiment is to obtain an estimate of the variance, replication also serves to introduce interspersion (Hurlbert, 1984). In a spatial context, this means that treatment effects are measured at several different, randomly selected locations in the experimental area. Without replication, there is no way to distinguish between legitimate treatment effects and differences in the response variable due to differences in the experimental substrate. Therefore, the method described in this section must be used with extreme caution in the comparison of treatments. If there is no interest in the comparison of two treatments, the method may be considered as a two-plot uniformity trial in which the estimate of n_e, together with an indication of the spatial consistency of the sample variance, may be useful in its own right as an indication of the effect of autocorrelation on the effective sample size.

16.4.2 Calculation of the Effective Sample Size

The method of estimation of n_e was discussed by Plant (2007) and is similar to that described in Section 11.2.3. It is based on an extension of the method of Clifford and Richardson (1985) and Clifford et al. (1989) described in Section 11.2.2. A spatial error model for the comparison of two means associated with autocorrelated variables $Y_1(x,y) = Y_{1j}$ and $Y_2(x,y) = Y_{2j}$ may be written as

$$Y_{ij} = \mu_i + \eta_{ij}, \, i = 1, 2, \, j = 1, n$$

$$\eta_{ij} = \lambda \Sigma_k w_{jk} \eta_{ik} + \varepsilon_{ij},$$

(16.9)

where i indexes the factor levels, j indexes the samples, λ is a scalar measuring the strength of autocorrelation of the errors, the w_{ik} are elements of the spatial weights matrix describing the connectivity among the samples, and the ε_{ij} are independent, identically distributed normal random variables with mean 0 and variance σ^2. We will assume that the autocorrelation coefficient λ is the same for both Y_1 and Y_2.

The null hypothesis of equality of means of Y_1 and Y_2 may be tested using Student's t statistic, $t = (\overline{Y}_1 - \overline{Y}_2)/s_{12}$, where s_{12} is the square root of $s_{12}^2 = (s_1^2 + s_2^2)/2$, with $s_1^2 = \Sigma_j (Y_{1j} - \overline{Y}_1)^2/(n-1)$ and similarly for s_2^2. If the values of Y_{1j} and Y_{2j} are spatially independent, then the sampling distribution of the t statistic under the null hypothesis of no difference between means has a variance of $\sigma_t^2 = D/(D-2)$ where $D = 2n-2$ is the number of degrees of freedom. In the case that Y_{1j} and Y_{2j} are each spatially autocorrelated, if an independent estimate $\hat{\sigma}_t^2$ of the sampling variance is available, then this can in principle be used as was done in Section 11.2 to estimate the effective sample size n_e by inverting the variance equation and substituting for n.

Unfortunately, the resulting estimate, $\hat{n}_e = (2\hat{\sigma}_t^2 - 1)/(\hat{\sigma}_t^2 - 1)$, has very poor numerical properties because $\hat{\sigma}_t^2$ appears in both the numerator and the denominator. A better estimator for n_e is obtainable by using the point biserial correlation coefficient (Tate, 1954). This statistic is computed by arranging the vectors Y_1 and Y_2 into a single column vector Y. The point biserial correlation coefficient r_w is then the correlation coefficient between Y and a vector Z whose components corresponding to Y_1 are 1 and corresponding to Y_2 are 0 (i.e., $Z = [1...1 \, 0...0]'$). When the components of $Y_1(x,y)$ and $Y_2(x,y)$ are independent the variance of the sampling distribution of r_w is $\sigma_r^2 = 1/2n$ (Tate, 1954). Therefore, given an independent estimate $\hat{\sigma}_r^2$ of the variance of r_w an estimate of the effective sample size can be obtained as

$$\hat{n}_e = \frac{1}{2\hat{\sigma}_r^2}.$$

(16.10)

As described in Section 11.4.2, one can obtain an estimate using either a block bootstrap or a parametric bootstrap method (Efron and Tibshirani, 1993; Davison and Hinkley, 1997). The bootstrap estimate $\hat{\sigma}_r^2$ is plugged into Equation 16.10 to generate $\hat{n}_e$. The t statistic for the test of the null hypothesis $r_w = 0$ is

$$t = \frac{r_w(2n-2)^{1/2}}{1-r_w^2},$$ (16.11)

with $2n-2$ degrees of freedom. The corrected t statistic for the test of the null hypothesis $r_w = 0$ may then be obtained by replacing n with $\hat{n}_e$ in Equation 16.11 to obtain

$$t_{corr} = \frac{r_w(2\hat{n}_e-2)^{1/2}}{1-r_w^2}$$ (16.12)

with $2\hat{n}_e-2$ degrees of freedom.

As usual, we will first develop a simulation using a simple artificial data set. First, we generate two autocorrelated variables Y_1 and Y_2 on a 14 by 14 lattice and remove the outer two layers to produce a 10 by 10 lattice.

```
> library(spdep)
> lambda <- 0.6
> nlist.14 <- cell2nb(14, 14)
> IrWinv.14 <- invIrM(nlist.14, lambda)
> nlist <- cell2nb(10, 10)
> W <- nb2listw(nlist)
```

Next, we create a function `para.samp()` that generates a bootstrap replication of the point biserial correlation coefficient r_w.

```
> para.samp <- function(Y1, Y2, IrWinv1.mod, IrWinv2.mod){
+    Y1.samp <- sample(Y1, length(Y1), replace = TRUE)
+    Y1.boot <- IrWinv1.mod %*% Y1.samp
+    Y2.samp <- sample(Y2, length(Y2), replace = TRUE)
+    Y2.boot <- IrWinv2.mod %*% Y2.samp
+    Z <- c(rep(1,100), rep(0,100))
+    Y <- c(Y1.boot, Y2.boot)
+    r.hat <- cor(Y,Z)
+ }
```

Next, we generate the two variables Y_1 and Y_2.

```
> set.seed(123)
> Y1.plus <- IrWinv.14 %*% rnorm(14^2)
> Y1 <- matrix(Y1.plus, nrow = 14, byrow = TRUE)[3:12,3:12]
> Y2.plus <- IrWinv.14 %*% rnorm(14^2)
> Y2 <- matrix(Y2.plus, nrow = 14, byrow = TRUE)[3:12,3:12]
```

The uncorrected t-test yields the following.

```
> print(t.test(Y1, Y2, "two.sided")$p.value, digits = 3)
[1] 0.0632
```

Next, we fit the data to a spatial error model (Equation 16.10) and extract the residuals and the values of λ.

```
> # Fit a spatial error to the data
> Y1.mod <- errorsarlm(as.vector(Y1) ~ 1, listw = W)
> e1.hat <- residuals(Y1.mod)
> print(lambda1.hat <- Y1.mod$lambda, digits = 2)
lambda
   0.39
> Y2.mod <- errorsarlm(as.vector(Y2) ~ 1, listw = W)
> e2.hat <- residuals(Y2.mod)
> print(lambda2.hat <- Y2.mod$lambda, digits = 2)
lambda
   0.47
```

Now we generate 100 bootstrap resamples of the data, compute the point biserial correlation coefficients, and use the plug-in principle to compute the variance estimate $\hat{\sigma}_r^2$

```
> IrWinv1.mod <- invIrM(nlist, lambda1.hat)
> IrWinv2.mod <- invIrM(nlist, lambda2.hat)
> U <- replicate(200, para.samp(e1.hat, e2.hat, IrWinv1.mod,
+     IrWinv2.mod))
> print(sigmar.hat <- var(U), digits = 3)
[1] 0.0125
```

By way of comparison, the uncorrected value of the variance is $1/200 = 0.005$. Finally, we use Equations 16.10 and 16.12 to compute the corrected t statistic.

```
> print(ne.hat <- 1 / (2*sigmar.hat), digits = 3)
[1] 39.9
> Z <- c(rep(1,100), rep(0,100))
> Y <- c(Y1, Y2)
> rw <- cor(Y, Z)
> t.corr <- rw*sqrt(2*ne.hat - 2) / (1 - rw^2)
> print(p.corr <- 2 * (1 - pt(q = abs(t.corr), df = 2*ne.hat - 2)),
+     digits = 3)
[1] 0.241
```

The corrected p value is considerably higher than the uncorrected one. In Exercise 16.4, you are asked to test the method in a Monte Carlo simulation.

A spatial lag model of the form

$$Y_{ij} = \mu_i + \rho\Sigma_k w_{ik}Y_{ik} + \varepsilon_{ij}, \ i = 1, 2, \ j = 1, n \tag{16.13}$$

may for certain data be more appropriate than the spatial error model. This can be used as well (Exercise 16.5).

16.4.3 Application to Field Data

As an example of the method, we will test the null hypothesis that the mean values of *SoilpH* in Field 4.2 are equal on the east and west sides of the field (Figure 16.6). This is a bit contrived, but not entirely so. Recursive partitioning picked soil acidity as a

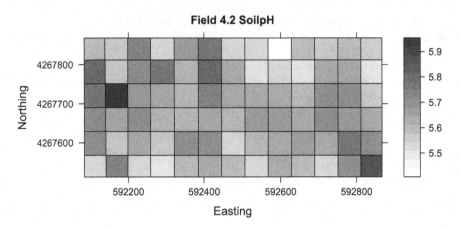

FIGURE 16.6
Thematic map of soil pH in Field 4.2.

variable associated with yield in Field 4.2 (Exercise 9.9), but higher yields are associated with more acidic soils, which is the opposite of what would be expected with wheat. There is a strong east-west yield trend in the 1996 data due to weed influence, and it is of interest to know whether soil pH might tend to be higher in the western part of the field, which could create a spurious association of soil pH with yield due to their common (and in the case of soil pH, probably coincidental) association with high weeds.

There are 13 sample locations in the east-west direction in the field, so we will eliminate the middle column and divide the field into two halves of 36 locations each.

```
> sort(unique(data.Set4.2$Easting))
 [1] 592107 592109 592110 592111 592112 592113 592168
 [8] 592229 592290 592354 592415 592476 592537 592598
[15] 592659 592720 592781 592842
> data.Set4.2W <- data.Set4.2[which(data.Set4.2$Easting < 592450),]
> data.Set4.2E <- data.Set4.2[which(data.Set4.2$Easting > 592500),]
```

For some reason, the Eastings of the westernmost sample locations are not exactly aligned (this is also evident in the slight irregularity of the Thiessen polygons on the west edge in Figure 16.6). Therefore, we have to be a bit careful in establishing the value of d2 for the function dnearneigh() and not make it too small.

```
> Y1 <- data.Set4.2W$SoilpH
> coordinates(data.Set4.2W) <- c("Easting", "Northing")
> nlist.1 <- dnearneigh(data.Set4.2W, d1 = 0, d2 = 70)
> W.1 <- nb2listw(nlist.1, style = "W")
> Y2 <- data.Set4.2E$SoilpH
> coordinates(data.Set4.2E) <- c("Easting", "Northing")
> nlist.2 <- dnearneigh(data.Set4.2E, d1 = 0, d2 = 70)
> W.2 <- nb2listw(nlist.1, style = "W")
```

Now we are ready to do the analysis. First, we will check the means and to an ordinary *t*-test.

```
> mean(data.Set4.2W$SoilpH)
[1] 5.651111
> mean(data.Set4.2E$SoilpH)
[1] 5.615833
> t.test(data.Set4.2W$SoilpH,data.Set4.2E$SoilpH, "two.sided")
        Welch Two Sample t-test
data: data.Set4.2W$SoilpH and data.Set4.2E$SoilpH
t = 1.6709, df = 68.797, p-value = 0.09928
```

The soil is very slightly more acidic on the west side, and the difference is not significant at the $\alpha = 0.05$ level and almost certainly not important. Let's follow up and compare the outcome of this test with the test that incorporates spatial autocorrelation. We first create a function para.samp()almost identical to the one in Section 16.4.2. Then we set up the spatial error models in the same way as in Section 16.4.2. Finally, we generate the bootstrap resample values, calculate the variance, and estimate the effective sample size and corrected p value.

```
> set.seed(123)
> U <- replicate(200, para.samp(e1.hat, e2.hat, IrWinv1.mod,
+     IrWinv2.mod))
> print(sigmar.hat <- var(U), digits = 3)
[1] 0.0366
> print(ne.hat <- 1 / (2*sigmar.hat), digits = 3)
[1] 13.7
> Z <- c(rep(1,length(Y1)), rep(0,length(Y2)))
> Y <- c(Y1, Y2)
> rw <- cor(Y, Z)
> t.corr <- rw*sqrt(2*ne.hat - 2) / (1 - rw^2)
> print(p.corr <- 2 * (1 - pt(q = abs(t.corr), df = 2*ne.hat - 2)),
+     digits = 3)
[1] 0.315
```

The effective sample size is about one-third the total sample size. Not surprisingly, the corrected p value is much higher than the uncorrected value. The comparison of p values provides an indication of the impact of spatial autocorrelation on the data analysis.

This example does not involve a treatment but rather measures differences in the substrate itself, so there is less concern with the interspersion issue. An example of a situation in which the interspersion effect would be important is if we had applied a compound such as gypsum to, say, the west side and measured the differences in pH after this application. In this case, without other information it would be impossible to distinguish differences in soil pH due to the gypsum from those present prior to the application.

16.5 Further Reading

Kutner et al. (2005) is a good source for a basic discussion of the analysis of variance, although any other basic linear modeling text will serve equally well. Gomez and Gomez (1984) has a very extensive discussion of both complete and incomplete block designs. Stroup (2002) provides a very cogent discussion of the pros and cons of the traditional "design based" approach versus the more recent "model based" approach to the incorporation of spatial effects into replicated experiments. Brownie et al. (1993) provide a very readable review and critique of the various methods and of the assumptions that underlie them. Griffith (1978) provides further discussion. The method of dealing with unreplicated experiments with two treatments discussed in Section 16.4 has some ideas in common with *before and after control intervention* (BACI) analysis (Thomas et al., 1978). Discussions of this include Smith et al. (1993) and Evans and Coote (1993).

With the exception of variety trials, the checkerboard layout common in statistical analyses of uniformity trials is uncommon in actual field experiments, at least with row crops. Many agronomists prefer to make the plots within the blocks long and narrow. This has both agronomic and statistical advantages. Agronomically, it is generally easier to apply the same treatment to an entire row. Also, the plots encompass as much of the block variability as possible within each plot (Little and Hills, 1978, p. 285; Gomez and Gomez, 1984, p. 500). Besag (1991) provides a good discussion of agricultural plots as well as a statistical method designed for the analysis of experiments with long, narrow plots.

Exercises

16.1   Repeat using a completely randomized design (CRD) the analysis done in Section 16.3.1 for an RCB design and compute the value of *EMP* for this design.

16.2   The correlated errors method described in Section 16.3.4 requires an estimate of the parameters of the variogram model for the errors. Test this method for sensitivity to the values of this variogram model.

16.3   Apply the completely random design, the randomized complete block, and the Papadakis method to the population data used in Chapter 5 to test various sampling schemes for the data of Field 4.2.

16.4   Repeat the analysis of Section 16.3.4 using a GAM to predict the trend surface.

16.5   Carry out a Monte Carlo simulation of the method discussed in Section 16.4.2 applied to the artificial data set used in that section. Compute the experimental error rate as a function of the parameter λ.

16.6   Using a spatial lag model rather than a spatial error model, repeat the analysis of Section 16.4.3.

17

Assembling Conclusions

17.1 Introduction

Formulating conclusions is the most important part of the analysis of a data set, and the most subjective. To interpret the results of the analysis, the investigator must apply his or her knowledge of the biophysics of the system, and as a result different people may legitimately look at the same statistical results and reach different conclusions. To the extent that the statistics support prior assumptions they may be regarded as an affirmation of these assumptions, but to the extent that they conflict with them, one must decide how to respond to this conflict. The analysis of each of the four data sets that has occupied this book has been a simulation of an actual research project. The objectives that were established in Chapter 1 are not actual current research objectives. Rather, the objectives were established, based on the research objectives of the original projects, for expository purposes, as questions that could be could legitimately be posed prior to the collection of the data and then addressed by the analysis of that data. Nevertheless, the fact that different people may use the same data set to reach entirely different conclusions still applies. For this reason, my goal in this chapter is not to convince you that my conclusions are correct, but rather to present an example of how one might approach the highly personal process of developing an interpretation of the implications of their analytical results. Others may come to different conclusions based on the same data, and the intent of this chapter is not to advocate for one particular approach but rather to provide an example of an approach, some parts of which you may find appropriate and some parts of which you may choose to ignore.

17.2 Data Set 1

The objective of the analysis of Data Set 1 is to test the predictive capacity of the California Wildlife Habitat Relationships (CWHR) model of habitat suitability for the yellow-billed cuckoo. The model involves four variables: patch area, patch width, floodplain age distribution as a surrogate for vegetation species distribution, and the ratio of area of tall vegetation (>20 m) to total area (Table 7.1). The latter two variables enter into the model in a non-monotonic fashion in which an intermediate value is optimal. A fifth variable, patch distance to water, is accounted for by excluding patches farther than 100 m from the river.

The data set does not allow us to distinguish effectively between patch area and patch width as limiting factors in determining habitat suitability, but a comparison of contingency tables in which one or the other was excluded indicates that for this data set, patch area plays a more critical role. In the exercises of Chapter 7, further contingency table tests with various combinations of the variables excluded were carried out. The results of the contingency table analysis indicate that a model including only *PatchArea* and *AgeRatio* suitability scores generates the same contingency table as the model with the full set of four variables. The table contains five true positives, twelve true negatives, two misclassified positives, and one misclassified negative. Among the single variable models, the contingency table of the model with only *AgeRatio* is closest to this. The Fisher test was used in Section 11.2 to test the null hypothesis that the model is not related to the presence-absence data. The test generated a *p* value of about 0.01 uncorrected and 0.07 when corrected for spatial autocorrelation.

Logistic and zero-inflated Poisson regression models were constructed for the system in Sections 8.3.3 and 8.3.4. These models also indicated that the most important predictors in the model are *PatchArea* and *AgeRatio*, although the manner in which these enter the model differed depending on the model (logistic vs. Poisson). Because of the relatively small number of data records, no attempt was made to incorporate spatial autocorrelation directly into these models.

Let's take a closer look at the habitat score data. After re-running the code from Section 7.2 and Exercises 7.2 and 7.3 we generate the data frame Set1.corrected, which contains the data use to carry out the analyses in the subsequent chapters. Here are the data, ordered by observation point ID number.

```
> d.f <- data.frame(with(Set1.corrected, cbind(PatchID,
+    AreaScore, AgeScore, HeightScore, obsID, HSIPred, PresAbs)))
> d.f[order(d.f$obsID),]
```

	PatchID	AreaScore	AgeScore	HeightScore	obsID	HSIPred	PresAbs
16	175	0.33	0.00	0.00	2	0	0
15	170	0.00	0.00	0.00	3	0	0
14	164	0.33	0.66	1.00	4	1	1
13	155	1.00	0.66	0.66	5	1	1
12	130	0.33	0.00	0.33	6	0	0
11	122	0.66	0.66	0.66	7	1	0
10	116	1.00	0.00	0.00	8	0	0
19	224	0.00	0.00	0.33	9	0	0
9	106	0.00	0.33	0.33	10	0	0
8	100	0.66	0.00	0.00	11	0	1
7	97	0.00	0.00	0.66	12	0	0
6	86	0.00	0.00	0.66	13	0	0
20	400	0.66	0.00	0.66	14	0	0
18	210	0.00	0.33	0.33	15	0	0
4	60	0.66	0.00	0.66	16	0	0
5	63	1.00	0.33	0.66	17	1	1
17	209	0.00	0.33	0.33	18	0	1
3	31	0.33	0.33	0.33	19	1	1
2	22	0.00	0.00	0.00	20	0	0
1	19	1.00	0.33	1.00	21	1	1

Of the 20 locations, there are none in which *HeightScore* is 0 and no other habitat score is 0. Indeed, every patch with *HeightScore* = 0 also has *AgeScore* = 0. The data set, therefore, cannot tell us anything about the capacity of *HeightScore* to limit patch suitability. One patch, patch 209, has *AreaScore* = 0 and positive values for both *AgeScore* and *HeightScore*, and is a false negative. This is misleading, however, because it happens that patch 209 is a very narrow patch classified as lacustrine that is in the center of a much larger patch classified as riverine (Exercise 7.4). In conclusion, as with patch width, we cannot provide any solid evidence for or against the predictive power of vegetation height in the model.

Our objective is to use a data set to test a classification of data based on an already specified model, as opposed to generating a new model. There is a statistic available, Cohen's kappa (Cohen, 1960; Lo and Yeung, 2007, p. 121), that is specifically designed for this objective. This statistic, which was briefly discussed in Section 15.4.1, accounts for the fact that, when comparing a predicted classification with the actual observed classification, some of observations will be correctly classified by chance. Cohen's kappa corrects for this chance agreement. It is defined as follows. Consider again the site classification matrix.

```
> UA <- with(Set1.corrected, which(HSIPred == 0 & PresAbs == 0))
> UP <- with(Set1.corrected, which(HSIPred == 0 & PresAbs == 1))
> SA <- with(Set1.corrected, which(HSIPred == 1 & PresAbs == 0))
> SP <- with(Set1.corrected, which(HSIPred == 1 & PresAbs == 1))
> print(cont.table <- matrix(c(length(SP),length(SA),
+     length(UP),length(UA)), nrow = 2, byrow = TRUE,
+     dimnames = list(c("Suit.", "Unsuit."),c("Pres.", "Abs."))))
        Pres. Abs.
Suit.       5    1
Unsuit.     2   12
```

This matrix describes the level of agreement between the predicted classification of sites of the CWHR model and the actual observed classification of sites. If it were a pure diagonal matrix, then all of the sites would be correctly classified. The off-diagonal elements are the number of misclassified sites. In this context, the matrix is called a *confusion matrix*. Let P_0 be the total portion of the sites that are correctly classified. Let P_c be the portion of the sites that would be correctly classified by chance if the classification were random. In terms of the marginal sums defined in Section 11.3.1, $P_c = n_{m1}n_{im} + n_{2m}n_{m2}$. The kappa statistic is given by

$$\kappa = \frac{P_0 - P_c}{1 - P_c}. \tag{17.1}$$

If all of the data are correctly classified then $\kappa = 1$, and if the classification is no better than random, then $\kappa = 0$. There are numerous packages available that contain a function to compute Cohen's kappa. We will use the function Kappa.test() of the fmsb package (Nakazawa, 2017).

```
> library(fmsb)
> print(kappa.stat <- Kappa.test(cont.table))
$Result

        Estimate Cohen's kappa statistics and test the null hypothesis
that the extent of agreement is same as random (kappa=0)
data:   cont.table
Z = 2.6127, p-value = 0.004491
95 percent confidence interval:
 0.3034305 1.0147513
sample estimates:
[1] 0.6590909
$Judgement
[1] "Substantial agreement"
```

Cohen (1960) showed that the kappa statistic satisfies an asymptotic normality property that permits it to be tested against the null hypothesis of random classification. As with the analysis of contingency table data, however, the small size of Data Set 1 limits the applicability of the results. We can carry out a permutation test similar to that used in Section 11.3.2 with the function `fisher.test()`. First, we carry out the test without restricted randomization.

```
> set.seed(123)
> sample.blocks <- function() sample(Set1.corrected$PresAbs)
> U <- replicate(1999,calc.kappa())
> print(obs.kappa <- Kappa.test(cont.table)$Result$estimate)
[1] 0.6590909
> print(p <- sum(u >= obs.kappa - 0.001) / 2000)
[1] 0.0075
```

Next, we carry out a restricted randomization using the blocks in Figure 11.4. Most of the code is identical to that used in Section 11.3.2 and is not shown here. The function `calc.kappa()` replaces `calc.fisher()`, and the function `sample.blocks()` from Section 11.3.2 replaces the function in the code sequence above.

```
> calc.kappa <- function(){
+    PA <- sample.blocks()
+    UA <- sum(Set1.corrected$HSIPred == 0 & PA == 0)
+    UP <- sum(Set1.corrected$HSIPred == 0 & PA == 1)
+    SA <- sum(Set1.corrected$HSIPred == 1 & PA == 0)
+    SP <- sum(Set1.corrected$HSIPred == 1 & PA == 1)
+    n <- matrix(c(SP, SA, UP, UA), nrow = 2, byrow = TRUE)
+    kappa.stat <- Kappa.test(n)$Result$estimate
+ }
```

Now we run the test.

```
> set.seed(123)
> u <- replicate(1999,calc.kappa())
> print(p <- sum(u >= obs.kappa - 0.001) / 2000)
[1] 0.0725
```

Taking spatial autocorrelation into account, the *p* value for the kappa statistic is about the same as that for the chi-squared statistic computed in Section 11.3.2. This is not terribly surprising since most of the sites are correctly classified.

The second issue raised above is that the model we are testing is intended to predict habitat suitability, not habitat occupancy. Therefore, an error of bird presence in a habitat patch classified as unsuitable would be considered more serious than an error of bird absence in a habitat patch classified as suitable. The remote sensing and GIS literature deals with this issue through the concept of *producer's accuracy* and *user's accuracy* (Lo and Yeung, 2007, p. 120). In the case of our data set, there is only one suitable site with an absence, and one of the sites with a misclassified presence (patch 209) has been accounted for. Therefore, this is not such an important issue for us.

Let's now consider the results of the generalized linear modeling. Because of the unreliable nature of the abundance data, we will focus on the logistic model. As with the contingency table model, the variables *PatchArea* and *AgeRatio* were picked out as most important. Here is a brief summary of the results:

```
> Set1.logArea <- glm(PresAbs ~ PatchArea,
+       data = Set1.norm1, family = binomial)
> AIC(Set1.logArea)
[1] 25.14631
> Set1.logAge <- glm(PresAbs ~ AgeScore,
+       data = Set1.norm1, family = binomial)
> AIC(Set1.logAge)
[1] 23.79338
> Set1.logAreaAge <- glm(PresAbs ~ PatchArea + AgeScore,
+       data = Set1.norm1, family = binomial)
> anova(Set1.logAge, Set1.logAreaAge, test = "Chisq")
Analysis of Deviance Table
Model 1: PresAbs ~ AgeScore
Model 2: PresAbs ~ PatchArea + AgeScore
  Resid. Df Resid. Dev Df Deviance P(>|Chi|)
1        18     19.793
2        17     17.086  1   2.7069   0.09991.
> summary(Set1.logAreaAge)
Coefficients:
            Estimate Std. Error z value Pr(>|z|)
(Intercept)  -1.7308     0.8444  -2.050   0.0404*
PatchArea     0.9876     0.6570   1.503   0.1328
AgeScore      4.8061     2.7064   1.776   0.0758.
AIC: 23.086
```

The single most important variable in the logistic regression model is *AgeScore*, and the model that includes both this and *PatchArea* is perhaps very slightly better. This is basically in concurrence with the contingency table analysis.

Figure 17.1 is a plot of the data values of *PresAbs* vs. *PatchArea*. The open circles are sites where *AgeScore* = 0, and the black circles are sites where *AgeScore* > 0. Also shown are plots of the fit of the model Set1.logAreaAge for *AgeScore* = 0 and for *AgeScore* = 1. The model appears to have been heavily influenced by the data record with *PresAbs* = 1, *AgeScore* = 0, and *PatchArea* $\cong$ 0.5. Overall, however, the logistic regression model does a reasonable job of fitting the data, with the exception that it appears to overestimate the probability of a patch being occupied when the patch area is high but the age score is low.

In summary, the data provide some support for the CWHR model. The relatively small sample size limits the power of the analysis, and this effect is exacerbated by spatial autocorrelation. To the extent that the data support the model, they also support the notion that the patch suitability is limited by its least suitable aspect. The data do not provide sufficient variability to distinguish the impact of patch width from that of patch area, nor do they provide evidence regarding the importance of the ratio of area of tall vegetation to total area. The one anomalous data record that cannot be explained is patch 100. This is a large patch with uniformly high floodplain age and a relative abundance of tall vegetation, indicating the strong possibility that it is dominated by oaks. The fact that there are four recorded sightings indicates that the chance of a false positive is low. This remains an unsolved mystery.

There is an epilogue to this story. Girvetz and Greco (2009) analyzed a similar data set consisting of 102 sites surveyed along a longer stretch of the Sacramento River. A major improvement in the analysis was to use the *Patch Morph* algorithm (Girvetz and Greco, 2007) to incorporate small isolated areas of different vegetation type into the habitat patches. This provided a better model of the patches as they are perceived by the cuckoos.

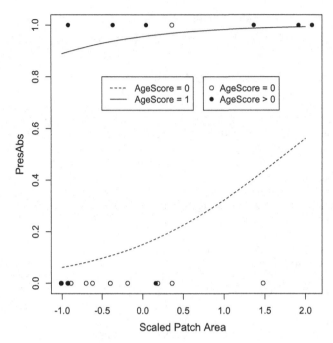

FIGURE 17.1
Plots of yellow-billed cuckoo presence/absence vs. scaled patch area and age score. The curves are predicted values based on the generalized linear model, and the circles are data values.

Girvetz and Greco (2009) used the CWHR model, logistic regression, and classification tree analysis to determine the factors having the greatest influence on patch suitability. Similarly to our analysis, they found that the total area of cottonwood forest within the patch was the most important factor in determining patch suitability. Their larger study area is bisected by a state highway and, surprisingly, they found a large reduction in patch occupancy on the north side of this highway. A legitimate inference from the data would be that the highway serves as some sort of barrier. This is surprising because, as Girvetz and Greco (2009) point out, the birds, having flown all the way from South America, should not find the crossing of a two-lane highway to be a major challenge. Taken together with our own anomalous patch 100 this shows that, despite our best efforts, the yellow-billed cuckoos have not yet revealed all of their secrets.

17.3 Data Set 2

Data Set 2 consists of observations of the presence or absence of blue oak trees together with measurements and calculated estimates of explanatory variables at 4,101 locations in California. Each of these locations was identified as containing at least one species of oak at the time it was sampled. The initial objective was to determine suitability characteristics of the individual sites for blue oak. In Section 12.6, we added a second related but different objective: to determine the characteristics of a geographic region, as defined by elevation and spatial proximity, that affect the average fraction of sites in the region containing a blue oak. We will refer to these as the pointwise and the regional analyses, respectively.

The data set naturally subdivides into four mountain ranges: the Sierra Nevada to the east, the Coast Range to the west, the Klamath Mountains to the north, and the Transverse Range to the south. We analyzed data from the Sierra Nevada and the Coast Range, leaving the Klamath Mountains and the Transverse Range for verification purposes.

The Sierra Nevada has a fairly simple topography, with a gradually increasing elevation as one moves from west to east. Blue oak prevalence declines as elevation increases (Figure 7.7), and there is a strong indication that mean annual precipitation, represented by the variable *Precip*, plays a prominent role in determining blue oak presence. This conclusion is supported by exploratory analysis (Figure 7.12), by the pointwise general linear model analysis under the assumption of independent data records (Section 8.4.2), and by the regional analysis in which spatial autocorrelation is accounted for (Section 12.6). The second explanatory variable in the pointwise model is *MAT*. In the Sierra Nevada, however, *MAT* is highly correlated with *Precip*, and it is unclear whether the estimated *MAT* effect is due to *MAT* or represents a quadratic effect of *Precip*. In the regional Sierra Nevada subset, the primary and secondary roles of *Precip* and *MAT* are exchanged. In both analyses, the soil-related variables *Permeab* and *AWCAvg* are also in the model. The variable *SolRad* is also in the pointwise model, and *PM400* is in the regional model.

Possibly because of its more complex topography, in the Coast Range subset the correlations among the explanatory variables are generally lower. This may help to disentangle the relationships a bit. We first focus on the issue of whether the blue oaks are responding to temperature. Regression tree analysis (Figure 9.10, Section 9.3.3) and contingency table analysis (Exercises 11.2 and 11.3) suggested that high mean July temperatures, low mean January temperatures, or both may favor blue oaks. Figure 17.2a shows a scatterplot of *JuMean* vs. *Precip* in the Coast Range subset, in which blue oak presence and absence is also shown. There appears to be a fairly strong boundary line effect (Abbott et al., 1970; Webb, 1972) of *JuMean*. A similar graph is obtained using *JuMax*. The use of the boundary line model represents an implicit acceptance of Leibig's law of the minimum (Loomis and Connor, 1992, p. 55). This law, as used by agronomists, asserts that yield is limited by the scarcest resource. As interpreted by ecologists, it states that species distribution will be controlled by the environmental factor for which the organism has the lowest range of adaptability (Krebs, 1994, p. 40), or what Bartholomew (1958) calls the *ecological tolerance*.

Figure 17.2a could also be interpreted as indicating that mean annual precipitation influences the boundary line effect of mean July temperature, with lower values of *Precip* corresponding to lover boundary values of *JuMean*. Figure 17.2b shows a plot of *JaMean* vs. *Precip* in the same format. There appears to be a similar boundary line effect, but in the opposite direction: blue oak presence is favored by values of *JaMean* below about 9.2. Figure 17.3a and b show maps indicating the locations that lie on the "wrong" side of the boundary lines in Figure 17.2a and b, respectively. Somewhat surprisingly, there is a strong negative correlation ($r = -0.57$) between *JaMean* and *JuMean* in the Coast Range. Figure 17.4 shows a scatterplot of the two variables identifying the values for data records lying on the "wrong" side of the boundary lines and with $QUDO = 1$. There may be some small indication that mean January temperature is more decisive than mean July temperature, since there are more anomalous data records of the latter type, but the evidence is not very strong. What is the effect of *JaMean* and *JuMean* in the Sierra Nevada? There are very few sites at which the either the value of *JaMean* is above 9.2 or the value of *JuMean* is below 19.8. Therefore, as with precipitation in the Coast Range, the data of the Sierra Nevada subset do not provide a sufficient range to determine the effect of temperature. Neither the Sierra Nevada subset nor the Coast Range subset of the pointwise data provided any convincing evidence for or against the effect of the soil related variables.

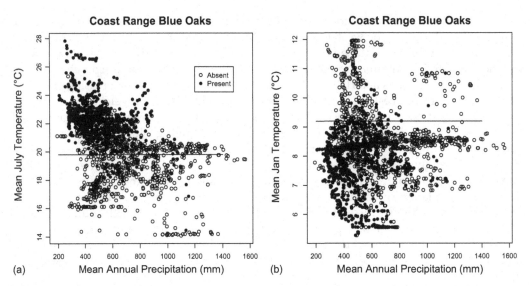

FIGURE 17.2
Scatterplots of (a) mean July temperature and (b) mean January temperature vs. mean annual precipitation. The filled circles indicate locations where a blue oak is present.

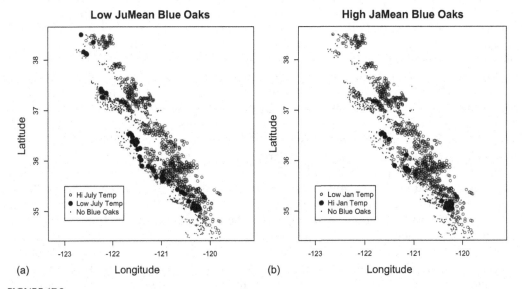

FIGURE 17.3
"Anomalous" locations in the Coast Range. The large filled circles indicate locations where a blue oak is present and: (a) the daily mean July temperature is less than 19.8°C, (b) the daily mean January temperature is greater than 9.2°C.

As a final check, we will test the models on the Klamath Range and Transverse Range data sets, which were not included in the model development. Figure 17.5 shows the results. Unfortunately, neither of these two mountain ranges has a sufficient range of values to truly test the models. The data are not, however, inconsistent with the notion that there is a boundary line temperature effect and that the location of the boundary is influenced by precipitation.

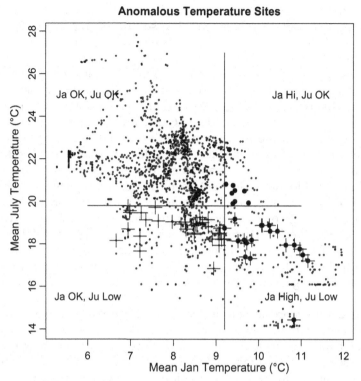

FIGURE 17.4

Scatterplot of daily mean January and July temperatures. The crosses indicate "anomalous" (i.e., low mean) July temperature sites containing a blue oak, and the filled circles indicate anomalous" (i.e., high mean) January temperature sites containing a blue oak.

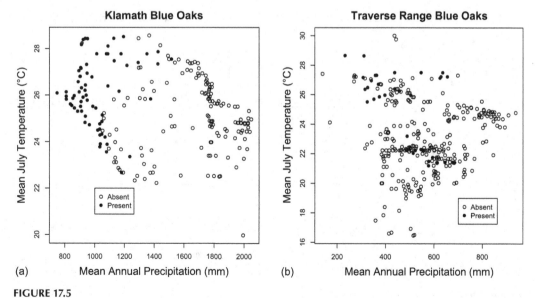

FIGURE 17.5

Scatterplots of mean July temperature vs. mean annual precipitation for (a) the Klamath Range and (b) the Transverse Ranges. The filled circles indicate locations where a blue oak is present.

We turn now to the regional data. The mixed model results in Section 12.6 identified soil permeability *Permeab*, length of growing season over 32°F *GS32*, and potential evapotranspiration *PE* as the principle explanatory variables in the Coast Range, and mean annual temperature *MAT*, mean annual precipitation *Precip*, and average available water capacity *AWCAvg* as the principle explanatory variables in the Sierra Nevada. Let's construct scatterplots of some of the regional data to see if we can get a handle on what the data are trying to say. Figure 17.6a shows a scatterplot of *QUDO* vs. *Permeab* for the Sierra Nevada and the Coast Range. The latter data set shows a boundary line effect with the maximum fraction of oak-occupied sites declining as the soil permeability increases. The Sierra Nevada only displays this effect to a much lesser extent. It is possible that this is a real effect, and that permeability is interacting with some other variable that differs between the Coast Range and the Sierra Nevada. It is also possible that permeability is serving as a surrogate for some other variable. The logical suspect is, of course, precipitation. There is indeed a positive correlation between *Permeab* and *Precip* ($r = 0.59$). Figure 17.6b shows a plot of *QUDO* vs. *Precip* for the regional data. There is again a negative relationship, but something seems to be interacting with *Precip*. Once again, we think of a logical suspect: something related to temperature. Playing around with scatterplots indicates that *JuMean* is a good possibility.

In summary, the evidence for the influence of precipitation is incontrovertible. The high level of autocorrelation among the variables makes further conclusions more or less speculative, but it appears that temperature, possibly either summer high temperatures or winter low temperatures, also limit the range of blue oaks. The xerophilic nature of blue oaks lends ecophysiological credence to the idea that the trees prefer hot summers, but regions that have hot summers tend to have cool winters, so it is hard to tease apart these effects with data. There are indications, both in the pointwise data and in the regional data, of an interaction between precipitation and temperature effects. The correlations between the data make it difficult to positively identify the effect of soil permeability.

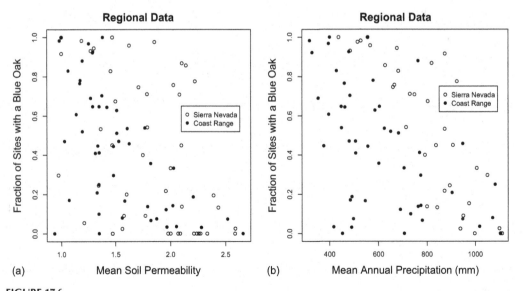

FIGURE 17.6
Plots of the fraction of sites in the regional data having a blue oak vs. (a) permeability and (b) mean annual precipitation for the Sierra Nevada and the Coast Range.

The initial exploratory analysis in Chapter 7 indicated that blue oak prevalence declined with increasing permeability, in opposition to the generally accepted properties of blue oak. Permeability and precipitation are highly correlated, however, and this may confound the results of the analysis.

17.4 Data Set 3

There is a wide range of values of yields among the rice farmers who participated in the study that led to Data Set 3. The ultimate goal in the analysis of the data set was to determine the management actions that those farmers take that would provide the greatest increase in their yield. No assumptions were made about either the economic value of the yield or the cost to the farmer of the management actions. As with Data Set 2, this data set and objective provides us with the opportunity to study a data analysis problem that involves multiple spatial scales. Two important early findings were that there was no significant effect of the season (i.e., the year of the study), and no significant rice year effect (i.e., of the position in the crop rotation). Based on these findings, data were pooled for the analysis. Much of the data analysis, especially the recursive partitioning in Chapter 9, examined the data at the landscape level, identifying the management variables that most influenced yield over this extent. The improvement in yield, however, must take place at the field level. In between these, there are differences in patterns at the regional level.

The landscape level recursive partitioning analysis in Section 9.3.4 identified the management variables N, *Irrig*, and *DPL* as playing an important role in the tree relating yield to management variables. When both management and exogenous variables are included in the model, nitrogen rate N and soil clay content *Clay* are identified as most important. An alternative approach was to identify the exogenous and management factors that most distinguished each farmer. The primary distinguishing characteristics of the farmers obtaining the highest yields in their respective regions were *DPL*, *Irrig*, and N.

The potential confounding factor is the quality of the land on which each farmer grows rice. Farmers with worse soil conditions might not improve their yield by mimicking the practices of farmers who get higher yield on better soil. Therefore, we attempted to determine the extent to which climatic or environmental factors played a role in determining the differences in rice yield observed between farmers. In other areas, clay-related quantities such as cation exchange capacity provide the best predictors of rice yield (Casanova et al., 1999; Williams, 2010, p. 5). Most of our analyses supported the conclusion that for the fields in our study, management variables played a more important role than field conditions (i.e., exogenous variables). The primary distinguishing characteristic at the landscape level between the lower yielding central region fields and the higher yielding northern and southern region fields is the high silt content in the central region. Figure 17.7 shows a soil texture triangle (Singer and Munns, 1996, p. 24) made with the function soil.texture() from the plotrix package (Lemon, 2006). Within region, however, there is no indication that silt content influences yield (Figure 7.20). This was cited in Section 11.5.2 as an example of the ecological fallacy. Normally, coarser textured soils require a higher level of applied nitrogen than high clay soils (Williams, 2010, p. 34), but farmers in the central region tended to apply fertilizer N at lower rates than farmers in the north and south.

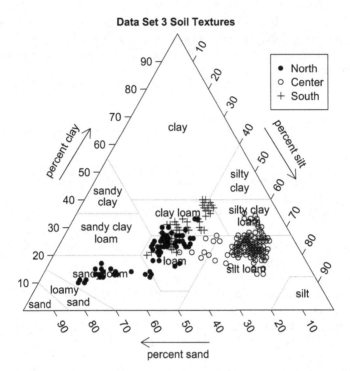

FIGURE 17.7
Soil texture triangle showing the soil textures of the fields in Data Set 3.

Our final analysis will try to predict the effect on yield that would occur if the farmers in the central region adopted the management actions of the northern and southern farmers. Here we must again be careful of running afoul of the ecological fallacy. The recursive partitioning analyses indicate that *DPL*, *N*, and *Irrig* are the most important management variables distinguishing high yield farmers from low-yield farmers. This conclusion is based on a landscape level analysis, but if we are going to compare farmers and regions, then we must accept this limitation. Gelman and Hill (2007, p. 252) point out that, just as the individual level model tends to understate regional variation, the regional means model tends to overstate it. We will try as much as possible, however, to predict the effect of modifications in farmer management practices using field scale analyses.

Our strategy will be the following. For each of the inputs *N*, *Irrig*, and *DPL*, we will carry out a regression analysis similar to that of Section 12.3 and determine the regression coefficients on a field by field basis for those fields at which the input was measured at more than one rate. We will then use these regression coefficients to predict the yield that would be achieved in each field in the central region if the farmer applied the input (*N*, *Irrig*, or *DPL*) at the rate deemed optimal by the analysis. Because of the scarcity of the data, we will not take spatial autocorrelation into account in the model. Because we are interested only in the specific fields, we will treat both the input and the field as a fixed effect, and therefore we will use ordinary least squares.

For nitrogen fertilization rate *N*, only fields 5, 11, 12, 13, and 16 contain measurements at different rates. We will use the function `lmList()` from the `nlme` package (Pinhiero et al., 2011) to obtain the regression coefficients.

```
> data.Set3N <- data.Set3[which(data.Set3$Field %in%
+     c(5,11,12,13,16)),]
> data.lis <- lmList(Yield ~ N | Field, data = data.Set3N)
> print(coef(data.lis), digits = 3)
   (Intercept)       N
5         37.5 126.1
11      -323.7 104.4
12       662.9 106.8
13      1351.1  86.6
16      3759.4  56.2
> b0.N <- coef(data.lis)[,1]
> b1.N <- coef(data.lis)[,2]
```

Now we compute the current yield and the predicted change in yield if *N* is applied at the maximal rate.

```
> print(Nmax <- max(data.Set3$N))
[1] 62.8
> print(Yield.predN <- b0.N + b1.N * Nmax, digits = 4)
[1] 7959 6233 7368 6789 7288
> print(Yield. N <- tapply(data.Set3N$Yield, data.Set3N$Field, mean),
+     digits = 4)
   5   11   12   13   16
6193 4771 6461 6053 7064
> print(delta. Y <- Yield.predN - Yield. N, digits = 3)
   5   11   12   13   16
1766 1462  908  736  224
```

We can carry out a similar analysis for *Irrig* and *DPL*. The code is analogous to that for *N*, and is not displayed. Here are the corresponding changes in yield.

```
> print(delta. YI <- Yield.predI - Yield. I, digits = 3)
       1        3        4        5        6        7        8        9       10
  154.20  1308.42   511.13  2873.80    -4.67   844.17  2811.60  2275.33  -448.35
      11       12       13       14       15       16
 1363.78  1083.96  1065.00   395.56  -216.50     9.90
> print(delta. YD <- Yield.predD - Yield. D, digits = 3)
    3     4     5    10    11    12    16
 -254  3419  3953  2827  5011 -2500   494
```

It is difficult to draw general conclusions, especially given the high variability in the predicted response to planting date. If we assume that there are no scheduling constraints on the growers that would prevent them from moving up their planting dates, then it would appear that this would be the lowest cost change that would improve their yield. Unless the cost of fertilizer becomes very high, increased nitrogen rate should also be economically justified. Given the high cost associated with land moving, changes in irrigation effectiveness within the field appear to be the least justified.

17.5 Data Set 4

The two fields that make up Data Set 4 were each sampled on a grid in 1995 at the start of the study, and yield data were collected in each of four growing seasons after that. Both fields were planted to a crop rotation that began with winter wheat in the first season and coursed to tomatoes in the second season. Field 4.1 was planted to beans and sunflower in years three and four, while Field 4.2 was planted to sunflower and corn (maize) during those years. The southern part of Field 4.1 (which is, alas, probably the most interesting part) was removed from the study after the second year. Both fields are laser leveled, so, although the topography prior to leveling almost surely influenced the current soil patterns, the current topography has no influence. The fields are fully irrigated, and there is virtually no rainfall (or even clouds) in the summer (cf. Figure 15.14b), so the last three years were not influenced by rainfall patterns. Temperature also does not appear to play much of a role, as the years were very consistent (Figure 15.14a).

The primary objective is to develop a methodology for determining the factors that influence spatial patterns of yield. There is no implication that the same patterns of influence will be present in other fields, but one might hope that the methodology can be applied more broadly. Field 4.2 is only about two kilometers away from Field 4.1, and is farmed by the same person, so that we might also hope that the scope of inference can be extended to Field 4.2. Although some of the exercises have involved graphical exploration and parameter estimation of Field 4.2, none of the model development has used information from this field.

Our model construction was founded on the assumption, based on several lines of evidence, that soil texture plays an important role in influencing yield in Field 4.1. Figure 17.8 shows the clay content over the field. There is one data record that is probably an outlier. Because of their high level of association with other variables, *Sand* and *SoilTOC* were not used in the analysis. *SoilK* and *SoilTN* are also highly associated (Figure 7.25), and both are associated to some degree with *SoilP*, with the association being stronger in the southern portion of the field (Exercise 8.3). There was no indication that any part of the field suffered from a nitrogen or potassium deficiency, but the northern and southern parts of the field had soil phosphorous levels below those recommended by University of California guidelines for wheat production (Figure 7.28a).

In Section 8.3, we developed several candidate models based on ordinary least squares regression. The linear regression models developed in Section 8.3 and later analyzed in Chapter 13 are the following:

```
> model.1 = lm(Yield ~ Clay + Silt + SoilpH + SoilTN + SoilK +
+      SoilP + Disease + Weeds + I(Clay*SoilP) + I(Clay*SoilK) +
+      I(Clay*SoilpH) + I(Clay*SoilTN), data = data.Set4.1)
> model.2 <- lm(Yield ~ Clay + Silt + SoilTN +
+      Weeds + I(Clay*SoilTN), data = data.Set4.1)
> model.3 <- lm(Yield ~ Clay + SoilpH + SoilP + Weeds +
+      I(Clay*SoilP) + I(Clay*SoilpH), data = data.Set4.1)
> model.4 <- lm(Yield ~ Clay + SoilpH + SoilP +
+      Weeds + I(Clay*SoilP) + Weeds, data = data.Set4.1)
> model.5 <- lm(Yield ~ Clay + SoilP + I(Clay*SoilP) +
+      Weeds, data = data.Set4.1)
```

Field 4.1 Clay Content

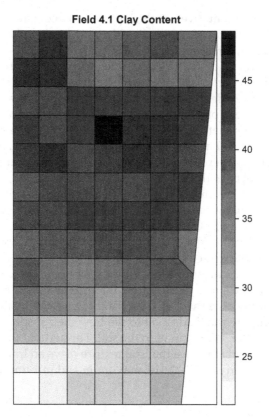

FIGURE 17.8
Thematic map of clay content in Field 4.1.

The first model, model.1, includes everything, the second model, model.2, was selected by backward selection, and the last model, model.5, is the model picked out by Mallow's C_p analysis. A real study would also have included exploration via forward and bidirectional selection. None of the models were strongly affected (i.e., there was no change in the significance level of any of the coefficients) by the inclusion of spatial autocorrelation in the model (Exercise 13.3a). These models identify clay content, soil pH, soil P, weeds, and soil TN as potentially important contributors to yield. Of these, the negative influence of weeds in certain parts of the field is easiest to distinguish. There is an interaction indicated between clay content and soil P, as well as possibly soil pH. This is also evident in the scatterplot matrices of the two regions of the field. Although inclusion of an interaction between *SoilP* and *SoilpH* would be biochemically justifiable, there is no evidence that such a term would be significant in a regression model (Exercise 8.6a).

The classification tree analysis in Chapter 15 of the spatiotemporal data clusters indicated that clay content was important throughout the four years (Figure 15.15a). The areas of higher clay content, coupled with either low soil P or high weed levels in 1996, had the lowest yield. Yields in the lower and higher yielding area over the four years did, however, tend to become more uniform (Figure 15.2), and by the fourth year, soil P level had apparently ceased to become as dominant in determining yield. This will be addressed below. There was some indication that soil pH and soil K content were associated with yield in the high clay region in 1999, with areas of lower pH and higher K being associated with a higher yield.

The evidence points to a tentative conclusion that clay content is a dominant factor in yield, that soil phosphorous content and possibly soil pH also play a role, and that weed level is important in certain areas and years, particularly in the first year along the eastern boundary of the field and in a small region along the south central part of the western boundary. In the northern part of the field, where clay content is high, aeration stress presumably limits yield. The evidence indicates that soil phosphorous level is also limiting in the far north. In the southern part of the field, the negative association between yield and soil phosphorous indicates that the crop is mining this immobile nutrient. In the southwestern corner, where the soil acidity is closer to neutral, presumably increasing P availability, the yield is highest. There is some indication that disease and either soil potassium level or soil nitrogen level or both may play a role in limiting yield, but the evidence is not conclusive. In this context, it is important to recall a comment that was made in Chapter 9 regarding the difference between regression methods and partitioning methods. This is that regression methods summarize process over the geographical extent of the region being studied, while partitioning methods as the partitioning advances increasingly restrict the analysis to separate geographical areas.

Now let's examine the data from Field 4.2 and see how it relates to these conclusions. Although the two fields are similar in location, management, and presumably properties, it would be naïve to simply plug the data from Field 4.2 into the model and see what comes out. First, we must compare the fields a bit to see how their response to the various yield limiting factors might differ. We begin with the soil. Throughout the analyses of the previous chapters, we have found evidence that soil texture plays an important role in Field 4.1. What is the role of texture in Field 4.2? The maximum and minimum values of *Clay* in the two fields are almost identical.

```
> range(data.Set4.1$Clay)
[1] 23.20 46.91
> range(data.Set4.2$Clay)
[1] 23.2 46.1
```

The distributions, however, are very different.

```
> stem(data.Set4.1$Clay)        > stem(data.Set4.2$Clay)
  22 | 279                        22 | 2
  24 | 68                         24 | 016
  26 | 559934578                  26 | 06
  28 | 0338                       28 | 50144
  30 | 5                          30 | 022788334455556777899
  32 | 2                          32 | 11238813679
  34 | 483456                     34 | 13338804557
  36 | 245670011                  36 | 03552689
  38 | 1145678935                 38 | 4689911258
  40 | 1225567801244555999        40 | 4493
  42 | 44456677790355678          42 | 9
  44 | 0135                       44 |
  46 | 9                          46 | 1
```

The *Clay* distribution in Field 4.1 is bimodal, whereas in Field 4.2 it is unimodal and positively skewed. Also, wheat yield in the 1996 harvest is much lower in the high clay area of Field 4.1 than it is in Field 4.2.

There is also a considerable difference in the range of *SoilP* levels between the fields.

```
> range(data.Set4.1$SoilP)
[1]  2.48 18.14
> range(data.Set4.2$SoilP)
[1]  13.5 52.3
```

There is nowhere in Field 4.2 even remotely close to the P application threshold of 6 ppm. The one variable among those in which we are interested with similar distributions between the two fields is *SoilpH*.

```
> stem(data.Set4.1$SoilpH)             stem(data.Set4.2$SoilpH)
  The decimal point is 1 digit(s)        The decimal point is 1 digit(s)
to the left of the |                   to the left of the |
   55 | 6                                 54 | 4
   56 | 001                               54 | 9
   56 | 555889                            55 | 000111234444
   57 | 000013344444                      55 | 5567777888899
   57 | 5555556777788999999               56 | 000000122222233333444
   58 | 00222                             56 | 55566788889
   58 | 44444677888899                    57 | 111111234
   59 | 11112                             57 | 55689
   59 | 5555667888889                     58 | 1
   60 | 0223                              58 | 7
   60 | 499                               59 | 2
   61 | 2
```

Field 4.2 is somewhat more acidic than Field 4.1.

During the actual study, standing water was observed in the north end of Field 4.1, which is a clear indication that the high level of association of clay content with yield was due to aeration stress, which was caused by poor drainage. There is only a very small isolated area of high clay content in Field 4.2, and the entire field is composed of well-drained soils (Andrews, 1972), so aeration stress should not be not a problem. Similarly, the high soil P levels in Field 4.2 likely preclude a relationship of yield with soil P. Therefore, we must conclude that the model of yield response to explanatory variables developed with the data of Field 4.1 cannot be extended to Field 4.2. Unfortunately, we were not privy to all of the cooperator's management decisions over the years, so we cannot tell whether the increased uniformity of yield is due to changed management practices. It may have also been due to the slightly higher temperatures in later years, as well as to the greater control that the farmer was able to exert over summer crops. A classification tree for the two clusters in Field 4.2 splits first (unsurprisingly) on *Weeds* $\geq$ 4.5 and then (surprisingly) on *SoilpH* <= 5.665. This is surprising because the more neutral soils are in the lower yielding cluster. Acidic soils are more commonly associated with reduced wheat yields. The area of reduced wheat yield associated with low soil pH is at the eastern end of the field. Unlike Field 4.1, it is unlikely that there is any substantial effect of acidity on P availability because P levels are so high. A scatterplot of wheat yield against soil pH reveals no trend, so it is likely that the association of these two quantities identified by the classification tree is spurious. There is definitely a pattern of lower wheat yields in the eastern end of the field, but this is not associated with anything measured in our study.

In the actual event, after he saw our first year's results, the cooperating farmer performed his own phosphate trial in the northern part of Field 4.1. He found that there was indeed a substantial yield response to added phosphate fertilizer. Increased P application in later years may have contributed to the increased uniformity. In that sense, the methodology worked as it was supposed to in Field 4.1. Field 4.2 turned out to be fairly uniform in its properties and is probably one of those fields for which the null hypothesis of precision agriculture (Whelan and McBratney, 2000) is valid, and uniform application of inputs is appropriate in this field.

17.6 Conclusions

The most important general conclusion that can be drawn from this text is that one should never rely on a single method to analyze any data set, spatial or otherwise. For each of the data sets we studied, every method applied to that data set provided some information, and no one method was sufficient by itself to provide all the information we needed to reach a conclusion. Data analysis is a nonlinear process. This is often not apparent in the publication of the results, and indeed the linear order of the chapters of this book may serve to reinforce the incorrect idea that data analysis proceeds from sampling to exploration to confirmation to conclusions. This is not the way a data set should be analyzed. Exploration should motivate modeling, and modeling should motivate further exploration. This is why you have to live with a data set for a long time to really come to know it.

Data from an observational study can never be relied on by themselves to produce a model of cause and effect. The best that can be accomplished with such data is to produce alternatives that can be parsed using biophysical arguments or controlled experiments. Sometimes, however, observational studies provide the only data that can be obtained. In the case of the Data Set 1, the yellow-billed cuckoo is an endangered species with a very small remaining habitat, and so it is doubtful that any replicated experiment could be undertaken that would contribute much to understanding the bird's behavior. In the case of Data Set 2, there have been numerous, highly informative replicated experiments that helped to understand blue oak seedling and sapling development. The distribution of mature blue oaks is more difficult to study in this way because trees of this species are extremely slow growing, and can remain in the seedling or sapling stage for decades before maturing (Phillips et al., 2007). This is a good example of a problem in which experimental data and observational study data must be combined to extend the scope and interpretation of the experiments.

Chamberlain (2008), in an editor's message that every ecologist (and every reviewer of ecological papers) should read, exhorts scientists not to "sacrifice biology on the altar of statistics." A research project always starts with a scientific question. Data are analyzed in a search for patterns that may help to provide an answer. One question (but not the only question) to ask about these patterns is how likely they are to have appeared by chance. This question is addressed via a probabilistic model. One must always keep in mind, however, that the most important objective is *not* to estimate the probability that the pattern in the data occurred by chance. It is to answer the scientific question that motivated the data collection in the first place. The problem with the use of a probabilistic model is that it lends an aura of credibility to the result. If the data do not satisfy the assumptions of the model, then this credibility may not be deserved. A more mathematically sophisticated model provides a greater aura of credibility and therefore a greater risk if its credibility is undeserved. A simpler model, on the other hand, is easier to understand and use but often has more restrictive underlying assumptions that are likely to be incorrect. For this reason, in selecting the most appropriate statistical model, the data analyst must balance the advantages of the simpler model against those of the more complex one. To do this correctly requires an understanding of the theory behind both models. In the words of Benjamin Bolker (2008, p. 220), "you have to learn all the rules, but also know when to bend them."

Appendix A: Review of Mathematical Concepts

A.1 Matrix Theory and Linear Algebra

A.1.1 Matrix Algebra

A *matrix* is an array of elements arranged in rows and columns. For example, the matrix

$$A = \begin{bmatrix} 3 & 17 \\ 21 & 0 \\ 4 & 16 \end{bmatrix} \tag{A.1}$$

has three rows and two columns and is referred to as a 3×2 matrix. The symbol a_{ij} is used to denote the element in row i and column j, so that a general 3×2 matrix may be written

$$A = \begin{bmatrix} a_{11} & a_{12} \\ a_{21} & a_{22} \\ a_{31} & a_{32} \end{bmatrix}. \tag{A.2}$$

A vector with n elements (or components) may be either an $n \times 1$ matrix, which is a *column vector*, or a $1 \times n$ matrix, which is a *row vector*. A matrix with the same number of rows and columns is called a *square matrix*.

The product of multiplication of a matrix A and a scalar c is defined as the matrix whose elements are ca_{ij}. The sum of two matrices A and B is defined only if they have the same number of rows and columns, in which case the elements of $A + B$ are $a_{ij} + b_{ij}$. The product of two matrices AB is defined only if B has the same number of rows as A has columns. In this case the elements of AB are obtained by adding the products across the rows of A and down the columns of B. For example, suppose

$$A = \begin{bmatrix} a_{11} & a_{12} \\ a_{21} & a_{22} \end{bmatrix}, B = \begin{bmatrix} b_{11} & b_{12} \\ b_{21} & b_{22} \end{bmatrix}. \tag{A.3}$$

Then

$$AB = \begin{bmatrix} a_{11}b_{11} + a_{12}b_{21} & a_{11}b_{12} + a_{12}b_{22} \\ a_{21}b_{11} + a_{22}b_{21} & a_{21}b_{12} + a_{22}b_{22} \end{bmatrix}. \tag{A.4}$$

In general, $AB \neq BA$. To see this, make up a pair of 2×2 matrices whose elements are the numbers 1–8 and compute the products.

The *identity matrix* is the matrix I that satisfies $AI = IA = A$ for any square matrix

A. The 3×3 identity matrix, for example, is

$$I = \begin{bmatrix} 1 & 0 & 0 \\ 0 & 1 & 0 \\ 0 & 0 & 1 \end{bmatrix}. \tag{A.5}$$

If A is a square matrix, then the *inverse* of A, A^{-1}, if it exists, is the matrix that satisfies $AA^{-1} = A^{-1}A = I$. If the inverse of a matrix does not exist, the matrix is said to be *singular*. To determine whether a matrix A is singular, one computes a quantity called the *determinant*, denoted det A (Noble and Daniel, 1977, p. 198). The matrix A is singular if and only if det $A = 0$. The *transpose* of the matrix A whose elements are a_{ij} is the matrix A' whose elements are a_{ji}. A matrix is *symmetric* if $A' = A$, i.e., if $a_{ij} = a_{ji}$ for all i and j.

A.1.2 Random Vectors and Matrices

A random vector or matrix is a vector or matrix that contains random variables. Let

$$Y = \begin{bmatrix} Y_1 \\ Y_2 \\ \cdots \\ Y_n \end{bmatrix}. \tag{A.6}$$

be a random vector. Then the expected value $E\{Y\}$ is given by

$$E\{Y\} = \begin{bmatrix} E\{Y_1\} \\ E\{Y_2\} \\ \cdots \\ E\{Y_n\} \end{bmatrix}, \tag{A.7}$$

and the variance-covariance matrix var$\{Y\}$ is given by

$$\text{var}\{Y\} = E\{(Y - E\{Y\})(Y - E\{Y\})'\}$$

$$= \begin{bmatrix} \text{var}\{Y_1\} & \text{cov}\{Y_1, Y_2\} & \cdots & \text{cov}\{Y_1, Y_n\} \\ \text{cov}\{Y_2, Y_1\} & \text{var}\{Y_2\} & \cdots & \text{cov}\{Y_2, Y_n\} \\ \cdots & \cdots & \cdots & \cdots \\ \text{cov}\{Y_n, Y_1\} & \text{cov}\{Y_n, Y_2\} & \cdots & \text{var}\{Y_n\} \end{bmatrix}. \tag{A.8}$$

Since $\text{cov}(Y_i, Y_j) = \text{cov}\{Y_j, Y_i\}$, the matrix var$\{Y\}$ is symmetric.

Suppose $W = AY$ where Y is a random vector and A is a fixed matrix. Then the expected value and variance-covariance matrix of W are given by

$$E\{W\} = E\{AY\} = AE\{Y\}$$

$$\text{var}\{W\} = \text{var}\{AY\} = A \, \text{var}\{Y\}A'. \tag{A.9}$$

These same relationships hold for the sample mean, sample variance, and sample covariance.

Now, suppose that instead of a single data field Y, we have a set of k data fields denoted Y_j, $i = 1,...,k$. Each data field is composed of n data records denoted Y_{ij}, $i = 1,...,n$. The set of all the data is represented by a matrix Y whose columns are the data fields Y_j and whose rows are the data records, that is,

$$Y = \begin{bmatrix} Y_{11} & Y_{12} & \cdots & Y_{1k} \\ Y_{21} & Y_{22} & \cdots & Y_{2k} \\ \cdots & \cdots & \cdots & \cdots \\ Y_{n1} & Y_{n2} & \cdots & Y_{nk} \end{bmatrix}. \tag{A.10}$$

We can write the sample variance-covariance matrix S_Y as

$$S_Y = \begin{bmatrix} \text{var}\{Y_1\} & \text{cov}\{Y_1, Y_2\} & \cdots & \text{cov}\{Y_1, Y_k\} \\ \text{cov}\{Y_2, Y_1\} & \text{var}\{Y_2\} & \cdots & \text{cov}\{Y_2, Y_k\} \\ \cdots & \cdots & \cdots & \cdots \\ \text{cov}\{Y_k, Y_1\} & \text{cov}\{Y_k, Y_2\} & \cdots & \text{var}\{Y_k\} \end{bmatrix}, \tag{A.11}$$

where the variances and covariances are now interpreted as sample statistics rather than population parameters. As an example, suppose that a data set consists of two data fields (say, mean annual rainfall and mean annual temperature) denoted by Y_1 and Y_2, and suppose that each quantity is measured at 30 locations. Suppose further that Y_1 is a normally distributed random variable and Y_2 is related to Y_1 according to the equation $Y_2 = 2Y_1 + 1.3 + \varepsilon$, where ε is normally distributed. The following code generates artificial data representing these data fields (please ignore the fact that the numerical values don't make any sense as measures of rainfall and temperature) and centers them so that their sample means are 0 (this is done for use in Section A.1.3). It then computes the sample variances and covariance and the covariance matrix in Equation A.11.

```
> set.seed(123)
> # Generate the data
> Y1 <- rnorm(30)
> Y2 <- 2 * Y1 + 1.3 * rnorm(30)
> # Center the variables
> Y1 <- Y1 - mean(Y1)
> Y2 <- Y2 - mean(Y2)
> var(Y1)
[1] 0.9624212
> var(Y2)
[1] 4.357569
> cov(Y1,Y2)
[1] 1.757146
> Y <- cbind(Y1,Y2)
> var(Y)
           Y1        Y2
Y1 0.9624212 1.757146
Y2 1.7571458 4.357569
```

A.1.3 Eigenvectors, Eigenvalues, and Projections

In a two-dimensional Cartesian coordinate system, every point in a region is described by a pair of coordinates that we can call x and y. Similarly, in a three-dimensional system we can add a third coordinate z. This notion of coordinates can be extended to general vectors through the use of *coordinate vectors*. In order to visualize this concept, we will illustrate it with the data fields Y_1 and Y_2 created in the previous subsection. Figure A.1a shows a scatterplot of the two data fields.

In order to avoid confusion, with too many symbols containing a Y, we will denote the ith data record by P_i. For example, data record 11, denoted P_{11}, is shown as a black dot in Figure A.1.a. This record has components (to three decimal places) $Y_{11,1} = 1.271$, $Y_{11,2} = 1.407$. Also shown in the scatterplot are two unit vectors u_1 and u_2. These unit vectors have coordinates $(1,0)$ and $(0,1)$, respectively. If we think of the data record P_{11} as a vector in this coordinate plane (shown in Figure A.1a as the arrow whose tip is at $Y_{11,1} = 1.271$, $Y_{11,2} = 1.407$), then this vector can be written in terms of the unit vectors u_1 and u_2 as (Figure A.1a)

$$P_{11} = 1.271u_1 + 1.407u_2. \tag{A.12}$$

Thus, there are two ways to interpret a vector. The first is as an array with one column, as in Equation A.6, and the second is as an arrow in a *vector space* like that of Figure A.1a. In this example, the vector as an array has two elements and is written

$$P_{11} = \begin{bmatrix} 1.271 \\ 1.407 \end{bmatrix}, \tag{A.13}$$

while the vector as an arrow is written in the form of Equation A.12. In the interpretation of vectors as arrows, the sum of two vectors is the arrow obtained by placing the tail of one arrow in the summand on the head of the other. That the representation of P_{11} in the form

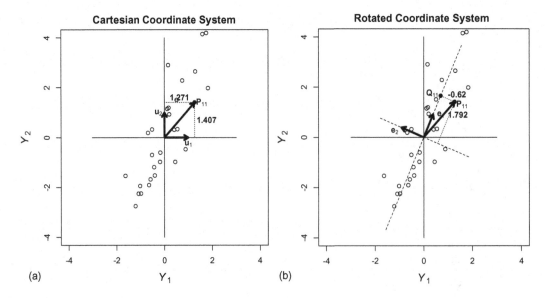

FIGURE A.1
(a) A vector expressed in Cartesian coordinates; (b) the same vector in a rotated coordinate system.

of Equation A.12 is equivalent to the representation in the form of Equation A.13 follows from the fact that we can write

$$u_1 = \begin{bmatrix} 1 \\ 0 \end{bmatrix}, u_2 = \begin{bmatrix} 0 \\ 1 \end{bmatrix}, \tag{A.14}$$

in which case Equation A.13 is obtained from Equation A.12 by a combination of multiplication by a scalar and matrix addition. The dual interpretation of a vector as an array or as an arrow in a vector space does not depend on the number of elements in the vector. If the vector as an array has more than three elements, then we can no longer visualize the corresponding vector space in which the vector as an arrow lives, but as an abstract mathematical entity it is equally valid.

We can rotate the unit vectors u_1 and u_2 as shown in Figure A.1b and define the vector P_{11} (or any other vector) in terms of a new rotated coordinate system, which is denoted e_1 and e_2 in the figure. We will continue to work in two dimensions; but all of our results generalize readily to three or more dimensions. Suppose we wish to represent the vector P_{11} in terms of the rotated coordinate system composed of e_1 and e_2 as $P_{11} = a_1 e_1 + a_2 e_2$. It turns out that if the coordinate vectors e_1 and e_2 are obtained by rotating through an angle θ counterclockwise from the original coordinate vectors u_1 and u_2, then the values of a_1 and a_2 are obtained by multiplying the array in Equation A.13 by a *rotation matrix* that is expressed as (Noble and Daniel, 1977, p. 282).

$$W = \begin{bmatrix} \cos\theta & \sin\theta \\ -\sin\theta & \cos\theta \end{bmatrix}, \tag{A.15}$$

For example, the coordinate vectors e_1 and e_2 in Figure A.1b are obtained from u_1 and u_2 by rotating through an angle of approximately $\theta = 67°$ (the reason for choosing this particular value for θ will be explained later). We have $\cos\theta = 0.3906338$ and $\sin\theta = 0.9205462$. Therefore, the vector P_{11} is expressed as follows:

```
> W
            [,1]           [,2]
[1,]   0.3906338    0.9205462
[2,]  -0.9205462    0.3906338
> print(P11 <- Y[11,])
       Y1            Y2
1.271186  1.407412
> W %*% P11
            [,1]
[1,]   1.7921559
[2,]  -0.6204022
```

Thus, $P_{11} = 1.792 e_1 - 0.62 e_2$. This is shown in Figure A.1b.

Because the vector P_{11} represents a data record in a sample composed of the data fields Y_1 and Y_2, there is a particular rotation that has special significance. This rotation has to do with the sample variance-covariance matrix. The variance-covariance matrix S_Y of the sample in this example is defined in Equation A.11 and is computed in the code given at the end of Section A.1.2. Like all variance-covariance matrices, it is symmetric. Suppose we wish to compute two new variables, denoted C_1 and C_2, that are linear combinations of Y_1 and Y_2 but are uncorrelated. That is, we would like to determine coefficients a_{ij} such that

$$C_{i1} = a_{11}Y_{i1} + a_{12}Y_{i2}$$
$$C_{i2} = a_{21}Y_{i1} + a_{22}Y_{i2}, i = 1, 2, \ldots, n,$$
(A.16)

so that the covariance between the vectors C_1 and C_2 is zero. In matrix theoretical terms, this says that the variance-covariance matrix S_C of the matrix C whose columns are these two new variables is a diagonal matrix. If we consider any data record $P_i = [Y_{i1} \, Y_{i2}]'$ (remember that the prime denotes transpose, and that P_i is a column vector) as a vector in the array sense, then we can write Equation A.16 as

$$C_i = AP_i, i = 1, \ldots, n$$
(A.17)

where A is a 2×2 matrix and C_i is a vector. For example, if for P_i we use the value P_{11} from our artificial data set, then we could write

$$\begin{bmatrix} C_{11,1} \\ C_{11,2} \end{bmatrix} = \begin{bmatrix} a_{11} & a_{12} \\ a_{21} & a_{22} \end{bmatrix} \begin{bmatrix} Y_{11,1} \\ Y_{11,2} \end{bmatrix} = \begin{bmatrix} a_{11} & a_{12} \\ a_{21} & a_{22} \end{bmatrix} \begin{bmatrix} 1.271 \\ 1.407 \end{bmatrix}.$$
(A.18)

One has to be a little careful of the subscripts here, because they have different uses. The subscript 11 of P_{11} and the $Y_{11,i}$ indicates that this is the eleventh data record. The subscripts 1 and 2 of $Y_{11,1}$ and $Y_{11,2}$ indicate that these are the first and second values of the eleventh data record (for example, they might represent the centered values of mean annual rainfall and mean annual temperature of the eleventh data record). The subscripts i and j of the elements a_{ij} represent the position of the element in the matrix A.

Returning to our problem, we want to determine a matrix A such that the variance-covariance matrix S_C of the data set expressed in the new variables C_1 and C_2 given by Equation A.16, or equivalently by Equation A.17, is a diagonal matrix. It turns out (Noble and Daniel, 1977, p. 272) that if a matrix A satisfying Equation A.17 exists such that the covariance matrix S_C is a diagonal matrix, then the matrices S_C, S_Y, and A are related by the equation

$$S_C = A'S_Y A.$$
(A.19)

It further turns out that because the variance-covariance matrix S_Y is symmetric, the matrix A satisfying Equation A.19 is guaranteed to exist. It further turns out that the matrix A satisfies

$$A^{-1} = A',$$
(A.20)

that is, its inverse is equal to its transpose. The properties represented in Equations A.19 and A.20 are a consequence of the symmetry of S_Y, and since all variance-covariance matrices are symmetric, these properties hold for any data set, no matter how many variables it contains.

The diagonal elements of S_C in Equation A.19 are called the *eigenvalues* of S_Y, and the columns of A are called the *eigenvectors*. *Eigen* is a German word that can be translated as "particular." "Eigenvalue" and "eigenvector" are hybrid German-English words that are usually rendered in English as "characteristic value" and "characteristic vector." In general, if there is a nonsingular matrix A such that the matrices B and D satisfy

$$D = A^{-1}BA, \qquad (A.21)$$

then B and D are said to be *similar*. If B is any matrix and D is a diagonal matrix, then the diagonal elements of D are the eigenvalues of B. Similarity is an equivalence relation, so any two matrices that are similar to each other have the same eigenvalues (Noble and Daniel, 1977, p. 279).

Continuing with the example begun earlier in this section, the R function `eigen()` computes the eigenvalues and eigenvectors of a matrix.

```
> print(SY <- var(P))
           Y1          Y2
Y1  0.9624212  1.757146
Y2  1.7571458  4.357569
> print(eigen.SY <- eigen(SY))
$values
[1]  5.1032143  0.2167763
$vectors
            [,1]          [,2]
[1,]  0.3906338  -0.9205462
[2,]  0.9205462   0.3906338
> A <- eigen.SY$vectors
> print (Sc <- t(A) %*% SY %*% A)
             [,1]           [,2]
[1,]   5.103214e+00  -1.053845e-16
[2,]  -7.859110e-17   2.167763e-01
```

Within the roundoff error of the computation, the matrix S_C is a diagonal matrix whose elements are the eigenvalues of S_Y.

The graphical interpretation shown in Figure A.1b demonstrates the connection of eigenvalues and eigenvectors to the rotation of coordinate systems discussed above. Applying the matrix multiplication in Equation A.19 is equivalent to rotating the coordinate system from the solid axes to the dashed axes, so that the unit vectors in this new coordinate system are the eigenvectors e_1 and e_2. The matrix A in Equation A.19 is the same as the rotation matrix W in Equation A.15 and corresponds to a rotation of 67°. The following shows the first column of the rotation matrix in that equation.

```
> W <- A
> acos(W[1,1]) * 180 / pi
[1] 67.00606
> asin(W[2,1]) * 180 / pi
[1] -67.00606
```

The values of data record 11 in this new coordinate system are given by equation (A.16).

```
> print(P11 <- Y[11,])
       Y1          Y2
1.271186  1.407412
> print(C <- A %*% P11)
```

```
          [,1]
[1,]   1.7921559
[2,]  -0.6204022
```

This is displayed in Figure A.1b, where the vector P_{11} is written in terms of the unit vectors e_1 and e_2 as $P_{11} = 1.792e_1 - 0.620e_2$.

Figure A.1b appears to indicate that the axis provided by the eigenvector e_1 has the property that among all rotations of the axes, it is the linear combination of the two variables having the maximum variance. This is in fact the case. The variable C_1 has the maximum variance among the possible rotations. Moreover, it turns out that the first eigenvalue, whose value to three decimal places is 4.859, is the variance of C_1. The variable C_2 is uncorrelated with C_1, and its variance is given by the second eigenvalue, which is 0.135. The sum of these variances is equal to the sum of the variances of the original variables Y_1 and Y_2 as shown here.

```
> with(eigen.SY, values[1] + values[2])
[1] 5.319991
> with(Y, var(Y1) + var(Y2))
[1] 5.319991
```

Now suppose we want to study a simplified version of the data set $\{Y_1, Y_2\}$ that consists of only one variable. It makes sense to use a variable that lies along the e_1 axis in Figure A.1b, since this would incorporate the greatest amount of variation (i.e., of information) from the original pair of variables Y_1 and Y_2. Consider the point P_{11} above. The point on the line of the e_1 axis that lies closest to P_{11} is the point Q_{11} shown in FigureA.1b. This point is called the *orthogonal projection*, or simply the *projection* of P_{11} onto e_1.

A.2 Linear Regression

A.2.1 Basic Regression Theory

The objective of regression is to predict the value of a random variable Y (the *response variable*) given one or more variables (which may or may not be random) denoted $X_1, X_2, ..., X_{p-1}$. The X_j are variously called *predictors*, *carriers*, and *explanatory variables*. The reason we write $p-1$ instead of just p will become evident later. Since the objective of the analysis in this book is generally to gain an increased understanding of process rather than to do prediction, we will use the last term of the three above for the X_j. Each explanatory variable is a vector that may take on more than one value X_{ij}, *viz.*, $X_{i1}, X_{i2}, ... X_{i,p-1}$, $i = 1, ..., n$. Each explanatory variable represents a data field, and each value represents one record (i.e., one measurement) of that data field.

We begin by discussing the case in which there is only one explanatory variable, denoted X, taking on values $X_1, X_2, ..., X_n$. This case is called *simple linear regression*. The explanatory variable X may be either a random variable, or it may take on fixed, predetermined values. The latter case is called the *Gaussian case*, and when this case applies, X is called a *mathematical variable.* We will illustrate the difference between a mathematical variable and a random variable using three examples.

> *Example 1.* An agronomist conducts an experiment in which fertilizer is applied to corn at the rate of 0, 50, 100, and 150 kg ha^{-1}. The explanatory variable X is fertilizer rate and the response Y is corn yield.
>
> *Example 2.* The same agronomist interviews 20 farmers and asks them how much fertilizer they apply and what their yield is. X and Y play the same roles. Although X is not "controlled" by the experimenter, it is selected by the farmer.
>
> *Example 3.* Our indefatigable agronomist randomly selects 20 corn fields and measures both the total accumulated precipitation during the growing season and the corn yield in each field. Here the explanatory variable X is annual precipitation and the response Y again is the yield.

Example 1 is clearly a Gaussian case, in which X is a mathematical variable. Example 3 is clearly a case in which X is a random variable. In example 2, X may look like a random variable, but actually it is a mathematical variable. Although the agronomist did not select the fertilization rates, the farmers did select them, and therefore they are not random.

It is easiest to develop regression theory for the Gaussian case. We assume a relationship between X and Y of the form

$$Y_i = \beta_0 + \beta_1 X_i + \varepsilon_i, \ i = 1, ..., n \tag{A.22}$$

where the β_i are parameters, the X_i are known constants and the ε_i are random variables (the "error" terms). The assumptions made in classical simple linear regression are

1. The ε_i are uncorrelated
2. $E\{\varepsilon_i\} = 0$
3. $\text{var}\{\varepsilon_i\} = \sigma^2$ is constant for all i
4. The ε_i are normally distributed.

Assumption 4 is not necessary for the results derived in this section to be valid. It will be needed in the later sections, when we discuss hypothesis testing.

Given a set of points (X_i, Y_i), $i = 1, ..., n$, we want to fit a straight line

$$\hat{Y}_i = b_0 + b_1 X_i \tag{A.23}$$

such that the quantity $Q(b_0, b_1) = \sum_{i=1}^{n} (Y_i - \hat{Y}_i)^2$ is minimized. The quantities b_0 and b_1 called the *regression coefficients*, are the estimators of β_0 and β_1, respectively. Differentiating Q with respect to the b_i and setting the derivatives equal to zero yields, after some algebra (Kutner et al., 2005, p. 17), the *normal equations,*

$$nb_0 + b_1 \Sigma X_i = \Sigma Y_i$$

$$b_0 \Sigma X_i + b_1 \Sigma X_i^2 = \Sigma X_i Y \tag{A.24}$$

These equations can then be solved for b_0 and b_1 to yield

$$b_1 = \frac{\Sigma(X_i - \bar{X})(Y_i - \bar{Y})}{\Sigma(X_i - \bar{X})^2}$$

$$b_0 = \bar{Y} - b_1\bar{X} = \frac{1}{n}(\Sigma Y_i - b_1\Sigma X_i)$$

(A.25)

An estimator b_i of the parameter β_i is *unbiased* if $E\{b_i\} = \beta_i$. The Gauss-Markov theorem (Kutner et al., 2005, p. 18) states that the estimator $\check{Y}_i$ of Equation A.23, where b_0 and b_1 satisfy Equation A.25, is the best linear unbiased estimator (or BLUE), in the sense that it has minimum variance among all unbiased linear estimators. The derivation of Equation A.24 assumes that X is a mathematical variable (the Gaussian case). Suppose X is a random variable. If both X and Y are normally distributed (the so-called *bivariate normal* case), then it turns out (Sprent, 1969, p. 8) that one may write $E\{Y \mid X = X_i\} = \mu_Y + \frac{\rho\sigma_Y}{\sigma_X}(X_i - \mu_X)$. As a result, some students have the impression that the linear regression model with random X is only valid for bivariate normal data. That is not true (Dawes and Corrigan, 1974). Theil (1971, p. 102) provides a lucid discussion of linear regression when X is a random variable. The results are summarized by Kutner et al. (2005, p. 83) as follows. The statistical results associated with the regression model discussed here are mathematically valid for random X and Y if the following two conditions are met:

1. The conditional distributions of the Y_i, given X_i, are normal and independent with mean $\beta_0 + \beta_1 X_i$ and constant variance σ^2.
2. The X_i are independent random variables whose probability distribution does not depend on β_0, β_1, or σ^2.

The exploitation of the relationship between regression and correlation does require bivariate normal data. Of course, to say that a regression model is mathematically valid does not mean that it is ecologically appropriate. Theil (1971, p. 155) gives an excellent short discussion of some of the issues associated with interpretation of regression models, and Venables (2000) provides an excellent discussion in more detail.

Given a regression model of the form (Equation A.23), we define the *error* e_i by

$$e_i = Y_i - \hat{Y}_i = Y_i - (b_0 + b_1 X_i).$$

(A.26)

The *error sum of squares, SSE,* the *regression sum of squares, SSR,* and the *total sum of squares, SST* are then given by

$$SSE = \Sigma(Y_i - \hat{Y}_i)^2$$

$$SSR = \Sigma(\hat{Y}_i - \bar{Y})^2$$

$$SST = \Sigma(Y_i - \bar{Y})^2.$$

(A.27)

These sums of squares satisfy an *orthogonality relationship* (Kutner et al., 2005, p. 65)

$$SST = SSE + SSR.$$

(A.28)

The *coefficient of determination,* which is generally denoted r^2 for simple linear regression, is defined by

$$r^2 = \frac{SSR}{SST}.$$

(A.29)

From Equations A.27 and A.29, it follows that $0 \le r^2 \le 1$ and a high value of r^2 is considered to imply a "good fit," that is, a high level of linear association between X and Y. The *mean squared error, MSE,* is defined for simple linear regression by

$$MSE = \frac{SSE}{n-2}.$$

(A.30)

It turns out (Kutner et al., 2005, p. 68) that the MSE is an unbiased estimator of the error variance σ^2, that is,

$$E\{MSE\} = \sigma^2.$$

(A.31)

A.2.2 The Matrix Representation of Linear Regression

The equations characterizing linear regression can be expressed very cleanly in matrix notation. Let

$$Y = \begin{bmatrix} Y_1 \\ \dots \\ Y_n \end{bmatrix}, \; \beta = \begin{bmatrix} \beta_0 \\ \beta_1 \end{bmatrix}, \; X = \begin{bmatrix} 1 & X_1 \\ \dots & \dots \\ 1 & X_n \end{bmatrix}, \; \varepsilon = \begin{bmatrix} \varepsilon_1 \\ \dots \\ \varepsilon_n \end{bmatrix}.$$

(A.32)

The matrix X is called the *design matrix.* With the definitions of Equation A.32, Equation A.22 becomes

$$Y = X\beta + \varepsilon,$$

(A.33)

and Equations A.25, the normal equations, become

$$X'Xb = X'Y.$$

(A.34)

Therefore, assuming $(X'X)^{-1}$ exists, Equations A.25 can be written

$$b = (X'X)^{-1}X'Y.$$

(A.35)

Finally, if we let $\hat{Y} = Xb$, then we have

$$\hat{Y} = Xb = X(X'X)^{-1}X'Y \equiv HY.$$

(A.36)

Because it puts a hat on Y, the matrix H is called the *hat matrix.*

Now that we have matrix notation, it is a very simple manner to extend the theory to cover multiple regression. Consider the multiple regression model

$$Y_i = \beta_0 + \beta_1 X_{i1} + \beta_2 X_{i2} + \dots + \beta_{p-1} X_{i,p-1} + \varepsilon_i, \; i = 1, \dots, n$$

(A.37)

If we replace Equation A.32 with

$$Y = \begin{bmatrix} Y_1 \\ ... \\ Y_n \end{bmatrix}, \beta = \begin{bmatrix} \beta_0 \\ \beta_1 \\ ... \\ \beta_{p-1} \end{bmatrix}, X = \begin{bmatrix} 1 & X_{11} & ... & X_{1,p-1} \\ 1 & X_{21} & ... & X_{2,p-1} \\ ... & ... & ... & ... \\ 1 & X_{n1} & ... & X_{n,p-1} \end{bmatrix}, \varepsilon = \begin{bmatrix} \varepsilon_1 \\ ... \\ \varepsilon_n \end{bmatrix}, \quad (A.38)$$

then all of the matrix regression equations still hold without modification. By enumerating the index of the explanatory variables from 1 to $p-1$, we have p regression coefficients, counting β_0. In the artificial rainfall-temperature data set caricatured in Section A.1, the data from the point P_{11} would be entered into the design matrix X as $X_{11,1}$ and $X_{11,2}$. It is important to add parenthetically that the term "linear regression" means that the regression is linear in the regression coefficients β_i, not necessarily in the explanatory variables X_i. Any of the explanatory variables in Equation A.37 may consist of powers or products of other explanatory variables.

Although the regression model represented by Equations A.32 through A.37 provides the best linear unbiased estimates b_i, given the data, of the regression coefficients β_i in Equation A.33, there may be problems with the data that cause it to be a poor representation of reality. Some of these problems are discussed in Chapter 8. One problem, which we discuss in Chapter 6, occurs when either one or more of the explanatory variables X or the response variable Y take on extreme values. A number of statistical measures have been devised to identify data records that have extreme values, and these are discussed in the next subsection.

A.2.3 Regression Diagnostics

One can always compute regression coefficients using Equation A.35, but the quality of the regression model as a representation of reality relies on the quality of the data as a representation of that same reality. Data are never perfect, however, and therefore some measures are required that provide an indication of whether the departures from perfection of a particular data set are sufficient to seriously reduce the usefulness of the regression model.

Although it is a bit of an oversimplification, we will in this subsection refer to a data record (X_i, Y_i) as an *outlier* if it has an extreme Y value, as a *leverage point* if it has an extreme X value (even if the corresponding Y value is consistent with the regression), and as an *influence point* if it is either an outlier or a leverage point. Outliers are discussed more fully in Section 6.2.1. Diagnostic statistics that identify influence points are called *influence measures*. One of the simplest influence measures is the set of diagonals of the hat matrix H. From Equation A.37, $H = X(X'X)^{-1}X'$, so it only depends on X and not on Y. Thus, it cannot detect outliers (unusual Y values), but it can detect leverage points (unusual X values). It turns out (Kutner et al., 2005, p. 392) that the ith diagonal h_{ii} of the hat matrix provides a measure of the influence of X_i on the regression. It further turns out that the h_{ii} satisfy $\Sigma h_{ii} = p$, so a "typical" value of h_{ii} should be roughly equal to p/n. The R function influence. measures() identifies data records with an h_{ii} value greater than $3p/n$ as influential.

In addition to the hat matrix, the matrix $C = (X'X)^{-1}$ is also useful in regression diagnostics. This stems from the fact that the variance-covariance matrix of the vector b of regression coefficients is given by var$\{b\} = \sigma^2 C$ (Kutner et al., 2005, p. 227). The matrix C is used in two diagnostic statistics, each of which is computed using a similar procedure. This is to delete on a record-by-record basis each data record, compute a statistic using the data set with the record deleted, and compare this to the same statistic computed with the full data set. The notation for a quantity computed with record i deleted is a subscript with the i in parentheses. Thus, for example $X_{(i)}$ denotes the design matrix X obtained by deleting data record i.

One diagnostic statistic computed from the matrix C is the *covariance ratio*, denoted *COVRATIO* (Belsley et al., p. 22). This is defined as

$$COVRATIO_i = \frac{MSE \det[(X'X)^{-1}]}{MSE_{(i)} \det[(X'_{(i)}X_{(i)})^{-1}]},$$ (A.39)

where $MSE_{(i)}$ is the value of MSE computed when the ith data record is deleted. As with other diagnostic measures, it turns out that the covariance ratio can be computed using a much simpler formula than that of the definition (Belsley et al., p. 22). The R function `influence.measures()` identifies data records with a covariance ratio value greater than $3p/n$ as influential.

The *DFBETAS* diagnostic statistic also makes use of the C matrix. It measures the influence of each data record on each regression coefficient (Kutner et al., 2005, p. 404). Let b_k denote the kth regression coefficient, $k = 0, \ldots, p - 1$, and let $b_{k(i)}$ denote the value of b_k when the ith data record is deleted. Let c_{kk} be the kth diagonal element of the matrix $(X'X)^{-1}$. Then the variance of b_k is $\text{var}\{b_k\} = \sigma^2 c_{kk}$ (Kutner et al. 2005, p. 404). This motivates the following definition for the *DFBETAS* statistic for the effect of data record i on regression coefficient b_k:

$$DFBETAS_{k(i)} = \frac{b_k - b_{k(i)}}{\sqrt{MSE_{(i)} c_{kk}}},$$ (A.40)

The R function `influence.measures()` identifies data records with a *DFBETAS* value greater than 1 as influential.

The *DFFITS* statistic measures the effect that data record i has on the fitted value $\hat{Y}_i$. It turns out that the variance of $\hat{Y}_i$ is $\text{var}\{\hat{Y}_i\} = \sigma^2 h_{ii}$, where h_{ii} is the ith diagonal of the hat matrix. This motivates the definition of $DFFITS_i$ as

$$DFFITS_i = \frac{\hat{Y}_i - \hat{Y}_{i(i)}}{\sqrt{MSE_{(i)} h_{ii}}},$$ (A.41)

where, as usual $\hat{Y}_{i(i)}$ is the value of $\hat{Y}_i$ computed when the ith data record is deleted. The R function `influence.measures()` identifies data records with a *DFFITS* whose absolute value is greater than $3\sqrt{p/(n-p)}$ as influential.

The *DFFITS* statistic measures the effect of data record i on the value of $\hat{Y}_i$. By contrast, the Cook's distance measures the effect of data record i on the entire regression model. The Cook's distance is defined as (Kutner et al., 2005, p. 402)

$$D_i = \frac{\sum_{j=1}^{n} (\hat{Y}_{ji} - \hat{Y}_{j(i)})^2}{pMSE}.$$ (A.42)

The R function `influence.measures()` identifies as influential data records with a Cook's distance for which the percentile of the distribution of the F percentile with p and $n-p$ degrees of freedom has a value greater than 0.5.

To illustrate the calculation influence measures, we will generate two normally distributed random variables according to the following code.

```
> set.seed(123)
> X1 <- rnorm(30)
> X2 <- 2 * X1 + 1.3 * rnorm(30)
```

Figure A.2a shows a plot of Y vs. X together with the least squares regression line. Applying the function influence.measures() to this regression yields the following.

```
> YX.lm <- lm(Y ~ X)
> im.lm <- influence.measures(YX.lm)
> im.lm$is.inf
    dfb.1_ dfb.X dffit cov.r cook.d hat
1   FALSE FALSE FALSE FALSE FALSE FALSE
2   FALSE FALSE FALSE FALSE FALSE FALSE
3   FALSE FALSE FALSE FALSE FALSE FALSE
4   FALSE FALSE FALSE FALSE FALSE FALSE
5   FALSE FALSE FALSE FALSE FALSE FALSE
6   FALSE FALSE FALSE FALSE FALSE FALSE
7   FALSE FALSE FALSE FALSE FALSE FALSE
8   FALSE FALSE FALSE FALSE FALSE FALSE
9   FALSE FALSE FALSE FALSE FALSE FALSE
10  FALSE FALSE FALSE FALSE FALSE FALSE
11  FALSE FALSE FALSE FALSE FALSE FALSE
12  FALSE FALSE FALSE FALSE FALSE FALSE
13  FALSE FALSE FALSE FALSE FALSE FALSE
14  FALSE FALSE FALSE FALSE FALSE FALSE
15  FALSE FALSE FALSE FALSE FALSE FALSE
16  FALSE FALSE FALSE FALSE FALSE FALSE
17  FALSE FALSE FALSE FALSE FALSE FALSE
18  FALSE FALSE FALSE  TRUE FALSE FALSE
19  FALSE FALSE FALSE FALSE FALSE FALSE
20  FALSE FALSE FALSE FALSE FALSE FALSE
```

Point 18, which lies very close to the regression line, is identified as a potential influence point by the covariance ratio, which indicates that deleting this point results in a noticeable change in the variance-covariance matrix of the regression coefficients. This (presumably) false positive shows that not every data record identified as suspicious should be automatically treated as an influence point.

As a first example of an influence point, we create an outlier, but one whose X value is close to $\bar{X}$. Here is the code.

```
> set.seed(123)
> X <- rnorm(20)
> Y <- X + 0.5 * rnorm(20)
> Y[2] <- Y[2] + 2
> YX.lm <- lm(Y ~ X)
> im.lm <- influence.measures(YX.lm)
> im.lm$is.inf
    dfb.1_ dfb.X dffit cov.r cook.d hat
1   FALSE FALSE FALSE FALSE FALSE FALSE
2    TRUE FALSE  TRUE  TRUE FALSE FALSE
```

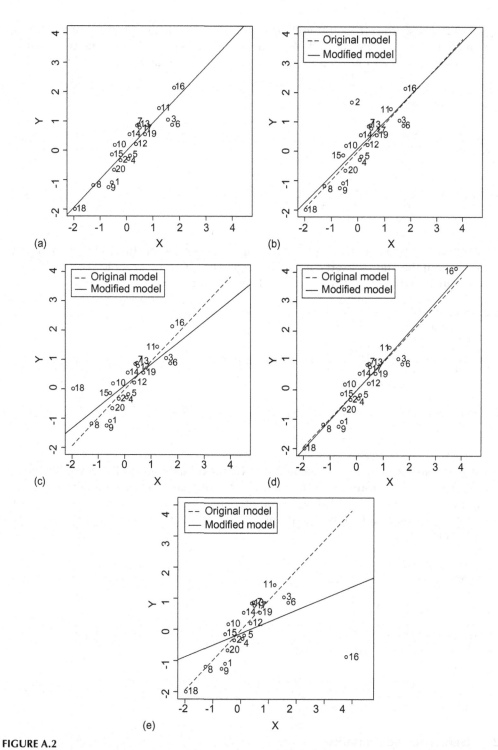

FIGURE A.2

A series of regression lines showing the effects of influence points: (a) no influence points; (b) an outlier near the median of the explanatory variables; (c) an influence point; (d) a point that is extreme in both the explanatory and response variables but consistent with the regression; (e) an outlier that is at an extreme value of the explanatory variables.

Figure A.2b shows the effect of this change: point 2 is clearly an extreme Y value, but the effect on the regression line is small. When a Y value corresponding to a more extreme X value is changed by the same amount, the effect is more dramatic (Figure A.2c).

```
> set.seed(123)
> X <- rnorm(20)
> Y <- X + 0.5 * rnorm(20)
> Y[18] <- Y[18] + 2
> YX.lm <- lm(Y ~ X)
> im.lm <- influence.measures(YX.lm)
> im.lm$is.inf
     dfb.1_ dfb.X dffit cov.r cook.d hat
       *     *     * DELETED    *     *     *
18 TRUE TRUE TRUE TRUE TRUE FALSE
```

This illustrates the reason for the term "leverage." As with a mechanical system, a point farther out from the middle exerts more "leverage" than a point closer to the middle. This is further illustrated in Figure A.2d, which is generated by the following code:

```
> set.seed(123)
> X <- rnorm(20)
> X[16] <- X[16] + 2
> Y <- X + 0.5 * rnorm(20)
> YX.lm <- lm(Y ~ X)
> im.lm <- influence.measures(YX.lm)
> im.lm$is.inf
     dfb.1_ dfb.X dffit cov.r cook.d hat
       *     *     * DELETED    *     *     *
16 FALSE TRUE TRUE TRUE FALSE TRUE
```

Point 16 is consistent with the regression model (the adjustment to the X value is made *before* the Y value is computed), so this point does not change the regression line much, although it is identified as an influence point. The alteration created by a data record that is both an outlier and a leverage point has the greatest impact of all of the changes (Figure A.2e).

```
> set.seed(123)
> X <- rnorm(20)
> Y <- X + 0.5 * rnorm(20)
> X[16] <- X[16] + 2
> Y[16] <- Y[16] - 3
> YX.lm <- lm(Y ~ X)
> im.lm <- influence.measures(YX.lm)
> im.lm$is.inf
     dfb.1_ dfb.X dffit cov.r cook.d hat
       *     *     *  DELETD    *     *     *
16 FALSE TRUE TRUE TRUE TRUE TRUE
```

A.2.4 Regression and Causality

Finally, it is important to note that a high level of linear association between X and Y does not imply a causal relationship between X and Y. This should be evident if the roles of X and Y in Example 2 in Section A.2.1 above are reversed. One would not expect that the fact that a strong linear relationship between the explanatory variable corn yield X and annual rainfall Y implies that the corn yield causes the rainfall. One of the more amusing examples of the misuse of correlation and regression to infer causality is the Theory of the Stork. There are a number of data sets, among them those of Box et al. (1978. p.8) and Höfer et al. (2004), that show a high level of linear association between the stork population in a given area and the birthrate in that area.

A.3 Nested Models and the General Linear Test

The *general linear test* is a means of testing a given null hypothesis involving the ordinary least squares regression model (Kutner et al. 2005, p. 72; Searle, 1971, p. 110). Suppose for example that, given a model of the form of Equation A.22, one wishes to test the null hypothesis $H_0 : \beta_1 = 0$ against the alternative hypothesis $H_a : \beta_1 \neq 0$. One regards Equation A.22 as the *full* model, denoted by the subscript F, and develops a model called the *restricted* model, denoted by the subscript R, in which the null hypothesis is satisfied. In our example the restricted model is

$$Y_i = \beta_0 + \varepsilon_i. \tag{A.43}$$

The restricted model of Equation A.43 is said to be *nested* in the full model of Equation A.22 because the model of Equation A.43 can be derived from the full model of Equation A.22 by restricting one or more of the parameters of the full model to certain values. In this particular example, the value of β_1 is restricted to $\beta_1 = 0$.

To carry out the general linear test, one computes the sums of squares of errors for both models and then computes the statistic

$$G = \frac{SSE_R - SSE_F}{df_R - df_F} \div \frac{SSE_F}{df_f}, \tag{A.44}$$

where SSE_i and df_i are the sums of squares of errors and the degrees of freedom of model i, $i = R, F$. If the null hypothesis is true, then the statistic G has an F distribution with $df_r - df_F$ and df_F degrees of freedom, and thus one can compute a p value based on this distribution. However, the statistic G has an F distribution only if the Y_i are independent (Kutner et al. 2005, p. 699). A consequence of the nonzero covariance structure of spatially autocorrelated errors is therefore that the general linear test on the sums of squares can no longer be used to test H_0. An alternative means of carrying out this test, the maximum likelihood method, remains valid when the response variables are not independent. This will be discussed in Section A.5.

A.4 The Method of Lagrange Multipliers

The Lagrange multiplier method is used to solve constrained optimization problems, that is, problems in which the objective is to find the maximum or minimum value of a scalar function $f(Y_1, Y_2, ..., Y_n)$ where the x_i are also required to satisfy some constraint of the form $g(Y_1, Y_2, ..., Y_n) = 0$. For simplicity, we will restrict the explanation to the case of two independent variables Y_1 and Y_2. The extension to the more general case is straightforward. We will only deal with the problem of finding *necessary* conditions for the maximum or minimum, and, to keep the explanation short, we will refer to this as "solving" the problem. The basic idea of the Lagrange multiplier is as follows (Sokolnikoff and Redheffer, 1966, p. 342). Suppose $f(Y_1, Y_2)$ is a scalar function of two variables whose derivatives in both variables exist everywhere, and suppose one wishes to find the values of Y_1 and Y_2 that maximize this function. A necessary (but not sufficient) condition that the values of Y_1 and Y_2 occur at a maximum of f is that they satisfy the equations

$$\frac{\partial f}{\partial Y_1} = 0$$

$$\frac{\partial f}{\partial Y_2} = 0. \tag{A.45}$$

Now suppose that we impose the condition that Y_1 and Y_2 maximize the function $f(Y_1, Y_2)$ subject to a constraint of the form $g(Y_1, Y_2) = 0$. To solve this, one can define a new function $F(Y_1, Y_2, \psi) = f(Y_1, Y_2) + \psi g(Y_1, Y_2)$. The term ψ is called a *Lagrange multiplier*. Since we require $g(Y_1, Y_2) = 0$, any values Y_1 and Y_2 that maximize $f(Y_1, Y_2)$ must also maximize $F(Y_1, Y_2, \psi)$. But $F(Y_1, Y_2, \psi)$ does not have any constraints on it; instead, it has the extra variable ψ. Therefore, the problem of finding necessary condition for a maximum of $f(Y_1, Y_2)$ subject to the constraint $g(Y_1, Y_2) = 0$ is equivalent to solving the unconstrained problem in three variables

$$\frac{\partial F}{\partial Y_1} = \frac{\partial f}{\partial Y_1} + \psi \frac{\partial g}{\partial Y_1} = 0$$

$$\frac{\partial F}{\partial Y_2} = \frac{\partial f}{\partial Y_2} + \psi \frac{\partial g}{\partial Y_2} = 0 \tag{A.46}$$

$$\frac{\partial F}{\partial \psi} = g(Y_1, Y_2) = 0.$$

Note that the constraint itself is recovered in the last equation. The problem of finding necessary conditions for a minimum of f subject to a constraint is solved in the same way since Equation A.46 is also a necessary constraint for a minimum. Thus, Lagrange multipliers can be used to convert a constrained optimization problem, which generally cannot be solved by simply setting the derivatives to zero, into an unconstrained optimization problem in one extra variable.

A.5 The Maximum Likelihood Method

A.5.1 The Likelihood Function

The method of maximum likelihood was worked out by R.A. Fisher over a period of years between 1912 and 1922 (Aldrich, 1997). Suppose there are a set of observations $Y_1, ... Y_n$, each having the same probability density $f(Y, \theta)$, where θ is a parameter or vector of parameters such as the expected value μ and/or variance σ^2. The joint probability distribution, when considered as a function of θ, is called the *likelihood function*, and written as

$$L(\theta|Y) = f(\theta, Y_1, ..., Y_n). \tag{A.47}$$

The values of the observations $Y_1, ..., Y_n$ are considered to be parameters of the function L.

As a very simple example, we will work out the likelihood function for the problem of estimating the mean and variance of a data set, so that the model is $Y_i = \mu + \varepsilon_i$, where the ε_i are independent, identically distributed random variables satisfying $\varepsilon_i \sim N(0, \sigma^2)$. Thus, in this case $\theta = (\mu, \sigma^2)$. At the very beginning, we run into a problem. This is that we wish to compute the likelihood function $L(\theta|Y)$, expressed in terms of the random variable Y, but the random variable whose distribution we know is ε, the error term. We therefore need to transform the variables of the probability distribution f in Equation A.47 from the ε variables to the Y variables. In the present case this is easy. The mean μ is fixed, and the ε_i are the only random variables in the model. Therefore, the Y_i have the same distribution as the ε_i, and the Y_i are also independent and normally distributed with mean μ and variance σ^2. The fact that the Y_i are independent permits us to write the likelihood as a product of normal density functions, but if they were not independent (as they are not in the spatial autocorrelation case), then in order to write the likelihood as a product we would have to transform the variables from the Y_i into the ε_i.

The probability density function for the random variable ε_i is

$$f(\varepsilon_i) = \frac{1}{\sqrt{2\pi\sigma^2}} \exp(-\varepsilon_i^2 / 2\sigma^2). \tag{A.48}$$

Since the ε_i are independent, the joint density function $f(\varepsilon_1, ..., \varepsilon_n)$ is given by

$$f(\varepsilon) = \prod_{i=1}^{n} f(\varepsilon_i, 0, \sigma^2). \tag{A.49}$$

where $f(\varepsilon_i, 0, \sigma^2)$ is the normal density function with mean 0 and variance σ^2. To express the likelihood in terms of Y, we substitute $\varepsilon_i = Y_i - \mu$ to get

$$L(\mu, \sigma^2 | Y) = f(Y | \mu, \sigma^2)$$

$$= \prod_{i=1}^{n} f(Y_i, \mu, \sigma^2) \tag{A.50}$$

$$= \frac{1}{(2\pi\sigma^2)^{n/2}} \exp(-\Sigma(Y_i - \mu)^2 / 2\sigma^2).$$

The fundamental principle of the maximum likelihood method is that estimates of the parameters of the model can be obtained by maximizing the likelihood function (Equation A.50). The likelihood function itself has a complex form. However, in the case of the normally distributed random variables in Equation A.50, the task of finding the maximum is simplified by maximizing the *log* likelihood, defined as

$$l(\theta | Y) = \log L(\theta | Y). \tag{A.51}$$

Since the logarithm is a monotonic function, l and L will attain their maxima at the same value of θ.

We now derive the log likelihood estimates for the parameters of our model $Y_i = \mu + \varepsilon_i$. The log likelihood function of Equation A.51 is (Theil, 1971, p. 90)

$$l(\mu, \sigma^2 \mid Y) = -\frac{n}{2}\log 2\pi - \frac{n}{2}\log \sigma^2 - \frac{1}{2\sigma^2}\sum_{i=1}^{n}(Y_i - \mu - \alpha_i)^2, \tag{A.52}$$

and therefore

$$\frac{\partial l}{\partial \mu} = \frac{1}{\sigma^2}\sum_{i=1}^{n}(Y_i - \mu) \tag{A.53}$$

$$\frac{\partial l}{\partial \sigma^2} = -\frac{n/2}{\sigma^2} + \frac{\Sigma(Y_i - \mu)}{2\sigma^4}.$$

Solving the first equation for μ yields $\mu = \bar{Y}$. Substituting this into the second equation yields

$$\sigma^2_{ML} = \frac{1}{n}\sum(Y_i - \bar{Y})^2 \tag{A.54}$$

Equation A.54 illustrates an important fact, (Kutner, 2005, p. 34), namely, that the maximum likelihood estimator of σ^2 is biased. Kendall and Stuart (1979, p. 90) show that the least squares estimate of σ^2 is the unbiased estimate

$$\hat{\sigma}^2 = \frac{1}{n-1}\sum_{i=1}^{n}(Y_i - \bar{Y})^2 \tag{A.55}$$

The fact that the maximum likelihood method yields a biased estimate for σ^2 is a problem that will occasionally have to be dealt with. The great advantage of the method is that we can apply it in situations where the method of least squares breaks down. One of these situations occurs in the mixed-model formulation discussed in Chapter 12, in which the values of the response variable Y_{ij} are correlated. A second situation where the method can be used is the spatial regression case discussed in Chapter 13, in which the error terms ε_{ij} are autocorrelated.

A.5.2 The Likelihood Ratio Test

In the hypothesis tests involving the mixed model discussed in Chapter 12, in which the response variables Y_{ij} are correlated, the general linear test statistic defined by Equation A.44 does not have an F distribution, and so the test cannot be easily applied. Therefore, a different test is required, and the likelihood ratio test is frequently used. Consider first the case where the parameter θ under consideration is a scalar rather than a vector (e.g., suppose $\theta = \mu$). Suppose the null hypothesis is $H_0 : \theta = \theta_0$ and the alternative is $H_a : \theta \neq \theta_0$. It turns out that under very general conditions the likelihood function $L(\theta \mid Y)$ is asymptotically normally distributed, that is, as the number n of observations increases the distribution of $L(\theta \mid Y)$ approaches normality (Theil 1971, p. 393). Let $\hat{\theta}$ be the maximum likelihood estimator of θ based on the observed data. For example, in the case $\theta = \mu$ the maximum likelihood estimator is the sample mean $\bar{Y}$. The likelihood ratio statistic

$$G \equiv -2\log(L(\theta)/L(\hat{\theta})) \tag{A.56}$$

is asymptotically distributed as χ_1^2, a chi square with one degree of freedom. Thus, the likelihood ratio test of the null hypothesis $H_0 : \theta = \theta_0$ consists of calculating G and comparing it with the percentiles of the chi-square probability distribution. This is an example of a likelihood ratio test. The test is only valid asymptotically, and McCulloch et al. (2008, p. 33) give an example in which the distribution of G is quite far from chi square for relatively moderate values of the sample size n. Nevertheless, the likelihood ratio test is very widely used.

More generally, suppose θ is a vector of parameters and we wish to compare a full model with a restricted model. For example, consider the ANCOVA model discussed in Chapter 12. In this case the full model is given by

$$Y_{ij} = \mu + \alpha_i + \beta(X_{ij} - \bar{X}_i) + \varepsilon_{ij}, \tag{A.57}$$

where the α_i are random effects and μ and β are fixed effects, and the restricted model is given by restricting $\beta = 0$ to obtain

$$Y_{ij} = \mu + \alpha_i + \varepsilon_{ij} \tag{A.58}$$

Note that the restricted model is nested in the full model, since it is obtained by restricting a parameter of the full model. Suppose in general that there are k_F parameters in the full model and k_R in the restricted model $(k_F > k_R)$. In our example, $k_F = 3$ and $k_R = 2$. Let L_F be the estimated value of the maximum likelihood of the full model, and let L_R of the restricted model. Then asymptotically in n the quantity

$$G = -2\log(L_F / L_R) = 2(l_F - l_R) \tag{A.59}$$

has a chi-square distribution with $(k_F - k_R)$ degrees of freedom (Kutner et al., 2005, p. 580; Theil, 1971, p. 396). In our case, $l_F = l(\mu, \alpha, \beta, \sigma^2 \mid Y)$, $l_R = l(\mu, \alpha, \sigma^2 \mid Y)$, and $k_F - k_R = 1$.

The likelihood ratio test is implemented by computing the maxima of the log likelihood statistics l_F and l_R and subtracting them. The computation is carried out by determining the values $\theta_i *$ of θ that maximize $l_i(\theta \mid z)$ for $i = R$ and F. If θ is a vector of k parameters (e.g., if $\theta = [\mu, \alpha_1, ..., \alpha_n, \beta, \sigma^2]'$, then $k = n + 3$), then to maximize $l_i(\theta \mid z)$ one must solve the equations

$$\frac{\partial l_i}{\partial \theta_j} = 0, \ i = R, S, \ j = 1, ..., k \tag{A.60}$$

In simple cases, these equations can be solved analytically. In more complicated cases, however, an approximate solution must be computed numerically.

One type of algorithm for the solution of problems such as these is called *hill climbing*, because if the values of the function $l_i(\theta \mid z)$ are considered as a surface or "hill" in the space of the variables $\theta_1, ..., \theta_k$, then the solution $\partial l_i / \partial \theta_j = 0$ is at the top of the hill (Figure A.3). Since the computer can only evaluate the function at specific values of $\theta_1, ..., \theta_k$, finding the solution is like using an altimeter to climb a hill in a thick fog. You know how high you

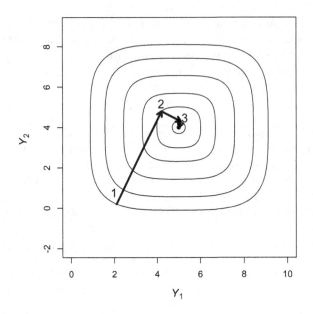

FIGURE A.3
An illustration of the Newton-Raphson hill climbing procedure.

are, but you can't see the quickest way to the top. The most common solution algorithm for a hill climbing problem is some form of the *Newton-Raphson* algorithm (Press et al., 1986, p. 269). To see how algorithms of this form work, imagine that you are at point 1 on the side of the likelihood hill in parameter space as shown in Figure A.3. You are in a thick fog so that you cannot see anything, but you are equipped with a compass and an altimeter. A good procedure to get to the top of the hill would be to take a small step in four perpendicular directions and based on this to estimate the direction of steepest ascent of the hill. This direction is called the *gradient*. Suppose you estimate that the direction of steepest ascent is along the dotted line passing through point 1. You move along this line, using your compass to maintain direction, counting paces and stopping every few paces to read your altimeter. Ultimately you will reach a crest, and the next altimeter reading will be less than the last. At this point, you stop, compute the gradient direction from this point, and repeat the process. As you get closer and closer to the top of the hill, the distance you travel before you reach a point where you change direction will decrease, until (after the third iteration in the figure) it becomes very small. When this distance is smaller than some predetermined value, you decide that you have reached the top of the hill and stop. If the likelihood function has the simple form shown in Figure A.3, then this iterative process is sure to converge, but if the function has a more complex form, then the algorithm may "get lost" and fail to converge. This occasionally happens when one attempts to compute the maximum likelihood. In addition, the computation of the gradient can also run into numerical difficulties.

A.5.3 Application of Maximum Likelihood to Linear Regression

The application of the theory of maximum likelihood to linear regression is a straightforward extension of the procedure discussed in Sections A.5.1 and A.5.2. We develop it here for the simple linear regression model (Equation A.22) based on the discussion in Kutner et al. (2005, p. 27). By analogy with Equation A.50, the likelihood function for the simple regression model, assuming independent, normally distributed errors with constant variance, is

$$L(\beta_0, \beta_1, \sigma^2 \mid Y) = \frac{1}{(2\pi\sigma^2)^{n/2}} \exp(-\sum_{i=1}^{n} \Sigma(Y_i - \beta_0 - \beta_1 X_i)^2 / 2\sigma^2). \tag{A.61}$$

The log likelihood is therefore

$$l(\beta_0, \beta_1, \sigma^2 \mid Y) = -\frac{n}{2}\ln 2\pi - \frac{n}{2}\ln \sigma^2 - \frac{1}{2\sigma^2}\sum_{i=1}^{n}(Y_i - \beta_0 - \beta_1 X_i)^2. \tag{A.62}$$

Taking partial derivatives with respect to the variables gives

$$\frac{\partial l}{\partial \beta_0} = \frac{1}{\sigma^2}\sum_{1}^{n}(Y_i - \beta_0 - \beta_1 X_i) = 0$$

$$\frac{\partial l}{\partial \beta_1} = \frac{1}{\sigma^2}\sum_{1}^{n}X_i(Y_i - \beta_0 - \beta_1 X_i) = 0 \tag{A.63}$$

$$\frac{\partial l}{\partial \sigma^2} = -\frac{n}{2\sigma^2} + \frac{1}{2\sigma^4}\sum_{1}^{n}(Y_i - \beta_0 - \beta_1 X_i)^2 = 0.$$

The solution to these equations is

$$\sum_{1}^{n}(Y_i - \hat{\beta}_0 - \hat{\beta}_1 X_i) = 0$$

$$\sum_{1}^{n}X_i(Y_i - \hat{\beta}_0 - \hat{\beta}_1 X_i) = 0 \tag{A.64}$$

$$\sum_{1}^{n}(Y_i - \hat{\beta}_0 - \hat{\beta}_1 X_i)^2 = n\hat{\sigma}^2.$$

The first two of these equations are the same as the normal Equations A.25 and indicate that the values $\hat{\beta}_i$ obtained by the method of maximum likelihood are equal to those obtained by least squares. The third equation indicates that the maximum likelihood estimate of the error variance is again biased, $\hat{\sigma}^2 = (n-1)s^2/n$.

The testing of a null hypothesis such as $H_0 : \beta_1 = 0$ via the likelihood ratio test is a straightforward extension of the method developed in Section A.5.2.

A.5.4 Maximum Likelihood and Restricted Maximum Likelihood

There is one further complication that must be discussed: the use of the *restricted maximum likelihood* (REML) method vs. the ordinary maximum likelihood (ML) method. As discussed above (Equation A.54), the maximum likelihood estimate of the variance of the simple mean estimation problem is biased, and indeed is $(n-1)/n$ times the unbiased estimate and therefore always biased downward. This tendency of the maximum likeli-hood method to underestimate the variance persists in the application to the mixed model. The REML is an alternative method (Pinhiero and Bates, 2000, p. 75) that provides bet-ter variance estimates, and also has some computational advantages. The REML works by splitting the solution of the model into two stages. In the first stage, the fixed terms, for example, the terms μ and β in the full model of Equation 12.6, and the term μ of the restricted model of Equation 12.5, are in effect removed from the model by assuming that they have a particularly simple probability distribution and integrating them over this distribution, which removes them from the calculation. The random effects estimates are then computed by solving the maximum likelihood equations under this simplify-ing assumption. In the second stage, these estimates of the random effects parameters are plugged back into the model and used to compute the estimates of the fixed effects parameters. While the REML method generally computes improved variance estimates, it cannot be used in a likelihood ratio test of a fixed-effect term. This is because the full and restricted models are computed using different variance structures and are therefore not comparable (Pinhiero and Bates, 2000, p. 76). Thus, we cannot use the REML to test the null hypothesis $\beta = 0$ in the model of Equation 12.6, because it is a fixed effect and must instead use the full maximum likelihood method.

A.6 Change of Variables of a Probability Density

In maximizing the likelihood function, moving from Equation A.48 to Equation A.50 requires a change of variables. Viewed as a probability density function f of the random variables ε_i rather than a likelihood function L of the parameters of the model, f depends on n variables ε_i. In order to get some insight for how to transform this to a function of the Y_i, we consider a very simple case, in which there is only one variable Y, and thus in Equation A.48 ε is a scalar. However, we will consider a more general form for the relation between Y and ε. The following argument is taken from Johnston (1972, p. 374). Assume X is a mathematical variable and let

$$Y = X\beta + g(\varepsilon). \tag{A.65}$$

Suppose ε has a probability density function $\varphi(\varepsilon)$. Our task is to express the likelihood function L in terms of Y rather than ε while maintaining its properties as a probability density function. Suppose g is monotonic in some region (i.e., some range of values of ε, so that g^{-1} exists in that region, and let $h = g^{-1}$, so that $\varepsilon = h(Y)$. Suppose that the relation between Y and ε in this region is shown in Figure A.4. Whenever ε lies in the region $\Delta\varepsilon$ shown in the figure, Y will lie in the corresponding region ΔY. Thus, we can write the following probability equation:

$$\Pr\{Y \text{ lies in } \Delta\Psi\} = \Pr\{\varepsilon \text{ lies in } \Delta\varepsilon\}. \tag{A.66}$$

We have

$$\psi(Y)\Delta Y = \phi(\varepsilon)\Delta\varepsilon. \tag{A.67}$$

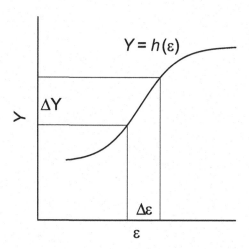

FIGURE A.4
Functional relationship between Y and ε.

where $\psi(Y)$ is the probability density expressed in terms of Y. Therefore,

$$\psi(Y) = \phi(\varepsilon)\frac{\Delta\varepsilon}{\Delta Y}. \tag{A.68}$$

Taking the limit for an infinitesimally small Δ leads to

$$\psi(Y) = \phi(\varepsilon)\frac{d\varepsilon}{dY} = \phi(h(Y))h'(Y). \tag{A.69}$$

where $h' = dh/dY$. In the figure, $h(Y)$ is an increasing function, so that h' is positive. If h' were negative, the same argument would hold except that, since the probability density must be positive, we would have $\psi(Y) = -\varphi(h(Y))h'(Y)$. Since we always multiply ϕ by a positive value, we can write the general equation

$$\psi(Y) = \varphi(h(Y))\big|h'(Y)\big|. \tag{A.70}$$

Now suppose ε and Y are vector functions with n components each. In this case, the formula is very similar (Hoel, 1971, p. 378). Suppose again that $\varepsilon = h(Y)$ locally, where now $h(Y)$ is an n dimensional function of the n dimensional vector Y. Then Equation A.70 becomes

$$\psi(\eta) = \varphi(h(Y))\big|J\big|, \tag{A.71}$$

where J is called the *Jacobian determinant* and is given by

$$J = \det \begin{bmatrix} \dfrac{\partial h_1}{\partial Y_1} & \dfrac{\partial h_1}{\partial Y_2} & \dfrac{\partial h_1}{\partial Y_n} \\[2mm] \dfrac{\partial h_2}{\partial Y_1} & \dfrac{\partial h_2}{\partial Y_2} & \dfrac{\partial h_2}{\partial Y_n} \\[4mm] \dfrac{\partial h_1}{\partial Y_n} & \dfrac{\partial h_2}{\partial Y_n} & \dfrac{\partial h_n}{\partial Y_n} \end{bmatrix}. \tag{A.72}$$

Appendix B: The Data Sets

B.1 Data Set 1

B.1.1 General Description

The information in this section is taken from Greco (1999). The area selected for the study is located on the Sacramento River between river-mile 196, at Pine Creek Wildlife Area, and river-mile 219, near the southern tip of Woodson Bridge State Recreation Area. The primary criteria that led to the selection of this particular location were: (1) it was a river sector with minimal levee influence that permitted fluvial geomorphological processes and riparian forest successional processes to function, (2) it had adequate amounts and distribution of public lands needed for access to riparian forest stands, (3) historical aerial photographic data were available, (4) the location contained two gauging stations with historical flow data, and (5) there were adequate yellow-billed cuckoo survey data.

B.1.2 Land Cover Mapping from 1997 Aerial Photography

The land cover mapping scheme for this project is a modified version of the California Wildlife Habitat Relationships (CWHR) system's land classification (Mayer and Laudenslayer 1988). The three habitat variables mapped were: (1) land type, (2) vegetation height class, and (3) vegetation canopy cover class. A set of original aerial photographs of the study site was obtained from the California Department of Water Resources (CDWR) for the purpose of cartographic and photogrammetric interpretation of land cover types. The set consisted of six flight lines of true-color stereo-pair aerial photographs dated 14-May 1997 at a representative fraction (RF) scale of 1:12,000. Land cover mapping and interpretation was conducted using a minimum mapping unit (MMU) resolution of 20 meters.

The CWHR system defines tree habitats based on measurements of tree trunk diameter at breast-height (DBH) or crown width to determine the height class of tree habitats (Mayer and Laudenslayer 1988). Since these parameters are not easily measured from aerial photographs at a photographic scale of 1:12,000, the parameter "tree height" was selected as a measure of tree size. In this study, tree height classes were determined using visual estimates of relative tree height from a stereoscope and applying height strata classification breaks at 6 meters and 20 meters. Vegetation patches were estimated for height by delineating zones of similarly sized or nearly homogeneous stands. Only vegetation land types were classified for height categories, the remainder of the land types (e.g., water, developed areas, etc.) were assumed to be in the "low" category.

Canopy cover was visually estimated from the aerial photographs using a standard photo interpretation tree stocking density scale (Wenger 1984) that was photo-enlarged for use with 1:12,000 photographs. Boundaries of regions derived from photointerpretation were drawn on clear acetate sheets. These were scanned, imported into Arc/Info (ESRI, Redlands, CA), and converted to a polygon coverage. The polygon coverages were then transformed to geographic coordinates using control points stored in a Universal Transverse Mercator (UTM), Zone 10 projection. The individually interpreted maps (one from each photograph in each set) were then assembled into a composite database of the whole study reach using a manual edge matching method that minimized positional error between the vector pairs. The composite database is included in Data Set 1 as the polygon shapefile *landcover.shp*. The coordinate system is UTM Zone 10, WGS 84. In addition to the *FID* and *Shape* fields, the attribute table contains the following data fields:

Area: The polygon area in m²

Perimeter: The polygon perimeter in m

VegType: Land cover type according to the following code (the bold types are those that may serve as yellow-billed cuckoo habitat): annual grassland (*a*), cropland (*c*), urban/developed (*d*), **freshwater emergent wetland (*f*)**, gravel bar (*g*), **lacustrine (non-flowing) (*l*)**, orchard (*o*), riverine (flowing) (*r*), **riparian (*v*)**, valley oak woodland (*w*)

HeightClass: Tree height in three categories: low (*l*), less than 6m; medium (*m*), 6–20m; and high (*h*), greater than 20m.

CoverClass: Cover class in three categories: sparse (*s*), 10–24 percent; open, 25–39 percent; moderate (*m*), 40–59 percent; dense (*d*), 60–100 percent; open (0), open water.

Data were ground field verified as described in Greco (1999).

B.1.3   Habitat Patches

The polygon shapefile *habitatpatches.shp* disaggregates the land cover data into patches as defined in the CWHR model (see Section 7.3). The file was created from the file *landcover.shp* through a series of GIS operations in ArcGIS that replicated the original operations carried out by Greco (1999). Only potential cuckoo habitat patches are included in the shapefile. These consist of polygons of land cover class freshwater emergent wetland (*f*), lacustrine (non-flowing) (*l*), or riparian (*v*) that are either within 100m of the river or directly adjacent to a polygon within 100m of the river. The steps used to create the file were as follows.

- **Step 1.** Add *landcover.shp*.
- **Step 2.** Create a shapefile as a 100m buffer of water. Select by location all polygons in *landcover.shp* that intersect this buffer. Then select all polygons in *landcover.shp* that touch the boundary of a selected polygon. Next select *VegType* = "f," "l," or "v." Then export this selection as the shapefile *riprval100m.shp*.
- **Step 3.** Use the ArcGIS *Dissolve* tool to dissolve the shapefile created in Step 2 on *VegType*.
- **Step 4.** Use the ArcGIS *Explode* tool to separate polygons into distinct entities. This does not create a new file.
- **Step 5.** Calculate area and perimeter of polygons in the shapefile.
- **Step 6.** Define a data field PatchID. Save this file as *habitatpatches.shp*.

One of the variables in the CWHR model is *patch width*. This was estimated in ArcGIS by converting the polygon shapefile *habitatpatches.shp* to raster and using the Raster Calculator command *zonalgeometry* to compute the thickness of each patch. The patch width was taken to be twice the thickness. This grid was then converted to vector and joined to the attribute table of *habitatpatches.shp* via an attribute join.

B.1.4 The Creation of the Full Set of Explanatory Variables

The shapefiles *landcover.shp* and *habitatpatches.shp* are not strictly necessary for the analysis of the data set and are included to aid in data exploration. The full set of explanatory variables is contained in the shapefile *set1data.shp*. This file was constructed in ArcGIS from the files *landcover.shp*, *habitatpatches.shp*, and a polygon shapefile named *floodplnage.shp*. This last file contained floodplain age estimated by the method described by Greco et al. (2007). The age categories are (in years) 10, 19, 31, 45, 59, 93, 101, and 102. Each category except the last represents the estimated number of years, based on interpretation of historical aerial images and maps, since the patch was last under water. The last category, 102, indicates patches that have not been under water in the last 101 years, which is as far back in time as accurate historical maps exist.

The shapefile *set1data.shp* was created as follows.

- **Step 1.** Union *landcover.shp* and *habitatpatches.shp*.
- **Step 2.** Eliminate redundant data fields and export the shapefile.
- **Step 3.** Intersect *floodplnage.shp* with the shapefile. Eliminate redundant data fields.
- **Step 4.** Save resulting file as *set1data.shp*.

The polygons in the file *set1data.shp* are thus subdivided according to habitat patch (which is defined by vegetation type), age class, and height class. Each polygon of *set1data.shp* is characterized by a unique combination of these three factors. The attribute data fields of *set1data.shp* include *PatchID*, *AgeID*, and *HtClID*, which are the ID number of the polygon determining each of these quantities, *PatchArea*, *AgeArea*, and *SzClArea*, which are the areas of each of these polygons, *PatchWidth*, *VegType*, *Age*, and *HeightClass*, which are the variables defining each polygon.

It happened that two separate habitat patches were given the same *PatchID* value (100) in the original data set. As luck would have it, each of these contained a cuckoo observation point. Based on the values of *HtClID*, one of these patches was reassigned the *PatchID* value 400.

B.1.5 Yellow-Billed Cuckoo Observation Points

The file *obspts.csv* contains the yellow-billed cuckoo observation data. This is an aggregation of observations collected during eleven years over a period between 1978 and 1992 (Gaines, 1974; Gaines and Laymon, 1984; Laymon and Halterman, 1987; Halterman, 1991, Greco, 2002). The survey includes a total of 21 locations within the study area. The five surveys each followed a similar protocol in which the presence or absence of the yellow-billed cuckoo is tested by visiting the site, playing a tape recording of the bird's *kowlp* call, and observing whether or not there is a response (Gaines, 1974). The standard procedure (Greco, 1999, p. 232) is to play the call five to ten times at each observation point, although the exact procedure varied somewhat. The call is audible for a distance of about 300 meters (Laymon and Halterman, 1987). Failure to elicit a response is taken to imply that no cuckoos are present at that location. The response may be simply a bird call and may not include an actual sighting. The sites are generally visited by boat and are often impenetrable, in which case the call is played from the boat. The original coordinates of two of the observation points (point 4 and point 19) in the data set were located in the river itself. These were taken to be the location of the boat at the time the call was played. They were adjusted to place the location in the nearest point on land.

B.1.6 The Data Files

The data describing Data Set 1 are housed in three entities: two ESRI format shapefiles and one comma separated variable file. The ESRI shapefiles actually consist of three disk files each the description of the contents of the files is given in Section 7.2. Assuming that the setwd() function has been used to set the working directory to the data directory, the data are loaded using the following R statements.

The data from Data Set 1 are housed in the *Set1* subdirectory of the *Data* directory. The shapefile consisting of the files *set1data.shp*, *set1data.shx*, set1data.*dbf* contains the data describing the explanatory variables. The statements to load this data set as a spatial features file are:

```
> library(sf)
> data.Set1.sf <- st_read("set1\\set1data.shp")
> st_crs(data.Set1.sf) <- "+proj=utm +zone=10 +ellps=WGS84"
```

To convert his to a SpatialPolygonsDataFrame object, we use the coercion process.

```
> library(maptools)
> data.Set1.sp <- as(data.Set1.sf, "Spatial")
```

The comma separated variable file *obspts.csv* contains the data associated with the yellow-billed cuckoo observations. The statements to load this data set, convert it to a SpatialPointsDataFrame object, and assign coordinates are:

```
> library(maptools)
> data.Set1.obs <- read.csv("set1\\obspts.csv", header = TRUE)
> coordinates(data.Set1.obs) <- c("Easting", "Northing")
> proj4string(data.Set1.obs) <- CRS("+proj=utm +zone=10 +ellps=WGS84")
```

The data may also be converted from a data frame into an sf object.

```
> library(sf)
> data.Set1.obs.sf <- st_as_sf(data.Set1.obs, coords = c("Easting",
+     "Northing"))
```

Two other data sets are provided. The shapefile consisting of the files *habitatpatches.shp*, *habitatpatches.shx*, and *habitatpatches.dbf* contains the habitat patches data. The statements to load this data set are:

```
> library(maptools)
> data.Set1.patches.sf <- st_read("set1\\habitatpatches.shp")
> st_crs(data.Set1.patches.sf) <- "+proj=utm +zone=10 +ellps=WGS84"
> data.Set1.patches.sp <- as(data.Set1.patches.sf, "Spatial")
```

The shapefile consisting of the files *landcover.shp, landcover.shx,* and *landcover.dbf* contains the land cover class data. The statements to load this data set are:

```
> library(maptools)
> data.Set1.landcover.sf <- st_read("set1\\landcover.shp")
> st_crs(data.Set1.landcover.sf) <- "+proj=utm +zone=10 +ellps=WGS84"
> data.Set1.landcover.sp <- as(data.Set1.landcover.sf, "Spatial")
```

B.2 Data Set 2

Data Set 2 is based on 4,101 plots from a historic dataset, the Vegetation Type Map (VTM) survey of California (Wieslander, 1935). This survey was conducted in the 1920s and 1930s. Field crews gathered data from more than 8,000 plots in the coastal ranges from San Francisco bay to the Mexican border and a large part of the Sierra Nevada. Allen and coworkers entered approximately 4,300 VTM records (all the plots featuring a *Quercus* species) into a database and analyzed them to construct a classification system for California's oak woodlands (Allen et al. 1991). Evett (1994, p. 27) added geographic coordinates and estimated climatic data to the database and built models of the environmental niche for six oak species using the GLM-based direct gradient analysis approach of Austin et al. (1990).

The VTM survey plots were 0.081 hectare rectangles chosen to be representative of the vegetation map units being delineated in the field on topographic maps (Wieslander, 1935). Vegetation information consisted of the number of stems per diameter class of overstory tree species and the percent cover of each understory plant species. Original environmental information included plot location, elevation, slope, aspect, soil surface texture, and parent material (Evett, 1994, p. 22). Temperature, length of growing season, and evapotranspiration data for each plot were derived by Evett (1994) from a series of county level isoline maps drawn by C. R. Elford, the California state climatologist, in the 1960s and 1970s. Vayssières et al. (2000) derived precipitation data from an isohyetal map of thirty-year annual average precipitation (1950 to 1980) drawn from approximately 4,100 stations by the California Department of Water Resources. Average available soil water capacity was computed from the State Soil Geographic (STATSGO) GIS soils database for California. Both precipitation and soil map data were at 1/250,000 scale. Values for individual plots were interpolated and/or extracted using the GIS program Arc/Info (ESRI, Redlands, CA). Potential cloudless solar radiation on a tilted surface (in MJ/m^2) was computed as a function of latitude, altitude, slope, and aspect using a FORTRAN program written by T. Rumsey (personal communication).

The data from Data Set 2 are housed in the *Set2* subdirectory of the *Data* directory. The data describing Data Set 2 are housed in the comma separated variable file *set2data.csv.* Assuming that the setwd() function has been used to set the working directory to the data directory, the data are loaded using the following R statement.

```
> data.Set2 <- read.csv("set2\\set2data.csv", header = TRUE)
```

A complete list of the quantities contained in Data Set 2 and the symbols that represent them is given in Table 7.2.

Because the locations in Data Set 2 span two UTM zones, longitude and latitude are generally used to represent spatial coordinates.

B.3 Data Set 3

A total of 16 fields, each planted to rice (*Oryza sativa* L.) were sampled during the course of three growing seasons, 2002–03, 2003–04, and 2004–05. The fields were located in three separate regions in northeastern Uruguay, latitude 32.8° to 33.8° S, longitude 53.7° to 54.5° W. A total of 323 data locations were sampled, 125 in 2002–03, 108 in 2003–04, and 100 in 2004–05 (Figure 1.8). Sample sites were selected by eye, with the intent of obtaining a data set that appropriately represented field variability. Field size varied between 30 to 50 ha. On average there was approximately one sample site per 2.5 ha. Sample sites were georeferenced with a hand held Garmin global positioning system (Garmin International, Oalthe, KS). In fields that were sampled in multiple years, sample sites were at the same location in each year. In addition to longitude, latitude, and a number identifying the field, a total of 19 attribute data items were recorded at each point (Table 7.3). Four of the data fields, emergence (*Emer*), weed level (*Weeds*), level of weed control (*Cont*), and irrigation effectiveness (*Irrig*) were scored on an ordinal scale of 1 to 5 based on expert judgment. Planting date (*DPL*) was recorded on a scale of 1 to 5 based on half-month increments from 1 October, so that 1 represented 1 October through 15 October, 2 represented 16 October through 31 October, etc. The approximate date of 50 percent flowering (*D50*) was recorded as the date the field was visited estimated to be that closest to the true date of 50 percent flowering. Prior to planting, at each sample site in each field, four soil cores (0–30 cm depth) were extracted from a circular region of about 8 m radius such that one core lay in each of the four quadrants. Sand (*Sand*), silt (*Silt*), and clay (*Clay*) content were measured. Soil pH (*pH*), organic matter (*Corg*), phosphorus (Bray) (*SoilP*), and potassium (*SoilK*) were determined at the Instituto Nacional de Investigación Agropecuaria (INIA) La Estanzulea Soils Laboratory.

A sample plot of size 2.5 × 3m was hand harvested at each location and threshed in an experimental combine. Yields were obtained by hand harvesting, standardized to 14 percent moisture content, and these standardized yields (*Yield*) and harvest moisture were recorded, as well as variety (*Var*). Climatic data were recorded at the INIA El Paso de la Laguna Rice Experiment Station near Treinta y Tres. Daily average temperature, rainfall and total hours of sun were summarized by period, where each month is divided into three periods of length approximately ten days. Period 1 contains the first third of October, Period 2 the second, and so on until Period 21, which contains the last third of April.

The data from Data Set 3 are housed in the *Set3* subdirectory of the *Data* directory. The data describing Data Set 3 are housed in the comma separated variable file *set3data.csv*. Assuming that the setwd() function has been used to set the working directory to the data directory, the data are loaded using the following R statement.

```
> data.Set3 <- read.csv("set2\\set3data.csv", header = TRUE)
```

B.4 Data Set 4

B.4.1 General Description

The site at which the study took place consists of two commercial fields located in Winters, CA, latitude 38°32′ N, longitude 121°58′ W. We identify the fields as Fields 1 and 2. They are 32 and 31 ha in area, respectively. Soils are predominantly alluvial clay loams, silty clay loams, and silty clays. Mean annual precipitation is 55.0 cm, almost all of which occurs during the winter. The fields, which had been laser leveled, were graded to a uniform slope for furrow irrigation. Spring wheat (*Triticum aestivum* L. cv. "Express") was drill seeded in December, 1995 into raised beds with furrows spaced 1.52 m apart. The furrows provided surface drainage (the fields do not have any other drainage system installed) during periods of heavy rain that occurred in January, 1996 and were used for a single irrigation in April, 1996 at anthesis. Normal commercial practices, including pest management and supplemental fertilization, were followed in producing the crop. Precipitation during the winter of 1995–96 was 82.2 cm, or 150 percent of normal. The wheat crop was harvested in June 1996.

Aerial photographs were taken on December 16, 1995, March 8, 1996, and May 4, 1996 from an altitude of approximately 1000 m using 70 mm Kodak Aerochrome II 2443 color infrared film. Positive images were scanned using an Agfa Arcus II scanner at 600 dpi. This corresponded to a resolution of approximately 1.8 m. Point sampling was carried out in 1995 on a 61 m square grid in each field. There were 86 and 78 points sampled in Fields 1 and 2, respectively. At each sample point, 25 soil cores (0–15 cm depth) were extracted prior to planting from the tops of the beds in a 25 m^2 area and composited. Sand, silt, and clay content were measured. Soil pH, nitrogen, phosphorus, potassium, and organic carbon levels were determined using standard methods of the University of California Division of Agriculture and Natural Resources Analytical Laboratory (Anonymous, 1998). At anthesis, visual determinations were made at each grid point of stand density and weed and disease infestation intensity, each of which were rated on a scale of 1 to 5. Locations of sample points were determined using a Trimble ProXL® GPS (Trimble Navigation, Sunnyvale, CA), which was corrected by post-processing.

In the second year, both fields were planted to tomato (*Solanum lycopersicum* L.). In the third year, Field 1 was planted to bean (*Phaseolus vulgaris* L.) and Field 2 was planted to sunflower (*Helianthus annuus* L.), and in the fourth year Field 1 was planted to sunflower and Field 2 was planted to corn (*Zea mays* L.). No soil or plant sampling was carried out after the first year. Grain crops were harvested with a combine equipped with an Ag Leader®/GPS yield mapping system (Ag Leader Technology, Ames, Iowa), and tomatoes were harvested with an experimental tomato yield monitor (Pelletier and Upadhyaya, 1999). Yield data were collected once per second and in some years it was not possible to get full coverage.

The data from Data Set 4 are housed in the *Set4* subdirectory of the *Data* directory. Assuming that the setwd() function has been used to set the working directory to the data directory, the data are loaded using the following R statements.

B.4.2 Weather Data

Weather data recorded daily at the University of California, Davis, about 20 km from the fields, is housed in the comma separated variable file *dailyweatherdata.csv*. The file is loaded using the following statement.

```
> weather.data <- read.csv("Set4\\dailyweatherdata.csv", header = TRUE)
```

B.4.3 Field Data

The following describes the loading of data from Field 1. The statements for the Field 2 data are identical except for the obvious substitution of "2" for "1" in the appropriate places.

The data collected manually at the 86 sample sites in Field 1 are housed in the comma separated variable file *set4.196sample.csv*. This file is loaded using the following statement.

```
data.Set4.1 <- read.csv("Set4\\set4.196sample.csv", header = TRUE)
```

"Raw" yield monitor data are housed in comma separated variable files *set4.196wheatyield.csv, set4.197tomatoyield.csv, set4.198beanyield.csv,* and *set4.199sunfloweryield.csv.* The 1996 data are loaded using the following statement.

```
> data.Set4.1Yield96raw <-
+    read.csv("Set4\\Set4.196wheatyield.csv", header = TRUE)
```

Analogous statements are used for the other files.

Georeferenced image data for remotely sensed images acquired in December 1995, March 1996, and May 1996 are stored in the following file pairs: *set4.11295.tif* and *set4.11295.tfw; set4.10396.tif* and *set4.10396.tfw;* and *set4.10596.tif* and *set4.10596.tfw,* respectively. The data can be loaded using the function `readGDAL()` or `raster()`. The following statement are used to load these data using `readGDAL()`.

```
> library(rgdal)
> data.4.1.Dec <- readGDAL("set4\\set4.11295.tif")
> data.4.1.Mar <- readGDAL("set4\\Set4.10396.tif")
> data.4.1.May <- readGDAL("set4\\Set4.10596.tif")
```

The following statements may be used to load the data as a `raster` brick object.

```
> library(raster)
> data.4.1.Dec <- brick("set4\\set4.11295.tif")
> data.4.1.Mar <- brick("set4\\Set4.10396.tif")
> data.4.1.May <- brick("set4\\Set4.10596.tif")
```

In Section 6.4.2, the contents of the "raw" file *set4.196wheatyield.csv* are cleaned, and the result is stored in the comma separated variable file *set4.1yld96cleaned.csv* in the *created* subdirectory. Data from this file are loaded using the following statement:

```
> data.Yield4.1idw <-
+    read.csv("created\\set4.1yld96ptsidw.csv", header = TRUE)
```

Field 2 of Data Set 4 contains a data set of high density apparent electrical conductivity measurements. Data from these measurements are stored in the comma separated variable file *Set4.2EC.csv.* Data from this file are loaded using the following statement.

```
> data.Set4.2EC <- read.csv("Set4\\Set4.2EC.csv", header = TRUE)
```

Appendix C: An R Thesaurus

R is a functional programming language, and this provides the user with both its greatest power and its greatest challenge. It is powerful because when you want to carry out some task, there is a very good chance that someone else has written a function that will carry out this task and has left the function for you to use. It is challenging because you have to find the function and learn how to use it. Then you have to remember it. This is particularly a problem for people who don't use the program on a regular basis. When I was first learning R, I found it very useful to keep a record of tasks I had performed and the functions I had used to perform them. This motivated me to create such a list of tasks and functions in this book. A thesaurus is a book that you use when you have an idea and you want to find a word to fit it, and this appendix provides something you can use when you have a task and you want to find a function to perform it. For that reason, I call this an "R Thesaurus." The thesaurus is divided into major categories of operations and subcategories within these categories. The task and function are listed, along with the package in which the function is found. If no package is listed, then the function is in the base package. The R file and the line number within the file identify one location of this function. If the function is discussed in the text, then this is the reference implementation; otherwise, it is generally the first implementation. There may be other implementations of the function in other files as well. To find these, a good text search app is useful. The one I like is Agent Ransack (https://www.mythicsoft.com/agentransack/).

C.1 Data Input and Output

Operation	Function	Package	File	Line
Read an attribute data file	read.csv()		2.4.1	2
Read a GeoTiff file	readGDAL()	Rgdal	1.1	8
	brick()	raster	2.4.3	81
Read an ESRI shapefile into an sf object	st_read()	sf	2.4.3	50
Write an sf object to an ESRI shapefile	st_write()	sf	2.4.3	46

C.2 Working with Objects

C.2.1 Attribute Data Objects and General Object Functions

Operation	Function	Package	File	Line
Convert a numeric object to a factor	as.factor()		2.4.1	18
Create an ordered factor with labels	ordered()		7.2	59
Display the names of the data fields of a data frame	names()		2.4.1	7

(Continued)

Operation	Function	Package	File	Line
Display only the first records of a data frame	head()		4.5.1	59
Display only the last records of a data frame	tail()		4.5.1	60
Display the class of an object	class()		2.2	5
Display the structure of an object	str()		2.4.1	3
Display the contents of a slot of and S4 object	slot()		2.4.3	62
Display only one element of an object with a long str() output	str()		5.3.4	130–131
Remove data records with missing values	is.na()		10.1	23
Construct a list	list()		2.4.1	21
Construct a function	function()		2.5	16
Convert a number to characters	as.character()		3.6.3	5
Paste character strings together	paste()		6.2.3	7
Convert a list to a vector	unlist()		2.4.3	68
Display the length of a vector	length()		2.4.3	25
Create a matrix	matrix()		2.2	54
Bind together vectors to form a matrix	cbind()		2.2	60
Assign names to rows and/or columns of a matrix	rownames() colnames()			
Construct a data frame	data.frame		2.4.1	28
Reference a data frame	with()		5.3.4	36
Add fields to a data frame	data.frame()		3.2.1	18–19
Select a subset of members of a data frame	subset()		7.5	129
Create a contingency table	table()		12.6.2	67
Generate a sequence of numbers	seq()		2.4.3	103
Coerce a sequence into a matrix	as.matrix()		10.1	10
Concatenate the columns of a matrix into a vector	stack()		3.2.1	196
List all of the unique values of an array	unique()		6.2.3	36
Center and scale the values of an array	scale()		7.2	212
Arrange the elements of an array in order of value of elements of another array	order()		6.2.3	59
Create an array of factors based on characters	factor(), as.character()		1.3.1	61
Coerce an object into a factor	as.factor()		2.4.1	18
Coerce an object into an integer	as.integer()		2.5	5
Associate labels with factors	factor()		8.4.2	101
Display the levels of a factor	levels()		7.2	57
Determine the array number of the smallest element in an array	which.min()		5.3.1	28
Join two data frames by merging elements with matching attribute values	merge()		7.2	167
Remove duplicate rows from a data frame	remove.dup. rows()	cwhmisc	9.3.4	76
Execute statements based on a condition	if() ifelse()		2.3.1 3.2.2	11 15

C.2.2 Spatial Data Objects

Operation	Function	Package	File	Line
Convert a data frame to an `sf` object	`st_as_sf()`	sf	2.4.3	9
Covert an `sf` object to an `sp` object	`as()`	sp	2.4.3	59
Covert an `sp` object to an `sf` object	`st_as_sf()`	sf	3.6.2	5,7
Convert a data frame to a `SpatialPointsDataFrame`	`coordinates()`	sp	2.4.3	73
Assign a projection to an `sf` object	`st_crs()`	sf	2.4.3	16
Assign a projection to an `sp` object	`proj4string()`	sp	2.4.3	74
Transform from one projection to another	`spTransform()`	rdgal	12.6.2	19
Create a grid of points	`expand.grid()`		2.4.3	103
Convert a `SpatialPointsDataFrame` to a `SpatialGridDataFrame`	`gridded()`	sp	4.6.1	35
Convert a band from an image to a `SpatialPointsDataFrame`	Several	maptools	1.1	77–82
Create an `sf` polygon object	Several	sf	2.4.3	37–43
Extract the coordinates of the label points of a `SpatialPolygons` object	`lapply()`		2.4.3	67
Extract the ID values of a `SpatialPolygons` object	`lapply()`		2.4.3 6.4.2	66 73
Digitize a region on the screen	`locator()`	sp	7.3	123
Select those points in a `SpatialPointsDataFrame` lying inside a `SpatialPolygons` object	`over()`	sp	5.3.4 7.5	121 43
Compute the means of an attribute value over points lying within a group of polygons	`over()`	sp	6.4.2 11.5.1	77 46
Resample from one grid to another	`raster()` `over()`	raster sp	6.4.3 6.4.2	17 77
Do a "polygon containing point" operation to create a data frame of attributes of polygons containing a set of points	`over()`	sp	7.2	120
Clip an interpolated grid to an irregular boundary	`over()`	sp	7.3	141
Convert a SpatialPoints object into a SpatialGrid object	`gridded()`	sp	4.6.1	35
Construct a regular lattice of polygons	`GridTopology()`, `as()`	sp	5.3.2	32
Create a set of Thiessen polygons using `spatstat` functions	Several	spatstat	3.6.1	41–44
Verify that attribute values match spatial locations correctly	Several		3.6.2	4–18
Create buffers around points	`disc()`	spatstat	6.4.2	36
Construct a neighbor list for a lattice	`cell2nb()`	spdep	3.2.1	9
Construct a neighbor list for `SpatialPolygons` or `SpatialPolygonsDataFrame` object	`poly2nb()`	spdep	3.6.3	4
Assign weights to a spatial weights matrix	`nb2listw()`	spdep	3.5.3	11
Assign weights to spatial weights matrix based on distance	`dnearneigh()`	spdep	3.6.1	13
Assign weights to spatial weights matrix based on number of neighbors	`knearneigh()`	spdep	3.6	23
Determine the cardinality (number of neighbors) of a point set	`card()`	spdep	14.5.2	13
Generate an autocorrelated data set	`invIrM()`	spdep	3.2.1	11
Bind two `SpatialPointsDataFrame` objects	`spRbind()`	sp	5.2.1	38
Crop a `raster` object	`crop()`	raster	7.4	72
Extract cell values from a raster layer	`extract()`	raster	5.7	38
Compute the slope and aspect of a raster layer	`slopeAspect()`	raster	7.4	73

C.3 Graphics

Graphics functions are discussed extensively in Section 2.6. The discussion, however, references figures in other chapters. The code to produce these figures is in the section containing the figure, rather than in Section 2.6. Also note that ggplot() can do many quick plotting operations much more easily.

C.3.1 Traditional Graphics, Attribute Data

Section 2.6.1 contains a discussion of traditional graphics for attribute data.

Operation	Function	Package	File	Line
Create a scatterplot	plot()		3.5.1	59
Create a line plot	plot()		3.2.1	46
Create a histogram	hist()		3.3	13
Create a histogram of two data sets	hist(), plot()		10.1	36
Create a hexagonal bin plot	hexbin()	hexbin	7.3	244
Create a boxplot	boxplot()		7.2	193
Create a barplot	barplot()		7.3	215
Create an aligned dot plot	matplot() plot()		7.2	218
Identify individual points in a scatterplot	identify()		6.3.3	57
Insert italic script	expression()		3.2.1	47
Insert a subscript	expression()		4.6.1	23
Make text bold	expression()		4.6.1	23
Insert a Greek character	expression()		3.5.1	59
Place a bar over a character	expression()		3.5.1	63
Insert a degree symbol	expression()		7.4	101
Examples of the use of expression() for complex formulas	expression()		3.2.1 4.6.1	53 44
Draw an arrow	arrows()		4.6.1	22
Insert a line of text	text()		3.2.1	51
Control the position of the text	text()		3.2.1	51
Define x and y limits of a function	plot()		4.6.1	41
Draw contours of a surface	contour()		12.6.1	58
Control plotting of axes	axis()		7.4	106
Plot an empty screen	plot()		7.1	1
Control the size of axis labels	plot()		3.2.1	45
Control the size of the margins	par()		3.2.1	45
Construct a perspective plot	persp()		3.2.1	76
Construct a circle	symbols()		3.5.2	41
Draw an ellipse	ellipse()	car	3.5.2	43

C.3.2 Traditional Graphics, Spatial Data

Section 2.6.2 contains a discussion of traditional graphics for spatial data.

Operation	Function	Package	File	Line
Define a gray scale	`grey()`		3.5.3	39
Display a GeoTiff file	`image()`		5.3.4	50
Display a `SpatialPointsDataFrame` over a GeoTiff image	`plot()`	`maptools`	1.1	17
Construct a bubble plot	`plot()`	`maptools`	3.2.1	163
Select colors for an image	`terrain.colors()`	`RColorBrewer`	1.1	26
Plot a map showing geographic coordinates on the axes	`title()`	`maptools`	1.1	21
Draw a map using data from `stateMapEnv`	Several	`maps, maptools`	1.3.1 7.3	5 25
Draw a map using data from `world`	Several	`maps, maptools`	1.3.3	4
Draw a scale bar on a map	`SpatialPolygonsRescale()`	`maptools`	1.3.1 7.3	37 39
Draw a north arrow on a map	`SpatialPolygonsRescale()`	`maptools`	1.3.1	28
Draw a small map inside a larger map	Several	`maptools`	1.3.1	40
Plot the value of an attribute by spatial location	`text()`,	`maptools`	4.5.3	23
Create a star plot map	`stars()`		7.5	172

C.3.3 Trellis Graphics, Attribute Data

Section 2.6.3 contains a discussion of trellis graphics for both attribute and spatial data.

Operation	Function	Package	File	Line
Create scatter or line plots	`xyplot()`	`lattice`	4.5.1 7.4	63 174
Create a scatterplot matrix	`splom()`	`lattice`	7.3	119
Print a `lattice` graph in gray scale	`trellis.device()`	`lattice`	3.5.3	40
Create a boxplot	`bwplot()`	`lattice`	7.4	27
Create a dotplot	`dotplot()`	`lattice`	8.3	22
Construct a trellis plot with two grouping variables	`Any trellis graphics function`	`lattice`	7.4	174
Adjust the parameters of a trellis graphics function	`Any trellis graphics function`	`lattice`	7.3	118

C.3.4 Trellis Graphics, Spatial Data

Section 2.6.3 contains a discussion of trellis graphics for both attribute and spatial data.

Operation	Function	Package	File	Line
Plot an interpolated surface	`spplot()`	sp	1.1	59
Plot a `SpatialPointsDataFrame` over an interpolated surface	`spplot()`	sp	1.2	20–23
Create a north arrow	`spplot()`	sp	1.3.1	65
Create a scale bar	`list()`	sp	1.3.1	67
Control the placement of the legend	`spplot()`	sp	15.2	135
Plot a thematic map of factor data (polygons)	`spplot()`	sp	7.2	65
Create a color map	`spplot()`	sp	1.3.1	80, 101
Create multiple panel plots	`spplot()`	sp	3.5.3	41
Overlay a point map on top of a polygon map.	`spplot()`	sp	7.2	65
Plot in color	`colorRampPalette()`	sp	3.5.3	48

C.4 Computations

C.4.1 Computations on Attribute Data

Operation	Function	Package	File	Line
Matrix multiplication	`%*%`		3.2.1	22
Compute $a \bmod(b)$	`%%`		6.3.3	112
Eliminate rows of a data frame not satisfying a criterion	`which()`		6.3.3	112
Iterate using a *for* loop	`for()`		2.3.1	3
Evaluate a conditional expression in a subscript	`"["`		2.3.1	21
Round a numeric value	`round()`		4.2.1	26
Generate a pseudorandom number seed	`set.seed()`		2.3.2	2
Generate a sequence of normally distributed random numbers	`rnorm()`		2.3.2	3
Generate a sequence of binomially distributed random numbers	`rbinom()`		3.2.2	6
Replicate a sequence	`rep()`		2.3.2	5
Sort a sequence of numbers	`sort()`		2.3.2	7
Test whether the elements of two arrays are all equal	`all.equal()`		3.6.2	10
Permute the elements of a sequence	`sample()`		3.4	29
Generate permutations of a sequence of numbers	`permutations()`	e1071	15.4.1	72

(Continued)

Operation	Function	Package	File	Line
Add a small random perturbation to each element of an array	`jitter()`		12.4	23
Carry out a Monte Carlo simulation by replicating an "experiment"	`replicate()`		3.3	9
Assign a value based on the outcome of a test	`ifelse()`		3.2.2	15
Apply a function to a matrix by row or by column	`apply()`		5.4	32
Apply a function to one sequence as indexed by another sequence	`tapply()`		2.3.2	10
Apply a function to a vector, creating a list	`lapply()`		4.3	8
Apply a function to a vector, creating a vector	`sapply()`		16.3.2	42
Apply a function to a data frame split by factors	`by()`		11.5.2	5
Aggregate the contents of a data frame	`aggregate()`		7.2	150
Average quantities over factors	`ave()`		9.3.4	66
Compute the distances between entries in a matrix	`vegdist()`	vegan	11.4.1	9
Carry out a *t*-test	`t.test()`		3.4	6
Test for independence in a contingency table using a chi-square test	`chisq.test()`		11.3.1	27
Test for independence in a contingency table using the Fisher test	`fisher.test()`		11.3.2	166
Compute a correlation coefficient	`cor()`		6.3.3	50
Test the significance of a correlation coefficient	`cor.test()`		11.2.1	10
Compute the variance	`var()`		3.3	11
Determine influence measures of a regression	`influence.measures()`		8.3	120
Plot a regression line	`abline()`		5.7	51
Carry out a bootstrap analysis	`boot()`	boot	10.2	50
Fit a linear regression	`lm()`		3.2.1	94
Fit a linear regression with higher order terms	`lm()`		3.2.1	125
Carry out an analysis of variance	`aov()`		16.2	79
Carry out an analysis of covariance	`aov()`		12.1	20
Test for homogeneity of variance of a linear regression	`bptest()` `leveneTest()`	lmtest car	8.3 11.2.4	129 10
Computing the AIC values resulting from dropping terms from a model	`drop1()`		8.3	61

(Continued)

Operation	Function	Package	File	Line
Computing the AIC values resulting from adding terms to a model	`add1()`		8.4.2	24
Construct added variable plots	`avPlots()`	`car`	8.2.2	68
Construct partial residual plots	`crPlots()`	`car`	8.3	124
Compute predicted values of a regression model	`predict()`		3.2.1	96
Compute separate linear regression models over a grouped variable	`lmList()`	`nlme`	12.1	15
Carry out *k*-means cluster analysis	`kmeans()`		12.6.1	7
Carry out a recursive partitioning analysis	Several	`rpart`	9.3.1	18
Carry out a random forest analysis	Several	`randomForest`	9.4.1	5–12
Compute a logistic regression	`glm()`		8.4.1	6
Fit a zero-inflated Poisson regression model	`zeroinfl()`	`pscl`	8.4.4	187
Mixed model analysis	`lme()`	`nlme`	12.3	5
Generalized additive model analysis	`gam()`	`mgcv`	9.2	79

C.4.2 Computations on Spatial and Spatiotemporal Data

Operation	Function	Package	File	Line
Compute a `gstat` variogram	`variogram`	`gstat`	4.6.1	39
Fit a `gstat` variogram model	`fit.variogram()`	`gstat`	4.6.1	40
Plot a `gstat` variogram and model	`plot()`	`gstat`	6.3.2	58
Calculate an experimental correlogram	`correlogram()`	`spatial`	11.2.2	24
Compute an IDW interpolation using `gstat`	`idw()`	`gstat`	6.3.1	86
Compute a kriging interpolation using `gstat`	`krige()`	`gstat`	6.3.2	80
Compute a co-kriging interpolation using `gstat`	Several	`gstat`	6.3.3	105
Construct a `gstat` object	`gstat()`	`gstat`	6.3.3	104
Fit data using median polish	`medpolish()`	`spdep`	3.2.1	103
Compute and test Moran's I	`moran()`	`spdep`	4.4	16
	`moran.test()`			17
	`moran.mc()`			26
Compute and test Geary's c	`geary.test()`	`spdep`	4.4	28
	`geary.mc()`			37
Construct a Moran correlogram	`sp.correlogram()`	`spdep`	4.5.1	16
Construct a Moran scatterplot	`moran.plot()`	`spdep`	4.5.2	14
Compute a geographically weighted regression	`gwr()`	`spgwr`	4.5.4	13
Construct a time sequence	`ISOdate()`		15.2.1	81
	`as.POSIXct()`		15.2.2	70
Construct an `stfdf` object	`STFDF()`	`spacetime`	15.2.1	89
Construct a spatiotemporal variogram	Several	`spacetime`	15.2.1	108

References

Abbott, A. J., G. R. Best, and R. A. Webb (1970). The relation of achene number to berry weight in strawberry fruit. *Journal of Horticultural Science* 45: 215–222.

Abramowitz, M., and I. A. Stegun (1964). *Handbook of Mathematical Functions*. Dover Publications, New York.

Agresti, A. (1996). *An Introduction to Categorical Data Analysis*. Wiley, New York.

Agresti, A. (2002). *Categorical Data Analysis*. Wiley, New York.

Akaike, H. (1974). A new look at the statistical model identification. *IEEE Transactions on Automatic Control* 19: 716–723.

Aldrich, J. (1997). R.A. Fisher and the making of maximum likelihood 1912–1922. *Statistical Science* 12: 162–176.

Alker, H. R. Jr. (1969). A typology of ecological fallacies. In Dogan, M. and Rokkan, S., Editors. *Quantiative Analysis and the Social Sciences* pp. 69–86. MIT Press, Cambridge, MA.

Allen, B. H., B. A. Holzman, and R. R. Evett (1991). A classification system for California's hardwood rangelands. *Hilgardia* 59: 1–45.

Allen, D. M. (1971). Mean square error of prediction as a criterion for selecting variables. *Technometrics* 13: 469–475.

Allen-Diaz, B. H., and B. A. Holtzman (1991). Blue oak communitites in California. *Madroño* 38: 80–95.

Alonso, M. (2008). Non-metric multidimensional scaling (NMDS) as a basis for a plant functional group classification and a Bayesian belief network formulation for California oak woodlands. Ph.D. Dissertation, University of California, Davis, CA.

Ames, W. F. (1977). *Numerical Methods for Partial Differential Equations*, 2nd edition. Academic Press, New York.

Anderson, K. M., D. C. Dumphy, and P. W. F. Wilson (1990). Management of data quality in a long-term epidemiologic study: The Framingham Heart Study. In Liepins, G. E. and Uppuluri, V. R. R., Editors. *Data Quality Control: Theory and Pragmatics* pp. 57–68. Marcel Dekker, New York.

Andrews, W. F., C. B. Goudey, G. J. Staidl, and L. A. Bates (1972). *Soil Survey of Yolo County, California*. USDA Soil Conservation Service, Washington, DC.

Andrienko, A., and G. Andrienko (1998). *Exploratory Analysis of Spatial and Temporal Data*. Springer, Berlin, Germany.

Anonymous. (2008). 2008 Regional Barley, Common Wheat and Triticale, and Durum Wheat Performance Tests in California. Agronomy Progress Report No. 296, October 2008. University of California, Davis, Davis, CA.

Anscombe, F. J. (1960). Rejection of outliers. *Technometrics*. 2: 123–166.

Anselin, L. (1988a). Lagrange multiplier test diagnostics for spatial dependence and spatial heterogeneity. *Geographical Analysis* 20: 1–17.

Anselin, L. (1988b). *Spatial Econometrics: Methods and Models*. Kluwer Academic Publishers, Dordrecht, the Netherlands.

Anselin, L. (1992). Spatial data analysis with GIS: An introduction to application in the social sciences, Technial report 92-10. National Center for Geographic Information and Analysis, Santa Barbara, CA.

Anselin, L. (1995). Local indicators of spatial association: LISA. *Geographical Analysis* 27: 93–115.

Anselin, L. (1996). The Moran scatterplot as an ESDA tool to assess local instability in spatial association. In Fischer, M., Scholten, H. J., and Unwin, D., Editors. *Spatial Analytical Perspectives in GIS* pp. 111–125. Taylor & Francis Group, London, UK.

Anselin, L. (2001). Spatial effects in econometric practice in environmental and resource economics. *American Journal of Agricultural Economics* 83: 705–710.

Anselin, L., and A. K. Bera (1998). Spatial dependence in linear regression models with an introduction to spatial econometrics. In Ullau, A. and Giles, D. E. A., Editors. *Handbook of Applied Economic Statistics* pp. 237–289. Marcel Dekker, New York.

Anselin, L., A. K. Bera, R. Florax, and M. J. Yoon (1996). Simple diagnostic tests for spatial dependence. *Regional Science and Urban Economics* 26: 77–104.

Anselin, L., and D. A. Griffith (1988). Do spatial effects really matter in regression analysis. *Papers of the Regional Science Association 65*: 11–34.

Anselin, L., and S. Rey (1991). Properties of tests for spatial dependence in linear regression models. *Geographical Analysis 23*: 112–130.

Anselin, L., I. Syabri, and Y. Kho (2006). GeoDa: An introduction to spatial data analysis. *Geographical Analysis 38*: 5–22.

Arlinghaus, S. L. (1995). *Practical Handbook of Spatial Statistics*. CRC Press, Boca Raton, FL.

Augustin, N. H., M. A. Mugglestone, and S. T. Buckland (1996). An autologistic model for the spatial distribution of wildlife. *Journal of Applied Ecology 33*: 339–347.

Austin, M. P., R. B. Cunningham, and P. M. Flemming (1984). New approaches to direct gradient analysis using environmental scalars and statistical curve-fitting procedures. *Vegetatio 55*: 11–27.

Austin, M. P., R. B. Cunningham, and R. Good (1983). Altitudinal distribution of several eucalypt species in southern New South Wales. *Australian Journal of Ecology 8*: 169–180.

Austin, M. P., A. O. Nicholls, and C. R. Margules (1990). Measurement of the realized qualitative niche: Environmental niches of five Eucalyptus species. *Ecological Monographs 60*: 167–177.

Baddeley, A., and R. Turner (2005). Spatstat: An R package for analyzing spatial point patterns. *Journal of Statistical Software 12*: 1–42.

Ball, S. T., and B. E. Frazier (1993). Evaluating the association between wheat yield and remotely-sensed data. *Cereal Research Communications 21*: 213–219.

Banerjee, S., B. P. Carlin, and A. E. Gelfand (2004). *Hierarchical Modeling and Analysis for Spatial Data*. Chapman & Hall/CRC Press, Boca Raton, FL.

Barbour, M. G., J. H. Burk, and W. D. Pitts (1987). *Terrestrial Plant Ecology*. Benjamin Cummings, Menlo Park, CA.

Barnett, V., and T. Lewis (1994). *Outliers in Statistical Data*. John Wiley & Sons, New York.

Bartels, C. P. A. (1979). Operational statistical methods for analysing spatial data. In Bartels, C. P. A., and Ketellapper, R. H., Editors. *Exploratory and Explanatory Statistical Analysis of Spatial Data* pp. 5–50. Kluwer Academic Publishers, Dordrecht, the Netherlands.

Bartels, C. P. A., and R. H. Ketellapper, Editors (1979). *Exploratory and Explanatory Analysis of Spatial Data*. Martinus Nijhoff Publishing, Boston, MA.

Bartholomew, G. A. (1958). The role of physiology in the distribution of terrestrial vertebrates. In Hubbs, C. L., Editor, *Zoogeography* pp. 81–95. American Association for the Advancement of Science, Washington, DC.

Bartlett, M. S. (1935). The effect of non-normality on the *t* distribution. *Proceedings of the Cambridge Philosophical Society 31*: 223–231.

Bartlett, M. S. (1975). *The Statistical Analysis of Spatial Pattern*. Chapman & Hall, London, UK.

Beale, C. M., J. J. Lennon, D. A. Elston, M. J. Brewer, and J. M. Yearsley (2007). Red herrings remain in geographical ecology: A reply to Hawkins et al. (2007). *Ecography 30*: 845–847.

Beck, A. D., S. W. Searcy, and J. P. Roades (2001). Yield data filtering techniques for improved map accuracy. *Applied Engineering in Agriculture 17*: 423–431.

Becker, R. A., J. M. Chambers, and A. R. E. Wilks (1988). *The New S Language: A Programming Environment for Data Analysis and Graphics*. Wadsworth & Brooks/Cole, Pacific Grove, CA.

Becker, R. A., A. R. Wilks, R. Brownrigg, and T. P. Minka (2016). Maps: Draw geographical maps. http://CRAN.R-project.org/package=maps.

Bellman, R. E. (1957). *Dynamic Programming*. Princeton University Press, Princeton, NJ.

Belsley, D. A., E. Kuh, and R. E. Welsch (1980). *Regression Diagnostics*. Wiley, New York.

Benedetti, R., and P. Rossini (1993). On the use of NDVI profiles as a tool for agricultural statistics: the case study of wheat yield estimate and forecast in Emilia Romagna. *Remote Sensing of the Environment 45*: 311–326.

Besag, J. (1991). Spatial statistics in the analysis of agricultural field experiments. In National Research Council, *Spatial Statistics and Digital Image Analysis* pp. 109–127. National Academy Press, Washington, DC.

Besag, J., and P. Clifford (1989). Generalized Monte Carlo significance tests. *Biometrika 76*: 633–642.

Besag, J. E. (1974). Spatial interaction and the statistical analysis of lattice systems. *Journal of the Royal Statistical Society, Series B 36*: 192–236.

Besag, J. E. (1983). Comment on "Nearest neighbour (NN) analysis of field experiments" by G.N. Wilkinson et al. *Journal of the Royal Statistical Society, Series B 45*: 180–183.

Besag, J. E., and R. Kempton (1986). Statistical analysis of field experiments using neighbouring plots. *Biometrics 42*: 231–251.

Biemer, P. P., and L. E. Lyberg (2003). *Introduction to Survey Quality*. Wiley, Hoboken, NJ.

Bierkens, M. F. P., P. A. Finke, and P. de Willigen (2000). *Upscaling and Downscaling Methods for Environmental Research*. Kluwer Academic Publishers, Dordrecht, the Netherlands.

Birrell, S. J., S. C. Borgelt, and K. A. Sudduth (1995). Crop yield mapping: Comparison of yield monitors and mapping techniques. In Robert, P. C., Rust, R. H., and Larson, W. E., Editors. *Proceedings of the Second International Conference on Precision Agriculture* pp. 15–32. American Society of Agronomy, Madison, WI.

Bivand, R. (with contributions by M. Altman, L. Anselin, R. Assunção, O. Berke, A. Bernat, G. Blanchet, E. Blankmeyer, M. Carvalho, B. Christensen, Y. Chun, C. Dormann, S. Dray, R. Halbersma, E. Krainski, N. Lewin-Koh, H. Li, J. Ma, G. Millo, W. Mueller, H. Ono, P. Peres-Neto, G. Piras, M. Reder, M. Tiefelsdorf and and D. Yu) (2011). spdep: Spatial dependence: Weighting schemes, statistics and models. http://CRAN.R-project.org/package=spdep.

Bivand, R., E. Pebesma, and V. Gómez-Rubio. (2013b). *Applied Spatial Data Analysis with R*, 2nd edition. Springer, New York.

Bivand, R., and C. Rundel (2017). rgeos: Interface to Geometry Engine: Open Source (GEOS). http://CRAN.R-project.org/package=rgeos.

Bivand, R., and N. Lewin-Koh (2017). Maptools: Tools for reading and handling spatial objects. R package version 0.9-2. https://CRAN.R-project.org/package=maptools.

Bivand, R., and D. Yu (2017). spgwr: Geographically weighted regression. R package version 06.31. https://CRAN.R-project.org/package=spgwr.

Bivand, R. S., and G. Piras (2015). Comparing implementations of estimation methods for spatial econometrics. *Journal of Statistical Software 63*(18): 1–36. http://www.jstatsoft.org/v63/i18/.

Bivand, R. S., J. Hauke, and T. Kossowski (2013a). Computing the Jacobian in Gaussian spatial autoregressive models: An illustrated comparison of available methods. *Geographical Analysis 45*(2): 150–179.

Blackmore, B. S., and C. J. Marshall (1996). Yield mapping: Errors and algorithms. In Robert, P. C., Rust, R. H., and Larson, W. E., Editors. *Proceedings of the Third International Conference on Precision Agriculture* pp. 403–416. American Society of Agronomy, Madison, WI.

Blackmore, S. and M. Moore (1999). Remedial correction of yield map data. *Precision Agriculture 1*: 53–66.

Bolker, B. M. (2008). *Ecological Models and Data in R*. Princeton University Press, Princeton, NJ.

Bolker, B. M., M. E. Brooks, C. J. Clark, S. W. Geange, J. R. Poulsen, M. H. H. Stevens, and J.-S. S. White (2009). Generalized linear mixed models: A practical guide for ecology and evolution. *Trends in Ecology and Evolution 24*: 127–135.

Bolthausen, E. (1982). On the central limit theorem for stationary mixing random fields. *Annals of Probability 10*: 1047–1050.

Bonham-Carter, G. F. (1994). *Geographic Information Systems for Geoscientists*. Pergamon, Kidlington, UK.

Box, G. E. P. (1976). Science and statistics. *Journal of the American Statistical Association 71*: 791–799.

Box, G. E. P. (1979). Robustness in the strategy of scientific model building. In Launer, R. L. and Wilkerson, G. N., Editors. *Robustness in Statistics* pp. 201–236. Academic Press, New York.

Box, G. E. P., and G. C. Tiao (1973). *Bayesian Inference in Statistical Analysis*. Addison-Wesley, Menlo Park, CA.

Box, G. E. P., and G. M. Jenkins (1976). *Time Series Analysis: Forecasting and Control*. Holden-Day, San Francisco, CA.

Box, G. E. P., W. G. Hunter, and J. S. Hunter (1978). *Statistics for Experimenters*. John Wiley & Sons, New York.

Boyd, D. W. (2001). *Systems Analysis and Modeling*. Academic Press, San Diego, CA.

Brady, N. C. (1974). *The Nature and Properties of Soils*. Macmillan, New York.

Braun, W. J. (2015). MPV: Data sets from Montgomery, Peck, and Vining's book. http://CRAN.R-project.org/package=MPV.

Bravington, M. B. (2013). debug: MVB's debugger for R. R package version 1.3.1. https://CRAN.R-project.org/package=debug

Breiman, L. (1996). Bagging predictors. *Machine Learning 24*: 123–140.

Breiman, L. (2001a). Random forests. *Machine Learning 45*: 5–32.

Breiman, L. (2001b). Statistical modeling: The two cultures. *Statistical Science 16*: 199–231.

Breiman, L., J. H. Friedman, R. A. Olshen, and C. J. Stone (1984). *Classification and Regression Trees*. Chapman & Hall, New York.

Brower, J. E., J. Zar, and C. von Ende (1990). *Field and Laboratory Methods for General Ecology*. Brown Publishers, Dubuque, IA.

Brownie, C., D. T. Bowman, and J. W. Burton (1993). Estimating spatial variation in analysis of data from yield trials: A comparison of methods. *Agronomy Journal 85*: 1244–1253.

Brus, D. J. (1994). Improving design-based estimation of spatial means by soil map stratification. a case study of phosphate saturation. *Geoderma 62*: 233–246.

Brus, D. J., and J. J. De Gruijter (1997). Random sampling or geostatistical modeling? Choosing between design-based and model-based sampling strategies for soil (with discussion). *Geoderma 80*: 1–44.

Bullock, D. G., and D. S. Anderson (1998). Evaluation of the Minolta SPAD-502 chlorophyll meter for nitrogen management in corn. *Journal of Plant Nutrition 21*: 741–755.

Burnham, K. P., and D. R. Anderson (1998). *Model Selection and Inference: A Practical Information-Theoretic Approach*. Springer, New York.

Burrough, P. A., and R. McDonnell (1998). *Principles of Geographical Information Systems*. Oxford University Press, Oxford, UK.

Buse, A. (1982). The likelihood ratio, Wald, and Lagrange multiplier tests: An expository note. *The American Statistician 36*: 153–157.

Cameletti, M. (2011). Stem: Spatio-temporal models in R. http://CRAN.R-project.org/package=Stem.

Cameron, A. C., and P. K. Trivedi (1998). *Regression Analysis of Count Data*. Cambridge University Press, Cambridge, UK.

Campbell, I. (2007). Chi-squared and Fisher-Irwin tests of two-by-two tables with small sample recommendations. *Statistics in Medicine 26*: 3661–3675.

Canty, A., and B. Ripley (2017). boot: Bootstrap R (S-Plus) Functions. R package version 1.3-19.

Cantrell, R. S., and C. Cosner (2003). *Spatial Ecology via Reaction-Diffusion Equations*. Wiley, Chichester, UK.

Capelli, L., S. Sironi, and R. Del Rosso (2014). Electronic noses for environmental monitoring applications. *Sensors 14*: 19979–20007.

Carl, G., and I. Kuhn (2007). Analyzing spatial autocorrelation in species distribution using Gaussian and logit models. *Ecological Modelling 207*: 159–170.

Carr, D. (2016). Ported by Nicholas Lewin-Koh, Martin Maechler and contains copies of lattice functions written by Deepayan Sarkar. hexbin: Hexagonal Binning Routines. R package version 1.27.1. https://CRAN.R-project.org/package=hexbin.

Casanova, D., J. Goudriaan, J. Bouma, and G. F. Epema (1999). Yield gap analysis in relation to soil properties in direct-seeded flooded rice. *Geoderma 91*: 191–216.

Case, A. C., and H. S. Rosen (1993). Budget spillovers and fiscal policy interdependence. *Journal of Public Economics 52*(3): 258–307.

Cassman, K. G., and R. E. Plant (1992). A model to predict crop response to applied fertilizer nutrients in heterogeneous fields. *Fertilizer Research 31*: 151–163.

Cattell, R. B. (1952). *Factor Analysis: An Introduction and Manual for the Psychologist and Social Scientist.* Harper, New York.

Cerioli, A. (1997). Tests of independence in 2 x 2 tables with spatial data. *Biometrics 53*: 619–628.

Chamberlain, M. J. (2008). Are we sacrificing biology for statistics? *Journal of Wildlife Management 72*: 1057–1058.

Chambers, J. M., W. S. Cleveland, B. Kleiner, and P. A. Tukey (1983). *Graphical Methods for Data Analysis.* Duxbury Press, Boston, MA.

Chang, K.-T. (2008). *Introduction to Geographic Information Systems.* McGraw-Hill, Boston, MA.

Chun, Y., and D. L. Griffith (2013). *Spatial Statistics & Geostatistics.* Sage Publications, London, UK.

Chou, Y.-H., R. A. Minnich, L. A. Salazar, J. D. Power, and R. J. Dezzani (1990). Spatial autocorrelation of wildfire distribution in the Idyllwild Quadrangle, San Jacinto Mountain, California. *Photogrammetric Engineering and Remote Sensing 56*: 1507–1513.

Christensen, R. (1991). *Linear Models for Multivariate, Time Series, and Spatial Data.* Springer Verlag, New York.

Clark, J. S., and A. E. Gelfand (2006). *Hierarchical Modelling for the Environmental Sciences.* Oxford University Press, Oxford, UK.

Clarke, K. C. (1990). *Analytical and Computer Cartography.* Prentice-Hall, Englewood Cliffs, NJ.

Clements, R. R. (1989). *Mathematical Modeling: A Case Study Approach.* Cambridge University Press, Cambridge, UK.

Cleveland, W. S. (1985). *The Elements of Graphing Data.* Wadsworth, Monterey, CA.

Cleveland, W. S. (1993). *Visualizing Data.* Hobart Press, Summit, NJ.

Cleveland, W. S., and S. J. Devlin (1988). Locally weighted regression: An approach to regression analysis by local fitting. *Journal of the American Statistical Association 83*: 596–610.

Cliff, A. D., and J. K. Ord (1973). *Spatial Autocorrelation.* Pion Limited, London, UK.

Cliff, A. D., and J. K. Ord (1981). *Spatial Processes: Models and Applications.* Pion Limited, London, UK.

Clifford, P., and S. Richardson (1985). Testing the association between two spatial processes. In Dudewicz, E. J., Plachky, D., and Sen, P. K., Editors. *Statistics and Decisions* pp. 155–160. R. Oldenbourg Verlag, Munich, Germany.

Clifford, P., S. Richardson, and D. Hemon (1989). Assessing the significance of the correlation between two spatial processes. *Biometrics 45*: 123–134.

Cochran, W. G. (1977). *Sampling Techniques.* Wiley, New York.

Cohen, J. (1960). A coefficient of agreement for nominal scales. *Educational and Psychological Measurement 20*: 37–46.

Cohen, P. R. (1985). *Heuristic Reasoning about Uncertainty: An Artificial Intelligence Approach.* Pitman Publishing, Boston, MA.

Colvin, T. S., and S. Arslan (2001). A review of yield reconstruction and sources of errors in yield maps. In Robert, P. C., Rust, R. H., and Larson, W. E., Editors. *Proceedings of the Fifth International Conference on Precision Agriculture* p. Unpaginated CD. American Society of Agronomy, Madison, WI.

Conard, S. G., R. L. MacDonals, and R. F. Holland (1977). Riparian vegetation and flora of the Sacramento Valley. In Sands, A., Editor. *Riparian Forests of California, their Ecology and Conservation* pp. 47–55. Publication 4101, Division of Agricultural Sciences, University of California, Berkeley, CA.

Congdon, P. (2003). *Applied Bayesian Modeling.* John Wiley & Sons, West Sussex, UK.

Congdon, P. (2010). *Applied Bayesian Hierarchical Methods.* CRC Press, Boca Raton, FL.

Cooperrider, A. Y. (1986). Habitat evaluation systems. In Boyd, R. J. and Stuart, H. R., Editors. *Inventory and Monitoring of Wildlife Habitat* pp. 757–776. US Department of the Interior, Bureau of Land Management, Denver, CO.

Corwin, D. L., and S. M. Lesch (2005). Apparent soil electrical conductivity measurements in agriculture. *Computers and Electronics in Agriculture 46*: 11–44.

Crawley, M. J. (2007). *The R Book*. Wiley, Chichester, UK.

Cressie, N. (1997). Change of support and the modifiable areal unit problem. *Geographical Systems 3*: 159–180.

Cressie, N., and C. K. Wikle (2011). *Statistics for Spatio-Temporal Data*. Wiley, New York.

Cressie, N. A. C. (1991). *Statistics for Spatial Data*. Wiley, New York.

Cunningham, R. B., and D. B. Lindenmeyer (2005). Modeling count data of rare species: Some statistical issues. *Ecology 86*: 1135–1142.

Dacey, M. F. (1965). A review of measures of contiguity for two and k-color maps. In Berry, B. L. J. and Marble, D. F., Editors. *Spatial Analysis: A Reader in Statistical Geography* pp. 479–495. Prentice-Hall, Englewood Cliffs, NJ.

Daniel, C. (1960). Locating outliers in factorial experiments. *Technometrics 2*: 149–156.

Davies, L., and U. Gather (1993). The identification of multiple outliers. *Journal of the American Statistical Association 88*: 782–792.

Davis, J. C. (1986). *Statistics and Data Analysis in Geology*. Wiley, New York.

Davison, A. C., and D. V. Hinkley (1997). *Bootstrap Methods and their Application*. Cambridge University Press, Cambridge, UK.

Davison, A. C., and E. J. Snell (1991). Residuals and diagnostics. In Hinkley, D. V., Reid, N., and Snell, E. J., Editors. *Statistical Theory and Modeling: In Honour of Sir David Cox, FRS* pp. 83–106. Chapman & Hall, London, UK.

Dawes, R. M., and B. Corrigan (1974). Linear models in decision making. *Psychological Bulletin 81*: 95–106.

De'ath, G. (2002). Multivariate regression trees: A new technique for modeling species-environment relationships. *Ecology 83*: 1105–1117.

De'ath, G., and K. E. Fabricius (2000). Classification and regression trees: A powerful yet simple technique for ecological data analysis. *Ecology 81*: 3178–3192.

de Smith, M. J., M. F. Goodchild, and P. A. Longley (2007). *Geospatial Analysis: A Comprehensive Guide to Principles, Techniques and Software Tools*. Matador, Leicester, UK.

Demers, M. N. (2009). *Fundamentals of Geographic Information Systems*. Wiley, New York.

Dempster, A. P. (1990). Construction and local computation aspects of network belief functions. In Oliver, R. M. and Smith, J. Q., Editors. *Influence Diagrams, Belief Nets and Decision Analysis* pp. 121–142. New York, Wiley.

Dennis, B., R. A. Desharnais, J. M. Cushing, and R. F. Costantino (1995). Nonlinear demographic dynamics: Mathematical models, statistical methods, and biological experiments. *Ecological Monographs 65*: 261–281.

Dettman, J. W. (1969). *Mathematics of Physics and Modern Engineering*. McGraw-Hill, New York.

Deutsch, C. V., and A. G. Journel (1992). *GSLIB Geostatistical Software Library and User's Guide*. Oxford University Press, New York.

Diggle, P. J. (1983). *Statistical Analysis of Spatial Point Patterns*. Academic Press, New York.

Diggle, P. J., P. Heagerty, K.-Y. Liang, and S. L. Zeger (2002). *Analysis of Longitudinal Data*. Oxford University Press, Oxford, UK.

Dimitriadou, E., K. Hornik, A. Weingessel, and F. Leisch (2017). e1071: Misc functions of the department of statistics, probability theory group (Formerly: E1071), TU Wien. R package version 1.6-8. https://CRAN.R-project.org/package=e1071.

Diniz-Filho, J. A. F., L. M. Bini, and B. A. Hawkins (2003). Spatial autocorrelation and red herrings in geographical ecology. *Global Ecology and Biogeography 12*(1): 53–64.

Diniz-Filho, J. A. F., B. A. Hawkins, L. M. Bini, P. De Marco, and T. M. Blackburn (2007). Are spatial regression methods a panacea or a Pandora's box? A reply to Beale et al. (2007). *Ecography 30*: 848–851.

Dobermann, A., and J. L. Ping (2004). Geostatistical integration of yield monitor data and remote sensing improves yield maps. *Agronomy Journal 96*: 285–297.

Donald, C. M., and J. Hamblin (1976). The biological yield and harvest index of cereals as agronomic and plant breeding criteria. In Norman, A. G., Editor. *Advances in Agronomy* pp. 361–411. Academic Press, San Diego, CA.

Dormann, C. F. (2007). Assessing the validity of autologistic regression. *Ecological Modeling 207*: 234–242.

Dormann, C. F., J. M. McPherson, M. B. Araújo, R. Bivand, J. Bolliger, G. Carl, R. G. Davies, et al. (2007). Methods to account for spatial autocorrelation in the analysis of species distributional data: a review. *Ecography 30*: 609–628.

Duncan, O. D., R. P. Cuzzort, and B. Duncan (1960). *Statistical Geography*. The Free Press, Glencoe, IL.

Dutilleul, P. (1993). Modifying the t test for assessing the correlation between two spatial processes. *Biometrics 49*: 305–314.

Edgington, E. S. (2007). *Randomization Tests*. Chapman & Hall/CRC, Boca Raton, FL.

Edwards, D. (2000). Data quality assurance. In Michener, W. K. and Brunt, J. W., Editors. *Ecological Data: Design, Management, and Processing* pp. 70–91. Blackwell Science, Oxford, UK.

Efron, B. (1979). Bootstrap methods: Another look at the jackknife. *Annals of Statistics 7*: 1–26.

Efron, B. (1986). Why isn't everyone a Bayesian? *The American Statistician 40*: 1–5.

Efron, B., and R. J. Tibshirani (1993). *An Introduction to the Bootstrap*. Chapman & Hall, New York.

Elliot, G., T. Rothenberg, and J. Stock (1996). Efficient tests for an autoregressive unit root. *Econometrica 64*: 155–181.

Emerson, J. D., and D. C. Hoaglin (1983). Analysis of two-way tables by medians. In Hoaglin, D. C., Mosteller, F., and Tukey, J. W., Editors. *Understanding Robust and Exploratory Data Analysis* pp. 166–210. Wiley, New York.

Engle, R. F. (1984). Wald, liklihood ratio, and Lagrange multiplier tests in econometrics. In Griliches, Z., and Intriligator, M. D., Editors. *Handbook of Econmetrics, Volume II* pp. 776–826. Elsevier Science Publishers, Amsterdam, the Netherlands.

Evans, J. C., and B. G. Coote (1993). Matching sampling designs and significance tests in environmental studies. *Environmetrics 4*: 413–437.

Evett, R. R. (1994). Determining environmental realized niches for six oak species in California through direct gradient analysis and ecological response surface modeling. Dissertation, University of California, Berkeley, CA.

Faraway, J. J. (2002). Practical regression and ANOVA using R. cran.r-project.org/doc/contrib/Faraway-PRA.pdf, The R Project.

Federer, W. T., and C. S. Schlottfeldt (1954). The use of covariance to control gradients in experiments. *Biometrics 10*: 282–290.

Ferraro, D. O., C. M. Ghersa, and D. E. Rivero (2012). Weed vegetation of sugarcane cropping systems of northern Argentina: Data-mining methods for assessing the environmental and management effects on species composition. *Weed Science 60*: 27–33.

Finley, A. O., S. Banerjee, and B. P. Carlin (2015). spBayes for large univariate and multivariate point-referenced spatio-temporal data models. *Journal of Statistical Software 63*(13): 1–28.

Fisher, R. A. (1922). On the interpretation of chi-square from contingency tables, and the calculation of P. *Journal of the Royal Statistical Society 85*: 87–94.

Fisher, R. A. (1935). *The Design of Experiments*. Oliver and Boyd, London, UK.

Fisher, R. A. (1936). The use of multiple measurements in taxonomic problems. *Annals of Eugenics 7*: 179–188.

Fortin, M. J., and M. R. T. Dale (2005). *Spatial Analysis: A Guide for Ecologists*. Cambridge University Press, Cambridge, UK.

Fortin, M.-J., and G. Jacquez (2000). Randomization tests and spatially autocorrelated data. *Bulletin of the Ecological Society of America 81*: 201–205.

Fortune, S. J. (1987). A sweepline algorithm for Voronoi diagrams. *Algorithmica 2*: 153–174.

Fotheringham, A. S., C. Brundson, and M. Charlton (2000). *Quantitative Geography*. Sage, London, UK.

Fotheringham, A. S., C. Brundson, and M. Charlton (2002). *Geographically Weighted Regression: The Analysis of Spatially Varying Relationships*. Wiley, Chichester, UK.

Fotheringham, A. S., M. Charlton, and C. Brunsdon (1997). Measuring spatial variations in relationships with geographically weighted regression. In Fischer, M. M. and Getis, A., Editors. *Recent Developments in Spatial Analysis* pp. 60–82. Springer-Verlag, Berlin, Germany.

Fotheringham, A. S., and P. A. Rogerson, Editors (2009). *The SAGE Handbook of Spatial Analysis*. Sage Publications, Los Angeles, CA.

Fox, J. (1991). *Regression Diagnostics*. Sage Publications, Newbury Park, CA.

Fox, J. (1997). *Applied Regression Analysis, Linear Models, and Related Methods*. Sage Publications, Thousand Oaks, CA.

Fox, J., and S. Weisberg (2011). *An R Companion to Applied Regression*. Sage Publications, Thousand Oaks, CA.

Freeman, D. H. Jr. (1987). *Applied Categorical Data Analysis*. Marecl Dekker, New York.

Gaines, D. A. (1974). Review of the status of the yellow-billed cuckoo in California: Sacramento Valley populations. *The Condor 76*: 204–209.

Gaines, D. A., and S. A. Laymon (1984). Decline, status, and preservation of the yellow-billed cuckoo in California. *Western Birds 15*: 49–80.

Geary, R. C. (1954). The contiguity ratio and statistical mapping. *The Incorporated Statistician 5*: 115–145.

Gehlke, C. E., and K. Biehl (1934). Certain effects of grouping upon the size of the correlation coefficient in census tract material. *Journal of the American Statistical Association, Supplement 29*: 169–170.

Gelb, A. (1974). *Applied Optimal Estimation*. MIT Press, Cambridge, MA.

Gelfand, A. E., A. Latimer, S. Wu, and J. A. Jr. Silander (2006). Building statistical models to analyze species distributions. In Clark, J. S. and Gelfand, A. E., Editors. *Hierarchical Modelling for the Environmental Sciences* pp. 77–97. Oxford University Press, Oxford, UK.

Gelman, A., J. B. Carlin, H. S. Stern, D. B. Dunson, A. Vehtari, and D. B. Rubin (2014). *Bayesian Data Analysis*, 3rd ed. CRC Press, Boca Raton, FL.

Gelman, A., and J. Hill (2007). *Data Analysis Using Regression and Multilevel/Hierarchical Models*. Cambridge University Press, Cambridge, UK.

Gelman, A., and D. Rubin (1992). Inference from iterative simulation using multiple sequences. *Statistical Science 7*: 457–511.

Geman, S., and D. Geman (1984). Stochastic relaxation, Gibbs distributions and the Bayesian restoration of images. *IEEE Transactions on Pattern Analysis and Machine Intelligence 6*: 721–741.

George, M. R., J. R. Brown, and W. J. Clawson (1992). Application of nonequilibrium ecology to management of Mediterranean grasslands. *Journal of Range Management 45*: 436–440.

German, S., and D. German (1984). Stochastic relaxation, Gibbs distributions and the Bayesian restoration of images. *IEEE Transactions on Pattern Analysis and Machine Intelligence 6*: 721–741.

Gething, P. A. (1993). Improving returns from nitrogen fertilizer: The potassium-nitrogen partnership. International Potash Institute Technical Report 13, Basel, Switzerland.

Getis, A., and J. Aldstadt (2004). Constructing the spatial weights matrix using a local statistic. *Geographical Analysis 36*: 90–104.

Getis, A., and J. K. Ord (1992). The analysis of spatial association by use of distance statistics. *Geographical Analysis 24*: 189–206.

Getis, A., and J. K. Ord (1996). Local spatial statistics: An overview. In Longley, P. and Batty, M., Editors. *Spatial Analysis: Modelling in a GIS Environment* pp. 261–277. GeoInformation International, Cambridge, UK.

Geweke, J. (1992). Evaluating the accuracy of sampling-based approaches to the calculation of posterior moments. In Berger, J. O., Bernardo, J. M., Dawid, A. P., and Smith, A. F. M., Editors. *Proceedings of the Fourth Valencia International Meeting on Bayesian Statistics* p. 169. Oxford University Press, Oxford, UK.

Gilks, W. R., S. Richardson, and D. J. Spiegelhalter, Editors. (2010). *Markov Chain Monte Carlo in Practice*. Chapman & Hall, Boca Raton, FL.

Girvetz, E. G. S. E. (2007). How to define a patch: A spatial model for hierarchically delineating organism-specific habitat patches. *Landscape Ecology 22*: 1131–1142.

Girvetz, E. G. S. E. (2009). Multi-scale predictive habitat suitability modeling based on hierarchically delineated patches: An example for yellow-billed cuckoos nesting in riparian forests, California, USA. *Landscape Ecology* 24: 1315–1329.

Goldberger, A. S. (1973). Structural equation models: An overview. In Goldberger, A. S. and Duncan, O. D., Editors. *Structural Equation Models in the Social Sciences* pp. 1–18. Seminar Press, New York.

Gomez, K. A., and A. A. Gomez (1984). *Statistical Procedures for Agricultural Research*. Wiley, New York.

Good, P. I. (2005). *Permutation, Parametric, and Bootstrap Tests of Hypotheses*. Springer, New York.

Good, P. I. (2001). *Resampling Methods*. Birkhauser, Boston, MA.

Goodchild, M. F. (1986). *Spatial Autocorrelation*. Geo Books, Norwich, UK.

Goodliffe, P. (2007). *Code Craft: The Practice of Writing Efficient Code*. No Starch Press, San Francisco, CA.

Goovaerts, P. (1997). *Geostatistics for Natural Resources Evaluation*. Oxford, UK, Oxford University Press.

Gotway, C. A., and W. W. Stroup (1997). A generalized linear model approach to spatial data analysis and prediction. *Journal of Agricultural, Biological, and Environmental Statistics* 2: 157–178.

Gotway, C. A., and R. D. Wolfinger (2003). Spatial prediction of counts and rates. *Satatistics in Medicine* 22: 1415–1432.

Gotway, C. A., and L. J. Young (2002). Combining incompatible spatial data. *Journal of the American Statistical Association* 97: 632–648.

Gotway, C. A., and L. J. Young (2007). A geostatistical approach to linking geographically aggregated data from different sources. *Journal of Computational and Graphical Statistics* 16: 115–135.

Gräler, B. (2016). gstat/demo/stkrige.R. https://github.com/edzer/gstat/blob/master/demo/stkrige.R.

Gräler B., E. J. Pebesma, and G. Heuelink (2016). Spatio-temporal interpolation using gstat. http://cran.r-project.org/web/packages/gstat. Accessed March 3, 2018.

Greco, S. E. (1999). Monitoring riparian landscape change and modeling habitat dynamics of the yellow-billed cuckoo on the Sacramento River, California. PhD Dissertation, University of California, Davis, CA.

Greco, S. E., A. K. Fremier, E. W. Larsen, and R. E. Plant (2007). A tool for tracking floodplain age land surface patterns on a large meandering river with applications for ecological planning and restoration design. *Landscape and Urban Planning* 81: 354–373.

Greco, S. E., R. E. Plant, and R. H. Barrett (2002). Geographic modeling of temporal variability in habitat quality of the yellow-billed cuckoo on the Sacramento River. In Scott, J. M., Heglund, P. J., and Morrison, M. L., Editors. *Predicting Species Occurrences: Issues of Accuracy and Scale* pp. 183–196. Island Press, Washington, DC.

Greenland, S., and H. Morgenstern (1989). Ecological bias, confounding, and effect modification. *International Journal of Epidemiology* 18: 269–274.

Griffin, T. W., J. P. Brown, and J. Lowenberg-DeBoer (2007). Yield monitor data analysis protocol: A primer in the management and analysis of precision agriculture data. http://www.purdue.edu/ssme, Purdue University Site Specific Management Center Publication.

Griffith, D. A. (1988). *Advanced Spatial Statistics*. Kluwer, Dordrecht, the Netherlands.

Griffith, D. A. (2005). Effective geographic sample size in the presence of spatial autocorrelation. *Annals of the Association of American Geographers* 95: 740–760.

Griffith, D. A. (1987). *Spatial Autocorrelation: A Primer*. Association of American Geographers, Washington, DC.

Griffith, D. A. (1978). A spatially adjusted ANOVA model. *Geographical Analysis* 10: 296–301.

Griffith, D. A., and L. J. Layne (1999). *A Casebook for Spatial Statistical Data Analysis*. Oxford University Press, Oxford, UK.

Griffith, D. A. (1989). *Spatial Statistics,: Past, Present, and Future*. Michigan Document Services, Ann Arbor, MI.

Gu, C. (2014). Smoothing Spline ANOVA Models: R Package gss. *Journal of Statistical Software* 58(5): 1–25. URL http://www.jstatsoft.org/v58/i05/.

Guyon, X. (1995). *Random Fields on a Network*. Springer-Verlag, New York.

Hadfield, J. D. (2010). MCMC methods for multi-response generalized linear mixed models: The MCMCglmm R package. *Journal of Statistical Software 33*: 1–22.

Haining, R. (1991). Bivariate correlation with spatial data. *Geographical Analysis 23*: 210–227.

Haining, R. (1988). Estimating spatial means with an application to remotely sensed data. *Communications in Statistics: Theory and Methods 17*: 573–597.

Haining, R. (1990). *Spatial Data Analysis in the Social and Environmental Sciences.* Cambridge University Press, Cambridge, UK.

Haining, R. (2003). *Spatial Data Analysis: Theory and Practice.* Cambridge University Press, Cambridge, UK.

Halterman, M. D. (1991). Distribution and habitat use of the yellow-billed cuckoo (Coccyzus americanus occidentalis) on the Sacramento River, California, 1987090. M.S. Thesis, California State University, Chico, CA.

Haneklaus, S., H. Lilienthal, E. Schnug, K. Panten, and E. Haveresch (2001). Routines for efficient yield mapping. In Robert, P. C., Rust, R. H., and Larson, W. E., Editors. *Proceedings of the Fifth International Conference on Precision Agriculture* p. Unpaginated CD. American Society of Agronomy, Madison, WI.

Harrell, F. E. (2001). *Regression Modeling Strategies.* Springer, New York.

Harrell, F. E. Jr. (2011). rms: Regression Modeling Strategies. http://CRAN.R-project.org/package=rms.

Hartling, J. M., L. E. Druffel, and F. J. Hilbing (1983). *Introduction to Computer Programming: a Problem-Solving Approach.* Digital Press, Bedford, MA.

Hastie, T. (2017). gam: Generalized Additive Models. R package version 1.14-4. https://CRAN.R-project.org/package=gam

Hastie, T., and R. Tibshirani (1986). Generalized additive models (with discussion) *Journal of Statistical Science 1*: 297–318.

Hastie, T., R. Tibshirani, and J. Friedman (2009). *The Elements of Statistical Learning: Data Mining, Inference, and Prediction.* Springer-Verlag, New York.

Hastings, W. K. (1970). Monte Carlo sampling methods using Markov chains and their applications. *Biometrika 57*: 97–109.

Hauselt, P., and R. E. Plant (2010). Spatial modeling of water use in California rice production. *The Professional Geographer 62*: 462–477.

Hawkins, B. A., J. A. F. Diniz-Filho, L. M. Bini, P. De Marco, and T. M. Blackburn (2007). Red herrings revisited: Spatial autocorrelation and parameter estimation in geographical ecology. *Ecography 30*: 375–384.

Hay, R. K. M. (1995). Harvest index: A review of its use in plant breeding and crop physiology. *Annals of Applied Biology 126*: 197–216.

Heiberger, R. M., and B. Holland (2004). *Statistical Analysis and Data Display.* Springer, New York.

Heidelberger, P., and P. Welch (1992). Simulation run length control in the presence of an initial transient. *Operations Research 31*: 1109–1144.

Henderson, C. R. (1982). Analysis of covariance in the mixed model: higher-level, nonhomogeneous, and random regressions. *Biometrics 38*: 623–640.

Henderson, H. V., and P. E. Velleman (1981). Building multiple regression models interactively. *Biometrics 37*: 391–411.

Hengl, T., E. van Loon, H. Sierdsema, and W. Bouten (2008). Advancing spatio-temporal analysis of ecological data: Examples in R. In Gervasi, O., Murgante, B., Lagana, A., Taniar, D., and Mun, Y., Editors. *Computational Science and its Applications: ICCSA 2008* pp. 692–707. Springer, Berlin/Heidelberg, Germany.

Hijmans, R. J. (2016). raster: Geographic Data Analysis and Modeling. R package version 2.5-8. https://CRAN.R-project.org/package=raster.

Hilborn, R., and M. Mangel (1997). *The Ecological Detective. Confronting Models with Data.* Princeton University Press, Princeton, NJ.

Hill, J., S. R. Roberts, D. M. Brandon, S. C. Scardaci, J. F. Williams, C. M. Wick, W. M. Canevari, and B. L. Weir (1992). *Rice Production in California*. Cooperative Extension, University of California, Oakland, CA.

Hill, M. O. (1979b). *DECORANA: A FORTRAN Program for Detrended Correspondence Analysis and Reciprocal Averaging*. Cornell University, Ithaca, NY.

Hill, M. O. (1979a). *TWINSPAN: A FORTRAN Program for Arranging Multivariate Data in an Ordered Two-Way Table by Classification of Individuals and Attributes*. Cornell University, Ithaca, NY.

Hines, R. J. O., and E. M. Carter (1993). Improved added variable and partial residual plots for the detection of influential observations in generalized linear models. *Journal of the Royal Statistical Society. Series C (Applied Statistics 42*: 3–20.

Hoel, P. G. (1971). *Introduction to Mathematical Statistics*. Wiley, New York.

Hoffmann, C. W. (2015). cwhmisc: CWH's functions for math, plotting, printing, statistics, strings, and tools. http://CRAN.R-project.org/package=cwhmisc.

Hope, A. C. A. (1968). A simplified Monte Carlo significance test procedure. *Journal of the Royal Statistical Society Series B 30*: 582–598.

Hunt, R. (1982). *Plant Growth Curves: The Functional Approach to Plant Growth Analysis*. University Park Press, Baltimore, MD.

Huntsinger, L., and J. W. Bartolome (1992). Ecological dynamics of Quercus dominated woodlands in California and southern Spain: A state-transition model. *Vegetatio 99–100*: 299–305.

Hurlbert, S. H. (1984). Psuedoreplication and the design of ecological field experiments. *Ecological Monographs 54*: 187–211.

Höfer, T., H. Przyrembel, and S. Verleger (2004). New evidence for the theory of the stork. *Paedoatric and Perinatal Epidemiology 18*: 88–92.

Isaaks, E. H., and R. M. Srivastava (1989). *An Introduction to Applied Geostatistics*. Oxford University Press, Oxford, UK.

Iverson, L. R. (2007). Adequate data of known accuracy are critical to advancing the field of landscape ecology. In Wu, J. and Hobbs, R., Editors. *Key Topics in Landscape Ecology* pp. 11–38. Cambridge University Press, Cambridge, UK.

Jackson, L. (1999). Nitrogen fertilization and grain protein content in California wheat. Fertilizer Research and Extension Program Report 97-0365: www.cdfa.ca.gov/is/docs/Jackson99.pdf.

James, G., D. Witten, T. Hastie, and R. Tibshirani (2013). *An Introduction to Statistical Learning*. Springer, New York.

Jenny, H. (1941). *Factors of Soil Formation*. McGraw-Hill, New York.

Jensen, F. V. (2001). *Bayesian Networks and Decision Graphs*. Springer, New York.

Jensen, H. A. (1947). A system for classifying vegetation in California. *California Fish and Game 33*: 199–266.

Jepson, W. L. (1910). *The Silva of California*. University of California Press, Berkeley, CA.

Jiang, P., Z. He, N. R. Kitchen, and K. A. Sudduth (2009). Bayesian analysis of within-field variability of corn yield using a spatial hierarchical model. *Precision Agriculture 10*: 111–127.

Johnston, C. (1998). *Geographic Information Systems in Ecology*. Blackwell Science, Oxford, UK.

Johnston, J. (1972). *Econometric Methods*. McGraw-Hill, New York.

Johnston, J., and J. DiNardo (1997). *Econometric Methods*, 4th edition. McGraw-Hill, New York.

Journel, A. G., and C. J. Huijbregts (1978). *Mining Geostatistics*. Academic Press, New York.

Juerschik, P., A. Giebel, and O. Wendroth (1999). Processing point data from combine harvesters for precision farming. In Stafford, J. V., Editor. *Precision Agriculture '99: Papers Presented at the Second European Conference on Precision Agriculture* pp. 297–318. Sheffield, Sheffield, UK.

Kanciruk, P., R. J. Olson, and R. A. McCord (1986). Data quality assurance in a research data base: the 1984 national water survey. In Michener, W. K., Editor. *Research Data Management in the Ecological Sciences* pp. 193–207. University of South Carolina Press, Columbia, SC.

Kanevski, M., M. Maignam. (2004). *Analysis and Modeling of Spatial Environmental Data*. Marcel Dekker, New York.

Kaya, A., and H.-Y. Fang (1997). Identification of contaminated soils by dielectric constant and electrical conductivity. *Journal of Environmental Engineering 123*: 169–177.

Keitt, T. H., R. Bivand, E. Pebesma, and B. Rowlingson (2017). rgdal: Bindings for the Geospatial Data Abstraction Library. http://www.R-project.org.

Kempthorne, O. (1952). *The Design and Analysis of Experiments*. Wiley, New York.

Kempthorne, O., and T. E. Doerfler (1969). The behavior of some significance tests under experimental randomization. *Biometrika 56*: 231–248.

Kendall, M., and A. Stuart (1979). *The Advanced Theory of Statistics*. Charles Griffin, London, UK.

Kendall, M., and J. K. Ord (1990). *Time Series*. Edward Arnold, Kent, Great Britain.

Kepner, R. A., R. Bainer, and E. L. Barger (1978). *Principles of Farm Machinery*. AVI Publishing, Westport, CT.

Kish, L. (1965). *Survey Sampling*. Wiley, New York.

Klingman, G. C. (1961). *Weed Control: As a Science*. Wiley, New York.

Koop, G. (2003). *Bayesian Econometrics*. Wiley, Chichester, UK.

Krebs, C. J. (1994). *Ecology: The Experimental Analysis of Distribution and Abundance*. HarperCollins, New York.

Krige, D. G. (1951). A statistical approach to some mine valuations and allied problems at the Witwatersrand. Master's thesis, University of Witwatersrand, Witwatersrand, South Africa.

Korner-Nievergelt, F. S. von Felten, T. Roth, B. Almasi, J. Guélat, and P. Korner-Nievergelt (2015). *Bayesian Data Analysis in Ecology Using Linear Models with R, BUGS, and Stan*. Elsevier, New York.

Kronmal, R. A. (1993). Spurious correlation and the fallacy of the ratio standard revisited. *Journal of the Royal Statistical Society Series A 156*: 379–392.

Kruskal, W. H. (1960). Some remarks on wild observations. *Technometrics 2*: 1–3.

Kutner, M. H., C. J. Nachtsheim, J. Neter, and W. Li (2005). *Applied Linear Statistical Models*. McGraw-Hill, Boston, MA.

Lacombe, D. J. (2004). Does econometric methodology matter? An analysis of public policy using spatial econometric techniques. *Geographical Analysis 36*: 105–118.

Lahiri, S. N., and J. Zhu (2006). Resampling methods for spatial regression models under a class of stochastic designs. *The Annals of Statistics 34*: 1774–1813.

Lambert, D. (1992). Zero-inflated poisson regression with an application to defects in manufacturing. *Technometrics 34*: 1–14.

Lambert, D. M., J. Lowenberg-DeBoer, and R. Bongiovanni (2004). A comparison of four spatial regression models for yield monitor data: A case study from Argentina. *Precision Agriculture 5*: 579–600.

Landwehr, J. M., D. Pregibon, and A. C. Shoemaker (1984). Graphical methods for assessing logistic regression models. *Journal of the American Statistical Association 79*: 61–71.

Langbein, L. I., and A. J. Lichtman (1978). *Ecological Inference*. Sage University Press, Beverly Hills, CA.

Lark, R.M. (1997). An empirical method for describing the joint effects of environmental and other variables on crop yield. *Communications in Applied Biology 131*: 141–159.

Lark, R. M., and R. Webster (2001). Changes in variance and correlation of soil properties with scale and location: Analysis using an adapted maximal overlap discrete wavelet transform. *European Journal of Soil Science 52*: 547–562.

Larsen, R. J., and M. L. Marx (1986). *An Introduction to Mathematical Statistics and its Applications*. Prentice-Hall, Englewood Cliffs, NJ.

Larsen, W. A., and S. J. McCleary (1972). The use of partial residual plots in regression analysis. *Technometrics 14*: 781–790.

Laymon, S. A., and M. D. Halterman (1987). Can the western subspecies of the yellow-billed cuckoo be saved from extinctin? *Western Birds 18*: 19–25.

Laymon, S. A., and M. D. Halterman (1989). A proposed habitat management plan for yellow-billed cuckoos in California. In D.L. Abell, Technical Coordinator. *Proceedings of the California Riparian Systems Conference: Protection, Management, and Restoration for the 1990's* pp. 272–277. USDA Forest Service, Pacific Southwest Forest and Range Experiment Station, Berkeley, CA.

Lee, J., J. Six, A. P. King, C. van Kessel, and D. E. Rolston (2005). Tillage and field scale controls on greenhouse gas emissions. *Journal of Environmental Quality 35*: 714–725.

Legendre, L., and P. Legendre (1998). *Numerical Ecology*. Elsevier, New York.

Leisch, F. (2004). S4 classes and methods. *UseR! 2004, Vienna, Austria* http://www.ci.tuwien.ac.at/Conferences/useR-2004/Keynotes/Leisch.pdf.

Lemon, J. (2006). Plotrix: A package in the red light district of R. *R News 6*: 8–12.

LeSage, J., and K. P. Pace (2009). *Introduction to Spatial Econometrics*. CRC Press, Boca Raton, FL.

Lesch, S. M., D. J. Strauss, and J. D. Rhoades (1995). Spatial prediction of soil salininty using electromagnetic induction techniques: 2: An efficient spatial sampling algorithm suitable for multiple linear regression model identification and estimation. *Water Resources Research 31*: 387–398.

Levy, R., and A. V. Crawford (2009). Incorporating substantive knowledge into regression via a Bayesian approach to modeling. *Multiple Linear Regression Viewpoints 35*: 4–9.

Lewin-Koh, N. J., and R. Bivand (2011). maptools: Tools for reading and handling spatial objects. *R Package Version 0.7-12*. URL http://www.R-project.org.

Lewis, G. N., and Randall. M. (1961). *Thermodynamics*. McGraw-Hill, New York.

Liang, K., and S. L. Zeger (1986). Longitudinal data analysis using generalized linear models. *Biometrika 73*: 13–22.

Liaw, A., and M. Wiener (2002). Classification and regression by randomForest. *R News 2/3*: 18–22.

Lichstein, J. W., T. R. Simons, S. A. Shriner, and K. E. Franzreb (2002). Spatial autocorrelation and autoregressive models in ecology. *Ecological Monographs 72*: 445–463.

Ligges, U. (2006). R help desk: Accessing the sources. *R News 6(4)*: 43–45.

Littell, R. C., W. W. Stroup, and R. J. Freund (2002). *SAS for Linear Models*. SAS Institute, Cary, NC.

Little, T. M., and F. J. Hills (1978). *Agricultural Experimentation*. John Wiley & Sons, New York.

Lloyd, C. D. (2010). *Spatial Data Analysis: An Introduction for GIS Users*. Oxford University Press, Oxford, UK.

Lo, C. P., and A. K. W. Yeung (2007). *Concepts and Techniques in Geographic Information Systems*. Pearson Prentice Hall, Upper Saddle River, NJ.

Lobell, D. B., J. I. Ortiz-Monasterio, G. P. Asner, R. L. Naylor, and W. P. Falcon (2005). Combining field surveys, remote sensing, and regression trees to understand yield variations in an irrigated wheat landscape. *Agronomy Journal 97*: 241–249.

Long, D. S. (1996). Spatial statistics for analysis of variance of agronomic field trials. In Arlinghaus, S. L., Editor. *Practical Handbook of Spatial Statistics* pp. 251–278. CRC Press, Boca Raton, FL.

Longley, P., and M. Batty, Editors (1996). *Spatial Analysis: Modelling in a GIS Environment*. GeoInformation International, Cambridge, UK.

Longley, P. A., M. F. Goodchild, D. J. Maguire, and D. W. Rhind (2001). *Geographic Information Systems and Science*. Wiley, New York.

Loomis, R. S., and D. J. Connor (1992). *Crop Ecology: Productivity and Management in Agricultural Systems*. Oxford University Press, Oxford, UK.

López-Sánchez, A., J. Schroeder, S. Roig, M. Sobral, and R. Dirzo (2014). Effects of cattle management on oak regeneration in northern Californian mediterranean oak woodlands. *PLoS One 9(8)*: e105472. doi:10.1371/journal.pone.0105472.

Lowenberg-DeBoer, J., and K. Erickson (2000). *Precision Farming Profitability*. Purdue University, West Lafayette, IN.

Lumley, T., and A. Miller (2017). leaps: Regression subset selection. http://CRAN.R-project.org/package=leaps.

Lunn, D. J., A. Thomas, N. Best, and D. Spiegelhalter (2000). WinBUGS, a Bayesian modelling framework: concepts, structure, and extensibility. *Statistics and Computing 10*: 325–337.

Mallows, C. L. (1973). Some comments on Cp. *Technometrics 15*: 661–676.

Manly, B. F. J. (1997). *Randomization, Bootstrap and Monte Carlo Methods in Biology*. Chapman & Hall, London, UK.

Mantel, N. (1967). The detection of disease clustering and a generalized regression approach. *Cancer Research 27*: 209–220.

Mantel, N., and R. S. Valand (1970). A technique of nonparametric multivariate analysis. *Biometrics* 547–558.

Marchesi, C. (2009). Regional analyses of constraints affecting rice production systems in California: Grain milling quality at harvest and evolution of herbicide resistant Echinochloa spp. PhD Dissertation, University of California, Davis, CA.

Markwell, J., J. C. Osterman, and J. L. Mitchell (1995). Calibration of the Minolta SPAD-502 leaf chlorophyll meter. *Photosynthesis Research 46*: 467–472.

Marques de Sá, J. P. (2007). *Applied Statistics Using SPSS, STATISTICA, MATLAB, and R*. Springer, Berlin, Germany.

Marschner, H. (1995). *Mineral Nutrition of Higher Plants*. Academic Press, London, UK.

Matheron, G. (1963). Principles of geostatistics. *Economic Geology 58*: 1246–1266.

Matloff, N. (1991). Statistical hypothesis testing: Problems and alternatives. *Environmental Entomology 20*: 1246–1250.

Mayer, K. E., and W. F. Laudenslayer, Jr. (1988). *A Guide to the Wildlife Habitats of California*. California Department of Forestry and Fire Protection, Sacramento, CA.

McBratney, A. B., R. Webster, and T. M. Burgess (1981). The design of optimal sampling schemes for local estimation and mapping of regionalized variables: I—Theory and method. *Computers and Geosciences 7*: 331–334.

McCarthy, M. A. (2007). *Bayesian Methods for Ecology*. Cambridge University Press, Cambridge, UK.

McClaran, M. P. (1986). *Age Structure of Quercus douglasii in Relation to Livestock Grazing and Fire*. Ph.D. Dissertation, University of California, Berkeley, CA.

McCullagh, P. (1983). Quasi-likelihood functions. *Annals of Statistics 11*: 59–67.

McClullagh, P., and J. A. Nelder (1989). *Generalized Linear Models*. Chapman & Hall, New York.

McCulloch, C. E., S. R. Searle, and J. M. Neuhaus (2008). *Generalized, Linear, and Mixed Models*. Wiley, Hoboken, NJ.

McDonald, P. M. (1990). *Quercus douglasii* - blue oak. In Burns, R. M. and Honkaka, B. H., technical coordinators. *Silvics of North America: 2. Hardwoods. Agricultural Handbook 654* US Department of Agriculture, Forest Service, Washington, DC.

Melsa, J. L., and D. L. Cohn (1978). *Decision and Estimation Theory*. McGraw-Hill, New York.

Mercer, W. B., and A. D. Hall (1911). The experimental error of field trials. *Journal of Agricultural Science 4*: 107–132.

Messing, S. A. (1992). The impact of European settlement on blue oak (Quercus douglasii) regeneration and recruitment in the Tehachapi Mountains, California. *Madroño 39*: 36–46.

Metropolis, N., Rosenbluth, A. W., Rosenbluth, M. N., Teller, A. H., and Teller, E. (1953). Equations of state calculations by fast computing machines. *Journal of Chemical Physics 21*: 1087–1091.

Miron, J. (1984). Spatial autocorrelation in regression analysis: A beginner's guide. In Gaile, G. L. and Willmott, C. J., Editors. *Spatial Statistics and Models* pp. 201–222. D. Reidel, Dordrecht, the Netherlands.

Miroshnikov, A., and E. Conlon (2014). parallelMCMCcombine: Methods for combining independent subset Markov chain Monte Carlo (MCMC) posterior samples to estimate a posterior density given the full data set. R package version 1.0. https://CRAN.R-project.org/package=parallelMCMCcombine.

Moellering, H., and W. Tobler (1972). Geographical variances. *Geographical Analysis 4*: 34–50.

Moran, P. A. P. (1948). The interpretation of statistical maps. *Journal of the Royal Statistical Society, Series B 10*: 243–251.

Mosteller, F., and J. W. Tukey (1977). *Data Analysis and Regression*. Addison-Wesley, Reading, MA.

Muller, K. E., and B. A. Fetterman (2002). *Regression and ANOVA: An Integrated Approach Using SAS Software*. SAS Institute, Cary, NC.

Murphy, D. P., E. Schnug, and S. Haneklaus (1995). Yield mapping: A guide to improved techniques and strategies. In Robert, P. C., Rust, R. H., and Larson, W. E., Editors. *Proceedings of the Second International Conference on Precision Agriculture* pp. 32–47. American Society of Agronomy, Madison, WI.

Murrell, P. (2006). *R Graphics*. CRC Press, Boca Raton, FL.

Nakazawa, M. (2017). *fmsb: Functions for Medical Statistics Book with some Demographic Data*. R package version 0.6.0. https://CRAN.R-project.org/package=fmsb.

Naus, J. I. (1975). *Data Quality Control and Editing*. Marcel Dekker, New York.

Nelder, J. A., and R. W. M. Wedderburn (1972). Generalized linear models. *Journal of the Royal Statistical Society Series A 135*: 370–384.

Neuwirth, E. (2014). RColorBrewer: ColorBrewer Palettes. R package version 1.1-2. https://CRAN.R-project.org/package=RColorBrewer.

Nicholls, A. O. (1989). How to make biological surveys go further with generalised linear models. *Biological Conservation 50*: 51–75.

Noble, B., and J. W. Daniel (1977). *Applied Linear Algebra*. Prentice-Hall, Englewood Cliffs, NJ.

Nolan, S. C., G. W. Haverland, T. W. Goddate, M. Green, D. C. Penny, J. A. Henriksen, and G. Lachapelle (1996). Building a yield map from geo-referenced harvest measurements. In Robert, P. C., Rust, R. H., and Larson, W. E., Editors. *Proceedings of the Third International Conference on Precision Agriculture* pp. 885–892. American Society of Agronomy, Madison, WI.

Ntzoufras, I. (2009). *Bayesian Modeling using WinBUGS*. Wiley, Hoboken, NJ.

O'Neill, R. V., D. L. DeAngelis, J. B. Waide, and T. F. H. Allen (1986). *A Hierarchical Concept of Ecosystems*. Princeton University Press, Princeton, NJ.

Odeh, I. O. A., A. J. Todd, J. Triantafilis, and A. B. McBratney (1998). Status and trends of soil salinity at different scales: The case for the irrigated cotton growing region of eastern Australia. *Nutrient Cycling in Agroecosystems 50*: 99–107.

Oden, N. L., and R. R. Sokal (1992). An investigation of three-matrix permutation tests. *Journal of Classification 9*: 275–290.

Okubo, A. (1980). *Diffusion and Ecological Problems: Mathematical Models*. Springer-Verlag, New York.

Olea, R. A. (1984). Sampling design optimization for spatial functions. *Mathematical Geology 16*: 369–392.

Openshaw, S. (1984). Ecological fallacies and the analysis of areal census data. *Environment and Planning A 16*: 17–31.

Openshaw, S., and P. Taylor (1979). A million or so correlation coefficients: Three experiments on the modifiable areal unit problem. In Wrigley, N., Editor. *Statistical Applications in the Spatial Sciences* pp. 127–144. Pion, London, UK.

Openshaw, S., and Taylor. P. J. (1981). The modifiable areal unit problem. In Wrigley, N. and Bennett, R. J., Editors. *Quantitative Geography: A British View* pp. 62–69. Rutledge & Kegan, London, UK.

Ord, J. K., and A. Getis (1995). Local spatial autocorrelation statistics: Distributional issues and an application. *Geographical Analysis 27*: 286–306.

Ord, K. (1975). Estimation methods for models of spatial interaction. *Journal of the American Statistical Association 70*: 120–126.

Oksanen, J., F. G. Blanchet, M. Friendly, R. Kindt, P. Legendre, D. McGlinn, P. R. Minchin, et al. (2017). vegan: Community Ecology Package. R package version 2.4-3. https://CRAN.R-project.org/package=vegan.

Oyana, T. J., and F. Margai (2015). *Spatial Analysis: Statistics, Visualization, and Computational Methods*. CRC Press, Boca Raton, FL.

Pannatier, Y. (1996). *Variowin: Software for Spatial Data Analysis in 2D*. Springer-Verlag, New York.

Papadakis, J. S. (1937). Statistical methods for field experiments (in French). *Bulletin De L'Institut De L'Amelioraton Des Plantes Thessaloniki*: 23.

Pavlik, B. M., P. C. Muick, S. G. Johnson, and M. Popper (1991). *Oaks of California*. Cachuma Press, Olivos, CA.

Pearl, J. (1986). On evidential reasoning in a hierarchy of hypotheses. *Artificial Intelligence 28*: 9–15.

Pebesma, E. J. (2004). Multivariable geostatistics in S: The gstat package. *Computers and Geosciences 30*: 683–691.

Pebesma, E. (2011). Spacetime: Classes and methods for spatio-temporal data. http://CRAN.R-project. org/package=spacetime.

Pebesma, E. (2012). Spacetime: Spatio-temporal data in R. *Journal of Statistical Software 51*: 7.

Pebesma. E. (2017). sf: Simple Features for R. R package version 0.5-3. https://CRAN.R-project.org/ package=sf.

Pebesma, E. (2018). Map overlay and spatial aggregation in sp. https://cran.r-project.org/web/ packages/sp/vignettes/over.pdf.

Pebesma, E. Z. (2001). *Gstat User's Manual*. Utrecht University, Utrecht, the Netherlands.

Pelletier, G., and S. K. Upadhyaya (1999). Development of a tomato load/yield monitor. *Computers and Electronics in Agriculture 23*: 103–118.

Phillips, R. L., N. K. McDougald, D. McCreary, and E. R. Atwill (2007). Blue oak seedling age influences growth and mortality. *California Agriculture 61*: 11–15.

Pielou, E. C. (1977). *Mathematical Ecology*. Wiley-Interscience, New York.

Pinhiero, J. C., and D. M. Bates (2000). *Mixed-Effects Models in S and S-PLUS*. Springer, New York.

Pinheiro, J., Bates, D., DebRoy, S., Sarkar, D., and R Core Team (2017). nlme: Linear and Nonlinear Mixed Effects Models. R package version 3.1-131. https://CRAN.R-project.org/package=nlme>.

Plant, R. E. (2007). Comparison of means of spatial data in unreplicated field trials. *Agronomy Journal 99*: 481–488.

Plant, R. E., and R. T. Cunningham (1991). Analyses of the dispersal of sterile Mediterranean fruit flies (Diptera, Tephritidae) released from a point source. *Environmental Entomology 20*: 1493–1503.

Plant, R. E., A. Mermer, G. S. Pettygrove, M. P. Vayssières, J. A. Young, R. O. Miller, L. F. Jackson, R. F. Denison, and K. Phelps (1999). Factors underlying grain yield spatial variability in three irrigated wheat fields. *Transactions of the ASAE 42*: 1187–1202.

Plant, R. E., and N. S. Stone (1991). *Knowledge-Based Systems in Agriculture*. McGraw-Hill, New York.

Plant, R. E., M. P. Vayssières, S. E. Greco, M. R. George, and T. E. Adams (1999). A qualitative spatial model of hardwood rangeland state-and-transition dynamics. *Journal of Range Management 52*: 50–58.

Plant, R. E., and L. T. Wilson (1985). A Bayesian method for sequential sampling and forecasting in agricultural pest management. *Biometrics 41*: 203–214.

Plummer, M., N. Best, K. Cowles, and K. Vines (2011). coda: Output analysis and diagnostics for MCMC. http://CRAN.R-project.org/package=coda.

Pocknee, S., B. C. Boydell, H. M. Green, D. J. Waters, and C. K. Kvein (1996). Directed soil sampling. In Robert, P. C., Rust, R. H., and Larson, W. E., Editors. *Proceedings of the Third International Conference on Precision Agriculture* pp. 159–168. American Society of Agronomy, Madison, WI.

Post, E. (1983). Real programmers don't use Pascal. *Datamation 29*: 1983.

Pregibon, D. (1985). Link tests. In Kotz, S. and Johnson, N. L., Editors. *Encyclopedia of Statistical Sciences* pp. 82–85. Wiley, New York.

Press, W. H., B. P. Flannery, S. A. Teukolsky, and W. T. Vetterling (1986). *Numerical Recipes*. Cambridge University Press, Cambridge, UK.

R Development Core Team (2017). *R: A Language and Environment for Statistical Computing*. R Foundation for Statistical Computing, Vienna, Austria.

Raftery, A. E., and S. M. Lewis (1992). Comment: One Long Run with Diagnostics: Implementation Strategies for Markov Chain Monte Carlo. *Statistical Science 7*: 493–497.

Raiffa, H., and R. Schlaifer (1961). *Applied Statistical Decision Theory*. Harvard Business School, Boston, MA.

Ramsey, F. L. S. D. W. (2002). *The Statistical Sleuth*. Brooks/Cole, Belmont, CA.

Ravindran, A., K. M. Ragsdell, and G. V. Reklaitis (2006). *Engineering Optimization Methods and Applications*. Wiley, New York.

Renka, R. J., A. Gebhardt, S. Eglen, S. Zuyev, and D. White (2011). tripack: Triangulation of irregularly spaced data. URL http://www.R-project.org.

Ribeiro, P. J., Jr and P. J. Diggle (2016). geoR: Analysis of Geostatistical Data. R package version 1.7-5.2. https://CRAN.R-project.org/package=geoR

Rice, E. L. (1984). *Allelopathy*. Academic Press, New York.

Richter, S. H., J. P. Garner, and H. Wurbel (2009). Environmental standardization: Cure or cause of poor reproducibility in animal experiments? *Nature Methods 6*: 257–261.

Ripley, B. (1987). *Stochastic Simulation*. Wiley, New York.

Ripley, B. D. (1977). Modelling spatial patterns. *Journal of the Royal Statistical Society, Series B 39*: 172–192.

Ripley, B. D. (1981). *Spatial Statistics*. Wiley, New York.

Rizzo, M. L. (2008). *Statistical Computing with R*. Chapman & Hall, Boca Raton, FL.

Robinson, C., and R. E. Schumacker (2009). Interaction effects: Centering, variance inflation factor, and interpretation issues. *Multiple Linear Regression Viewpoints 35*: 6–11.

Robinson, T. P., and G. Metternicht (2005). Comparing the performance of techniques to improve the quality of yield maps. *Agricultural Systems 85*: 19–41.

Robinson, W. S. (1950). Ecological correlations and the behavior of individuals. *American Sociological Review 15*: 351–357.

Roel, A., H. Firpo, and R. E. Plant (2007). Why do some farmers get higher yields? Multivariate analysis of a group of Uruguayan rice farmers. *Computers and Electronics in Agriculture 58*: 78–92.

Roel, A., and R. E. Plant (2004a). Factors underlying rice yield variability in two California fields. *Agronomy Journal 96*: 1481–1494.

Roel, A., and R. E. Plant (2004b). Spatiotemporal analysis of rice yield variability in two California fields. *Agronomy Journal 96*: 77–90.

Romesburg, H. C. (1985). Exploring, confirming, and randomization tests. *Computers and Geosciences 11*: 19–37.

Rossiter, D. G. (2007). Technical Note: Co-kriging with the gstat package of the R environment for statistical computing. http://www.itc.nl/~rossiter/teach/R/R_ck.pdf.

Rubinstein, R. Y. (1981). *Simulation and the Monte Carlo Method*. Wiley, New York.

Russo, D. (1984). Design of an optimal sampling network for estimating the variogram. *Soil Science Society of America Journal 48*: 708–716.

Rutherford, A. (2001). Introduction to ANOVA and ANCOVA: A GLM Approach. Sage, London, UK.

Sarkar, D. (2008). *Lattice: Multivariate Data Visualization with R*. Springer, New York.

Schabenberger, O., and C. A. Gotway (2005). *Statistical Methods for Spatial Data Analysis*. Chapman & Hall, Boca Raton, FL.

Schabenberger, O., and F. J. Pierce (2002). *Contemporary Statistical Models for the Plant and Soil Sciences*. CRC Press, Boca Raton, FL.

Scheiner, S. M., and J. Gurevitch, Editors (2001). *Design and Analysis of Ecological Experiments*. Oxford University Press, Oxford, UK.

Schwarz, G. E. (1978). Estimating the dimension of a model. *Annals of Statistics 6*: 461–464.

Searle, S. R. (1971). *Linear Models*. Wiley, New York.

Searle, S. R. (1987). *Linear Models for Unbalanced Data*. Wiley, New York.

Searle, S. R., G. Casella, and McCulloch (1992). *Variance Components*. Wiley, New York.

Shanahan, J. F., J. S. Schepers, D. D. Francis, G. E. Varvel, W. W. Wilhelm, J. M. Tringe, M. R. Schlemmer, and D. J. Major (2001). Use of remote sensing imagery to estimate corn grain yield. *Agronomy Journal 93*: 583–589.

Shapiro, S. S., and M. B. Wilk (1965). An analysis of variance test for normality (complete samples). *Biometrika 52*: 591–611.

Sibley, D. (1988). *Spatial Applications of Exploratory Data analysis*. Geo Books, Norwich, UK.

Simbahan, G. C., A. Doberman, and J. L. Ping (2004). Screening yield monitor data improves grain yield maps. *Agronomy Journal 96*: 1091–1102.

Singer, M. J., and D. N. Munns (1996). *Soils: An Introduction*. Prentice Hall, Upper Saddle River, NJ.

Smith, B. J. (2007). boa: An R Package for MCMC output convergence assessment and posterior inference. *Journal of Statistical Software* 21: 1–37.

Smith, E. P., D. R. Orvos, and J. Cairns (1993). Impact assessment using the before-and-after-control-impact model: concerns and comments. *Canadian Journal of Fisheries and Aquatic Science* 50: 627–637.

Smouse, P. E., J. C. Long, and R. R. Sokal (1986). Multiple regression and correlation extensions of the Mantel test of matric correspondence. *Systematic Zoology* 35: 627–632.

Snedecor, G. W., and W. G. Cochran (1989). *Statistical Methods*. Iowa State University Press, Ames, IA.

Sokal, R. R., and N. L. Oden (1978). Spatial autocorrelation in biology: 1. Methodology. *Biological Journal of the Linnean Society* 10: 199–228.

Sokal, R. R., N. Oden, B. A. Thomson, and J. Kim (1993). Testing for regional differences in means: distinguishing inherent from spurious spatial autocorrelation by restricted randomization. *Geographical Analysis* 25: 199–210.

Sokal, R. R., N. L. Oden, B. A. Thomson, and J. Kim (1993). Testing for regional differences in means: distinguishing inherent from spurious spatial autocorrelation by restricted randomization. *Geographical Analysis* 25: 199–210.

Sokal, R. R., and F. J. Rohlf (1981). *Biometry*. W.H. Freeman, San Francisco, CA.

Sokolnikoff, I. S., and R. M. Redheffer (1966). *Mathematics of Physics and Modern Engineering*. McGraw-Hill, New York.

Spector, P. (1994). *An Introduction to S and S-Plus*. Duxbury Press, Belmont, CA.

Spiegelhalter, D., A. Thomas, N. Best, and D. Lunn (2003). *WinBUGS User Manual, Version 1.4*. http://www.mrc-bsu.cam.ac.uk/bugs.

Sprent, P. (1969). *Models in Regression and Related Topics*. Methuen, London, UK.

Sprent, P. (1998). *Data Driven Statistical Methods*. Chapman & Hall, London, UK.

Standiford, R., N. McDougald, W. Frost, and R. Phillips (1997). Factors influencing the probability of oak regeneration on southern Sierra Nevada woodlands in California. *Madroño* 44: 170–183.

Stanish, W. M., and N. Taylor (1983). Estimation of the intraclass correlation coefficient for the analysis of covariance model. *The American Statistician* 37: 221–224.

Steel, R. G. D., J. H. Torrie, and D. A. Dickey (1997). *Principles and Procedures of Statistics: A Biometrical Approach*. McGraw-Hill, New York.

Stein, A., F. van der Meer, and B. Gorte, Editors (1999). *Spatial Statistics for Remote Sensing*. Kluwer Academic Publishers, Dordrecht, the Netherlands.

Stevens, S. S. (1946). On the theory of scales of measurement. *Science* 103: 677–680.

Stram, D. O., and J. W. Lee (1994). Variance components testing in the longitudinal mixed effects models. *Biometrics* 50: 1171–1177.

Street, R. L. (1973). *The Analysis and Solution of Partial Differential Equations*. Brooks/Cole, Monterey, CA.

Stroup, W. W. (2002). Power analysis based on spatial effects mixed modes: A tool for comparing design and analysis strategies in the presence of spatial variability. *Journal of Agricultural, Biological, and Environmental Statistics* 7: 491–511.

Stroup, W. W., P. S. Baenziger, and D. K. Mulitze (1994). Removing spatial variation from wheat yield trials: a comparison of methods. *Crop Science* 86: 62–66.

Student (1914). The elimination of spurious correlation due to position in time or space. *Biometrika* 10: 179–180.

Sturtz, S., U. Ligges, and A. Gelman (2005). R2WinBUGS: A package for running WinBUGS from R. *Journal of Statistical Software* 12: 1–16.

Sudduth, K. A., S. T. Drummond, and D. B. Myers (2013). Yield Editor 2.0: software for automated removal of yield map errors. http://extension.missouri.edu/sare/documents/asabeyieldeditor2012.pdf.

Sudduth, K. A., N. R. Kitchen, W. J. Wiebold, W. D. Batchelor, G. A. Bollero, D. E. Clay, H. L. Palm, F. J. Pierce, R. T. Schuler, and K. D. Thelen (2005). Relating apparent electrical conductivity to soil properties across the north-central USA. *Computers and Electronics in Agriculture* 46: 263–283.

Tamura, R. N., L. A. Nelson, and G. C. Naderman (1988). An investigation of the validity and usefulness of trend analysis for field plot data. *Agronomy Journal 80*: 712–718.

Tate, R. F. (1954). Correlation between a discrete and a continuous variable. Point-biserial correlation. *Annals of Mathematical Statistics 25*: 603–607.

Theil, H. (1971). *Principles of Econometrics*. Wiley, New York.

Therneau, T. M., B. Atkinson, and B. Ripley (2011). rpart: Recursive Partitioning. http://CRAN.R-project.org/package=rpart.

Therneau, T. M., and E. J. Atkinson. 1997. An introduction to recursive partitioning using the RPART routines. Biostatistics Technical Report 61. Mayo Clinic, Rochester, MN.

Thomas, A., B. O'Hara, U. Ligges, and S. Sturtz (2006). Making BUGS open. *R News 6*: 12–17.

Thomas, A., N. Best, D. Lund, R. Arnold, and D. Spiegelhalter. (2004). GeoBUGS User Manual. http://www.mrc-bsu.cam.ac.uk/bugs/winbugs/geobugs12manual.pdf.

Thomas, J. M., J. A. Mahaffey, K. L. Gore, and D. G. Watson (1978). Statistical methods used to assess biological impact at nuclear power plants. *Journal of Environmental Management 7*: 269–290.

Thompson, J. F., and R. G. Mutters (2006). Effect of weather and rice moisture at harvest on milling quality of California medium grain rice. *Transactions of the ASABE 49*: 435–440.

Thompson, L. A. (2009). R (and S-PLUS) manual to accompany Agresti's Categorical Data Analysis (2002). https://Home.Comcast.Net/~Lthompson221/Splusdiscrete2.Pdf.

Thylen, L., and D. P. L. Murphy (1996). The control of errors in momentary yield data from combine harvesters. *Journal of Agricultural Engineering Research 64*: 271.

Tiefelsdorf, M. (2000). *Modelling Spatial Processes*. Springer-Verlag, Berlin, Germany.

Tiefelsdorf, M. G., D. A. Griffith, and B. Boots. (1999). A variance-stabilizing coding scheme for spatial link matrices. *Environment and Planning A 31*: 165–180.

Tobler, W. (1970). A computer movie simulating urban growth in the Detroit region. *Economic Geography 46*: 234–240.

Triantafilis, J., A. I. Huckel, and I. O. A. Odeh (2001). Comparison of statistical prediction methods for estimating field-scale clay content using different combinations of ancillary variables. *Soil Science 166*: 415–427.

Tucker, C. J. (1979). Red and photographic infrared linear combinations for monitoring vegetation. *Remote Sensing of Environment 8*: 127–150.

Tufte, E. R. (1983). *The Visual Display of Quantitative Information*. Graphics Press, Cheshire, CT.

Tufte, E. R. (1990). *Envisioning Information*. Graphics Press, Cheshire, CT.

Tukey, J. W. (1977). *Exploratory Data Analysis*. Addison-Wesley, Reading, MA.

Tukey, J. W. (1986). Sunset salvo. *The American Statistician 40*: 72–76.

Turner, R. (2011). deldir: Delaunay Triangulation and Dirichlet (Voronoi) Tessellation. http://CRAN.R-project.org/package=deldir.

Tyler, C. M., B. Kuhn, and E. W. Davis (2006). Demography and recruitment limitations of three oak species in California. *Quarterly Review of Biology 81*: 127–152.

University of California. (2006). Small grains production manual. Report 8028. Division of Agriculture and Natural Resources, Oakland, CA.

Unwin, D. (1975). *An Introduction to Trend Surface Analysis*. Geo Abstracts, Norwich, UK.

Upton, G. J. G., and B. Fingleton (1985). *Spatial Data Analysis by Example, vol. 1: Point Pattern and Quantitative Data*. Wiley, New York.

Upton, G. J. G., and B. Fingleton (1989). *Spatial Data Analysis by Example, vol. 2: Categorical and Directional Data*. Wiley, New York.

Valliant, R., J. A. Dever, and F. Kreuter (2013). *Practical Tools for Designing and Weighting Survey Samples*. Springer, New York.

Valliant, R., A. H. Dorfman, and R. M. Royall (2000). *Finite Population Sampling and Inference: A Prediction Approach*. Wiley, New York.

van Groenigen, J. W., W. Siderius, and A. Stein (1999). Constrained optimization of soil sampling for minimization of the kriging variance. *Geoderma 87*: 239–259.

van Groenigen, J. W., and A. Stein (1998). Constrained optimization of spatial sampoing using continuous simulated annealing. *Journal of Environmental Quality 27*: 1078–1086.

Vayssières, M. P., R. E. Plant, and B. H. Allen-Diaz (2000). Classification trees: An alternative non-parametric approach for predicting species distributions. *Journal of Vegetation Science 11*: 679–694.

Venables, W. N. (2000). Exegeses on linear models. Paper Presented to the S-PLUS User's Conference. Washington, DC, 8–9th October, 1998: www.stats.ox.ac.uk/pub/MASS3/Exegeses.pdf.

Venables, W. N., and C. M. Dichmont (2004). A generalized linear model for catch allocation: An example from Australia's northern prawn industry. *Fisheries Research 70*: 409–426.

Venables, W. N., and B. D. Ripley (2000). *S Programming*. Springer-Verlag, New York.

Venables, W. N., and B. D. Ripley (2002). *Modern Applied Statistics with S*. Springer-Verlag, New York.

Venables, W. N., D. M. Smith, and R Development Core Team (2008). *R: A Language and Environment for Statistical Computing*. R Foundation for Statistical Computing, Vienna, Austria.

ver Hoef, J. M., and N. Cressie (1993). Multivariable spatial prediction. *Mathematical Geology 25*: 219–240.

ver Hoef, J. M., and N. Cressie (2001). Spatial statistics. Analysis of field experiments. In Scheiner, S. M. and Gurevitch, J., Editors. *Design and Analysis of Ecological Experiments* pp. 289–307. Oxford University Press, Oxford, UK.

Veronesi, F. (2015). Spatio-temporal kriging in R. http://r-video-tutorial.blogspot.ch/2015/08/spatio-temporal-kriging-in-r.html.

Wackerly, D. D., W. Mendenhall, and R. L. Scheaffer (2002). *Mathematical Statistics with Applications*. Duxbury, Pacific Grove, CA.

Waddell, K. L., and T. M. Barrett (2005). *Oak Woodlands and other Hardwood Forests of California, 1990s*. Pacific Northwest Research Station, US Department of Agriculture Resource Bulletin PNW-RB-245, Portland, OR.

Wall, M. M. (2004). A close look at the spatial structure implied by the CAR and SAR models. *Journal of Statistical Planning and Inference 121*: 311–324.

Waller, L. A., and C. A. Gotway (2004). *Applied Spatial Statistics for Public Health Data*. Wiley-Interscience, New York.

Walters, C. (1986). *Adaptive Management of Renewable Resources*. McGraw-Hill, New York.

Walvoort, D. J. J., D. J. Brus, and J. J. Gruijter (2010). An R package for spatial coverage sampling and random sampling from compact geographical strata by k-means. *Computers and Geosciences 36*: 1261–1267.

Wang, P. C. (1985). Adding a variable in generalized linear models. *Technometrics 27*: 273–276.

Warton, D. I. (2005). Many zeros does not mean zero inflation: Comparing the goodness-of-fit of parametric models to multivariate abundance data. *Envirometrics 16*: 275–289.

Webb, R. A. (1972). Use of the boundary line in the analysis of biological data. *Journal of Horticultural Science 47*: 309–319.

Webster, R. (1997). Regression and functional relations. *European Journal of Soil Science 48*: 557–566.

Webster, R., and M. A. Oliver (1990). *Statistical Methods in Soil and Land Resource Survey*. Oxford University Press, Oxford, UK.

Webster, R., and M. A. Oliver (1992). Sample adequately to estimate variograms of soil properties. *Journal of Soil Science 43*: 177–192.

Webster, R., and M. A. Oliver (2001). *Geostatistics for Environmental Scientists*. Wiley, New York.

Wedderburn, R. W. M. (1974). Quasi-likelihood functions, generalized linear models and the Gauss-Newton method. *Biometrika 61*: 439–447.

Wenger, K. F. (1984). *Forestry Handbook*. John Wiley & Sons, New York.

Westoby, M., B. Walker, and I. Noy-Meir (1989). Opportunistic management for rangelands not at equilibrium. *Journal of Range Management 42*: 266–274.

Whalley, R. D. B. (1994). State and transition models for rangelands. 1. Successional theory and vegetation change. *Tropical Grasslands 28*: 195–205.

Whelan, B. M., and A. B. McBratney (2000). The "null hypothesis" of precision agriculture management. *Precision Agriculture 2*: 265–279.

Whittle, P. (1954). On stationary processes in the plane. *Biometrika 41*: 434–449.

Wickhan, H. (2009). *ggplot2: Elegant Graphics for Data Analysis*. Springer, New York.

Wickham, H., and G. Grolemund (2017). *R For Data Science*. O'Reilly Media, Sebastopol, CA.

Wieslander, A. E. (1935). A vegetation type map of California. *Madroño 3*: 140–144.

Wiles, L., and M. Brodahl (2004). Exploratory data analysis to identify factors influencing spatial distributions of weed seed banks. *Weed Science 52*: 936–947.

Wilkinson, G. N., S. R. Eckert, T. W. Hancock, and O. Mayo (1983). Nearest neighbour (NN) analysis of field experiments. *Journal of the Royal Statistical Society, Series B 45*: 151–178.

Wilkinson, L. (2005). *The Grammar of Graphics*. Springer, New York.

Williams, J. F. (2010). *Rice Nutrient Management in California*. University of California Division of Agriculture and Natural Resources Publication 3516, Oakland, CA.

Wilson, A. D., D. G. Lester, and C. S. Oberle (2004). Development of conductive polymer analysis for the rapid detection and identification of phytopathogenic microbes. *Phytopathology 94*: 419–431.

Winfree, A. (1977). Spatial and temporal organization in the Zhabotinsky reaction. *Advances in Biological and Medical Physics 16*: 115–136.

Wollenhaupt, N. C., D. J. Mulla, and N. C. Gotway Crawford (1997). Soil sampling and interpolation techniques for mapping spatial variability of soil properties. In Pierce, F. J. and Sadler, E. J., Editors. *The State of Site Specific Management for Agriculture* pp. 19–54. American Society of Agronomy, Madison, WI.

Wollenhaupt, N. C., R. P. Wolkowski, and M. K. Clayton (1994). Mapping soil test phosphorous and potassium for variable-rate fertilizer application. *Journal of Production Agriculture 7*: 441–448.

Wong, D. W. S. (1995). Aggregation effects in geo-referenced data. In Arlinghaus, S. L., Griffith, D. A., Arlinghaus, W. C., Drake, W. D., and Nystuen, J. D., Editors. *Practical Handbook of Spatial Statistics* pp. 83–106. CRC Press, Boca Raton, FL.

Wood, A. T. A., and G. Chan (1994). Simulation of stationary Gaussian process in [0,1]^d. *Journal of Computational and Graphical Statistics 3*: 409–432.

Wood, S. N. (2006). *Generalized Additive Models: An Introduction with R*. CRC Press, Boca Raton, FL.

Wooldridge, J. M. (2013). *Introductory Econometrics, a Modern Approach*. South-Western, Mason, OH.

Wu, M., and Y. Zuo (2008). Trimmed and Winsorized standard deviations based on a scaled deviation. *Journal of Nonparametric Statistics 20*: 319–325.

Yee, T. W., and N. D. Mitchell (1991). Generalized additive models in plant ecology. *Journal of Vegetation Science 2*: 587–602.

Yule, G. U. (1926). Why do we sometimes get nonsense-correlations between time-series?—A study in sampling and the nature of time-series. *Journal of the Royal Statistical Society 89*: 1–63.

Zalom, F. G., P. B. Goodell, L. T. Wilson, W. W. Barnett, and W. J. Bentley. 1983. Degree-Days: The Calculation and Use of Heat Units in Pest Management. Leaflet 21373. Division of Agriculture and Natural Resources, University of California, Berkeley, CA.

Zavaleta, E. S., K. B. Hulvey, and B. Fulfrost (2007). Regional patterns of recruitment success and failure in two endemic California oaks. *Diversity and Distributions 13*: 735–745.

Zeileis A., and G. Grothendieck (2005). zoo: S3 infrastructure for regular and irregular time series. *Journal of Statistical Software 14*: 1–27. doi:10.18637/jss.v014.i06.

Zeileis, A., and T. Hothorn (2002). Diagnostic checking in regression relationships. *R News 2*: 7–10.

Zeileis, A., C. Kleiger, and S. Jackman (2008). Regression models for count data in R. *Journal of Statistical Software 27*: 108.

Zhang, M. H., P. Hendley, D. Drost, M. O'Niell, and S. Ustin (1999). Corn and soybean yield indicators using remotely sensed vegetation index. In Robert, P. C., Rust, R. H., and Lawson, W. E., Editors. *Proceedings of the Fourth International Conference on Precision Agriculture* pp. 1475–1481. American Society of Agronomy, Madison, WI.

Zhu, J., and G. D. Morgan (2004). Comparisons of spatial variables over subregions using a block bootstrap. *Journal of Agricultural, Biological, and Environmental Statistics 9*: 91–104.

Zimmerman, D. L., and D. A. Harville (1991). A random field approach to the analysis of field-plot experiments and other spatial experiments. *Biometrics* 47: 223–239.

Zuur, A. F., E. N. Ieno, and G. M. Smith (2007). *Analyzing Ecological Data*. Springer, New York.

Zuur, A. F., E. N. Ieno, N. J. Walker, and A. A. S. G. M. Saveliev (2009). *Mixed Effects Models and Extensions in Ecology with R*. Springer, New York.

Index

Note: Page numbers in italic and bold refer to figures and tables respectively.

Printed in the United States
by Baker & Taylor Publisher Services